Design Analysis in Rock Mechanics
3rd Edition

Design Analysis in Rock Mechanics
3rd Edition

William G. Pariseau

Malcolm McKinnon Endowed Chair, Emeritus
Department of Mining Engineering,
University of Utah, Salt Lake City, Utah, USA

CRC Press is an imprint of the
Taylor & Francis Group, an **informa** business
A BALKEMA BOOK

Front: Open pit gold mine, Nevada, USA, 2001

Back: The author in a small prospect adit in the north Cascade Mountains of Washington, USA

Published by:
CRC Press/Balkema
Schipholweg 107C, 2316 XC Leiden, The Netherlands

CRC Press/Balkema is an imprint of the Taylor & Francis Group, an informa business

ISBN-13: 978-1-138-02958-3 (hbk)
ISBN-13: 978-1-03-265239-9 (pbk)
ISBN-13: 978-1-315-20615-8 (ebk)

DOI: 10.1201/9781315206158

Typeset in Times New Roman by
Integra Software Services Private Ltd

Publisher's Note
The publisher has gone to great lengths to ensure the quality of this reprint but points out that some imperfections in the original copies may be apparent.

Visit the Taylor & Francis Web site at
http://www.taylorandfrancis.com

and the CRC Press Web site at
http://www.crcpress.com

Library of Congress Cataloging in Publication Data

Applied for

A Solutions Manual is available exclusively to lecturers adopting this publication for their courses. To gain access to the instructor resources for this title, please visit:
http://routledgetextbooks.com/textbooks/instructor_downloads/

Contents

Preface

This 3rd edition of *Design Analysis in Rock Mechanics* continues with the philosophy followed in the first two editions: There are important design problems in rock mechanics that can be addressed on the basis of *mechanics of materials*, a course often taught during the third year of undergraduate engineering study. The usual prerequisite of statics, dynamics, fluid mechanics, mathematics, chemistry and physics are assumed. Familiarity with the concept of stress concentration is expected. Studies of soil mechanics, geology and mining methods in civil, geological and mining engineering are helpful but not required. Again, a careful reading by students who were offered bounty points for identifying "typos" was helpful in clarifying the text. Units continue to be a mix of metric and imperial. What is new in this edition is a chapter on design of foundations on jointed rock. Design guidelines in the form of load – settlement (force –displacement) plots for stiff footings or slabs are presented. Nonlinearity and heterogeneity are often present in case of jointed rock and thus these guidelines are necessarily based on numerical results, specifically, on results obtained using the popular finite element method. Example problems illustrate application of guideline plot information; problems at the end of the chapter are available for home study and detailed solutions are available in a companion 3rd edition *Solution Manual.*

William G. Pariseau
Salt Lake City, Utah
December, 2016

Preface to the second edition

This second edition benefits from five years of experience with the first edition and valuable feedback from users of the first edition including students, faculty, and practicing engineers. This edition differs from the first with the addition of an important chapter on dynamic phenomena including discussion of rock bursts and coal bumps. More examples have been added to each chapter to further illustrate application of design principles and procedures developed in the text and more problems for home study have been added at the end of each of the nine chapters in the text. Several appendices have been expanded; some reduced. All of the text and companion *Solution Manual* have been edited with the intention of correcting mistakes, although there is no guarantee of perfection. Use of metric and imperial units is continued. The author remains convinced that (1) there are important design problems associated with excavations in rock that can be addressed from a background in mechanics of materials and a basic understanding of stress, strain, elasticity, and strength, and (2) understanding is a prerequisite for responsible use of computer programs intended for excavation design whether for surface or underground excavations.

William G. Pariseau
Salt Lake City
May, 2011

Preface to the first edition

A preface gives the author an opportunity to explain his work: what the objective is, what level of understanding is assumed, why some material is included and some not, and so forth. Like most experienced practitioners and teachers of rock mechanics, I find that there is a lack of suitable textbooks for undergraduate courses and consequently I rely on personal notes and handouts. There are a number of *reference* books available that cover rock mechanics in varying degrees of depth and breadth, but most are best suited for supporting graduate study. None have a satisfactory array of worked out example problems, a selection of problems for homework assignment at the end of each chapter and a key containing detailed solutions as is customary in undergraduate textbooks.

There is an opinion that rock mechanics is better left for advanced study, so any undergraduate course should be preparation for graduate study and should therefore emphasize the more scientific features of rock mechanics. My view is that there are important, practical engineering problems in rock mechanics that should and can be addressed at the undergraduate level. An important example is rock slope stability. Hence, the title of this book is *Design Analysis in Rock Mechanics.* Emphasis is on application to practical problems. Although intended primarily for undergraduate study, first year graduate students, whose introduction to rock mechanics was one of engineering science, may find application to engineering design of some interest.

Of course, not all issues in rock mechanics can be addressed in a single book. The rich and varied area of numerical methods and computational rock mechanics must necessarily be left out. In any event, computer code usage should be learned only after a firm grasp of fundamentals is in hand. Numerical analyses provide much helpful guidance toward the solution of complex problems in rock mechanics, but becoming adept at keystrokes and program execution is no substitute for thoughtful analysis based on fundamental principles.

This book builds on mechanics of materials, a course that is required of almost all engineers. Familiarity with the concepts of stress, strain, and elastic stress–strain relations for axially loaded members, for torsion of circular bars and for beams is assumed. Torsion is not often encountered in rock mechanics design, but axial load and beam bending find important parallels with pillars in underground excavations and roof spans in stratified ground. These problems are essentially one-dimensional and may be handled analytically and quantitatively with help from a calculator.

The concept of stress concentration about a hole is also introduced in the first course in mechanics of materials. Application here, for example, is to circular mine shafts and rectangular tunnels. These two-dimensional problems (plane strain) require the ability to transform the state of stress and strain to a rotated set of axes. The important concepts of

principal stresses and strains are implied in any two-dimensional analysis as is maximum shear stress. Rotation of the reference axes is also needed to compute normal and shear stresses on an inclined joint in a pillar. Specialized three-dimensional features of stress and strain may also be handled.

Design applications addressed in the text fall into three broad categories: surface excavations, underground excavations, and subsidence. The concept of safety factor is introduced early as an empirical index to design adequacy and is carried through the various sections of the text. Justification is found in a simple analysis using Newton's second law of motion to show that a factor of safety less than 1 implies acceleration. Thus, the general objective of rock mechanics, which is to compute the motion of a mass of interest, is replaced by the practical objective of computing a suitably defined safety factor that serves the particular problem at hand. In this regard, a distinction is made between global and local safety factors that measure nearness to structural collapse and nearness to the elastic limit at a point, respectively.

Surface excavations pose questions of slope stability which is perhaps the most important problem area in rock mechanics because of the economic impact slope angle has on surface mine operations. Of course, highway cuts and dam abutments in rock are also of considerable importance for obvious reasons. Mine operations produce enormous quantities of waste rock from stripping and milling that lead to soil-like slope stability problems, which are also addressed in the text.

Underground excavations pose a greater variety of design problems. Safety of shafts and tunnels is paramount, and is addressed with respect to intact rock and joint failure mechanisms by building on the concept of stress concentration about holes. Engineered reinforcement and support considerations are often needed to insure safety of these lifelines to the underground. Discussion of tunnel support also offers an opportunity to introduce rock mass classification systems for engineering purposes. Although not a substitute for engineering design, familiarization with rock classification schemes is important to understanding much of the ongoing discussion on practical design analysis in the technical literature.

Stratified ground poses questions of safe roof spans and pillar sizes for natural primary support and what other installed support may be required for adequate ground control. Hard rock excavations range over a large set of mining methods and geometries from horizontal room and pillar excavations to stopes (excavations) in narrow, steeply dipping veins to large open stopes, and to caving methods. Backfill and cable bolting are common to all except caving methods which often require robust combination support systems. Backfill study presents an opportunity to consider soil size classification for engineering behavior and seepage in porous media as well as a review of the concept of effective stress.

Subsidence that links the surface to the underground is divided into two parts: chimneys over hardrock excavations and troughs over softrock (e.g. coal) excavations. The goal is to understand the forces and geometries of both with forecasting surface motion as an objective.

Sparing use of computer programs is made in the belief that understanding should precede computer usage. Teaching key strokes is avoided, but at the same time students are referred to some menu-driven course downloads from the Department Website and to other commercial programs licensed to our Department. Use of these programs is mainly to check analytical solutions that require some computations with a hand calculator or a spreadsheet. These programs also help to familiarize students with possibilities for interactions with consultants after graduation, but are better deferred to more advanced courses.

Some comments should perhaps be made concerning the matter of references and bibliography. First, this book is neither an academic treatise nor a reference work for the use of research investigators, but rather a text for undergraduates. Controversial issues under active research scrutiny about which there is considerable doubt have been avoided. Consequently, much of the material and concepts presented are in the broad area of common knowledge. Second, only publications that the author is personally acquainted with are cited. This narrows the selection of references to those primarily in the areas of interest to the author and of necessity neglects a vast amount of literature of possible relevance to rock mechanics. But again, the purpose of this book is to aid in the development of a working knowledge of some, but not all, important rock mechanics problems, an understanding of their main physical features, and how to carry out a useful design analysis for engineering action.

Finally, a rather long review of fundamentals was included in an earlier draft of this text, but after some debate, it was omitted in favor of an immediate development of design analysis. Some instructors may elect to supplement the current material with a review of stress, principal stresses, stress change, strain, strain rates, Hooke's law and other material models, and so forth. However, such an addition is likely to require shortening of other material presented in the text that now leads to a full three-credit hour semester course of 15 weeks with allowance for several examinations along the way. Consideration is also being given to the addition of chapters on foundations and rock dynamics. The brief mention of bearing capacity within the context of slope stability hardly does justice to this important subject. Mention is also made of seismic loading, again, within the context of slope stability, but certainly a more systematic development of elementary wave mechanics would be helpful to understanding effects of blasting on stability of surface and underground openings, for example.

William G. Pariseau
Salt Lake City

Acknowledgments

Financial support of the Malcolm N. McKinnon endowed chair in the Department of Mining Engineering at the University of Utah contributed to the expense of this work, especially toward the lengthy process of drafting and redrafting the many figures essential to the understanding of the concepts developed in the text. In this regard, there remains a long-standing, unmet need for undergraduate texts in mining engineering. The relatively small size of the discipline and market place publishing economics militate against textbook development for undergraduates. A professor will find compensation for time spent much more advantageous in consulting and research. However, sharing discovery and knowledge is at the heart of teaching that textbooks hopefully make more effective. Again, considerable thanks are extended to the support provided by the Malcolm McKinnon endowed chair.

Acknowledgment is also made for use of figures from publications after other authors cited in the text.

Finally, appreciation is expressed to the many students who have persevered in their introduction to design analysis in rock mechanics using chalk board notes, handouts, slides, videos, and a variety of reference books but without an actual textbook.

William G. Pariseau

Chapter 1

Introduction

In a broad sense, what one attempts to do in rock engineering is to anticipate the motion of a proposed structure under a set of given conditions. The main design objective is to calculate displacements, and as a practical matter, to see whether the displacements are acceptable. Very often restrictions on displacements are implied rather than stated outright. This situation is almost always the case in elastic design where the displacements of the structure of interest are restricted only to the extent that they remain within the range of elastic behavior.

In rock mechanics, the "structure" of interest is simply the rock mass adjacent to a proposed excavation. The proposed excavation may be started at the surface, or it may be a deepening of an existing surface excavation, a start of a new underground excavation or an enlargement of an existing underground excavation. In any case, the excavation plan, if actually carried out, would cause changes in the forces acting in the neighborhood of the excavation and would therefore cause deformation of the adjacent rock mass. Of course, the rock mass may also move in response to other forces such as those associated with earthquakes, equipment vibration, blasting, construction of a nearby waste dump, filling or draining of an adjacent water reservoir, temperature changes, and so on.

Regardless of the specific identity of the forces acting, the associated motion must always be consistent with basic physical laws such as the conservation of mass and the balance of linear momentum. In this respect, rock is no different from other materials. Any motion must also be consistent with the purely geometrical aspects of translation, rotation, change in shape, and change in volume, that is, with kinematics. However, physical laws such as Newton's second law of motion (balance of linear momentum) and kinematics are generally not sufficient for the description of the motion of a deformable body. The number of unknowns generally exceeds the number of equations.

This mathematical indeterminancy may be removed by adding to the system as many additional equations as needed without introducing additional unknowns. The general nature of such equations becomes evident following an examination of the internal mechanical reaction of a material body to the externally applied forces. The concepts of stress and strain arise in such inquiry and the additional equations needed to complete the system are equations that relate stress to strain. Stress–strain relationships represent a specific statement concerning the nature or "constitution" of a material and are members of a general class of equations referred to as constitutive equations. Constitutive equations express material laws. Whereas physical laws and kinematics are common to all materials, constitutive equations serve to distinguish categories of material behavior. Hooke's law, for example, characterizes materials that respond elastically to load. A system of equations that describes the motion of a

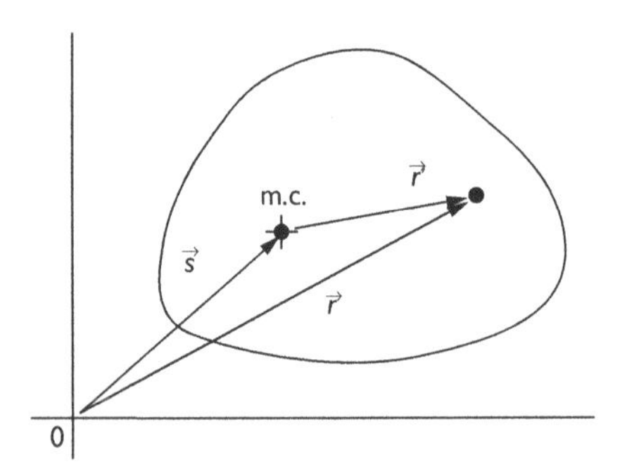

Figure 1.1 Position relative to the mass center of a body.

deformable body necessarily includes all three types of equations: physical laws, kinematics, and material laws.

In reality, a system describing the motion of a material body is only an approximation. Mathematical complexities often dictate additional simplification and idealization. Questions naturally arise as to what simplifications should be made and, once made, how well the idealized representation corresponds to reality. Questions of this type relate more to the art than to the science of engineering design and have no final answers. Experience can, of course, be a great aid in this regard, when such experience is informed by a clear understanding of the fundamental concepts.

Example 1.1 The mass center of a body is defined by

$$M\vec{s} = \int_V \vec{r}\mathrm{d}m$$

where $\vec{s}$ and $\vec{r}$ are vectors shown in Figure 1.1. Differentiation twice with respect to time gives the interesting result

$$M\ddot{\vec{s}} = \int_B \vec{a}\mathrm{d}m = \int_V \rho\vec{a}\mathrm{d}V = \dot{\vec{P}} = F$$

that shows that the mass center moves as if it were a particle accelerating according to Newton's second law (resultant of external forces = time rate of change of linear momentum). Thus, even though one cannot determine the acceleration everywhere in the body of interest at this juncture, there is a possibility of at least following the motion of the center of mass of the body. This fact remains true even if the body disintegrates.

Consider a mass M of rock in a landslide or avalanche and suppose that the external forces are: (i) the weight W of the slide mass and (ii) the contact force R acting between the slide mass and the parent rock mass from which the slide mass has become detached. The contact force R may be frictional, viscous, and displacement-dependent, that is, $R = R_0 + R_1\ \mathrm{d}s/\mathrm{d}t - R_2\ s$. R increases with speed but decreases with displacement of the mass center and resists the downhill component of weight $D = D_0$. According to the previous result

$$F = D - R = M\ddot{s}$$

Hence

$$M\ddot{s} + R_1\dot{s} - R_2 s = D_0 - R_0$$

describes the motion of the slide mass center. The coefficients in this equation may depend on time and position (except M). Over a short period of time, however, a reasonable assumption is that they are constant in which case the form of the solution is known. Reasonable initial conditions are that the slide mass is at rest or moving at a constant speed (steady creep).

The logic followed in Example 1.1 illustrates in a very compact form how one proceeds from physical laws (conservation of mass, Newton's second law) through problem simplification (look at mass center motion only, disregard deformation, disintegration, motion of individual elements) and material idealization (assumptions concerning resistance) to a mathematically tractable representation of the original problem (landslide dynamics). Of course, simplification is a relative notion. Here simplification means one has progressed from an essentially hopeless situation to a situation where useful information may be extracted. In this example, useful information might refer to estimation of slide mass travel in conjunction with zoning regulation for geologic hazard. The solution effort required may still be considerable. However, there may also be unexpected benefits. In this example, "triggering" of catastrophic landslides under load level fluctuations that were formerly safe becomes understandable in relatively simple physical terms.

1.1 A practical design objective

A tacit assumption in rock mechanics that is often made in the absence of inelastic behavior is that large displacements accompanying failure are precluded. Under these circumstances, design is essentially an analysis of safety and stability. A practical design objective is then to calculate a factor of safety appropriate for the problem at hand.

An appropriate factor of safety depends on the problem and is an empirical index to "safety" or "stability." Safety and stability are often used interchangeably in rock mechanics, although strictly speaking, they are not synonymous. Stability often connotes a possibility of fast failure or the onset of large displacements below the elastic limit. An example is strata buckling where kinks in thin laminations may form suddenly below the yield point. Safety typically relates to strength and nearness to the elastic limit or yield point. If forces are of primary concern, then a ratio of forces resisting motion to forces that tend to drive the motion is an appropriate safety factor. If rotation is of primary concern, then a ratio of resisting to driving moments would be an appropriate safety factor. When yielding at a point is of interest, then a ratio of "strength" to "stress" defines a useful safety factor when measures of "strength" and "stress" are well defined.

Example 1.2 Consider a rock mass high on a steep canyon wall that may pose a threat to facilities below. A reasonable index to stability is a factor of safety FS defined as a ratio of forces tending to drive the slide mass downhill to forces resisting the motion. Show that a safety factor greater than one implies safety.

Solution: By definition, FS = R/D. The mass center then moves according to

$$F = D - R = D(1 - \text{FS}) = M\ddot{s}$$

Hence, a safety factor less than one implies downhill acceleration, while a safety factor greater than one, implies stability (uphill acceleration is physically meaningless in this situation).

Example 1.3 Stress concentration about a vertical shaft results in a compressive stress of 8,650 psi (59.66 MPa) at the shaft wall where the rock mass has an unconfined compressive strength of 12,975 psi (89.48 MPa). Determine the shaft wall safety factor at the considered point.

Solution: An appropriate safety factor at the shaft wall is the ratio of strength to stress. Thus,

$$\begin{aligned} \text{FS}_\text{c} &= \frac{\text{Strength}}{\text{Stress}} \\ &= \frac{C_\text{o}}{\sigma_\text{c}} \\ &= \frac{12{,}975}{8{,}650} = \frac{89.48}{59.66} \\ \text{FS}_\text{c} &= 1.50 \end{aligned}$$

1.2 Problem solving

This text has been written from the point of view that whenever the main physical features of a problem are well known, as they are in the determination of tunnel support requirements, for example, then a strength of materials background should be sufficient for the development of quantitative analysis procedures. The emphasis throughout is upon the time-tested engineering approach to problem solving requiring (i) a brief but concise statement as to what is being sought; (ii) a listing of pertinent known data; (iii) a sketch of the "structure" for analysis showing in particular the applied loads and reactions; (iv) the equations and assumptions used; and (v) an outline of the major calculational steps taken in obtaining the desired results. Some of these steps may be combined as circumstances allow.

Example 1.4 A large array of square support pillars are formed by excavating rooms in a horizontal stratum 5 m (16.4 ft) thick at a depth of 300 m (984 ft). The pillars are 15 m (49.2 ft) on edge and are spaced on 22 m (72.2 ft) centers. Determine the average vertical stress in the pillars.

Solution: A large array implies that the pillars in the array are similar, so consideration of equilibrium of one pillar should reveal the relationship of forces acting at equilibrium. The pillars are the materials that remain after the rooms have been excavated, and must carry the weight of the overburden that, prior to excavation, was supported by all materials in the seam.

1. Sketch the geometry of the problem in plan view and vertical section.
2. Apply force equilibrium in the vertical direction. Thus, $W = F_\text{p}$ as shown in the sketch where W = (specific weight)(volume) and F_p = (average vertical pillar stress)(pillar area). One may estimate the overburden specific weight as, say, 24.8 kN/m^3 (158 pcf).
3. Do calculations. (24.8)(300)(22)(22) = S_p(15)(15), so S_p = 16 kN/m^2 (2,320 psi).

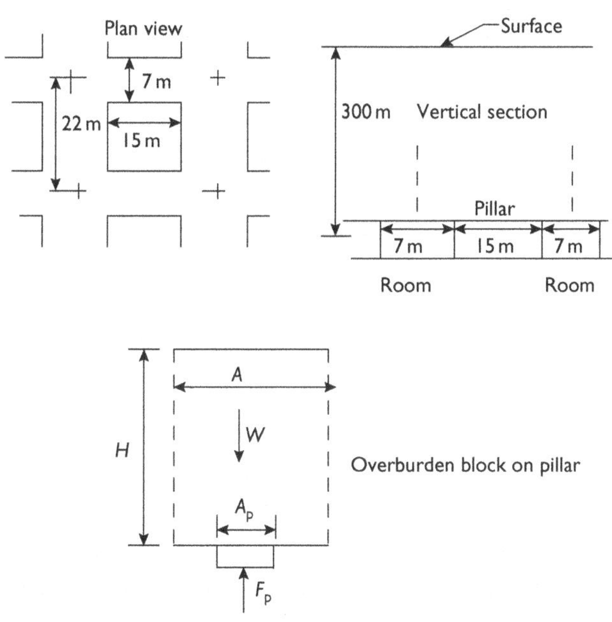

Sketches not to scale

1.3 Units

No one system of units is more "scientific" than another, although the SI system is advantageous in thermodynamic analysis compared with the use of pound mass and pound force units rather than slugs and pound force often used in mechanics. Why the switch to mass units in traditional engineering classes in the USA is a mystery and once a source of unnecessary confusion for many student engineers. Fortunately, many texts now use metric (SI) units and English engineering units. Both are also used in this text which, however, tilts toward the latter.

Both view Newton's second law of motion as $F = ma$ where F is the resultant of external forces (pound force, Newton), m is mass (slug, kilogram) and a is acceleration (feet/second^2, meter/second^2). A 1 pound force results when 1 slug is accelerated 1 ft/s^2. Thus, 1 lbf = slug-ft/s^2. (A slug is a 32.174 pound mass and 1 lbf = 1 lbm-ft/s^2.) A 1 N force results when 1 kg is accelerated 1 m/s^2, so 1 N = kg-m/s^2. Sometimes both units are given with one in parentheses following the other (Example 1.4). Tables of data may be given in either system. Sometimes conversion factors are given with a table of data when both units are not presented, but not always. Some useful conversion factors are given in Table 1.1.

Example 1.5 An estimate of preexcavation vertical stress is 1 psi per foot of depth. This estimate is based on an assumed overburden specific weight of 144 pcf. An improved estimate would use 158 pcf or 1.1 psi/ft, that is, $S_v = 1.1h$ where S_v is the vertical stress in psi and h is depth in feet. Modify this last equation to give S_v in kN/m^2 with h in meters.

***Solution*:** The estimate in detail is:
$S_v\,(\text{kN/m}^2) = 1.1(\text{psi/ft})\;\;6894.9(\text{N/m}^2/\text{psi})\;\;3.281(\text{ft/m})\;\;10^{-3}\;\;(\text{kN/N})\;\;h(\text{m})$, that is, $S_v\,(\text{kN/m}^2) = 24.9h(\text{m})$.

Table 1.1 Conversion factors[a]

Feet (ft)[b]	0.3048	Meters (m)
Inches (in.)	2.54	Centimeters (cm)
Meters (m)	3.2808	Feet (ft)
Centimeters (cm)	0.3937	Inches (in.)
Pound force (lbf)	4.4482	Newton (N)[c]
Newton (N)	0.2248	Pound force (lbf)
lbf/square inch (psi)	6894.8	Newton/square meter
Newton/square meter	1.4504 (10^{-4})	lbf/square inch (psi)
lbf/ft^3 (pcf)	157.09	N/m^3
N/m^3	6.366 (10^{-3})	lbf/ft^3 (pcf)

Notes

a Multiply units on the left by the number in the middle column to obtain units on the right.

b Abbreviations for units are not terminated with a period with the exception of inches.

c N/m^2 = 1 Pascal, 1 kN/m^2 = 1 kPa, 1 MN/m^2 = 1 MPa, k = kilo (10^3), M = mega (10^6). Also: 1 bar = 14.504 psi = 100 kPa, 1 atm = 14.7 psi.

Alternatively, from Table 1.1 and the given data (158 pcf)(157.09) = 24.8 (10^3) N/m^3 = 24.8 kPa/m within truncation and roundoff error.

Example 1.6 Given the specific gravity (SG) of a rock sample as 2.67, determine the specific weight γ in lbf/ft^3 and kN/m^3.

***Solution*:** By definition, SG is the ratio of a given mass to the mass of water occupying the same volume at the same temperature. This definition is thus the ratio of mass density of the given material ρ to the mass density of water ρ_w at the given temperature. Thus, SG = ρ/ρ_w. Also by definition specific weight γ is the weight of a unit volume of material, and weight is the force of gravity acting on the given mass, so that $\gamma = \rho g$. Hence, $\gamma = \rho g$ = (SG)(ρ_w)g = (SG)(γ_w) where the last term is the specific weight of water which is 62.43 lbf/ft^3. Thus, γ = 2.67(62.43) = 167 lbf/ft^3 and from the conversion factor in Table 1.1 γ = (167 lbf/ft^3)(157.09)(10^{-3}) = 26.2 kN/m^3.

Example 1.7 Estimate the range of tensile strength (T_o) of Bedford limestone between one standard deviation limits (one-sigma limits).

***Solution*:** Examination of Table B.5 in Appendix B lists the average tensile strength of Bedford limestone as 7.5 MPa (1,080 psi) and the standard deviation as 3.6 MPa (520 psi). The one-sigma limits are therefore 7.5–3.6 and 7.5+3.6 or 3.9 (560 psi) and 11.1 MPa (1,600 psi), respectively (within rounding).

Example 1.8 Estimate the unconfined compressive strength (C_o) of quartzite.

***Solution*:** According to Table B.7, strengths of two quartzites listed are 143.7 and 112.4 MPa from tests on 20 and 50 samples, respectively. A simple average is 128.1 MPa (18,570 psi). A weighted average is [(143.7)(20)+(112.4)(50)]/(20+50) = 121.3 MPa (17,590 psi), a slightly lower value.

Example 1.9 Estimate joint normal and shear stiffness in shale.

***Solution*:** Table B.11 in Appendix B shows ranges of 0.31–3.72(10^5 psi/in.) and 0.44 – 3.5(10^5 psi/in.) for normal and shear stiffness, respectively. Mid-range values of normal and shear stiffness are 2.02(10^5 psi/in.) and 1.97(10^5 psi/in.), respectively.

Example 1.10 Joint strength is often characterized by a cohesion (c_j) with units of stress and a friction angle (φ_j). Estimate c_j and φ_j for joints in granite. Table B.12 in Appendix B gives a range of 40–160 psi for c_j and $27° - 37°$ for φ_j. Mid-range estimates are thus 100 psi and 32° for c_j and φ_j, respectively.

1.4 Background information

Background information of interest certainly includes the technical literature in rock mechanics and data on rock properties pertinent to design analysis.

Rock mechanics literature

Background literature includes texts that develop fundamentals of strength of materials, mechanics of materials, elasticity theory, and continuum mechanics. Also included are reference works in rock mechanics. An annotated list of references, which the author has personally consulted on occasion, is given in Appendix A. There is also a list of references cited in the text.

Mechanical proper ties of rock

Rock properties data necessary for engineering design analysis may be found in numerous technical articles and handbooks. Some observations on laboratory rock properties data are of interest at this stage. The most important rock properties for engineering design are elastic moduli and strengths of intact rock and joint stiffnesses and strengths. Other properties such as thermal conductivity may be important in special cases, of course. A discussion of intact rock, rock joints and composite elastic moduli, and strengths and failure criteria is given in Appendix B.

Example 1.11 Given the two-dimensional state of stress where compression is positive:

$$\sigma_{xx} = 1,450 \text{ psi } (10 \text{ MPa}),\ \sigma_{yy} = 2,900 \text{ psi } (20 \text{ MPa}),\ \tau_{xy} = 725 \text{ psi } (5 \text{ MPa}).$$

Find the principal stresses and directions.

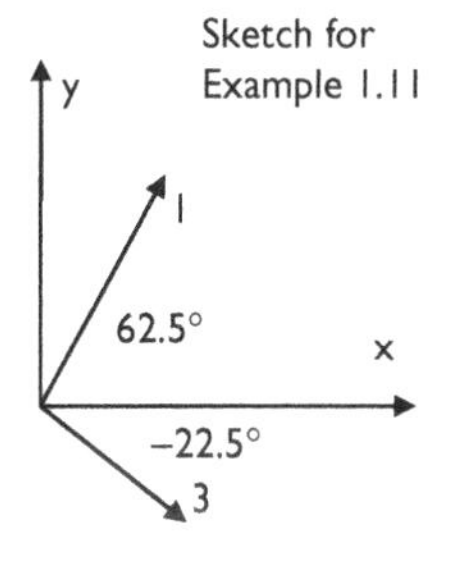

Sketch for Example 1.11

Solution: According to formula D.11 in Appendix D,

$$\tan(2\theta) = \frac{\tau_{xy}}{(1/2)(\sigma_{xx} - \sigma_{yy})}$$
$$= (2)(725)/(1450 - 2900)$$
$$\tan(2\theta) = -1 \quad \therefore 2\theta = -45^\circ, +135^\circ$$
$$\theta = -22.5^\circ, +62.5^\circ$$

Which of the two solutions gives the maximum or major principal stress can be determined by back substitution into the equations of transformation that allow rotation of the reference axes at a considered point. These equations are the first three of D.1. In fact, the correct choice is $\theta = 62.5°$. Another way of making the proper choice is to chose the angle that is nearest the largest given normal stress. In this example, σ_{yy} is the largest given normal stress and the angle $\theta = -22.5^o$ measured clock-wise is nearest the y-axis.

According to formula D.12, that major and minor principal stresses are

$$\left.\begin{matrix}\sigma_1\\ \sigma_3\end{matrix}\right\} = (1/2/)(\sigma_{xx} + \sigma_{yy}) \pm \sqrt{[(\sigma_{xx} - \sigma_{yy})/2]^2 + (\tau_{xy})^2}$$

$$\left.\begin{matrix}\sigma_1\\ \sigma_3\end{matrix}\right\} = (1/2/)(1450 + 2900) \pm \sqrt{[(1450 - 2900)/2]^2 + (725)^2}$$

$$\left.\begin{matrix}\sigma_1\\ \sigma_3\end{matrix}\right\} = \left.\begin{matrix}3{,}200\\ 1{,}150\end{matrix}\right\} \text{psi}, \qquad \left.\begin{matrix}22.07\\ 7.93\end{matrix}\right\} \text{Mpa}$$

As a check on the calculations, one notes that the sum of the normal stresses is independent of the angle of rotation. This sum is 1,450+2,900 = 4,350 = 3,200+1,150 or 10+20 = 30 = 22.07+7.93.

Example 1.12 Consider a three-dimensional state of stress relative to xyz coordinates where x = east, y = north, and z = up in a rock mass with a Young's modulus of 12.4 GPa and a Poisson's ratio of 0.20. Thus, in MPa with compression positive,

$$\sigma_{xx} = 12.50, \quad \sigma_{yy} = 16.3, \quad \sigma_{zz} = 25.0, \quad \tau_{xy} = -3.4, \quad \tau_{yz} = 0, \quad \tau_{zx} = 0$$

Find the strains relative to xyz coordinates assuming the state of stress is in the linearly elastic range of deformation and the rock mass is isotropic.

Solution: The relationship between stress and strain is simply Hooke's law. Thus, from Appendix D, one has the equations D.27 that for the normal strains give

$$\varepsilon_{xx} = (1/12.4)(10^{-9})[12.50 - (0.2)(16.3) - (0.2)(25.0)](10^6) = 0.442(10^{-3})$$
$$\varepsilon_{yy} = (1/12.4)(10^{-9})[16.3 - (0.2)(25.0) - (0.2)(12.5)](10^6) = 0.710(10^{-3})$$
$$\varepsilon_{zz} = (1/12.4)(10^{-9})[25.0 - (0.2)(12.5) - (0.2)(16.3)](10^6) = 1.552(10^{-3})$$

The shear strains require the shear modulus G that by the formula

$$G = E/[2(1+\nu)] = 12.4/[2(1+0.2)] = 5.167 \text{ GPa}$$

Hence, the shear strains are

$$\gamma_{xy} = (1/5.167)(10^{-9})(-3.4)(10^6) = -0.6581(10^3)$$

$$\gamma_{yz} = \gamma_{zx} = 0$$

The z-direction shear strains are zero because the corresponding shear stresses are zero.

Example 1.13 Consider a strain rosette mounted on a traction-free surface with three gauges A, B, and C that are separated by 45° angles as shown in the sketch. Suppose the A-gauge coincides with the x-axis, so the C-gauge coincides with the y-axis. Derive a formula for the engineering shear strain γ_{xy}. Recall $\gamma_{xy} = 2\varepsilon_{xy}$ where ε_{xy} is "mathematical" shear strain.

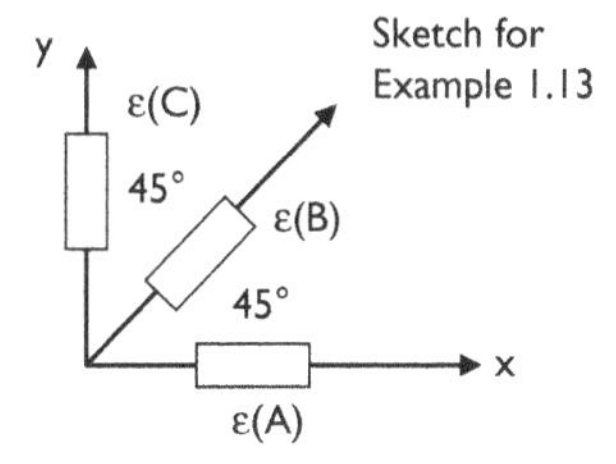

Sketch for Example 1.13

***Solution*:** According to the transformation of strain formulas D.17 in Appendix D

$$\varepsilon(45) = (1/2)(\varepsilon_{xx} + \varepsilon_{yy}) + (1/2)(\varepsilon_{xx} - \varepsilon_{yy})\cos(2\theta) + \varepsilon_{xy}\sin 2(\theta)$$

In this layout, $\theta = 0$, so in view of the A- and C-gauge positions, one has

$$\varepsilon(45) = (1/2)(\varepsilon(0) + \varepsilon(90)) + (1/2)(\varepsilon(0) - \varepsilon(90)\cos(0) + \varepsilon_{xy}\sin 2(0) \therefore$$

$$\gamma_{xy} = -[\varepsilon(0) + \varepsilon(90) - 2\varepsilon(45)]$$

Example 1.14 A three-gauge strain rosette with gauge angles 0-45-90 degrees is attached to a rock outcrop. Overcoring the gauge relieves the stress at the site and yields strains readings:

$$\varepsilon(0) = 1{,}200, \quad \varepsilon(45) = 2{,}400, \quad \varepsilon(90) = -750$$

where tension is positive and the readings are in *micro-strains* (e.g., micro-inches/inch, micro-meters/meter). The rock is isotropic with Young's modulus and Poisson's ratio of 10 GPa and 0.20, respectively. Determine the state of stress and strain at the rosette site.

***Solution*:** First attach a set of reference axes to the gauge site. A convenient choice is to align an x-axis with the 0-gauge and a y-axis with the 90-gauge. The z-axis is then perpendicular to the rosette. Because the rosette is attached to a load-free surface, the z-direction stresses are automatically zero. Thus, $\sigma_{zz} = 0$, $\tau_{zx} = \tau_{xz} = 0$, $\tau_{zy} = \tau_{yz} = 0$. Because the rock is isotropic, zero shear stress implies a corresponding zero shear strain. For example, from Hooke's law $\gamma_{xz} = \tau_{xz}/G$ where G is the rock shear modulus. Thus, the shear strains $\gamma_{xz} = \gamma_{zx} = 0$ and $\gamma_{yz} = \gamma_{zy} = 0$.

The task now is to compute the remaining unknowns, σ_{xx}, σ_{yy}, τ_{xy}, ε_{xx}, ε_{yy}, γ_{xy}, and ε_{zz}. The two normal strains are given directly by $\varepsilon_{xx} = \varepsilon(0) = 1{,}200(10^{-6})$, $\varepsilon_{yy} = \varepsilon(90) = -750(10^{-6})$.

According to formula D.18, the remaining shear strain is given by

$$\gamma_{xy} = -[\varepsilon(0) + \varepsilon(90) - 2\varepsilon(45)]$$
$$= -[1,200 + (-750) - 2(2,400)](10^6)$$
$$\gamma_{xy} = 4,350(10^{-6})$$

Hooke's law may now be used to compute the remaining stresses. Thus for the shear stress,

$$\tau_{xy} = G\gamma_{xy}$$
$$= \{E/[2(1+\nu)]\}(4,350)(10^{-6})$$
$$= \{10(10^9)/[2(1+0.2)]\}(4,350)(10^{-6})$$
$$\tau_{xy} = 18.125 \text{ Mpa}$$

For the two normal stresses remaining

$$\sigma_{xx} = \left(\frac{E}{1-\nu^2}\right)(\varepsilon_{xx} + \nu\varepsilon_{yy})$$
$$= [10(10^9)/(1-0.2^2)][1,200 + 0.2(-750)](10^{-6})$$
$$\sigma_{xx} = 10.94 \text{ Mpa}$$

and

$$\sigma_{yy} = \left(\frac{E}{1-\nu^2}\right)(\varepsilon_{yy} + \nu\varepsilon_{xx})$$
$$= [10(10^9)/(1-0.2^2)][-750 + 0.2(1,200)](10^{-6})$$
$$\sigma_{yy} = -5.31 \text{ Mpa}$$

Finally, the normal strain in the z-direction from Hooke's law is

$$\varepsilon_{zz} = (-\nu/E)(\sigma_{xx} + \sigma_{yy})$$
$$= (-0.2/10)(10^{-9})(10.94 + (-5.31))(10^{+6})$$
$$\varepsilon_{zz} = 113(10^{-6})$$

In summary:

$$\begin{array}{ll}
\varepsilon_{xx} = 1,200(10^{-6}) & \sigma_{xx} = 10.94 \text{ MPa} \\
\varepsilon_{yy} = -750(10^{-6}) & \sigma_{yy} = -5.31 \text{ MPa} \\
\varepsilon_{zz} = 113(10^{-6}) & \sigma_{zz} = 0 \text{ MPa} \\
\gamma_{xy} = 4,350(10^{-6}) & \tau_{xy} = 18.13 \text{ MPa} \\
\gamma_{yz} = 0\tau_{yz} & = 0 \text{ MPa} \\
\gamma_{zx} = 0\tau_{zx} & = 0 \text{ MPa}
\end{array}$$

Comment: The strains are relieved from the overcoring operation and are equal but opposite in sign to the strains *in situ*. The stresses *in situ* are also opposite in sense to those associated with the relieved, that is, measured strains. For example, the x-direction stress *in situ* is a negative (compressive) 10.94 MPa. Tension is considered positive in this rosette data reduction problem because strain meters typically indicate tension as a positive quantity.

Example 1.15 Consider the three-dimensional state of stress relative to xyz coordinates where x=east, y=north, and z=up. Thus in MPa with compression positive,

$$\sigma_{xx} = 12.50, \quad \sigma_{yy} = 16.3, \quad \sigma_{zz} = 25.0, \quad \tau_{xy} = -3.4, \quad \tau_{yz} = 0, \quad \tau_{zx} = 0$$

Find the magnitude and directions of the major, intermediate and minor principal stresses.

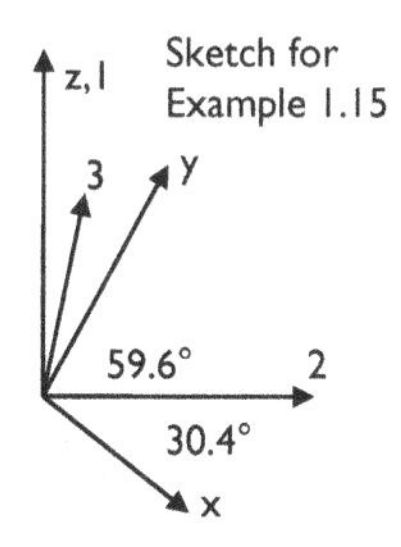

Solution: Because the z-direction shear stresses are zero, the surface perpendicular to the z-axis is a principal plane, that is, the xy plane is a principal plane. The z-direction normal stress is therefore a principal stress, although which one is unknown at this stage. The other two principal stresses are in the xy plane. The directions are given by

$$\begin{aligned} \tan(2\theta) &= \frac{\tau_{xy}}{(1/2)(\sigma_{xx} - \sigma_{yy})} \\ &= (2)(-3.4)/(12.5 - 16.3) \\ \tan(2\theta) &= 1.7895 \quad 2\theta = 60.8^0, \ 240.8^0 \\ \theta &= 30.4^0, \ 120.4^0 \end{aligned}$$

The direction of the maximum stress in the xy plane is 30.4 degrees counter-clockwise from the y-axis. The magnitudes of the maximum and minimum stresses in the xy plane are

$$\left.\begin{matrix}\sigma_1 \\ \sigma_3\end{matrix}\right\} = (1/2/)(\sigma_{xx} + \sigma_{yy}) \pm \sqrt{[(\sigma_{xx} - \sigma_{yy})/2]^2 + (\tau_{xy})^2}$$

$$\left.\begin{matrix}\sigma_1 \\ \sigma_3\end{matrix}\right\} = (1/2/)(12.5 + 16.3) \pm \sqrt{[(12.5 - 16.3)/2]^2 + (-3.4)^2}$$

$$\left.\begin{matrix}\sigma_1 \\ \sigma_3\end{matrix}\right\} = \left.\begin{matrix}18.3 \\ 10.5\end{matrix}\right\} \text{MPa}$$

After reordering the three principal stresses in three-dimensions, the major, intermediate and minor principal stresses are $\sigma_1 = 25.0$, $\sigma_2 = 18.3$, $\sigma_3 = 10.5$ MPa, respectively. The major principal stress is vertical; the *azimuth* (clockwise angle from north) of the intermediate principal is 59.6° (ENE) or 59.6° counter-clockwise from the x-axis, and the *azimuth* of the minor principal stress is −30.4° (NNE).

Example 1.16 Consider the state of stress given in Example 1.15. Find the resultant normal and shear stress on a plane with dip direction $\alpha = 45°$ and dip $\delta = 60°$. Dip is measured positive down, while dip direction corresponds to a horizontal arrow pointing in the direction of the dip.

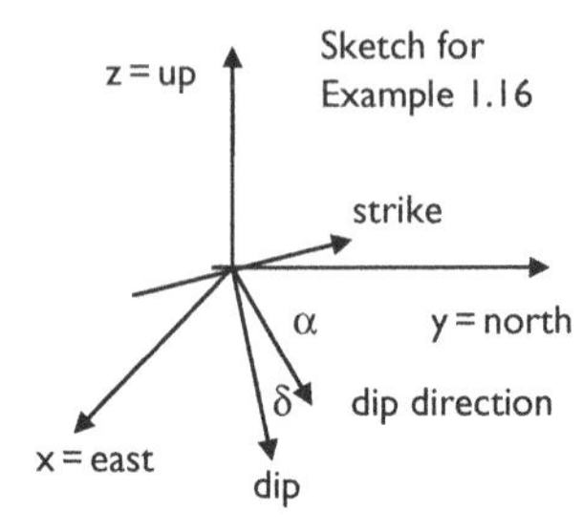

***Solution*:** This problem requires the use of formulas in Appendix D from the section on **Normal and shear stress on a plane**. The first step is to compute the direction cosines *cx, cy, cz* from the given dip and dip direction. Thus,

$$\begin{aligned}
&\cos(cx) = \sin(\delta)\sin(\alpha), && \cos(cy) = \sin(\delta)\cos(\alpha), && \cos(cz) = \cos(\delta)\\
&\cos(cx) = \sin(60)\sin(45), && \cos(cy) = \sin(60)\cos(45), && \cos(cz) = \cos(60)\\
&\cos(cx) = (\sqrt{3}/2)(1/\sqrt{2}), && \cos(cy) = (\sqrt{3}/2)(1/\sqrt{2}), && \cos(cz) = 1/2\\
&\cos(cx) = 0.61237, && \cos(cy) = 0.61237, && \cos(cz) = 0.5000
\end{aligned}$$

The next step is to compute the components of the stress vector. Thus in MPa,

$$\begin{aligned}
T_x &= (12.5)(0.61237) + (-3.4)(0.61237) + (0)(0.5) = 5.573\\
T_y &= (-3.4)(0.61237) + (16.3)(0.61237) + (0)(0.5) = 7.90\\
T_z &= (0)(0.61237) + (0)(0.61237) + (25.0)(0.5) \quad = 12.500
\end{aligned}$$

The resultant stress vector is

$$T = \sqrt{(5.573)^2 + (7.90)^2 + (12.5)^2} = 15.80 \text{ MPa}$$

The normal component is given by the projection of T onto the normal direction. Thus,

$$\begin{aligned}
N &= T_x\cos(cx) + T_y\cos(cy) + T_z\cos(cz)\\
&= (5.57)(0.61237) + (7.90)(0.61237) + (12.5)(0.5)\\
N &= 14.5 \text{ MPa}
\end{aligned}$$

The shear component is then

$$\begin{aligned}
S &= \sqrt{T^2 - N^2}\\
&= \sqrt{(15.8)^2 - (14.5)^2}\\
S &= 6.28 \text{ MPa}
\end{aligned}$$

Example 1.17 Suppose the plane in Example 1.16 is a clean joint that lacks cohesion and has an associated friction angle of 29°. Determine is there is a danger of shear slip on the joint.

***Solution*:** Shear slip is possible if the shear stress τ (stress) exceeds shear strength. The shear stress from Example 1.16 is 6.28 MPa (911 psi). Shear strength is computed as the product on coefficient of friction by normal stress, that is, τ (strength) = (σ)tan(φ). Hence, τ (strength) = (14.5)tan(29) = 8.04 MPa (1,165 psi). Because shear strength exceeds shear stress, joint slip is not expected. One could also compute safety factor fs = τ (strength)/ τ (stress) = 8.04/6.28 = 1.28 indicating safety or no-slip.

Example 1.18 Consider the state of stress in isotropic, elastic ground at depth z caused by gravity loading only. (a) Compute the ratio of horizontal stress to vertical stress (σ_h / σ_v) and the ratio of horizontal strain to vertical strain ($\varepsilon_h / \varepsilon_v$). (b) Then specify these ratios for three cases: (1) the hydrostatic case, (2) the case of no lateral restraint, and (3) the case of complete lateral restraint. The hydrostatic case is an allusion to pressure in a fluid at rest. Lateral restraint refers to an image of a column of rock of height z and unit cross-sectional area and the motion of the sides of the column under application of gravity.

***Solution*:** Stresses and strains are related through Hooke's law and because there is no distinction in horizontal directions under gravity loading, the horizontal stresses and strains are equal. The vertical stress may be computed by considering equilibrium of a rock column of height z and unit cross-sectional area. Thus, W=F where W=column weight and F is the vertical reaction at the column bottom; W= γ V where γ =specific weight of rock and V=column volume (V= zA= z1). Also F= σ_vA= σ_v 1. Hence, $\sigma_v = \gamma_z$. This estimate of vertical stress is often referred to as the "overburden" load or the "lithostatic" stress where γ is the average specific weight of the rock mass overlying the considered point. If γ is given as 144 (lbf/ft^3) and z is in ft, then σ_v =144 lbf/ft^2 per foot of depth, that is, *1 psi per foot of depth*. This estimate is a handy but rough estimate of stress at depth. In SI units, this estimate is *22.6 kPa/m*. Implicit in this estimate is that the vertical and horizontal stresses are principal stresses; shear stresses are absent in the chosen coordinate system (x=east, y=north, z=up). According to Hooke' law with v=Poisson's ratio and E =Young's modulus

$$\varepsilon_{xx} = (1/E)(\sigma_{xx} - v\sigma_{yy} - v\sigma_{zz})$$
$$\varepsilon_{yy} = (1/E)(\sigma_{yy} - v\sigma_{zz} - v\sigma_{xx})$$
$$\varepsilon_{zz} = (1/E)(\sigma_{zz} - v\sigma_{xx} - v\sigma_{yy})$$

that reduces to

$$\varepsilon_{hh} = (1/E)[(\sigma_{hh}(1 - v) - v\sigma_{vv})]$$
$$\varepsilon_{vv} = (1/E)(\sigma_{vv} - 2v\sigma_{hh})$$

Thus,

$$(\varepsilon_{hh}/\varepsilon_{vv}) = (\sigma_{hh}(1 - v) - v\sigma_{vv}]/(\sigma_{vv} - 2v\sigma_{hh}),$$

therefore with the ratios

$$K(\varepsilon) = (\varepsilon_{hh}/\varepsilon_{vv}) \text{ and } K(\sigma) = (\sigma_{hh}/\sigma_{vv})$$

$$K(\varepsilon) = [(1-\nu)K(\sigma) - \nu]/[K(\sigma) - 2\nu] \text{ and}$$
$$K(\sigma) = [K(\varepsilon) + \nu]/[(1-\nu) + 2\nu K(\varepsilon)]$$

Case	*K(ε)*	*K(σ)*
1. Hydrostatic	1	1
2. No lateral restraint	$-\nu$	0
3. Complete restraint	0	$\nu/(1-\nu)$

The last case (3) is often used to estimate the horizontal stress in the gravity field. For example, with $\nu = 0.25$, a reasonable value, then $\sigma_h = (1/3)\sigma_v$.

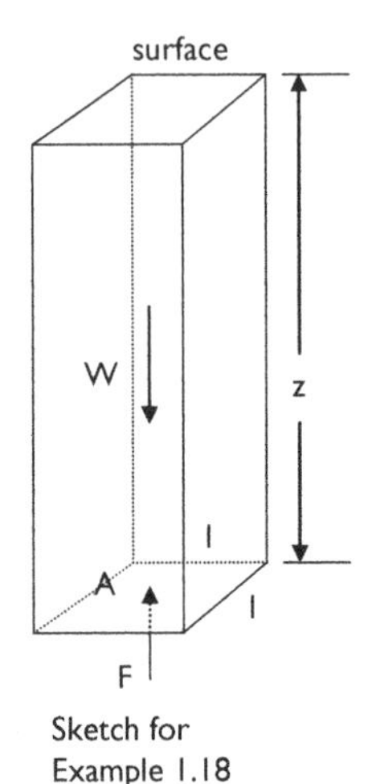

Sketch for Example 1.18

1.5 Problems

Basics

1.1 Identify the three major categories of equations that form the "recipe" for calculating rock mass motion. Give an example of each in equation form.

1.2 If ρ is mass density, explain why or why not the conservation of mass may be expressed as

$$\dot{M} = \int_V \dot{\rho} \mathrm{d}V$$

for a body of mass M occupying volume V.

1.3 Name the two types of external forces recognized in mechanics and give an example of each.

1.4 A huge boulder is inadvertently cast high into the air during an open pit mine blast. Using the definition of mass center and Newton's second law of motion, show that the center of mass of the boulder travels the same path even if it disintegrates into 1,001 pieces of various sizes and shapes.

1.5 Suppose a static factor of safety is defined as the ratio of resisting to driving forces, that is, FS = R/D. Show that a factor of safety less than one implies acceleration is impending.

1.6 Given a cylindrical sandstone laboratory test specimen of diameter D and height L subjected to a vertical load F (force), find (a) the forces N and T acting normal and tangential to a surface inclined at an angle θ to the vertical and (b) stresses, that is, the forces per unit area, σ and τ on the inclined surface in terms of the applied stress σ_o (F/A), where A is the area acted on by F. Illustrate with sketches. Note: The inclined surface is elliptical and has an area πab where a and b are the semi-axes of the ellipse.

1.7 Given a rectangular prism of edge length D and height L subjected to a vertical load F (force) applied to the prism ends and a horizontal pressure p applied to the sides of the prism, find (a) the forces N and T acting normal and tangential to a surface inclined at an angle θ to the vertical, measured in a plane parallel to a side of the prism and (b) stresses, that is, the forces per unit area, σ and τ on the inclined surface in terms of the applied stress σ_o (F/A), and pressure p where A is the area acted on by F. Illustrate with sketches.

1.8 Is the following equation a physical law, kinematic relationship, or material law?

$$Ms = \int_V r\mathrm{d}m$$

Here M = mass in volume V, dm = a mass element, r = position vector of dm, and s is the position of the mass center.

1.9 Is the following equation a physical law, kinematic relationship, or material law?

$$\dot{\sigma} = \frac{\partial \sigma}{\partial t}$$

1.10 Darcy's law for fluid flow in porous media relates the seepage velocity v to the hydraulic gradient h through a constant k (hydraulic conductivity). Thus,

$$v = kh$$

Is this relationship a physical law, kinematic relation, or material law?

Review of stress

1.11 Consider a "two-dimensional" state of stress in the x–y plane characterized by:

$$\sigma_{xx} = 2,500 \quad \sigma_{yy} = 5,200 \quad \tau_{xy} = 3,700$$

where units are psi and tension is positive. Find: the magnitude and direction of the principal stresses and illustrate with a sketch.

1.12 Given the stress state in Problem 1.11, find: the magnitude and direction of the maximum shear stress and illustrate with a sketch.

1.13 Consider a "two-dimensional" state of stress in the x–y plane characterized by:

$$\sigma_{xx} = 17.24 \quad \sigma_{yy} = 35.86 \quad \tau_{xy} = 25.52$$

where units are MPa and tension is positive. Find: the magnitude and direction of the principal stresses and illustrate with a sketch.

1.14 Given the stress state in Problem 1.13, find: the magnitude and direction of the maximum shear stress and illustrate with a sketch.

1.15 Suppose that

$$\sigma_{xx} = 2{,}500 \quad \sigma_{yy} = 5{,}200 \quad \tau_{xy} = 3{,}700$$

and the z-direction shear stresses (τ_{zx}, τ_{zy}) are zero, while the z-direction normal stress (σ_{zz}) is 4,000 psi. Find: the major, intermediate, and minor principal stresses.

1.16 With reference to Problem 1.15, find the magnitude of the maximum shear stress and show by sketch the orientation of the plane it acts on.

1.17 Suppose that

$$\sigma_{xx} = 17.24 \quad \sigma_{yy} = 35.86 \quad \tau_{xy} = 25.52$$

in MPa and the z-direction shear stresses (τ_{zx}, τ_{zy}) are zero, while the z-direction normal stress (σ_{zz}) is 27.59 MPa. Find: the major, intermediate, and minor principal stresses.

1.18 With reference to Problem 1.17, find the magnitude of the maximum shear stress and show by sketch the orientation of the plane it acts on.

1.19 Consider a "two-dimensional" state of stress in the x–y plane characterized by:

$$\sigma_{xx} = 5{,}200 \quad \sigma_{yy} = 2{,}500 \quad \tau_{xy} = -3{,}700$$

where units are psi and tension is positive. Find: the magnitude and direction of the principal stresses and illustrate with a sketch.

1.20 Consider a "two-dimensional" state of stress in the x–y plane characterized by:

$$\sigma_{xx} = 35.86 \quad \sigma_{yy} = 17.24 \quad \tau_{xy} = -25.52$$

where units are MPa and tension is positive. Find: the magnitude and direction of the principal stresses and illustrate with a sketch.

1.21 Consider a three-dimensional state of stress characterized by:

$$\sigma_{xx} = 3{,}000 \quad \sigma_{yy} = 2{,}000 \quad \tau_{xy} = 0.$$
$$\sigma_{zz} = 4{,}000 \quad \tau_{zx} = 0. \quad \tau_{yz} = 2{,}500$$

where units are psi and compression is positive. Find: the magnitude and direction of the principal stresses and illustrate with a sketch.

1.22 Given the stress state in Problem 1.21, find the state of stress relative to axes $abc(\sigma_{aa}, \ldots, \tau_{ca})$ rotated 30° counterclockwise about the z-axis.

1.23 Consider a three-dimensional state of stress characterized by:

$$\sigma_{xx} = 20.69 \quad \sigma_{yy} = 13.79 \quad \tau_{xy} = 0.0$$
$$\sigma_{zz} = 27.59 \quad \tau_{zx} = 0.0 \quad \tau_{yz} = 17.24$$

where units are MPa and compression is positive. Find: the magnitude and direction of the principal stresses and illustrate with a sketch.

1.24 Given the stress state in Problem 1.23, find the state of stress relative to axes $abc(\sigma_{aa}, \ldots, \tau_{ca})$ rotated 30° counterclockwise about the z-axis.

1.25 Show in two dimensions that the mean normal stress $\sigma_m = (\sigma_{xx} + \sigma_{yy})/2$ is invariant under a rotation of the reference axes and is equal to one-half the sum of the major and minor principal stresses in the considered plane. In the context of this problem, invariant means independent of the magnitude of the rotation angle.

1.26 Show in two dimensions that the maximum shear stress $\tau_m = \{[(\sigma_{xx} - \sigma_{yy})/2]^2 + (\tau_{xy})^2\}^{(1/2)}$ is invariant under a rotation of the reference axes and is equal to one-half the difference of the major and minor principal stresses in the considered plane.

1.27 Consider an NX-size drill core (2–1/8 in., 5.40 cm diameter) with an L/D ratio (length to diameter) of two under an axial load of 35,466 pounds (158.89 kN). Find:

1 The state of stress within the core sample relative to a cylindrical system of coordinates ($r\theta z$).
2 The state of stress relative to an abc set of axes associated with a plane that bears due north and dips 60° to the east. The c-axis is normal to this plane, the b-axis is parallel to strike and the a-axis is down dip. Use the equations of transformation.
3 Show by a direct resolution of force components and calculation of area that the normal stress (σ_{cc}) on the plane in part (2) is the same as that given by the equations of transformation.
4 Is the shear stress (τ_{ac}) acting on the plane in part (2) the same using forces and areas as that given by the equations of transformation? Show why or why not.

Review of strain and elasticity

1.28 Strain measurements are made on a wide, flat bench in a dimension stone quarry using a 0-45-90 rosette. The 0-gauge is oriented N60E. Specific weight of the stone (a granite) is 162 pcf (25.6 kN/m^3); Young's modulus $E = 12.7(10)^6$ psi (87.59 GPa) and Poisson's ratio $v = 0.27$. Tension is considered positive. Measured strains in microunits per unit are:

$$\epsilon(0) = -1480, \quad \epsilon(45) = -300, \quad \epsilon(90) = -2760$$

Find (where x = east, y = north, and z = up):

1 the strains $\epsilon_{xx}, \epsilon_{yy}, \epsilon_{xy}$ (tensorial shear strain);
2 the stresses $\sigma_{zz}, \tau_{yz}, \tau_{zx}$ (in psi or MPa);
3 the stresses σ_{xx}, σ_{yy}, and τ_{yx} (in psi or MPa);
4 the strain ϵ_{zz} (microinches per inch or mm/m);
5 the strains ϵ_{yz} and ϵ_{zx} (tensorial shear strains);
6 the direction of the true principal stresses $\sigma_1, \sigma_2, \sigma_3$, where $\sigma_1 \geq \sigma_2 \geq \sigma_3$ (tension is positive) and sketch;
7 the magnitudes of the principal stresses $\sigma_1, \sigma_2, \sigma_3$; where $\sigma_1 \geq \sigma_2 \geq \sigma_3$ (tension is positive);
8 the magnitudes of the principal strains $\epsilon_1, \epsilon_2, \epsilon_3$;
9 the directions of the principal strains; and sketch;
10 the change ϵ_v in volume per unit volume that would occur if the stresses were entirely relieved, that is reduced to zero, for example, by overcoring the rosette.

1.29 Given the stresses in psi:

$$\sigma_{xx} = 1{,}500 \quad \sigma_{yy} = -2{,}000 \quad \sigma_{zz} = 3{,}500$$
$$\tau_{xy} = 600 \quad \tau_{yz} = -300 \quad \tau_{zx} = -500$$

where x = east, y = north, z = up, compression is positive, find: the secondary principal stresses in the zx-plane and sketch.

1.30 Given the stresses in MPa:

$$\sigma_{xx} = 10.35 \quad \sigma_{yy} = -13.79 \quad \sigma_{zz} = 24.14$$
$$\tau_{xy} = 4.14 \quad \tau_{yz} = -2.07 \quad \tau_{zx} = -3.45$$

where x = east, y = north, z = up, compression is positive, find: the secondary principal stresses in the zx-plane and sketch.

1.31 Given the strains:

$$\epsilon_{xx} = 2{,}000 \quad \epsilon_{yy} = 3{,}000 \quad \epsilon_{zz} = 4{,}500$$
$$\gamma_{xy} = -200 \quad \gamma_{yz} = 300 \quad \gamma_{zx} = 225$$

where x = east, y = north, z = up, compression is positive, the units are microinches per inch, Young's modulus $E = 5.0(10)^6$ psi, and shear modulus $G = 2.0(10)^6$ psi, find the corresponding stresses for a linear, homogeneous, isotropic elastic response.

1.32 Consider a cylindrical test specimen under a confining pressure of 3,000 psi and an axial stress of 3,000 psi with compression positive, so that in cylindrical coordinates

$$\sigma_{zz} = 3{,}000 \quad \sigma_{rr} = 3{,}000 \quad \sigma_{\theta\theta} = 3{,}000$$
$$\tau_{rz} = 0.0 \quad \tau_{z\theta} = 0.0 \quad \tau_{\theta r} = 0.0$$

Find:

1 $\epsilon_{rr}, \epsilon_{zz}, \epsilon_{\theta\theta}, \gamma_{rz}, \gamma_{z\theta}, \gamma_{\theta r}$;
2 the axial stress σ_{zz} required to maintain a zero axial strain;
3 the strain energy and strain energy density, if the test specimen is an NX core with a height to diameter ratio of two. Note: $E = 2.4 \times 10^6$ psi, $v = 0.20$, and the sample is isotropic.

1.33 Given the strains:

$$\epsilon_{xx} = 2{,}000 \quad \epsilon_{yy} = 3{,}000 \quad \epsilon_{zz} = 4{,}500$$
$$\gamma_{xy} = -200 \quad \tau_{yz} = 300 \quad \gamma_{zx} = 225$$

where x = east, y = north, z = up, compression is positive, the units are microinches per inch, Young's modulus E = 34.48 GPa, and shear modulus G = 13.79 GPa psi, find the corresponding stresses for a linear, homogeneous, isotropic elastic response.

1.34 Consider a cylindrical test specimen under a confining pressure of 20.69 MPa and an axial stress of 20.69 MPa with compression positive, so that in cylindrical coordinates

$$\sigma_{zz} = 20.69 \quad \sigma_{rr} = 20.69 \quad \sigma_{\theta\theta} = 20.69$$

$\tau_{rz} = 0.0 \qquad \tau_{z\theta} = 0.0 \qquad \tau_{\theta r} = 0.0$

Find:

1 $\epsilon_{rr}, \epsilon_{zz}, \epsilon_{\theta\theta}, \gamma_{rz}, \gamma_{z\theta}, \gamma_{\theta r}$;
2 the axial stress σ_{zz} required to maintain a zero axial strain;
3 the strain energy and strain energy density, if the test specimen is an NX core with a height to diameter ratio of two. Note: $E = 16.55$ GPa, $v = 0.20$, and the sample is isotropic.

1.35 Show that under complete lateral restraint and gravity loading only, that the vertical normal stress in a homogeneous, isotropic linearly elastic earth at a depth z measured from the surface with compression positive is γz and that the horizontal normal stresses are equal to $v/(1 - v)$ times the vertical stress where v is Poisson's ratio.

1.36 Suppose an NX-size test cylinder with an L/D ratio of two has a Young's modulus of 10 million psi, a Poisson's ratio of 0.35 and fails in uniaxial compressive at 0.1% strain. Find:

1 the axial load (kN) and stress (unconfined compressive strength, psi) at failure;
2 the relative displacement of the cylinder ends at failure (in inches).

1.37 Suppose an NX-size test cylinder with an L/D ratio of two has a Young's modulus of 68.97 GPa, a Poisson's ratio of 0.35, and fails in uniaxial compressive at 0.1% strain. Find:

1 the axial load (kN) and stress (unconfined compressive strength, MPa) at failure;
2 the relative displacement of the cylinder ends at failure (in mm).

1.38 Consider gravity loading under complete lateral restraint in flat, stratified ground where each stratum is homogeneous, isotropic, and linearly elastic. Assume compression is positive, v is Poisson's ratio, z is depth, and γ is *average* specific weight to any particular depth. A 250 ft thick, water-bearing sandstone is encountered at a depth of 1,300 ft. Water pressure at the top of the sandstone is 80 psi. Estimate the total and effective stresses at the center of the sandstone.

1.39 Consider gravity loading under complete lateral restraint in flat, stratified ground where each stratum is homogeneous, isotropic, and linearly elastic. Assume compression is positive, v is Poisson's ratio, z is depth, and γ is *average* specific weight to any particular depth. A 76.2 m thick, water-bearing sandstone is encountered at a depth of 396 m. Water pressure at the top of the sandstone is 552 kPa. Estimate the total and effective stresses at the center of the sandstone.

1.40 Consider gravity loading only under complete lateral restraint in flat strata with properties given in Table 1.2. Vertical stress at the top of the geologic column given in Table 1.2 is 1,000 psi. Compression is positive, γ is specific weight, h is thickness, E is Young's modulus, and G is shear modulus. Find the horizontal stress at the bottom of the sandstone and the top of the limestone.

1.41 Consider gravity loading only under complete lateral restraint in flat strata with properties given in Table 1.3. Vertical stress at the top of the geologic column given in Table 1.3 is 6.9 MPa. Compression is positive, SG is specific gravity, h is thickness, E is Young's modulus, and G is shear modulus. Find the horizontal stress at the bottom of the sandstone and the top of the limestone.

Table 1.2 Strata properties for Problem 1.40

Property stratum	γ *(pcf)*	h *(ft)*	E *(10^6 psi)*	G *(10^6 psi)*
Shale	134	120	4.52	1.83
Coal	95	25	0.35	0.13
Sandstone	148	50	3.83	1.53
Limestone	157	74	5.72	2.40
Mudstone	151	133	7.41	2.79

Table 1.3 Data for Problem 1.41

Property stratum	SG *(–)*	h *(m)*	E *(GPa)*	G *(GPa)*
Shale	2.15	36.6	31.2	12.6
Coal	1.52	7.6	2.41	0.9
Sandstone	2.37	15.2	26.4	10.6
Limestone	2.52	22.6	39.4	16.6
Mudstone	2.42	40.5	51.1	19.2

1.42 A stress measurement made in a deep borehole at a depth of 500 m (1,640 ft) indicates a horizontal stress that is three times the vertical stress and that the vertical stress is consistent with an estimate based on overburden weight alone, that is, 25 kPa per meter of depth (1.1 psi per ft of depth). Estimate the portions of horizontal stress associated with weight alone and with other forces.

Additional problems

1.43 Consider an isotropic elastic solid under plane strain conditions, so $\varepsilon_{zz} = \gamma_{yz} = \gamma_{zx} = 0$. Show that the directions of principal strain and stress coincide.

1.44 Show that the readings of a 4-gauge strain rosette with gauges at 0, 45, 90, and 135 degrees from the 0-gauge provide a fast, convenient check on experimental data from the rosette because the sum of readings from the 0- and 90-gauges should equal the sum of readings from the 45- and 135-gauges.

1.45 Explain how the data reduction procedure for a three-gauge 0–45-90 rosette is changed when attached to a surface under fluid pressure p from the case when the rosette is attached to a traction-free surface. The material is isotropic with Young's modulus E, shear modulus G, and Poisson's ratio ν.

Chapter 2

Slope stability

Stability of slopes is an important engineering topic because of economic impact on mining and civil enterprises. Failures may be costly in the extreme and have tragic personal consequences. Steep slopes are favorable to the economics of surface mines, while low slope angles favor stability. The trade-off between these two trends almost always results in some slope failures in large, open pit mines. In civil works such as highway cuts, slope failures cannot be tolerated because of the threat to public safety.

Slopes may be classified in a number of ways, for example, as rock slopes or soil slopes. While description of the two share fundamental features such as physical laws, methods of analysis are generally different. Of course, weathered rock slopes grade into soil slopes, so there may not always be a clear distinction between the two and some consideration must be given to the most appropriate analysis method. Slopes may also be classified according to an expected mode of failure, say, by translation or rotation. High strength rock slope failures initially are often rigid-body like translational motions, while low strength soil slopes often begin to fail in rigid-body like rotations. Continued failure generally results in disintegration of the slide mass in any case.

Intact rock tested in the laboratory typically has unconfined compressive strength of the order of 10–20 MPa (14,500–30,000 psi). Surface mine depth is of the order of 300 m (1,000 ft). Even after allowance for stress concentration at the toe of a surface mine slope, stress is unlikely to exceed strength of intact rock. For this reason, structural discontinuities that are generally present in rock masses, especially rock masses that host ore bodies, are important to rock slope stability. Faults, joints, bedding planes, fractures ("joints" for brevity), and contacts between different rock types, are essential considerations in stability analysis of rock slopes. Translational sliding along joints is the most common form of rock slope failure; slopes in soils tend to fail by rotation, although in densely jointed rock masses rotational failure is certainly possible. Mine waste dumps, tailings dams, earthen embankments, road fills, storage piles of broken rock, aggregate, and so forth, are soil-like in mechanical behavior, are usually much weaker than rock slopes, and also tend to fail by rotation. In any case, major determinants of slope stability are slope height and angle. Water is another factor of major importance to slope stability analyses. All are subject to some engineering control.

A primary objective of slope stability analysis is estimation of a factor of safety for the considered slope and slide mass. An intuitive formulation of a safety factor appropriate to translational sliding is the ratio of resisting to driving forces acting parallel to the direction of translation as shown in Figure 2.1. Thus, for translational slides,

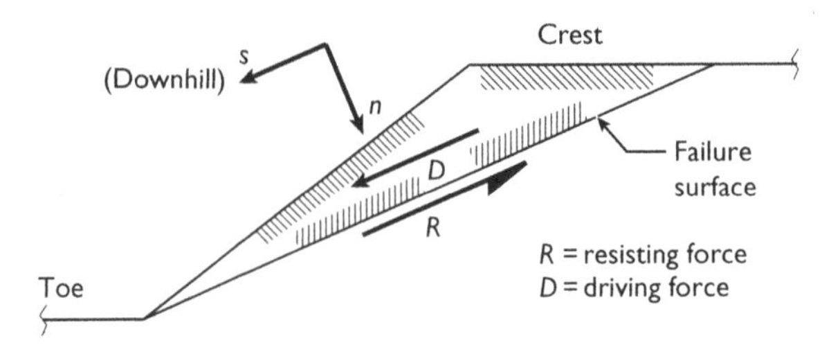

Figure 2.1 Resisting and driving forces during translational slope failure.

$$\mathrm{FS} = \frac{R}{D} \tag{2.1}$$

where R and D are the total resisting and driving forces. In this regard, a sign convention is needed that determines whether a force is driving or resisting. The convention is simple: forces that act downhill in the direction of the potential slide are driving forces; uphill forces are resisting. Forces directed uphill should not be subtracted from downhill driving forces, nor should downhill forces reduce resisting forces. Otherwise a different value of safety factor results. Normal forces acting into the slope are also positive.

The reason for this uphill–downhill force convention and safety factor definition is associated with slide mass motion or acceleration. Motion impends if the resultant of external forces is greater than zero. In this case, according to Newton's second law of motion,

$$F = \dot{P} = ma$$

where F, dP/dt, m, and a are resultant of external forces, time rate of change of linear momentum, slide mass, and acceleration of the slide mass center, respectively. The resultant of external forces may be decomposed into downhill and uphill forces based on the sign convention adopted in the safety factor formula, that is, $F = D - R$. With this decomposition and the expression for safety factor, one has for acceleration

$$D(1 - \mathrm{FS}) = ma$$

Clearly, a safety factor less than one indicates a positive, downhill acceleration. A safety factor greater than one indicates a negative acceleration that for a slide mass at rest is physically meaningless; the slide mass would remain at rest. If the slide mass were moving downhill, then a negative acceleration would indicate a reduction in velocity. Conceivably, a slide mass could be moving uphill after gaining momentum from a previous downhill motion. However, the factor of safety concept is not intended for *slide dynamics*; such questions are better addressed directly through Newton's second law.

The factor of safety defined here is a *global* factor of safety in contrast to a *local* factor of safety (FS) that is defined as the ratio of strength to stress at a point. A local factor of safety less than one indicates the elastic limit may be exceeded, yielding, and failure ensue, at the considered *point*. A global safety factor less than one implies yielding and failure over an extended failure surface and is indicative of *collapse*.

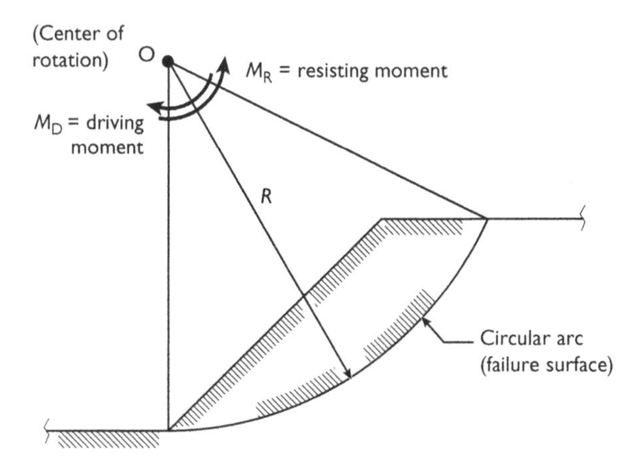

Figure 2.2 Resisting and driving moments during conventional rotational slope failure.

In the case of rotational sliding, as shown in Figure 2.2,

$$M = \dot{H}$$

$$\mathrm{FS} = \frac{M_\mathrm{R}}{M_\mathrm{D}} \tag{2.2}$$

$$M_\mathrm{D}(1 - \mathrm{FS}) = \dot{H}$$

where M, dH/dt, M_R, M_D are resultant of external moments, time rate of change of angular momentum, resisting moment total, and driving moment total, respectively. A partition of external moments into driving and resisting moments that depends on direction of action is made in the equation of motion. Resisting moments are then eliminated from the equation of motion using the safety factor definition. Again, a safety factor less than one indicates potential angular acceleration of the slide mass from rest, while a safety factor greater than one indicates stability.

Other definitions of safety factor are possible, but without adherence to a sign convention that is consistent with driving and resisting directions, interpretation with respect to acceleration cannot be justified. The physical implication is then uncertain and usefulness in doubt.

What constitutes an adequate safety factor is often a matter of engineering judgment, but may also be required by construction codes. In surface mining and cut rock slopes, safety factors are usually near one. Soil-like slopes such as waste rock dumps may require a safety factor of 1.5 or so. In any case, if an estimated safety factor is too low for the problem at hand, then *remedial measures* are required. There are always two choices for improving a factor of safety (1) increase resistance or (2) decrease driving forces or moments.

A change in parameters of an analysis dictates a recalculation of the safety factor, of course. As design changes are explored, the burden of calculation grows and with it justification for automated analysis using computer programs. Some programs, complete with graphical displays, are available from software firms that often offer training in program use. Other programs are proprietary, while still others are available as freeware. But regardless of source, intelligent, effective use of computer programs for slope stability analysis and design requires an understanding of basic slope mechanics and safety factor definitions.

2.1 Translational rock slope failures

Two important types of translational rock slope failures are (1) *planar block slides* and (2) *wedge slides*. The term "slide" is somewhat misleading because it seems to imply frictional sliding of a rigid block down an inclined plane. Although there are similarities in analysis, the mechanism of rock mass motion during a slope failure is complex and far from that of simple "sliding."

Planar block slides

Planar block slides are possible when "joints" dip into a surface excavation and the excavation is long relative to height. Figure 2.3 shows a potential planar block slide in cross section that has a possible tension crack near the crest of the slope. A Mohr–Coulomb (MC) strength criterion is reasonable for many rock and soil slopes. With the assumption of MC, the planar block slide safety factor is

$$FS_p = \frac{N'\tan(\phi) + C + T_b}{D} \tag{2.3}$$

where ϕ is the rock mass angle of internal friction. This simple formula masks several practical difficulties that include determining a likely failure surface and the strength properties of the considered surface. These difficulties must be resolved before proceeding with the computation of resisting and driving forces.

Generally, geologic structure and possibly previous slide histories in the vicinity of a considered slide site provide guidance to likely failure surfaces, while strength properties are estimated through a combination of rock and joint properties testing and site examination. The former leads to friction angle and cohesion for joints (ϕ_j, c_j) and intact rock between joints (ϕ_r, c_r). The latter not only provides statistical data for joint orientation and spacing, but importantly, estimates of joint persistence. Persistence p is defined as the ratio of joint area A_j in a joint plane to total area A composed of joint area plus area of intact rock bridges A_r between joint segments ($A = A_j + A_r$), that is, $p = A_j/A$. A simple weighted average for rock mass properties may be obtained from:

$$\begin{aligned} \tan(\phi) &= (1-p)\tan(\phi_r) + (p)\tan(\phi_j) \\ c &= (1-p)c_r + (p)c_j \end{aligned} \tag{2.4}$$

If the failure surface is entirely joint, then $p = 1$. In this case, slope failure is very much like a rock block sliding down an inclined plane. A special case occurs when one assumes that intact rock and joint friction angles are equal and joint cohesion is negligible. This case leads to the well-known Terzaghi model of jointed rock mass strength where $\phi = \phi_r = \phi_j$ and $c = (1 - p)c_r = (A_r/A)c_r$. The range of rock mass friction angles is relatively large; as a rough guide $25° < \phi < 55°$. Cohesion estimates are more difficult because of the dependency on intact rock cohesion and joint persistence. Joints of large areal extent are likely to

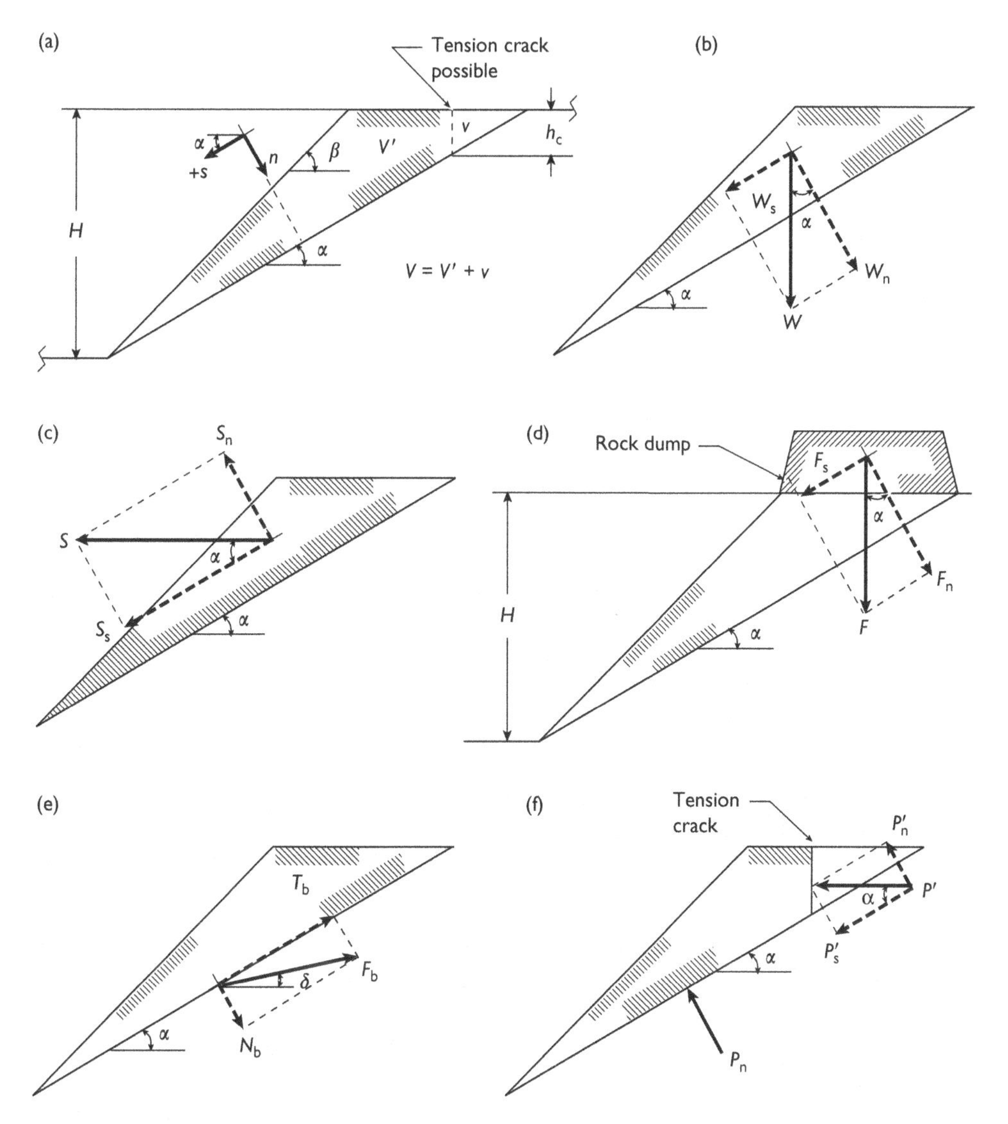

Figure 2.3 Planar block slide volumes and forces: (a) volumes, (b) weight, (c) seismic–dynamic force, (d) surcharge force, (e) bolt forces, and (f) water forces.

have low or negligible cohesion, while cohesion of intact rock is highly variable. As a rough guide and in consideration of the Terzaghi model, rock mass cohesion may be an order of magnitude less than intact rock cohesion.

Example 2.1 Consider a rock mass where field mapping indicates joint persistence is 0.87 and laboratory tests on intact rock and joint samples show that $\phi_r = 54°$, $c_r = 2,750$ psi, $\phi_j = 32°$, $c_j = 25$ psi. Estimate the rock mass cohesion and angle of internal friction.

***Solution*:** By definition, persistence is the ratio of joint area to area of joints and intact rock bridges in the joint plane, and according to (2.4)

$$\tan(\phi) = (1 - p)\tan(\phi_r) + (p)\tan(\phi_j)$$

$$= (1 - 0.87)\tan(54) + 0.87\tan(32)$$

$$\tan(\phi) = 0.7226$$

$$\therefore$$

$$\phi = 36^\circ$$

$$c = (1 - 0.87)2,750 + 0.87(25) = 379 \text{ psi}$$

Also shown in Figure 2.3 are various forces to be considered in analysis. Weight is always present, so there is always a driving force, but other forces may be absent. Driving forces include the downhill components of:

1 slide mass weight W_s;
2 seismic force S_s;
3 surcharge force F_s;
4 water force P'_s.

The total driving force is

$$D = W_s + S_s + F_s + P'_s \tag{2.5}$$

If there is no tension crack at the crest, then no downhill water force acts, that is, $P'_s = 0$. If no surcharge is present, then $F_s = 0$, and if seismic activity is not a concern, then $S_s = 0$ also. Generally, slope failures are gravity-driven and the downhill component of weight is most important. Equilibrium in the downhill direction requires resisting forces to be at least equal to D, that is, a safety factor of at least one.

Transient forces associated with blasting could be added to the list of forces to be considered. Some measure of the acceleration induced by a blast would then be needed in order to quantify the effect on the safety factor. One way of doing so is to treat a blast transient in the same manner as a seismic load. In effect then, blast transients are subsumed here under seismic load.

Resisting forces include uphill components of frictional resistance mobilized by normal loads, cohesive resistance, and direct resistance from reinforcement (bolting):

1 slide mass weight W_n;
2 seismic force S_n;
3 surcharge force F_n;
4 reinforcement normal force N_b;
5 reinforcement tangential force T_b;
6 cohesive force C;
7 water forces P_n, P'_n.

Normal forces are summed algebraically to obtain the net normal force available to mobilize frictional resistance to sliding. This net normal force is, in fact, the *effective* normal force and may be obtained from an equilibrium analysis of forces acting in the normal direction. Figure 2.3 shows the positive normal direction which is perpendicular to the potential slope failure surface (not the slope face) and directed into the slope. The positive shear direction is down slope and parallel to the potential failure surface. In the normal direction, equilibrium requires

$$N' = W_n - S_n + F_n + N_b - P_n - P'_n \tag{2.6}$$

where N' is the normal reaction of the rock mass below the failure surface to the slide mass above.

Computation of the various forces that enter the safety factor calculation begins with the most important force which is weight of the slide mass W. Although this calculation is straightforward and is simply the product of specific weight γ times slide volume V, that is, $W = \gamma V$, calculation of the volume using slope height and angles that define the failure surface and slope face is helpful in bringing these variable into an explicit relationship. Thus,

$$V = \left(\frac{H^2 b}{2}\right)[\cot(\alpha) - \cot(\beta)] \tag{2.7}$$

where H = slope height (vertical distance from toe to crest), b = breadth of slope (arbitrary distance perpendicular to the cross-section), α = failure surface angle measured from the horizontal, β = slope angle (face angle) measured from the horizontal. Generally, the slope angle enters the factor of safety calculation only through slide mass volume. Breadth is often taken to be one distance unit, say, one meter or foot, but it may also be convenient to use breadth as the horizontal spacing between fans of bolts when used to reinforce a slope. Face angle β is under some engineering control. The failure surface angle α is usually dictated by unfavorably oriented geologic structure, for example, bedding plans that dip into the excavation.

Example 2.2 Consider a planar block slide without a tension crack and loaded by gravity only. Slope face angle β is 45°; slope height H is 600 ft. A potential failure surface is inclined 35° (α) from the horizontal. Cohesion c of the failure surface is estimated to be 300 psi, while the friction angle ϕ is estimated to be 30°. Specific weight γ of the block is 150 pcf. Determine the slope safety factor FS when the water table is below the slope toe.

***Solution*:** Sketch and then draw a free body diagram of the slide mass.

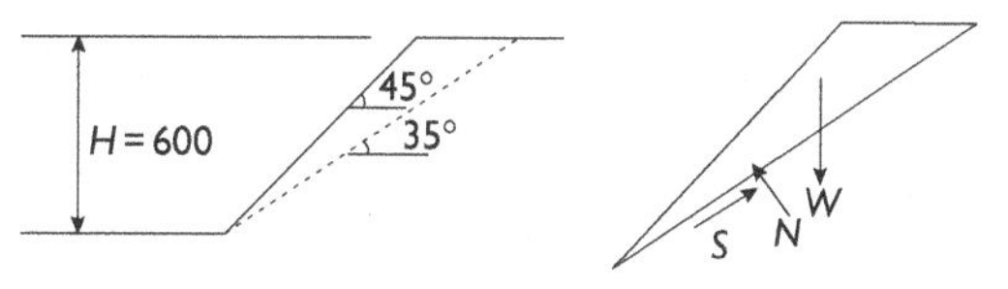

Sketch Free body diagram

By definition

$FS = R/D$,	R = forces resisting,	D = forces driving,
$R = N\tan(\phi) + C$,	N = normal force,	$C = cA$, cohesive force,
	$N = W\cos(\alpha)$,	$A = H/\sin(\alpha)$,
	$W = \gamma V$,	V = volume.

One then has

$$
\begin{aligned}
V &= \frac{bH^2}{2}[\cot(\alpha) - \cot(\beta)] \\
&= \frac{(1)(600)^2}{2}[1.4281 - 1.0000] \\
V &= 7.707(10^4)\text{ft}^3 \\
W &= \gamma V \\
&= 150(7.707)(10^4) \\
W &= 11.57(10^6)\text{lbf}
\end{aligned}
$$

and

$$
\begin{aligned}
N &= 11.56(10^6)\cos(35^\circ) \\
N &= 9.469(10^6)\text{ lbf} \\
C &= (300)(12)\left[\frac{(600)(12)}{\sin(35^\circ)}\right] \\
C &= 45.190(10^6)\text{ lbf} \\
D &= W\sin(\alpha) \\
&= 11.56(10^6)\sin(35^\circ) \\
D &= 6.631(10^6)\text{ lbf} \\
\text{FS} &= \frac{N\tan(\phi) + C}{D} \\
&= \frac{9.469(10^6)\tan(30^\circ) + 45.190(10^6)}{6.631(10^6)} \\
&= \frac{5.467 + 45.190}{6.63} \\
\text{FS} &= 7.64
\end{aligned}
$$

In these calculations, the slide mass is considered to be one foot thick into the page, that is, $b = 1$ ft. N and S are normal and shear reactions, that is, they replace the effect of the material removed from the material remaining in the free body diagram. N is obtained from analysis of equilibrium in the normal direction and is simply the normal component of weight. The downhill or driving force D is the tangential component of weight. The tangential or shear reaction S does not enter the safety factor calculation but would be equal to D in order to satisfy equilibrium in the downhill direction.

The safety factor is high because of the cohesion which makes a contribution to the resisting force that is about nine times that of the frictional resistance. If the cohesion were only one-tenth of the given value, the safety factor would be about 1.5, but if there were no cohesion, the slide mass would fail. Generally speaking, a small amount of cohesion on a potential failure surface makes a large difference in the safety factor of a planar block slide and should be given careful consideration in a design analysis.

When a tension crack is present a small volume "correction" is needed. This volume is

$$v = \left(\frac{h_c^2 b}{2}\right) \cot(\alpha) \tag{2.8}$$

and should be subtracted from V when a tension crack is present at a depth h_c as shown in Figure 2.3.

Seismic force S is an inertia force that is considered to act horizontally and directed away from the slope into the excavation. Magnitude of S is related to weight by a seismic coefficient a_s that is a decimal fraction of the acceleration of gravity g. This coefficient depends on the seismic zone where the excavation is located and ranges from 0.0 to 0.2. Thus,

$$S = ma = \left(\frac{W}{g}\right) a_s g = W a_s \tag{2.9}$$

where m and a are slide mass and acceleration, respectively.

Example 2.3 Consider the slope shown in the sketch and further suppose a seismic force should be considered in a zone where the seismic coefficient is 0.10. Determine the maximum face angle in consideration of this seismic loading.

Solution: The free body diagram now has a horizontal load S that represents seismic loading in a quasistatic manner.

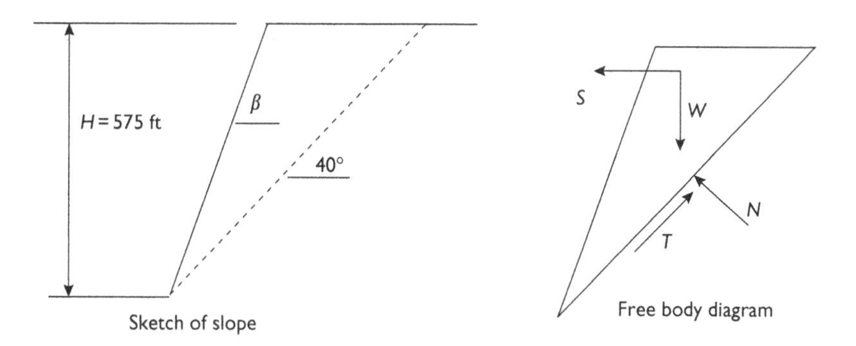

The face angle β occurs only in the weight W, but $S = ma$ where m = slide mass and a = acceleration, that is, $S = (W/g)(0.1\ g) = 0.1\ W$. Resolution of forces normal and tangential to the failure surface leads to the resisting forces R and the driving forces D. Thus,

$$R = [W\cos(\alpha) - 0.1W\sin(\alpha)]\tan(\phi) + C$$

$$D = W\sin(\alpha) + 0.1W\cos(\alpha)$$

The safety factor is one, so $R = D$ and

$$W[\sin(40) + (0.1)\cos(40) - \cos(40)\tan(32) + (0.1)\sin(40)\tan(32)] = 2.576(10^6)\text{ lbf}$$

$$\therefore$$

$$W = 9.172(10^6)\text{ lbf}$$

$$\therefore$$

$$\cot(\beta) = \cot(40) - \frac{9.172(10^6)}{(1/2)(158)(1)(575)^2} = 0.8406$$

$$\therefore$$

$$\beta = 49.9^\circ$$

In this case, a seismic design results in a significant reduction in slope angle.

The force F caused by a surcharge, for example, from a rock dump near the crest, is simply the weight of the dump. The dump may extend beyond the point where the slide surface intersects the slope crest, but only the portion of the dump bearing on the crest loads the slide mass and contributes to the surcharge force. Other sources of surcharge are possible, for example, from equipment traveling near the crest. If so, then equipment weight would constitute a surcharge.

Example 2.4 Given the possible planar block slide shown in the sketch and the associated free body diagram with $\alpha = 35°, \beta = 45°, c = 30$ psi, $\phi = 28°, \gamma = 156, H = 400$ ft, determine the slope safety factor without a surcharge and with a surcharge from a stockpile of broken rock 50 ft high with a specific weight of 96 pcf.

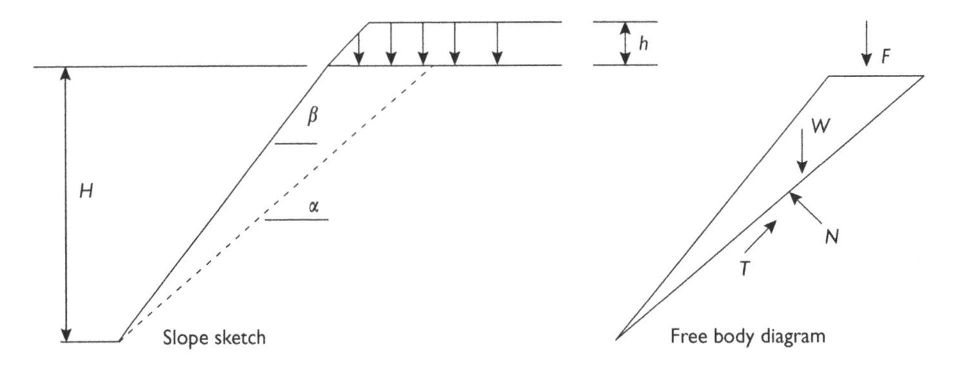

***Solution*:** (a) No surcharge.

$$W = \left(\frac{1}{2}\right) b\gamma H^2 \left[\cot(\alpha) - \cot(\beta)\right] = \left(\frac{1}{2}\right)(1)(156)(400)^2 \left[\frac{1}{\tan(35)} - \frac{1}{\tan(45)}\right]$$

$$W = 5.343(10^6)\ \text{lbf}$$

$$\text{FS} = \frac{5.343(10^6)\cos(35)\tan(28) + (30)(144)(1)(400/\sin 35)}{5.343(10^6)\sin(35)}$$

$$\text{FS} = 1.74$$

(b) With a surcharge.
The surcharge is estimated as a uniformly distributed vertical load f caused by weight of the stockpile material. Thus, $f = (96)(50) = 4800$ psf. The force associated with this load is the product of load times area of action. Although the load extends past the failure surface intercept at the crest, only the portion above the failure surface acts on the slide mass. This area is A given by $(b)(H)[(\cot(\alpha) - \cot(\beta)] = (1)(400)[1/\tan(35) - 1/\tan(45)]$, that is, $A = 171.3$ sqft. The surcharge $F = (171.3)(4, 800) = 0.822(10)^6$ lbf and has components acting normal and tangential to the failure surface, F_n and F_s, respectively. These are

$$F_n = 0.822(10^6)\cos(35) = 0.673(10^6)$$

$$F_s = 0.822(10^6)\sin(35) = 0.472(10^6)$$

The safety factor with surcharge is

$$\text{FS} = \frac{5.339(10^6) + 0.673(10^6)\tan(28)}{3.064(10^6) + 0.472(10^6)}$$

$$= \frac{5.697}{3.536}$$

$$\text{FS} = 1.61$$

Thus, the surcharge in this case is detrimental to stability. However, this result may not always be the case because the surcharge adds to the resisting and driving forces. Mathematically, there would seem to be a possibility that the resistance will increase more than the driving force increase to an extent that the slope safety factor actually increases.

Bolting force F_b is a total bolting force in the slide section that has breadth b. This force is the sum of forces from all bolts installed in the slope that intersect the considered failure surface. If each bolt hole has force f_b and there are n holes in rows spaced b ft apart horizontally, then the total bolt force $F_b = nf_b$, provided all holes have the same inclination angle δ, as shown in Figure 2.3.

Water forces act perpendicular to the surface of contact. Figure 2.3 shows the water force P_n acting on the inclined portion of a failure surface. This water force is purely normal and has no downhill component, of course. When a tension crack is present and the water table is above the bottom of the tension crack, then a horizontal water force P' acts on the tension crack. This force has normal and tangential components, as shown in Figure 2.3. The normal components of water forces are *uplift* forces that act in the negative normal direction and reduce the normal force transmitted across the failure surface. Again, the net normal force transmitted is the effective normal force and is simply the reaction N'.

Frictional resistance is mobilized by the effective normal force and is

$$R_f = N' \tan(\phi) \tag{2.10}$$

In principle, the effective normal force could be negative in the presence of sufficient uplift force from water and possibly from seismic or dynamic loading. If N' is negative, then frictional resistance is zero.

Cohesive resistance is the product of rock mass cohesion c and surface area A of the inclined failure plane. If a tension crack is present, cohesion is destroyed and because of the separate surfaces defining the crack, no frictional resistance is possible. Thus, cohesive resistance is simply

$$C = cA \tag{2.11}$$

where $A = bH / \sin(\alpha)$ in the absence of a tension crack. When a tension crack is present, then $A = b(H - h_c)/ \sin(\alpha)$, as seen from the geometry in Figure 2.3. Uplift forces that lead to a negative N' are also likely to destroy cohesion along the inclined portion of the failure surface with the net consequence of eliminating all natural resistance to failure. Some form of reinforcement would then be necessary for stability.

Water force P_n requires knowledge of the distribution of water pressure over the wetted portion of the failure surface. In principle, then

$$P_n = \int_{A'} p\mathrm{d}A$$

where A' is the wet portion of the failure surface and p is position-dependent water pressure. A reasonable assumption is that p does not vary significantly across the breadth of the failure surface, but varies along the failure surface, so

$$P_n = \int_L pb\mathrm{d}l$$

where l is measured along the failure surface and L is the wetted length; b is a constant. In view of the planar nature of the slide surface,

$$P_n = \int_{z'} pb\left(\frac{\mathrm{d}z}{\sin(\alpha)}\right)$$

where z is depth below the uppermost wet point on the failure surface and z' is the vertical projection of the wetted length of the failure surface as shown in Figure 2.4. Unfortunately, the distribution of $p(z)$ cannot be accurately determined without a detailed seepage analysis, so progress now depends on making some reasonable approximation for $p(z)$.

One simple approximation is to suppose water pressure increases linearly with depth to the half-depth point between water table and slope toe, as shown in Figure 2.5, and then decreases linearly to the toe. A reasonable estimate of water pressure increase is then $p = \gamma_w z$ where γ_w is specific weight of water and z is depth below the water table to the half-depth point. Below the half-depth point, $p(z) = \gamma_w (z_0 - z)$ where z_0 is the vertical distance between

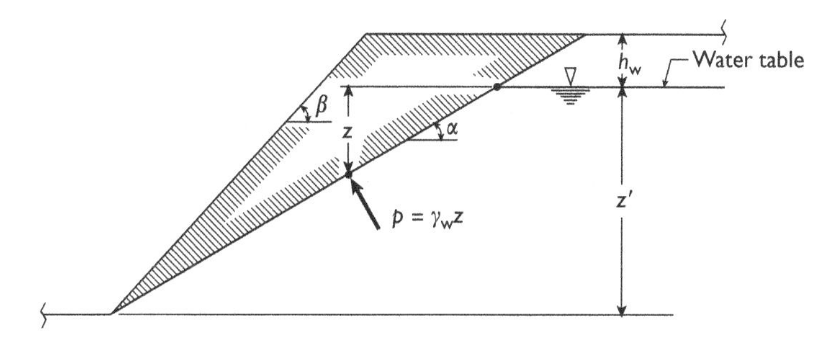

Figure 2.4 Basis for water pressure estimation.

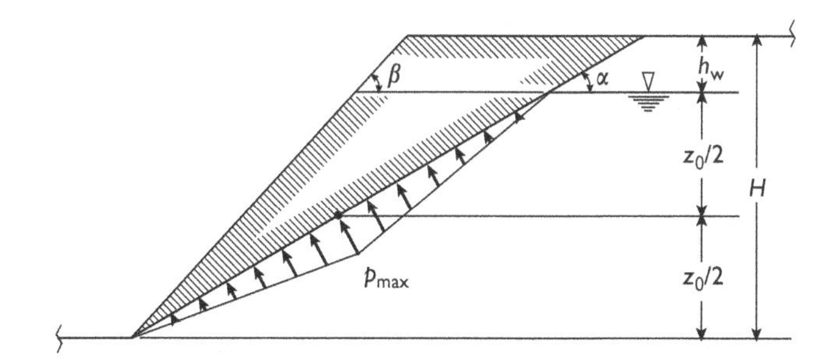

Figure 2.5 Approximation of water pressure distribution using half-depth point.

the water table and slope toe. Because the pressure distribution is linear and distributed over a rectangular surface, the average pressure is just one-half the maximum and

$$P_n = \int_{A'} p\mathrm{d}A = \bar{p}A' = \left[\gamma_w\left(\frac{z_0}{2}\right)\left(\frac{1}{2}\right)\right]\left[b\left(\frac{z_0}{2}\right)\left(\frac{2}{\sin(\alpha)}\right)\right] \tag{2.12}$$

where the first term in brackets on the right is the average pressure and the second term is wetted area in contact with the two pressure triangles shown in Figure 2.5.

However, an analysis of the geometry shown in Figure 2.6 indicates that the distance from the half-depth point to the top of the water table is not always $z_0/2$. The reason is that the water table is defined by the surface of zero water pressure and this line is assumed to follow the slope from the intersection point of the face with the horizontal water table line down to the toe, as shown in Figure 2.6. In Figure 2.6(a), the depth from the water table to the half-depth point is $z_0/2$ and the maximum pressure is $\gamma_w z_0/2$, but in Figure 2.6(b), the depth is y_w and the maximum pressure is $\gamma_w y_w$. If $\tan(\beta) \geq 2\tan(\alpha)$, the full half-depth $z_0/2$ may be used to calculate the maximum and therefore the average water pressure, otherwise the smaller distance y_w should be used. This value is then given by

$$y_w = \left(\frac{z_0}{2}\right)\left[\left(\frac{\tan(\beta)}{\tan(\alpha)}\right) - 1\right] \tag{2.13}$$

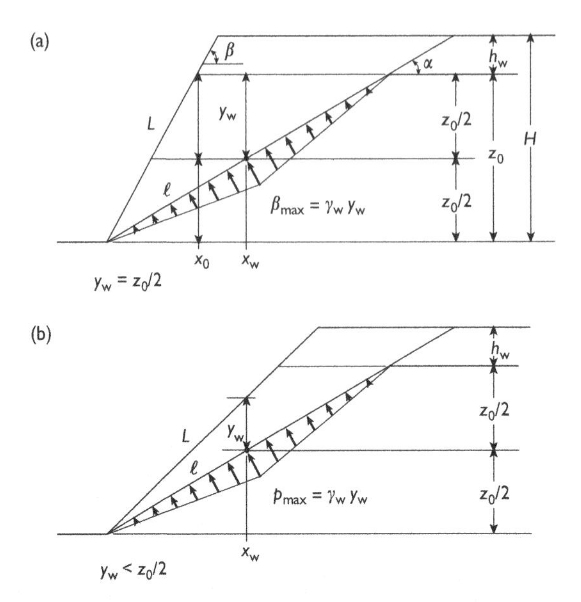

Figure 2.6 Modified half-depth pressure maximum.

which is less than $z_0/2$, provided $\tan(\beta) \leq 2\tan(\alpha)$. The face angle is always greater than the failure plane angle, so $\tan(\beta) > \tan(\alpha)$ and the term in brackets is always positive. When y_w is used, the water force on the failure plane is

$$P_n = \left[\gamma_w y_w \left(\frac{1}{2}\right)\right] \left[b\left(\frac{z_0}{2}\right)\left(\frac{2}{\sin(\alpha)}\right)\right] \tag{2.14}$$

where the first term in brackets is now the average pressure, while the second remains the wetted area of the failure plane.

Example 2.5 Consider the planar block data in Example 2.2 and further suppose the water table is at the crest of the slide. Determine the slide mass safety factor under this "wet" condition.

***Solution*:** All the data in Example 2.2 apply, but now the effective normal force $N' = N - P$ must be calculated where P is the force of water acting normal to the slide surface as shown in the free body diagram.

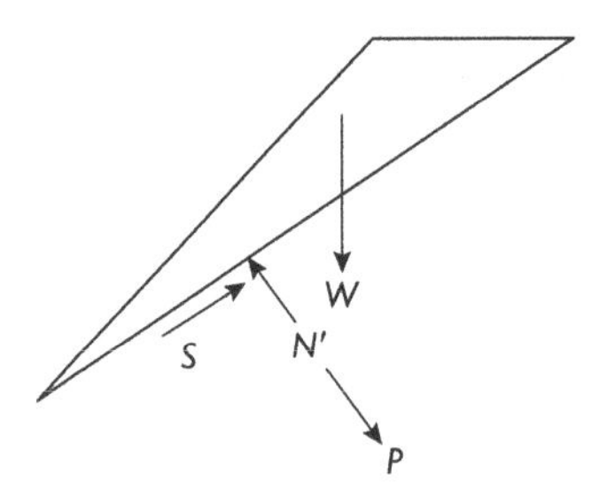

Calculation of the water force requires knowledge of the pressure distribution over the wet slide surface area A. One may approximate this distribution by assuming a linear increase of pressure with depth to $Z/2$ from 0, and then a linear decrease to zero at the toe, as shown in the water pressure diagram. Here Z is the height of the water table above the toe of the slide.

The average pressure over a triangular distribution illustrated in the text is just one-half the maximum pressure. Thus, over a triangle of pressure

$$\bar{p} = \frac{p(\max)}{2} = \frac{\gamma(\text{water})h}{2}$$

$$\Sigma F_{\text{n}} = 0$$

$$0 = W\cos(\alpha) - N' - P$$

$\therefore$ Water pressure distribution $N' = W\cos(\alpha) - P$

$$P = \int_A p\mathrm{d}A = \bar{p}A$$

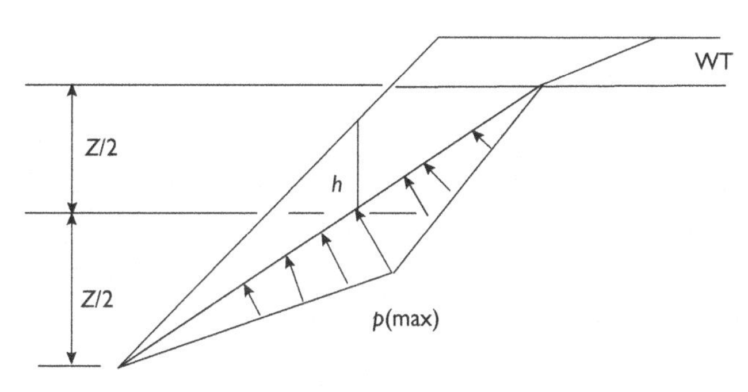

Water pressure distribution

What depth h should be used to calculate $p(\max)$ depends on the angles α and β. Examination of the slide mass geometry of the water pressure distribution shows that $h = y_w$ (2.13) should be used if $2\tan(\alpha) > \tan(\beta)$, then $h = (Z/2)[(\tan(\beta)/\tan(\alpha)) - 1]$, else one should use $h = Z/2$. The water table in the sketch of water pressure distribution is indicated by WT. In this example where the water table is at the crest of the slide, $Z = H$, where H is slope height, otherwise Z is slope height less water table depth. (The sketch shows the case where the water table is below the crest.)

$$h = (600/2)[-1 + \tan(45°)/\tan(35°)] = 128.4 \text{ ft}$$

$$p(\max) = (62.4)(128.4) = 8015 \text{ lbf/ft}^2$$

$$\bar{p} = 8015/2 = 4008 \text{ lbf/ft}^2$$

$$P = (4,008)(600/\sin(35°)(1) = 4.192(10^6) \text{ lbf}$$

$$N' = 9.469(10^6) - 4.192(10^6) = 5.277(10^6)$$

$$R = 5.277(10^6)\tan(30°) + 45.190(10^6) = 48.24(10^6) \text{ lbf}$$

$$\text{FS} = \frac{48.24(10^6)}{6.63(10^6)} = 7.28$$

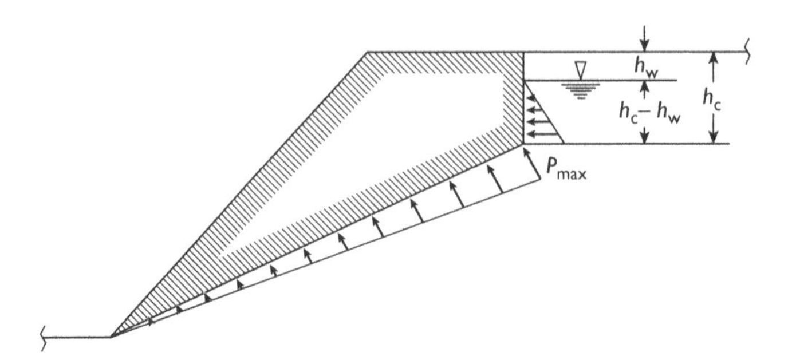

Figure 2.7 Alternative pressure distribution when a tension crack is present.

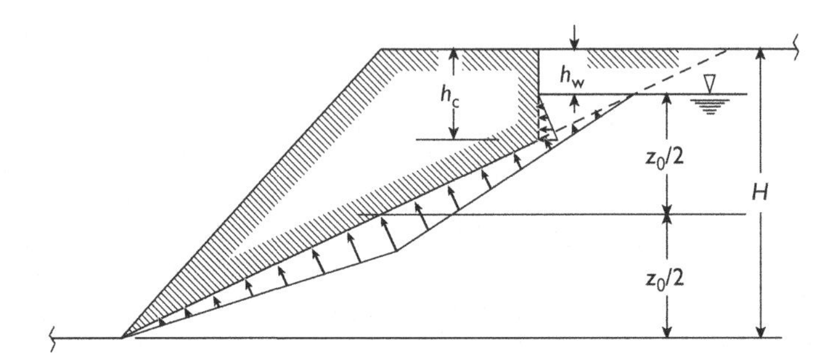

Figure 2.8 Half-depth water pressure estimation with tension crack.

In this case of a high water table, the effect is small. The main reason is the relatively high cohesion with respect to slope height.

When a tension crack is present and the water table is higher than the bottom of the crack, a different water pressure distribution may suffice. This distribution is shown in Figure 2.7 where the pressure increases linearly with depth to a maximum at the crack bottom and then decreases linearly to the slope toe. A difficulty with this distribution is the potential for a serious *underestimation* of the water force acting on the failure plane. This easily occurs if the tension crack is very shallow or if the water table is near the bottom of the crack. Both situations result in very small maximum water pressure and correspondingly small water forces. In the extreme, one might suppose the water table at the tension crack being at the bottom. If the crack is at the top and the crack has a vanishingly small depth, the result is the same–zero water force.

There is no reason why the half-depth approximation to water pressure distribution cannot be used in the presence of a tension crack, as shown in Figure 2.8. In this case, a small correction to the water force is needed. The water force on the inclined surface is then

$$P''_n = P_n - \left[\gamma_w\left(\frac{h_c - h_w}{2}\right)\right]\left[\frac{(h_c - h_w)}{\sin(\alpha)}\right] b \tag{2.15}$$

where the first term in brackets on the right is the average pressure between the water table and bottom of the tension crack, while the second term in brackets is the area of the inclined failure surface that is cut off by the tension crack.

The horizontal force of water acting on the wetted portion of the (vertical) tension crack is simply

$$P'_n = \left[\gamma_w\left(\frac{h_c - h_w}{2}\right)\right](h_c - h_w)b \tag{2.16}$$

These results could also be obtained by direct integration of the assumed pressure distribution between the proper limits measured down from the water table.

Safety factor improvement

Improvement of the safety factor for a planar block slide may be accomplished in two distinct ways: by increasing the resistance R and by decreasing the driving force D. Increasing resistance occurs when (1) the normal force mobilizing frictional resistance is increased, (2) cohesive resistance is increased, and (3) when reinforcement is installed. Decreasing weight decreases the downhill components of weight and seismic force; decreasing surcharge force and water force in a tension crack also decreases the total driving force. Decreasing weight also decreases the normal component of weight, and therefore the associated frictional resistance. Hence, some interaction between resisting and driving forces is inevitable with the consequence that the net effect on the factor of safety is unclear without an analysis.

However, closer examination of the safety factor formula shows that decreasing water forces always increases the safety factor. The numerator R increases while the denominator D decreases with reduction in water forces. In this context, "drainage" means depressurization, a reduction in water pressure, and does not imply desaturation with complete removal of water. Indeed, drainage is the most important method of improving slope stability. The effect of drawing down the water table (drainage) is readily quantified in the safety factor formula. Thus, where only gravity acts and no tension crack is present

$$\mathrm{FS} = \frac{(W_n - P_n)\tan(\phi) + C}{W_s} \tag{2.17}$$

Example 2.6 Consider the slope shown in the sketch.

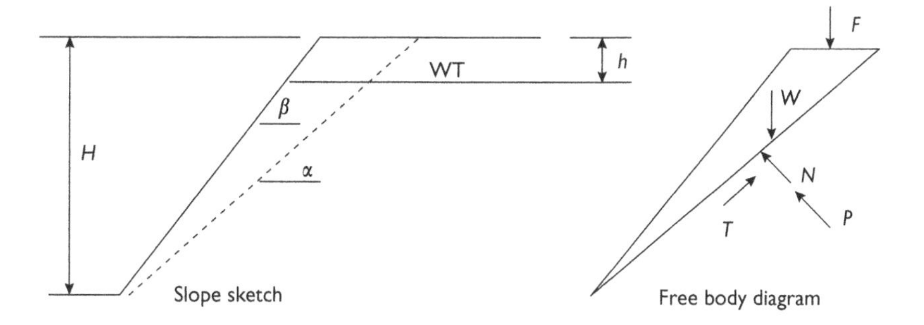

One method of improving a slope safety factor in wet ground is drainage. The objective of drainage is to reduce water forces that are always destabilizing. The maximum effect is achieved when the water table is drawn down below the toe of the slope. However, in some cases this action is not practical, and only a partial reduction in water forces is possible.

Given a possible planar block slide with the associated free body diagram where $\alpha = 30°, \beta = 45°, c = 20$ psi, $\phi = 35°, \gamma = 156, H = 1000$ ft, $h = 100$ ft, determine the slope safety factor (a) dry with the water table below the slope toe, (b) wet with the water table 100 ft below the crest, and (c) with the water table pushed back 150 ft into the slope by horizontal drains.

Solution: (a) *Dry slope*:

$$W = (1/2)(158)(1000)^2\left(\frac{1}{\tan(30)} - \frac{1}{\tan(45)}\right) = 57.10(10^6)\text{ lbf}$$

$$W_n = W\cos(30) = N$$

$$N = 57.10(10^6)\cos(30) = 49.45(10^6)\text{ lbf}$$

$$W_s = W\sin(30) = D$$

$$D = 57.10(10^6)\sin(30) = 28.55(10^6)\text{ lbf}$$

$$\begin{aligned} C &= cA = cbL \\ &= (20)(144)(1)[1000/\sin(30)] \end{aligned}$$

$$C = 5.76(10^6)\text{ lbf}$$

$$\begin{aligned} \text{FS} &= \frac{R}{D} = \frac{N\tan(\phi) + C}{D} \\ &= \frac{49.45(10^6)\tan(35) + 5.76(10^6)}{28.55(10^6)} \end{aligned}$$

$$\text{FS}(dry) = 1.415$$

(b) *Wet slope*: The new quantity needed to compute the slope safety factor in this case is the water force P acting normal to the potential failure surface.

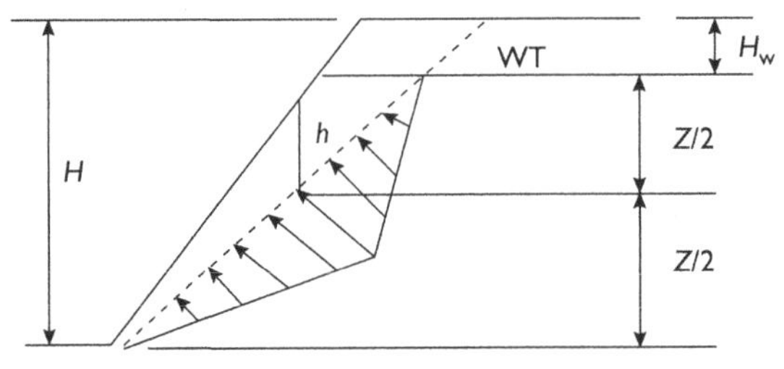

Water pressure distribution before drainage sketch

The calculation is

$$P = \bar{p}A = [p(\max)/2](b)L_w$$
$$= [p(\max)/2](b)(H - H_w)/\sin(\alpha)$$
$$P = (62.4)(h)(1/2)(1)(1000 - 100)/\sin(30)$$

The water head h is given by

$$h = \left(\frac{Z}{2}\right)\left[-1 + \frac{\tan(\beta)}{\tan(\alpha)}\right]$$
$$= (900/2)[-1 + \tan(45)/\tan(30)]$$
$$h = 329.4 \text{ ft}$$

Thus, $P = 18.50(10^6)$ lbf and so $N' = [49.45 - 18.50](10^6) = 30.9(10^6)$ lbf. Hence,

$$\text{FS} = \frac{30.9\tan(35) + 5.76}{28.55} = 0.96$$

which illustrates the detrimental effect of water pressure on slope stability.
(c) *Partially drained slope*: This case requires relocation of the water table 150 ft behind the face as shown in the sketch.

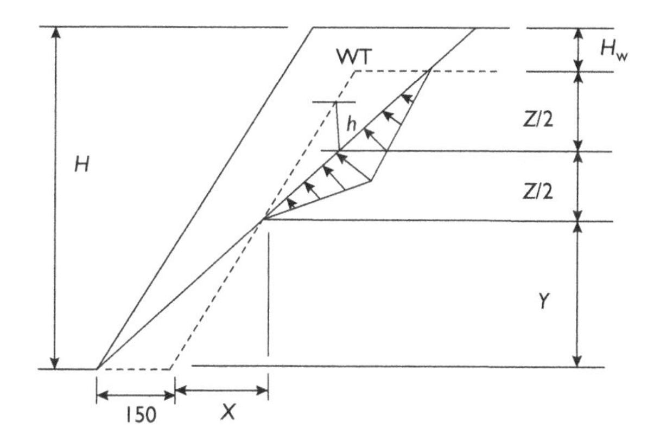

Sketch of water pressure distribution after partial drainage of the slope face.

The water table is given by the dashed line and is parallel to the face. Again, the change is in the water force. The form of the calculation is the same as in the wet case, but the geometry of the water pressure distribution has changed. Preliminary calculations of X and Y are necessary to determine Z and h which are required for the water force calculation. Thus, from the geometry of the sketch,

$$\tan(30) = \frac{Y}{150 + X}; \quad \tan(45) = \frac{Y}{X}$$

that may be solved for X and Y. The results are $X = 204.9$ ft and $Y = 204.9$ ft. Again, from the sketch geometry and given data, $Z = 1000 - 100 - 204.9 = 695.1$ ft. Water head h in this case is given by

$$h = (695.1/2)[-1 + \tan(45)/\tan(30)] = 254.3 \text{ ft}$$

The water force is

$$P = (62.4)(254.3)(1/2)(1)(695.1/\sin(30) = 11.04(10^6)) \text{ lbf}$$

and the effective normal is therefore

$$N' = [49.45 - 11.04](10^6) = 38.41(10^6) \text{ lbf}$$

The factor of safety in this partially drained case is then

$$\text{FS} = \frac{38.41 \tan(35) + 5.76}{28.55} = 1.144$$

which shows the effectiveness of the proposed horizontal drainage scheme in this slope design.

Although drainage implies removal of water, it does not imply removal of all water. As a practical matter, a slope is likely to be saturated even after drainage. Indeed, removal of all water to bring the degree of saturation to zero is generally not practical nor necessary. What is needed is the depressurization of the slope which may be accomplished without desaturation.

Improving the safety factor of a dry slope is usually done by decreasing the *slope angle*, that is, by reducing the inclination of the face angle β. Consider a simple, but common (dry) case where only gravity acts. The safety factor is then given by

$$\text{FS} = \frac{\tan(\phi)}{\tan(\alpha)} + \frac{2c}{\gamma H \sin^2(\alpha)[\cot(\alpha) - \cot(\beta)]} \tag{2.18}$$

that clearly shows that decreasing the slope angle β increases the safety factor, other parameters remaining the same. This simple case also indicates that an increasing slope height decreases the slope safety factor. Indeed, a plot of FS as a function of the reciprocal slope height H^{-1} is a straight line, as shown in Figure 2.9. If the safety factor intercept of this line is greater than one, then no critical depth exists at which slope failure would occur.

A gravity only safety factor expression is sometimes written as

$$\text{FS} = \text{FS}_\phi + \text{FS}_\text{c} \tag{2.19}$$

where the first safety factor on the right is a "factor of safety with respect to friction," while the second term is a "factor of safety with respect to cohesion." The overall safety factor is greater than one whenever the failure surface angle is less than the friction angle, that is, FS > 1, provided $\phi > \alpha$, even in the absence of cohesion. This observation allows for rapid screening of a problem. If the friction safety factor is greater than one, then so is the overall slope safety factor. However, any improvement would necessarily involve the factor of safety with respect to cohesion.

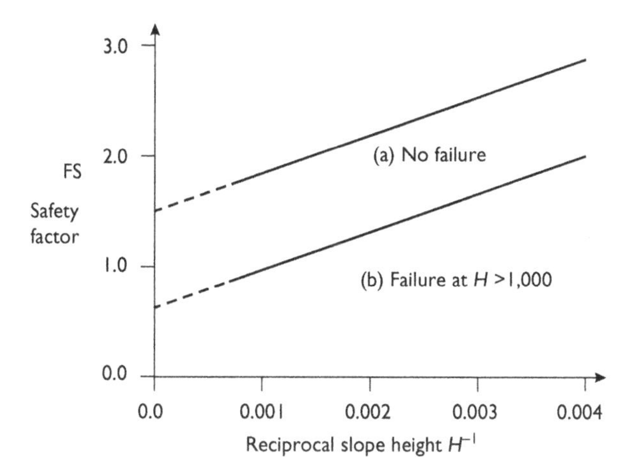

Figure 2.9 Planar gravity slide factor of safety as a function of reciprocal slope height.

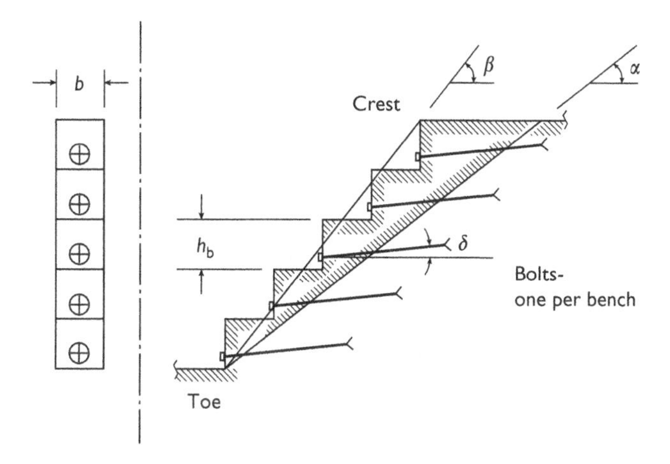

Figure 2.10 Cable bolted planar block slide.

When reinforcement with tensioned bolts is considered, the factor of safety may be written in the form

$$\mathrm{FS} = \mathrm{FS_o} + \Delta\mathrm{FS_b} \tag{2.20}$$

where the first term on the right is the unbolted slope safety factor and the second term on the right is the improvement obtained by bolting. The improvement is

$$\Delta\mathrm{FS_b} = \frac{N_\mathrm{b}\tan(\phi) + T_\mathrm{b}}{D} = \frac{nf_\mathrm{b}\,[\sin(\alpha-\delta)\tan(\phi) + \cos(\alpha-\delta)]}{D} \tag{2.21}$$

where all n bolts in the slope slab of thickness b have the same bolting angle and force f_b, as shown in Figure 2.10, so the total slope bolting force is F_b. The dimension b is also the horizontal row spacing of bolts. Vertical spacing is likely to be bench height in open pit mines, although under special circumstances more than one bolt may be installed per bench.

The optimum bolting angle obtained by maximizing the bolting improvement is

$$\delta_{opt} = \alpha - \phi \tag{2.22}$$

which is positive for failure surface angle greater than the friction angle and negative when less. A positive angle is measured counterclockwise from the horizontal (x-axis). If the failure surface angle is less than the friction angle, then the unbolted safety factor must be greater than one and bolting may not be needed. Otherwise the optimum bolting angle is positive. From a rock mechanics view then, bolt holes should be drilled uphill. Uphill holes have the advantage of being easier to flush clean. However, cable bolting of surface mine slopes involves relatively large drill holes that accommodate many cables, unlike underground cable bolting where one or two cables per hole is common. Cables or "strands" are usually 1/2-in. in nominal diameter, have an area of 0.1438 sqin., a unit strength of 250 ksi and thus an ultimate strength of 36,000 lbf. Higher strengths may be used. Six to sixteen or more 1/2-in. strands may be installed per hole. The weight of an assemblage becomes considerable and may prove impossible to manually move into an up-hole. For this reason, down-holes, say, at −5°, are preferred for cable bolting surface excavations. Even then, a pulley installed at hole bottoms may be required to aid installation. Spacing is also considerably greater in cable bolted surface mine slopes compared with underground cable bolt spacing. Holes must always be drilled deep enough to intersect the failure plane, provide for adequate grouted anchorage length, and for some sludge settling in the hole bottom.

Anchorage length is determined by bond strength and must be sufficient to support the ultimate tensile load the hole assemblage can sustain. In this regard, the entire hole length may not be grouted, especially if pretensioning is done. In this event, cable exposed over the ungrouted, open-hole length should be protected against corrosion. A simple equilibrium analysis that equates tensile strength to bond strength shows that anchorage length L_a is about 0.6 ft of hole per 1/2-in. strand of 36,000 lbf steel when a bond strength of 450 psi is assumed. The bond considered is between rock and grout. A 10-strand assemblage in a 4-in. diameter hole would require about 6 ft of anchorage length. Because progressive failure of the bond is possible, additional length should be provided. Higher strength strands would require additional anchorage lengths. Thus, about one to two feet of anchorage length per strand may be needed. In this regard, specialists in cable bolting surface excavations recommend pullout tests to determine what anchorage length is satisfactory at a given site because rock strength also affects bolting effectiveness.

Grouted bolt anchorage fills the bolt hole and thus acts as a *dowel* whether tensioned or not. The assemblage of steel and cement in the hole provides a direct resistance to shear along the potential failure surface. A dowel effect may be incorporated into a factor of safety estimate by simply adding the associated shear resistance to the other resisting forces being considered in an analysis. This resistance is a combination of grout and steel shear strength and respective areas, and is also likely to be small and have little effect on a slope safety factor. For example, a 6-in. diameter hole containing 24 cables would add less than 200,000 pounds to the resisting forces and when distributed over an area defined by even a close bolting spacing of 10 ft × 20 ft, results in less than 8 psi shear resistance. This rough estimate does suggest that in rather weak rock masses, a combination of closely spaced, grouted, and tensioned bolts may be an effective control measure.

Quantitative analysis of actual surface mine bolting cases shows that bolting provides very little improvement in the unbolted safety factor at bolt spacings of the order of bench height, say, 50 ft more or less. Horizontal spacing can be much less, but decreased spacing comes at increased cost. Generally, bolting open-pit mine rock slopes is not economical except in unusual circumstances, for example, where an in-pit crusher requires protection. In civil works, public safety often justifies the cost of maintaining stable slopes using bolting methods. These methods include cable bolting and use of solid bars and individual wires, although cables are the usual choice.

Example 2.7 Consider a cable-bolted slope where a potential failure may occur in the form of a planar block slide and further suppose the slope is loaded by gravity only. Determine the optimum bolting angle for the cable bolts assuming all bolts are tensioned to the same force in the pattern shown in the sketch. The bolting angle η is positive counterclockwise; total bolt tension is T_b, vertical bolt spacing is h, and horizontal spacing is b units into the page.

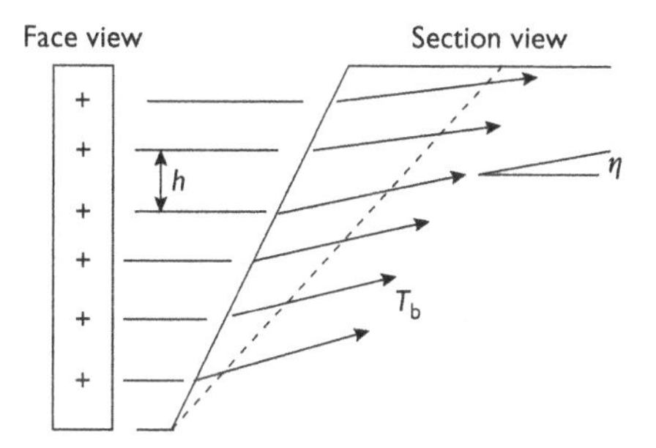

Sketch for slope bolting analysis

***Solution*:** A free body diagram showing the forces and directions is drawn first. The total force exerted by all bolts in a row is the number of bolts times the tensile force in each bolt, $T_b = nf_b$ where f_b is the force per bolt.

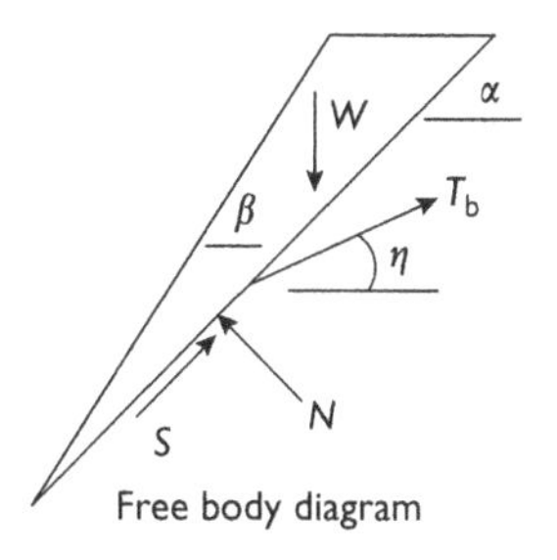

Free body diagram

Assuming Mohr–Coulomb strength for the potential failure surface with cohesion c and angle of friction ϕ, the resisting forces, driving forces, and safety factor are

$$R = W_n \tan(\phi) + C + N_b \tan(\phi) + S_b$$

$$D = W_s$$

$$\text{FS} = \frac{W_n \tan(\phi) + C + N_b \tan(\phi) + S_b}{W_s}$$

$$\text{FS} = \text{FS}_o + \Delta\text{FS}$$

where W_n, W_s, N_b, S_b, FS_o, and ΔFS are normal component of weight, tangential component of weight, normal component of the bolt force, tangential component of the bolt force, factor of safety without bolting, and change in the factor of safety caused by bolting. Downhill forces are driving forces; uphill forces are resisting forces.

The improvement in the slope safety factor by bolting is

$$\Delta\text{FS} = \frac{T_b \sin(\alpha - \eta)\tan(\phi) + T_b \cos(\alpha - \eta)}{W_s} = \frac{T_b \cos(\alpha - \eta - \phi)}{W_s \cos(\phi)}$$

A maximum of this last expression as a function of the bolting angle may be found in the usual way or simply by noting that the cosine is maximum when the argument is zero. Thus

$$\eta(\text{opt}) = \alpha - \phi$$

which indicates the bolts should make an angle of $\phi°$ to the failure surface. A negative bolting angle is indicated when the friction angle is greater than the angle the failure surface makes with the horizontal, that is, when $\phi > \alpha$. However, in this case bolting would not be necessary because frictional resistance alone would give a safety factor greater than one. Cohesion would further increase this safety factor. Although from the mechanics view, bolting should be uphill, difficulty in installing multiple strands of flexible cable uphill tends to be impractical. Most installations are downhill at about 5°.

At the optimum bolt angle, the total bolting force must be a substantial fraction of the driving force to make a significant contribution to the slope safety factor. This is unlikely to be the case where bench height largely determines bolt spacing in the vertical direction and in the horizontal direction when bolting on a square pattern.

Wedge failures

Wedge failures are mainly translational slides that occur when joint planes combine to form a rock block that may slide down the line of intersection of the two joint planes. Figure 2.11 shows a wedge failure of this type. Also shown in Figure 2.11 is another type of wedge failure. A rock block that forms a "wedge" (actually, a tetrahedron) which may slide along one joint plane only or the other and may also rotate about an axis perpendicular to one joint plane or the other and may even tumble forward out of the face of the slope. There are, in fact, three translational and three rotational modes of wedge failure. However, sliding down the line of intersection is the most important wedge failure mode in surface excavations.

The basic mechanics of wedge failures are the same as for planar block slides, but the geometry of wedge failure is decidedly more complicated. A factor of safety with respect to resisting and driving forces for a wedge failure shown in Figure 2.11 is

$$\text{FS}_w = \frac{R}{D} = \frac{R_A + R_B}{D} \tag{2.23}$$

where the subscripts A and B refer to the joint planes shown in Figure 2.11. A distinction is needed because each joint set may have different frictional and cohesive properties and

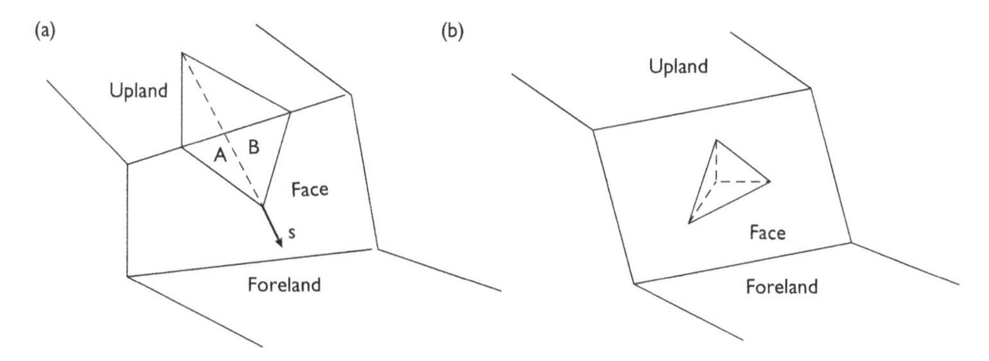

Figure 2.11 Two types of wedge failures in a surface excavation: (a) surface wedge, and (b) face wedge.

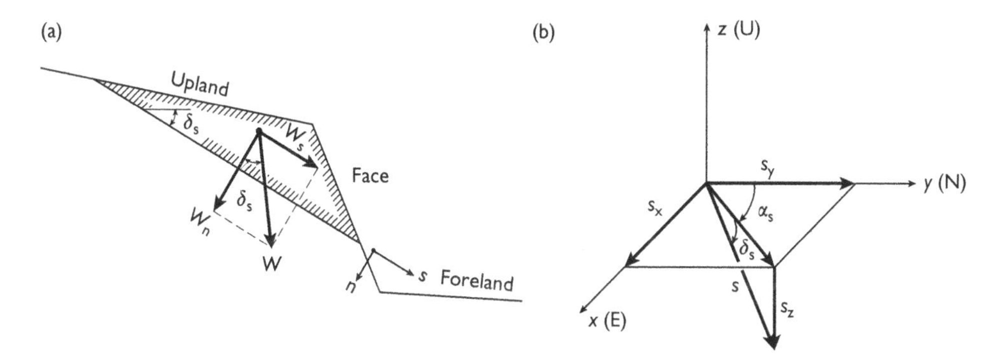

Figure 2.12 Dip of the line of intersection δ_s.

differing persistence, as well. An assumption of 100% persistence is often justified and leads to a conservative or low estimate of the factor of safety. The reason is that wedge failures tend to be relatively small, of the order of bench height in open pit mines or about 50 ft more or less. At this scale, persistence may be locally high, that is, 100%, even though over a large distance or scale, the average persistence is less.

The driving force down the line of intersection of the joint planes is the downhill component of weight when gravity only is considered. In this case,

$$D = W_s = W \sin(\delta_s) \tag{2.24}$$

where W is the rock wedge weight and δ_s is the dip of the line of intersection, as shown in Figure 2.12. The downhill direction is the positive s direction in Figure 2.12. Calculation of the driving force thus requires calculation of weight and intersection line dip. Weight is the product of specific weight γ and wedge volume V, which is simple to calculate in principle but requires considerable effort in practice.

Dip direction is somewhat easier and begins with specification of joint plane orientation by the angle of dip δ, which is measured from the horizontal down and ranges between 0 and 90°, and the dip direction α which is the azimuth of a horizontal line from which the dip is measured. Figure 2.13 shows the relationship between compass coordinates where x = east,

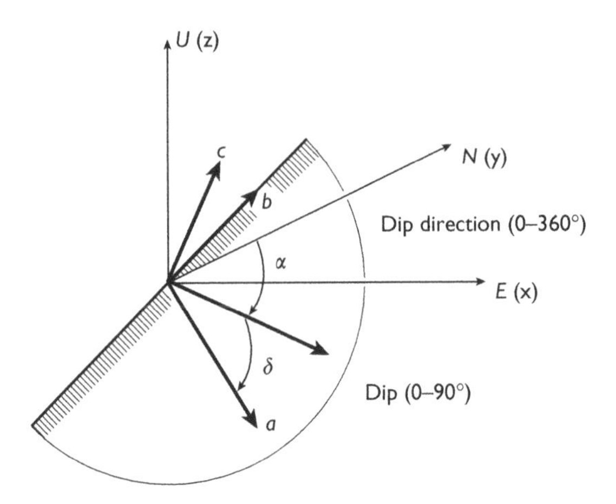

Figure 2.13 Compass coordinates, joint plane coordinates.

Table 2.1 Directions cosines between compass and joint plane coordinate directions

Compass direction joint plane direction	*x (east)*	*y (north)*	*z (up)*
a (dip)	cos(δ) sin(α)	cos(δ) cos(α)	−sin(δ)
b (strike)	−cos(α)	sin(α)	0
c (normal)	sin(δ) sin(α)	sin(δ) cos(α)	cos(δ)

Notes: α = dip direction (+cw from north, 0–360°), δ = dip (+down, 0–90°).

y = north, z = up, and joint plane coordinates where the coordinate directions are a = dip, b = strike, and c = normal. The strike direction α' is the dip direction less 90°; the dip direction is an azimuth that is measured positive in a clockwise direction form north; if measured counterclockwise, then the dip direction is negative. Directions cosines that relate compass to joint plane coordinates are given in Table 2.1.

Example 2.8 Show that the direction cosines in Table 2.1 define a mutually orthonormal triple for joint plane coordinates.

***Solution*:** Direction cosines are components of unit vectors pointing from an origin in the coordinate directions. Lengths of these vectors must be one and they must be orthogonal. Thus,

$$
\begin{aligned}
|a| &= \sqrt{\vec{a}\cdot\vec{a}} \\
&= \sqrt{a_x^2 + a_y^2 + a_z^2} \\
&= \left\{[\cos(\delta)\,\sin(\alpha)]^2 + [\cos(\delta)\,\cos(\alpha)]^2 + [-\sin(\delta)]^2\right\}^{1/2}
\end{aligned}
$$

$$= \{[\cos(\delta)]^2 + [\sin(\alpha)]^2\}^{1/2}$$

$$|a| = 1$$

$$|b| = \sqrt{b_x^2 + b_y^2 + b_z^2}$$

$$= \{[-\cos(\alpha)]^2 + [\sin(\alpha)]^2 + 0\}^{1/2}$$

$$|b| = 1$$

$$|c| = \sqrt{c_x^2 + c_y^2 + c_z^2}$$

$$= \{[\sin(\delta)\ \sin(\alpha)]^2 + [\sin(\delta)\ \cos(\alpha)]^2 + [\cos(\delta)]^2\}^{1/2}$$

$$= \{[\sin(\delta)]^2 + [\cos(\alpha)]^2\}^{1/2}$$

$$|c| = 1$$

where the dot or inner product is indicated in the first of the above equations. These results show the directions cosines are indeed components of unit vectors. The dot product may also be used to show orthogonality, while the cross product of two vectors should give the third. Thus,

$$\vec{a} \cdot \vec{b} = 0, \quad \vec{b} \cdot \vec{c} = 0, \quad \vec{c} \cdot \vec{a} = 0$$

$$\vec{a} \times \vec{b} = c, \quad \vec{b} \times \vec{c} = \vec{a}, \quad \vec{c} \times \vec{a} = \vec{b}$$

For example,

$$\vec{a} \cdot \vec{b} = [\cos(\delta)\ \sin(\alpha)(-\cos(\alpha)) + [\cos(\delta)\ \cos(\alpha)\ (\sin(\alpha)) + [-\sin(\alpha)(0)] = 0$$

$$\vec{a} \times \vec{b} = \vec{c}$$

$$= [\cos(\delta)\ \cos(\alpha)(0) - (-\sin(\delta))(\sin(\alpha)],$$

$$[(-\cos(\alpha)(-\sin(\delta)) - \cos(\delta)\sin(\alpha)(0)],$$

$$[\cos(\delta)\ \sin(\alpha)\ \sin(\alpha) - (-\cos(\alpha))\ \cos(\delta)\ \cos(\alpha)]$$

$$= [\sin(\delta)\ \sin(\alpha)],\ [\sin(\delta)\ \cos(\alpha)],\ [\cos(\delta)]$$

$$\vec{a} \times \vec{b} = \vec{c}$$

Example 2.9 Given the transformation from compass to joint coordinates, find the angle of intersection between vectors **b** and **c**.

Solution: By definition the dot product is $\vec{b} \cdot \vec{c} = |b||c| \cos(\theta)$ and as the considered vectors are unit vectors, one has

$$\begin{aligned} \cos(\theta) &= b_x c_x + b_y c_y + b_z c_z \\ &= -\cos(\alpha) \sin(\delta) \sin(\alpha) + \sin(\alpha) \sin(\delta) \cos(\alpha) + (0) \cos(\delta) \\ &= 0 \end{aligned}$$

$\therefore$

$$\theta = \frac{\pi}{2}$$

which agrees with the requirement for orthogonality.

Example 2.10 Given a transformation to joint coordinates from compass coordinates outlined in Example 2.8, find the area subtended by vectors **a** and **b**.

Solution: Recall the geometric interpretation of the cross product illustrated in the sketch.

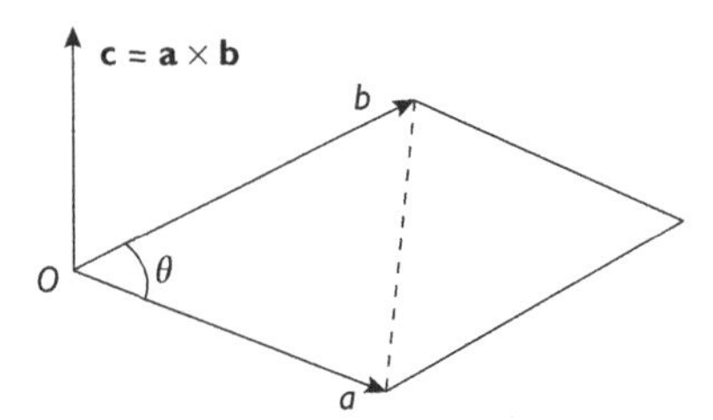

Sketch of vector cross product geometry

By definition the cross product is $\vec{a} \times \vec{b} = |a||b| \sin(\theta) = (\text{base})(\text{height})$ where base and height refer to the parallelogram defined by the considered vectors. The result is just the area A of the parallelogram. In this problem, the vectors are unit orthogonal vectors, so the parallelogram is just a square with edges of unit length. The area of the square is one, of course. The area of either triangle formed by the diagonal between the vector ends is just one-half the cross product.

The direction of a line of intersection between two planes is specified by the direction cosines of the line. These direction cosines may be found by forming the vector product (cross product) of the normal vectors of the considered joint planes. The resulting vector is perpendicular to both normal vectors and thus is parallel to both planes and must then be parallel to the line of intersection. A view looking up the line of intersection of joint planes A and B is shown in Figure 2.14.

If the normal vectors to planes A and B have components $(a_x\ a_y\ a_z)$ and $(b_x\ b_y\ b_z)$, then the direction numbers of the line of intersection are $(S_x\ S_y\ S_z)$ given by the 2×2 determinants

$$S_x = \begin{vmatrix} a_y & a_z \\ b_y & b_z \end{vmatrix}, \quad S_y = -\begin{vmatrix} a_x & a_z \\ b_x & b_z \end{vmatrix}, \quad S_z = \begin{vmatrix} a_x & a_y \\ b_x & b_y \end{vmatrix} \tag{2.25}$$

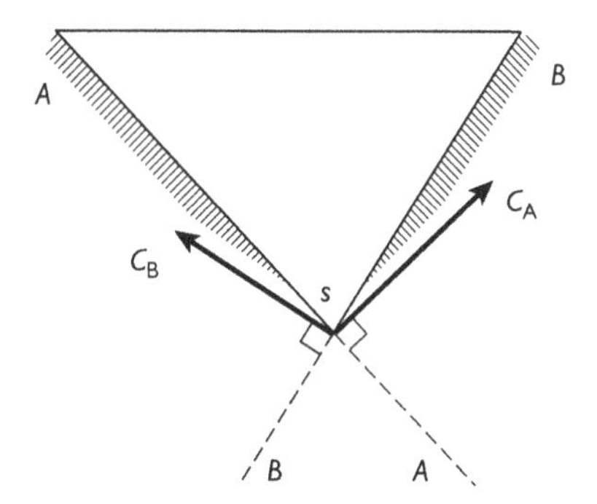

Figure 2.14 View up the line of joint plane intersection.

The direction cosines $(s_x\ s_y\ s_z)$ are

$$s_x = \frac{S_x}{S}, \quad s_y = \frac{S_y}{S}, \quad s_z = \frac{S_z}{S} \tag{2.26}$$

where S is the magnitude

$$S = \sqrt{S_x^2 + S_y^2 + S_z^2} \tag{2.27}$$

In consideration of the right-hand rule for the vector product, the line of intersection vector points downhill, as shown in Figure 2.12(a). The dip direction and dip of the line of intersection may be obtained from

$$\begin{aligned} \sin(\alpha_s) &= \frac{s_x}{\sqrt{s_x^2 + s_y^2}} \\ \sin(\delta_s) &= \frac{-s_z}{\sqrt{s_x^2 + s_y^2 + s_z^2}} \end{aligned} \tag{2.28}$$

as shown in Figure 2.12(b).

Example 2.11 Consider a potential wedge failure by sliding down the line of intersections between two joints that belong to joint sets A and B. Dip and dip direction of A-joints are 45° and 0°, respectively, and for B-joints 45° and 90°, respectively. Thus A-joints strike due west and B-joints strike due north. Determine the dip and dip direction of the line of intersection.

***Solution*:** A sketch of the given data shows the situation where α_s and δ_s are the dip direction (azimuth) and dip of the line of intersection formed by two joints from sets A and B.

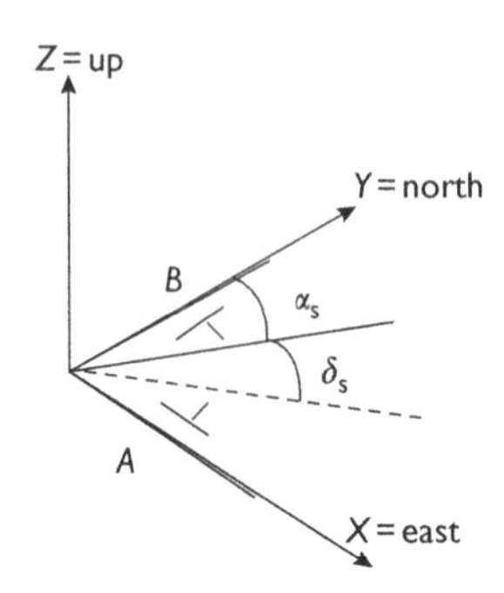

Sketch for joint set line of intersection

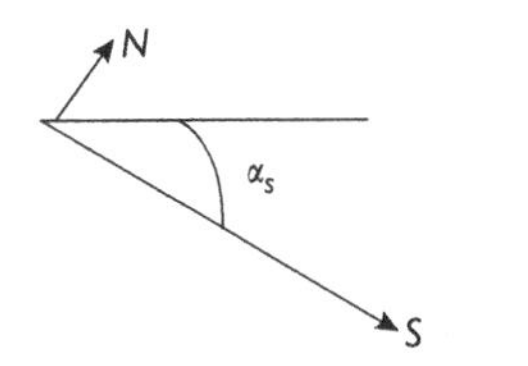

(a) True slope of the line of intersection

(b) Looking up the line of intersection

Views that show the true slope of the line of intersection and one looking up the line of intersection are helpful. The line of intersection is seen as a point when looking up the line. Construction of the normals to the line of intersection from the dip and dip direction of the joint planes can be done with the aid of the formulas in Table 2.1. Thus,

$$\vec{c}_A = (0,\ 1/\sqrt{2},\ 1/\sqrt{2}),\quad \vec{c}_B = (1/\sqrt{2},\ 0,\ 1/\sqrt{2})$$

Examination of the geometry of the situation shows that the cross product of the normal vectors is a vector **S** parallel to the line of intersection. Thus,

$$\vec{c}_A \times \vec{c}_B = \vec{S} = (1/2,\ 1/2,\ -1/2)$$

which has a length of $\sqrt{3}/4$. Direction cosines of **S** are then $(1/\sqrt{3},\ 1/\sqrt{3},\ -1/\sqrt{3})$, which is a unit vector pointing down the line of intersection of the two joint planes. These direction cosines define a space diagonal with a negative (dipping) slope. An angle with cosine $1/\sqrt{3}$ is 54.7°. An angle with cosine $-1/\sqrt{3}$ is 125.3°; this is the angle between the positive z-axis and the vector **S**. The dip is measured from the horizontal, so the dip of the line of intersection is just +35.3° (down as always for dip).

The azimuth of the line of intersection is measured clockwise from north to the dip direction in a horizontal plane. A formula for the sine of this angle is

$$\sin(\alpha_s) = \frac{s_x}{\sqrt{(s_x)^2 + (s_y)^2}} = \frac{(1/2)}{\sqrt{(1/2)^2 + (1/2)^2}} = 1/\sqrt{2}$$

$$\therefore \alpha_s = 45^\circ$$

which is intuitively seen as the correct result in this case.

A similar analysis applies to the intersection lines between face and upland and between foreland and face, L_{fu} and L_{fl}, respectively, as shown in Figure 2.15. As before, the analysis begins with determination of the normal vectors to the pair of planes considered, each with direction cosines that may be determined from Table 2.1 after specification of the dip and dip direction of the upland, face, and foreland. In this regard, forelands and uplands are often flat, so the dip directions α are indeterminate and may be assumed due north as a matter of convenience.

Other lines of intersection occur between joints, face, and upland, as shown in Figure 2.16. These lines are: L_{Af}, L_{fB}, L_{Bu}, and L_{uA}, where the order of subscripts corresponds to the order of vectors in the vector product. For example, L_{Af} corresponds to the vector product between the normal vectors to joint plane A and the face, that is, $c_A \times c_f$. If the order is reversed, then the direction of the result is reversed. Again, the direction of the vector product is the direction of the unit vector formed by the direction cosines ($s_x\ s_y\ s_z$) of the considered line.

Line length determination requires introduction of distance between points that define the wedge. Two lengths that are intuitively important are (1) the vertical height H of the wedge measured from the point of joint intersection d at the face and a point a on the slope crest intersected by joint plan A, as shown in Figure 2.16, and (2) length along the upland from the crest point a to the point of joint plane intersection c on the upland, shown in Figure 2.16. Given H, the dip direction α of L_{Af} (ad) and dip δ, the length of the line L_{Af} is $H / \sin \delta$. The length of L_{uA} may be measured directly in the field.

Associated with the introduction of distance is the calculation of joint surface areas and wedge volumes that are needed for safety factor calculations. These calculations may be done

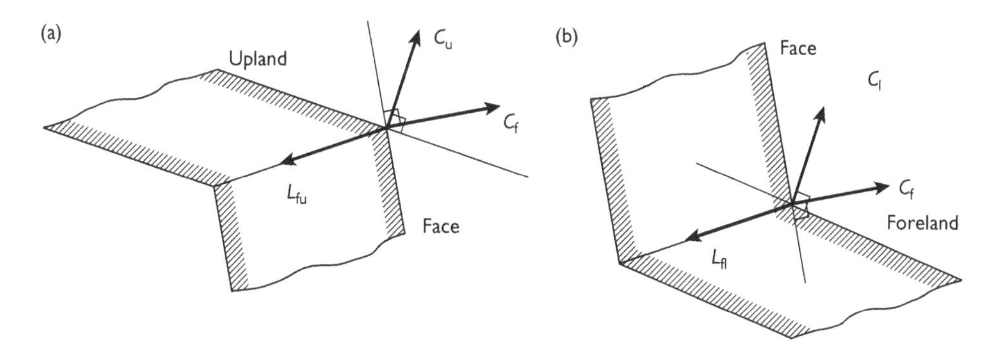

Figure 2.15 Intersection vectors between: (a) face and upland, and (b) foreland and face.

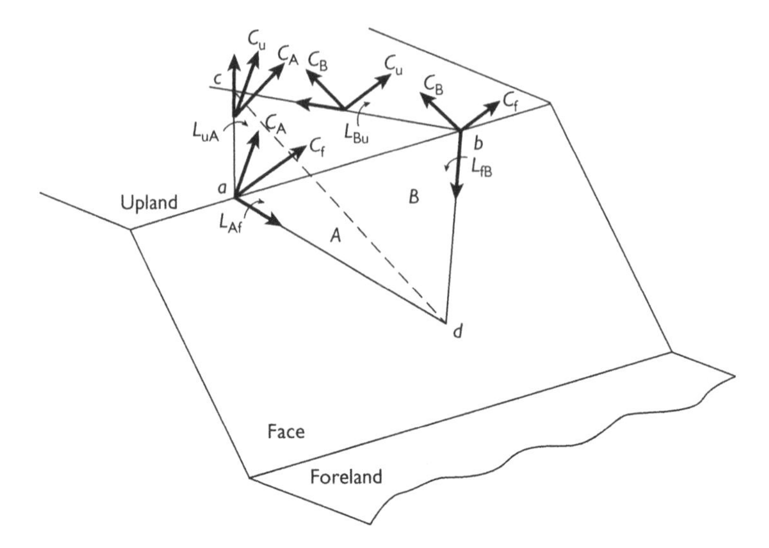

Figure 2.16 Lines of intersection formed by face, upland, and joint plane intersections.

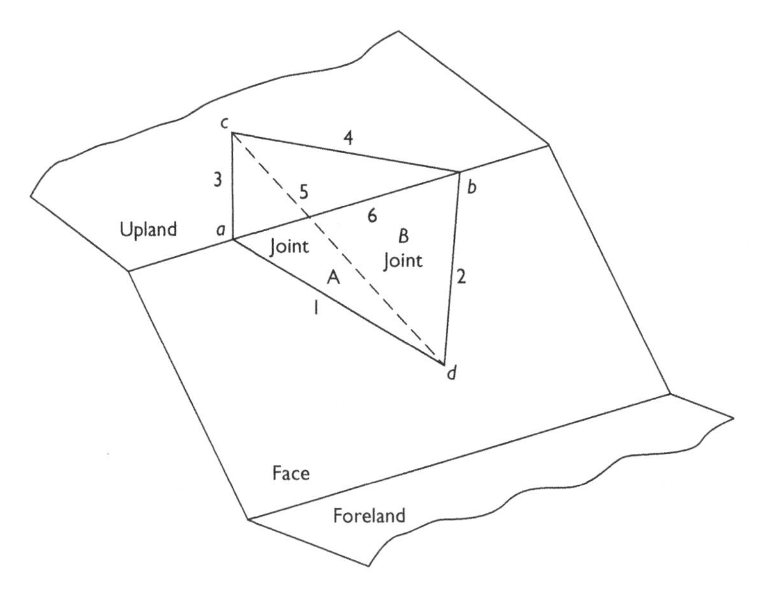

Figure 2.17 Line numbering for wedge analysis.

systematically with cross products, dot products, the sine law, and the cosine law for triangles. A relabeled diagram of the wedge from Figure 2.16 helps to simplify the notation and is shown in Figure 2.17 where lines are numbered. For example, line 6 connects points a and b. Numerical evaluation of a cross product requires knowledge of the angle between the two vectors involved. This angle may be obtained from the scalar or dot product of the considered vectors. For example, the cosine of the angle between normal vectors to joint plane A and the foreland is $\cos(c_A, c_f) = c_A \cdot c_f / |c_A||c_f|$ where the vertical bars indicate vector length or magnitude.

The area of joint plane A may be obtained from the cross product of vectors connecting points a and d and points a and c. Thus, $A_A = (1/2)\, L_1 L_3 \sin(\theta_{13})$ where the angle between lines 1 and 3 is obtained from the dot product $\cos(\theta_{13}) = S_1 \cdot S_3$. The unit vectors S_1 and S_3 are line of intersection vectors between face and joint plane A and between upland and joint plane A, respectively; they are directed away from point a toward d and c. Interestingly, the result of this vector product is a vector normal to joint plane A. All the line of intersection vectors are known in principle, so any and all angles between lines may be computed.

Length of the line of intersection $L5$ and other angles of joint plane A may be found using the cosine law (1) and sine law (2). With reference to Figure 2.18,

1 $$L_5^2 = L_1^2 + L_3^2 - 2L_1L_3\cos(\theta_{13})$$

2 $$\frac{\sin(\theta_{53})}{L_1} = \frac{\sin(\theta_{15})}{L_3} = \frac{\sin(\theta_{31})}{L_5}$$

where the order of subscripts on angles is not important. The first equation may be solved for the unknown line length L_5. The second equation may then be solved for the other two unknown angles, if desired. For joint plane B, lengths L_2 and L_4 may be obtained from the sine law. Thus,

$$\frac{\sin(\theta_{24})}{L_5} = \frac{\sin(\theta_{45})}{L_2} = \frac{\sin(\theta_{52})}{L_4}$$

where the angles are known from intersection line dot products. For example, $\cos(\theta_{24}) = s_2 \cdot s_4$. The area of joint plane B is then

$$A_B = \left(\frac{1}{2}\right) L_2 L_4 \sin(\theta_{24})$$

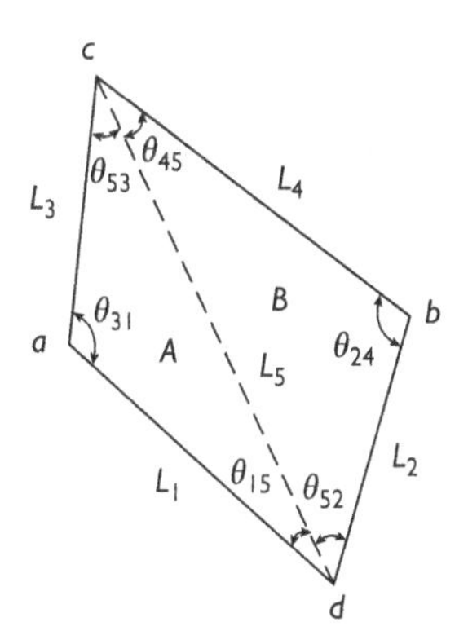

Figure 2.18 Joint planes A and B with lengths and angles shown.

The volume of the tetrahedron may be obtained from the triple product of vectors emanating from point c. Thus $V = (1/6)\,(S_4 x S_3 \cdot S_5)$ and the weight $W = \gamma\, V$. The vectors S_4, S_3, and S_5 are known from previous calculations of line lengths, dip directions, and dips.

Example 2.12 Calculate the volume of a tetrahedron with vertices at (0,0,0), (1,0,0), (0,1,0), (0,0,1) as shown in the sketch.

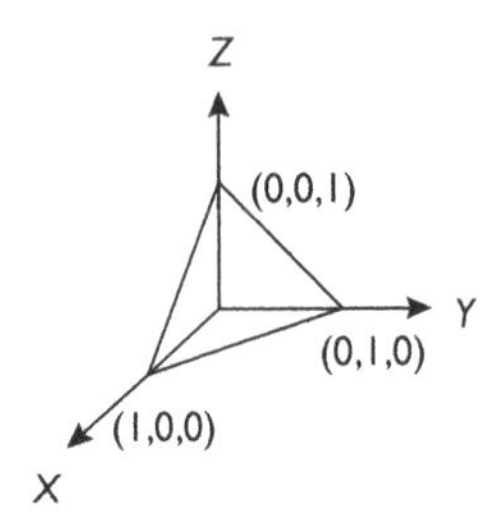

Sketch for calculating the volume of a tetrahedron

***Solution*:** A formula cited here without proof states that the volume $V = (1/6)(\mathbf{abc})$ where **abc** is the vector triple product, that is,

$$V = \left(\frac{1}{6}\right)\vec{a} \cdot (\vec{b} \times \vec{c}) = \left(\frac{1}{6}\right)\vec{a} \cdot \vec{d}$$

$$\vec{d} = \vec{b} \times \vec{c}$$

$$= \begin{vmatrix} x & y & z \\ 0 & 1 & 0 \\ 0 & 0 & 1 \end{vmatrix} \text{ (symbolically)}$$

$$\vec{d} = (1,\ 0,\ 0)$$

$$V = \left(\frac{1}{6}\right)\ [(1)(1) + (0)(0) + (0)(0)] = \left(\frac{1}{6}\right)$$

which is just one-sixth the volume of a unit cube defined by vectors **a**, **b**, and **c**.

A complication occurs when a third joint set is present or a tension crack forms, as shown in Figure 2.19 where a triangular crack that has sides 7, 8, and 9 and corners e, f, and g cuts off a portion of the original wedge. The "tension" crack plane has an orientation defined by a dip direction, and dip (α_c, δ_c), and a position defined by the distance d_c between points a and e. The distance l_3 is readily obtained as the difference between the distance from a to c measured in the field previously and d_c. The lengths of all lines in the triangles on the joint planes cut off by the tension crack can be determined using the sine law, once the angles between the lines of intersection formed by the tension crack, joint planes, and upland are calculated. This information is obtained by calculating the dip directions and dips of the lines 7 and 8 formed by the intersection of the tension crack with the two joint planes. Figure 2.20 shows the notation for such calculations. Thus for joint plane A,

$$\frac{\sin(\theta_{75})}{l_3} = \frac{\sin(\theta_{37})}{l_5} = \frac{\sin(\theta_{53})}{l_7}$$

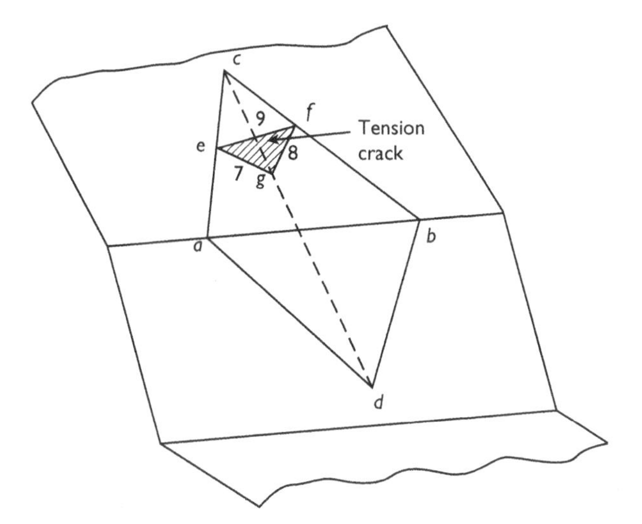

Figure 2.19 Wedge with a tension crack.

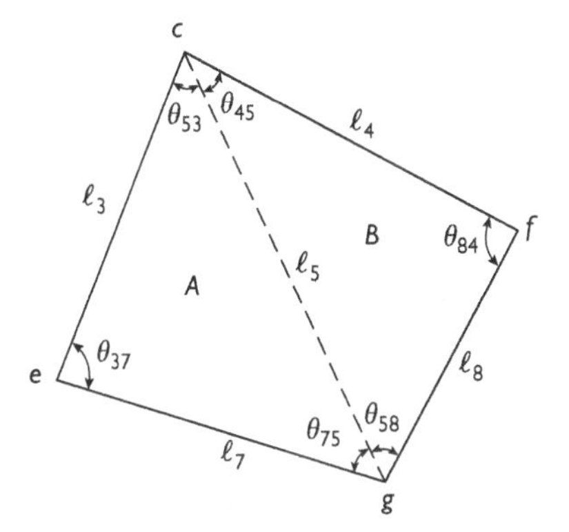

Figure 2.20 Joint plane triangles cut off by a tension crack.

A similar set of equations applies to joint plane B that share the common line segment l_5. The volume of the cutoff tetrahedron v is

$$v = \left(\frac{1}{6}\right) (S'_4 x S'_3 \cdot S'_5)$$

where the prime indicates analogous vectors for calculating the total wedge volume with the same directions but with the proper magnitudes or cutoff lengths shown in Figure 2.20. The cutoff wedge volume $V' = V - v$ and the weight now is $W = \gamma\, V'$. There is a possibility of a tension crack forming on the face, but this possibility and others of rather problematic geometry are not considered here.

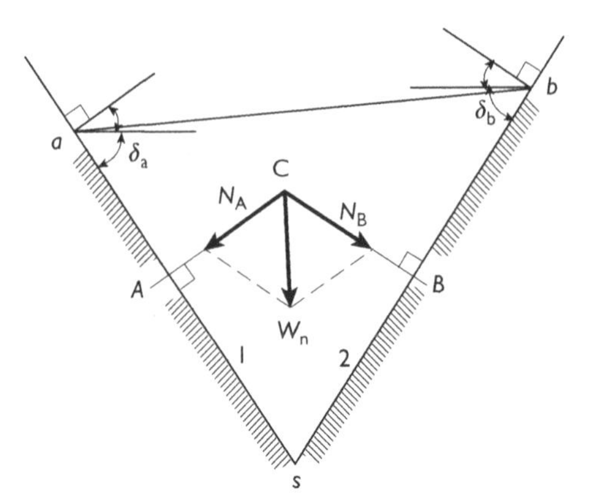

Figure 2.21 Normal forces resolved on joint planes *A* and *B* in a section perpendicular to the line of intersection.

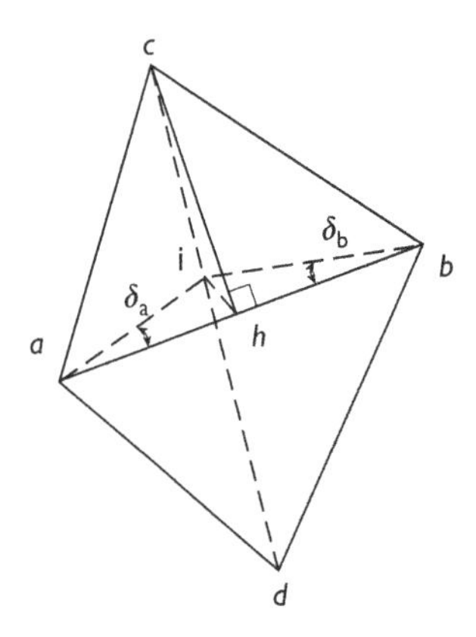

Figure 2.22 Construction of a plane normal to the line of intersection.

While the driving force is simply the downhill component of weight, normal forces acting on the joints require further analysis. Figure 2.21 shows a resolution of normal forces acting on the joint planes. The plane containing the normal component of weight is perpendicular to the line of intersection and to the dip direction as shown in Figure 2.22. The angles δ_a and δ_b are given by

$$\begin{aligned} \tan(\delta_a) &= \sin(\delta_s)\tan(\alpha_s - \alpha_a) \\ \tan(\delta_b) &= \sin(\delta_s)\tan(\alpha_b - \alpha_s) \end{aligned} \tag{2.29}$$

where subscripts *s, a*, and *b* refer to the line of intersection and dip directions of joint planes *A* and *B*, respectively. The line *hi* in Figure 2.22 is perpendicular to the line of intersection of the joint planes. Note: the line *ab* in Figure 2.22 is *not* the wedge crest. Careful inspection of Figure 2.22 shows that the angles δ_a and δ_b are *not* dips of the joint planes.

Equilibrium in the x and y directions in the plane of analysis requires

$$\begin{aligned} W_{\mathrm{n}} &= N_a \cos(\delta_a) + N_b \cos(\delta_b) \\ 0 &= N_a \sin(\delta_a) + N_b \sin(\delta_b) \end{aligned} \tag{2.30}$$

Hence

$$\begin{aligned} N_a &= \frac{W_{\mathrm{n}} \sin(\delta_b)}{\sin(\delta_a + \delta_b)} \\ N_b &= \frac{W_{\mathrm{n}} \sin(\delta_a)}{\sin(\delta_a + \delta_b)} \end{aligned} \tag{2.31}$$

In case of water infiltrated joints, the total normal forces become the sum of effective normals forces and water forces, that is, in wet ground

$$\begin{aligned} N'_a &= N_a - P_a \\ N'_b &= N_b - P_b \end{aligned} \tag{2.32}$$

As usual, the effective normal forces mobilize frictional resistance. Thus, in the absence of a tension crack and forces other than weight and water, resistances are

$$\begin{aligned} R_a &= N'_a \tan(\phi_a) + c_a A_A \\ R_b &= N'_b \tan(\phi_b) + c_b A_B \end{aligned} \tag{2.33}$$

An interesting special case occurs when the rock mass is dry, cohesion is lacking on both joint planes and friction angles are equal. In this case

$$\mathrm{FS}_{\mathrm{w}} = \left(\frac{W \cos(\delta_s) \tan(\phi)}{W \sin(\delta_s)} \right) \left[\frac{\sin(\delta_b) + \sin(\delta_a)}{\sin(\delta_a + \delta_b)} \right] \tag{2.34}$$

The term in parenthesis is a safety factor for a dry, cohesionless planar block slide FS_{p}, while the term in brackets is never less than one. Hence, in this special case

$$\mathrm{FS}_{\mathrm{w}} \geq \mathrm{FS}_{\mathrm{p}} \tag{2.35}$$

Thus it seems that simply "folding" the failure surface of a planar block slide into a "winged" failure surface increases the resistance to frictional sliding.

Calculation of water forces is done by integration of pressure distributions over wetted areas of joint planes and tension crack, when present. As before, an estimate of the pressure distribution is needed before the calculations can be done. The worst case occurs when the water table follows the ground surface and so is coincident with the upland. Consider joint plane A shown in Figure 2.23 where the water pressure is assumed to vary linearly up to a depth equal to one-half the vertical distance between points c and d on the line of intersection. The water pyramid has a weight given by

$$W_{\mathrm{w}} = \gamma_{\mathrm{w}} A_{\mathrm{A}} \left(\frac{1}{3} h \right) = \gamma_{\mathrm{w}} A_{\mathrm{A}} \left(\frac{1}{3} \right) \left[\frac{1}{2} L_5 \sin(\delta_{\mathrm{s}}) \right] = P_A \tag{2.36}$$

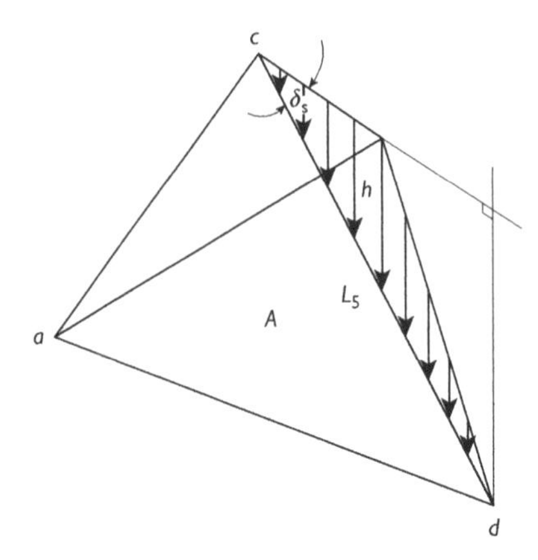

Figure 2.23 Water pyramid on joint plane *A*.

which is just the force of water acting normal to joint plane A. For joint plane B

$$W_w = \gamma_w A_B\left(\frac{1}{3}h\right) = \gamma_w A_B\left(\frac{1}{3}\right)\left[\frac{1}{2}L_5\sin(\delta_s)\right] = P_B \tag{2.37}$$

If a tension crack is present, then these forces need to be reduced by the forces cut off by the crack. These forces are

$$\begin{aligned} P'_A &= \gamma_w A'_A\left(\frac{1}{3}\right)\left[\frac{1}{2}l_5\sin(\delta_s)\right] \\ P'_B &= \gamma_w A'_B\left(\frac{1}{3}\right)\left[\frac{1}{2}l_5\sin(\delta_s)\right] \end{aligned} \tag{2.38}$$

where the prime on areas signifies cutoff areas of the joint planes. The water force acting normal to a tension crack is also given by a "pyramid" formula. Thus,

$$W_w = \gamma_w A_c\left(\frac{1}{3}h_c\right) = \gamma_w A_c\left(\frac{1}{3}\right)\left[\frac{1}{2}l_5\sin(\delta_s)\right] = P_c \tag{2.39}$$

which generally has uplift and downhill components because of the tension crack dip. Average pressure over any surface is just weight divided by area. Thus, the average water pressure acting on a triangle is just one-third the maximum in comparison with one-half maximum for a rectangle.

If the vertical projection of the half-depth point intersects the face, so $\tan(\beta) < 2\tan(\delta_s)$ where β is the face angle, then a different maximum pressure should be used, as was the case for planar block slides. In this case the vertical distance should be

$$y_w = \left[\frac{1}{2}L_5\sin(\delta_s)\right]\left[\frac{\tan(\beta)}{\tan(\delta_s)} - 1\right] \tag{2.40}$$

and the maximum pressure is $\gamma_w y_w$.

Table 2.2 Dip and directions, cohesion, and friction angles for a wedge

Property plane	*Dip dir. (α,°)*	*Dip (δ,°)*	*Cohesion (c, psf)*	*Friction angle (φ,°)*
A-joints	−30	60	1440	29
B-joints	120	60	720	23
Face	45	75	—	—
Foreland	45	0	—	—
Upland	45	0	—	—
Tension crack	45	80	0	35

Example 2.13 Data for a potential wedge failure by sliding down the line of intersection are given in Table 2.2

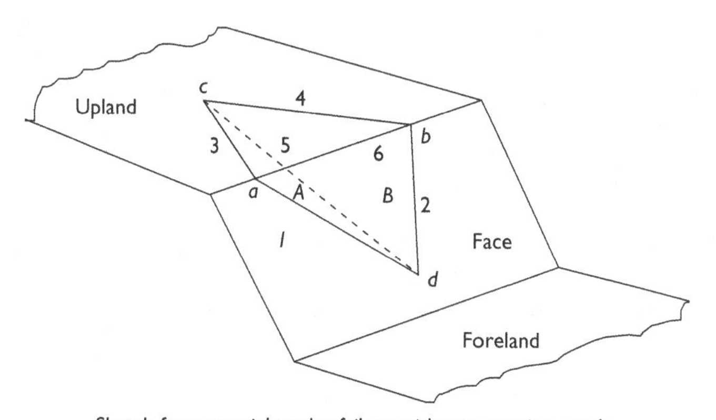

Sketch for potential wedge failure without a tension crack

Assume the tension crack is not present, then determine dip and dip direction of the line of intersection of joints A and B.

***Solution*:** First determine the joint plane normals $\mathbf{n}_A$ and $\mathbf{n}_B$. From Table 2.1, $\mathbf{n}_A$ = sin(60)sin(−30), sin(60)cos(−30), cos(60) $= (-\sqrt{3}/4),\ (3/4),\ (1/2)$, $\mathbf{n}_B$ = sin(60) sin(120), sin(60)cos(120), cos(60) $= (3/4),\ (-\sqrt{3}/4),\ (1/2)$ which are seen to be unit vectors as they should be.Next form the cross product $\mathbf{n}_A \times \mathbf{n}_B$,

$$\vec{n}_A \times \vec{n}_B = \begin{vmatrix} x & y & z \\ -\sqrt{3}/4 & 3/4 & 1/2 \\ 3/4 & -\sqrt{3}/4 & 1/2 \end{vmatrix}$$

$$= \left(\frac{3+\sqrt{3}}{8}\right),\ \left(\frac{3+\sqrt{3}}{8}\right),\ \left(-\frac{3}{8}\right)$$

$$\vec{n}_A \times \vec{n}_B = (0.5915,\ \ 0.5915,\ \ -0.3750)$$

The length of a vector **S** pointing down the line of intersection from this result is 0.9167, so the direction cosines of **S** are the direction numbers divided by the length, that is,

(0.6452, 0.6452, −0.4091) which constitute a vector of unit length. An angle with cosine −0.4091 is 114.1°. Hence the dip of the line of intersection $\delta_s = 24.1°$.

The azimuth of the line of intersection is, by inspection, 45°. Also,

$$\tan(\alpha_s) = \frac{s_x}{s_y} = \frac{0.6452}{0.6452} = 1$$

which confirms the intuitive result that $\alpha_s = 45°$.

Example 2.14 Given the data in Example 2.13 but without a tension crack, determine the length of Line 1, if the vertical distance between points *a* and *d* is 68 ft.

***Solution*:** The lines in the sketch in Example 2.13 are all lines of intersection between the various planes *A, B, F* (face), and *U* (upland). In particular, Line 1 is the intersection of joint plane *A* and the face plane *F*. The cross product of the normals to these planes results in a vector pointing along Line 1. The components of this vector allow for the determination of dip and dip direction of the line. The solution procedure is similar to that for the line of intersection of the joint planes in Example 2.13.

The normal vectors from the formulas in Table 2.1 are

$$\mathbf{n}_A = \sin(60)\sin(-30),\ \sin(60)\cos(-30),\ \cos(60) = (-0.4330),\ (0.7500),\ (0.5000)$$

$$\mathbf{n}_F = \sin(75)\sin(45),\ \sin(75)\cos(45),\ \cos(75) = (0.6830),\ (0.6830),\ (0.2588)$$

$$\vec{n}_A \times \vec{n}_F = \begin{vmatrix} x & y & z \\ -0.4330 & 0.7500 & 0.5000 \\ 0.6830 & 0.6830 & 0.2588 \end{vmatrix}$$

$$\therefore\ \vec{n}_A \times \vec{n}_F = (-0.1474,\ 0.4536,\ -0.8080)$$

The length of this vector is 0.93826, so the direction cosines of the resulting vector **S1** are (−0.1568, 0.4834, −0.8611). An angle whose cosine is −0.8611 is 149.4°, so the dip of this line is $\theta_1 = 59.4°$, as shown in the sketch.

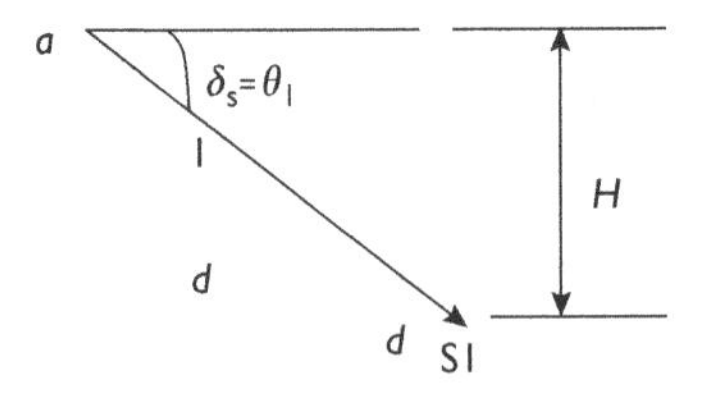

Sketch for finding the length of Line 1

The length of Line 1 is $L1 = H / \sin(\theta_1) = 68/0.8611 = 79.0$ ft.

Example 2.15 Consider the wedge data given in Table 2.2, Example 2.13, and the normal vectors

$$\mathbf{n}_A = \sin(60)\sin(-30),\ \sin(60)\cos(-30),\ \cos(60) = (-0.4330),\ (0.7500),\ (0.5000)$$

$$\mathbf{n}_B = \sin(60)\sin(120),\ \sin(60)\cos(120),\ \cos(60) = (3/4),\ (-\sqrt{3}/4),\ (1/2)$$

$$\mathbf{n}_F = \sin(75)\sin(45),\ \sin(75)\cos(45),\ \cos(75) = (0.6830),\ (0.6830),\ (0.2588)$$

developed from the formulas in Table 2.1. Cross products of $\mathbf{n}_A$ and $\mathbf{n}_B$ lead to a unit vector pointing down the line of intersection **S5** = (0.6452, 0.6452, −0.4091); $\mathbf{n}_A$ and $\mathbf{n}_F$ lead to a unit vector pointing down Line 1 which is the intersection of the face and joint plane A, that is **S1** = (−0.1568, 0.4834, −0.8611). Both lines of intersection vectors are unit vectors. Find the angles between Line 1 and Line 5, between Line 1 and Line 3, and between Line 3 and Line 5.

***Solution*:** The required angle may be obtained from the dot product between **S1** and **S5**. Thus, $\cos(\theta_{15})$ = **S1** · **S5** = (−0.1568)(0.6452) + (0.4834)(0.6452) + (−0.8611)(−0.4091) = 0.5616, so $\theta_{15} = 55.8°$.

The angles θ_{13} and θ_{35} require a vector associated with Line 3. Line 3 is formed by the intersection of the upland and joint plane A, so a cross product of normals to U and A is needed. The normal to A is known, and using the formula for a normal to a plane, the normal to U is

$\mathbf{n}_U$ = sin(0)sin(45), sin(0)cos(45), cos(0) = (0.0, 0.0, 1.0) which is intuitively correct as a vertical, upward pointing vector of unit length.

The cross product $\mathbf{n}_U$ x $\mathbf{n}_A$ = **S3** =(−0.7500, −0.4330, 0.0000) which points from point a to point c. The direction cosines of **S3** = (−0.8660,− 0.5000, 0.0000). Both the x- and y-components are negative, while the z-component of **S3** is zero, so the vector points into the third quadrant. The angles are −150° (or +210°) from the x-axis, and 120° from the y-axis. This result could also be obtained from the dip direction of A and the fact that the upland is flat, so the azimuth of Line 3 is known (and has zero dip).

The angle θ_{13} can be obtained with the dot product. Thus, $\cos(\theta_{13})$ = (−0.8660)(−0.1568) + (−0.5000)(0.4834) + (0)(−0.8611) = −0.1059. The negative sign indicates an angle greater than 90°. Thus θ_{13} =96.1°.

The third angle of the A joint plane triangle is $\theta_{35} = 180° - \theta_{13} - \theta_{15}$ = 180−96.1−55.8 = 28.1°. This result should be checked by using the dot product of vectors **S3** and **S5**. However, for this calculation, the direction of **S3** should be reversed. Thus, −**S3** · **S5** = $\cos(\theta_{35})$ = (0.8660, 0.5000, 0.0000) · (0.6452, 0.6452, −0.4091) = 0.8813. An angle with cosine 0.8813 is 28.2°. Thus, θ_{35} = 28.2° to within roundoff accuracy.

Example 2.16 Consider the wedge data given in Table 2.2, and results from previous examples:

$$\mathbf{S1} = (-0.1568,\ 0.4834,\ -0.8611),$$

$$\mathbf{S3} = (-0.8660,\ -0.5000,\ 0.0000),$$

$$\mathbf{S5} = (0.6452,\ 0.6452,\ -0.4091),$$

which are unit vectors with direction cosines in parentheses and

$$\theta_{15} = 55.8^\circ,$$
$$\theta_{13} = 96.0^\circ,$$
$$\theta_{35} = 28.2^\circ,$$

which are the angles between sides of the A joint plane triangle where $L1$ = 79.0 ft and roundoff error is compensated so the angles total 180°. Determine the area of joint plane A.

Solution: Use the cross product to determine the area. In this case we may choose from

$$\frac{L1}{\sin(\theta_{35})} = \frac{L3}{\sin(\theta_{15})} = \frac{L5}{\sin(\theta_{13})}$$

$$L3 = \frac{79.0}{\sin(28.2)}\sin(55.8) = 138.3 \text{ ft}$$

$$L3 = \frac{79.0}{\sin(28.2)}\sin(96.0) = 166.3 \text{ ft}$$

several combinations. For example, **S1** × **S3** = 2A. Thus, $2A = |\vec{S}1||\vec{S}3|\sin(\theta_{13})$. Here, one views the line vectors having the direction of the unit vectors along $L1$ and $L3$, but also having an actual physical length. In case of **S**1, the length (magnitude) is 79.0 ft. Before proceeding, then one must determine the length of $L3$. A calculation using the sine law serves the purpose. Hence, 2A =(79.0)(138.3)sin(96.0) = 10, 865, and A = 5, 433 ft^2 (joint plane A).

Example 2.17 Consider the wedge data given in Table 2.2, and results from previous examples:

$$\mathbf{S1} = (-0.1568, 0.4834, -0.8611),$$
$$\mathbf{S3} = (-0.8660, -0.5000, 0.0000),$$
$$\mathbf{S5} = (0.6452, 0.6452, -0.4091),$$

which are unit vectors with direction cosines in parentheses and

$$\theta_{15} = 55.8^\circ,$$
$$\theta_{13} = 96.0^\circ,$$
$$\theta_{35} = 28.2^\circ,$$

which are the angles between sides of the A joint plane triangle where $L1$ = 79.0 ft, $L3$ = 138.3 ft, and joint plane area A is 5433 ft^2, and roundoff error is compensated so the angles total 180°. Determine the volume of the wedge.

Solution: The volume of the wedge is the volume of the tetrahedron that forms the wedge. Tetrahedron volumes are given by the vector triple product, that is, V = (1/6)**abc** where the vectors have their origin at one of the vertices of the tetrahedron and have lengths equal to corresponding edge lengths. The triple product may be written as $\mathbf{a} \cdot \mathbf{b} \times \mathbf{c}$. In this problem one may identify vectors **b** and **c** as **S**1 and **S**3; **a** is then identified as a vector associated with Line 6 **S**6, all of which have lengths equal to the respective line lengths. They are not unit vectors for this calculation. The cross product part of the triple product is just twice the A joint plane area, so $V = (1/6)L_6\,(2A)\cos(\theta)$ where L is the magnitude of **a** or length of Line 6 and θ is the angle between the normal to A and $L6$. The product $L_6\cos(\theta)$ is the altitude of the "pyramid" with base $A, V = (1/3)hA$ which is just the formula for volume of a pyramid. Thus, the problem requires knowing **S**6 (direction cosines and length).

The required information may be obtained in several ways. Because of the flat upland (horizontal) in this particular case, one may proceed intuitively by first determining the angles in the triangle that forms the top of the wedge in the upland. The sine law then yields lengths of the two unknown edges from the trace length of Line 3 computed earlier. In the general case, one must resolve angles and lengths of the face facet of the wedge.

In this case, examination of the geometry of the upland facet shown in the sketch results in $\theta_{34} = 30°$, $\theta_{36} = 75°$, $\theta_{46} = 75°$.

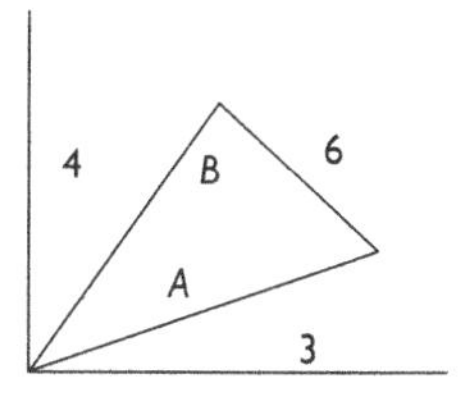

According to the sine law

$$\frac{L_3}{\sin(\theta_{46})} = \frac{L_4}{\sin(\theta_{36})} = \frac{L_6}{\sin(\theta_{34})}$$

$\therefore$

$$L_4 = \frac{L_3}{\sin(\theta_{46})}\sin(\theta_{36}) = \frac{138.3}{\sin(75)}\sin(75) = 138.3 \text{ ft}$$

$$L_6 = \frac{L_3}{\sin(\theta_{46})}\sin(\theta_{34}) = \frac{138.3}{\sin(75)}\sin(30) = 71.6 \text{ ft}$$

Also by inspection of the sketch and in consideration of the given data, the direction cosines of **S6** are $(-1/\sqrt{2},\ 1/\sqrt{2},\ 0)$, a unit vector. When associated with $L6$ length, **S**6 (−50.6, 50.6, 0). The normal to joint plane A is given by the formulas in Table 2.1. Thus, $\mathbf{n}_A$ = [sin(60)sin(330), sin(600)cos(330), cos(60)] = (−0.4330, 0.7500, 0.5). The cosine of the angle between the normal to joint plane A and $L6$ is the dot product, that is, $\cos(\theta)$= (−0.4330)(−0.7071) + (0.7500)(0.7071) (0.5)(0.0) = 0.8365. Hence θ = 33.23°. This angle is just the θ needed to complete the volume calculation. Thus, V = (1/3)cos(33.23)(71.6)(5433) = 108,464 ft^3.

Example 2.18 Consider the wedge data in Example 2.13 and further suppose no tension crack is present but a water table exists at the surface. The water table coincides with the slope face. Use an approximation that the water pressure increases linearly with depth to a point half the distance to the toe of the wedge and then decreases linearly to the toe. Determine the water force on joint plane A that has area 5,433 ft^2.

Solution: The water force P is just the integral of water pressure p over the area A. The assumed water pressure distribution down the line of intersection is shown in the sketch.

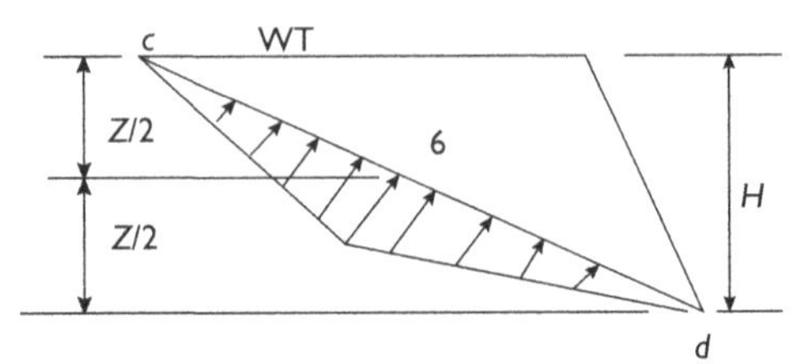

Water pressure distribution down the line of intersection

In this case, the maximum water pressure $p(\max) = \gamma_w (Z/2) = (62.4)(68/2) = 2{,}122$ lbf/ft^2. The water pressure distribution over the joint plane A must meet the requirements of zero pressure along the traces of the joint plane on the upland and on the face.

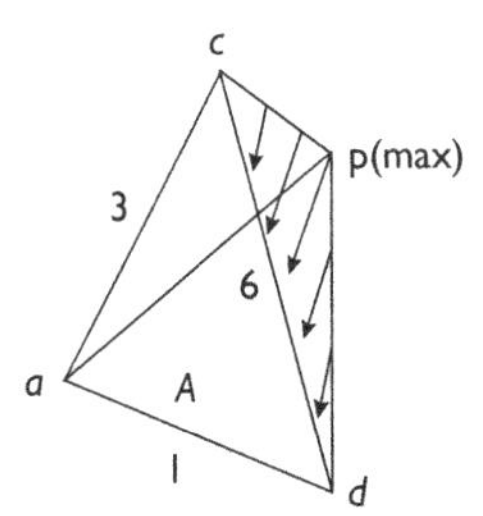

Water pressure distribution on joint plane A

An integration of this water pressure distribution gives a general formula for the water force. Thus, $P = p(\max)A/3$ where A is the area of the wet plane surface. The average water pressure over a triangular area is seen to be just one-third the maximum pressure. In this particular case, the water force $P = (2{,}122/3)(5{,}433) = 3.84(10^6)$ lbf on joint plane A.

Example 2.19 The wedge in Example 2.8 has volume $V = 108{,}464$ ft^3. With a specific weight of 158 lbf/ft^3, the weight is $17.14(10^6)$ lbf. Given the line of intersection dip direction $\alpha_s = 45°$ and dip $\delta_s = 24.1°$, determine the normal forces acting on joint planes A and B.

Solution: Normal forces must satisfy equilibrium; forces are illustrated in the sketches.

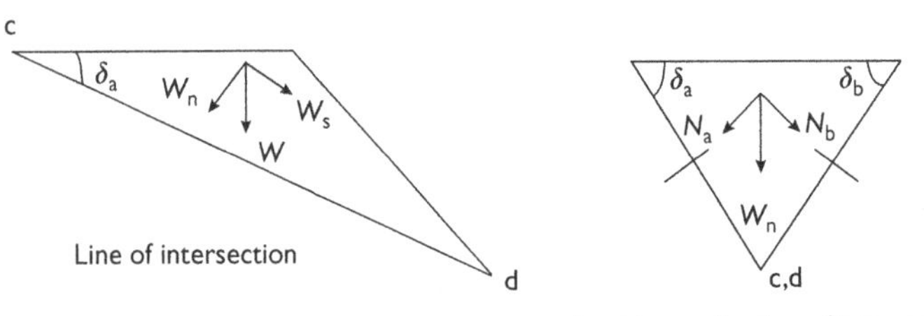

Sketches of forces needed for normal force determination

The normal and tangential components of weight are

$$W_n = W\cos(\delta_s) = 17.14(10^6)\cos(24.1) = 15.64(10^6)\text{ lbf}$$
$$W_n = W\sin(\delta_s) = 17.14(10^6)\sin(24.1) = 7.000(10^6)\text{ lbf}$$

Equilibrium requires

$$W_n = N_a\cos(\delta_a) + N_b\cos(\delta_b)$$
$$0 = N_a\sin(\delta_a) + N_b\sin(\delta_b)$$

Hence, the normal forces are

$$N_a = 15.64(10^6)\left[\frac{\sin(\delta_b)}{\sin(\delta_b+\delta_a)}\right] = 15.64(10^6)\frac{\sin(56.7)}{\sin(113.4)} = 14.24(10^6)$$
$$N_b = 15.64(10^6)\left[\frac{\sin(\delta_a)}{\sin(\delta_b+\delta_a)}\right] = 15.64(10^6)\frac{\sin(56.7)}{\sin(113.4)} = 14.24(10^6)$$

Determination of the angles δ_a and δ_b was done with the aid of the formulas from the text. These forces are equal as they should be in view of the symmetry of the wedge data. Thus,

$$\tan(\delta_a) = \sin(\delta_s)\tan(\alpha_s - \alpha_a) = \sin(24.1)\tan(45-(-30)) = 1.525$$
$$\tan(\delta_b) = \sin(\delta_s)\tan(\alpha_b - \alpha_s) = \sin(24.1)\tan(120-45) = 1.525$$
$$\therefore$$
$$\delta_a = \delta_b = 56.7^\circ$$

Example 2.20 Given the wedge data from Example 2.13 where the downhill component of wedge weight is $7.000(10^6)$ lbf, the normal forces acting on joint planes A and B are $14.24(10^6)$ lbf, and each joint plane area is 5,433 ft^2, determine the wedge factor of safety (a) dry and (b) wet with water force $P = 3.84(10^6)$ lbf.

***Solution*:** By definition, the factor of safety is

$$\text{FS} = \frac{R}{D}$$
$$= \frac{R_A + R_B}{W_s}$$
$$= \frac{N'_a\tan(\phi_A) + c_A A_A + N'_b\tan(\phi_B) + c_B A_B}{W_s}$$
$$= \frac{14.24(10^6)\tan(29) + 1440(5433) + 14.24(10^6)\tan(23) + 720(5433)}{7.000(10^6)}$$
$$\text{FS} = 3.67(\text{dry})$$
$$= \frac{[14.24(10^6) - 3.84(10^6)]\tan(29) + 1440(5433) + [14.24(10^6) - 3.84(10^6)]\tan(23) + 720(5433)}{7.0(10^6)}$$
$$\text{FS} = 3.13(\text{wet})$$

2.2 Rotational slope failures

Two important types of rotational slope failures are (1) a *conventional* reverse rotation along a surface that is often approximated as a circular arc transecting a soil-like material and (2) a forward rotation associated with *toppling* of rock blocks. Although such failures may initially have a rigid-body character, slide masses deform considerably and often disintegrate as motion continues beyond the stage of incipient failure. Soil slopes may be natural slopes, cut slopes for roads or built-up slopes such as waste dumps, leach heaps, tailings dams, canal banks, water retention dams, and so forth.

Stability analysis of soil slopes shares basic principles with slopes in jointed rock masses, but the expected mode of failure is different. Rotational failure is the norm. Figure 2.24 illustrates a rotational failure that occurs along a segment of a circular arc. A safety factor with respect to moments is appropriate. Thus,

$$FS = \frac{M_R}{M_D} \tag{2.41}$$

which is simply the ratio of resisting to driving moments.

With reference to Figure 2.24, moment equilibrium about the center of rotation O requires

$$\int_V r\sin(\alpha)\gamma \mathrm{d}V = \int_A R\tau \mathrm{d}A \tag{2.42}$$

where r is the radius to a typical volume element in the slide mass, γ is specific weight of slide mass material, R is the radius of the "slip" circle, V is volume of the slide mass, and A is the area of the failure surface which is the length of the circular arc multiplied by thickness of the section, b. Evaluation of the integral on the right clearly requires knowledge of the distribution of stress over the failure surface.

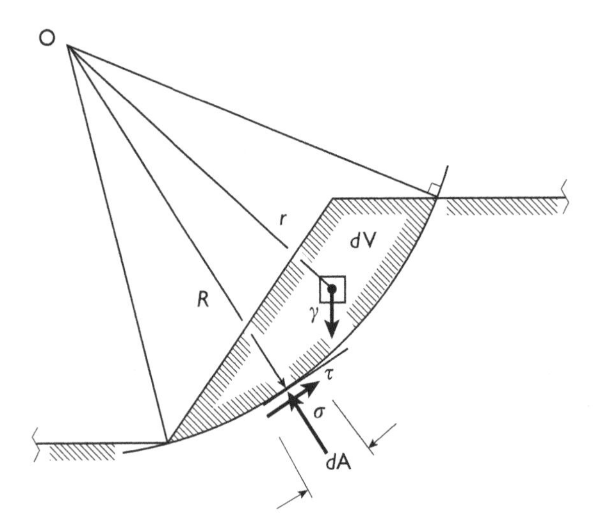

Figure 2.24 Moments from surface and body forces for a slope failure along a circular arc.

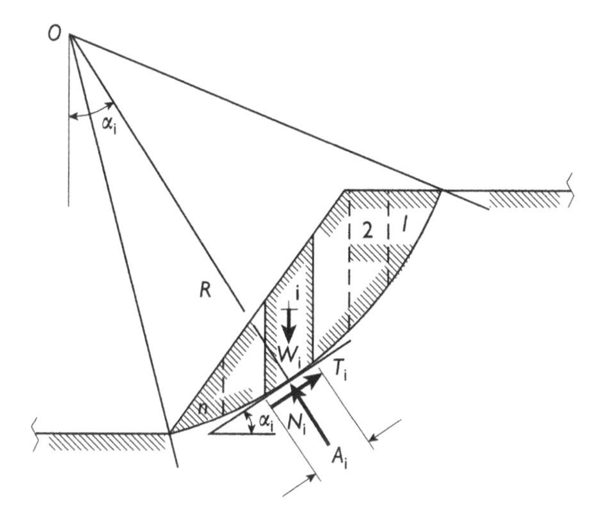

Figure 2.25 Circular arc slope failure and subdivision into slices. Slices are numbered 1, 2, ..., *n*.

The *method of slices* is a way of approximating the overall equilibrium requirement of the slip circle mass. There are many varieties of the method; an important beginning is the simplified Bishop method. Figure 2.25 shows the same slope divided into vertical slices. Replacement of integration by summation results in the approximation,

$$\sum_{1}^{n} R\sin(\alpha_i)W_i = \sum_{1}^{n} RT_i(\text{stress}) = \sum_{1}^{n} RT_i(\text{strength})/\text{fs}_i \tag{2.43}$$

where a local safety factor fs is introduced for each slice i as the ratio of strength to stress and there are n slices in the slope. If the slice safety factors are all the same, that is, $\text{fs}_i = \text{constant} = \text{FS}$, then assuming a Mohr–Coulomb failure criterion

$$\text{FS} = \frac{\sum_1^n RT_i(\text{strength})}{\sum_1^n R\sin(\alpha_i)W_i} = \frac{\sum_1^n N_i' \tan(\phi_i) + c_iA_i}{\sum_1^n \sin(\alpha_i)W_i} \tag{2.44}$$

where the prime denotes effective normal force and radius R is a constant. Terms in the denominator are just "downhill" or driving components of slice weights $W_s(i)$.

Effective normal slice forces are estimated from an approximate slice equilibrium analysis. Figure 2.26 shows forces acting on a typical slice. With neglect of the side forces, the total normal force is simply the normal component of slice weight and the effective normal force is the normal component of weight less the water force. Thus, the slope safety factor is now

$$\text{FS} = \frac{\sum_i [(W_n - P)\tan(\phi) + cA]}{\sum_i W_s} \tag{2.45}$$

where all quantities after the summation signs may *vary* from slice to slice. This feature of the method of slices allows for considerable complexity to be taken into account during a stability analysis.

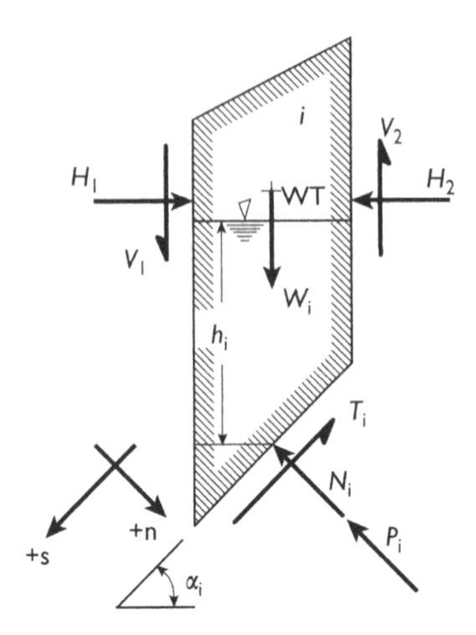

Figure 2.26 Slice free body diagram.

Example 2.21 Consider a 45° slope, 150 ft high with a potential circular arc failure surface that is a quarter-circle with moment center shown in the sketch. Suppose $\phi = 25°$, $c = 1{,}440$ psf, $\gamma = 100$ pcf, and the water table is below the slope toe. Determine the slope safety using just five slices of equal width (30 ft). Number the slices beginning at the toe, so slice 5 is at the crest.

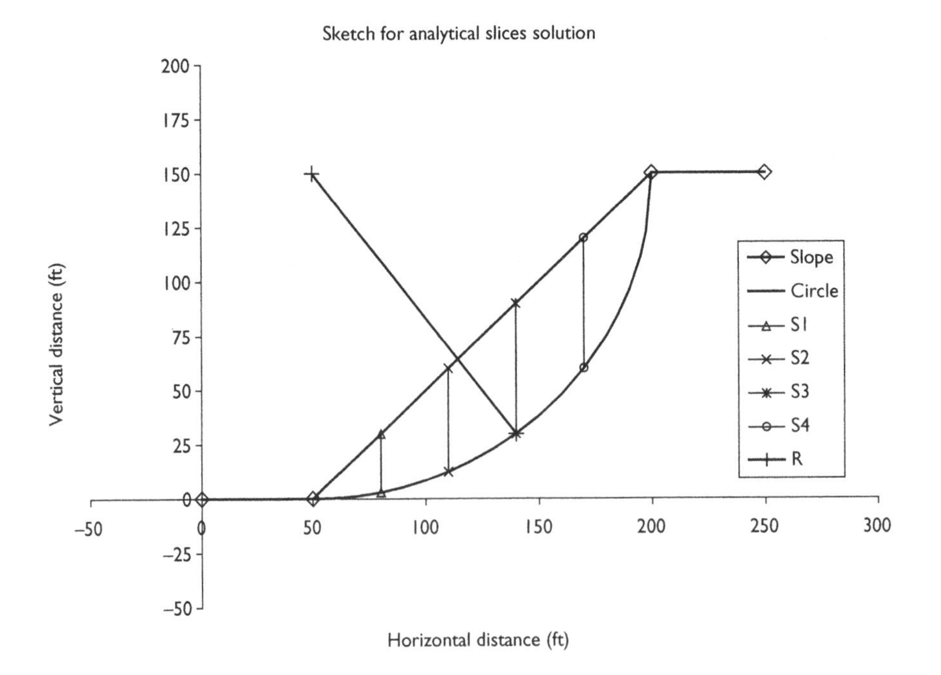

Solution: There are several avenues leading to a solution. One avenue is graphical where a carefully drawn diagram to a suitable scale allows for measurements that need to be made from the diagram. Another is to do calculations by hand. A third and, perhaps, the most efficient is to use a spreadsheet guided by analytical expressions for the slope face and slip circle.

Table 2.3 presents slice geometry data from an evaluation of analytical expressions, while volume and weight data are given in Table 2.4.

The angle of inclination of a slice base from the horizontal (α) is calculated from the radius R and the horizontal distance to the midplane of the slice. For example, in case of slice 3, sin (α) = (1/2)(60 + 90)/150 = 0.5 and $\alpha = 30°$. The chord length of a slice base can be computed from slice width and angle. In case of slice 3, L = 30/cos(30°) =34.64 ft.

The factor of safety with respect to moments for the considered failure surface is

$$\text{FS} = \frac{(150)(211,783 + 298,186)}{(150)(325,154)} = 1.568$$

A computer code result gives 1.536 for simplified Bishop and 1.584 for Janbu methods where 25 slices were used in the computations. The difference in the coarse, 5-slice calculation and the finer 25-slice is about 2% using the simplified Bishop method. The Janbu method is somewhat optimistic compared with the simplified Bishop method in this example with a safety factor that is higher by about 3% according to computer code results. The difference is less when compared with the example, the spreadsheet result, which is about 1% lower than the Janbu result. All three results show a safety greater than 1.5 and would indicate a safe, although rather steep slope.

Water forces are calculated as the product of average pressure and wetted area. Thus, for each slice below the water table

$$P_i = \bar{p}A_i = \gamma_w h_i A_i \tag{2.46}$$

where h_i is the *average head* of water in the ith slice, $A_i = bl_i$ is the area of the base of the ith slice, b is breadth, and l_i is slope length. The average head may be approximated by the vertical distance from water table to midpoint of the slice bottom shown in Figure 2.26.

Example 2.22 Consider the data in Example 2.21 and further suppose a water table is present at the crest and follows the face to the toe of the bank. Determine the safety factor of the given slip circle under this wet condition, again using five slices.

Solution: After adding a column for water force to the given data, one has Table 2.5 data.

Water forces P are calculated using the formula $P = \gamma\ h(\text{ave})Lb$. For example, in case of slice 3, P = (62.4)(53.74)(34.64)(1) = 116,161 lbf.

Table 2.3 Slice points, side lengths, and average height

Slice	*x (ft)*	*y (ft)*	*h (ft)*	*h (ave ft)*
	0.00	0.00	0.00	
1				13.48
	30.00	3.03	26.97	
2				37.22
	60.00	12.52	47.78	
3				53.74
	90.00	30.00	60.00	
4				60.00
	120.00	60.00	60.00	
5				30.00
	150.00	150.00	0.00	

Origin is at the slice toe.
Slice width Δx = 30 ft.
Length of side h is distance between slope face and slip circle.
Average length is the mean of the side lengths, the height of the midplane.

Table 2.4 Slice volume, weight, and force data

Slice	*Volume* (ft^3)	*Weight* (*lbf*)	*Angle* (°)	W_n (*lbf*)	W_s (*lbf*)	*Ntan* (ϕ) (*lbf*)	*cLb* (*lbf*)
1	404.54	40,454	5.74	40,251	4,045	18,769	43,418
2	1,116.70	111,670	17.46	106,526	33,501	49,674	45,286
3	1,612.16	161,216	30.00	139,617	80,608	65,105	49,883
4	1,800.00	180,000	44.43	128,546	126,000	59,942	60,492
5	900.00	90,000	64.16	39,230	81,000	18,293	99,108
Total	5, 833.4	583,340			325,154	211,783	298,186

Slices are b = 1 ft thick into the page.
Volume = h(ave)$b\Delta x$.
$W_n = W\cos(\alpha)$.
$W_s = W\sin(\alpha)$.

Table 2.5 Slice volume, weight, and force data (wet case)

Slice	*Volume* (ft^3)	*Weight* (*lbf*)	*Angle* (°)	W_n (*lbf*)	*P* (*lbf*)	W_s (*lbf*)	$N'tan(\phi)$ (*lbf*)	*cLb* (*lbf*)
1	404.54	40,454	5.74	40,251	25,371	4,045	6,939	43,418
2	1,116.70	111,670	17.46	106,526	73,047	33,501	15,612	45,286
3	1,612.16	161,216	30.00	139,617	116,161	80,608	10,938	49,883
4	1,800.00	180,000	44.43	128,546	157,290	126,000	−13,399	60,492
5	900.00	90,000	64.16	39,230	128,840	81,000	−41,786	99,108
Total	5,833.4	583,340				325,154	−21,696	298,186

Slices are b =1 ft thick into the page. Volume= h(ave)$b\Delta x$. $W_n = W\cos(\alpha)$. $W_s = W\sin(\alpha)$.

The factor of safety with respect to moments for the considered failure surface is

$$\mathrm{FS} = \frac{(150)(-21,696 + 298,186)}{(150)(325,154)} = 0.850$$

A computer code gives 0.840 for simplified Bishop and 0.800 for Janbu methods where 25 slices were used in the computations. All results indicate an unsafe slope.

Although one example is not proof, the simplified Bishop method is generally optimistic in cases of high water tables. However, all results cited indicate the need for remedial measures, one of which is, obviously, drainage.

The presence of negative effective normal forces indicates water forces in excess of the normal component of slice weight. This situation is physically acceptable (and may occur on steeply inclined slices) provided cohesion is adequate to support the indicated tension (negative force). Otherwise a negative shear strength would be implied in which case slice resistance should be set to zero.

Seismic, surcharge, and bolting forces may be included in a stability analysis by the method of slices. The seismic load on the ith slice $S_i = W_i a_s$ which is directed horizontally toward the slope face. Surcharge forces F_j act only on slices j supporting the surcharge and act vertically. Bolt forces are applied to slice bottoms. Resolution of these forces into normal (and tangential) components then modifies the effective normal force through the simplified normal equilibrium equation. However, the moment arms are more easily calculated for the total forces. Figure 2.27 shows seismic, surcharge, and bolting forces acting on a slice and the moment arms associated with each force. Thus,

$$\begin{aligned} M_S(i) &= R_c(i)\cos(\alpha_{ic})S(i) \\ M_F(i) &= R\sin(\alpha_i)F(i) = RF_s \\ M_B(i) &= RT_b(i) \end{aligned} \tag{2.47}$$

where moments associated with seismic and surcharge forces are driving, while the bolting moments associated with the tangential component of bolting forces are resisting. Subscript c refers to the slice center, as shown in Figure 2.27 and T_B is the tangential bolting force component: $T_B = F_B\cos(\alpha - \delta)$ where δ is the bolting angle, which may be negative if the bolt hole is down.

The factor of safety with respect to moments with seismic surcharge and bolt forces is

$$\mathrm{FS} = \frac{\sum_i N'\tan(\phi) + cA}{\sum_i W_s + (M_S/R) + F_s} + \frac{\sum_i F_B[\cos(\alpha - \delta - \phi)][1/\cos(\phi)]}{\sum_i W_s + (M_S/R) + F_s} \tag{2.48}$$

where $N' = W_n - P - S_n + F_n$. If the effective normal force is negative (tensile), then frictional resistance is nil and the corresponding contribution to the summation is zero. Not all terms may be present in every slice. For example, if surcharge or bolting forces are absent, then no contributions from the considered slice are made to the summations. If the bottom of a slice is above the water table, no water force is present in the expression for effective normal force

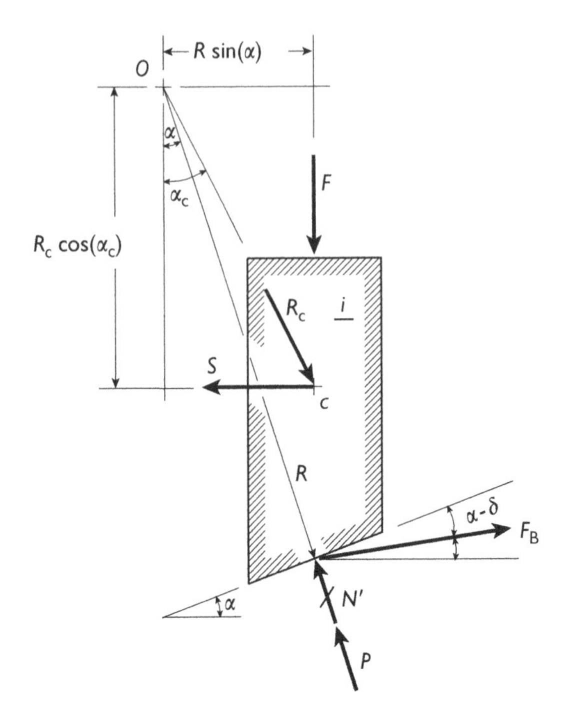

Figure 2.27 Seismic, surcharge, and bolting forces acting on the *i*th slice. Point *c* is slice center.

for that slice. Again, cohesion and angle of internal friction may vary from slice to slice. The second term on the right is associated with bolting and shows the safety factor improvement that may be obtained by bolt reinforcement. The presence of the slip circle radius R in this expression is a reminder that the mode of failure is rotational and the safety factor is with respect to moments.

In some cases the slip circle may dip below the toe of the slope as shown in Figure 2.28. Inspection of Figure 2.28 shows that tangential components of weight to the left of the bottom of the slip circle (O') oppose rotation and are therefore associated with resisting moments rather than driving moments. These resisting moments need to be added to the other resisting moments rather than to the driving moments. Symbolically then with only weight forces considered,

$$\mathrm{FS} = \frac{\sum_i M_R + \sum_j RW_s}{\sum_{i-j} RW_s} \tag{2.49}$$

where there are i total slices in the slide mass, j resisting slices to the left of the slip circle bottom, and $i-j$ driving slices to the right.

Example 2.23 In some cases, the slip circle may dip below the toe of a slope, as shown in the sketch. In such cases, some slices have an inclination of the tangential component of

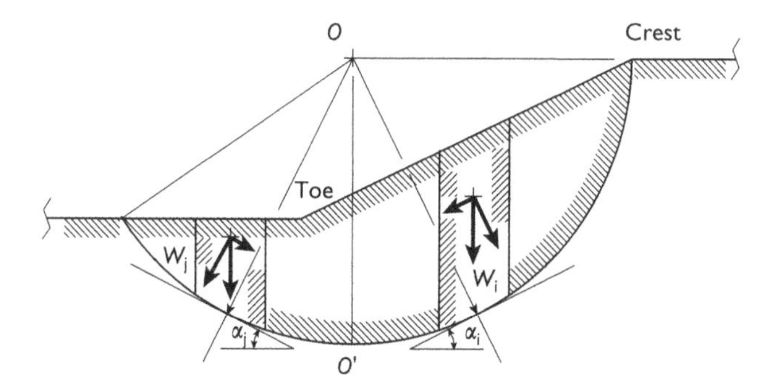

Figure 2.28 Slip circle dips below slope toe.

weight that opposes rotation of the slide mass. Modify the safety factor expression to accomodate the case where the slip circle dips below the toe.

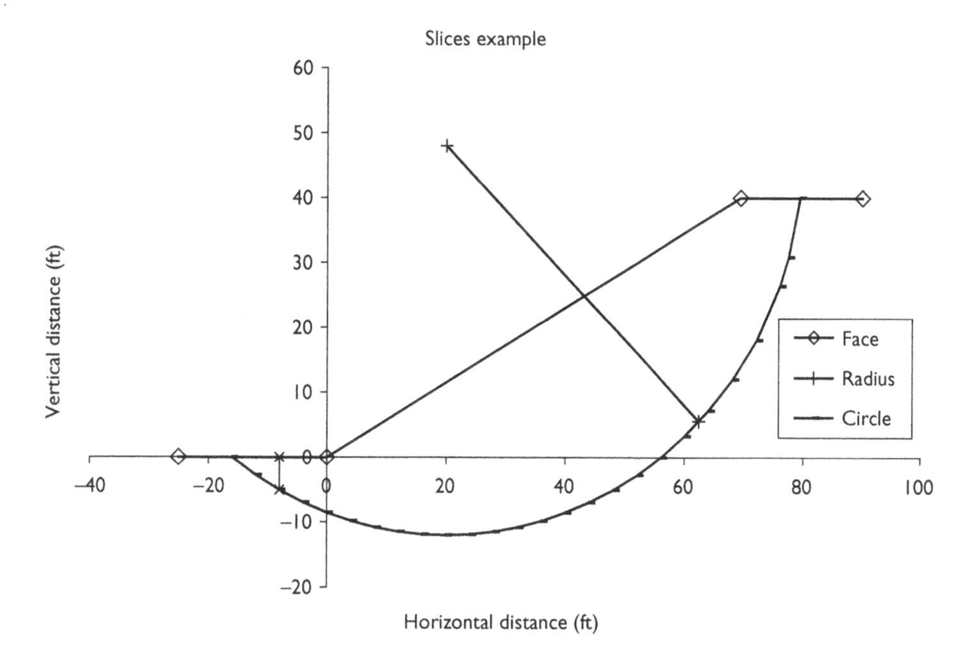

***Solution*:** The safety factor equation for the case where the slip circle does not dip below the toe is

$$\text{FS} = \frac{M_R}{M_D} = \frac{\sum_1^n [N_i' \tan(\phi) + C_i]R}{\sum_1^n W_i \sin(\alpha_i)R}$$

where W = weight, α = inclination of slice bottom chord from the horizontal, R = circle radius, C = cohesive force, N' = effective normal force, and ϕ = angle of internal friction. The number of slices is n. Modification to allow for the case when the slip circle dips below the toe requires examination of the moments resisting and driving, M_R, M_D, respectively. After distinguishing between slices with tangential components of weight opposing and driving the motion, one has

$$FS = \frac{\sum_1^n [N_i' \tan(\phi_i) + C_i]R + \sum_1^m RW_i \sin(\alpha_i)}{\sum_{m+1}^n RW_i \sin(\alpha_i)}$$

where m is the number of slices that have tangential components of weight that oppose the motion. These slices contribute to resisting moments. Slices $m + 1$ to n have tangential components of weight that tend to drive the motion.

An important question in slip circle stability analyses is whether the considered slip circle is the most dangerous one that has the lowest factor of safety. Consideration of a different potential slip circle would result in a factor of safety that could be greater or less than the safety factor from the first analysis. Clearly, a search is necessary to discover the slip circle with the lowest safety factor. A new slip circle could obviously be generated by simply increasing the slip circle radius R from the same center of rotation. A change in the center of rotation using the original slip circle radius would also generate a new slip circle. As a search is conducted, there may be multiple minimums because there is no guarantee that two different slip circles will not have the same minimum safety factor. For example, there may be a small slip circle entirely in the slope face that has a low safety factor and another slip circle well into the slope that also has a low safety factor. Thus, although a simple hand calculation using a limited number of slices, say, five or six, may suffice for a single analysis, multiple analyses are needed to find the critical slip circle that is associated with a minimum slope safety factor. Moreover, the failure surface may not be circular, so a modified method of slices is required that allows for noncircular slip surfaces.

When multiple analysis or parametric design is done in the presence of different soil types, water tables and so forth, the computational burden increases considerably and, as a practical matter, a computer program is needed for such analyses. There are a number of such programs in use that allow for a variety of forces, water tables including perched water and noncircular slip surfaces that can be used to great advantage once the fundamentals are well understood.

Remedial measures

Measures to improve safety against rotational slope failure must increase the net resisting moment or decrease the net driving moment. A base case where weight provides the driving moment has an associated factor of safety given by

$$FS = \frac{\sum (W \cos(\alpha) - P) \tan(\phi) + cA}{\sum W \sin(\alpha)}$$

where a tacit assumption is made that the slip circle does not dip below the slope toe and summation is over all slices. However, c and ϕ may vary from slice to slice.

Reduction of water forces by depressurization (drainage) is clearly beneficial. Complete drainage would eliminate the negative term in the safety factor expression. This action increases resisting moments and is the method of choice to improve the slope safety factor in wet ground.

Inspection of the last term on the right shows that reduction of weight W increases the contribution of cohesion to the factor of safety and is beneficial. The resisting moment associated with friction is also affected by weight. Thus, reduction of weight decreases both resisting and driving moments, so the situation is unclear. However, if one considers a single slice safety factor with respect to forces or moments for a circular slip surface with neglect of side forces as before, then

$$\mathrm{fs}_i = \frac{W_i \cos(\alpha_i)\tan(\phi_i) + c_i A_i}{W_i \sin(\alpha_i)}$$

Reduction of slice weight Wi clearly increases this slice safety factor. The suggestion is that the slope safety factor FS is also increased by reducing weight.

Weight can be reduced in two ways (1) by decreasing the slope face angle β, and (2) by removing material from the crest. The first applies only to slices that have tops along the slope face, while the second applies to all elements that have driving moments caused by weight. Relieving the crest of weight, as shown in Figure 2.29, effectively reduces the slope angle when the angle is the inclination of a line drawn from toe to crest of slope. Slope height remains unchanged. Crest relief is readily quantified after recalculating weights of the affected slices only.

Another method of improving the safety factor is to buttress the toe. This method is shown in Figure 2.30. The effect of the toe buttress may be computed after assuming an extended failure surface that requires the toe to be pushed horizontally by the slide mass during impending rotation along the considered slip circle. Frictional and cohesive resisting forces are determined and then applied to the toe where the contribution to the net resisting moment is calculated. Thus,

$$M_T = R\cos(\alpha_T)[W_T \tan(\phi_T) + c_T A_T] \tag{2.50}$$

where the subscript T means "toe," W_T is weight of the buttress, ϕ_T is the friction angle between buttress material and base or toe material the buttress contacts, c_T is cohesion

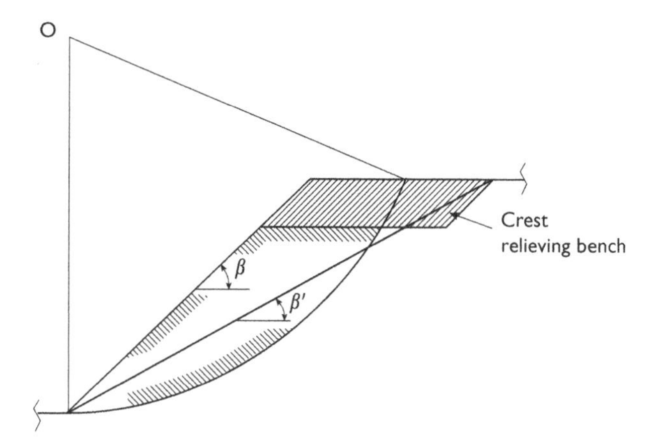

Figure 2.29 Crest relief.

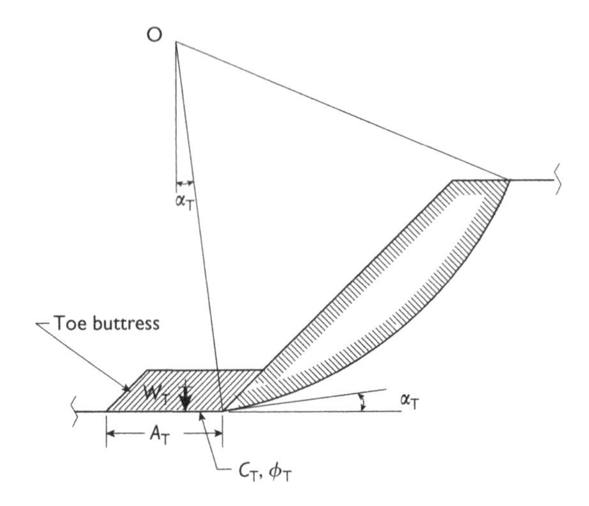

Figure 2.30 Toe buttress for safety factor improvement.

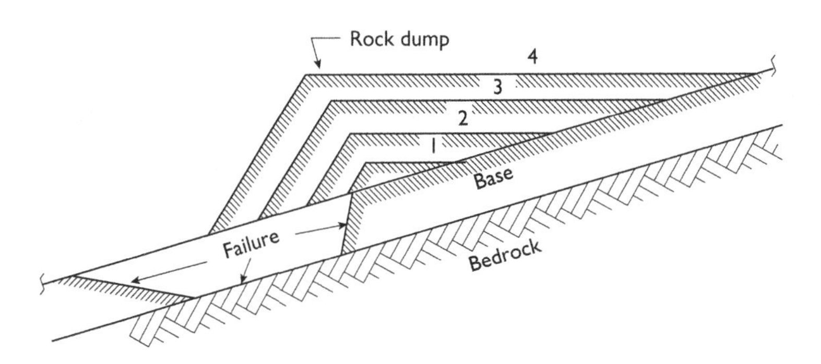

Figure 2.31 Evolving waste dump and potential base failure.

between buttress and base, and A_T is the area of the buttress bottom. This buttress moment is added to the net resisting moment in the safety factor expression. With reference to Figure 2.30, a portion of the buttress extends up the slope face and loads slices below. This weight effect could be introduced into the slice weight calculation directly or perhaps as a surcharge. If so, then the toe weight should be reduced accordingly.

Example 2.24 Consider a circular failure surface with a toe buttress as shown in Figure 2.30. Show that the factor of safety with respect to moments can be decomposed into two parts, one part being the factor of safety without a toe buttress and a second part being the improvement obtained with a toe buttress.

***Solution*:** The safety without a toe buttress is given by equation 2.4. Thus, $FS = M_R/M_D$. With a toe buttress, an additional resisting moment is present according to equation 2.5. Hence,

$$FS = \{M_R + R\cos(\alpha_T)[W_T\tan(\phi_T) + c_T A_T]\}/M_D$$
$$= M_R/M_D + \{R\cos(\alpha_T)[W_T\tan(\phi_T) + c_T A_T]\}/M_D$$
$$= FS_o + \{R\cos(\alpha_T)[W_T\tan(\phi_T) + c_T A_T]/M_D\}$$
$$FS = FS_o + \Delta FS$$

where the terms on the right in the last equation match the terms in the expression above.

Example 2.25 Consider a circular failure surface that intersects a tension crack as shown in the sketch. Suppose the slope is dry, then show that the expression for the slope factor of safety with respect to moments may be computed by simply omitting a slice from the summations involved.

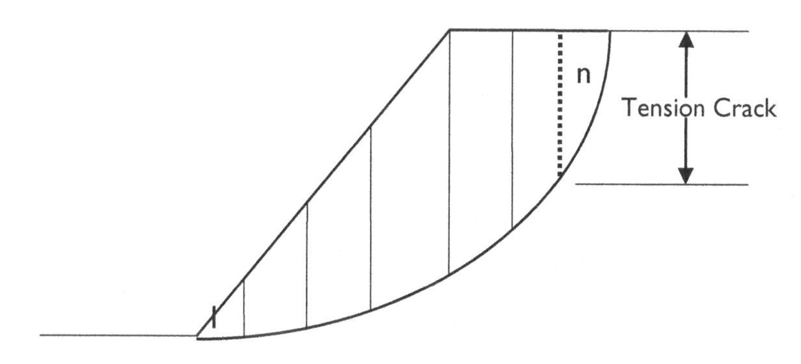

Sketch for Example 2.25

***Solution*:** Inspection of the sketch shows the circular part of the failure surface passing through the bottom of the tension crack. Because the tension crack also is part of the total failure surface, only the material to the left of the tension crack constitutes the slide mass. Hence by simply omitting the nth slice in summations over the slices indicated and therefore leaving the slice to the right of the tension crack out of the safety factor calculation, one can compute the safety in the usual manner.

Note: If the circular failure surface extends through the tension crack, the consequence is the same. Should the failure surface pass below the tension crack, then all slices should be included in the safety factor summations.

Base failures

An important departure from failure over a segment of a circular arc is base failure. Base failure occurs in foundation material below a slope or other "surcharge," for example, as shown in Figure 2.31 where a waste dump is formed. Successive additions to the dump eventually cause the superincumbent load to exceed the bearing capacity of the base. If the base material above bedrock is relatively thin, failure may follow the contact between base and bedrock, as illustrated in Figure 2.31. A failure mechanism then forms that allows for large displacements with very little increase in load; collapse impends.

Bearing capacity is related to base material cohesion, angle of internal friction, and to the contact properties characterized by a different cohesion (adhesion) and friction angle. Water

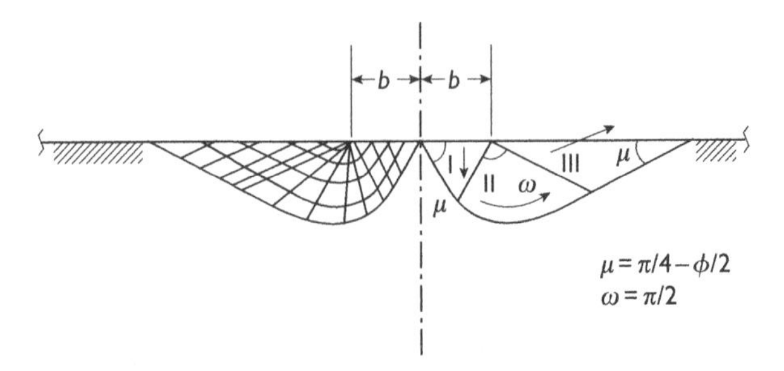

Figure 2.32 Limit equilibrium diagram for simple base failure.

seepage through waste dump and base material further complicate the situation, especially if collection ditches and drains are present. Base failures may also form below tailings dams whether placed on flat ground, hillsides, or in valleys. Any base failure obviously poses a threat to failure of the waste dump or tailings dam above.

A limiting equilibrium solution is rather complex even in an idealized two-dimensional case shown in Figure 2.32 where the length of the base is long relative to the width, $2b$. The reason is the superincumbent load tends to penetrate base material pushing it downward immediately below the load, shearing the material radially at the corner point, and pushing adjacent base material aside. This combination of failure mechanisms is indicated schematically in Figure 2.32. An analytical solution exists for base material that is considered weightless. Thus,

$$\sigma_u = \left(\frac{c}{\tan(\phi)}\right)\left\{\left(\frac{1+\sin(\phi)}{1-\sin(\phi)}\right)\exp[\pi\tan(\phi)] - 1\right\} \tag{2.51}$$

which is the major principal stress acting normal to the (weightless) base material and is the "ultimate" bearing capacity.

Assumption of weightlessness allows the analytical solution to go forward, but there is an associated cost. This solution indicates that a cohesionless sand has zero bearing capacity contrary to reality. Neglect of weight is the cause of this deficiency. Several adjustments have been suggested in the form of adding a pseudocohesion to the formula that is estimated from the effect of weight as a confining pressure on cohesionless material strength. Thus cohesion c in the above formula should be replaced by

$$c' = c + \left(\frac{1}{2}\right)\gamma b\tan(\phi)\sqrt{\frac{1+\sin(\phi)}{1-\sin(\phi)}} \tag{2.52}$$

where γ is specific weight of base material and b is the half-width of the loaded area that is assumed to be long relative to width (two-dimensional view). Analysis of base failures falls naturally into the important category of foundation design analysis that is beyond the scope of this discussion. The main point is to avoid taking base stability for granted in the analysis of embankments of any type.

Example 2.26 Consider a soil-like material with cohesion $c = 1$ MPa, angle of internal friction $\varphi = 30°$, and specific weight $\gamma = 22.5$ kN/m^3. (a) Determine the ultimate bearing capacity of this material, (b) estimate the ultimate bearing capacity, if the cohesion $c = 0$ and the width of the region of load application is 10 m.

***Solution*:** According to equation 2.51,

$$\begin{aligned}
\sigma_u &= \left(\frac{c}{\tan(\phi)}\right)\left\{\left(\frac{1+\sin(\phi)}{1-\sin(\phi)}\right)\exp[\pi\tan(\phi)] - 1\right\} \\
&= \left(\frac{1}{\tan(30)}\right)\left\{\left(\frac{1+\sin(30)}{1-\sin(30)}\right)\exp[\pi\tan(30)] - 1\right\} \\
&= (1.7321)\{(3.0)(6.1337) - 1.0\} \\
\sigma_u &= 30.14 \text{ MPa } (4{,}370 \text{ psi}) \quad \text{(a)}
\end{aligned}$$

In case of a cohesionless, sand-like material, equation 2.52 is used to estimate a pseudo-cohesion for use in equation 2.51. Thus,

$$\begin{aligned}
c' &= c + (1/2)b\gamma\tan(\phi)\sqrt{(1+\sin(\phi))/(1-\sin(\phi))} \\
&= 1(10^3) + (1/2)(5)(22.5)(0.5774)\sqrt{3} \\
c' &= 56.25 \text{ kPa} \\
\sigma_u &= (56.25)(1.7321)\{(3.0)(6.1337) - 1\} \\
\sigma_u &= 1.7 \text{ MPa (b)}
\end{aligned}$$

Example 2.27 A 160,000 lbf load per foot of length is applied to a foundation that is long compared with width $2b$. The material has a cohesion $c = 14.5$ psi, angle of internal friction of $\varphi = 30°$, and specific weight 143.2 lbf/ft^3. Determine the minimum width of foundation needed for a safety factor of 2.0 or greater.

***Solution*:** A safety factor in this situation may be defined as the ratio of resisting to driving forces. Thus, $FS = F_R/F_D$ that must be equal to or greater than 2.0, that is, $FS \geq 2.0$. Therefore, $F_R \geq 2F_D$. The resisting force is given by $F_R = (2b)(\sigma_u)$ where the last term is given by equation 2.51. Accordingly,

$$\begin{aligned}
\sigma_u &= \left(\frac{c}{\tan(\phi)}\right)\left\{\left(\frac{1+\sin(\phi)}{1-\sin(\phi)}\right)\exp[\pi\tan(\phi)] - 1\right\} \\
&= \left(\frac{14.5}{\tan(30)}\right)\left\{\left(\frac{1+\sin(30)}{1-\sin(30)}\right)\exp[\pi\tan(30)] - 1\right\} \\
&= (14.5)(1.7321)\{(3.0)(6.1337) - 1\} \\
\sigma_u &= 437 \text{ psi } (3.01 \text{ MPa})
\end{aligned}$$

The safety factor requirement gives

$$\begin{aligned}
2b &\geq 2(160000)/(387)(144) \\
2b &\geq 5.74 \text{ ft } (1.75 \text{ m})
\end{aligned}$$

Example 2.28 Suppose a material is clay-like and has cohesion c but lacks frictional strength ($\phi = 0°$). Develop a formula for the ultimate bearing capacity(σ_u).

***Solution*:** The formula 2.51 appear indeterminate in this case. However, by plotting the formula for reasonable angles, say, to 45 degrees, and then for small angles, say, less than 5 degrees, or even to a fraction of a degree, one may estimate the bearing capacity for zero degrees as shown in the plot. The exact value is 5.14 ($2+\pi$) for a unit cohesion (c =1). Thus, for $\phi = 0°$, $\sigma_u = 5.14\ c$.

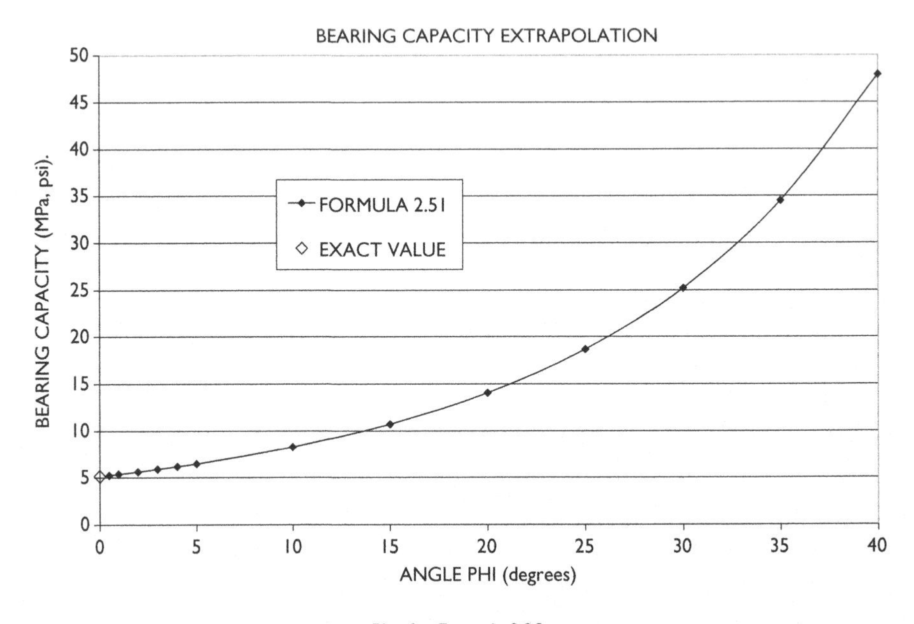

Plot for Example 2.28

Toppling failures

The simplest type of toppling failure is a kinematic instability that occurs when conditions lead to forward rotation, tumbling, of a relatively high, narrow rock block as shown in Figure 2.33. The block in Figure 2.33 may "fail" in several ways (1) sliding down the inclined plane supporting the block; (2) toppling by falling forward, and (3) sliding and toppling. A simple equilibrium analysis leads to stability conditions for each mode of failure. Considering the block in frictional contact with the plane below, the no-slip condition is FS $\geq$ 1. Thus, $\tan(\alpha) < \tan(\phi)$ insures stability against sliding. Moment equilibrium shows that when $\tan(\alpha) < (b/h)$ toppling does not occur. Combinations of both conditions satisfied, one or the other satisfied, and neither satisfied are shown in Figure 2.34 where b/h is plotted as a function of α.

A second form of toppling appears when rather continuous joints dip into a slope as shown in Figure 2.35(a) where blocks formed by intersecting joints form gaps near the slope face and lean downslope one against the other. An equilibrium analyses of a series of idealized blocks is possible that requires force and moment equilibrium in consideration of frictional contact between blocks.

An alternative view is that the slope face is deforming in shear, shown in Figure 2.35(b). Shear resistance along the steeply dipping continuous joints is much smaller than shear

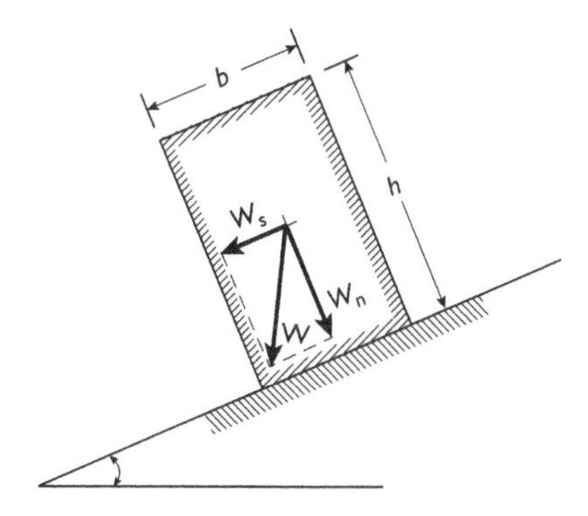

Figure 2.33 Simple toppling and sliding schematic.

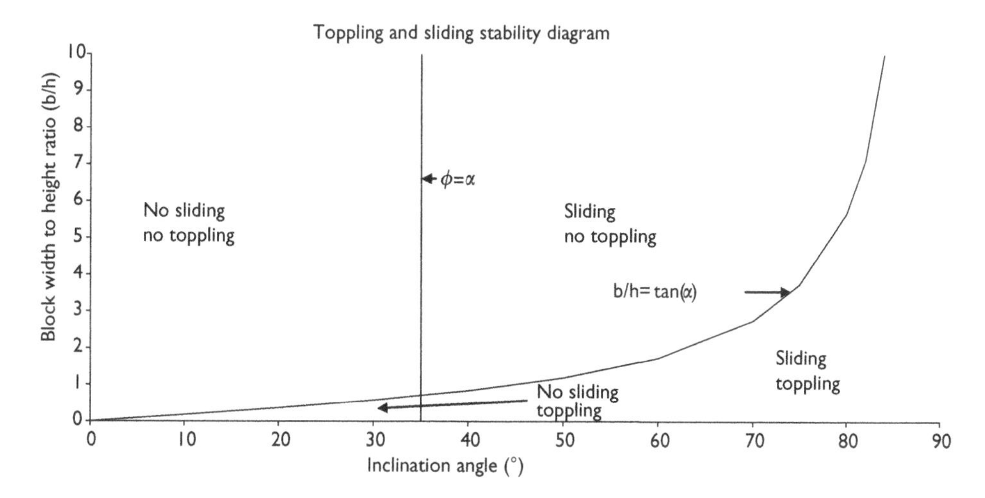

Figure 2.34 Toppling and sliding of a block on an inclined plane (Figure 2.33).

resistance down the slope and across the joints. As excavation proceeds, this anisotropy of rock mass strength leads to a stepped face and small reverse slopes formed by blocks jutting upward from the face.

Example 2.29 A 60m high (h=60) slab 10m (b=10) wide has separated from a cliff-forming sandstone. Water fills the crack behind the slab and tends to push the slab away from the parent sandstone with resulting stress distributions shown in the sketch. Derive an expression from equilibrium considerations that prevents toppling of the slab.

***Solution*:** Draw a free body diagram as shown in (b) of the sketch. Summation of forces in the vertical and horizontal direction show that W=N and T=P where W=slab weight, N=normal reaction at the slab bottom, T=horizontal reaction at the slab bottom, P=water force. The triangular distribution of the water pressure is shown in (a) of the sketch. The resultant force acts through the centroid of the distribution at h/3 from the slab bottom. The same is true of the normal stress distribution, but for generality, the resultant normal force N is assumed to act at x from the middle of the slab bottom. Summation of moments about the

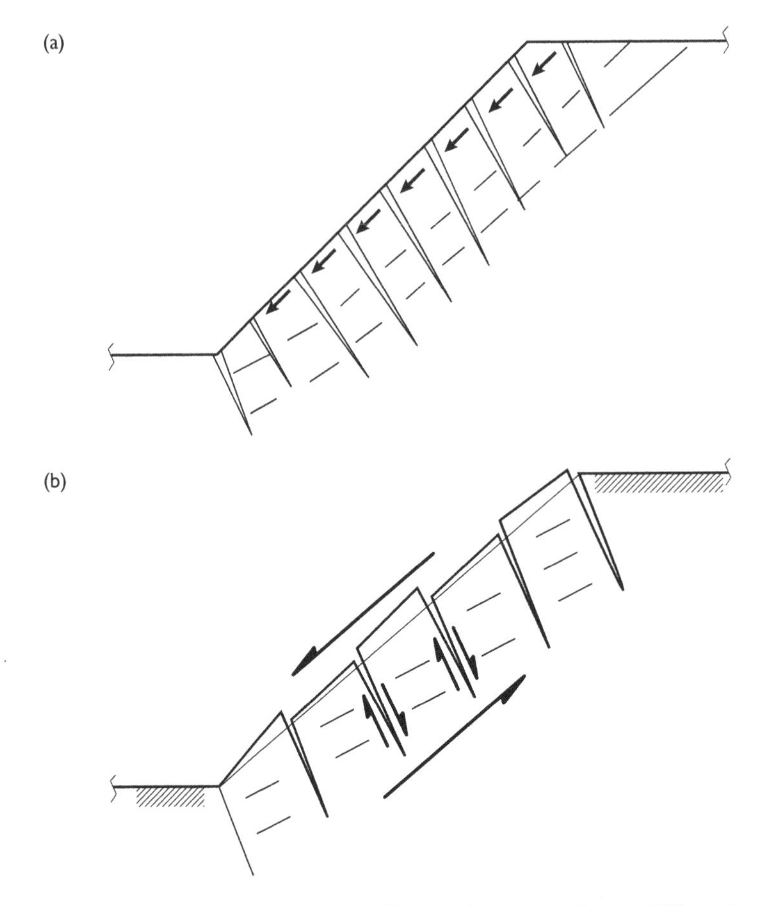

Figure 2.35 Alternative "toppling" interpretation (a) joints dipping into face and (b) results of shearing of an anisotropic rock mass near face.

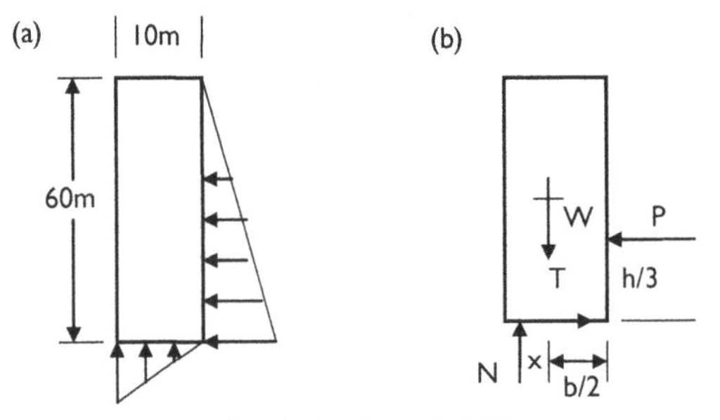

Sketch for Example 2.29

center of the block of unit thickness into the page gives

$$0 = Nx + P(h/2 - h/3) - T(h/2)$$
$$= Nx - P(h/3)$$

$$= \gamma_r hb(1)x - \bar{p}h(1)(h/3)$$
$$= \gamma_r hbx - (p_{\max}\, h/2)(h/3)$$
$$= \gamma_r hbx - (\gamma_w h)(h/2)(h/3) \therefore$$
$$x = (\gamma_w h^2/6)/(\gamma_r b)$$

where γ is specific weight and subscripts "r" and "w" refer to rock and water. Inspection of the sketch shows that$x \leq (b/2) - (b/3) = b/6$. Hence $x = (\gamma_w\, h^2\, /6)/(\gamma_r\, b) \leq (b/6)$. The condition for no-toppling of the slab is therefore

$$(h/b)^2 \leq (\gamma_r/\gamma_w)$$

One has $(70/10) \leq \sqrt{}24.5/9.80 = 2.5$, so the slab would topple were the given conditions realized.

Example 2.30 Suppose the slab in Example 2.29 experiences an ice force P uniformly distributed from the bottom of the slab to a height y as shown in the sketch. In this case the distance x is maximum and corresponds to a triangular distribution of normal stress on the slab bottom. Determine the maximum height that ice can form without toppling the slab.

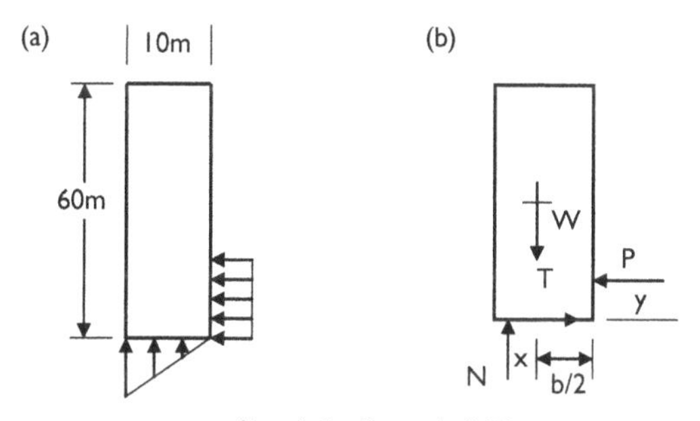

Sketch for Example 2.30

***Solution*:** Equilibrium of forces again requires W=N and P=T. Equilibrium of moments requires

$$0 = Nx - Py$$
$$= Wx - py(1)y$$
$$= \gamma_r bh(1)(b/6) - py(1)y/2$$

where p is the ice pressure assumed to be uniformly distributed as shown in the sketch. Solving for the maximum ice height y gives

$$y = \sqrt{\gamma_r hb^2/3p}$$

In this case $y = \sqrt{(24.5)(10^3)(70)(10)(10)/3p}$ m. An estimate of the ice pressure p is needed to complete a quantitative estimate. In the absence of an ice pressure estimate, one may bound the ice pressure by maximizing the ice height. Thus, $p = [(24.5)(10^3)(70)(10)^2]/[(3)(70)^2]$, that is, $p = 11.67$ kPa.

An ice pressure less than 11.67 kPa will not topple the slab. However, a greater ice pressure distributed over a lower height may also be safe. Alternatively, if ice forms to a height y behind the slab, then a pressure $p \le (\gamma_r b^2 h)/(3y^2)$ will not topple the slab with $y \le h$.

Example 2.31 Consider the motion of a slide mass center down hill across a valley floor and part way up the opposite slope as shown in the sketch. Show that the slope of a line between the starting and stopping points of the slide mass center is $\tan(\varphi)$, where φ is a friction angle.

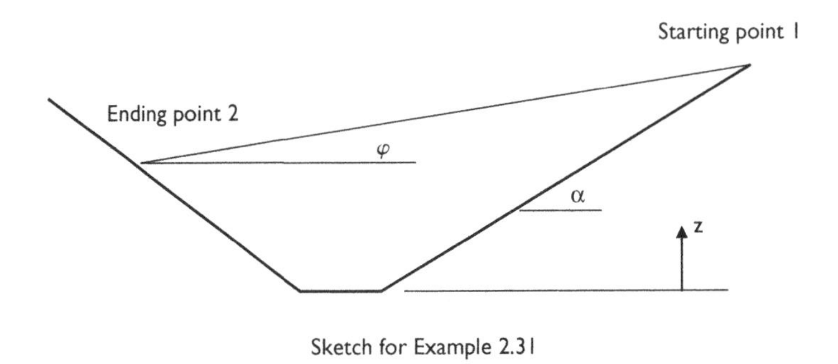

Sketch for Example 2.31

***Solution*:** A mechanical energy balance states that the work (W) done on the slide mass equals the energy change (ΔE) that in turn is the sum of changes in kinetic (ΔKE) and potential energies (ΔPE). Thus,

$$\Delta E = \Delta KE + \Delta PE = W$$
$$(1/2)m(v_2^2 - v_1^2) + W(z_2 - z_1) = F(s_2 - s_1)$$

where v, z, and s are velocity, elevation and distance along the mass center path. The slide mass starts from rest at point 1 and comes to a stop at point 2, so the starting and stopping velocities are zero. The resisting force is frictional only because cohesion is considered destroyed during sliding. Hence, $F = N\tan(\varphi) = W\cos(\alpha)\tan(\varphi)$ where N is the force normal to the slope that has inclination α. The energy balance is now

$$W(z_2 - z_1) = W\cos(\alpha)\tan(\varphi)(s_2 - s_1) \therefore$$
$$\tan(\varphi) = (z_2 - z_1)/(h_2 - h_1)$$

where h is horizontal distance, that is, $h = (s)\cos(\alpha)$.

Example 2.32 With reference to Example 2.31, show that when a line drawn from the starting point of a slide mass center at an angle φ from the horizontal intersects a hillside, the slide mass comes to rest at that point.

Solution: A sketch of the situation indicates possible peaks and valleys along the slide route. The energy balance is

$$\Delta E = \Delta KE + \Delta PE = W$$
$$(1/2)m(v_2^2 - v_1^2) + W(z_2 - z_1) = F(s_2 - s_1)$$
$$(1/2)m(v_2^2 - v_1^2) + W(z_2 - z_1) = W\cos(\alpha)\tan(\varphi)(s_2 - s_1)$$
$$(1/2)m(v_2^2 - v_1^2) = -W(z_2 - z_1) + W\tan(\varphi)(h_2 - h_1)$$
$$= -W(z_2 - z_1) + W(z_2 - z_1)\therefore$$
$$(1/2)m(v_2^2 - v_1^2) = 0$$

which shows the velocity at point 2 is zero (the slide mass starts from rest).

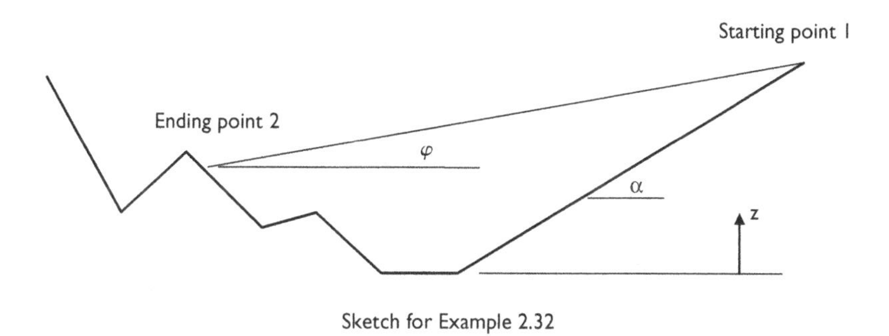

Sketch for Example 2.32

2.3 Problems

Planar block slides

Work according to homework assignment standards. Problems 2.1–2.4 are algebraic, as you will see. Problems 2.5–2.9 are quantitative. Download or run the program PLANAR or similar program (e.g., the commercially available RocPlane from Rocscience, Inc.) from the "net" and use it to check your answers to Problems 2.5–2.9 inclusive. Resolve or explain any discrepancies.

2.1 Derive an expression for the factor of safety FS for a planar block slide with a tension crack behind the crest. The slope height is H and has a face angle β measured from the horizontal to the slope face; the failure surface is inclined at α degrees to the horizontal. Tension crack depth is h. Only the force of gravity acts on the slide mass; specific weight is γ. A Mohr–Coulomb failure criterion applies to the failure surface (cohesion is c, angle of internal friction is ϕ). Note: The slide mass extends b units into the plane of the page. Show schematically a plot of FS (y-axis, range from 0.0 to

4.0) as a function of the reciprocal of the slope height H^{-1} (x-axis, range from 0.0 to 0.001) and discuss briefly the form and significance of the plot.

2.2 Modify the expression for the safety factor from Problem 2.1 to include the effect of a water force P acting on the inclined failure surface. Note that the water table is below the bottom of the tension crack.

2.3 With reference to Problem 2.2, suppose the water pressure p increases linearly with depth (according to $p = \gamma_w z$ where z = depth below water table and γ_w = specific weight of water) from the water table elevation down to a point one-half the vertical distance to the slide toe and then decreases linearly from the halfway point to the toe.

2.4 Consider a seismic load effect S that acts horizontally through the slide mass center with an acceleration a_s given as a decimal fraction a_o of the acceleration of gravity g ($a_s = a_o g$, typically $a_o = 0.05$ to 0.15, depends on seismic zone). Modify the safety factor expression from Problem 1 to include seismic load effect S.

2.5 Given the planar block slide data shown in the sketch where a uniformly distributed surcharge σ is applied to the slope crest over an area (bl), first find the slope safety factor without a surcharge and then find the magnitude of the surcharge possible for a slope safety factor of 1.1 against translational sliding. Note that $b = 25$ ft (7.62 m).

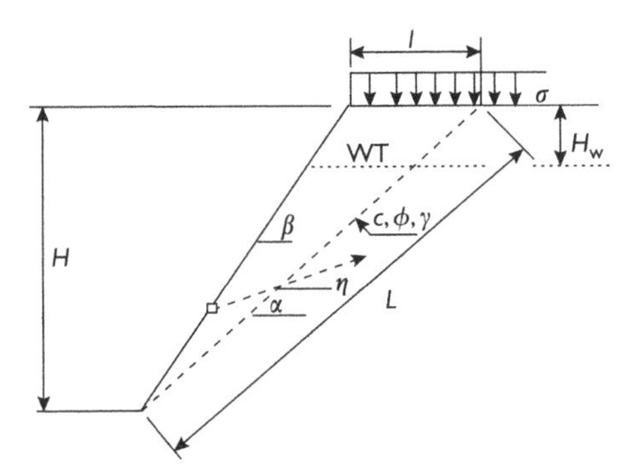

Sketch for Problems 2.5–2.9 Data: c = 50 psi (345 kPa), γ = 156 pcf (24.68 kN/m3), $\alpha = 40°$, $\beta = 50°$, $\phi = 29°$, H =500 ft (152 m), Hw =depth of water table below crest, b = distance into the page, η = bolting angle (+ up).

2.6 Consider the planar block slide in the sketch. If no surcharge is present, find the maximum depth H of excavation possible before failure impends.

2.7 Suppose the slope in the sketch is cable bolted. No surcharge is present. Bolt spacing is 50 ft (15.2 m) in the vertical direction and 25 ft (7.6 m) in the horizontal direction. Bolts assemblages are composed of 12 strands of Type 270 cable (495,600 lbf, ultimate strength or 2.22 MN) and are installed in down holes (5°). Design tension is 60% of the ultimate strength. Find the safety factor obtained by bolting and therefore the improvement in the safety factor obtained (difference between the bolted and unbolted slope safety factors).

2.8 Consider the planar block slide in sketch without surcharge, seismic load, and bolt reinforcement and suppose the water table rises to the top of the slide. Find the safety factor of the "flooded" slide mass.

2.9 Suppose the cohesion of the slide mass shown in the sketch decreases to zero and no surcharge, seismic load, or water is present. Determine whether the slide mass will accelerate, and if so, the magnitude and direction of the acceleration of the slide mass center.

2.10 Some data for a possible planar block slide are given in the sketch. If $\alpha = 35°$, $\beta = 45°$, and $H = 475$ ft (145 m), what cohesion c (psf, kPa) is needed to give a safety factor of at least 1.5?

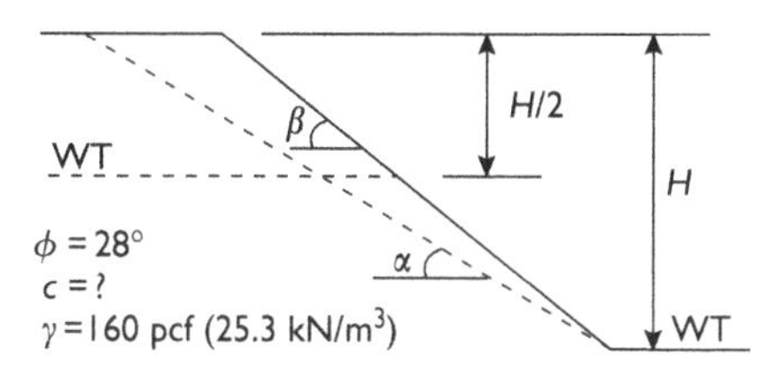

2.11 Consider the planar block slide in the sketch. (a) Show algebraically that reduction of the slide mass volume from V_o by V_1 by excavating a relieving bench near the crest necessarily increases the safety factor, provided the water table is lowered below the toe. (b) Show that placing a berm of weight W_1 at the toe of the slide, where cohesion c_1 and friction angle ϕ_1 are mobilized at the berm bottom, necessarily increases the safety factor, other factors remaining the same.

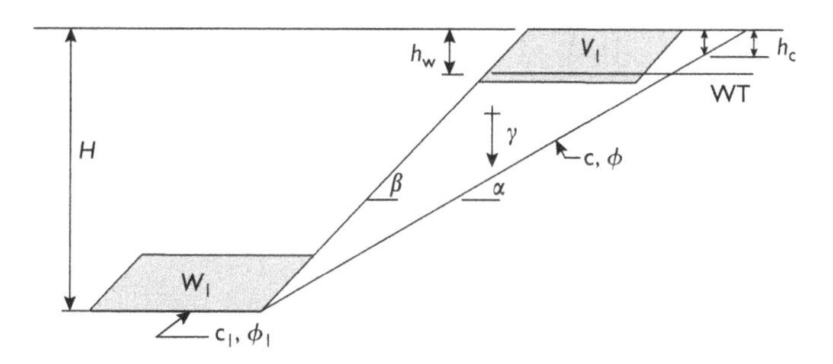

Sketch for Problems 2.11–2.12. Schematic and notation for Problems 2.11, 2.12. WT = water table, h_w =water table depth, h_c =tension crack depth. Water table is fixed and follows the slope face down to the toe.

2.12 With reference to the sketch, if no tension crack is present, $\alpha = 29°$, $\gamma = 156$ pcf (24.7 kN/m^3), joint persistence = 0.87, no crest relieving bench and no toe berm are being considered, determine (a) the maximum slope height H when $\beta = 50°$ and the water table is at the crest, and (b) when the slope is completely depressurized. Rock and joint properties are given in Table 2.6.

2.13 A generic diagram of a slope in a jointed rock mass that is threatened by a planar block slide is shown, in the sketch. Although not shown, bench height is 55 ft (16.76 m).

Table 2.6 Properties for Problem 2.12

Property material	$\phi°$	*c psf/Mpa*
Rock	32.0	64,800/3.12
Joint	25.0	1,620/0.078

Given that Mohr–Coulomb failure criteria apply, the clay-filled joints constitute 93% of the potential shear failure surface, no tension cracks have yet appeared and:

1 slope height $H = ?$ ft (m);
2 failure surface angle $\alpha = 32°$;
3 slope angle $\beta = 49°$;
4 friction angle (rock) $\phi_r = 38°$
5 cohesion (rock) $c_r = 1000$ psi (6.90 MPa);
6 friction angle (joint) $\phi_j = 27°$;
7 cohesion (joint) $c_j = 10$ psi (0.069 MPa);
8 specific weight $\gamma = 158$ pcf (25.0 kN/m^3);
9 tension crack depth $h_c = 0.0$ ft (m);
10 water table depth $h_w = 0.0$ ft (m);
11 seismic coefficient $a_o = 0.00$;
12 surcharge $s = 0.0$ psf (kPa)

Find the maximum pit depth possible without drainage.

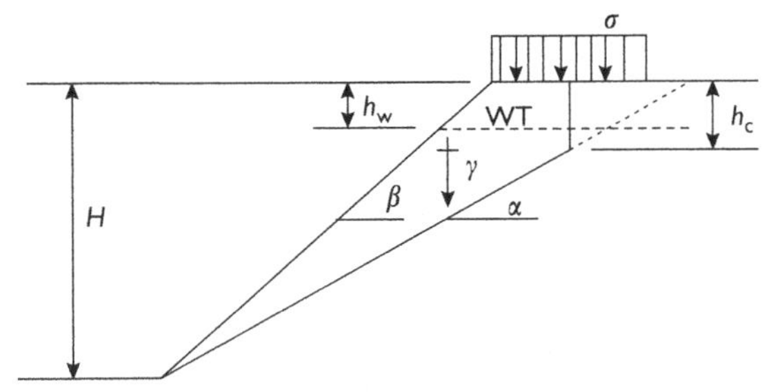

Sketch and notation for Problems 2.13, 2.14, 2.16

2.14 A generic diagram of a slope in a jointed rock mass that is threatened by a planar block slide is shown in the sketch. A minimum safety factor of 1.05 is required. Although not shown bench height is 60 ft (18.3 m). Determine if FS =1.05 is possible, given that Mohr–Coulomb failure criteria apply, the joints constitute 79% of the potential shear failure surface, slide block weight is $1.351(10^7)$ lbf/ft (198.6 MN per meter) of thickness and:

1 slope height $H = 540$ ft (165 m);
2 failure surface angle $\alpha = 32°$;
3 slope angle $\beta = 45°$;
4 friction angle (rock) $\phi_r = 33°$;
5 cohesion (rock) $c_r = 2870$ psi (19.8 MPa);
6 friction angle (joint) $\phi_j = 28°$;

7	cohesion (joint)	c_j = 10.0 psi (0.069 MPa);
8	specific weight	γ = 158 pcf (25.0 kN/m^3);
9	tension crack depth	h_c = 50.0 ft (15.2 m);
10	water table depth	h_w = 60.0 ft (18.3 m);
11	seismic coefficient	a_o = 0.15;
12	surcharge	σ = 0.0 psf (0.0 kPa).

(assume the maximum pressure occurs at $z_o/2$ for this problem). If FS < 1.05, will drainage allow the safety factor objective to be achieved?

2.15 Given the following planar block slide data:

1	slope height	H = 613 ft (187.8 m);
2	failure surface angle	α = 34°;
3	slope angle	β = ?°;
4	friction angle	ϕ = 30°;
5	cohesion	c = 1440 psf (0.069 MPa);
6	specific weight	γ = 162 pcf (26.63 kN/m^3).

Mohr–Coulomb failure criterion applies
Find the maximum slope angle β possible, when the water table is drawn down below the toe of the slope. Note: No tension crack forms.

2.16 A generic diagram of a slope in a jointed rock mass that is threatened by a planar block slide is shown in the sketch. With neglect of any seismic load, find the slope height (pit

1	slope height	H = ? ft (m);
2	failure surface angle	α = 37°;
3	slope angle	β = 48°;
4	friction angle (rock)	ϕ_r = 33°;
5	cohesion (rock)	c_r = 2580 psi (17.79 MPa);
6	friction angle (joint)	ϕ_j = 33°;
7	cohesion (joint)	c_j = 0.0 psi (0.0 MPa);
8	specific weight	γ = 162 pcf (25.63 kN/m^3);
9	tension crack depth	h_c = 0.0 ft (m);
10	water table depth	h_w = 0.0 ft (m);
11	seismic coefficient	a_o = 0.15;
12	surcharge	σ = 0.0 psf (kPa).

depth) possible with a safety factor of 1.15, given that Mohr–Coulomb failure criteria apply, the joints constitute 86% of the potential shear failure surface and:

2.17 With reference to the sketch of the potential slope failure shown in the sketch, find:

(a) the factor of safety of a cable bolted slope when the water table is drawn down 100 ft (30.5 m), bench height is 40 ft (12.2 m) (vertical bolt spacing), horizontal bolt spacing is 20 ft (6.1 m), the bolting angle is 5° down, and bolt loading is 700 kips (3.14 MN) per hole;

(b) the factor of safety of the same slope but without bolts when the water table is drawn down below the toe.

The design trade-off here is between drainage and bolting. Any reasons for preferring one or the other?

Note: A tension crack 37 ft (11.3 m) deep is observed at the crest behind the face. The slope is 320 ft (97.5 m) high, failure surface angle is 32°, slope angle is 40°, specific weight is 158 pcf (25.0 kN/m3), cohesion is 1440 psf (0.069 MPa), and the slide surface friction angle is 28°.

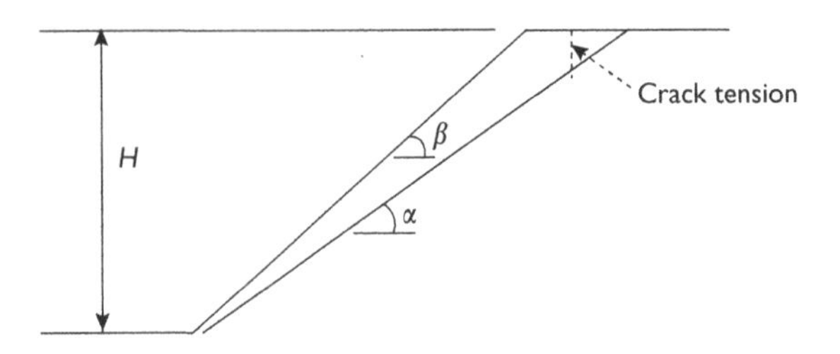

Wedge failures

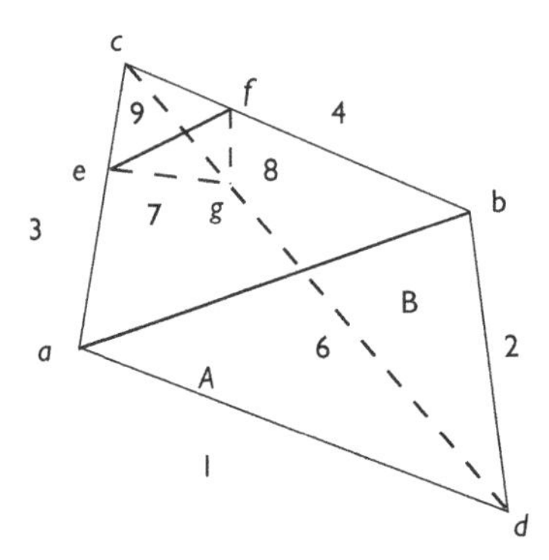

Sketch and generic wedge notation for Problems 2.18–2.22

2.18 Given the data in Table 2.7, determine the direction angles of the normal vectors to joint planes *A* and *B*.

2.19 Calculate the dip and dip direction of the line of intersection of joint planes *A* and *B*, then sketch the result in compass coordinates.

2.20 Given that the vertical distance between points *a* and *d* is 120 ft (36.6 m), determine the areas of joint planes *A* and *B*, assuming no tension crack is present.

2.21 Use computer programs to verify the area data in Table 2.7 with a tension crack present and a rock specific weight of 158 pcf (25.0 kN/m^3). What is the wedge safety factor when the water table is below the toe of the slope? Note: Tension crack offset (*ae*) is 90 ft (27.4 m).

2.22 With reference to the generic wedge in the sketch and the specific data in Table 2.8 if the safety factor with respect to sliding down the line of intersection is greater than 1.10, then safety is assured. The water table is at the crest of the slope (but may be lowered to meet the safety criterion), vertical distance between *a* and *g* is 85 ft (25.9 m), point *e* is 48 ft (14.6 m) from point *a*, rock specific weight is 158 pcf (25.0 kN/m^3) and active wedge volume is 3,735 yd^3 (2,856 m^3). What range of face orientations (dip directions) is of concern in an open pit mine where benches 95 ft (28.96 m) high are planned?

2.23 Two joints K1 and K2 are mapped in the vicinity of a proposed surface mine. Joints in set K1 have dip directions of 110° and dips of 38°; joints in set K2 have dip directions of 147° and dips of 42°. Determine if potential wedge failures may form, and if so, what the dip direction and dip of the lines of intersection would be, and the range of

Table 2.7 Data for Problem 2.18[a]

Parameter plane	*Dip dir. (°)*	*Dip (°)*	*Cohesion psf (kPa)*	*Friction angle (°)*	*Direction angles of normals* α, β, γ *(°)*	*Area ft² (m²)*
Joint *A*	0	60	1,080 (51.7)	32		8,072 (750)
Joint *B*	90	60	1,640 (78.5)	37		8,072 (750)
Face	45	85	—	—		
Upland	45	5	—	—		
Tension crack	45	75	0	39		1,312 (122)

Notes
a Wedge volume is 11,683 yd^3 (8,932 m^3).

Table 2.8 Data for Problem 2.22

Parameter plane	*Dip dir. (°)*	*Dip (°)*	*Cohesion psf (kPa)*	*Friction angle (°)*	*Direction angles of normals* α, β, γ *(°)*	*Area ft² (m²)*
Joint *A*	90	45	1,800 (86.2)	26		3,318 (308)
Joint *B*	180	45	1,080 (51.7)	32		338 (31.4)
Face	135	75	—	—		
Upland	0	0	—	—		
Tension crack	148	68	0	48		7,144 (664)

Table 2.9 Data for Problem 2.24

Parameter plane	*Dip dir. (°)*	*Dip (°)*	*Cohesion psf (kPa)*	*Friction angle (°)*	*Direction angles of normals* α, β, γ *(°)*	*Area ft² (m²)*
Joint *A*	30	60	2,800 (134)	26		
Joint *B*	120	60	5,280 (253)	32		
Face	75	85	—	—		
Upland	65	5	—	—		
Tension crack	55	75	0	39		

bench face azimuths that should be examined more closely for slope stability. Sketch the data and results.

Note: Formula for a joint plane normal (n_x n_y n_z): (sin(δ) sin(α), sin(δ) cos(α), cos(δ)). Also, the components of a unit normal vector to a joint in set K1 are (0.5785, −0.2106, 0.7880).

2.24 With reference to the generic wedge shown in the sketch and the data in Table 2.9, the water table is below the toe, vertical distance between *A* and *O* is 85 ft (25.9 m), point *T* is 40 ft from point *A*, rock specific weight is 158 pcf (25.0 kN/m^3), and active wedge volume is 3,850 yd^3 (2,944 m^3). (a) Find the dip direction (azimuth) and dip of the line of intersection of joint planes *A* and *B*. (b) Find the range of intersection line azimuths that preclude sliding down the line of intersection. (c) Find the length of the line of intersection between the face and plane *A*. Note: The foreland is flat.

Table 2.10 Data for Problem 2.25

Parameter plane	*Dip dir. (°)*	*Dip (°)*	*Cohesion psf (kPa)*	*Friction angle (°)*	*Direction angles of normals* α, β, γ (°)			*Area ft² (m²)*
Joint *A*	60	60	1,440 (68.97)	35	41.4	64.4	60.0	10,550 (980)
Joint *B*	210	60	720 (34.48)	28	115.7	138.6	60.0	10,642 (989)
Face	135	70	—	—	48.4	131.6	70.0	—
Upland	0	0	—	—	90.0	90.0	0.0	—
Tension crack	118	58	0	48	41.5	113.5	58.0	310 (28.8)

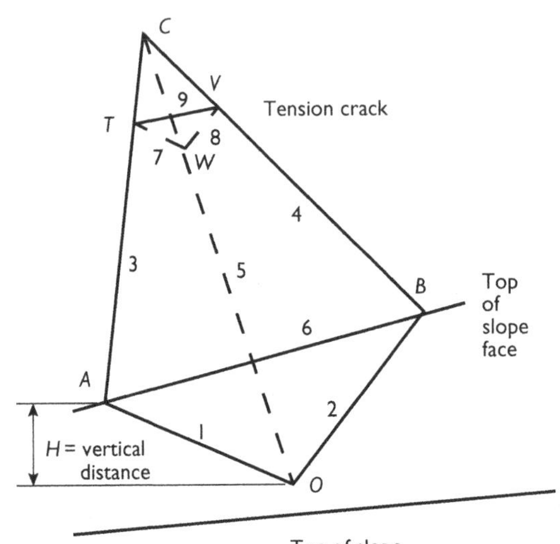

Sketch and notation for Problems 2.24–2.27

2.25 With reference to the wedge shown in the sketch and the Table 2.10 data, if the safety factor with respect to sliding down the line of intersection is greater than 1.10, then safety is assured. Can this objective be met? Yes, no, maybe? Explain. Note: The water table is below the toe of the slope, vertical distance between A and O is 100 ft, point T is 150 ft from point A and wedge volume is 308,000 ft³.

2.26 Given the wedge data in Table 2.11, if the vertical distance between crest and the point where the line of intersection between the joint planes intersects the face is 120 ft (36.6 m), find the dip direction, dip, and length of the line formed by the intersection between the joint plane *A* and the face *F*. Note: This is line 1 in the sketch. Upland is flat.

2.27 With reference to the generic wedge shown in the sketch and the data in Table 2.12, the water table is below the toe, vertical distance between *A* and *O* is 25.9 m (85 ft), point *T* is 12.2 m (40 ft) from point *A*, rock specific gravity is 2.46, and active wedge volume is 2,944 m³ (3,850 yd³). (a) Find the dip direction (azimuth) and dip of the line of intersection of joint planes *A* and *B*. (b) Find the range of intersection line azimuths that preclude sliding down the line of intersection. (c) Find the length of the line of intersection between the face and plane *A*. Note: The foreland is flat.

Table 2.11 Data for Problem 2.26

Angle surface	*Dip dir. (°)*	*Dip (°)*
Joint plane A	0	45
Joint plane B	120	60
Face F	90	45
Upland U	90	45

Table 2.12 Data for Problem 2.27

Parameter plane	*Dip dir. (°)*	*Dip (°)*	*Cohesion kPa (psf)*	*Friction angle (°)*	*Direction angles of normals* α, β, γ *(°)*	*Area* m^2 *(*ft^2*)*
Joint A	30	60	134 (2,800)	26		
Joint B	120	60	253 (5,280)	32		
Face	75	85	—	—		
Upland	65	5	—	—		
Tension crack	55	75	0	39		

Table 2.13 Slice widths in ft

Slice	*1*	*2*	*3*	*4*	*5*	*6*	*7*	*8*	*9*
Width									
ft	8	8	16	16	16	12	9.28	8	2.19
m	2.44	2.44	4.88	4.88	4.88	3.66	2.83	2.44	0.67

Rotational slides

2.28 Consider a bank 40 ft (12.2 m) high with a slope of 30° that may fail by rotation on a slip circle of radius 60 ft (18.3 m) with center at (20, 48 ft) or (6.10, 14.63 m) relative to coordinates in ft with origin at the bank toe. In this case, the clip circle dips below the toe. Cohesion and friction angle are 720 psf (34.5 kPa) and 18°. Table 2.13 gives widths of nine slices to be used for analysis. Numbering proceeds from the left near the toe to the right at the slope crest. Specific weight is 100 pcf.(15.8 kN/m^3).

(a) Use a spreadsheet to analyze the slope stability and to show that the factor of safety is 2.164. Verify the reliability of this result using the computer program *Slide* or similar method of slices program. Attach a print of your spreadsheet and also attach prints of the slope and slip circle from the Interpreter and the Infoviewer of Slide. Note: The water table is below the bottom of the slip circle, that is, dry case.

(b) Repeat the analysis using your spreadsheet and then verify with *Slide* or equivalent using a water table that coincides with the toe-slope face-crest profile. What are the safety factors from your spreadsheet and from Slide? Again, attach prints of your spreadsheet and two prints from Slide.

2.29 With reference to the sketch and tables, find the safety factors for the four cases posed using the method of slices. Include all important calculations. If you use a spreadsheet, include suitable data prints.

Complete a table for each of the conditions following:

1. With water table (WT) 1 and no relieving benches
2. With WT 2 and no relieving benches
3. With WT 2 and a 140 × 60 ft (42.7 × 18.3 m) relieving bench at the crest
4. With WT 1, a 140 × 60 ft relieving bench at the crest, and a relieving Bench 2 at the toe required to give a factor of safety of 1.05. Compute the length L of Bench 2 in this case.

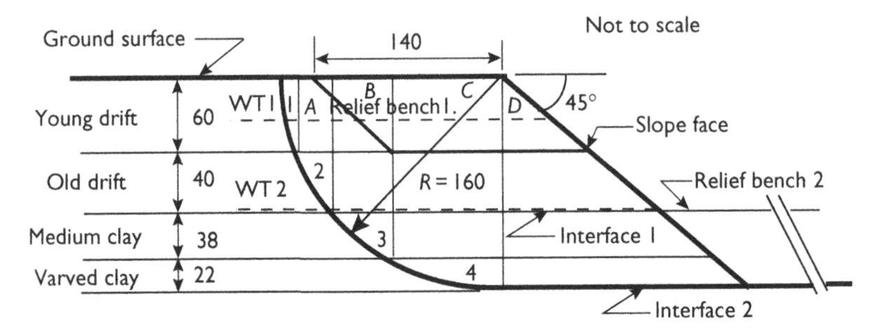

Sketch for Problem 2.29. Units are ft.

Table for Problem 2.29

Section	*Weight*	α	W_s	W_n	P	$W'n$	$\tan\phi$	R	c	L	cL

Factor of safety = ________________________

Notes: Units are feet-ft (m).

Water table 1 (WT1) is 30 ft (9.14 m) below the ground surface.

Water table 2 (WT2) is 100 ft (30.5 m) below the ground surface.

Areas ft^2 (m^2):

Parts of relief bench 1: slices:

$A = 118$ (10.96)	$1 = 470$ (43.7)
$B = 1.682$ (156.3)	$2 = 1{,}937$ (180)
$C = 4{,}800$ (445.9)	$3 = 5{,}451$ (506.4)
$D = 1{,}800$ (167.2)	$4 = 12{,}246$ (1,137.7)

Properties[a] :

Property material	*Cohesion c psf (kPa)*	*Friction angle ϕ (°)*	*Specific weight γ pcf (kN/m^3)*
Young drift	—	36	110 (17.4)
Old drift	—	33	125 (19.8)
Interface I	800 (38.3)	—	—
Medium clay	2,300 (110.2)	—	125 (19.8)
Varved clay	2,800 (134.1)	—	105 (16.6)
Interface I	700 (33.5)	—	
Iron formation	—	—	—

Notes

a A blank means zero (0). Drift is cohesionless, clay and interfaces are frictionless. The difference between dry and wet specific weight is neglected in this problem.

2.30 With reference to the sketch illustrating a rotational slide being considered for a method of slices stability analysis, show that the inclusion of seismic forces as quasistatic horizontal forces results in the expression

$$\mathrm{FS} = \frac{M_R}{M_D} = \frac{\sum_i^n [N_i' \tan(\phi) + C_i]R}{\sum_i^n [W_i \sin(\alpha_i) + S_i(r_i/R)\cos(\beta_i)]R}$$

and present a formula for the effective normal force in this expression. Illustrate with a slice force diagram and be sure to state important assumptions.

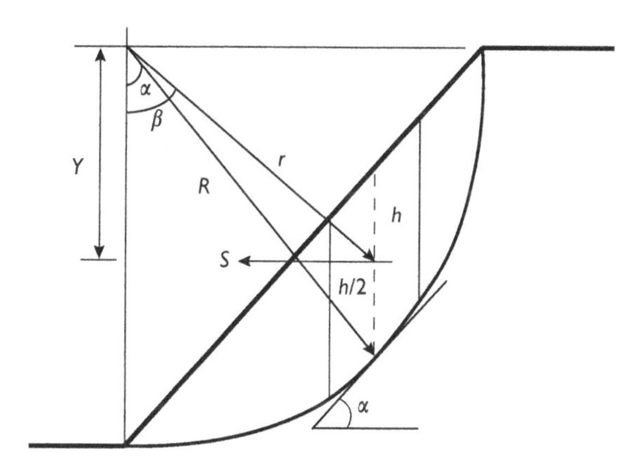

Sketch for analysis of seismic slice force S

2.31 Given the potential one-foot thick slope failure along the circular arc (R = 300 ft, 91.44 m) shown in the sketch, the data in Table 2.14, a slope height = 120 ft (36.6 m); slope angle = 29°, specific weight = 95 pcf (15.0 kN/m^3), cohesion = 367 psf (17.6 kPa), and angle of internal friction= 16°, find:

1 seismic force on slice 7, if the seismic coefficient is 0.15;
2 water force at the bottom of slice 7;

3 force safety factor of slice 7 alone taking into account the seismic and water forces;
4 safety factor for the considered slip surface with no seismic or water forces acting.

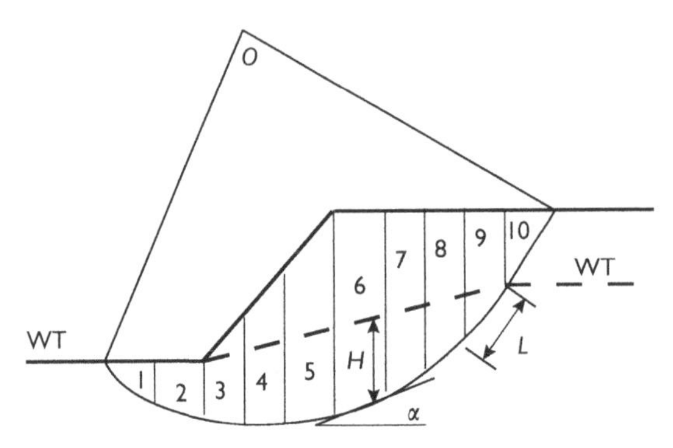

Sketch and circular arc schematic for Problems 2.31–2.33.

2.32 Given the potential one-foot thick slope failure along the circular arc (R = 300 ft, 91.44 m) shown in the sketch and the data in Table 2.14, find the safety factor possible with drawdown of the water table below the failure surface. Slope height is 120 ft (36.6 m); slope angle is 29° and specific weight is about 95 pcf (15 kN/m3). Cohesion is 367 psf (17.6 kPa) and the angle of internal friction of the material is 16°.

2.33 Given the potential one-foot thick slope failure along the circular arc shown in the sketch and the data in Table 2.14, find the safety factor for slice 2 and for slice 7, then show algebraically the safety factor for the slope. Slope height is 120 ft (36.6 m),

Table 2.14 Data for Problems 2.31–2.33. Note: Alphas 1–5 are negatives.

Par. slice	*Weight lbf 10^5 (MN)*	α(°)	*L ft (m)*	*H ft (m)*	*W_n lbf 10^5 (MN)*	*P lbf 10^5 (MN)*	*N′ lbf 10^5 (MN)*	*W_s lbf 10^5 (MN)*
1	1.283 (0.575)	40	60 (18.29)	30 (9.14)	0.983 (0.440)			0.825 (0.370)
2	2.778 (1.245)	28	30 (9.14)	52 (15.85)	2.453 (1.100)			1.304 (0.584)
3	4.489 (2.011)	24	30 (9.14)	75 (22.86)	4.101 (1.837)			1.826 (0.818)
4	6.413 (2.873)	13	30 (9.14)	90 (27.43)	6.248 (2.800)			1.443 (0.646)
5	10.90 (4.58)	3	56 (17.18)	97 (29.57)	10.885 (4.877)			0.570 (0.255)
6	11.33 (5.076)	10	60 (18.29)	97 (29.57)	11.158 (4.999)			1.976 (0.885)
7	10.29 (4.610)	24	75 (22.86)	82 (24.99)	9.400 (4.2110)			4.185 (1.875)
8	5.985 (2.681)	38	60 (18.29)	60 (18.29)	4.716 (2.113)			3.629 (1.626)
9	2.779 (1.245)	45	56 (17.18)	15 (4.57)	1.965 (0.880)			1.965 (0.880)
10	1.283 (0.575)	66	83 (25.30)	—	0.799 (0.358)			1.172 (0.525)

slope angle is 29°, and specific weight is about 95 pcf (15 kN/m^3). Cohesion is 367 psf (17.6 kPa) and the angle of internal friction of the material is 16°.

2.34 Given the circular failure illustrated in the sketch, find algebraically the factor of safety. Assume that the material properties and geometry of slope and slices are known.

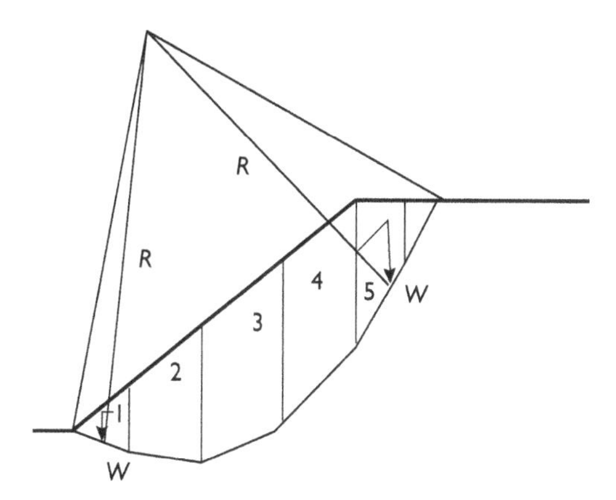

Dynamics, toppling

2.35 Consider the possibility of a rock slide starting at point A and ending at B, as shown in the sketch, and suppose the slide is driven by gravity and resisted by friction. Derive expressions for acceleration, velocity, and distance moved down slope by the mass center of the slide over any segment of constant inclination from the horizontal. Be sure to identify all terms and assumptions.

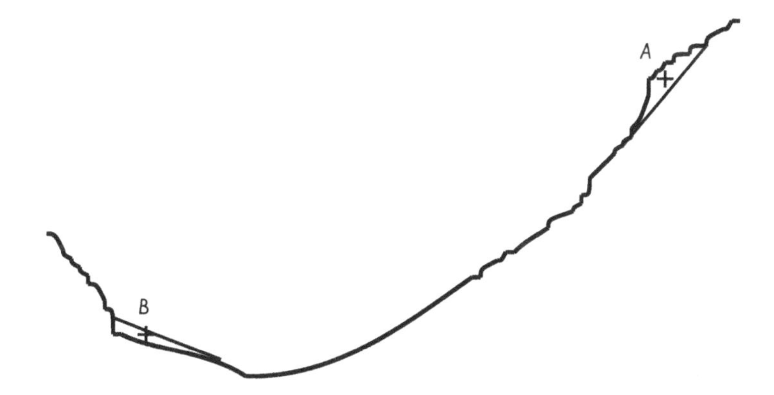

2.36 Given the slope profile in vertical section shown in the sketch and an angle of friction $\phi = 15°$, if the mass center of the slide is initially at 1, determine the height of the mass center after sliding. This is the distance of 5 above the valley floor.

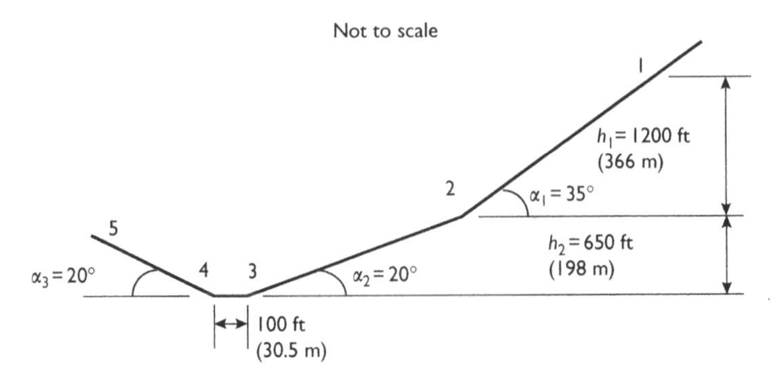

2.37 Toppling of tall blocks is possible when one corner tends to lift off from the base supporting the block, as shown in the sketch. Show that the block is stable provided $\tan(\alpha) < (1/3)\tan(\beta)$ where $\tan(\beta) = b/h$ when the ground reaction is a triangular distribution. Note: Weight W is vertical; the base is inclined $\alpha°$ from the horizontal.

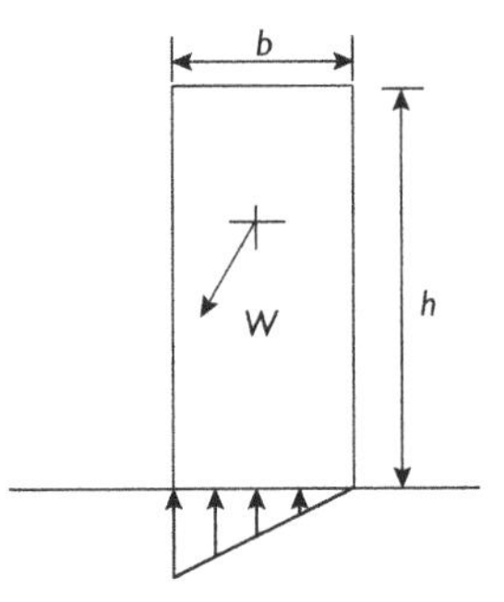

2.38 A rock block with a square base rests on a 28° slope. The block base plane has a friction angle of 32°. There is no adhesion between block and base plane. Find the base dimensions necessary to just prevent toppling.

Additional problems

2.39 Consider a "highwall" in flat sedimentary strata as shown in the sketch. Three sets of joints are present. One set, J1, is composed of bedding plane joints. The other two sets, J2 and J3, are at right angles to the bedding and to each other as is often the case. The vertical joints J2 and J3 allow vertical separation of a large block of ground. For each meter of block into the page, estimate the acceleration (from blasting) that will just move the block forward as it slips along the clay interface. Note: the cohesion of the clay interface is 100 kPa; the friction angle is zero.

2.40 Suppose the cohesion of the failure surface in the sketch is diminished with each blast by fracturing of the interface, so the persistence p, defined as the ratio of facture area to total area, decreases by steps with each blast. Further suppose blasting occurs twice a day and the slope moves downhill 0.1 inch per blast. Is there a day when the slope continues to accelerate after a blast? Note: cohesion and angle of friction of the failure

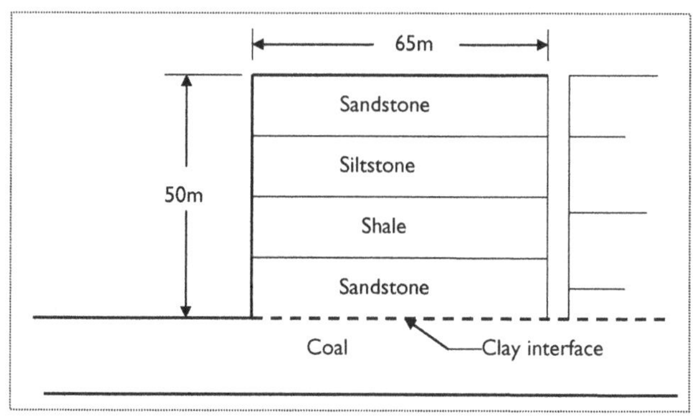

Sketch for Problem 2.39

surface are (900 psi, 40°). The fractured portion of the failure surface has cohesion and friction angle (0 psi, 30°). Specific weight is 150 pcf.

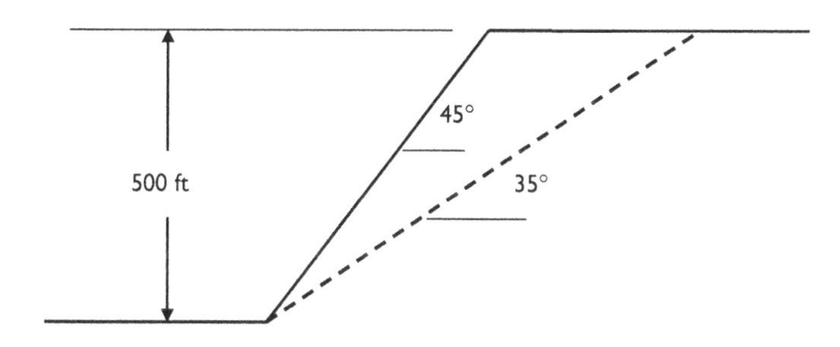

Sketch for Problem 2.40

2.41 Bolting reinforcement of a potential wedge failure is anticipated. Bolts will be installed perpendicular to the face of the wedge as indicated in the sketch. All bolts will be

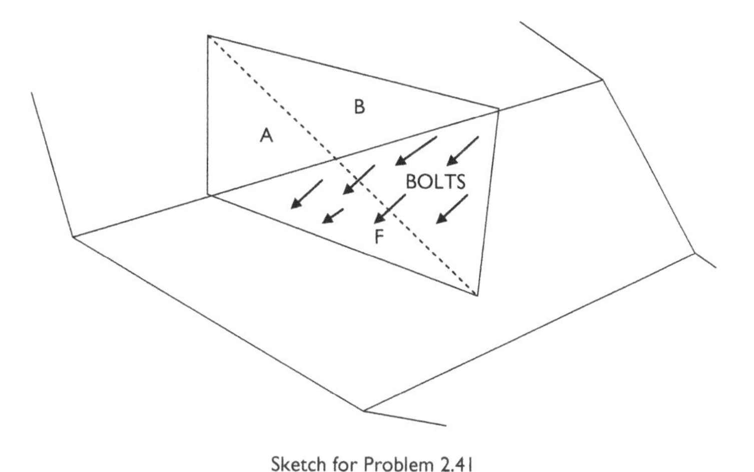

Sketch for Problem 2.41

tensioned the same amount and installed on a square pattern SxS m. Derive a formula for the wedge factor of safety with respect to sliding down the line of intersection of joints A and B that includes the effect of bolting. Assume there are m_A and m_B bolts intersecting planes A and B.

2.42 Consider a potential wedge failure with seismic load W_s illustrated in the associated sketch. The seismic force is estimated as wedge weight W times a seismic coefficient a_s appropriate for the region. The seismic force acts horizontally through the center of the wedge and has an azimuth equal to the dip direction of the line of intersection. Modify a wedge safety factor formula to include such a seismic force.

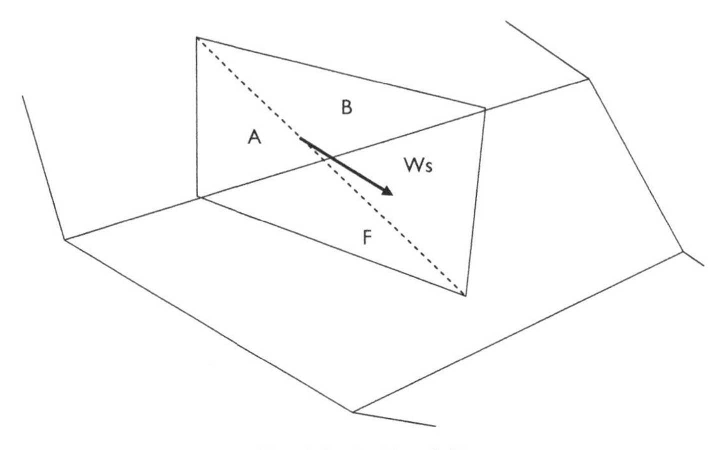

Sketch for Problem 2.42

2.43 Consider a rotational slide in the form of a spherical segment as shown in the sketches. Outline a "method of columns" analogous to the method of slices that accounts for three-dimensional effects not present in the two-dimensional analysis using the method of slices.
(After Chen, 1981)

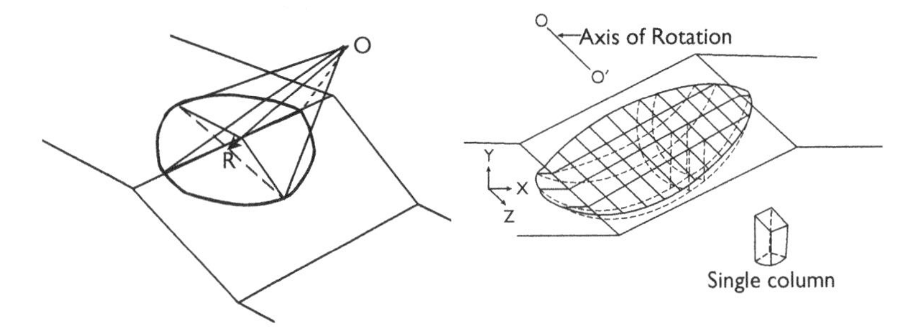

Sketch for Problem 2.43

2.44 Consider a rotational failure as shown in the sketch with the water table at the crest. The water pressure on the circular slip surface is given by $p = \gamma_w z$, that is, by the product of water specific weight times water depth. The pressure distribution will show an increase from zero at the crest to a maximum and then a decrease to zero at the toe of the slope (a). An impervious seal is applied to the slope, but no drain holes are provided (b). Derive a formula that shows the effect of the seal on the slope safety factor

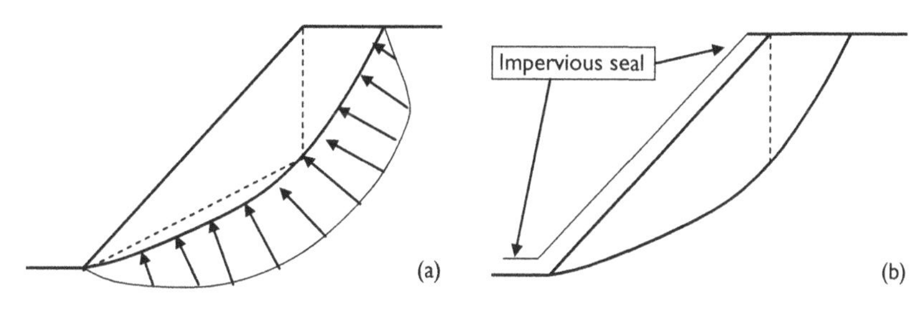

Sketch for Problem 2.44

2.45 An unstable slope poses a threat to a housing development below. The height to the unstable mass is H=185 ft and the slope angle α =30 degrees. The effective angle of sliding friction is φ. (a) Estimate the "run-out" distance the mass center moves from the slope toe before coming to rest. (b) Estimate a barrier height required to stop the slide mass before traveling the entire run-out distance.

Chapter 3

Shafts

Shafts are vital access ways for many underground mines. They provide passageways for personnel, materials, and ventilation air. Vertical shafts are the norm, but inclines are also used. Shafts must be engineered for a long, stable design life; they must be safe at all points along the shaft route. Where the rock mass is strong and the *in situ* stress is low, an unlined shaft may be acceptable, for example, a ventilation shaft, but usually some provision is needed for preventing loose rock from cascading down the shaft. Even small rock fragments can do great harm after falling from considerable height. Rock bolts, steel strapping, wire mesh, screen of some type, lacing, and lagging are commonly used to prevent minor falls of rock in shafts and are usually installed during shaft sinking. A relatively thin concrete liner serves the same purpose and also provides a smooth surface that reduces air friction loss in ventilation shafts.

If the safety factor of the unlined, unsupported shaft wall is acceptable, then no additional support is needed beyond that provided by shaft framing which may be wood or steel, although steel is the most commonly used material today. Shaft framing partitions the shaft space into compartments for hoisting personnel and material, and for utilities (electrical power, compressed air, drill water, and pump water) as well as for ladders that allow passage of personnel. Spacing of shaft framing is variable but generally of the order of feet or meters. Ends of shaft framing are usually placed in notches excavated in the shaft walls (bearing sets), but may also rest on bearing plates (ledges) bolted to the walls.

3.1 Single unlined naturally supported shafts

Calculation of a safety factor (FS) for an unlined shaft wall is straightforward in principal, but requires attention to geological details. Thus,

$$\begin{aligned} \text{FS} &= \frac{\text{Strength}}{\text{Stress}} \\ \text{FS} &= \frac{C_o}{\sigma_c} \\ \text{FS} &= \frac{T_o}{\sigma_t} \end{aligned} \tag{3.1}$$

where C_o and T_o are the unconfined compressive and tensile strengths of the shaft wall rock, and σ_c and σ_t are the peak compressive and tensile stresses at the shaft wall. If tension or compression is absent, then the associated safety factor is not an issue. These safety factors use stress and strength at a point and therefore are *local*. However, when the most highly stressed point at a shaft wall is within the *elastic limit*, then all other points at the shaft wall have higher safety factors and thus overall or *global* safety is assured. Safety factors of 2 to 4 in compression and 4 to 8 in tension have been suggested (Obert *et al.*, 1960, chapter 3).

Although a safety factor calculation as such is simple in the extreme, determining the requisite strength and stress requires thoughtful attention to details of rock mass strength and stress and how they may be influenced by shaft route geology. The safety factor approach implies that the rock mass response to load is initially elastic, but that the elastic range of deformation is limited by rock mass strength. Within the elastic range of stress, shaft wall displacements remain small. These conditions may not always be met. For example, an important exception occurs when a shaft route penetrates bedded salt that behaves much like a highly viscous fluid. Shales may also exhibit a significant time-dependent component of displacement and lead to "squeezing" ground. Densely jointed rock masses may tend to flow because of joint alteration mineralogy and inelasticity. The presence of acid mine waters, saline solutions, and liquid and gaseous hydrocarbons may further complicate a rock mass response to stresses induced by shaft excavation. Nevertheless, elastic behavior is the usual expectation and is the model followed in the absence of field evidence to the contrary.

Shaft wall stress concentration

In elastic ground, the peak stresses act tangentially at the shaft wall and may be obtained from stress concentration factors K_c and K_t for compression and tension, respectively. Thus,

$$\begin{aligned} \sigma_c &= K_c S_1 \\ \sigma_t &= K_t S_1 \end{aligned} \tag{3.2}$$

where S_1 is the major principal stress before excavation. If a stress concentration factor is negative, then the associated stress is opposite in sense to S_1. Stress concentration factors are obtained from solutions to problems in the mathematical theory of elasticity and are simply the ratios of peak to reference stresses. The usual reference stress in rock mechanics is the preexcavation or applied major principal stress. Usually compression is considered positive.

Peak stresses occur at hole boundaries and depend on:

1. hole shape
2. aspect ratio
3. principal stress ratio M
4. orientation.

Common shapes are rectangular and circular. New shafts are quite likely to be circular and for this reason the circular shaft is important. Older shafts are likely to be rectangular, and so extension of an existing shaft may involve a rectangular shape. Elliptical sections are possible but rare, although a circle may be considered a special ellipse. Aspect ratio refers to the ratio of section dimensions, for example, "width" W_o to "height" H_o. The principal stress ratio M is the ratio of minor to major premining stress S_3/S_1. Orientation is determined by the angle the long axis of the opening makes with the major principal premining stress.

Unlined circular shafts

A model of an unlined shaft with circular cross section is a circular hole excavated in a slab of unit thickness that is loaded by uniformly distributed stresses over the sides of the slab. Relative displacement between slab faces is zero (plane strain). In linearly elastic, homogeneous, isotropic rock, the circumferential (tangential) normal stress at the shaft wall is

$$\sigma_t = (S_1 + S_3) - 2(S_1 - S_3)\cos(2\theta) \tag{3.3}$$

where S_1 and S_3 are the major and minor applied (preshaft) principal stresses and θ is the angle from the direction of S_1 to the point of action of σ_t (see, for example, Obert and Duvall, 1967, chapter 3) Circumferential stress concentration K at the shaft wall is then

$$K = (1 + M) - 2(1 - M)\cos(2\theta) \tag{3.4}$$

where M is the ratio of preshaft minor to major principal stress and θ is a counterclockwise angle from S_1 to the point of stress concentration at the shaft wall. Figure 3.1 illustrates a compressive stress concentration at the wall of a circular shaft according to Equations (3.3) and (3.4).

Equation (3.4) is similar in each quadrant, so only the first quadrant needs to be examined. Maximum and minimum values of K occur at 0° and 90° in this quadrant and have values $K_{max} = (3 - M)$ and $K_{min} = (-1 + 3M)$. By definition, $M < 1$. Although this restriction allows for very large but negative values of M, a range between -1 and $+1$ is reasonable, while a range between 0 and $+1$ includes most practical cases. Table 3.1 describes the state of stress for selected values of M. The first case in Table 3.1 (pure shear or "diagonal tension") is unlikely, while the last case is highly improbable in nature, but is included for the sake of completeness.

Figure 3.2(a) is a plot of (3.4) for a range of M values and shows the vanishing of tension at the shaft wall for $M \geq 1/3$. This result suggests that tensile failure of an unlined shaft is often not a consideration. All curves have a common point of $K = 2$ at $\theta = 60°$. Figure 3.2(b) is a

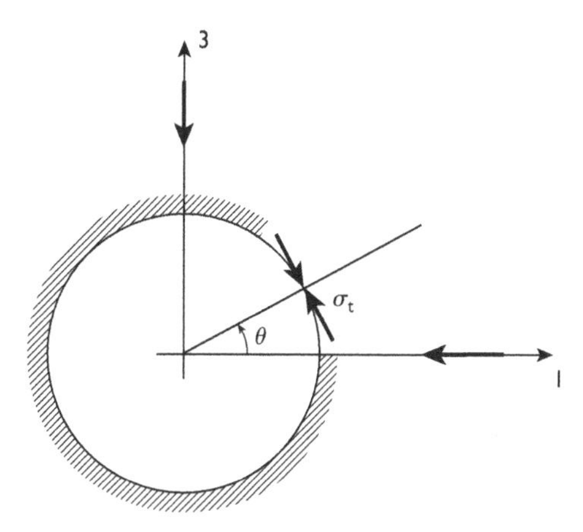

Figure 3.1 Compressive stress concentration at the wall of a circular shaft.

Table 3.1 Principal stresses for selected M values[a]

Case M	*Principal stresses*	*Physical description*
−1	$S_3 = -S_1$	Pure shear
0	$S_3 = 0$	Uniaxial compression
+1	$S_3 = +S_1$	Hydrostatic compression
− ∞	$S_1 = 0$	Uniaxial tension

Note
a Compression is considered positive.

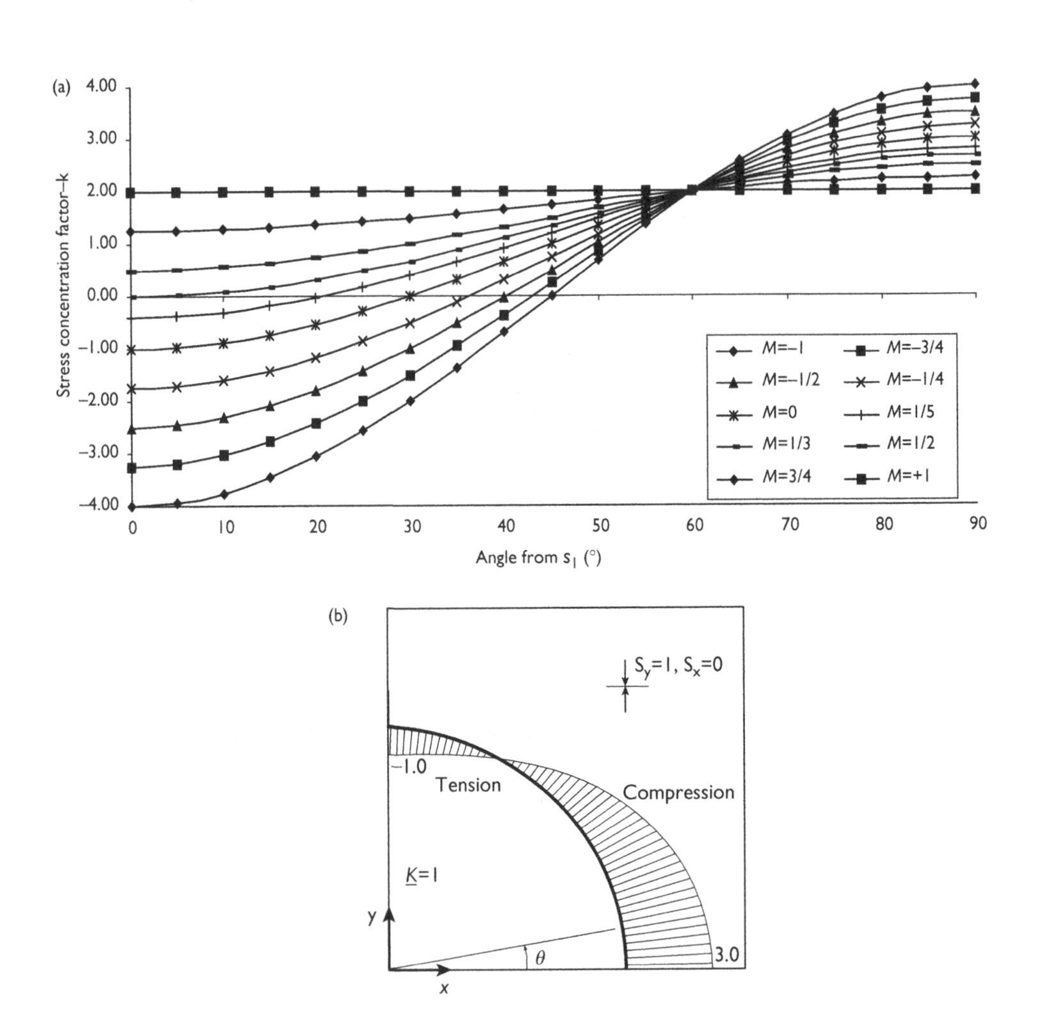

Figure 3.2 (a) Stress concentration at the wall of a circular shaft at 5° intervals for a range of principal preshaft stress ratios (*M* values) and (b) graphical representation of stress concentration about a circular hole.

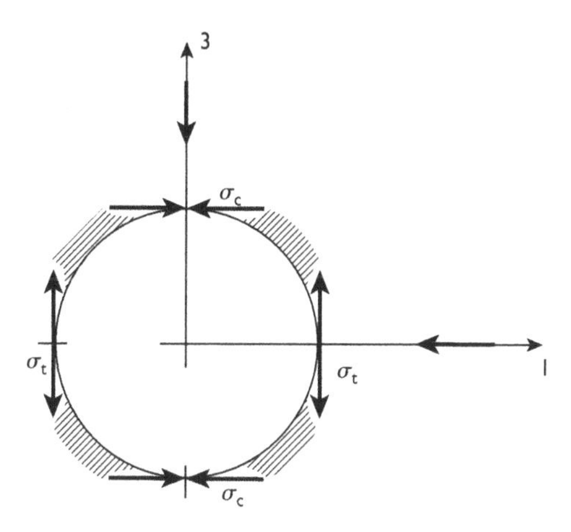

Figure 3.3 Peak tensile and compressive stress concentration about a circular hole.

graphical representation of (3.4) for $M = 0$ (uniaxial compression). Stress concentration magnitude is proportional to radial line length; tension (negative) is plotted inside the excavation boundary, while compression (positive) is plotted outside. The scale bar under $K = 1$ shows a stress concentration of one.

The peak compressive stress concentration K_c always occurs parallel to the direction of S_1 at ±90° and ranges from 2.0 to 3.0, as a practical matter. When $M > 1/3$, the minimum stress concentration factor is compression and only safety with respect to compressive stress failure is of concern. However, when tension is present ($M < 1/3$), the peak tensile stress concentration factor K_t occurs perpendicular to S_1 at 0° (and 180°) and ranges from – 1 to nil, as a practical matter. Figure 3.3 shows the location and orientation of these peak compressive and tensile stress concentrations.

Consider a hypothetical case of a vertical, circular shaft sunk in a preshaft stress field attributable to gravity alone. The vertical preshaft stress is then unit weight of rock times depth ($S_v = \gamma H$); the horizontal stress in any direction is some fraction of the vertical stress. Assume compass coordinates, so x = east, y = north and z = up, then in the x- and y-directions, the horizontal stresses S_H and S_h are some fraction of the vertical stress, that is, $S_H = S_h = K_o S_v$. In plan view, $S_H = S_h = S_1 = S_3$, $M = 1$ (hydrostatic case) and the shaft wall is in a uniform state of compression. The compressive stress concentration factor $K_c = 2$, hence $FS = FS_h = C_o/2K_o S_v$ in plan view (horizontal cross section). In vertical section, a vertical compressive stress acts at the shaft wall that is equal to the preshaft vertical stress. Thus, in vertical section, $S_v = S_1$ and $S_3 = 0$, but there is no stress concentration of the vertical stress. In this view, the unlined shaft wall safety factor is simply $FS = FS_v = C_o/S_v$. In a gravity-only initial stress field; FS_v is often less than FS_h and therefore governs the design, but both safety factors should always be computed.

Example 3.1 A vertical, unlined circular shaft is sunk in a preexcavation stress field caused by gravity alone. Estimate the *peak stresses* in horizontal and vertical sections as functions of depth. Note: The ratio of horizontal to vertical stress is 1/3.

Solution: Peak stress may be obtained from stress concentrations and reference stresses. There are no tensile stresses. In plan view, the stress concentration factor is 2 and in section is 1.

However, the reference stress in plan view is the horizontal preexcavation stress, while in section, it is the vertical preexcavation stress. Thus, the peak compression in plan view is

$$\begin{aligned}\sigma_c &= 2S_h \\ &= 2K_o S_v \\ &= (2)(1/3)\gamma h \\ \sigma_c &= (2/3)(\gamma/144)h\end{aligned}$$

where stress σ_c is in psi, specific weight γ is in pcf, and depth h is in ft. The peak stress in psi in this case is about 2/3 times the depth in ft.

In vertical section, the preexcavation and post-excavation stresses are equal, so the post-stress compression at the shaft wall $\sigma_v = \gamma\, h$ or about 1 psi per foot of depth.

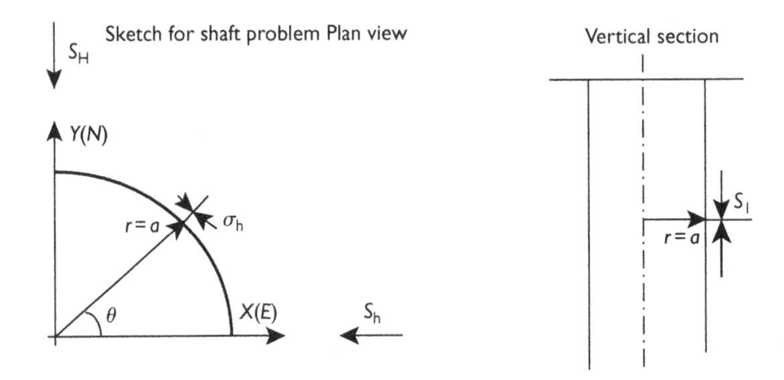

Example 3.2 A vertical, unlined circular shaft is sunk in a preexcavation stress field caused by gravity alone. Estimate the *peak stress concentrations* in horizontal and vertical sections as functions of depth.

Solution: In a gravity stress field, the vertical preexcavation normal stress is unit weight of rock times depth, so $S_v = \gamma\, h$ where γ is unit weight of rock and h is depth from the surface. Compression is considered positive. The unit weight is not given, so an estimate is necessary for a numerical calculation. Horizontal normal stresses, S_h and S_H, say, in the east and north directions (compass coordinates, z is positive up) are equal in a gravity stress field and are some fraction of the vertical stress. Thus, $S_H = S_h = K_o S_v$ where K_o is some decimal fraction. An estimate of K_o can be obtained by considering the gravity field to be applied to an elastic rock mass under complete lateral restraint. This assumption implies zero lateral strains. In consideration of Hooke's law

$$\begin{aligned}E\varepsilon_h &= 0 \\ &= S_h - \nu S_H - \nu S_v \\ 0 &= (1-\nu)S_h - \nu S_v \\ \therefore & \\ S_h = S_H &= \left(\frac{\nu}{1-\nu}\right)S_v\end{aligned}$$

A reasonable value for Poisson's ratio is 0.25, so that $K_o = 1/3$.

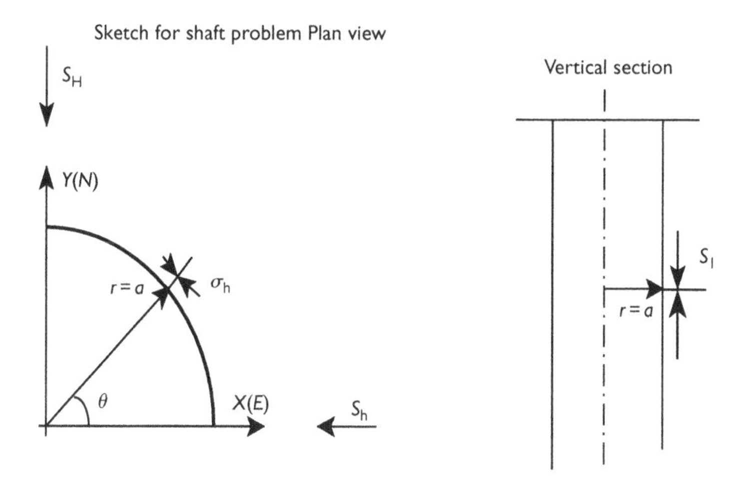

A formula for stress concentration about a circular hole can be obtained from the expression for the tangential stress. Thus, at the shaft wall of radius a

$$\begin{aligned}\sigma_\theta &= \left(\frac{S_h + S_H}{2}\right)(2) - \left(\frac{S_h - S_H}{2}\right)(4)\cos(2\theta)\\ &= 2S_h\\ &= 2K_o S_v\\ \sigma_\theta &= (2)(1/3)\gamma h\end{aligned}$$

as shown in the sketch.

The compressive stress concentration K_c in a horizontal plane (plan view) is by definition the ratio of peak stress to reference stress which is the preexcavation major principal stress *in the considered view*, that is, S_h. Hence, $K_c = 2$ in plan view and is independent of depth.

In vertical section, the postshaft vertical stress is equal to the preshaft vertical stress. Hence the stress concentration factor (in compression) $K_{cv} = S_l/S_v = 1$ and is also independent of depth.

There are no tensile stresses in either view, so tensile stress concentration need not be considered.

Example 3.3 A vertical circular shaft is proposed in a region of high horizontal stress. Measurements along the shaft route indicate the preshaft principal stresses in psi (compression positive) are given by

$$S_v = 1.1(\gamma/144)h, S_h = 100 + 1.5(\gamma/144)h, S_H = 500 + 2.2(\gamma/144)h$$

(h is depth in feet).

Determine the strengths (C_o, T_o) needed for safety factors with respect to compression and tension of 1.5 and 2.5, respectively, at 3,750 ft, with $\gamma = 160$ pcf.

Solution: A sketch illustrates the situation in plan view. In this view, the tangential stress is given by

$$\sigma_\theta = (1/2)(S_H + S_h)(2) - (1/2)(S_H - S_h)(4)\cos(2\theta)$$

The extreme values of this function with respect to θ are at zero at 0° and 90°. The extreme values are $-S_H + 3S_h(\theta = 0°)$ and $3S_H - S_h(\theta = 90°)$. Because $S_H > S_h$, the first is the minor stress which is tensile when $3S_h < S_H$; the second is the major compression.

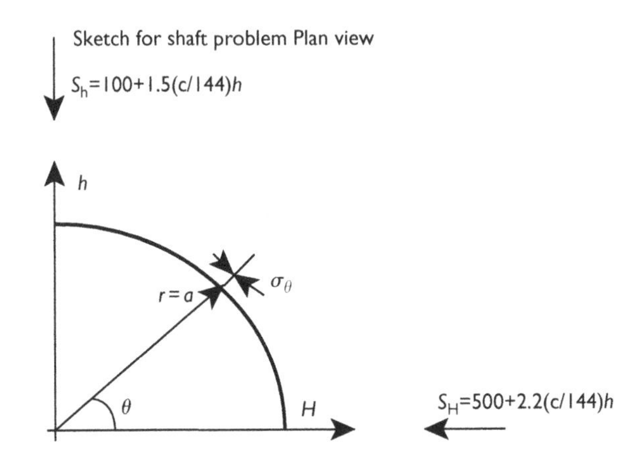

Thus,

$$\sigma(0) = -[500 + 2.2(160/144)3{,}750] + (3)[100 + 1.5(160/144)3{,}750] = 9{,}383 \text{ psi}$$
$$\sigma(90) = (3)[500 + 2.2(160/144)3{,}750] - [100 + 1.5(160/144)3{,}750] = 22{,}650$$

By definition $FS_c = C_o/\sigma_c$ and $FS_t = T_o/\sigma_t$. No tension is present, so tensile strength is not a factor. The compressive strength needed is (1.5)(22,650) = 33,975 psi, a relatively high value even for intact core tested in the laboratory. Reinforcement and support would be needed before reaching the considered depth. In vertical section $\sigma = S_v = 1.1(160/144)(3{,}750) = 4{,}583$ psi which is considerably less than the peak compression in plan view.

Unlined elliptical shafts

An elliptical shape introduces two additional considerations for unlined shaft safety factor calculations: aspect ratio and orientation. In case of elliptical shafts, when the axes of the ellipse coincide with the preshaft principal directions, if a is the semi-axis parallel to S_1 and b is parallel to S_3, then the stress concentration tangential to the shaft wall is given by

$$K = \frac{(1-k^2)(1-M) + 2k(1+M) - (1-M)(1+k)^2\cos(2\alpha)}{(1+k^2) - (1-k^2)\cos(2\alpha)} \tag{3.5}$$

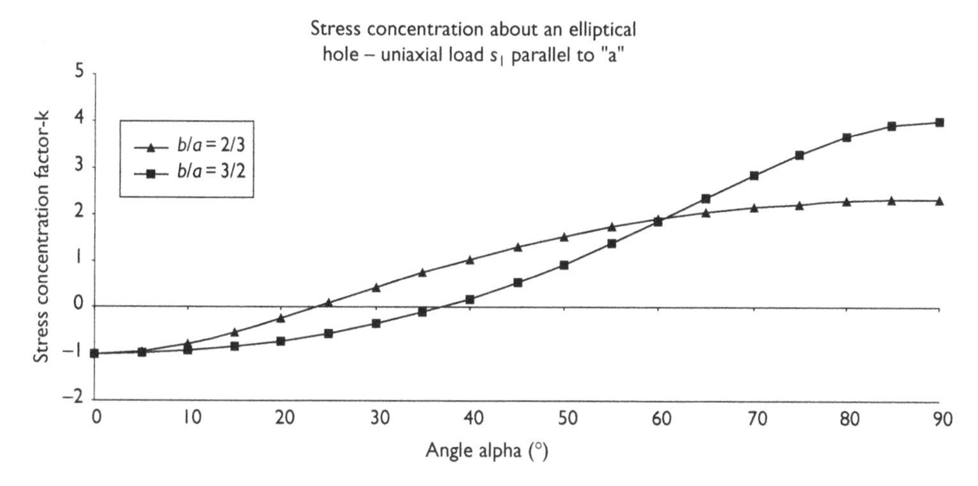

Figure 3.4 Stress concentration about an elliptical shaft with semi-axes ratios of 2/3 and 3/2 under uniaxial stress $(M = 0)S_1$ applied parallel to the semi-axis a.

where k is the aspect ratio b/a, which may be greater or less than one (see, e.g. Jaeger and Cook, 1969). The angle α is related to the polar angle by $\tan(\theta) = (b/a)\tan(\alpha)$ where angles are measured counterclockwise from the direction of S_1. In the case of a circular shaft, $k = 1$ and (3.5) reduces to (3.4). Figure 3.4 shows how K varies with α for two selected values of k (2/3 and 3/2). Inspection of Figure 3.4 suggests that peak stress concentrations occur at the ends of the semi-axes, that is, at the sides and ends of an ellipse ($\theta = \alpha = \pm 90°$, $\theta = \alpha = 0°$ and 180°) as in the case of a circular shaft. Indeed, this is the case. In fact, stress concentrations at the ends of an ellipse semi-axes with S_1 parallel to semi-axes a are

$$
\begin{aligned}
K_a &= -1 + M\left(\frac{2}{k} + 1\right) \\
K_b &= (1 + 2k) - M
\end{aligned}
\tag{3.6}
$$

which reduce to the circular case when $k = 1$. Figure 3.5(a) shows the meaning of stress concentration factors K_a and K_b when the long axis of an ellipse is parallel to $S_1 (k < 1)$; Figure 3.5(b) shows the situation when the long axis is perpendicular to $S_1 (k > 1)$. In case $M = 0$, the tension under S_1 at the ends of the a semi-axis is equal to S_1 (but opposite in sign) regardless of k, while the compression at the ends of the b semi-axes is much greater when the long axis of the ellipse is perpendicular to $S_1 (k > 1)$. Figures 3.5(c) and (d) show graphically two uniaxial cases: (c) when the applied load is parallel to the long axis of an ellipse with semi-axes ratio $b/a = 2/3$ and (d) when the applied load is perpendicular to the long axis of an ellipse with semi-axes ratio $b/a = 3/2$. These cases are those plotted in Figure 3.4. If both stress concentration factors are positive, then tension is not a factor.

Consider an elliptical shaft with a semi-axes ratio $k = 1/3$ in a preshaft stress field characterized by $M = 1/4$. The long axes of the shaft cross section is parallel to S_1 and according to (3.6), $K_a = +0.75(+3/4)$, $K_b = +1.42(+1 + 5/12)$, and tension is absent. If the long

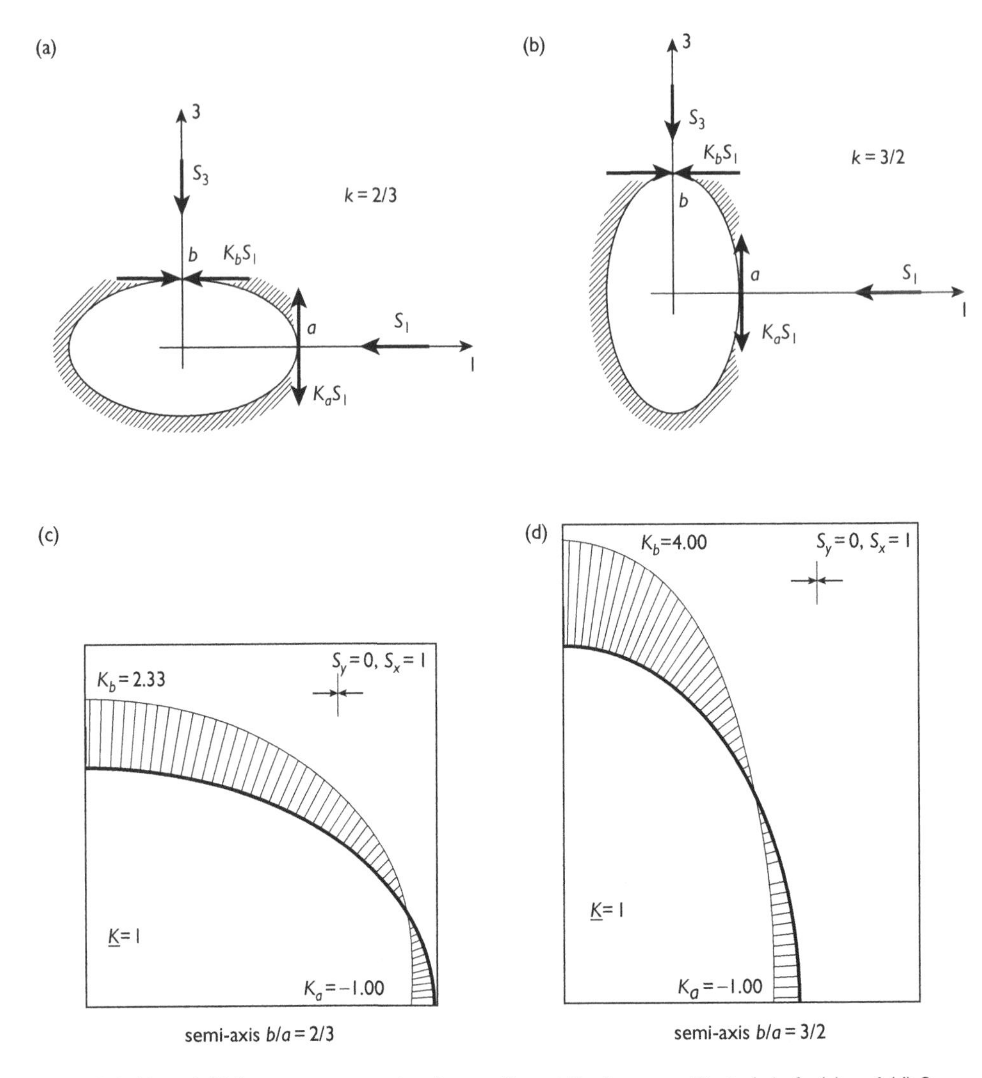

Figure 3.5 (a) and (b) Stress concentration factors K_a and K_b about an elliptical shaft. (c) and (d) Stress concentration about two uniaxially loaded ellipses.

axis is perpendicular to S_1, so $k = 3$, then $K_a = -0.59(-7/12)$ and $K_b = +6.75(+6 + 3/4)$. These numbers illustrate a *rule of thumb* that states *the most favorable orientation of an elliptical excavation is with the long axis parallel to the major preshaft compression.*

Example 3.4 An elliptical cross section may be advantageous in a stress field where the major and minor preexcavation stresses S_1 and S_3 are aligned with the geometric axes of the section. Show that the preferred orientation is with the long axis of the ellipse parallel to S_1.

Solution: The peak stress concentrations at the wall of an elliptical section occur at the ends of the geometric axes. The peak stresses are given by the formulas

$$\sigma_a = K_a S_1 = -S_1 + (1 + 2/k)S_3 = S_1[(1 + 2/k)M - 1]$$

$$\sigma_b = K_b S_1 = -S_3 + (1 + 2k)S_1 = S_1[(1 + 2k) - M]$$

where the meaning of terms is shown in the sketch, $k = b/a$ and $M = S_3/S_1$.

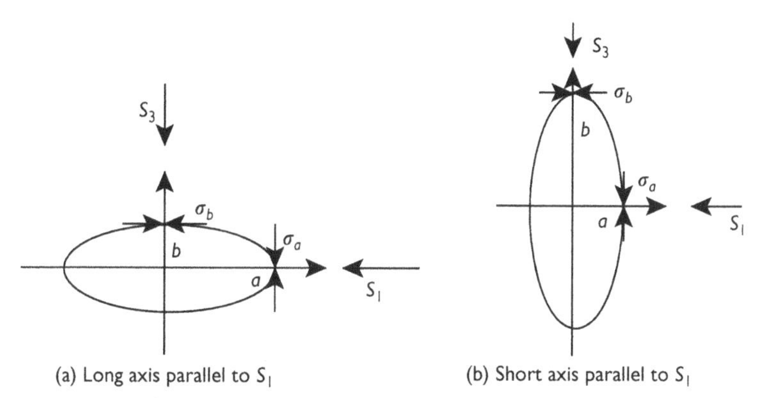

(a) Long axis parallel to S_1 (b) Short axis parallel to S_1

Sketch illustrating the meaning of ellipse formula terms

In the uniaxial case, $M = 0$, so when the long axis is parallel to S_1, $k < 1$, $K_a = -1$ and $K_b = 1 + 2k$, and when the short axis is parallel to S_1, $k = k' > 1$, $K_a = -1$, as before, and $K_b = 1 + 2k'$, which is a larger number than the first case. These observations suggest that when the long axis is parallel to S_1, the peak compression will be minimal. A change in axis orientation corresponds to using the reciprocal of k. If k is the aspect ratio when the long axis is parallel to the x-axis (Case 1), then use of the reciprocal $k' = 1/k$ corresponds to the case where the long axis is normal to the x-axis (Case 2). Examination of the stress concentration factors for points a and b shows that $K_b' > K_b$ and $K_a > K_a'$ always and that $K_b > K_a$ when $k > M$. With this proviso, the stress concentration factors are ordered: $K_b' > K_b > K_a > K_a'$. Thus, stress concentrations in Case 1 are bounded by those in Case 2, provided $k > M$.

For a given value of preexcavation stress ratio M, one may choose an ellipse aspect ratio k that meets this condition. The best orientation is then with the long axis parallel to the major compression before excavation. High aspect ratio ellipses are unlikely to be considered in practice, so the condition $k > M$ is not too restrictive. For example, given $M = 1/3$, then for $k = 1/2, K_b' = 1 + 2(2) - 1/3 = 4.67, K_b = 1 + 2(1/2) = 2, K_a = [1 + 2/(1/2)](1/3) - 1] = 0.67, K_a' = -0.33$. Thus, when the long axis of the ellipse is oriented parallel to S_1, the stress concentrations at b and a are 2 and 0.67, respectively. If the long axis were normal to S_1, the stress concentrations would be 4.67 and −0.33, respectively, where the negative sign indicates a tension (at a).

Consider a problem of selecting the best elliptical shape for a given *in situ* stress field. The problem is to determine the optimum k for a given M. Optimum from the ground control view means lowest stress concentrations. Suppose the given stress state is hydrostatic, so $M = 1$, then $K_a = 2/k$ and $K_b = 2k$ according to (3.6). Both stress concentration factors are positive. Thus tension is not a factor. Optimum or least compressive stress concentration can now be obtained by reducing the larger of K_a or K_b. Ideally, one has neither $K_a > K_b$ nor $K_b > K_a$. Hence, the optimum occurs when $K_a = K_b$, that is, when $k = 1$. Thus, a circular shape, a special case of an ellipse, is optimum when the preshaft stress state is hydrostatic. This situation occurs for vertical shafts in ground where the stress state is caused by gravity alone.

Example 3.5 A vertical shaft of elliptical section is being considered in a region of high horizontal stress. Measurements along the shaft route indicate the preshaft principal stresses in psi (compression positive) are given by $S_v = 1.1h$, $S_h = 100 + 1.5h$, $S_H = 500 + 2.2h$ (h is depth in feet). Determine the optimum orientation and aspect ratio of the section.

***Solution*:** The peak stress concentrations at the wall of an elliptical section occur at the ends of the geometric axes. The peak stresses are given by the formulas

$$\sigma_a = K_a S_1 = -S_1 + (1 + 2/k)S_3 = S_1[(1 + 2/k)M - 1]$$

$$\sigma_b = K_b S_1 = -S_3 + (1 + 2k)S_1 = S_1[(1 + 2k) - M]$$

where the meaning of terms is shown in the sketch, $k = b/a$ and $M = S_3/S_1$.

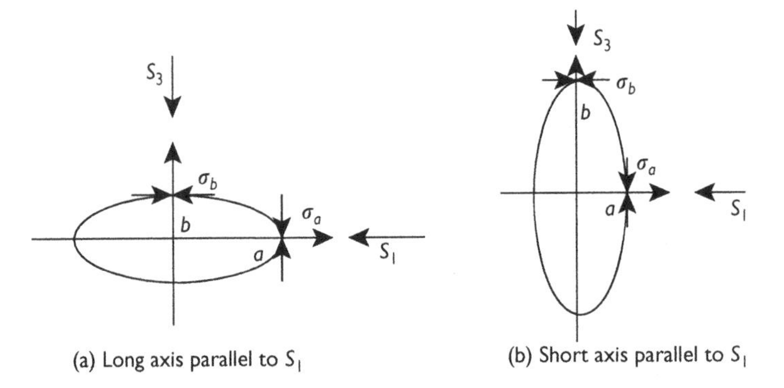

(a) Long axis parallel to S_1 (b) Short axis parallel to S_1

Sketch illustrating the meaning of ellipse formula terms

An optimum orientation would be one that reduces stress concentration to a minimum, so the stress concentration at b is no greater or less than at a and tension is absent. Thus,

$$K_a = K_b \geq 0$$

$$\therefore$$

$$(1 + 2/k)M - 1 = (1 + 2k) - M$$

$$\therefore$$

$$k = M = \frac{100 + 1.5h}{500 + 2.2h}$$

so the aspect ratio varies with depth, but is equal to the preexcavation stress ratio at any particular depth. At the surface $k = 0.200$ and at great depth $k = 1.5/2.2 = 0.682$.

The best orientation is with the long axis a parallel to S_H. In this orientation, with $k = M$, $K_a = K_b = 1 + M$ and indeed the wall of the ellipse is in a uniform compression. At the surface, the stress concentration is 1.2; at great depth the stress concentration is 1.682.

With the short axis parallel to S_H, $K_a' = 2M^2 + M$ and $K_b' = 1 - M + 2/M$, which is greater than $1 + M$ and therefore less favorable at any depth.

If a circular section were considered, the stress concentrations would be $-1 + 3M$ and $3 - M$ or -0.4 and 2.8 at the surface; 1.046 and 2.318 at great depth. The elliptical section shows a considerable advantage over the circular section in this stress field. However, other considerations such as difficulty in excavation to the proper shape would need to be considered before final selection of the shaft shape.

In vertical section, the shaft wall stress is equal to the preshaft stress and stress concentration is one. This value is less than stress concentration in plan view of the best orientation and aspect ratio (1+M) for $M = k > 0$. Hence, the plan view section governs design.

Example 3.6 Suppose a vertical shaft of elliptical section 14 ft × 21 ft is being considered for sinking to a depth of 4,350 ft. The horizontal stress in the north–south direction is estimated to be twice the vertical stress caused by rock mass weight, while the east–west stress is estimated to be equal to the "overburden" stress. Unconfined compressive strength of rock is 22,000 psi, as measured in the laboratory; tensile strength is 2,200 psi. Determine shaft wall safety factors as functions of depth with the section oriented in a way that minimizes compressive stress concentration.

***Solution*:** Shaft wall safety factors with respect to compression and tension, by definition, are

$$\mathrm{FS_c} = \frac{C_o}{\sigma_c}, \quad \mathrm{FS_t} = \frac{T_o}{\sigma_t}$$

In case of an ellipse, the peak stresses at the ends of the semi-axes are:

$$\sigma_a = K_a S_1 = -S_1 + (1 + 2/k)S_3 = S_1[(1 + 2/k)M - 1]$$

$$\sigma_b = K_b S_1 = -S_3 + (1 + 2k)S_1 = S_1[(1 + 2k) - M]$$

$$k = 14/21 = 2/3 \leq 1, \quad M = S_{EW}/S_{NS} = 1/2,$$

$$\sigma_a = K_a S_1 = -S_1 + (1 + 2/k)S_3 = S_1[(1 + 2/(2/3))(1/2) - 1] = S_1(1.00)$$

$$\sigma_b = K_b S_1 = -S_3 + (1 + 2k)S_1 = S_1[(1 + 2(2/3)) - 1/2] = S_1(1.83)$$

The maximum and minimum stress concentration factors are both positive, so no tension is present. With respect to compressive stress, the shaft wall safety factor as a function of depth is

$$\mathrm{FS_c} = \frac{C_o}{1.83S_1}$$

$$= \frac{22,000}{1.83(2)(158/144)h}$$

$$\mathrm{FS_c} = \frac{5,478}{h}$$

where h is in feet and the specific weight of the rock mass is estimated to be 158 pcf. The long axis of the section is in a north–south alignment.

The safety factor in vertical section is

$$FS_{cv} = \frac{22,000}{(158/144)h} = \frac{20,050}{h}$$

When the ellipse axes do not coincide with the preshaft principal axes, an orientation effect is introduced into the stress concentration formula (3.6). Thus,

$$K = \frac{(1-k^2)(1-M)\cos(2\beta) + 2k(1+M) - (1-M)(1+k)^2\cos(2\beta - 2\alpha)}{(1+k^2) - (1-k^2)\cos(2\alpha)} \quad (3.7)$$

where β is an inclination angle measured counterclockwise from S_1 to a as shown in Figure 3.6(a). Alternatively, the applied stresses include a shear stress relative to $x - y$ axes coincident with the $a - b$ semi-axes, as shown in Figure 3.6(b). Principal stresses and directions seen in the shaft cross section are given by the usual equations. Thus,

$$\left.\begin{matrix} S_1 \\ S_3 \end{matrix}\right\} = \frac{S_{xx} + S_{yy}}{2} \pm \left[\left(\frac{S_{xx} - S_{yy}}{2}\right)^2 + (T_{xy})^2\right]^{1/2}$$

$$\tan(2\beta) = \left[\frac{T_{xy}}{(S_{xx} - S_{yy})/2}\right] \quad (3.8)$$

where S_{xx}, S_{yy}, T_{xy} are the preshaft stresses in the shaft cross section and β is measured counterclockwise from the x-axis. There are two solutions for β; care must be exercised in selecting the correct one. One method is to consider that the major principal stress will be "nearest" to the largest given normal stress and then to measure the angle obtained from a calculator using the arctan function from the direction of the largest given normal stress using

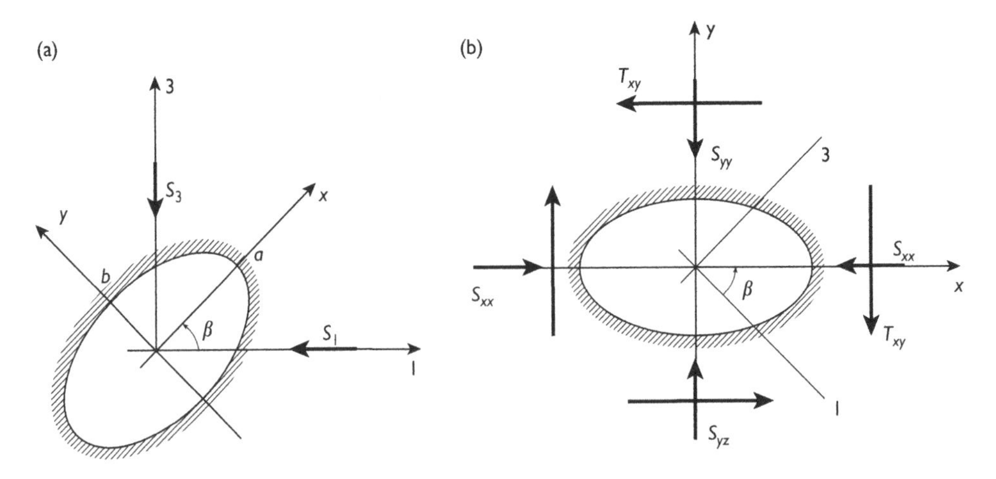

Figure 3.6 Orientation of an elliptical shaft when shear stress is present.

the sign convention that counterclockwise is positive. For example, suppose stress measurements show $S_{xx} = 1{,}500$, $S_{yy} = 2{,}500$, $T_{xy} = 866$ in unspecified units. According to Equation (3.8), $S_1 = 3{,}000$, $S_3 = 1{,}000$, $\beta = -30°$. Hence, the direction of S_1 is at a 30° clockwise angle from the y-axis (60° counterclockwise from the x-axis).

An interesting exception to the rule of peak stresses occurring at the semi-axes ends occurs when tension is present at the walls of an inclined ellipse. Figure 3.7(a) shows stress concentration at the walls of an elliptical shaft with semi-axes ratio 1/4 under uniaxial compression ($M = 0$) as a function of polar angle θ (measured counterclockwise from the a semi-axis) for inclinations up to 90°. In the worst case shown in Figure 3.7(a), $\beta = 30°$ and the peak tensile stress concentration is about -1.5 which occurs about 8° away from the a semi-axis. This tension is 50% greater than the frequent peak tension of -1. The peak tension and angle away from the semi-axis increase with decreasing aspect ratio. A similar compressive stress concentration effect is also evident in Figure 3.7(a). As the ellipse is rotated from being parallel to $S_1 (\beta = 0°)$ and having the most favorable orientation perpendicular to $S_1 (\beta = 90°)$, which is the least favorable orientation, the compressive stress concentration factor increases from 1.5 to 9.0. The peak tensile stress concentration ranges from -1.0 to about -1.5 at $\beta = 30°$ and back to -1.0. Figures 3.7(b) and (c) show graphically the distribution of stress concentration in this case ($\beta = 30°$) and in the case when $\beta = 60°$. However, these effects are not of much practical importance for shafts because the elliptical shape itself is not very practical and uniaxial stress *in situ* is rare, while tension is greatly reduced under biaxial stress.

Example 3.7 Consider a vertical shaft with an elliptical cross section that has a semi-axis ratio $k = 1/3$ and is excavated in a stress field characterized by a principal stress ratio $M = 1/4$ in plan view. The section of interest is at a depth of 1,200 ft where the vertical stress is 1,350 psi. Intact rock strengths are: $C_o = 13{,}500$ psi and $T_o = 1{,}250$ psi. Suppose the a-axis is inclined 30° to S_1 and that the xy compass directions coincide with the principal 1-3 directions. Determine shaft wall safety factors with respect to compression and tension.

***Solution*:** According to Equation (3.7), (with $M = 1/4$, $k = 1/3$, $\beta = 30°$, $\alpha = 0°, 30°$), stress concentration factors at the ends of the ellipse are: $K_a = 9/4$ and $K_b = 11/12$. Thus, tension is absent and peak compression is at a. If $S_1 = 1{,}200$, estimated at 1 psi/ft of depth, then $\sigma_t = 2{,}700$ psi at the ends of the a semi-axes and 1,100 psi at the ends of the b semi-axes. The shaft wall safety factor with respect to compression is

$$\mathrm{FS_c} = \frac{13{,}500}{(9/4)(1{,}200)} = 5.00$$

$$\mathrm{FS_v} = \frac{13{,}500}{1{,}350} = 10.00$$

where the first calculation is in plan view and the second is in vertical section. In this example, safety is determined in horizontal section.

Unlined rectangular shafts

In the case of rectangular shapes, rounding of corners is an important consideration. Amathematical sharp corner leads to infinite stress at a corner point. A rounded corner, characterized by a specified radius of curvature, results in finite stress. The greater the

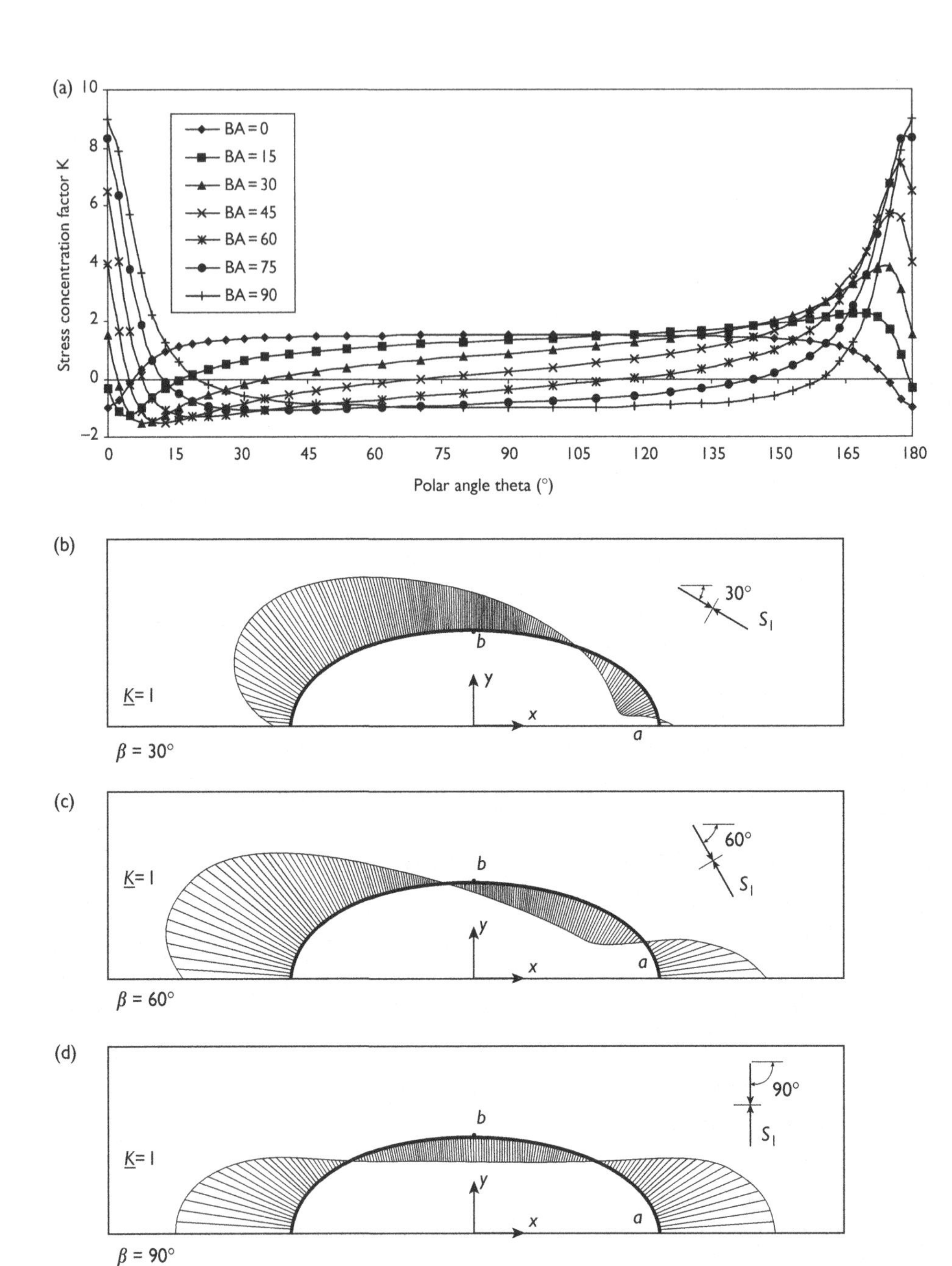

Figure 3.7 (a) Stress concentration at the wall of an inclined elliptical shaft under uniaxial stress ($M = 0$). BA = inclination angle of the a semi-axis to S_1, $b/a = 0.25$. Semi-axes ratio $k = 1/4$. (b), (c), and (d) stress concentration about an ellipse of semi-axes ratio $b/a = 1/2$. β = *ccw* angle from S_1 to semi-axis a. Tension inside, compression outside section.

rounding, the smaller the stress concentration. The maximum corner radius possible is one-half the least dimension of the section and results in an ovaloid. As with ellipses, one may characterize rectangular shafts by an aspect ratio k where $k = b/a$, a is the semi-axes parallel to S_1 and b is parallel to S_3. A range of k between 1/8 and 8 more than covers the range of practical rectangular shaft dimensions.

A *new* phenomenon occurs in stress concentration about rectangular shafts. Unlike elliptical shaft shapes, rectangular and ovaloidal sections experience peak compressive stress concentration near corners, not at the end of the semi-axes. However, when tension is present, peak tension still occurs at the ends of a semi-axis, that is, at mid-sides; which sides (short or long) depends on orientation with respect to the preshaft principal axes. These observations are made with the understanding that the preexcavation principal stress directions are parallel to the semi-axes of the section.

Simple mathematical expressions for stress concentration about rectangular shafts are not available as they were for circular and elliptical shaft sections. Although stress concentration about rectangular sections in homogeneous, isotropic, linearly elastic ground also depends on the ratio of preshaft principal stresses, aspect ratio, and orientation, useful stress concentration formulas are lengthy and complex in derivation. Numerical evaluation is required that introduces a degree of approximation to symbolic but exact results from the mathematical theory of elasticity.

Figure 3.8 shows two cases of stress concentration about a rectangular shaft with an aspect ratio of 1/2 and 2. Data in Figures 3.8(a) and (b) were obtained numerically from an approximate analyses (Heller *et al.*, 1958). Inspection of Figure 3.8 shows that indeed the peak compressive stress concentration always occurs at the corners, regardless of preshaft principal stress ratio M, and that peak tension, when present, occurs at the ends of a semi-axis. Figure 3.8 also suggests that for likely preshaft principal stress ratios ($M > 0$), tension is relatively constant until the considered point nears the corner where tension rapidly diminishes and stress concentration rises sharply to peak compression. Comparisons of peak compressive stress concentration (compression positive) suggests that the most favorable orientation of a rectangular section is with the long axis parallel to the major preshaft principal stress S_1, as was the case with an elliptical shaft section, for the practical range of $M > 0$. In this regard, the long axis is parallel to S_1 for $k < 1$ and is perpendicular to S_1 when $k > 1$. Examination of many such plots shows that this rule of thumb does indeed apply to rectangular shafts when the choice is whether to orient the long axis of a section parallel or perpendicular to the major preshaft compression.

Maximum and minimum stress concentration about rectangular sections with corners rounded to 1/12th the short dimension are shown in Figure 3.9 for a range of preexcavation principal stress ratios. Figure 3.9(a) shows an *almost linear increase in peak compressive stress* (maximum stress) concentration when the semi-axis ratio $k > 1$ (long axis perpendicular to S_1), that is, when unfavorably orientated. The case $M = 1/3$ falls in the center of the cluster of almost parallel lines to the right of $b/a = 1$ and has the equation $K_c = 4.06 + 1.19[b/a - 1]$. Extrapolation to $b/a = 20$ gives $K_c = 26.7$. At this aspect ratio the difference between $M = 0$ and $M = 1$ from $M = 1/3$ is about 4% less and 8% more, respectively.

Figure 3.9(b) shows an *almost constant tensile stress* (minimum) concentration when the semi-axis ratio $k > 1$; the minimum stress concentration is tension for values of

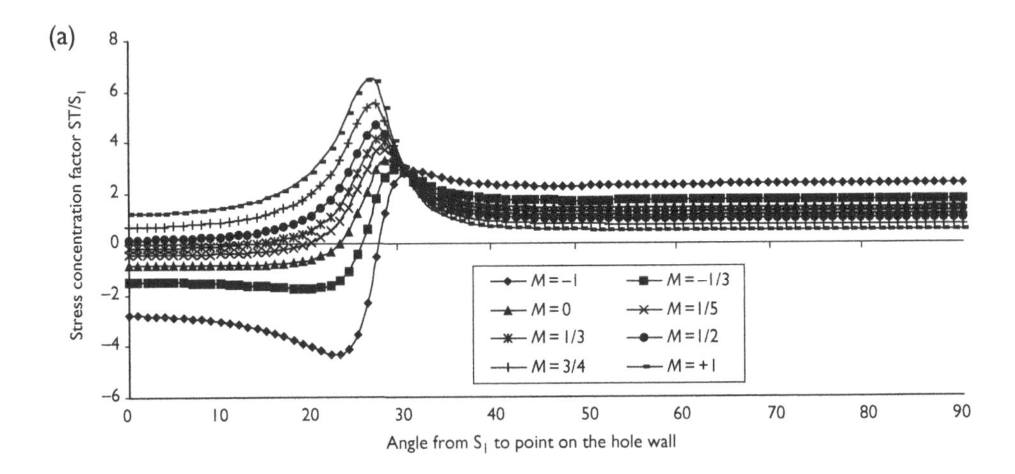

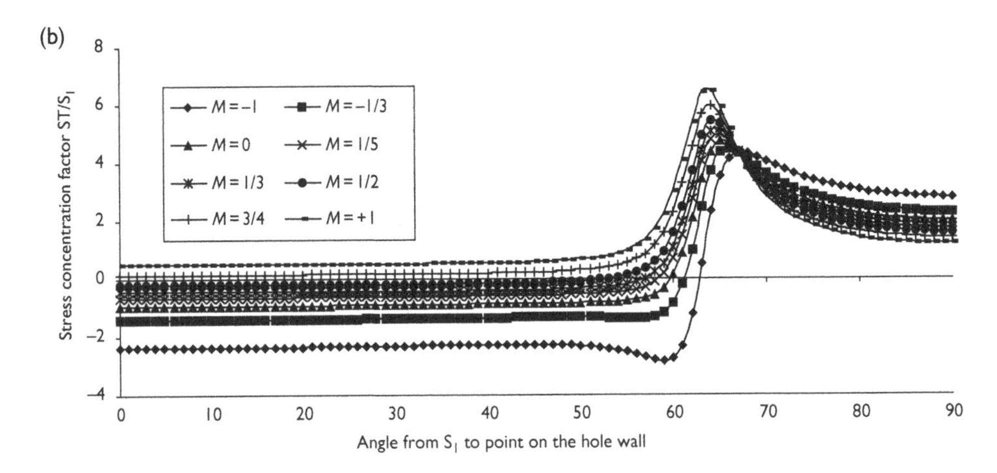

Figure 3.8 Examples of stress concentration about a rectangular shaft (a) S_1 is parallel to the long axis *a* and $k = 1/2$ (b) S_1 is perpendicular to the long axis *b* and $k = 2$, for various preshaft principal stress ratios *M*.

$M < 3/4$. For $M = 0$, an extreme value, as a practical matter, the peak tensile stress concentration is about -1. For a more likely value of $M = 1/3$, the peak tension is about -0.5. When $k < 1$ (favorable orientation), peak tension is even less in the practical range of $M > 0$.

The data in Figure 3.9 are summarized in an abbreviated tabular form in Table 3.2 for *M* in the practical range of zero to one. Generally the maximum stress concentration occurs at the corner of the section, while the minimum stress concentration occurs at a midside. Two cases are presented: Case 1 where the major compression acts parallel to the long axis of the section ($k < 1$) and Case 2 where the major compression acts perpendicular to the long axis

($k = k' > 1$), as illustrated in the sketch. In both cases, tension when present, occurs "under the load" at the midside normal to S_1.

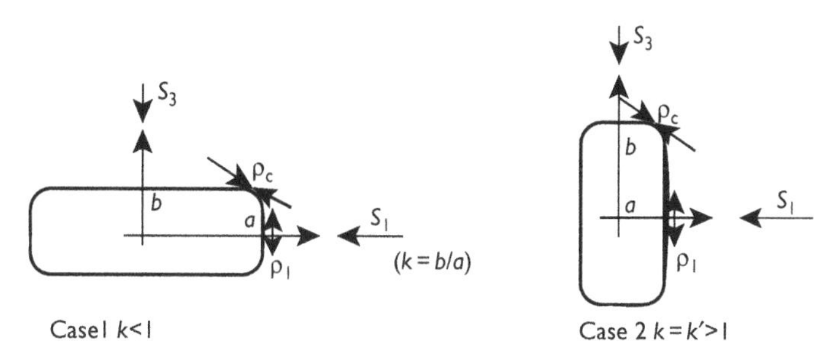

Sketch for rectangular opening stress concentration

Comparisons of peak compressive and tensile stress concentrations at the walls of elliptical and rectangular sections are presented in Table 3.2(b), respectively, for a practical value of preexcavation principal stress ratio ($M = 1/3$) and two values of semi-axis ratios ($k = 1/2$ and 2). Table 3.2(b) data show the superiority of the ellipse when in the favorable orientation with the long axis of the section parallel to the major principal stress before excavation. The tension is small for the rectangle and nil for the ellipse, so tension is not an important factor. However, the difference in compressive stress concentration is considerable, more than double in the case of the rectangle. When the long axis of the considered section is perpendicular to the preexcavation principal stress, the differences are much less and the ellipse shows only a slight advantage in compression and tension. In this case, the rectangle would be favored as a practical matter because of greater ease of construction.

Another comparison can be readily made in the case of hydrostatic preexcavation stress ($M = 1$). In this case there are no orientation effects on peak stresses. Tension is absent from both sections and is not a factor. Peak compressive stress concentration for the ellipse is 4.00 and for the rectangle is 6.46 (circle, 2.0).

Figure 3.9(c) shows the variation of maximum stress concentration about a rectangle as a function of aspect ratio for M in the range [0, 3/4]. The regression model is a multilinear function of aspect ratio k, reciprocal $1/k$ and preexcavation principal stress ratio M. The constants in the regression function were determined using all the data in Table 3.2(a), and for this reason, the fit at $M = 0$ is best. For this fit: $a_1 = 2.00$, $a_2 = 1.22$, $a_3 = 0.51$, and $a_4 = 2.28$. Changing the range of the regression analysis would enable one to obtain better fits at other M values. For example, Figure 3.9(d) shows a fit using M in the practical range [0, 1].

Example 3.8 Compare the peak compressive and tensile stress concentrations for elliptical and rectangular shaft sections when the sections are 10×20 ft and are oriented in the most favorable way and the preexcavation ratio of minor to major principal stress is 1/3. If the preexcavation major principal stress is 2,500 psi in plan view and compression is positive, determine the peak shaft wall compressive stresses for the two sections in both orientations.

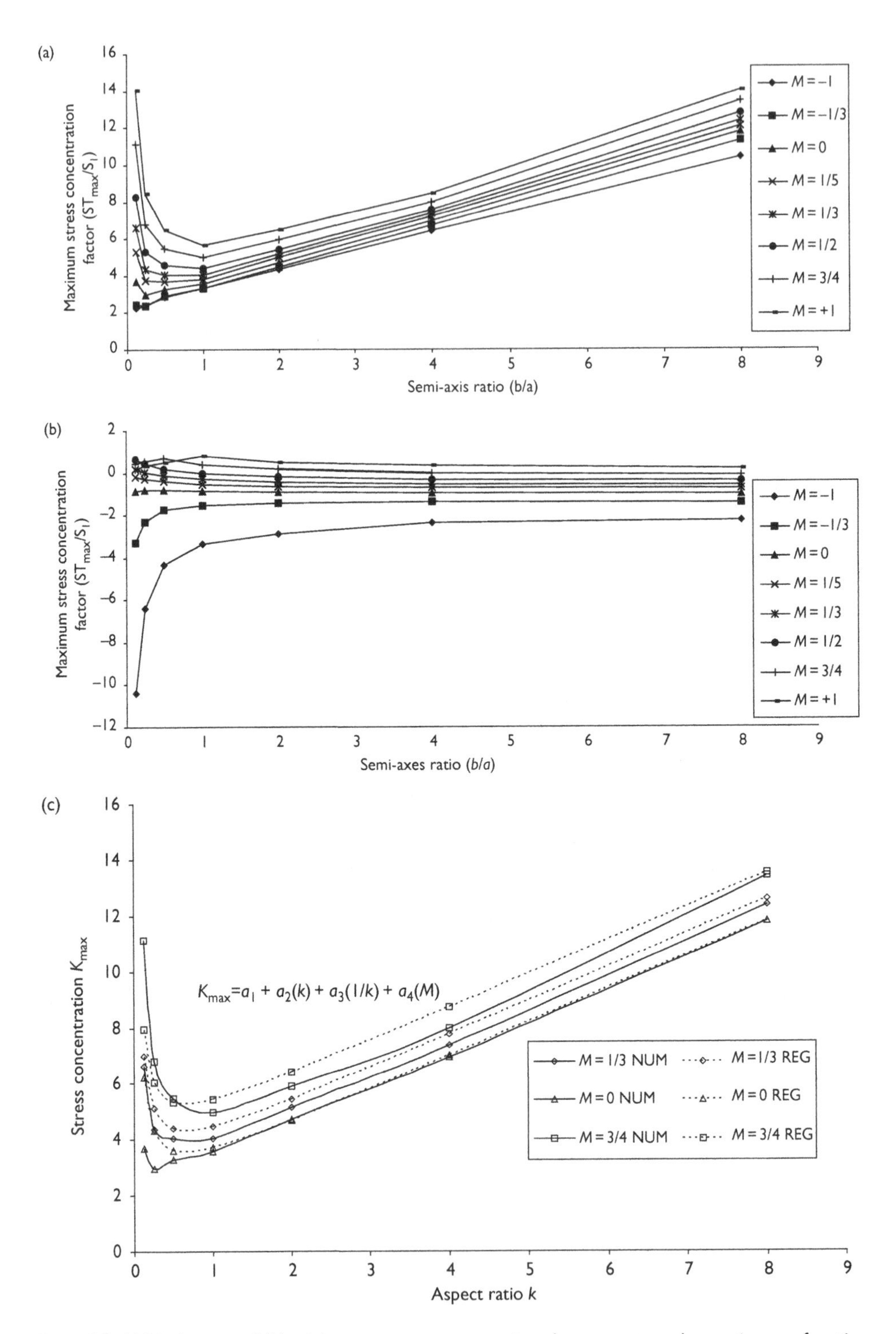

Figure 3.9 (a) Maximum and (b) minimum stress concentration about a rectangular section as a function of aspect ratio k for various preexcavation principal stress ratios (M), (c) maximum stress concentration about a rectangle as a function of aspect ratio. Solid line from numerical analysis. Dotted line from regression analysis using the equation model shown in the plot, (d) using M in the range [0,1]. For this fit: M = 1/3 and a_1 = 0.69, a_2 = 1.27, a_3 = 0.68, a_4 = 3.83.

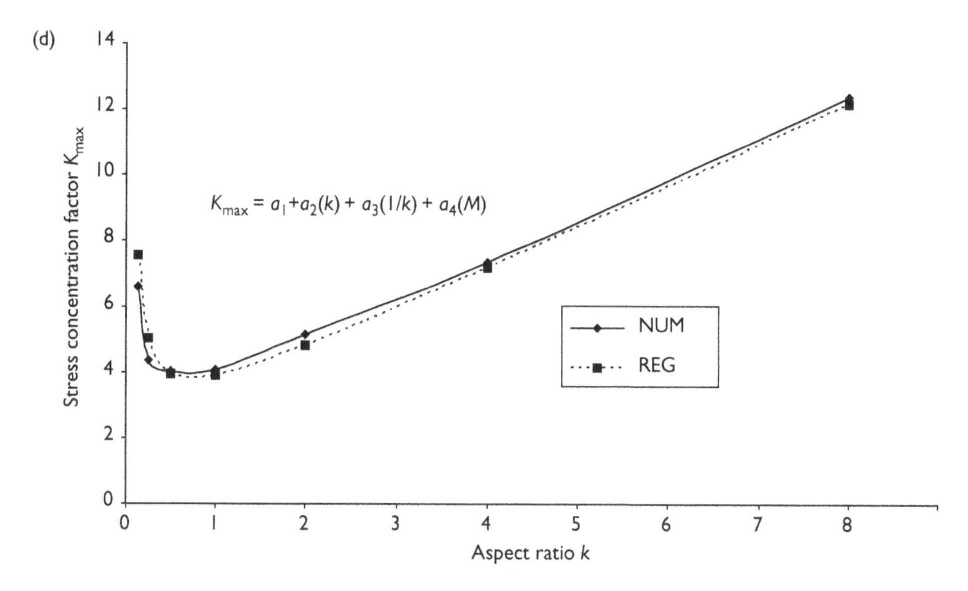

Figure 3.9 Continued.

Solution: The semi-axis ratio for both sections is either 1/2 or 2 depending on orientation, as shown in the sketch.

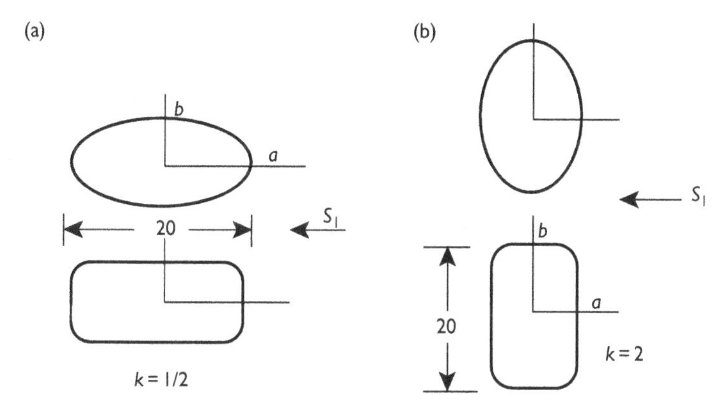

The most favorable orientation, according to a rule of thumb is to position the long axis of the section parallel to the major compression before excavation. This orientation is (a) in the sketch. In this orientation, the semi-axis ratio b/a is 1/2 for both sections. With $M = 1/3$, one may use data from Table 3.2(b) that gives a peak compressive stress concentration factor of 1.67 for the ellipse and 4.05 for the rectangle. Tension is absent about the ellipse, while the tensile stress concentration factor for the rectangle is 0.14.

In this orientation, peak compression for the ellipse is $\sigma_c = K_c S_1 = (1.67)(2{,}500) = 4{,}175$ psi and for the rectangle $\sigma_c = (4.05)(2{,}500) = 10{,}125$ psi.

In the adverse orientation, peak compression for the ellipse is $\sigma_c = K_c S_1 = (4.67)(2{,}500) = 11{,}675$ psi and for the rectangle $\sigma_c = (5.15)(2{,}500) = 12{,}875$ psi.

Table 3.2a Stress concentration factors about a rectangular section

k k′	M										
	0		1/5		1/3		1/2		3/4		1
	K_{max}	K_{min}	K_{max}	K_{min}	K_{max}	K_{min}	K_{max}	K_{min}	K_{max}	K_{min}	K_{max} K_{min}
1/8	3.70	−0.84	5.30	−0.18	6.62	0.20	8.28	0.64	11.1	0.42	14.04
8	11.8	−1.01	12.08	−0.77	12.4	−0.60	12.8	−0.40	13.4	−0.10	0.18
1/4	2.96	−0.81	3.72	−0.29	4.36	0.03	5.28	0.42	6.80	0.56	8.43
4	6.96	−0.96	7.21	−0.71	7.37	−0.53	7.58	−0.32	7.99	0.00	0.32
1/2	3.28	−0.82	3.70	−0.41	4.05	−0.14	4.57	0.19	5.48	0.69	6.46
2	4.72	−0.92	4.97	−0.64	5.15	−0.44	5.44	−0.20	5.92	0.16	0.52
1	3.58	−0.85	3.83	−0.52	4.06	−0.29	4.39	−0.02	4.97	0.40	5.65 0.80

Notes
M = principal stress ratio (S_3/S_1).
K = aspect ratio (b/a), S_1 is applied parallel to the a-axis (long axis of the section).
k′ = 1/k corresponding to application of S_1 parallel to the short axis of the section.
K(max) = maximum stress concentration factor.
K(min) = minimum stress concentration factor.

Table 3.2b Peak stress concentration factor comparison of elliptical and rectangular sections at M = 1/3

Section	*Case*			
	M = 1/3, k = 1/2		*M = 1/3, k = 2*	
	K_c	K_t	K_c	K_t
Ellipse	1.67	—	4.67	−0.33
Rectangle	4.05	−0.14	5.15	−0.44

An interesting orientation effect occurs in the case of rectangular sections when the preexcavation principal stress directions do not coincide with the section semi-axes. Let the a and b semi-axes coincide with x and y axes, respectively, and the *in situ* stresses be S_{xx}, S_{yy}, and T_{xy}, then the preexcavation principal stresses and directions are given by the usual Equations (3.8) where, as before, β is the angle between the x-axis and the direction of S_1.

Orientation effects on a rectangular section with semi-axes ratio of 1/8 are shown in Figure 3.10 for a range of principal stress ratios. Figure 3.10(a) shows that the maximum stress concentration factor K_{max} is compressive at all inclinations (compression positive) and that there is a *peak* K_{max} at about 35° for non-negative M. In the hydrostatic case there is no orientation effect, of course, and this result is also shown in Figure 3.10(a). Figure 3.10(b) shows that the minimum stress concentration factor K_{min} is generally tensile except for small inclinations at high principal stress ratios. There is a *peak* K_{min} at about 35° that increases with M. Again, there is no orientation effect in the hydrostatic case.

Figures 3.10(c), (d), and (e) show graphically stress concentration about a rectangular opening with semi-axes ratio of 0.5 ($b/a = 1/2$) when the a or x semi-axis is inclined at $\beta = 30°$, 60°, and 90° to S_1. The corners in these figures are rounded off to one-twelfth the short dimension b and $M = 0$ in all cases. Interestingly, peak tension and compression occur at the corners when $\beta = 30°$. When $\beta = 60°$, peak compression remains at a corner while tension peaks occur at a corner and at the end of the long semi-axes a. When $\beta = 90°$ and the preexcavation compression is perpendicular to the long dimension of the section, peak tension occurs at the centerline of the section, that is, at the end of the b semi-axis. Peak compression still occurs at the corner of the section.

These results call into question the rule of thumb that the most favorable orientation is with the long axis parallel to the preexcavation major principal stress. The orientation effect is to reduce the peak compressive stress while increasing the peak tension for practical M values

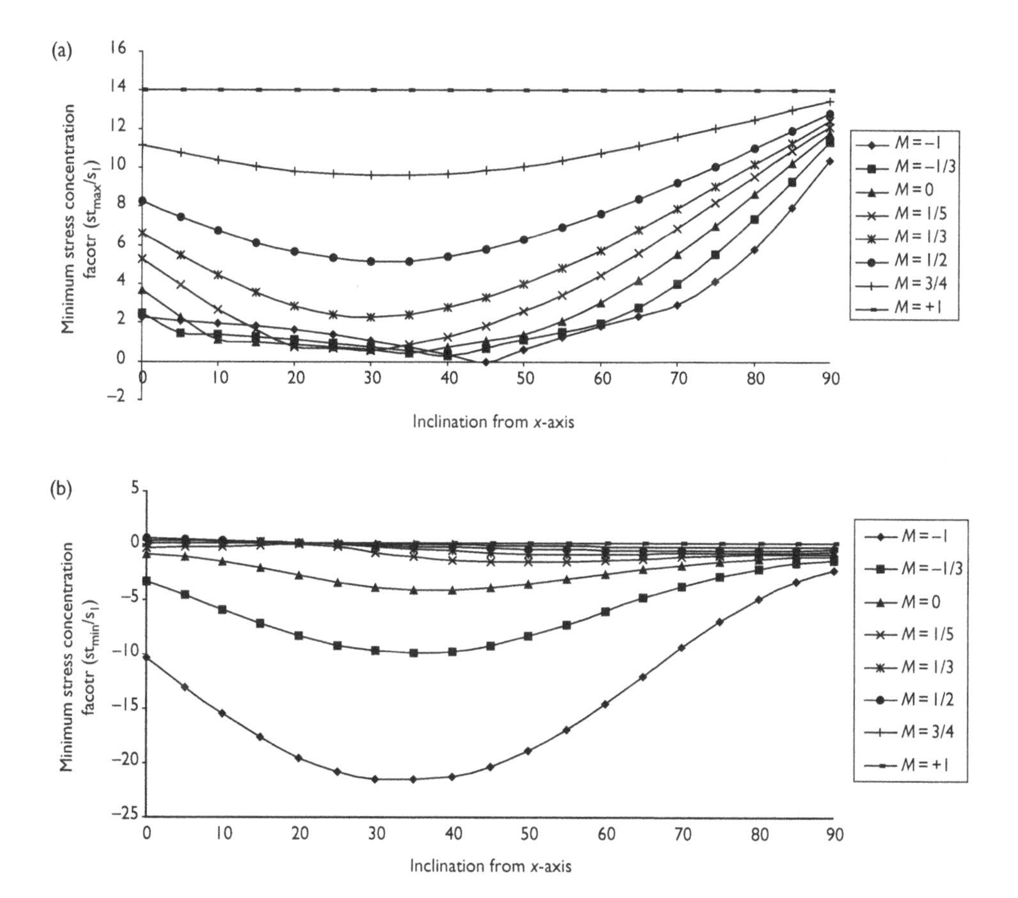

Figure 3.10 (a) and (b) Orientation effect on a rectangular section with semi-axis ratio $k = 1/8$. (c),(d), and (e) Stress concentration about a rectangle of semi-axis ratio $b/a = 1/2$. β = ccw angle from S_1 to semi-axis a.Tension inside, compression outside section. Corners rounded to 1/6th semi-axis b.

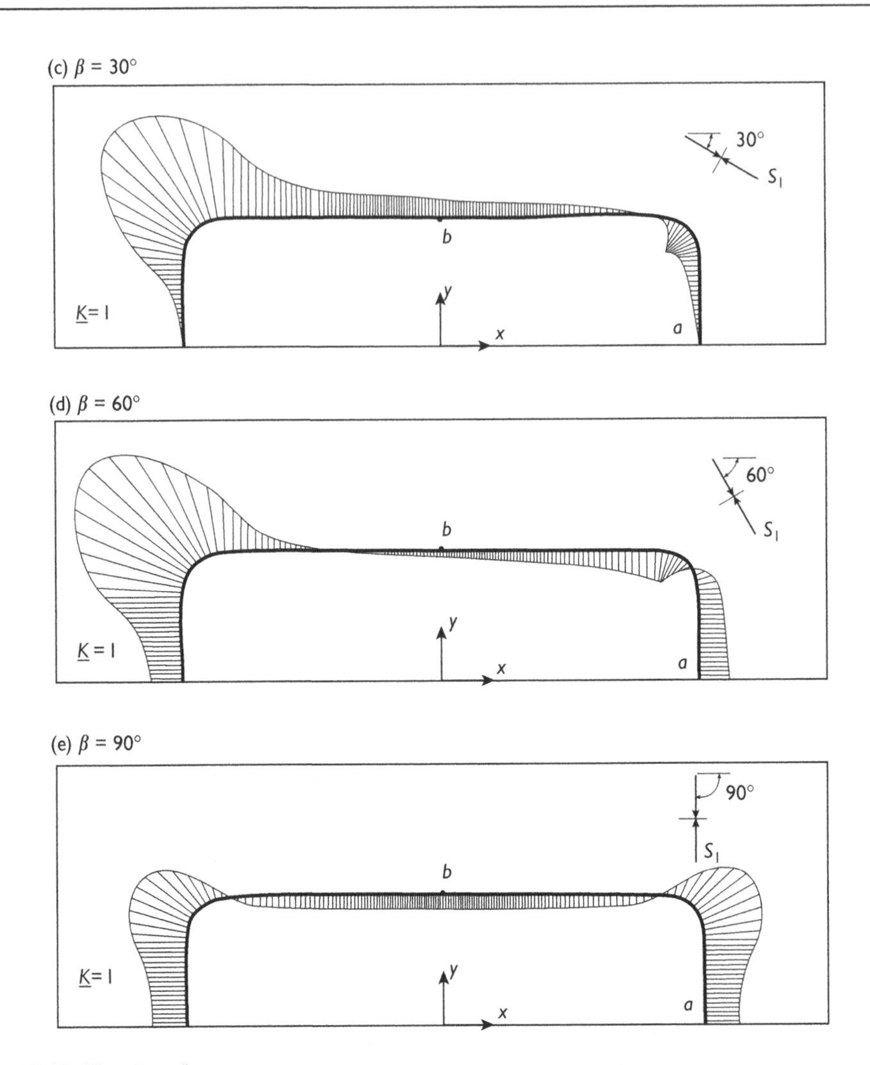

Figure 3.10 (Continued)

(positive M). For example, at $M = 1/3$, the peak compressive stress concentration is about 6.6 in favorable orientation (inclination is zero), while at an inclination of about 30°, the peak is only about 2.3. There is no tension in the favorable orientation (inclination is zero), while at 30°, a slight tension appears with a concentration of −0.08. Whether the compressive stress effect remains large at lesser and more practical semi-axes ratios is an open question at this juncture as is the question of the most favorable orientation of a rectangular shaft with respect to the *in situ* stresses.

Returning to the important practical case of a vertical shaft in a preexcavation stress field caused by gravity alone, one recalls that in horizontal section, the principal stresses are equal and therefore $M = 1$ (hydrostatic case). Hence there are no orientation effects and the question of the most favorable orientation is moot. The peak stresses can then be readily determined

Table 3.3 Peak compressive (K_c) stress concentration factors for rectangular sections[a] in a hydrostatic stress field (M = 1). Tension is absent

k =	1/8	1/4	1/2	1	2	4	8
K_c =	14.04	8.43	6.46	5.65	6.46	8.43	14.04

Note

a Corners rounded to 1/12 radius of least dimension.

from stress concentration factors in Figure 3.9. Data for this case are given in Table 3.3. These data are symmetric about $k = 1$ (square section) because there is no unique orientation with respect to principal directions. A simple straight line fit of Table 3.3 data for $k > 1$ leads to a $K_c = 5.65 + 1.2(b/a - 1)$ which allows for estimation of peak compressive stress concentration at higher aspect ratios (for $M = 1$).

A square is the best shaped rectangle in a hydrostatic stress field according to the data in Table 3.3. By comparison, the optimum elliptical shaft excavated in a hydrostatic stress field was a circle. Increasing corner sharpness increases peak stress; rounding decreases peak stress. In fact, rounding the corners of a square to maximum (½ least dimension) creates a special ovaloid. This ovaloid is actually a circle which has a peak stress concentration of 2.0. Reduction of peak stress by rounding corners is significant, by almost a factor of 3. One may speculate if corners are not deliberately rounded by design, then stress reduction will occur naturally in small, localized regions that yield under high stress at corners. There is also a practical trade-off between more usable cross-sectional area in a square and less stress concentration about a circular shaft and between a square and other rectangles in a gravity-only stress field.

In plan view (horizontal cross section), the safety factor FS associated with a square with 1/12 corner rounding is $FS_h = C_o/5.65K_oS_v$. In vertical section, a vertical compressive stress acts at the shaft wall that is equal to the preshaft vertical stress. Thus, in vertical section, $S_v = S_1$ and $S_3 = 0$. In this view, the unlined shaft wall safety factor is simply $FS = FS_v = C_o/S_v$. In a gravity-only initial stress field; FS_v is less than FS_h when $5.65K_o < 1$ or K_o about less than 1/5 which is seldom the case. Thus, the safety factor computed in plan view usually governs rectangular shaft section design, although both safety factors should always be computed.

Example 3.9 Stress measurements with respect to compass coordinates x = east, y = north, z = up show $\sigma_{xx} = 1{,}200$; $\sigma_{yy} = 2{,}350$; $\sigma_{zz} = 3{,}420$; $\tau_{xy} = -760$; $\tau_{yz} = 0$; $\tau_{zx} = 0$ where compression is positive. A 12 × 24 ft rectangular section is scheduled to pass through this region. What orientation of the shaft is best and what are the shaft wall safety factors in compression and tension, given unconfined compressive and tensile strengths of 18,500 and 1,250 psi, respectively?

***Solution*:** The most favorable orientation is one that induces the least stress concentration at the shaft wall, and according to a rule of thumb, is one with the long axis of the section parallel to major compression before excavation.

By inspection, the z-direction is a principal direction because of the vanishing z-direction shear stresses. Orientation of the major principal stress in the xy-plane is given by

$$\tan(2\theta) = \frac{\tau_{xy}}{(1/2)(\sigma_{xx} - \sigma_{yy})}$$

$$= \frac{-760}{(1/2)(1,200 - 2,350)}$$

$$\tan(2\theta) = 1.322$$

$\therefore$

$$\theta = 26, 116^{\circ}$$

The larger angle is the correct solution. One may intuitively see that this is correct by supposing the major principal stress is nearest to the largest given normal stress which is σ_{yy} in this case. Measuring 26°, positive counterclockwise from the y-axis gives the desired orientation, as shown in the sketch.

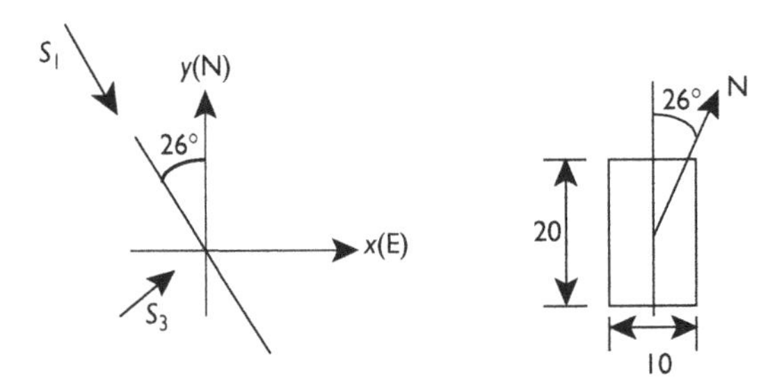

The major and minor principal stresses before excavation are (in psi):

$$S_1 = (1/2)(S_{xx} + S_{yy}) \pm [(1/4)(S_{xx} - S_{yy})^2 + (T_{xy})^2]^{1/2}$$

$$= (1/2)(1,200 + 2,350) \pm [(1/4)(1,200 - 2,350)^2 + (-760)^2]^{1/2}$$

$$= 1,775 \pm 938$$

$\therefore$

$$S_1 = 2,713$$

$$S_3 = 837$$

The principal stress ratio $M = 837/2, 713 = 0.31$, a value close to 1/3 which is associated with known stress concentration factors in Table 3.2(b). Thus, to a reasonable approximation, $K_c = 4.07$ and $K_t = 0.14$. Hence, $FS_c = 18,500/(4.07)(2,713) = 1.68$ and $FS_t = 1,250/0.14)(2,713) = 3.29$.

Inspection of Table 3.2(a) indicates that the compressive stress concentration would be somewhat less, while the tensile stress concentration would be somewhat more than the values used. The actual safety factors would be somewhat greater with respect to compression and somewhat less with respect to tension.

Example 3.10 Arectangular shaft is being considered for extension to 5,250 ft from 4,200 ft. The *in situ* stresses are given by

$$S_v = 1.1h$$
$$S_{NS} = 100 + 0.7h$$
$$S_{EW} = 1,000 + 0.5h$$

where h is depth in feet and stress is in psi. The shaft is 12 × 24 ft in section and is oriented in the most favorable direction. Unconfined compressive and tensile strengths are 18,000 and 1,200 psi, respectively. Determine the shaft wall safety factors as functions of depth to 5,250 ft from the surface.

Solution: In plan view, the EW stress is larger than the NS stress above 4,500 ft, so the long axis of the shaft is in the EW direction (most favorable) and the EW stress is the major compression in plan view. The plan view preexcavation principal stress ratio M varies with depth. Consequently, the peak stress concentration factors also vary with depth as do the shaft wall safety factors. A tabulation of data is shown in the table and a plot is given in the sketch. Table data were interpolated from Table 3.2(a) in the text and extended to 6,000 ft.

Depth ft	S_v *psi*	*SNS psi*	*SEW psi*	*M*	*k*	K_c	K_t	FS_c =	FS_t =
0	0	100	1,000	0.10	0.5	3.49	–0.62	5.16	1.95
500	550	450	1,250	0.36	0.5	4.13	–0.09	3.48	11.08
1,000	1,100	800	1,500	0.53	0.5	4.69	0.26	2.56	No tens
1,500	1,650	1,150	1,750	0.66	0.5	5.14	0.50	2.00	
2,000	2,200	1,500	2,000	0.75	0.5	5.48	0.69	1.64	
2,500	2,750	1,850	2,250	0.82	0.5	5.76	0.64	1.39	
3,000	3,300	2,200	2,500	0.88	0.5	5.99	0.60	1.20	
3,500	3,850	2,550	2,750	0.93	0.5	6.17	0.57	1.06	
4,000	4,400	2,900	3,000	0.97	0.5	6.33	0.54	0.95	
4,200	4,620	3,040	3,100	0.98	0.5	6.38	0.53	0.91	
4,500	4,950	3,250	3,250	1.00	0.5	6.46	0.52	0.86	
5,000	5,500	3,600	3,500	0.97	2	6.40	0.54	0.78	
5,100	5,610	3,670	3,550	0.97	2	6.39	0.47	0.77	
5,250	5,775	3,775	3,625	0.96	2	6.37	0.46	0.75	
5,500	6,050	3,950	3,750	0.95	2	6.35	0.45	0.72	
6,000	6,600	4,300	4,000	0.93	2	6.31	0.42	0.66	

Inspection of the table and figure data show that tension is not a concern, especially at depth. However, below 3,500 ft failure in compression becomes possible as the shaft wall safety factor decreases below 1.0. One may presume that the original shaft required support and further support will be needed with the extension to depth.

Shaft wall strengths

Strengths that enter shaft wall safety factor calculations limit the range of elastic deformation. Indeed, strength may be defined as stress at the elastic limit. A naturally supported shaft wall is unconfined; applicable strengths are therefore unconfined compressive and tensile

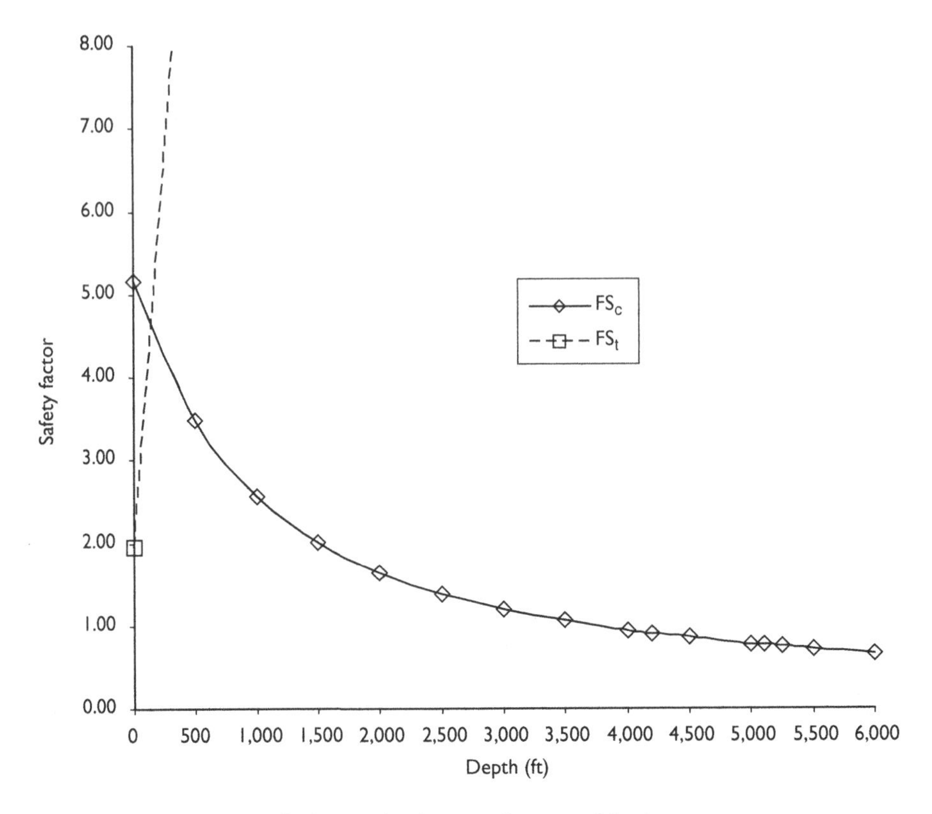

Shaft wall safety factors as functions of depth

strengths, C_o and T_o, respectively. These strengths should be consistent with the concept of a homogeneous material used in elastic analyses for stress and stress concentration. Homogeneity is easy to define in the context of mechanics and implies identical properties throughout a considered region, for example, same Young's modulus, same compressive strength at every point.

In this regard, a remarkable feature of two-dimensional stress analyses in isotropic elastic material obscures the question of homogeneity. This feature is the absence of elastic properties, such as Young's modulus and Poisson's ratio, in the distribution of stress. Consequently, associated stress concentrations about shaft walls are independent of elastic moduli. However, displacements induced by excavation are not independent of elastic moduli even in two-dimensional analyses, but are tacitly assumed to be small because of the expected elastic response.

Although easy to define, the concept of homogeneity raises important practical concerns and is central to much of theoretical and applied rock mechanics. Natural rock masses generally contain numerous cracks, fractures, joints, and faults ("joints" for brevity) that significantly alter the behavior of a rock mass from the mechanical response of intact rock. The issue of homogeneity can be simply posed by asking what should be done about the "joints." There are two distinct approaches to addressing this question: (1) address joint failure and failure of intact rock between separately, or (2) homogenize the rock mass using equivalent moduli and strengths and calculate safety factors based on these properties.

Calculation of stress in a jointed rock mass using the first approach is necessarily simplified by first considering the joints to be absent and then using the resulting stresses for analysis of intact rock and joint safety. This approach is homogenization in the extreme that is convenient, obviously inconsistent and therefore quite approximate, but still useful for the guidance provided. Safety factors for intact rock between joints can be calculated immediately following specification of C_o and T_o, preferably from laboratory measurements, but also possibly from estimates based on tabulated values of similar rock types. The latter specification is certainly cheaper but obviously much less certain than measured values. Safety factors for joints require consideration of joint orientation in addition to joint strength and the stress field at the shaft wall after excavation.

The well-known Mohr–Coulomb criterion is the most frequently used failure criterion for "joints" and is adopted here in the traditional form

$$\tau_j = \sigma_j \tan\left(\phi_j\right) + c_j \tag{3.9}$$

where τ_j, σ_j, ϕ_j, and c_j are joint *shear strength,* normal stress, friction angle, and cohesion, respectively, and the subscript j indicates joint. A joint safety factor FS_j is then

$$FS_j = \frac{\tau_j(\text{strength})}{\tau_j(\text{stress})} = \frac{\sigma_j \tan\left(\phi_j\right) + c_j}{\tau_j} \tag{3.10}$$

which is simply the ratio of shear strength to shear stress. Joint properties, ϕ_j and c_j are usually measured in direct shear test apparatus.

Joint normal and shear stresses must be calculated from the post-excavation stress state in consideration of joint orientation. The calculations are straightforward but require several steps:

1. Rotate the post-excavation stress state at a point on the shaft wall from $x'\ y'\ z'$ reference axes to compass coordinates xyz (x = east, y = north, z = up) about the vertical z axis with $z' = z$. Stresses relative to the prime system are: $\sigma_{x'x'} = 0$, $\sigma_{y'y'} = \sigma_t = KS_1$, $\sigma_{z'z'} = S_v$ where the stress concentration factor K is obtained from (3.4) in case of circular shafts and shear stresses are nil.
2. Compute direction cosines of the joint normal from the dip direction α and dip δ of the considered joint relative to compass coordinates xyz. These direction cosines are $n_x = \sin(\delta)\cos(\alpha)$, $n_y = \sin(\delta)\sin(\alpha)$, $n_z = \cos(\delta)$.
3. Compute joint tractions $T_x = \sigma_{xx}n_x + \tau_{xy}n_y + \tau_{xz}n_z$, $T_y = \tau_{xy}n_x + \sigma_{yy}n_y + \tau_{yz}n_z$, $T_z = \tau_{xz}n_x + \tau_{yz}n_y + \sigma_{zz}n_z$.
4. Compute magnitude $T = [(T_x)^2 + (T_y)^2 + (T_z)^2]^{1/2}$.
5. Compute $\sigma_j = N = T_x n_x + T_y n_y + T_z n_z$.
6. Compute $\tau_j = [T^2 - N^2]^{1/2}$.

The first step may be executed from formulas for rotation of reference axes about the given z-axis. In symbolic form this rotation is

$$\begin{bmatrix} \sigma_{xx} & \tau_{xy} & \tau_{xz} \\ \tau_{xy} & \sigma_{yy} & \tau_{yz} \\ \tau_{xz} & \tau_{yz} & \sigma_{zz} \end{bmatrix} = \begin{bmatrix} \cos(\theta) & \sin(\theta) & 0 \\ -\sin(\theta) & \cos(\theta) & 0 \\ 0 & 0 & 1 \end{bmatrix} [\sigma'][R]^{\mathrm{t}} \tag{3.11}$$

where $[\sigma']$ is a 3×3 array of stresses in the prime system, θ is the angle of rotation (positive counterclockwise) about the z' -axis from x' to x, and $[R]^{\mathrm{t}}$ is the transpose of the rotation matrix that precedes $[\sigma']$. The array $[\sigma']$ is diagonal because of the absence of shear stresses at the shaft wall. Results of (3.11) are the stresses in compass coordinates:

$$\begin{aligned} \sigma_{xx} &= \sigma_{\mathrm{t}}\left(\frac{1-\cos(2\theta)}{2}\right), \quad \tau_{xy} = \sigma_{\mathrm{t}}\left(\frac{\sin(2\theta)}{2}\right) \\ \sigma_{yy} &= \sigma_{\mathrm{t}}\left(\frac{1+\cos(2\theta)}{2}\right), \quad \tau_{yz} = 0 \\ \sigma_{zz} &= S_{\mathrm{v}} \quad \tau_{xz} = 0 \end{aligned} \tag{3.12}$$

where σ_{t} is the post-excavation normal stress acting tangentially to the shaft wall, as before.

The actual stress field that results from excavation in the presence of joints differs from the simplified view to an unknown degree and there is no certainty that joint safety factors computed from the assumed stress field are less than they would be in actuality. Thus, there is no assurance that a shaft wall joint safety factor greater than one guarantees joint safety in this simplified approach. However, low safety factors are unambiguously indicative of the need to consider installation of rock reinforcement and support as excavation continues.

Where cracks, fractures, and joints are too numerous to be treated individually, homogenization is indicated. Homogenization is a process that results in a heterogeneous sample volume responding *on average* as an *equivalent* homogeneous sample. Most equivalent rock properties models assume the sample is *representative* of the rock mass. A representative rock mass volume must contain so many joints that addition or elimination of a joint does not significantly affect the results of homogenization. Consider a cubical sample of edge length L containing a joint set with spacing S, then to be representative, $L >> S$, say, L should be an order of magnitude greater than S. At the same time, the homogenized sample volume must be much smaller than the size of the considered excavation for valid stress analyses. For example, the diameter of an unlined shaft D_{o} should satisfy the inequality $D_{\mathrm{o}} >> L$. Cracks and fractures spaced a few inches or centimeters indicate samples of a foot or somewhat less than a meter on edge. Shaft diameters of ten feet or meters are common. Homogenization using representative volumes is useful in such cases of relatively closely spaced joints. However, joint sets with joint spacings of a meter or more lead to much larger representative rock mass volumes. A meter joint spacing indicates a representative volume element L of ten meters and therefore unrealistically large shaft sizes D_{o} for applicability of the representative sample volume approach to equivalent rock mass properties. In such cases, a nonrepresentative volume element approach to equivalent properties is necessary, one that takes into account the unique geological structure in relative small volumes about the shaft wall.

Generally, joints reduce moduli and strengths from intact rock testing values. Scale factors that multiply intact rock elastic moduli and strengths to give rock mass moduli and strengths are often used for homogenization. For example, scale factors for Young's modulus R_E and unconfined compressive strength R_C are simply the ratios E_f/E_l and $(C_o)_f/(C_o)_l$, respectively, where subscripts f and l stand for field and laboratory scales. Clearly, such scale factors must be less than one and greater than zero, that is, they must be in the interval [0,1]. If no data are available to guide scale factor estimates, then one might simply choose the mid-point and reduce laboratory moduli (Young's modulus, E, shear modulus, G) and strengths (C_o and T_o) by 1/2. Other estimates may be based on pseudodimensional analyses. For example, one may suppose the strain to failure under uniaxial compressive stress is the same at the laboratory scale and at the scale of the rock mass, that is, $\varepsilon_f = C_o/E$. According to this concept, scale factors for moduli and strength are the same. Alternatively, one may suppose that under uniaxial compressive stress, the strain energy to failure is the same, so $(\sigma\varepsilon)_f = (C_o)^2/E$. Hence, the scale factor for moduli is the square of the scale factor for strength. These simple guides relate compressive strength and moduli scale factors but do not indicate how either are to be estimated. A similar analysis could be applied to tensile strength and related modulus.

Example 3.11 Given a reasonable 0.3% strain to failure and corresponding laboratory test values of 30,000 psi and 10(10^6^) psi for unconfined compressive strength and Young's modulus, respectively, determine rock mass compressive strength when displacement measurements in the field show Young's modulus to be 2.5(10^6^) psi. Use (a) strain to failure scaling and (b) energy to failure scaling.

***Solution*:**
(a) Strain to failure is

$$\varepsilon_f = (0.003) = [C_o/E]_l = [C_o/E]_m$$

$$\therefore$$

$$C_o(\text{rock mass}) = (0.003)(2.5)\left(10^6\right) = 7,500\,\text{psi}$$

(b) Energy to failure is

$$U = (1/2)[(C_o)^2/E]_l = (1/2)[(C_o)^2/E]_m$$
$$= (1/2)[3(10^4)]^2/10(10^6) = (9/2)10$$
$$U = 45$$

$$\therefore$$

$$(C_o)^2 = (45)(2)(2.5)\left(10^6\right)$$

$$\therefore$$

$$C_o(\text{rock mass}) = 15,000\,\text{psi}$$

In case (a) strength and modulus scale in direct proportion. In case (b), strength scales as the square root of the modulus scaling. In this example, Young's modulus for the rock mass was 1/4th the laboratory value, so the rock mass strength is 1/4 laboratory strength in case (a) and 1/2 laboratory strength in case (b), as the computations show.

The Terzaghi jointed rock mass strength model is one approach to estimating rock mass strength directly (Terzaghi, 1962). A strength scale factor can then be computed and subsequently a modulus scale factor using strain to failure or energy at failure guidelines. The Terzaghi model recognizes an important feature of rock masses and that is jointing is generally *discontinuous*. Shear failure along a plane of discontinuous jointing requires joint shear and shear of intact rock bridges between joint segments. If areas of intact rock and joints are A_r and A_j, then the total sample area $A = A_r + A_j$. The ratio of joint area to total area is *joint persistence p*. Thus, $p = A_j/A$ which ranges over the interval [0,1]. Shear strength along the joint is

$$\begin{aligned} \tau &= \tau_r(A_r/A) + \tau_j(A_j/A) \\ &= \tau_r(1-p) + \tau_j(p) \\ &= (1-p)[\sigma\tan(\phi) + c]_r + (p)[\sigma\tan(\phi) + c]_j \\ \tau &= \sigma\tan(\phi) + c \end{aligned} \tag{3.13}$$

where Mohr–Coulomb strength criteria are assumed for intact rock and joint which are indicated with subscripts r and j, respectively. Comparison of the last two equations in (3.13) shows that the rock mass angle of internal friction and cohesion are given by

$$\begin{aligned} \tan(\phi) &= (1-p)\tan(\phi_r) + (p)\tan\left(\phi_j\right) \\ c &= (1-p)c_r + (p)c_j \end{aligned} \tag{3.14}$$

which are persistence weighted averages. Equations (3.14) allow direct estimation of rock mass strengths using the Mohr–Coulomb strength relationships:

$$\left.\begin{matrix} C_o \\ T_o \end{matrix}\right\} = \frac{2(c)\cos(\phi)}{1 \mp \sin(\phi)}, \quad \sin(\phi) = \left(\frac{C_o/T_o - 1}{C_o/T_o + 1}\right), \quad c = \frac{\sqrt{C_o T_o}}{2} \tag{3.15}$$

that also apply to intact rock and joints. When the same strength scale factor is applied to unconfined compressive and tensile strengths, then the angle of internal friction remains unchanged between laboratory test values and rock mass values, while cohesion is scaled by the same strength scale factor. In fact, Terzaghi suggested equality of friction angles for joints and intact rock and that joints be considered cohesionless. Cohesionless joints are a limiting case that allows unconfined compressive and tensile strengths to approach zero while the ratio C_o/T_o remains a constant equal to $[1 + \sin(\phi)]/[1 - \sin(\phi)]$. Rock mass cohesion in this special case is given by the second equation of (3.14).

Example 3.12 Laboratory test data show that unconfined compressive and tensile strengths are 15,000 and 1,500 psi, respectively. Field measurements indicate a joint persistence of 0.75. Further laboratory testing shows that joint cohesion and friction angle are 8 psi and 25°, respectively. Estimate cohesion and angle of internal friction for intact rock tested in the laboratory, then determine rock mass values of cohesion, friction angle, unconfined compressive strength, and tensile strength, (c, ϕ, C_o, T_o).

***Solution*:** Assuming Mohr–Coulomb criteria,

$$\sin(\phi) = \frac{C_o - T_o}{C_o + T_o} = \frac{1,5000 - 1,500}{1,5000 + 1,500} = \frac{9}{11}$$
$$\therefore$$
$$\phi = 54.9^\circ$$
$$c = \frac{\sqrt{C_o T_o}}{2} = \sqrt{(1,5000)(1,500)}/2$$
$$\therefore$$
$$c = 2{,}372 \text{ psi}$$

for the intact rock from laboratory strength values.

For the rock mass

$$\tan(\phi) = (1 - 0.75)\tan(54.9) + 0.75\tan(25) = 0.7054$$
$$\therefore$$
$$\phi(\text{rock mass}) = 35.2^\circ$$
$$\therefore$$
$$c = (1 - 0.75)(2372) + 0.25(8) = 599$$
$$\therefore$$
$$c(\text{rock mass}) = 599 \text{ psi}$$

Also,

$$\left.\begin{matrix} C_o \\ T_o \end{matrix}\right\} = \frac{2c\cos(\phi)}{1 \mp \sin(\phi)} = \frac{2(2{,}372)\cos(35.2)}{1 \mp \sin(35.2)}$$
$$\therefore$$
$$\left.\begin{matrix} C_o \\ T_o \end{matrix}\right\}(\text{rock mass}) = \left.\begin{matrix} 9{,}152 \\ 2{,}459 \end{matrix}\right\}\text{psi}$$

A check on the ratio of unconfined compressive strength to tensile strength is

$$\frac{C_o}{T_o} = \frac{1 + \sin(\phi)}{1 - \sin(\phi)}$$
$$\therefore$$
$$\frac{9{,}152}{2{,}459} = 3.7218 = (?), \quad \frac{1 + \sin(35.2)}{1 - \sin(35.2)} = \frac{1.5764}{0.4236} = 3.7218(\text{checks})$$

The Terzaghi model implies directionally dependent shear strength for the rock mass because of the association with a joint plane of specific orientation and persistence measured over the considered joint plane. However, rock mass strengths computed from (3.15) imply isotropy, equal strengths in all directions. This inconsistency is sometimes addressed by adopting a *ubiquitous* joint model that assumes joints are seemingly everywhere, are fully persistent and of all orientations. This assumption also leads to a *microplane model*. Both concepts are invoked to explain and justify reducing laboratory values of elastic moduli and strengths to better represent rock mass behavior when joints are dominant and the role of intact rock between joints is negligible. In essence, these concepts

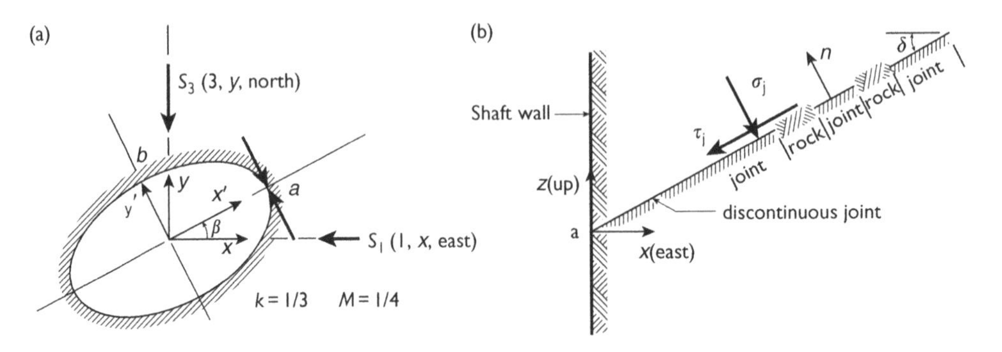

Figure 3.11 Elliptical shaft wall and joint geometry: (a) plan view and (b) vertical N–S section at *a*.

assign joint properties to the rock mass when site conditions justify the assumptions involved.

Example 3.13 Consider the data in Example 3.7 and a joint set with dip direction $\alpha = 0°$ and dip $\delta = 60°$ and suppose joint cohesion $c = 135$ psi and joint angle of friction $\phi = 35°$. Given these data and joint intersections of the shaft wall at the ends of the semi-axes, a and b, compute joint safety factors according to the steps outlined previously.

***Solution*:** The first step involves an angle of rotation from x' to x of $-30°$ as seen in Figure 3.11. According to (3.12) stresses at a relative to compass coordinates are

$$\sigma_{xx} = 2,700\left(\frac{1-0.5}{2}\right) = 675$$

$$\sigma_{yy} = 2,700\left(\frac{1+0.5}{2}\right) = 2,025$$

$$\sigma_{zz} = 1,350$$

$$\tau_{xy} = 2,700\left(\frac{-\sqrt{3}/2}{2}\right) = -1,170$$

Direction cosines from Step 2 are

$$n_x = \left(\frac{\sqrt{3}}{2}\right)(1) = 0.8667$$

$$n_y = \left(\frac{\sqrt{3}}{2}\right)(0) - 0.0$$

$$n_z = \left(\frac{1}{2}\right) = 0.5$$

Joint tractions are then

$$T_x = (0.8667)(675) + (0)(-1170) + (0.5)(0) = 585$$

$$T_y = (0.8667)(-1{,}170) + (0)(2{,}025) + (0.5)(0) = -1{,}013$$

$$T_z = (0.8667)(0) + (0)(0) + (0.5)(1{,}350) = 675$$

and magnitude $T = 1{,}351$. Joint normal stress from Step 5 is

$$\sigma_{\mathrm{j}} = N = (585)(0.8667) + (-1{,}013)(0) + (675)(0.5) = 845$$

and joint shear stress from Step 6 is

$$\tau_{\mathrm{j}} = [(1{,}351)^2 - (845)^2]^{1/2} = 1{,}054$$

The last two calculations enable an estimate of the joint safety factor. Thus,

$$\mathrm{FS_j} = \frac{(845)\tan(35^\circ) + 135}{1{,}054} = 0.69$$

which indicates that joint slip is highly probable as shaft excavation proceeds past the considered depth.

Example 3.14 Data from Examples 3.7 and 3.13 may be used to compute rock mass strengths following the Terzaghi jointed rock mass model. Use these data to estimate shaft wall safety factors.

***Solution*:** Tension is absent, so only safety with respect to compressive strength is of concern. For this site, persistence $p = 0.75$. From Equations (3.15), intact rock cohesion $c_{\mathrm{r}} = 2{,}134$ and angle of internal friction $\phi_{\mathrm{r}} = 54.9°$. Rock mass cohesion $c = 635$ and angle of internal friction $\phi = 41.4°$ from (3.14). Rock mass unconfined compressive strength $C_{\mathrm{o}} = 2{,}813$ from (3.15). Shaft wall safety factor FS = 2,813/2,700 = 1.04 in plan view and indicates marginal safety. In vertical section, FS = 2,813/1,350 = 2.08.

Joint persistence is measured over a representative area and is thus an average quantity. Locally, at the shaft wall, a joint segment is present or absent. In the worst case then, $p = 1.0$ and therefore the rock mass is assigned joint properties. Thus, $C_{\mathrm{o}} = 519$ from (3.15) and FS = 519/2,700 = 0.19 in plan view and 0.38 in vertical section. Both values indicate lack of safety and potential for failure if excavated without support provisions. If the joint is absent, then the FS's is associated with values for intact rock computed previously (5, 10), Example 3.7.

Similar analyses may be carried out for the opposite end of semi-axis a and for the ends of the b semi-axes. In this regard, there is a kinematic aspect of joint slip direction that should be considered. At the a point opposite to the one considered in detail, joint dip is into the rock mass rather than into the excavation and thus poses less of a threat to shaft wall failure even if the joint safety factor is less than one indicating slip.

3.2 Shaft wall support and liners

Rock bolting is a common method of preventing rock falls during shaft sinking. Rock bolts are steel rods that are secured in a drill hole on a spot basis, as needed in the judgment of the sinking crew, or systematically according to an engineered plan. In blocky ground, steel strapping may also be used in conjunction with bolts for surface support to prevent small rock falls between bolts from developing into larger wall failures. Other types of surface support may be used. Permanent support is often in the form of a concrete liner. Occasionally, in ground difficult to hold, smooth or corrugated steel lining and steel rings may be used either separately or in combination with concrete.

Shaft wall bolting

Rock bolts are secured in drill holes mechanically, with cement of some type, or frictionally. Figure 3.12 illustrates a variety of rock bolts. Bolting plans may be designed to reinforce

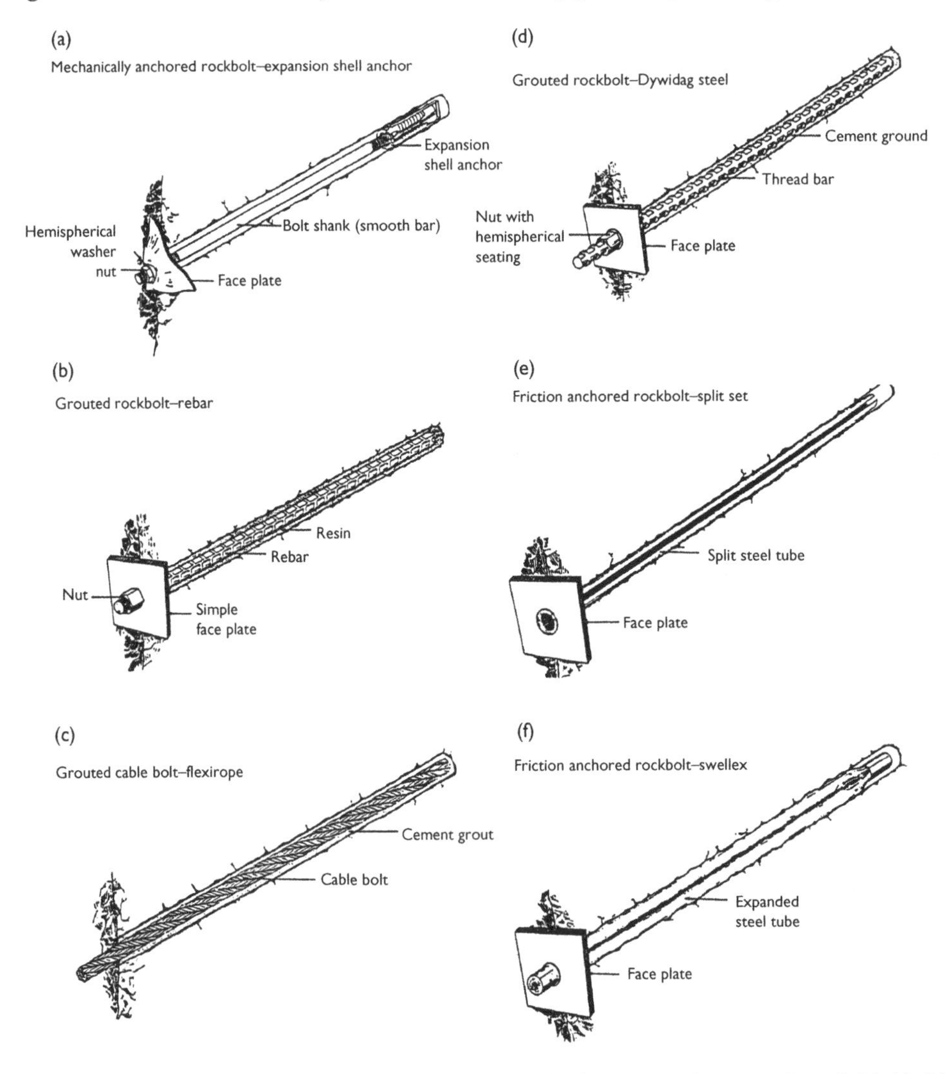

Figure 3.12 Several types of rock bolts (after Stillborg, 1986): (a) mechanical point anchored, (b), (c), (d) cemented anchorage, (e), (f) friction anchorage.

joints in a given joint set, to secure rock blocks formed by intersecting joints and to systematically reinforce the rock mass. Bolt length, diameter, steel grade, installation tension, and spacing are essential elements of such plans. Some data for bolt strengths according to steel grade are presented in Table 3.4. These data are part of the American Society for Testing and Materials (ASTM) Standard F432. Table 3.5 from the same standard gives additional bolt steel information.

Point anchored mechanical bolts are threaded at both ends. The end that goes into the drill hole first has a serrated wedge that is expanded against the drill hole wall as the bolt is spun into the hole. The wedge grips the rock and thus anchors the bolt in the hole. A small steel plate at the collar of the drill hole secures the other end of the bolt as a nut is tightened against the collar plate that bears on wall rock. Torque applied to the nut tensions the bolt that thus clamps and confines the rock mass along the bolt length. Bolts are generally tensioned to about 2/3 to 3/4 of the elastic limit (yield point). Associated bolt safety factors at installation time are therefore 3/2 to 4/3. Safety factors with respect to ultimate strength are higher, of course. Bolt spacing is generally in the range of two to eight feet; four feet is a common spacing; bolt spacing is also related to bolt length.

Bolt length may be more or less one-half shaft diameter. Eight foot long bolts could be used in a systematic plan for reinforcing a ten foot diameter shaft, while four foot long bolts could be used for spot bolting in a thirty foot diameter shaft. Bolt length combinations may also be used, short bolts for securing steel strapping, longer and larger bolts for reinforcement. The type of bolt plan dictates bolt length that in any case should be less than shaft diameter in consideration of the "1D" rule of thumb for extent of the zone of stress concentration about the shaft.

Consider a dipping joint set in the vicinity of a shaft wall and bolts passing through joints of the set as shown in Figure 3.13. The joint safety factor before bolting is given by Equation (3.10). Bolting reinforces the joints against slip directly through an uphill bolt force component and indirectly through a normal force that mobilizes frictional resistance to slip. The bolted joint safety factor $FS = FS_j + \Delta FS_j$. The improvement obtained by bolting is

$$\Delta FS_j = \frac{N_b \tan\left(\phi_j\right) + T_b}{\tau_j(\text{stress})A_j} \tag{3.16}$$

where N_b, T_b, and A_j are bolt normal force, bolt tangential force, and reinforced joint area, respectively. A reasonable estimate of A_j depends on bolt length L_b and spacings l_h and l_v in the horizontal and vertical directions, respectively. Bolt length must be sufficient to cross a joint and terminate in secure ground adequate for anchorage. The spacing product defines a bolt area of influence A_r half the distance to adjacent bolts, so $A_r = l_h l_v$. With reference to Figure 3.13, $A_j = A_r \cos(\beta)/sin(\delta - \beta)$ where δ is joint dip and β is the bolting angle shown in Figure 3.13. Other estimates of A_j could be formulated, for example, $A_j = L_b \cos(\delta - \beta)$ per foot of shaft perimeter or $A_j = L_b/\cos(\delta - \beta)$ per foot of shaft perimeter. Rearrangement of (3.16) in terms of bolt tensile force F_b is helpful. Thus,

$$\Delta FS_j = \frac{F_b \sin(\delta - \beta)\tan\left(\phi_j\right) + F_b \cos(\delta - \beta)}{\tau_j(\text{stress})A_j} \tag{3.17}$$

where F_b is the bolt force. Bolting angle β is a design variable and thus may be optimized for maximum safety factor improvement. Inspection of (3.17) shows that the optimum occurs

Table 3.4 Load support requirements of steel bolts, threaded bars, threaded deformed bars, and threaded slotted bars whose slot has been produced without metal Removal[a] N = Newton kN = kiloNewton I lb = 4.448 N or 0.004448 kN (adapted with permission, see ASTM in References)

Nominal diameter in.	*Thread stress area in.²· (cm²)*	*Grade*							
		55		60[b]		75		100	
		Yield	*Ultimate*	*Yield*	*Ultimate*	*Yield*	*Ultimate*	*Yield*	*Ultimate*
		lbs(kN)[c]		*lbs(kN)*[c]		*lbs(kN)*[c]		*lbs(kN)*[c]	
5/8	0.226	12,400	19,200	13,600	20,300	17,000	22,600	22,600	26,300
	(1.46)	(55.16)	(85.4)	(60.50)	(90.30)	(75.6)	(100.5)	(100.5)	(130.8)
3/4	0.334	18,400	28,400	20,000	30'100	25,100	33,400	33,400	41,800
	(2.65)	(81.8)	(126.3)	(89.0)	(133.9)	(111–6)	(148.6)	(146.6)	(193.5)
7/8	0.462	25,400	39,300	27,700	41,600	34,700	46,200	46,200	57,800
	(2.96)	(113.0)	(174.8)	(123.2)	(185.0)	(154.3)	(205.5)	(205.5)	(267.3)
I	0.606	33,300	51,500	36,400	54,500	45,500	60,600	60,600	75,800
	(3.91)	(148.1)	(229.0)	(161.9)	(242.4)	(202.4)	(269.5)	(269.5)	(350.5)
I 1/2	0.763	42,000	64,900	45,800	68,700	57,200	76,300	76,300	95,400
	(4.92)	(186.8)	(286.7)	(203.7)	(305.6)	(254.4)	(339.4)	(339.4)	(441.2)
I 1/4	0.969	53,300	82,400	58'100	87,200	72,700	96,900	96,900	121,100
	(6–25)	(237.1)	(366.5)	(258.4)	(387.9)	(323.4)	(431.0)	(431.0)	(560.4)
I 3/8	1.155	63,500	96,200	69,300	104,000	86,600	115,500	115,500	144,400
	(7.45)	(282.4)	(436.8)	(308.2)	(462.6)	(385.2)	(513.7)	(513.7)	(668.1)
I 1/2	1.405	77,300	I 19,400	84,300	126,500	105,400	140,500	140,500	175,600
	(9.06)	(343.8)	(531.1)	(375.0)	(562.7)	(468.8)	(624.9)	(624.9)	(814.6)

Notes

a Tests of bolts and threaded bars shall be performed using full-diameter products.

b Information for Gr 60 only applies to deformed bars.

c Required yield and tensile loads shown are calculated by multiplying thread stress areas times the yield point and tensile strength values shown in Table 3.4. Thread stress area is calculated from the mean root and pitch diameters of adrenal threads as follows:

$A_s = 0.7854(D - 0.9743/n)^2$

where:

A^2 = stress area, in.2

D = nominal diameter, in., and

n = number of threads per inch

$D^{5/8}$ in. Gr 40 products are not covered.

Table 3.5 Steel strength and grade[a]

Grade	*Yield point psi (MPa)*	*Tensile strength psi (MPa)*
40	40,000 (276)	70,000 (483)
55	55,000 (379)	85,500 (586)
60	60,000 (414)	90,000 (621)
75	75,000 (517)	100,000 (688)
100	100,000 (689)	125,000 (862)

Note
a After ASTM Standard F432.

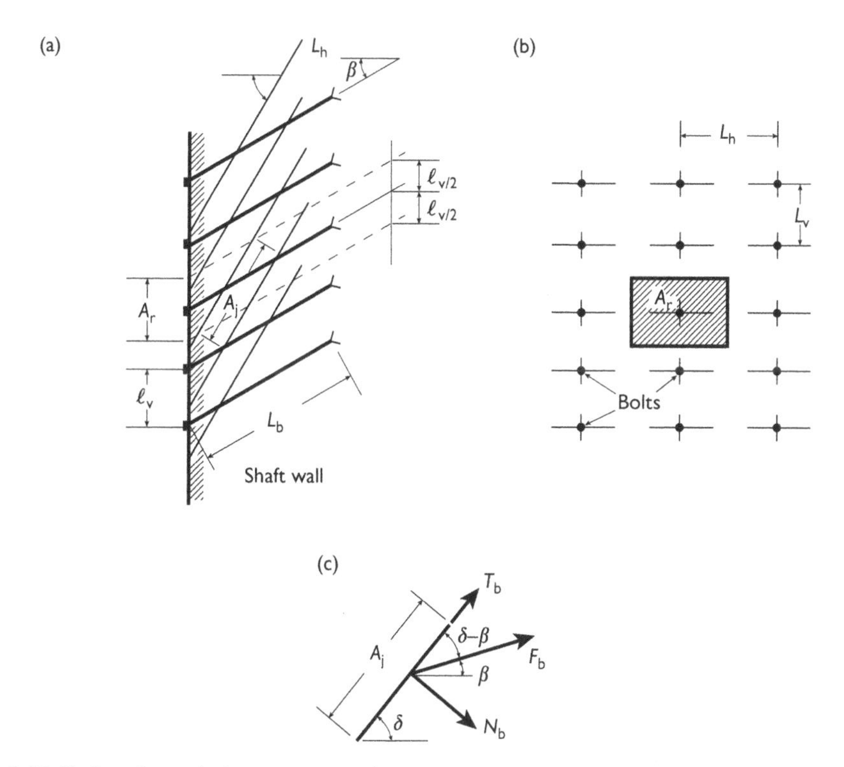

Figure 3.13 Shaft wall joint bolting geometry: (a) vertical section, (b) wall view, and (c) bolt force detail.

when the numerator is greatest, that is, when $\cos(\delta - \beta - \phi_j) = 1$, provided A_j is fixed. The optimum bolting angle $\beta = \delta - \phi_j$. Angling a bolt hole upwards has the advantage of producing a cleaner hole by allowing drill cuttings to flow freely from the hole. However, this advantage may be small compared with the disadvantage of orienting drill machines to drill upward rather than the more natural horizontal direction in a vertical shaft. In the horizontal direction, which is likely to be the case in practice, $\beta = 0$. If the joint is then vertical, $\delta = \pi/2$, and $\Delta FS_j = F_b \tan(\phi_j)/\tau_j(\text{stress})A_r$ as may be obtained directly.

The rock bolting safety factor improvement formula (3.17) masks a feature of bolting and that is bolting improvement is generally very small. Indeed, this is the case for much of rock support and reinforcement. The reason lies in the small "pressures" that support and reinforcement are capable of supplying to the rock mass.

Example 3.15 Consider a vertical shaft excavated in a stress field where $S_v = 1{,}350$, S_h (north) = 300 psi and S_H(east) = 1,200 psi and a joint set that strikes due north and dips 60° with cohesion $c_j = 135$ psi and friction angle $\phi_j = 35°$. For convenience, fix the x-axis due east parallel with S_H. Determine joint safety factors before excavation, after excavation and the improvement that a bolting plan could achieve.

Solution: Joint normal and shear stresses for this purpose may be obtained by rotation of the reference axes. Thus, before excavation

$$\sigma_j = \left(\frac{1{,}200 + 1{,}350}{2}\right) - \left(\frac{1{,}200 - 1{,}350}{2}\right)\cos(120°) = 1{,}313 \text{ psi}$$

$$\tau_j = -\left(\frac{1{,}200 - 1{,}350}{2}\right)\sin(120°) = 65 \text{ psi}$$

where the presence of the joint is assumed not to affect the given stress field. The joint safety factor is

$$\text{FS}_\text{j}(\text{before}) = \frac{1{,}313\tan(35°) + 135}{65} = 16.2$$

so joint slip or separation is not impending. After excavation

$$\sigma_j = \left(\frac{0 + 1{,}350}{2}\right) - \left(\frac{0 - 1{,}350}{2}\right)\cos(120°) = 1{,}013 \text{ psi}$$

$$\tau_j = -\left(\frac{0 - 1{,}350}{2}\right)\sin(120°) = 585 \text{ psi}$$

which shows some normal stress reduction but a significant increase in shear stress; both tend to reduce safety. The joint safety factor is now

$$\text{FS}_\text{j}(\text{after}) = \frac{1{,}013\tan(35°) + 135}{585} = 1.44$$

which is greatly reduced and probably too low for shaft wall safety.

If bolts are installed horizontally on a square pattern of three foot centers and tensioned to 18,500 pounds force, the safety factor improvement is

$$\Delta\text{FS}_\text{j} = \frac{18{,}500[\sin(60° - 0)\tan(35°) + \cos(60° - 0)]}{(585)(3)(3)\cos(0)/\sin(60°)} = 0.023$$

or less than 2% over the post-excavation joint safety factor. The small improvement in joint safety factor obtained by bolting suggests that bolting may not be an effective support method. This suggestion is definitely *not* the case. The example results do clearly indicate that stresses induced by bolting are quite small relative to stresses *in situ*. Indeed, bolts are effective when they prevent rock blocks from falling into an excavation and are essential

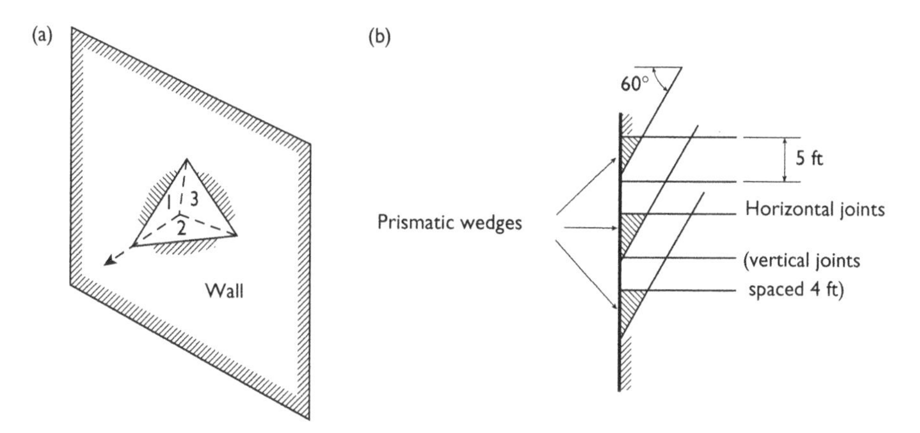

Figure 3.14 Potential wedge failures in a shaft wall: (a) tetrahedral wedge and (b) vertical section looking north.

to shaft safety. In this regard, slip at a point on a joint is not synonymous with rock block fall.

If a joint is the base of a rock block defined by other intersecting joints, then shaft safety requires the potential failure to be eliminated. However, not all rock blocks defined by intersecting joints, fractures, cracks, and bedding planes, pose a threat to safety. Only those blocks that are kinematically disposed to move into the excavation are of concern. For example, tetrahedral blocks formed by three intersecting joints that have a baseline of intersection dipping away from an excavation are safe. If the line of intersection dips into the excavation, as shown in Figure 3.14(a), then a safety analysis and possibly bolting reinforcement are indicated. A variety of block shapes may be formed by joints of different orientations and spacings and the excavation wall proper.

Example 3.16 Consider a wedge formed in a vertical shaft excavated where one joint set strikes due north and dips 60° west with uncertain cohesion c_j and friction angle $\phi_j = 35°$, a second set is vertical with spacing of four feet and dip direction due north, no cohesion and friction angle $\phi_j = 35°$ and a third is horizontal with spacing of five feet, no cohesion and nil friction angle. Fix the x-axis due east perpendicular to strike of the first joint, as shown in Figure 3.14(b). Determine whether the block will be a threat to excavation safety and, if so, the second question is then what bolting plan could eliminate the threat by improving a suitable factor of safety to some acceptable value.

***Solution*:** Arough first analysis may be based on a model of a block sliding down an inclined plane. This model neglects sides and top and because the inclination of the base is greater than the friction angle, sliding would occur under the weight of the block if cohesion were also neglected. Thus, further analysis is required. In view of the uncertain joint cohesion, one may determine what cohesion is necessary to achieve an acceptable safety factor, say, $FS_j = 3.5$. A force safety factor against sliding is

$$\begin{aligned}
\mathrm{FS_j} &= \frac{\text{resisting force}}{\text{driving force}} \\
&= \frac{R}{D} \\
&= \frac{W\cos(\delta)\tan\left(\phi_\mathrm{j}\right) + c_\mathrm{j}A_\mathrm{j}}{W\sin(\delta)} \\
3.5 &= \frac{(4{,}500)\cos(60°)\tan(35°) + c_\mathrm{j}(23)(144)}{(4{,}500)\sin(60°)}
\end{aligned}$$

where W is block weight and A_j is area of the block base (Volume = (4)(5)(5 tan(30)/2) = 28.87 cubic feet). Solving this last equation gives $c_j = 3.6$ psi, assuming a specific weight of rock of 156 pcf. Thus, a small amount of joint cohesion may be important.

But whether a reliable determination can be made is another question. Without cohesion, the safety factor is 0.40, and as a practical matter, bolting would probably be specified. For example, a horizontal bolt centered on the block face and is tensioned to 18,500 pounds improves the force safety factor against sliding by

$$\Delta\mathrm{FS_j} = \frac{18{,}500[\sin(60°)\tan(\phi) + \cos(60°)]}{4{,}500\,\sin(60°)} = 5.3$$

This result for a bolted rock block provides an entirely different view of joint reinforcement by bolting. Even if the rock block weight were tripled, the improvement in joint safety factor would, say, for the horizontal bolting plan, be 1.8.

In essence, a *dead weight* load model was adopted for the alternative joint reinforcement analysis, an approach that is conservative because of the neglect of physical features that would otherwise increase safety. In blocky ground formed by a multiplicity of joint intersections, block size and kinematics pose a formidable design challenge that may be met with specialized "key block" computer programs.

Consideration of gravity sliding of an individual block aides understanding how rock bolts reinforce a jointed rock mass, but when blocks are numerous and variable, a systematic bolting plan is indicated. An approach to understanding how pattern bolting may be effective is based on the influence of confining pressure on rock strength and how bolting generates a confining pressure at a shaft wall.

Compressive strength C_p under confining pressure p according to the Mohr–Coulomb strength criterion is

$$C_p = C_o + \left(\frac{C_o}{T_o}\right)p \tag{3.18}$$

The ratio of unconfined compressive to tensile strength for intact rock tested in the laboratory is usually in the range of 10–20, so a small confining pressure is greatly amplified and may contribute substantially to compressive strength. Intact, laboratory-scale strengths may be reduced or scaled to characterize a field-scale rock mass, perhaps by an order of magnitude. A 10,000 psi laboratory strength would be 1,000 psi in the field. Causes of such a reduction or scaling are the natural joints in the rock mass and cracks and fractures induced by excavation which are certainly formed by

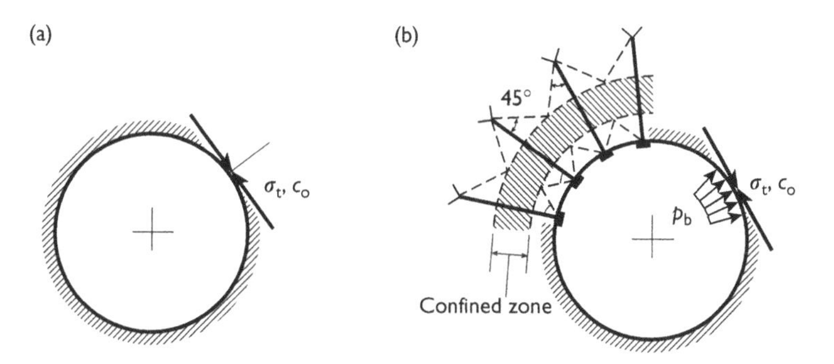

Figure 3.15 Radial shaft wall bolting plan: (a) before bolting and (b) after bolting.

blasting. However, the strength ratio would remain constant, provided that both strengths were reduced by the same scale factor.

Shaft wall safety factors (with respect to compression) before and after application of a confining pressure by bolting and the improvement achieved by bolting are therefore

$$\mathrm{FS_o} = \frac{C_\mathrm{o}}{\sigma_\mathrm{c}}, \ \ \mathrm{FS} = \frac{C_\mathrm{o} + (C_\mathrm{o}/T_\mathrm{o})p_\mathrm{b}}{\sigma_\mathrm{c}}, \ \ \Delta\mathrm{FS} = \frac{(C_\mathrm{o}/T_\mathrm{o})p_\mathrm{b}}{\sigma_\mathrm{c}} \tag{3.19}$$

where the confining pressure p_b is specifically attributed to bolt forces. Implicit in (3.19) is the assumption that the post-excavation stress σ_c concentrated at the shaft wall is unaffected by bolting. Bolting would tend to decrease this compression, so the assumption is conservative in the sense that the safety factor would be somewhat higher if the effect were easily quantified and taken into account. Figure 3.15 shows stress concentration in a circular shaft wall before bolting and a ring of bolts installed in the wall after excavation. Bolting pressure is a statically equivalent distributed normal stress given by $p_\mathrm{b} = F_\mathrm{b}/A_\mathrm{r}$ where A_r is the rock area influenced by a bolt. This area is simply defined by bolt spacing, that is, $A_\mathrm{r} = l_\mathrm{h}l_\mathrm{v}$, the product of horizontal by vertical bolt spacing on a shaft wall. Amodification to the equivalent distributed load view is to increase the area of distribution gradually from the collar and anchor points of force application. A 45° slope is often used for this purpose, as shown in Figure 3.15. The confining pressure in the clamped ring of rock is the same in either case.

Example 3.17 Consider bolts tensioned to 18,500 pounds force on three foot centers horizontally and four foot centers vertically, as shown in Figure 3.15, and suppose the tangential stress concentration at the shaft wall is 3,000 psi. Bolting pressure is about 15 psi. If intact rock unconfined compressive strength C_o = 12,000 psi and tensile strength T_o = 1,200 psi and both are scaled by a 1/4, then rock mass C_o = 3,000 psi. Determine safety factors before and after bolting and show the safety factor improvement achieved by bolting is about 0.05 (5%).

***Solution*:** By definition

$$\text{FS}_\text{o} = \frac{C_\text{o}}{\sigma_\text{c}} = \frac{3{,}000}{3{,}000} = 1.0$$

$$\text{FS} = \frac{C_\text{o} + (C_\text{o}/T_\text{o})p_b}{\sigma_\text{c}} = \frac{3{,}000 + (10)(18{,}500/9)(1/144)}{3{,}000} = 1.05$$

$$\Delta\text{FS}\% = (1.05 - 1.00)100 = 5\%$$

Again, in Example 3.17, the effects of bolting forces are small relative to rock mass stresses and thus bring into question whether the concept of bolt functioning by increasing strength with confining pressure is valid. The suggestion is that bolts may function according to a different physical mechanism, for example, by restraining rock blocks from falling and thus by preventing raveling of ground into a major failure. Indeed, bolts are generally considered temporary support of excavations designed for a long service life. They function mainly by preventing small block falls developing into large failures. Bolts may be used in combination with metal strapping, wire mesh, chain link, plastic webbing, or even membrane coatings for this purpose.

Bolts with distributed anchorage, bolts cemented into a hole, and friction bolts, function in much the same way as point anchored bolts that are tensioned during installation. Pretensioned bolts are *active* because they provide immediate resistance to rock mass displacement at the time of installation. Distributed anchorage bolts are ordinarily not tensioned during installation and are *passive* because they provide support only to the extent that the rock mass tensions bolts after installation. Both types support the rock mass by resisting displacement along the bolt hole. In addition, distributed anchorage bolts provide resistance to shearing motion across bolt holes because the hole is filled. This resistance is *dowel* action. Point anchored bolts have an empty annulus between bolt and hole that allows for easy displacement across bolt holes and thus are much less effective against shearing action. *Both types of bolts should be oriented to be stretched in tension by anticipated rock mass displacements.* Cable bolts in particular must be tensioned to be effective.

Circular shaft liners

Shaft liners are usually concrete that is formed to the shaft wall. Concrete may be used in combination with steel liner plate or steel rings which may be embedded in the concrete. Shaft walls may also be sprayed with cement (shotcrete, gunite) after bolting and screening. Liner plate and light segment corrugated steel may also be used in shafts. Concrete and shotcrete are brittle materials that may be reinforced by steel rebar (concrete) or by steel or glass fibers (shotcrete) to improve strength and resistance to fracture.

There are two systems for installing concrete shaft liners. For shafts of moderate depth, the liner is installed after the shaft is sunk. Concrete is poured from the bottom up using slip forms. In this case, shaft wall motion induced by mining is complete before the liner is installed assuming the rock mass responds elastically to excavation. Economics of deep shafts dictates installation of the liner concurrently with sinking. By keeping the last pour several shaft "diameters" above the shaft bottom, motion of the shaft walls after the liner is installed is slight and induces little stress in the liner as sinking continues, again, assuming the rock mass response is elastic. In either case, the liner experiences negligible loading from contact with the rock mass.

If the rock mass exhibits a pronounced time-dependent component of deformation and thus creeps significantly or behaves like a highly viscous fluid, then high stresses may eventually be induced in the liner. Indeed, as time passes, the full premining stress may come to bear on the liner. Under such circumstances, a rigid liner is a poor support choice. Support that accommodates yielding or flow of the rock mass would be a better choice.

Water is an important source of shaft liner load and represents an all around pressure on the liner after installation. During shaft sinking, a grout curtain is often used to dam the flow of water to the unlined shaft, but in time the curtain may deteriorate, fracture, and allow seepage to the liner. Liners should be designed for this eventuality.

A model for a *circular* shaft liner loaded with water pressure may be found in the solution to the problem of a hollow (thick-walled) cylinder loaded by radial "pressures" p_a and p_b acting on the inside of the liner at a radius a and at the outside of the liner at a radius b (unlined shaft diameter $D_o = 2b$). The cylinder is considered elastic with Young's modulus E and Poisson's ratio v and under plane strain conditions ($\varepsilon_{zz} = 0$). The problem is also one of axial symmetry. Stresses, strains, and displacements are therefore independent of z and θ. Under these conditions there is only the radial displacement u to be considered. In fact, $u = Ar + B/r$ (see, e.g. Love, 1944) and in cylindrical coordinates ($r\theta z$), the radial and circumferential strains are

$$\varepsilon_{rr} = A - \frac{B}{r^2}$$

$$\varepsilon_{\theta\theta} = A + \frac{B}{r^2}$$

According to Hooke's law under plane strain conditions, the radial and circumferential stresses are related to strains by

$$\sigma_{rr} = \frac{E}{(1+v)(1-2v)}[(1-v)\varepsilon_{rr} + v\varepsilon_{\theta\theta}]$$

$$\sigma_{\theta\theta} = \frac{E}{(1+v)(1-2v)}[(1-v)\varepsilon_{\theta\theta} + v\varepsilon_{rr}]$$

Hence,

$$\sigma_{rr} = \frac{E}{(1+v)(1-2v)}\left[A - (1-2v)\frac{B}{r^2}\right]$$

$$\sigma_{\theta\theta} = \frac{E}{(1+v)(1-2v)}\left[A + (1-2v)\frac{B}{r^2}\right]$$

Use of the boundary conditions on the radial stress at $r = a$ and $r = b$ allows for the determination of the constants A and B. Thus,

$$\frac{EA}{(1+v)(1-2v)} = \left(\frac{p_a a^2 - p_b b^2}{a^2 - b^2}\right)$$

$$\frac{EB}{(1+v)} = \left(\frac{p_a - p_b}{a^2 - b^2}\right)(a^2 b^2)$$

Hence,

$$\sigma_{rr} = \left(\frac{p_a a^2 - p_b b^2}{a^2 - b^2}\right) - \left(\frac{p_a - p_b}{a^2 - b^2}\right)\left(\frac{a^2 b^2}{r^2}\right)$$

$$\sigma_{\theta\theta} = \left(\frac{p_a a^2 - p_b b^2}{a^2 - b^2}\right) + \left(\frac{p_a - p_b}{a^2 - b^2}\right)\left(\frac{a^2 b^2}{r^2}\right)$$

$$u = \left(\frac{1+\nu}{E}\right)\left[(1-2\nu)\left(\frac{p_a a^2 - p_b b^2}{a^2 - b^2}\right)r + \left(\frac{p_a - p_b}{a^2 - b^2}\right)\left(\frac{a^2 b^2}{r}\right)\right]$$

where, if compression is positive, then a positive radial displacement is inward.

In the usual case, there is no internal pressure p_a acting on the liner, so

$$\sigma_{rr} = \left(\frac{p_b}{1 - (a/b)^2}\right)\left(1 - \frac{a^2}{r^2}\right)$$

$$\sigma_{\theta\theta} = \left(\frac{p_b}{1 - (a/b)^2}\right)\left(1 + \frac{a^2}{r^2}\right)$$

$$u = \left(\frac{1+\nu}{E}\right)\left[(1-2\nu)\left(\frac{p_b}{1 - (a/b)^2}\right)r + \left(\frac{p_b}{1 - (a/b)^2}\right)\left(\frac{a^2}{r}\right)\right]$$

A Mohr–Coulomb failure criterion is appropriate to concrete and includes steel as a special case. A convenient form in terms of principal stresses is

$$\left(\frac{\sigma_1 - \sigma_3}{2}\right) = \left(\frac{\sigma_1 + \sigma_3}{2}\right)\sin(\phi) + (c)\cos(\phi)$$

where c and ϕ are cohesion and angle of internal friction, respectively. In the hollow cylinder model of a shaft liner, the radial and circumferential stresses are principal stresses. Hence,

$$\left(\frac{p_b}{1 - (a/b)^2}\right)\left(\frac{a^2}{r^2}\right) = \left(\frac{p_b}{1 - (a/b)^2}\right)\sin(\phi) + (c)\cos(\phi)$$

that is first satisfied where r is the least, that is, at $r = a$. At greater r the left side of the failure criterion is too small to satisfy the equality. Thus, liner failure initiates at the inside of the liner when

$$p_b = \left[\frac{(c)\cos(\phi)}{1 - \sin(\phi)}\right]\left[1 - (a/b)^2\right]$$

The circumferential stress on the inside of the liner ($r = a$) is maximum. Thus,

$$\sigma_{\theta\theta}(a) = \frac{2p_b}{1 - (a/b)^2}$$

where the radial stress is zero. As b becomes indefinitely large, the stress approaches two times the applied stress which agrees with the case of a circular hole under hydrostatic stress. If p_b is a reference stress and K is a liner stress concentration factor, then

$$\sigma_{\theta\theta} = Kp_b, \quad K = \left[\frac{2}{1-(a/b)^2}\right]$$

The safety factor for the liner is given by

$$\mathrm{FS_c} = \frac{C_o}{\sigma_c} = \frac{C_o}{Kp_b} = C_o\left[\frac{1-(a/b)^2}{2p_b}\right] \tag{3.20}$$

If the liner thickness h is specified, so $h = b-a$, then the factor of safety may be calculated, given the liner strength C_o and applied stress p_b.

Alternatively, a liner thickness may be calculated for a given safety factor or, equivalently, for a given maximum allowable stress (C_o/FS). Two formulas are possible, one in terms of the inner liner radius a, the other in terms of the outer liner radius b which is also the radius of the unlined shaft wall. Thus, after solving the liner safety factor equation,

$$\begin{aligned} h &= b\left[1-\left(\frac{C_o - 2p_b\mathrm{FS}}{C_o}\right)^{(1/2)}\right] \\ h &= a\left[\left(\frac{C_o}{C_o - 2p_b\mathrm{FS}}\right)^{(1/2)} - 1\right] \end{aligned} \tag{3.21}$$

Either may be used; both follow directly from the liner safety factor formula. In consideration of the square root operation, the liner load can in no case exceed one-half the unconfined compressive strength of the liner material. Liner loads from water pressure are an order of magnitude less, although there are notable exceptions. In exceptional cases, a concrete liner alone may not suffice; combinations of steel and concrete may be needed to support very high water pressure.

Example 3.18 A circular concrete liner in a vertical shaft is subject to an external pressure of 90 psi. Concrete strength is 5,000 psi. If the inside diameter of the liner must be 18 ft, determine the liner thickness necessary to achieve a liner safety factor of 3.5.

***Solution*:** According the second equation of (3.21)

$$\begin{aligned} h &= \left(\frac{18}{2}\right)\left\{\left[\frac{5{,}000}{5{,}000-(2)(9)(3.5)}\right]^{1/2} - 1\right\} \\ &= 0.627 \text{ ft} \\ h &= 7.52 \text{ in.} \end{aligned}$$

This thickness is rather small and as a practical matter the actual liner would probably be one foot thick.

Example 3.19 A circular concrete liner in a vertical shaft is subject to an unknown external pressure that cause the liner to fail in compression on the inside of the liner wall where traces of vertical shear cracks are observed. Concrete strength is 5,000 psi. The inside diameter of the liner is 18 ft and the thickness is 7.52 in. Determine the liner load at failure.

Solution: At failure, the liner safety factor is 1.0; according to (3.20), this safety factor is

$$FS_c = C_o \left[\frac{1-(a/b)^2}{2p_b} \right]$$

$$\therefore$$

$$p_b = C_o \left[\frac{1-(a/b)^2}{2FS_c} \right]$$

$$= (5{,}000) \left\{ \frac{1-[9/(9+7.52/12.0)^2}{(2)(1)} \right\}$$

$$p_b = 315 \text{ psi}$$

The small thickness leads to a relatively small failure load. A one foot thick liner would require a pressure of 475 psi, about 50% greater, to cause failure.

Example 3.20 A 3 ft thick circular concrete shaft liner is designed for an allowable stress of 3,500 psi in compression. Outside diameter is 24 ft. Determine the thickness of a steel (A36) liner with a safety factor of 1.5 that would support the same load but allow for a greater inside diameter and therefore a greater useful cross-sectional area of the shaft.

Solution: The load the concrete liner is designed to support is

$$p_b = C_o \left[\frac{1-(a/b)^2}{2FS_c} \right] = (3{,}500) \left[\frac{1-(9/12)^2}{2} \right] = 766 \text{ psi}$$

According to the first equation of (3.21)

$$h = (12) \left[1 - \left(\frac{36{,}000 - 2(766)(1.5)}{36{,}000} \right)^{1/2} \right]$$

$$= 0.389 \text{ ft}$$

$$h = 4.67 \text{ in.}$$

The inside shaft diameter would increase from 21 ft to 23 ft. Whether the increase is warranted would require a cost analysis.

The *average* circumferential stress acting through the thickness of the liner may be computed by integrating the circumferential stress through the thickness of the liner and then dividing by the thickness. The result is

$$\sigma_{ave} = \frac{p_b b}{h}$$

which may also be obtained by substituting $a = b - h$ in the safety factor formula, then expanding the square and neglecting $(h/b)^2$ of the result on the grounds that the thickness is small compared with the radius, that is, $(b-a)/(b+a) << 1$. Figure 3.16 shows how the ratio of maximum to average circumferential stress varies with the ratio of thickness to outer radius. The average stress is within about 10% of the maximum stress up to a thickness of about 0.2 the outer radius (unlined shaft radius). A 20 ft diameter shaft lined with concrete 1 ft

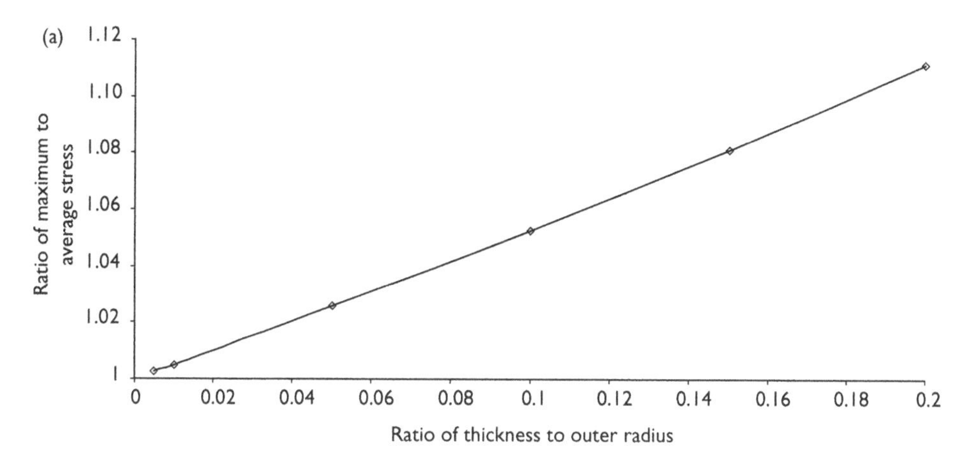

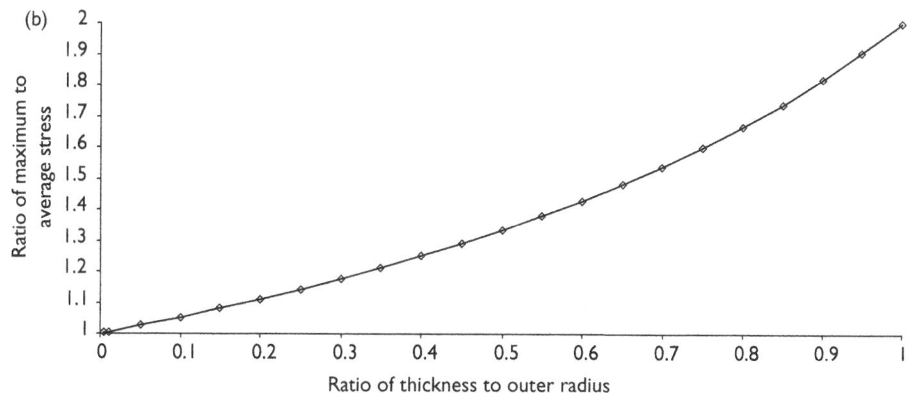

Figure 3.16 Ratio of maximum to average liner stress under external pressure as a function of the ratio of liner thickness to outer radius. (a) Small range and (b) large range.

thick could be designed using the thick-walled cylinder formula (exact solution) or using the simpler average stress. The maximum stress in this example is 10.52 p_b while the average is 10.00 p_b, a difference of about 5%.

Example 3.21 Suppose a 20 ft diameter shaft is sunk through an aquifer where the fluid pressure is 180 psi and a safety factor of 4.8 is desired using concrete with $C_o = 4{,}800$ psi. The required liner thickness according to (3.21) is

$$h = (20/2)\left[1 - \left(\frac{4{,}800 - (2)(180)(4.8)}{4{,}800}\right)^{1/2}\right] = 2.0 \text{ ft}$$

assuming the diameter given is the outside diameter. The finished inside shaft diameter would be 16 ft.

If the finished shaft diameter needed were 20 ft, then the liner thickness required would be

$$h = (20/2)\left[\left(\frac{4{,}800}{4{,}800 - (2)(180)(4.8)}\right)^{1/2} - 1\right] = 2.5 \text{ ft}$$

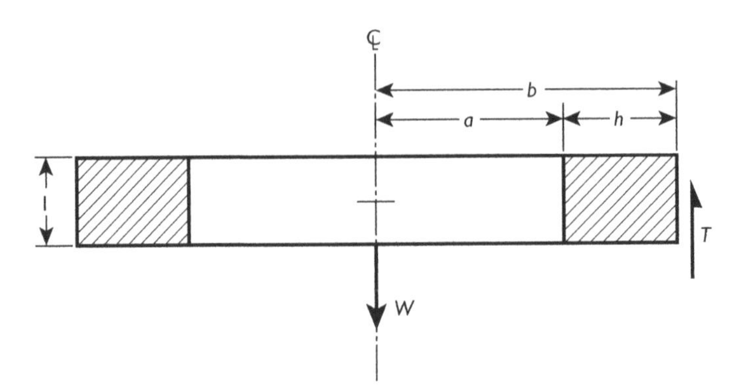

Figure 3.17 Shaft ring weight equilibrated.

If the liner were steel instead of concrete, and $C_o = 36{,}000$ psi, then by the same formula that leads to a 2.5 ft thickness of concrete, $h = 0.25$ ft or a 3.0 in. thickness of steel. The useful area of a shaft is, of course, defined by the inside diameter (20 ft).

In this example, the unlined shaft would need to be excavated to a diameter of 25 ft to allow for the required concrete liner thickness and finished diameter of 20 ft where the shaft route encounters the high water pressure.

Weight of a concrete liner also loads the liner, so some account of this load is of interest. The weight of a 20 ft diameter shaft liner, 1 ft thick, is the product of concrete specific weight and ring volume. With reference to Figure 3.17 that shows a ring in equilibrium with a shearing force acting on the perimeter of the ring at the concrete–rock contact with no change in vertical force from top to ring bottom, the shear force required for equilibrium is just

$$W = T$$

$$\gamma\pi\left(b^2 - a^2\right) = 2\pi b\tau$$

where τ is the average shear stress acting over a one foot run of liner. Solution for τ gives

$$\tau = \frac{\gamma}{2b}(b - a)(b + a)$$

$$\tau \approx \frac{\gamma}{2b}(h)(2b) = \gamma h$$

Thus, shear stress required to support the liner is about equal to the product of liner specific weight γ and liner thickness h. This product is roughly 1 psi per foot of liner thickness. A one foot thick liner would require about 1 psi of supporting shear stress at the liner–rock contact. In consideration of the roughness of the unlined shaft wall and the adhesion of concrete to rock, a shear stress of a few psi is easily supported. For this reason, the weight of a liner does not accumulate with depth. The load is distributed over the perimeter of the liner and does not develop a large vertical force over the bottom of the liner.

Example 3.22 A circular concrete shaft liner has an inside diameter of 20 ft and an outside diameter of 24 ft. Determine the bond strength between concrete and rock such that liner weight does not accumulate. Assume specific weight of concrete is 140 pcf.

Solution: A free body diagram of a section of the liner is shown in the sketch.

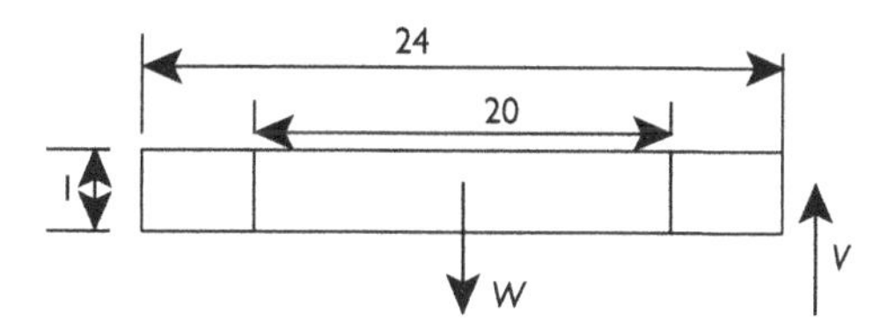

For weight not to accumulate, each vertical foot of liner must be supported in shear at the concrete–rock interface. Thus,

$$W = V = (140)(\pi/4)\left(24^2 - 20^2\right)(1) = \tau(\pi)(24)(1)$$

$$\therefore$$

$$\tau = 257 \text{ psf } (1.9 \text{ psi})$$

However, if a liner is decoupled from the rock and thus moves freely in the vertical direction, then liner weight does accumulate and must be supported by reactions at the bottom of the liner. Decoupling may be desirable in the presence of shaft wall motion or subsidence induced by mining or other activity such as drainage of an aquifer or fault.

If indeed the rock mass does not load a liner because post-excavation displacement is negligible, then liner thickness should be independent of shaft depth. Empirical evidence for this inference is shown in Figure 3.18 (Cornish, 1967) where depth is plotted as a function of the ratio of liner thickness to diameter (in./ft) for 43 shafts. Depth ranges to over 6,000 ft, but there is clearly no trend line that would indicate a correlation between depth and liner thickness to diameter ratio.

Figure 3.19 illustrates a computer simulation of concurrent shaft sinking and concrete lining using the popular finite element method (Pariseau, 1977). The shaft is circular and the analysis is axially symmetric with a preexcavation principal stress ratio of one ($M = 1$). Unlined shaft diameter was 22 ft and liner thickness was 1 ft; finished diameter was 20 ft. Results in Figure 3.19 show that total shaft wall displacement increases linearly with depth as does stress. However, liner displacement is small and independent of depth. Most of the shaft wall displacement occurred simultaneously with sinking rounds that were equal to 20 ft initially and then for computation efficiency increased to 40 ft. The liner was never installed closer than one shaft diameter to the shaft bottom and was never more than 60 ft from the shaft bottom. Keeping support at least one diameter from the shaft bottom not only defends support against residual wall displacement but also reduces damage from blasting.

Example 3.23 A 1-ft thick circular concrete shaft liner has an unconfined compressive strength of 3,500 psi, a Young's modulus of $3.5(10^6)$ psi, a Poisson's ratio of 0.20, and a specific weight of 140 pcf. Inside diameter is 20 ft. Determine the change in inside diameter when the load is sufficient to cause liner failure.

Solution: The radial displacement of the liner is given by

$$u(r) = \left(\frac{1+v}{E}\right)\left[(1-2v)\left(\frac{p_b}{1-(a/b)^2}\right)r + \left(\frac{p_b}{1-(a/b)^2}\right)\left(\frac{a^2}{r}\right)\right]$$

$$\therefore$$

$$u(a) = \left(\frac{1+0.20}{3.5(10^6)}\right)\left[\left(1-2(0.20)\right)\left(\frac{p_b}{1-(9/10)^2}\right)9 + \left(\frac{p_b}{1-(9/10)^2}\right)\left(\frac{9^2}{9}\right)\right]$$
$$= 2.6(10^{-5})p_b$$

At failure, the liner load is

$$p_b = (C_o/2)[1-(a/b)^2] = (3,500/2)[1-(9/10)^2] = 333\text{ psi}$$

$$\therefore$$

$$u(a) = 2.6(10^{-5})(333)(12) = 0.104\text{ in.}$$

The change in diameter is just twice the radial displacement, so $\Delta D = 0.208$ in. Although, seemingly small, accurate measurement during a monitoring program would not be difficult with a calibrated tape extensometer or similar device.

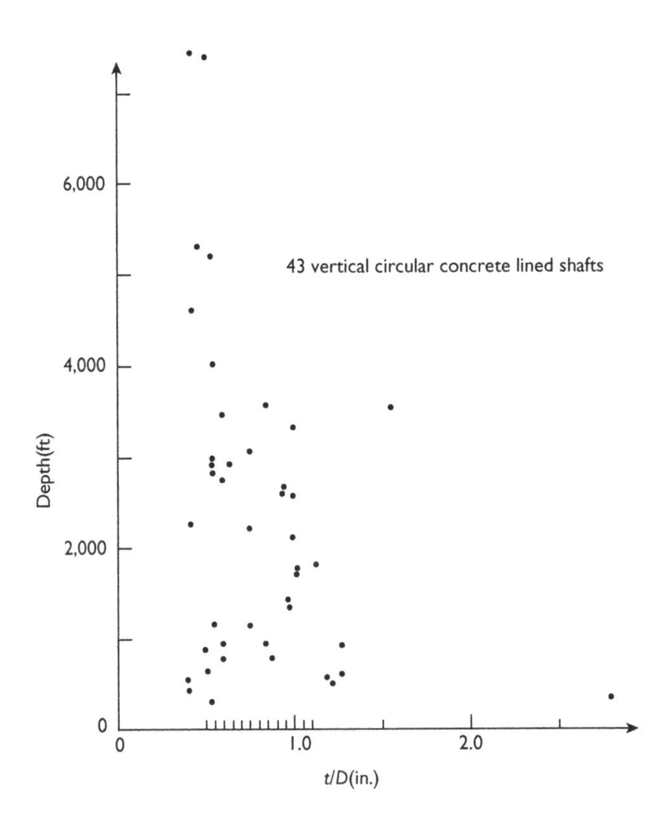

Figure 3.18 Shaft depth versus liner thickness to diameter ratio (after Cornish, 1967).

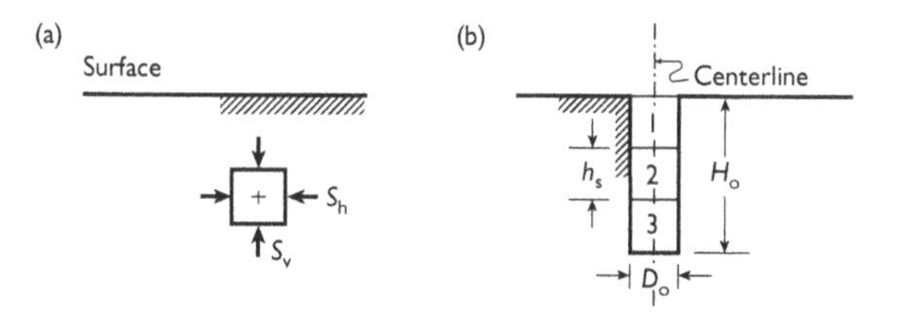

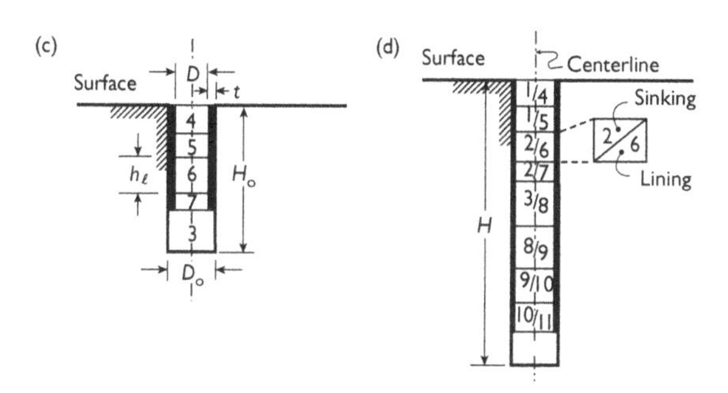

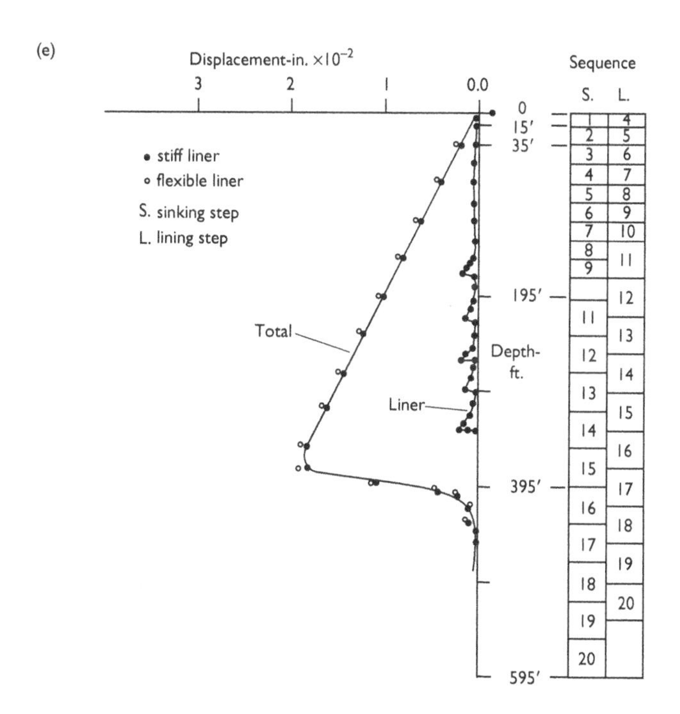

Figure 3.19 Schematic representation of shaft sinking and lining operations performed by the computer. Numbers identify the sequence of operations. (a) Generation of the premining stress field, (b) collaring of shaft and initial sinking to depth H_o in lifts of height h_s, (c) first installation of the liner in lifts of hight h_l, (d) shaft deepening to depth H with concurrent sinking and lining operation, and (e) shaft sinking and lining sequence. Also total shaft and liner wall displacements at shaft depth of 395 ft.

Note
Actual curves are based on more than 100 data points.

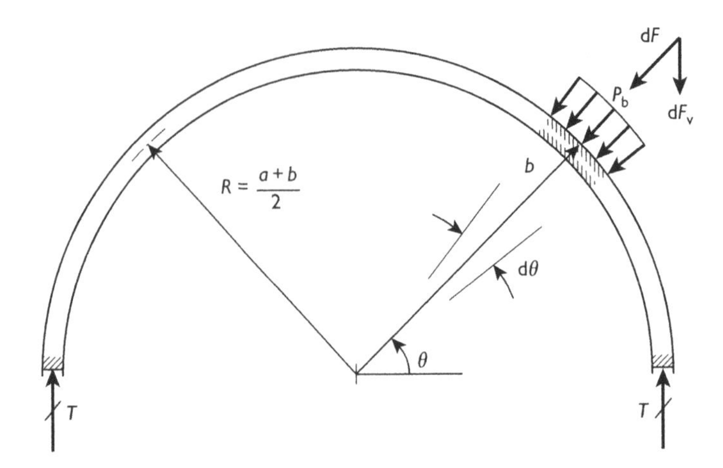

Figure 3.20 Thin, continuous ring equilibrium.

Circular steel rings

Integration of radial pressure over a semicircular shaft ring allows for estimation of the net horizontal force applied to the ring. This force must be equilibrated by internal reactions at the semi-ring ends as shown in Figure 3.20. The equilibrium requirement allows for calculation of the average stress acting through the ring thickness. Thus,

$$F_v = 2T$$

$$= \int_0^{\pi} dF_v$$

$$= \int_0^{\pi} \sin(\theta) b p_b S d\theta$$

$$F_v = 2bSp_b$$

where S is distance along the ring, which is assumed to be continuous and has cross section Sh. Hence, $T = bSp_b$ and the average stress is just $T/Sh = bp_b/h$ in this continuous liner ring.

If the ring is not continuous but discrete, a steel set, for example, then each ring supports the load one half the distance to the next set. Some load may be transmitted to a ring by lagging or lacing (wood, steel) between sets, but the largest load would be a line load P distributed over the outside of the ring with magnitude p_bS. In this case, the support load or pressure p_b is likely to be a rock pressure rather than a water load. Regardless of the identity of the load, if the spacing of ring sets is S, then ring support reaction T is

$$T = p_b RS \tag{3.22}$$

where the radius is now to the center of the ring which is of acceptable accuracy in consideration of ring radius and thickness. Clearly if the liner load were an internal pressure tending to expand the ring, the support reaction T would be a tension given by the same formula within the accuracy of practical geometry (thickness small compared with radius, say, a few inches compared with several feet).

A safety factor for steel ring sets in vertical shafts may be defined with respect to forces acting and resisting or simply as a ratio of strength to stress. Thus,

$$FS = \frac{C_s A_s}{T} = \frac{C_s}{T/A_s} = \frac{C_s}{\sigma_s} \tag{3.23}$$

where C_s and σ_s are steel compressive strength and stress, respectively.

Equations (3.22) and (3.23) allow one to estimate steel size (cross-sectional area) needed to support an anticipated rock pressure p_b in an unlined shaft of radius R (or b). Steel rings are curved beams shaped and sized to design specifications and are usually A36 steel that has an initial yield strength of about 36,000 psi. Steel area and ring spacing interact, so there is no unique solution to support design.

Example 3.24 Consider an unlined shaft 20 ft in diameter where the rock pressure is estimated to be 15 psi, a steel safety factor of 2.5 is specified, and the maximum allowable stress in a ring is 36,000/2.5 or 14,400 psi. Determine the steel area A_s required.

Solution: Although the spacing is not uniquely determined, an expression relating steel area to spacing may be obtained. Thus,

$$A_S = \frac{(15)(20/2)(144)S}{14{,}400} = 1.5S$$

where A_s is in square inches and spacing S is in feet.

A wide spacing clearly implies a heavier ring. A W or wide flange shape with an 8 in. flange width spaced on 6 ft centers with an area of about 9.0 square inches would weigh 31 pounds (W8×31) per lineal foot. Increasing spacing by 33% to 8 ft would increase weight proportionally and call for a W8×40 shape (*Manual of Steel Construction* (7th ed.) AISC, NY, 1970). This calculation points out once again, that seemingly small support loads, in this case, 15 psi, lead to sizeable support requirements and significant costs that must be considered in the overall design.

A wide spacing tends to lower costs by reducing the number of ring sets installed over a given length of excavation, while increasing cost and installation time because of greater difficulty in handling heavier steel. An important consideration in this regard is any tendency of the rock mass to ravel. Even falls of small rock pieces cannot be tolerated in a shaft, so lagging between rings is essential. Lagging may be wood, galvanized corrugated sheet metal, or some other structural material and may be placed between flanges or on top of the outside flange. Lagging is thin relative to length and experiences beam action under rock pressure. A common 3 in. thick piece of wood lagging 6 ft long under a distributed load of 15 psi experiences a tensile stress of 6480 psi which is well above design value for No. 1 Douglas Fir parallel to the grain. (*National Design Specification for Wood Construction Supplement* (1997 ed.) American Forest & Paper Association). Double lagging (6 in.) would reduce the bending stress by 1/4 at the very most, to 1,620 psi which is marginal. Smaller sets closer together may be required to allow thick lagging to safely restrain small rock pieces detached from the rock mass.

3.3 Multiple naturally supported shafts

Often several shafts are needed to meet requirements for ventilation, personnel, and materials handling. Efficient placement of surface and underground facilities favors closely spaced shafts, while wide spacing favors lower stress concentration caused by interaction between excavations. Multiple shafts also raise an important rock mechanics or safety question concerning the most favorable *shaft row orientation*. Two new rules of thumb provide guidance to these design questions.

The first new rule states that *excavations should be separated by at least one "diameter" to reduce interaction to an acceptable amount*. Separation refers to wall to wall distance. In case of circular shafts "diameter" is obviously the unlined shaft diameter. An acceptable amount of interaction occurs when stress concentration is reduced to about 10% of that associated with a single excavation. For example, by separating circular shafts by one diameter,"1-D," one hopes to reduce peak stress concentrations associated with a row of circular shafts to that of a single opening in the same preshaft stress field. In case of a single vertical shaft of circular cross section excavated in a stress field caused by gravity alone, circumferential stress concentration $K = 1 + (a/r)^2$. At 1-D away from the shaft wall $r = 3a$ and $K = 1.11$, so the stress concentration has decreased to within 11% of the preshaft value. In case of elliptical and rectangular shafts, "diameter" refers to a characteristic linear dimension of the section that reduces interaction to an acceptable amount. As a practical matter, the characteristic linear dimension of an ellipse or rectangular section for application of the 1-D rule is the *long* dimension of the opening.

The second new rule of thumb states that *the most favorable orientation of a row of identical excavations is with the row axis parallel to the major principal preexcavation compression*. Thus, the most favorable row orientation is the same as the most favorable orientation of a single opening.

Results from the mathematical theory of elasticity supply stress distributions and concentrations about multiple circular holes in a row. Results are in a variety of forms (analytical, numerical, computer code) and often require considerable additional work for practical assessment. Three reasonably assessable cases are of interest here: (1) two identical circular holes; (2) an infinite row of identical circular holes, and (3) two circular holes of different diameters.

Circular shafts in a row

Results from solutions to the problem of two circular holes side by side and an infinite row of circular holes allow one to evaluate the influence of one hole upon the other and to further assess the "1-D" rule of thumb for treating multiple identical openings as isolated, single openings. Figure 3.21 shows two identical circular holes side by side centered on the x-axis. Points of interest are holes sides (A and B), and top and bottom (C) where the stresses are at or near peak compressive and tensile values. Stress concentration at these points depends on the

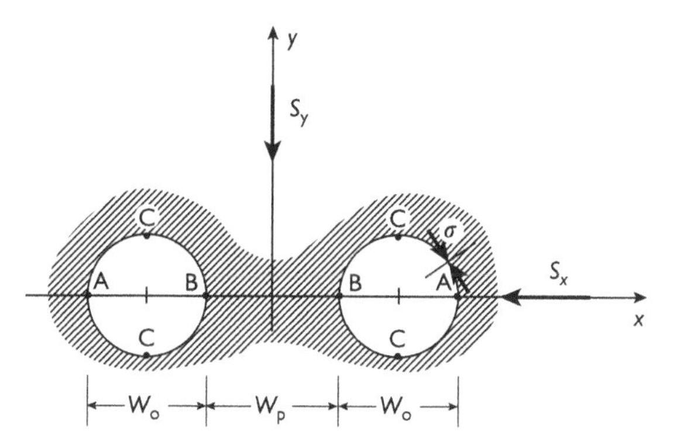

Figure 3.21 Two identical circular openings.

applied stresses (S_x, x = east and S_y, y = north) and on hole separation $W_o + W_p$ relative to hole size where W_o = width of opening (diameter) and W_p = width of pillar.

A plot of stress concentration at points A, B, and C as a function of the ratio Wp/Wo when a compression is applied *parallel* to the x-axis is shown in Figure 3.22(a). The reference stress $S_x = S_1$ and is the major principal stress (compression positive). When the holes are far apart, the stress concentration factors are −1.0 at points A and B and +3.0 at points C. When the holes are separated by one diameter (pillar width equals opening width, $W_p/W_o = 1$), the tension outside at A is reduced only slightly to about −0.9, while the tension inside at B is reduced to about −0.6. The compression at the top and bottom, point C, is reduced about 10% to 2.7 from 3.0. These results are in keeping with the rule of thumb stating holes more than one diameter apart experience only a slight interaction. As hole separation is reduced further from $W_p/W_o = 1$ and the holes nearly touch ($W_p/W_o = 0.2$), compression at C is reduced further. This effect occurs because one hole is in the "shadow" of the other when the major compression acts parallel to the axis of hole centers. The peak tension on the outside at A changes very little from −1.0 regardless of separation, while tension on the inside at B is reduced as separation is reduced.

Figure 3.22(b) shows the results when the major compression is *perpendicular* to the axis of hole centers. In this case $S_y = S_1$ and is the reference stress. The peak tension in this case occurs at the hole tops and bottoms, points C, and changes very little with hole separation. The peak compression occurs at B inside the holes and increases very rapidly from +3.0 as hole separation decreases below one diameter. Clearly, the most favorable orientation of the axis of hole centers in the case of two identical circular holes is parallel to the major compression when the choice is between parallel or perpendicular. The material between holes is shielded, shadowed, by the holes when the compression is parallel, but directly exposed when the applied compression is perpendicular to the axis of hole centers.

The two cases of uniaxial compression shown in Figure 3.22 may be combined under the principle of superposition. Table 3.6 presents the stress concentration factors for the two uniaxial cases. Thus in the hydrostatic compression case ($S_x = 1$, $S_y = 1$) with $W_p/W_o = 0.5$, stress concentration at pointsA, B, and C are 2.25, 2.89, and 1.74, respectively. Either applied

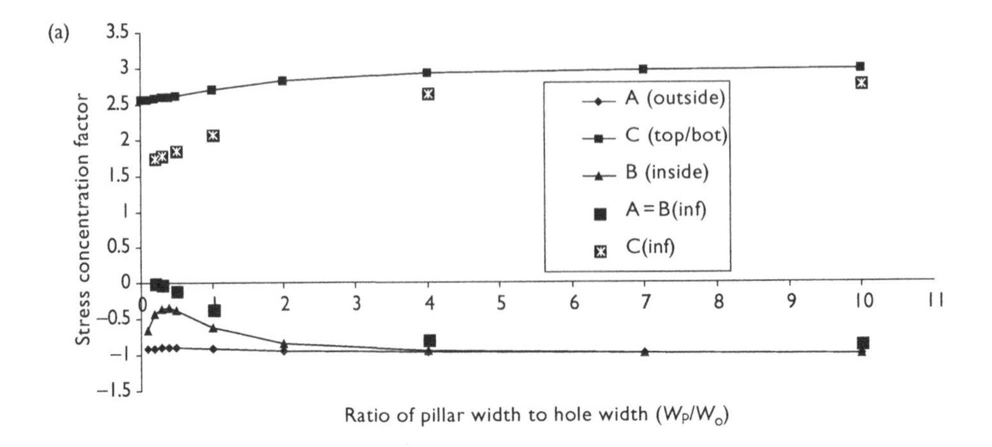

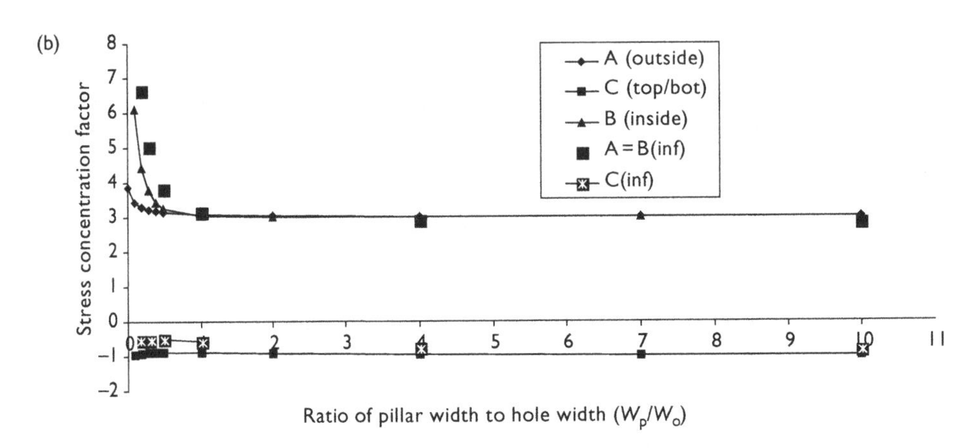

Figure 3.22 Stress concentration about two circular holes loading: (a) compression parallel to the row axis; $S_x = 1$, $S_y = 0$ and (b) compression perpendicular to row axis; $S_x = 0$, $S_y = 1$.

stress may be used as a reference stress. A single circular hole would experience a stress concentration of 2.0 at all points on the hole wall.

As another example of superposition, in case of a relatively low $S_y[S_y = (1/3)S_x]$ and $W_p/W_o = 0.5$, the stress concentrations at hole sides, top and bottom (points A, B, and C) are, 0.15, 2.33, and 0.71, respectively (reference stress is S_x). When S_y is high relative to $S_x[S_x = (1/3)S_y]$ and $W_p/W_o = 0.5$ as before, then stress concentrations at A, B, and C are 2.85, 3.14, and −0.01, respectively (reference stress is S_y). If one of the applied stresses is a tension, for example, S_y and $S_y = -S_x$, then stress concentrations for $W_p/W_o = 0.5$ at A, B, and C are −4.05, −3.64, and 3.51, respectively (S_x is the reference stress). If S_x is the tension, then the signs of the stress concentrations are reversed. If both of the applied stresses are tension, the stress concentration factors are the same as if both were compression.

Table 3.6 Stress concentration about two identical circular holes

W_p/W_o	S_x, S_y					
	$S_x = 1.0$ $S_y = 0.0$			$S_x = 0.0$ $S_y = 1.0$		
	A	C	B	A	C	B
0.0		2.569		3.869		
0.1	–0.918	2.568	–0.651	3.414	–0.958	6.106
0.2	–0.905	2.580	–0.423	3.289	–0.924	4.423
0.3	–0.899	2.593	–0.351	3.231	–0.905	3.768
0.4	–0.896	2.608	–0.346	3.185	–0.892	3.443
0.5	–0.896	2.623	–0.377	3.151	–0.884	3.264
1.0	–0.908	2.703	–0.609	3.066	–0.881	3.020
2.0	–0.94	2.825	–0.838	3.020	–0.92	2.992
4.0	–0.927	2.927	–0.948	3.004	–0.964	2.997
7.0	–0.987	2.970	–0.981	3.001	–0.985	2.998
10.0	–0.993	2.984	–0.991	3.000	–0.992	3.000

Example 3.25 Consider two vertical, circular shafts of the same diameter (7 m) in a preexcavation stress field defined by

$$S_{xx} = 2.5 \quad S_{yy} = 3.5 \quad S_{zz} = 10.0 \quad T_{yz} = 0.0 \quad T_{zx} = 0.0 \quad T_{xy} = -1.5$$

where units are MPa, x = east, y = north, z = up. Determine the best orientation of the shaft pair and the stress concentrations in plan view at points A, B, and C defined in Figure 3.21. Assume the shafts are separated by 1.4 m.

***Solution*:** According to rule of thumb, the row axis of multiple openings should be parallel to the major preexcavation compression. The direction of the preexcavation principal stress may be found by the solution of

$$\tan(2\alpha) = \frac{\tau_{xy}}{(1/2)(\sigma_{xx} - \sigma_{yy})}$$

$$= \frac{T_{xy}}{(1/2)(S_{xx} - S_{yy})}$$

$$= \frac{-1.5}{(1/2)(2.5 - 3.5)}$$

$$\tan(2\alpha) = 3.0$$

$$\therefore$$

$$2\alpha = 71.6°, \ 251.6°$$

$$\alpha = 35.8°, \ 125.8°$$

The proper choice of the two solutions may be made by measuring from the axis of the largest preexcavation normal stress, in this example, the y-axis, and observing the sign convention that positive angles are measured counterclockwise, as shown in the sketch.

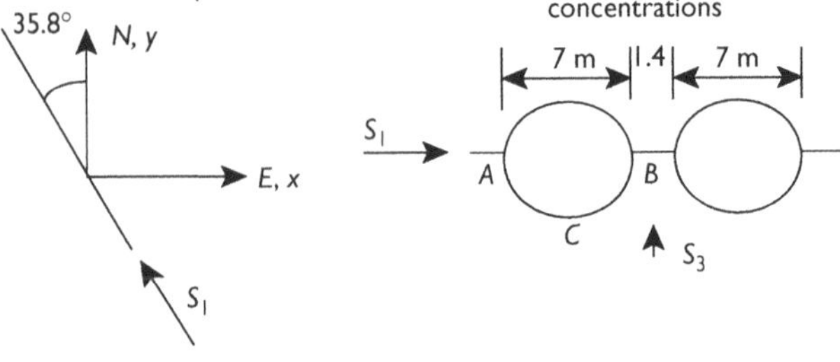

Numerical evaluation of stress concentrations at points A, B, and C shown in the sketch requires magnitudes of the preexcavation principal stresses. These are

$$\left.\begin{matrix} S_1 \\ S_3 \end{matrix}\right\} = \left(\frac{S_{xx}+S_{yy}}{2}\right) \pm \sqrt{\left(\frac{S_{xx}+S_{yy}}{2}\right)^2 + T_{xy}^2}$$

$$= \left(\frac{2.5+3.5}{2}\right) \pm \sqrt{\left(\frac{2.5+3.5}{2}\right)^2 + (-1.5)^2}$$

$$\left.\begin{matrix} S_1 \\ S_3 \end{matrix}\right\} = 3.0 \pm 1.58$$

$\therefore$

$$S_1 = 4.58$$
$$S_3 = 1.42$$

Accordingly, the ratio $M = 1.42/4.58 = 0.31$. The pillar width to opening width ratio $W_p/W_o = 1.4/7.0 = 0.2$. With S_1 and S_3 in the roles of S_x and S_y in Table 3.6, the stress concentrations are

$$K_A = -0.905 + 0.31(3.289) = 0.11$$
$$K_B = -0.423 + 0.31(4.423) = 0.95$$
$$K_C = 2.580 + 0.31(-0.924) = 2.29$$

There is no tension and the peak compressive stress concentration is 2.29. A single shaft would have a peak stress concentration of 2.69, about 14% higher in compression, and a tensile stress concentration of −0.07.

Example 3.26 Consider the data in Example 3.25 and suppose the major compression is perpendicular to the shaft pair axis rather than parallel. Determine the stress concentrations in this orientation.

***Solution*:** As before, the ratio $M = 1.42/4.58 = 0.31$. The pillar width to opening width ratio $W_p/W_o = 1.4/7.0 = 0.2$. With S_1 and S_3 in now in the roles of S_y and S_x in Table 3.6, the stress concentrations are

$$K_A = 3.289 + 0.31(-0.905) = 3.01$$
$$K_B = 4.423 + 0.31(-0.423) = 4.29$$
$$K_C = -0.924 + 0.31(2.58) = -0.124$$

Table 3.7 Stress concentration about an infinite row of identical circular holes

W_p/W_o	S_x, S_y			
	$S_x = 1.0$	$S_y = 0.0$	$S_x = 0.0$	$S_y = 1.0$
	A = B	C	A = B	C
0.2	–0.000	1.753	6.642	–0.535
0.3	–0.014	1.788	5.036	–0.518
0.5	–0.101	1.855	3.824	–0.495
1.0	–0.365	2.077	3.159	–0.555
4.0	–0.762	2.648	2.876	–0.795
10.0	–0.852	2.777	2.827	–0.851
∞	–1.000	3.000	3.000	–1.000

The tensile stress concentration factor has magnitude 0.124, which was absent in example 3.25. The peak compressive stress concentration is 4.29 which is 87% greater than when the shaft row is oriented parallel to the preexcavation principal stress (compression positive).

As more holes are added to the two-hole row, stress concentrations change. Eventually, the addition of one more identical circular hole results in a negligible change to holes in the interior of the row. When the row is of indefinite (infinite) extent, stress concentrations about each hole are repeated from hole to hole. Because of the periodicity of the array, stress concentration at points A and B are the same. Stress concentrations at the hole sides (A, B), top and bottom (C) are given in Table 3.7.

Comparison of two-hole row data with the infinite row data points in Figure 3.22 shows that loading *parallel* to the row axis leads to *lower* compressive and tensile stress concentrations in the infinite row case. This result indicates that the shadow effect is increased as holes are added to the row. In the case of loading *normal* to the row axis, compressive stress concentrations are generally *higher* while tensile stress concentrations are lower in the infinite row case. This effect is accentuated as the distance between holes is decreased. The generally lower stress concentrations under parallel loading lends further support to the rule of thumb that states the most favorable row orientation is parallel to the major principal stress (compression).

Superposition applies, so the results in Table 3.7 may be used to determine stress concentrations for combined loadings. For example, under hydrostatic compression ($S_x = S_y$) at $W_p/W_o = 1.0$, stress concentrations at A (and B) and C are 2.794 and 1.552, respectively. If $S_y = (1/3)S_x$, then stress concentration at A (and B) and C are both compressive, 3.037 and 0.137, respectively.

As the distance between holes ($W_o + W_p$) is decreased in the case of an infinite row of circular holes loaded *perpendicularly* to the line of hole centers, not only does the compressive stress concentration increase at the hole sides, but the average stress increases as well. Figure 3.23 shows the distribution of σ_{xx} (east stress) and σ_{yy} (north stress) along the x-axis between holes. Even though there is no applied stress in the east or x-direction, a stress develops that increases from zero towards the interior of the material between holes. The interior portion of the material between holes is confined by this x-direction normal stress buildup.

The average north (y-direction) pillar stress S_p may be calculated exactly from equilibrium of forces for holes in an infinite row. Thus, $S_p W_p = S_n(W_o + W_p)$, so $S_p = S_n/(1-R)$ where S_n is

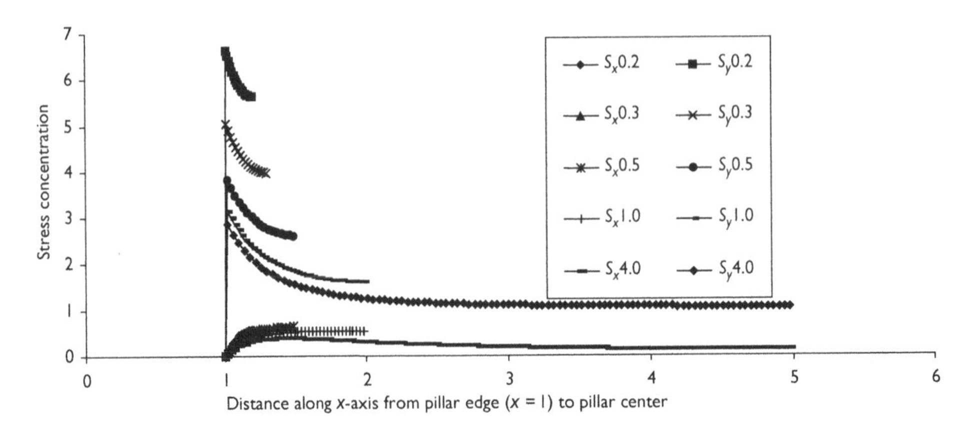

Figure 3.23 Stress concentration between two circular holes loaded perpendicularly to the line of centers (x = east = parallel to line of centers, y = north = perpendicular to line of centers). Plots start at pillar edges and end at pillar centers. Pillar to opening width ratio W_p/W_o ranges from 0.2 to 4.0.

the preexcavation normal stress acting in the north direction and $R = W_o/(W_o + W_p)$, ratio of opening width to hole spacing. In the cases shown in Figure 3.23, $W_p/W_o = 4$, 1, 0.5, 0.3, and 0.2, $S_p/S_n = 1.25$, 2.00, 3.00, 4.33, and 6.00, respectively. These averages may be compared with the corresponding peaks: 2.88, 3.16, 3.82, 5.03, and 6.64, respectively (Table 3.7). As the distance between holes is decreased both peak and average stress increase, while the difference decreases. At $W_p/W_o = 0.2$, the difference between peak and average stress is only about 10%.

Shaft pillar safety

A shaft pillar factor of safety based on *average pillar stress* S_p rather the peak stress is easy to compute and if too low indicates the need for more detailed analysis including determination of peak stresses. There is another type of shaft pillar about mine shafts. This second type of shaft pillar protects shafts from damage that may be induced by extracting ore too close to the shafts. Concern here is with pillars formed by multiple shafts. These pillars should be dimensioned to prevent damaging interaction between shafts. As always, geologic structure, "joints," need to be taken into account. The rock mass between shafts of equal size in an infinite row of shafts forms pillars that experience an average stress $S_p = S_n(W_o + W_p)/W_p$ where S_n is the horizontal preshaft stress acting normal to the line of centers and S_p is horizontal post-shaft stress (average). An area extraction ratio R may be defined as the ratio of the area excavated to a total area of excavation and pillar, that is, $R = W_o/(W_o + W_p)$, measured along the line of centers (horizontally) and one unit of distance vertically. The desired safety factor based on average compressive stress is then

$$\mathrm{FS}_{\text{ave}} = \frac{C_p}{S_p} = \frac{C_o(1-R)}{S_n} \tag{3.24}$$

where C_p is shaft wall compressive strength and C_o is rock mass unconfined compressive strength. If a size effect on strength is considered important, then C_o must include such an effect. Equation (3.24) may be used to evaluate a safety factor based on average pillar stress or to evaluate pillar size associated with a specified safety factor.

Example 3.27 Suppose a safety factor of 3.0 is desired in a rock mass with unconfined compressive strength of 4,800 psi where the preshaft stress is hydrostatic and 1,200 psi. A row of 25 ft diameter circular shafts is planned. Determine the shaft separation needed to meet the required safety factor.

Solution: By definition

$$FS_p = \frac{C_p}{S_p} = \frac{C_o}{S_n/(1-R)}$$

where no size effect on strength is assumed. This equation may be solved for $1-R$,

$$1-R = \frac{S_n(FS_p)}{C_o}$$

$$= \frac{(1,200)(3)}{4,800}$$

$$1-R = 0.75$$

The ratio R may also be computed from the shaft geometry, that is,

$$1-R = \frac{W_p}{W_o + W_p}$$

$$\therefore$$

$$0.75W_o = W_p(1-0.75)$$

$$\therefore$$

$$W_p = 3W_o = (3)(25) = 75 \text{ ft}$$

This pillar size separates the shafts of three diameters, so the shafts are effectively isolated and spaced on 100 ft centers. Single shafts would have a stress concentration of 2.0 in compression and tension would be absent. A safety factor based on peak stress at the shaft walls would be FS = C_o/K_cS_1 = 4,800/(2) (1,200) = 2.0. Thus, a design layout based on averages, although seemingly conservative at FS = 3.0, masks a danger of potentially unsafe peak stress conditions at excavation walls. In this case, the shaft wall safety factor is 2.0 and thus indicates a safe condition with respect to rock mass failure.

Because the safety factor based on average stress separates the shafts by more than one diameter, the spacing could be reduced to one diameter without significantly increasing peak stress. At a one diameter pillar width, W_p = 25 ft, R = 0.5, and FS_{ave} = (4,800)(0.5)/1,200 = 2.0. The safety factor at the shaft wall would be only slightly higher.

Example 3.28 According to the analysis summarized in Figure 3.23, the difference between peak and average stress concentration about two identical circular shafts is about 10% when W_p/W_o = 0.2. At this ratio, the peak stress concentration is 6.64; the average stress concentration is approximately 6.00. Estimate the rock mass strength needed under these circumstances to achieve a safety factor of 3 in a 1,200 psi hydrostatic stress field.

Solution: Because of the closeness of average to peak stress concentration, one may use either, therefore

$$\mathrm{FS_c} = \frac{C_o}{\sigma_c}$$

$$= \frac{C_o}{K_c S_1}$$

$$3 = \frac{C_o}{(6.64)(1{,}200)}$$

$$\therefore$$

$$C_o = 23{,}900 \text{ psi}$$

Use of unconfined compressive strength tends to underestimate the safety factor based on average stress. The reason is the buildup of horizontal stress in the pillars, as shown in Figure 3.23. At a pillar width equal to 1-D, Wo, the central portion of a pillar is confined by a stress almost equal to the preexcavation horizontal stress. If one adopts a Mohr–Coulomb rock mass failure criterion, then $C_p = C_o + (C_o/T_o)p$ where p=horizontal confining stress or "pressure." For example, even if p were only one-half the preexcavation stress in a hydrostatic stress field of 1,200 psi, the rock mass compressive strength $C_p = 4{,}800 + (10)(600) =$ 10,800 psi where a reasonable ratio of unconfined compressive to tensile strength of 10 is assumed and the unconfined compressive strength is 4,800 psi. In this case, other factors being equal, the pillar core has a safety factor that is more than double an estimate based on unconfined compressive strength.

A technically sound and economically viable shaft pillar defense strategy for circular shafts in a row allows for relatively close spacing (pillar width equal to shaft diameter), based on the extraction ratio Formula (3.24), with provision of shaft wall safety against peak stress failure. The latter provision almost certainly requires rock mass reinforcement that prevents falls of small rock and the start of progressive failure that could lead to pillar collapse. As in the case of single circular shafts, if the shaft wall is safe, then the rock mass beyond is also safe.

Formula (3.24) does not depend on shaft shape and may therefore be applied to other shaft sections in an infinite row of identical vertical shafts. Figure 3.24 illustrates before and after stress distributions normal to the line of centers of rectangular shafts. The x and y axes coincide with preshaft principal directions. Of course, the most favorable orientation would align the shaft section axis and row axis parallel to the major principal stress before excavation. In any case, the preshaft normal stress $S_n = S_y$ which is a principal stress. The post-shaft row pillar stress $S_p = S_n(1 - R)$ where R is the area extraction ratio $W_o/(W_o + W_p)$ and W_o is the section dimension *parallel* to the x-axis. Opening dimension in the y-direction, "height" of the opening is H_o which is also "height" of the pillar H_p. Of course, these distinctions were unnecessary in the case of circular shafts in a row.

Possibilities of failure along faults and major joints of considerable continuity in shaft pillars also need to be considered. Numerous small joints and fractures are considered in estimation of rock mass strength and corresponding safety factor calculations. Analysis of

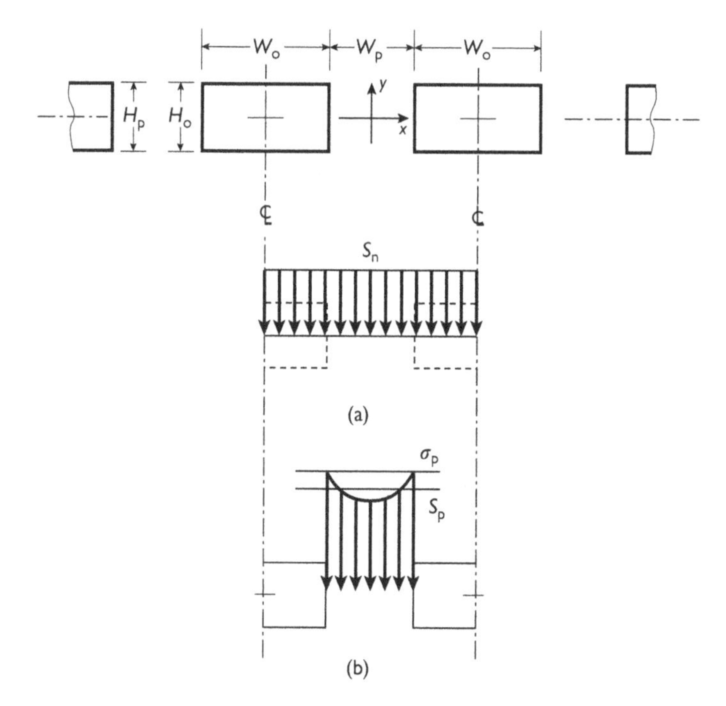

Figure 3.24 Pillar stress distribution normal to the line of centers before and after excavation: (a) before and (b) after.

pillar joint safety based on average stresses proceeds in much the same manner as wall joint safety discussed in the section on *Shaft Wall Strengths*. Here, "compass" coordinates, *xyz* are aligned with preshaft principal directions and with the understanding that the shaft row axis or line of centers is parallel to one of these principal directions. The best choice is with x parallel to S_1. Stresses with respect to *xyz* are: $S_{xx} = 0$, $S_{yy} = S_p$, and $S_{zz} = S_v$; shear stresses $T_{xy} = T_{yz} = T_{zx} = 0$, and the notation implies average stress. The next step is to compute the direction cosines of the normal to a joint plane of interest (n_x, n_y, n_z). Components of the stress vector (T_x, T_y, T_z) can then be calculated and finally the normal and shear stresses (S_j, T_j) acting on the joint plane may be obtained. The pillar joint safety factor, assuming Mohr–Coulomb failure, is then

$$\text{FS}_j = \frac{\sigma_j \tan\left(\phi_j\right) + c_j}{\tau_j} \tag{3.25}$$

Example 3.29 Consider a row of 18 ft diameter vertical shafts spaced on 45 ft centers at a depth of 3,200 ft where the vertical preexcavation stress is 3,600 psi and the horizontal stress is 1,800 psi. A small fault with dip direction 30° and dip 30° transects a potential shaft pillar. The gouge in the fault zone has an angle of internal friction of 28° and a cohesion of 20 psi. If the shaft excavation plan is realized, is there a danger of fault slip, that is, what is the post-excavation fault safety factor?

Solution: With the x-axis along the line of centers, the post-shaft stresses may be estimated as $S_{xx} = 0$, $S_{yy} = S_p$, $S_{zz} = S_v$, and $T_{xy} = T_{yz} = T_{zx} = 0$. The area extraction ratio $R = W_o/(W_o + W_p) = 18/45 = 0.4$, and $S_p = S_n/(1-R) = 1{,}800/(1 - 0.4) = 3{,}000$ psi. Direction cosines of the fault plane normal are:

$$n_x = \sin(\delta)\sin(\alpha) = \sin(60°)\sin(30°) = (\sqrt{3}/2)(1/2) = \sqrt{3}/4$$
$$n_y = \sin(\delta)\cos(\alpha) = \sin(60°)\cos(30°) = (\sqrt{3}/2)(\sqrt{3}/2) = 3/4$$
$$n_z = \cos(\delta) = \cos(60°) = 1/2$$

Tractions on the fault plane are then

$$T_x = (0)(\sqrt{3}/4) + (0)(3/4) + (0)(1/2) = 0$$
$$T_y = (0)(\sqrt{3}/4) + (3{,}000)(3/4) + (0)(1/2) = 2{,}250$$
$$T_z = (0)(\sqrt{3}/4) + (0)(3/4) + (3{,}600)(1/2) = 1{,}800$$

The normal stress on the fault plane is

$$N_j = (0)\left(\sqrt{3}/4\right) + (2{,}250)(3/4) + (1{,}800)(1/2) = 2{,}588$$

and the shear stress is

$$S_j = [(0)^2 + (2{,}250)^2 + (1{,}800)^2 - (2{,}588)^2]^{1/2} = 1{,}267$$

The fault plane factor of safety according to a Mohr-Coulomb strength criterion is then

$$\mathrm{FS_j} = \frac{(2{,}588)\tan(28°) + 20}{1{,}267} = 1.10$$

which is close to failure and would be cause for concern.

Of interest is the fault plane safety factor before excavation. From a similar analysis, $\mathrm{FS_j} = 1.56$, which indicates excavation would tend to induce fault plane shearing.

However, the pillar is 27 ft wide, more than one diameter, so consideration of confining pressure effects is reasonable. There are three effects, one that increases the normal stress after excavation, one that reduces the post-excavation shear stress, and one that increases strength. The strength effect increases compressive strength under confining pressure, but does not affect cohesion or angle of internal friction. If a cautious view of the confining pressure effects is taken by assuming the post-excavation normal stress in the x-direction in the pillar core is 900 psi, just one-half of the preexcavation horizontal stress, the effect is to increase the post-excavation normal stress to 2,334 psi, while reducing the shear stress to 889 psi. The fault plane safety factor after taking confining pressure into account is 1.42, which is somewhat less than the preexcavation safety factor but significantly greater than the unconfined safety factor calculation that lead to an estimate of 1.10. These results indicate that the core of the pillar would be safe with respect to fault plane slip. Slip at the shaft walls would still need to be considered in conjunction with loading by the circumferential stress concentrated at the shaft wall and the vertical stress, as in the case of a single shaft.

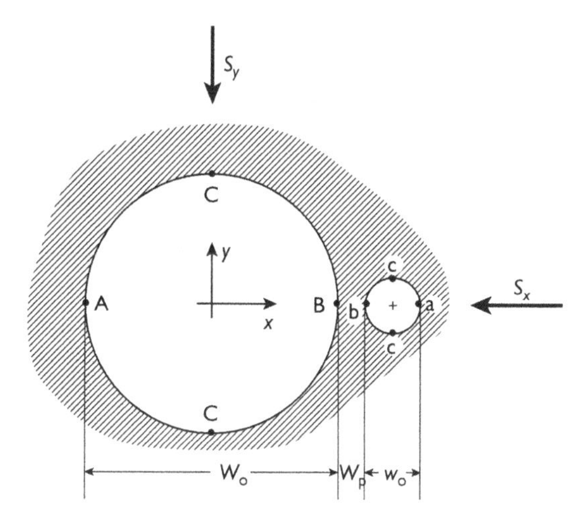

Figure 3.25 Circular excavations of different diameters.

Two circular shafts of different diameter

When two circular holes or shafts of *different* diameters are excavated in the same neighborhood, one may suppose that the large hole will significantly influence stress concentration near the small hole, while the small hole will have little influence on stress about the big hole. A practical example would involve a small diameter borehole for utilities excavated adjacent to a larger service shaft. Figure 3.25 illustrates the situation where W_o is width (diameter) of the large opening, w_o is the small opening width (diameter), and W_p is the width of the pillar between.

Peak stress concentration data for two diameter ratios W_o/w_o of 5 and 10 and for different pillar width to large hole width (diameter) W_p/W_o ratios are shown in Figure 3.26. The lower case legend with solid symbols pertains to the small hole; the upper case legend with open symbols pertain to the large hole. Loading is *parallel* to the line of hole centers in Figure 3.26 (a) and perpendicular in Figure 3.26(b). If either hole were isolated, then under compressive stress loading, peak compressive stress concentrations of 3.0 would occur at points C and c. Peak tensions of −1.0 would occur at A and B, and a and b (Figure 3.25). When loading is normal to the line of hole centers, the peak compression and tension would occur at points A, B, a, b, and at points C, c, respectively, provided the holes were isolated from each other. Data in Figure 3.26 show that the large diameter hole is largely unaffected by the small hole even when the separation or pillar width is less than the small hole diameter (about $W_p/W_o < 0.1$ in Figure 3.25). The large hole tends to shadow the small hole and reduces the peak small hole compression in the vicinity of points *c* with reduction in W_p/W_o. However, peak tension on the inside of the small hole at *b* tends to increase rapidly over −1.0 with close approach ($W_p/W_o <$ 0.1) of the small hole to the big hole. High tensile stress in the pillar could pose a serious threat to stability of both excavations.

Data for the case when the load is applied *perpendicular* to the line of hole centers is shown in Figure 3.26(b). Peak tensile stress concentrations about the big hole occur at points C and are much the same as if the hole were isolated, that is, $K_t = -1.0$. Peak compressive

(a)

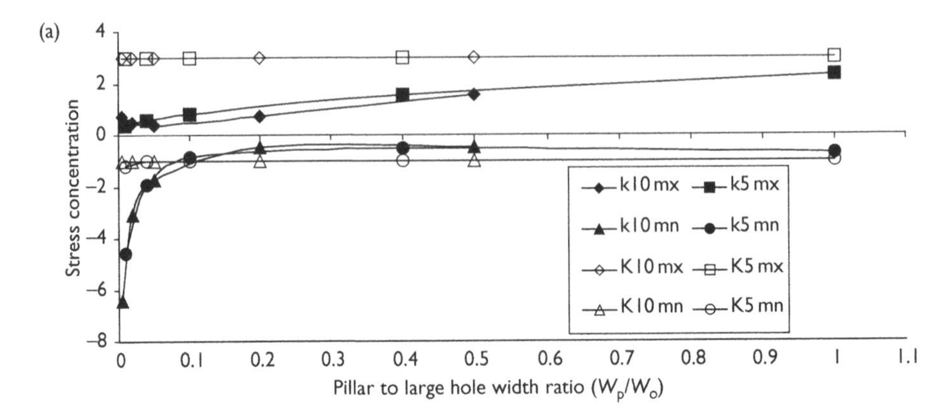

(b)

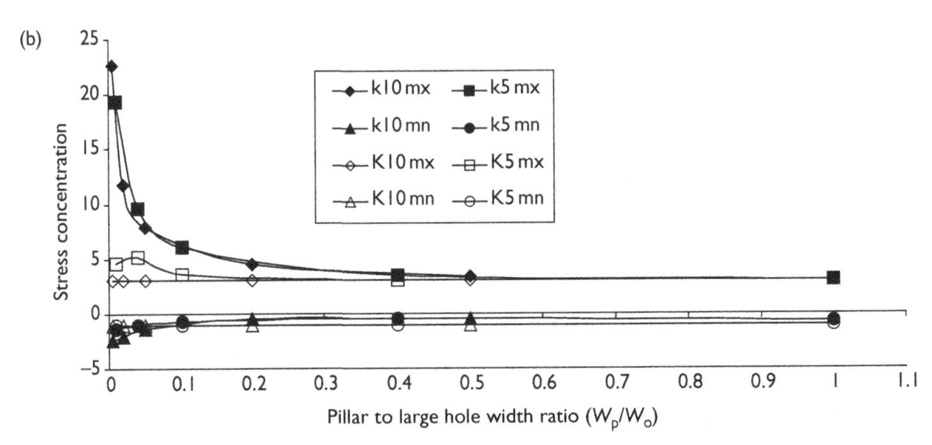

Figure 3.26 Maximum and minimum stress concentration about two circular holes of unequal diameters of 5 and 10. Lower case solid symbols pertain to the small hole. Upper case open symbols pertain to the large hole. Large hole diameter W_o is 5 and 10 times small hole diameter w_o. Rock distance between holes is W_p (after Haddon, 1967). (a) Parallel and (b) perpendicular to the line of centers.

stress concentration K_c is also largely unaffected by the small hole until the holes are quite close, $W_p/W_o < 1.0$. Peak compression at the wall of the large hole occurs on the inside at B. There is a hole size effect in that the case of the smaller big hole ($W_o/w_o = 5$), compressive stress increases to about 5.0 from 3.0 and then decreases with even closer approach of the small hole. This effect would be amplified as the big hole diameter is decreased. Indeed, when hole diameters are equal and $W_p/W_o = 0.1$, the compressive stress concentration is 6.11 (Table 3.6). Peak tension and compression at the wall of the small hole increase with decreasing separation. The increase accelerates with close approach as W_p/W_o decreases below 0.1, as seen in Figure 3.26(b).

The data for two circular holes of unequal diameter show that stress concentrations about the large hole are not significantly affected by the small hole, while stress concentration about the small hole is definitely influenced by the large hole. Intuitively, one might suppose that the

1-D rule applies using the large hole diameter and, in fact, if the two holes were separated by a pillar width $W_p = W_o$, then the data in Figure 3.26 make clear that the two holes are effectively isolated with negligible interaction. When isolated, maximum and minimum stress concentrations are 3.0 and −1.0. However, such separation is too conservative; a lesser separation is possible. Superposition applies and in practice, peak stress concentrations will be less than the uniaxial load case shown in Figure 3.26. Using the small hole diameter for the 1-D rule would be risky. A reasonable compromise and guide would be to separate circular holes of unequal diameter by $W_p = 0.5W_o$, when the large hole is more than five times the diameter of the small hole. Otherwise the 1-D rule should be applied using the large hole diameter in case of long life openings such as shafts. The rule of thumb for the most favorable orientation with the line of hole centers parallel to the row axis still applies as the data in Figure 3.26 indicate.

Consider a 25 ft diameter vertical shaft and 5 ft diameter borehole excavated in a hydrostatic stress field ($M = 1$) at a depth of 3,600 ft in a preexcavation stress field caused by gravity alone and where the horizontal stress is 1,200 psi and rock mass strength is 4,800 psi. If the holes are effectively isolated according to the 1-D rule (25 ft apart, $W_p = 25$) then $W_p/W_o = 1.0$. Tension is then absent and the peak compressive stress concentration is 2.0. The safety factor with respect to compression is then 2.0. Superposition of stress concentration data from Figure 3.26 allows for closer spacing. At $W_p/W_o = 0.4$, tension is still absent, while the peak compression stress concentration appears at the inside of the small hole and is almost 3.0. The holes are 10 ft apart ($W_p = 10$) and the associated safety factor is 1.33.

Elliptical shafts in a row

When the holes in a row are not circular, two orientation questions arise: (1) the best orientation for the individual holes and (2) the best orientation for the row of holes. The first question may be tentatively answered by applying a rule of thumb for isolated, single openings that states the best orientation is to align the long axis of the hole with the direction of the major principal compressive stress. Whether this applies for multiple noncircular holes in a row when interaction occurs is an open question. The second question calls into play the rule of thumb that states the most favorable row of holes orientation is with the row axis, the line of centers, also parallel to the major compression. Whether this rule applies to noncircular holes is also an open question. Analytical or closed form solution to this problem is not known. However, a purely numerical approach allows for quantitative evaluation.

A comparison of the analytical and numerical results in the case of two identical circular holes excavated in a hydrostatic stress field is presented in Figure 3.27. Numerical data in Figure 3.27 were obtained using the finite element method. The agreement between the two methods is excellent and indicates reliability of the finite element results. Data in Figure 3.27 may also be computed using superposition and Table 3.6 data for points A and B outside and inside the two holes.

Figure 3.28 shows four combinations of a pair of elliptical holes under uniaxial loading. The ratio of the major to the minor axes of the ellipses is two; the ratio of hole separation to hole width (W_p/W_o), both measured along the x-axis is 0.5, so interaction is expected. The results are summarized in Table 3.8. These limited results support the applicability of the

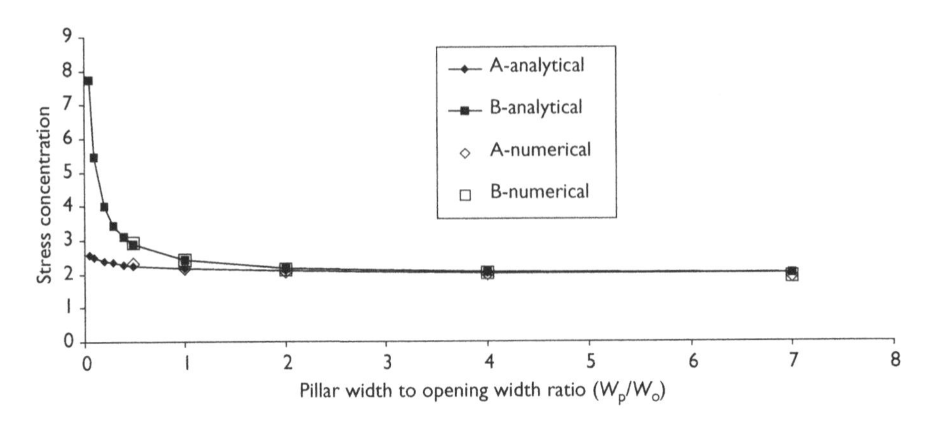

Figure 3.27 Comparison of analytical and numerical results and stress concentration in the case of two identical circular holes excavated in a hydrostatic stress field (M = 1).

combined rules of thumb for single openings and rows of openings that state the long axis of the hole and the row should be parallel to the major compression.

Inspection of Table 3.8 shows that the greatest compression (5.56) occurs inside ellipses (B) when the applied compressive stress is *normal* to the hole and row axes – combination (c) in Figure 3.28. The tension in this case (at C) is also very nearly equal to the greatest tension (−1.03). The smallest tension and compression (−0.64, 1.91, respectively) occur when the applied stress is *parallel* to the hole and row axes – combination (a) – at points B and C, respectively. For comparison, if the ellipses were widely separated and effectively isolated then loading parallel to the hole axis leads to compressive stresses of 2.0 at C and tensile stress of −1.0 at A and B. The noticeable reduction in tension at points B between the ellipse pair indicates a significant shadow effect in combination (a) of Table 3.8.

A single ellipse with an axis ratio of two under an applied stress acting *normal* to the long axis of the hole would experience stress concentrations of −1.0 at the hole top and bottom (C) and 5.0 at the hole sides. By comparison, the ellipse pair in case (c) shows a tension of −1.02 and compressions 5.16 and 5.56 outside and inside the holes (points A and B), respectively, and thus an interaction amounting to about a 10% increase in peak compression occurs in this unfavorable case.

Superposition applies, so one may use the data in Table 3.8 for combined S_x and S_y loading and $W_p/W\text{o} = 0.5$. For example in the hydrostatic case when $S_x = S_y$, and the ellipse axes are parallel to the x-axis, stress concentrations at A, B, and C are 4.33, 4.92, and 0.89, respectively. When the ellipse axes are parallel to the y-axis, stress concentrations at A, B, and C are 1.19, 2.00, and 3.23, respectively. In either case tensions are absent under compressive hydrostatic loading.

Table 3.9 shows stress concentrations for the four combinations of hole and infinite row axis orientation relative to the applied compression used in the two-ellipse cases (Table 3.8). The main trends are similar to the infinite circular hole row trends: (1) lower tensile and

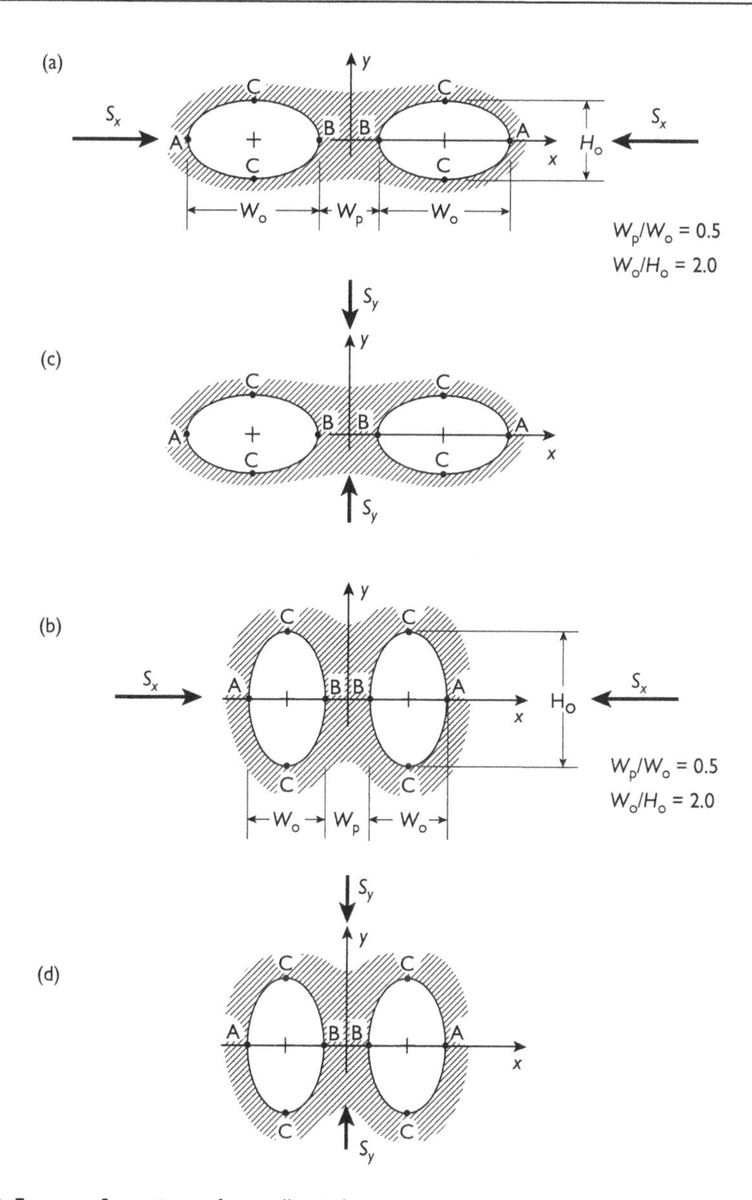

Figure 3.28 Four configurations of two elliptical excavations.

compressive stress concentration when the load is parallel to the row axis, and (2) lower tensile, but higher compressive stress concentrations when the load is normal to the row axis. Again the most favorable combination is with hole and row axes parallel to the major compression; the least favorable combination is with hole and row axes normal to the major compression.

Table 3.8 Stress concentration about two identical elliptical holes (Wp/Wo = 0.5)

Combination	S_x, S_y					
	$S_x = 1.0\ S_y = 0.0$			$S_x = 0.0\ S_y = 1.0$		
	A	C	B	A	C	B
(a) 1–1	–0.83	1.91	–0.64			
(b) 0–1	–1.03	4.09	–0.55			
(c) 0–0				5.16	–1.02	5.56
(d) 1–0				2.22	–0.86	2.55

Notes
(a) hole axis and row axis parallel to S1; (b) hole axis normal, row axis parallel; (c) hole axis and row axis normal; and (d) hole axis parallel, row axis normal.

Table 3.9 Stress concentration about an infinite row of identical elliptical holes

Combination	S_x, S_y			
	$S_x = 1.0$	$S_y = 0.0$	$S_x = 0.0$	$S_y = 1.0$
	A = B	C	A = B	C
(a) 1–1	–0.34	1.55		
(b) 0–1	–0.00	2.19		
(c) 0–0			5.58	–0.66
(d) 1–0			3.81	–0.31

Notes
(a) hole axis and row axis parallel to S_1; (b) hole axis normal, row axis parallel; (c) hole axis and row axis normal; and (d) hole axis parallel, row axis normal.

Rectangular shafts in a row

An elliptical section is useful for illustrating the concept of stress concentration about single and multiple holes and transition from analytical to numerical solutions. However, a rectangular section is a much more practical shape. Again, two related questions arise. One concerns orientation of individual holes; the second concerns orientation of a row of holes. Both orientations are with respect to the direction of the major preexcavation compression S_1. Four combinations of identical rectangular holes are shown in Figure 3.29; these combinations are the same used in analyses of elliptical holes. The long dimension of all holes is always twice the short dimension and the ratio of hole separation to hole width (W_p/W_o) is always 0.5 as measured along the x-axis. Thus, the openings are relatively close together and interact significantly, as was the case for elliptical holes.

Table 3.10 shows stress concentration factors for two cases of uniaxial loading of just *two* rectangular holes with different hole and row axis alignment relative to the load direction. Peak stress concentration occurs at corners in all cases. When loading is parallel to the x-axis (cases *a* and *b*), peak stress occurs on the outside corners (D_o); when loading is parallel to the y-axis (cases *c* and *d*), peak stress occurs on the inside corners (D_1). The most favorable case with respect to peak compressive stress concentration (under compressive loading) is case (*a*)

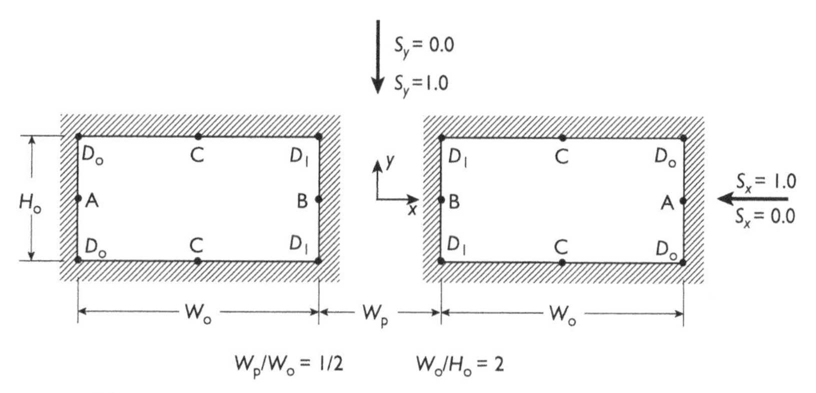

(a) hole and row axes parallel to $S_x = S_1$, $M = 0$

(c) hole axis and row axes normal to $S_y = S_1$, $M = 0$

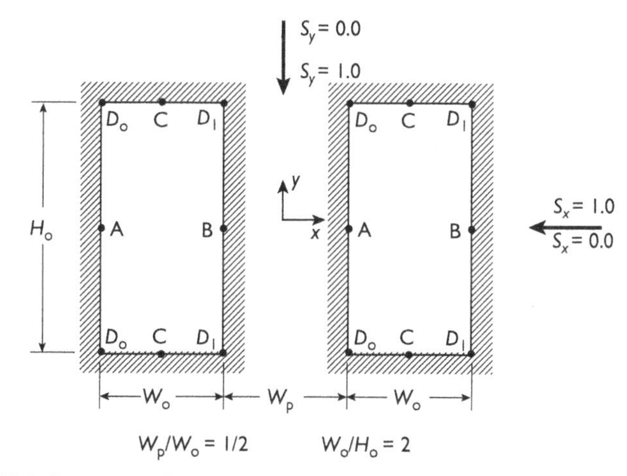

(b) hole axis normal row axis parallel to $S_x = S_1$, $M = 0$

(d) hole axis parallel row axis normal to $S_y = S_1$, $M = 0$

Figure 3.29 Four configurations of two identical rectangular excavations.

with hole and row axes parallel to the direction of the major preexcavation compression S_1 (S_x). The least favorable case is (*c*) with both axes normal to the applied stress. The same conclusions are obtained relative to tension: case (*a*) is most favorable; case (*c*) is least favorable.

Table 3.11 shows stress concentrations at the ends of hole semi-axes (A, B, C) and at hole corners (D_o, D_1) in four cases of an *infinite* row of identical rectangular holes. As in cases of two rectangular holes, peak compressive stress concentration always occurs at the corners of a rectangular section. Tension, when present, occurs "under the load," that is, at the ends of the semi-axis that is parallel to the major compression. Data in Table 3.11 show that loading *normal* to the row axis or line of centers (*c*) is quite unfavorable because of high compressive stress concentration at corners (D_o, D_1) and high tensile stress at midsides (C). Stress concentration at corners is greatly reduced when the row axis is oriented *parallel* to the major preexcavation compression; tension is also reduced.

Table 3.10 Stress concentration about two identical rectangular holes

Case	S_x, S_y									
	$S_x = 1.0$ $S_y = 0.0$					$S_x = 0.0$ $S_y = 1.0$				
	A	C	B	D_O	D_I	A	C	B	D_O	D_I
(a) 1 –1	−0.81	1.36	−0.42	2.84	2.01					
(b) 0–1	−1.02	2.04	−0.39	4.81	2.27					
(c) 0–0						2.50	−1.05	2.85	5.58	6.66
(d) 1-0						1.64	−0.90	2.07	3.32	3.40

Notes
(a) hole axis and row axis parallel to Sl; (b) hole axis normal, row axis parallel; (c) hole axis and row axis normal; and (d) hole axis parallel, row axis normal.

Table 3.11 Stress concentration about an infinite row of identical rectangular holes

Combination	S_x, S_y					
	$S_x = 1.0$ $S_y = 0.0$			$S_x = 0.0$ $S_y = 1.0$		
	A = B	C	$D_o = D_I$	A = B	C	$D_o = D_I$
(a) 1–1	−0.23	1.06	1.81			
(b) 0–1	0.00	1.07	1.57			
(c) 0–0				2.94	−0.84	7.30
(d) 1–0				3.00	−0.77	5.13

Notes
(a) hole axis and row axis parallel to S_1; (b) hole axis normal, row axis parallel; (c) hole axis and row axis normal; and (d) hole axis parallel, row axis normal.

The main trends as holes are added to the two rectangular hole cases are similar to those for elliptical holes: (1) lower tensile and compressive stress concentration when the load is *parallel* to the row axis (cases *a* and *b*) because of increased "shadow" effects and (2) lower tensile, but higher compressive stress concentrations when the load is *normal* to the row axis (cases *c* and *d*).

Interestingly, the condition of uniaxial loading (*b*), with the hole axis normal to the load direction and the row axis parallel to the load direction, is slightly more favorable than (*a*), which was more favorable for elliptical holes. In consideration of a more realistic *in situ* stress state than uniaxial ($M = 0$), for example $M = 1/3$ or $M = 1$, case (a) with both axes (hole and row) parallel to S_1, (*b*) is still more favorable. This trend continues at larger separations (greater W_p). To see these results, one may use superposition of data from Table 3.11 for midside points *A*, *B*, and *C* and by numerical calculation for corners (D_o and D_1). In case $M = 1/3$ with hole and row axes *parallel* to $S_1 = S_x$ and $S_3 = S_y = (1/3)S_1$, stress concentrations at *C* and *A(B)* and are 0.78 and 0.75, respectively. Tension is absent. Peak compressive stress concentration still occurs at the corners (D_o, D_1). However, direction and location of peak corner compression varies somewhat from case to case, so superposition does not apply at corners. A separate series of calculations leads to the data in Table 3.12 for $M = 1/3$. Results for hydrostatic case, $M = 1$, are given in Table 3.13.

Table 3.12 Stress concentration about an infinite row of identical rectangular holes

Combination	S_x, S_y					
	$S_x = 1$ $S_y = 1/3$			$S_x = 1/3$ $S_y = 1$		
	A = B	C	$D_O = D_I$	A = B	C	$D_O = D_I$
(a) 1–1	0.75	0.78	3.49			
(b) 0–1	1.00	0.81	2.69			
(c) 0–0				2.86	–0.49	7.62
(d) 1–0				3.00	–0.42	5.24

Notes
(a) hole axis and row axis parallel to SI; (b) hole axis normal, row axis parallel; (c) hole axis and row axis normal; and (d) hole axis parallel, row axis normal. $W_p/W_o = 0.5$

Table 3.13 Stress concentration about an infinite row of identical rectangular holes

Combination	S_x, S_y					
	$S_x = 1$ $S_y = 1$			$S_x = 1$ $S_y = 1$		
	A = B	C	$D_O = D_I$	A = B	C	$D_O = D_I$
(a) + (c)	2.71	0.22	8.20			
(b) + (d)				3.00	0.30	5.46

Notes
(a) hole axis and row axis parallel to SI; (b) hole axis normal, row axis parallel; (c) hole axis and row axis normal; and (d) hole axis parallel, row axis normal.
$W_p/W_o = 0.5$

Stress concentration data for an infinite row of rectangular holes thus indicates that *(b)* is more favorable than *(a)* with respect to peak compressive stress concentration at corners. However, the reverse was true with just two rectangular holes; *(a)* was most favorable. In the case, say, of five rectangular holes, stress concentration about the outside holes and associated pillars would be similar to stress concentration about just two holes. Stress concentration about the inside holes and pillars would more likely be similar to the infinite row of holes data. Because stress concentrations are higher in the two hole cases, especially at the outside corners, than in the infinite row of holes cases, the choice of configuration *(a)* or *(b)* is logically decided by the two hole case in favor of *(a)*. Thus, alignment of hole and row axes parallel to the major compression is the best choice, when the number of holes in the row is few, say, about five.

Example 3.30 Consider a relatively close spaced row of vertical, rectangular shafts such that $W_p = 0.5W_o$ with hole and row axis parallel. Further suppose the opening dimensions are 12 × 24 ft and therefore $H_o/W_o = 1/2$. These data correspond to the geometry of Figure 3.29 (a). The shaft route penetrates a weak zone at 3,600 ft where rock mass compressive and tensile strengths are 15,000 and 1,500 psi, respectively. If the preexcavation stress field is caused by gravity alone, estimate the shaft wall safety factors with respect to tension and compression.

***Solution*:** In plan view, the preexcavation gravity field shows equal stress, so $M = 1$. Either the results in Table 3.11 or Table 3.13 may be used for stress concentration values. From Table 3.13, a peak compressive stress concentration of 8.20 occurs at the corners (D); no tension is indicated. With respect to compression

$$FS_c = \frac{C_o}{K_c S_1} = \frac{15,000}{(8.20)(1,200)} = 1.52$$

which is low for shaft wall stability. In comparison, the safety factor at midside of the short edges (A, B) is 4.61. The midside points of the long edges (C) have even higher safety factors. Danger of large-scale failure is therefore not indicated.

Example 3.31 Stresses at a shaft wall are composed of two parts, a preshaft part and a change part induced by excavation of the shaft. Thus, $\sigma = \sigma^o + \Delta\sigma$, where σ, σ^o, and $\Delta\sigma$ post-shaft wall stress, preshaft stress, and stress change. In case of a vertical, circular shaft, the post-shaft wall circumferential stress $\sigma_t = (S_1 + S_3) - 2(S_1 - S_3)\cos(2\theta)$ as given by equation (3.3) and illustrated in Figure 3.1. Suppose the preshaft stress field with respect to compass coordinates (x=east, y=north, z=up) is $S_x = S_H = S_2$, $S_y = S_h = S_3$, and $S_z = S_v = S_1$. Shear stresses are nil. Compression is positive and the principal stresses in three dimensions are ordered such that $S_1 \geq S_2 \geq S_3$. Determine the stress changes induced by shaft excavation as seen in a plan view of the shaft.

***Solution*:** The normal stress perpendicular to the shaft wall (σ_r) is nil, so the stress change is equal in magnitude but opposite in sign to the preshaft stress. Similarly the shear stress in plan view ($\tau_{r\theta}$), also referred to cylindrical coordinates ($r\theta z$), is equal but opposite in sign to the preshaft stress. The circumferential stress change is the final stress less the preshaft stress. Thus,

$$\Delta\sigma_t = \Delta\sigma_\theta = (S_1 + S_3) - 2(S_1 - S_3)\cos(2\theta) - [(1/2)(S_1 + S_3) - (1/2)(S_1 - S_3)\cos(2\theta)]$$

where in plan view $S_1 = S_H$ and $S_3 = S_h$ and the term in [...] is the preshaft circumferential stress. The preshaft stress is obtained by rotation of reference axes about the given, vertical, z-axis. Hence, $\Delta\sigma_t = \Delta\sigma_\theta = (1/2)(S_1 + S_3) - (3/2)(S_1 - S_3)\cos(2\theta)]$. Summarizing, the induced stress changes are

$$\begin{aligned}
\Delta\sigma_r &= -[(1/2)(S_H + S_h) + (1/2)(S_H - S_h)\cos(2\theta)] \\
\Delta\sigma_\theta &= (1/2)(S_H + S_h) - (3/2)(S_H - S_h)\cos(2\theta) \\
\Delta\tau_{r\theta} &= (1/2)(S_H - S_h)\sin 2(\theta)
\end{aligned}$$

Example 3.32 Consider a circular shaft of radius a excavated in a preshaft stress field relative to compass coordinates (x=east, y=north, z=up), so $S_x = S_H = S_2$, $S_y = S_h = S_3$, and $S_z = S_v = S_1$. Shear stresses are nil as in Example 3.31. Further suppose the shaft axis is inclined with direction cosines $n_x = \cos(x, c) = \cos(\alpha)$, $n_y = \cos(y, c) = \cos(\beta)$, $n_z = \cos(z, c) = \cos(\gamma)$ where the direction angles of the shaft axis c are (α, β, γ). Find formulas for the preshaft stresses relative to a rotated set of axes (abc) where a is dip of a plane normal to the shaft axis c, and b is horizontal.

Solution: A plane normal to the shaft axis is analogous to a "joint" plane, so a table of direction cosines between axes (xyz) and (abc) is just that given in the text, that is, Table 2.1. Figure 2.13 shows the relationship between axes. Stresses in the shaft system (abc) may be computed from formulas given inAppendix D. Thus, from (D.4), $[\sigma(abc)] = [R][\sigma(xyz)][R]^t$ where the matrix $[R]$ is given by (D.6) which is essentially Table 2.1. The multiplication is lengthy but straightforward. The result in a compact form using the symbols in (D.2) for (D.6) and the given stress state is

$$\sigma_{aa} = S_H l_1^2 + S_h m_1^2 + S_v n_1^2$$

$$\sigma_{bb} = S_H l_2^2 + S_h m_2^2 + S_v n_2^2$$

$$\sigma_{cc} = S_H l_3^2 + S_h m_3^2 + S_v n_3^2$$

$$\tau_{ab} = S_H l_1 l_2 + S_h m_1 m_2 + S_v n_1 n_2$$

$$\tau_{bc} = S_H l_2 l_3 + S_h m_2 m_3 + S_v n_2 n_3$$

$$\tau_{ca} = S_H l_3 l_1 + S_h m_3 m_1 + S_v n_3 n_1$$

Note that the (xyz) directions are actually principal directions. For this reason no shear stresses are present in these formulas.

Example 3.33 Suppose the shaft axes in Example 3.32 has dip direction due north and dip of 60 degrees. Further suppose the stress state before shaft sinking is caused by gravity alone. Estimate the preshaft stresses at a depth of 300 m relative to an (abc) set of axes with a down the dip of the cross-section, b on strike, and c the shaft axis.

Solution: A reasonable estimate of the preshaft vertical stress is 25 kPa/m. The horizontal stresses are equal in a gravity field and are some fraction of the vertical stress, say, 1/3. Thus,

$$S_v = 7.5 \text{ MPa, and } S_H = S_h = 2.5 \text{ MPa.}$$

The cross-section of the shaft is in the role of a "joint" plane that is normal to the shaft axis. The dip of the cross-section is 30 degrees. A table of direction cosines is needed. Thus, with α =0 and δ=30 degrees,

system	x	y	z
a (dip)	l1 = cos(30)sin(0) l1 = 0	ml = cos(30)cos(0) m1 = √3/2	nl = –sin(30) nl = –1/2
b (strike)	l2 = –cos(0) l2 = – 1	m2 = sin(0) m2 = 0	n2 = 0
c (axis)	l3 = sin(30)sin(0) l3 = 0	m3 = sin(30)cos(0) m3 = 1/2	n3 = cos(30) n3 = √3/2

Using the formulas from the previous example:

$$
\begin{aligned}
\sigma_{aa} &= [(1/3)(0) + (1/3)(3/4) + (1/4)](7.5) \\
\sigma_{bb} &= [(1/3)(1) + (1/3)(0) + (0)](7.5) \\
\sigma_{cc} &= [(1/3)(0) + (1/3)(1/4) + (3/4)](7.5) \\
\tau_{ab} &= [(1/3)(0)(-1) + (1/3)(\sqrt{3}/2)(0) + (-1/2)(0)](7.5) \\
\tau_{bc} &= [(1/3)(-1)(0) + (1/3)(0)(1/2) + (0)(\sqrt{3}/2)](7.5) \\
\tau_{ca} &= [(1/3)(0)(0) + (1/3)(1/2)(\sqrt{3}/2) + (\sqrt{3}/2)(-1/2)](7.5)
\end{aligned}
$$

These evaluate to:

$$
\left.\begin{aligned}
\sigma_{aa} &= 3.75 \\
\sigma_{bb} &= 2.50 \\
\sigma_{cc} &= 6.25 \\
\tau_{ab} &= 0.00 \\
\tau_{bc} &= 0.00 \\
\tau_{ca} &= -2.17
\end{aligned}\right\} \quad MPa
$$

Example 3.34 Consider the results from Example 3.33 that give the preshaft stresses relative to a rectangle coordinate system (*abc*) after rotation from compass coordinates (*xyz*). The axis *c* is parallel to the shaft axis that dips 60 degrees with dip direction due north, *b* is horizontal, and *a* is inclined. Sketch the relationship between the axes and then compute algebraically the preshaft stresses relative to a cylindrical coordinate system attached to the shaft.

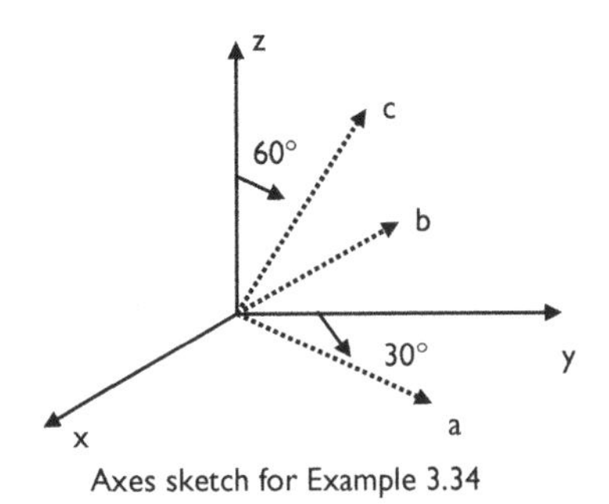

Axes sketch for Example 3.34

***Solution*:** The sketch with x=east, y=north, and z=up is looking down the shaft axes and gives a plan view of the circular shaft and the relationship between (*abc*) and cylindrical coordinates ($r\theta z$). Thus, Stresses in cylindrical coordinates are obtained by a rotation of reference axes from (*abc*) to ($r\theta z$) about the shaft axes. Thus, from (D.1) in Appendix D, with the necessary change in subscripts,

$$
\begin{aligned}
\sigma_{rr} &= (1/2)(\sigma_{aa} + \sigma_{bb}) + (1/2)(\sigma_{aa} - \sigma_{bb})\cos(2\theta) + \tau_{ab}\sin(2\theta) \\
\sigma_{\theta\theta} &= (1/2)(\sigma_{aa} + \sigma_{bb}) - (1/2)(\sigma_{aa} - \sigma_{bb})\cos(2\theta) - \tau_{ab}\sin(2\theta) \\
\tau_{r\theta} &= (-1/2)(\sigma_{aa} - \sigma_{bb})\sin(2\theta) + \tau_{ab}\cos(2\theta)
\end{aligned}
$$

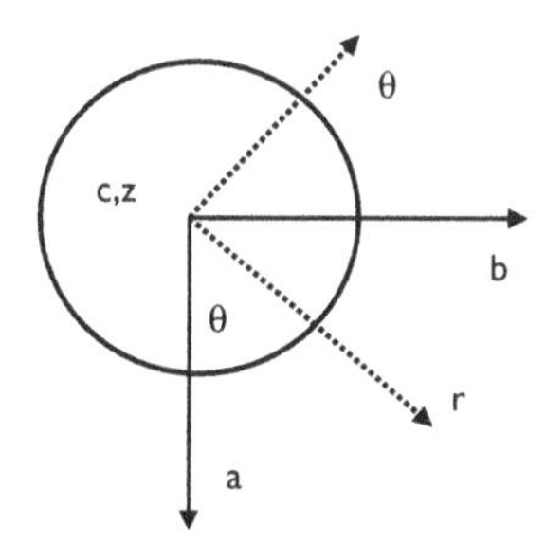

Shaft coordinates, rectangular and cylindrical.

$$\sigma_{zz} = \sigma_{cc}$$

$$\tau_{rz} = \tau_{ac}\cos(\theta) + \tau_{bc}\sin(\theta)$$

$$\tau_{\theta z} = -\tau_{ac}\sin(\theta) + \tau_{bc}\cos(\theta)$$

These are the preshaft stresses relative to shaft coordinates.

Example 3.35 Analysis of an inclined circular shaft shows that the stress changes induced by mining at the shaft wall are given by:

$$\Delta\sigma_r = -[(1/2)(S_r + S_\theta) + (1/2)(S_r - S_\theta)]$$

$$\Delta\sigma_\theta = [(1/2)(S_r + S_\theta) - (3/2)(S_r - S_\theta)]$$

$$\Delta\tau_{r\theta} = -T_{r\theta}$$

$$\Delta\sigma_{zz} = -(2v)(S_r - S_\theta)$$

$$\Delta\tau_{rz} = -T_{rz}$$

$$\Delta\tau_{\theta z} = T_{\theta z}$$

where the capital letters indicate preshaft stresses relative to the shaft coordinate system ($r\theta z$). These are the stresses given in Example 3.34 with a simple change in notation, e.g., $S_r = \sigma_{rr}$ and $T_{rz} = \tau_{rz}$and so on. Determine the post-shaft wall stresses in terms of the preshaft stresses.

Note: The stress change formulas are computed on the assumption that the shaft diameter is small compared with shaft depth and the section of interest is beyond the influence of the surface where the shaft is collared, say, several diameters.

***Solution*:** The r-stress changes are equal but opposite in sign to the preshaft stresses. Consequently, the total post-shaft stresses $\sigma_r = S_r + \Delta\sigma_r$, $\tau_{rz} = T_{rz} + \Delta\tau_{rz}$, and $\tau_{r\theta} = T_{r\theta} + \Delta\tau_{r\theta}$ are zero as, indeed, they should be in consideration of the traction-free shaft wall that is generated through excavation.

The remaining three post-shaft wall stresses by similar logic are

$$\sigma_\theta = S_\theta + [(1/2)(S_r + S_\theta) - (3/2)(S_r - S_\theta)]$$

$$\sigma_{zz} = S_z - (2v)(S_r - S_\theta)$$

$$\tau_{\theta z} = T_{\theta z} + T_{\theta z}$$

Alternatively, after collecting terms

$$\sigma_\theta = -S_r + 3S_\theta$$
$$\sigma_{zz} = S_z - (2v)(S_r - S_\theta)$$
$$\tau_{\theta z} = 2T_{\theta z}$$

Example 3.36 Consider a circular shaft excavated in a stress field caused by gravity only. Inclination of the shaft axis with respect to the vertical varies. Derive a procedure to evaluate the maximum and minimum shaft wall stress concentration factors for circumferential stress, K(max) and K(min) at any orientation from the vertical to the horizontal. In this case, use the vertical preshaft stress $S_v$as the reference stress.

Solution: The preshaft stress field is caused by gravity alone. Hence, the vertical and horizontal preshaft stresses have the form $S_v = \gamma H$, $S_h = S_H = K_o S_v$ where γ , H, and K_o are specific weight of rock, depth to a point of interest, and the ratio of horizontal to vertical preshaft stress (some number between 0 and 1, perhaps 1/3).

By definition, stress concentration $K = (stress)/(reference\ stress)$ and $K(\max) = \sigma(\max)/S_1$, $K(\min) = \sigma(\min)/S_1$where ordinarily $S_1$is the major principal stress *in the considered view.* However, in this example, the reference stress is given simply as S_v. The reason is S_1 in the cross-section view varies continuously with inclination while S_v is constant and in fact is the preshaft major principal stress in three dimensions.

From Example 3.35, the post-shaft wall stresses in terms of preshaft stresses are S_r, S_θ, S_z, T_{rz}, $T_{r\theta}$, $T_{\theta z}$ where S is a normal stress and T is a shear stress. These stresses are referred to a cylindrical coordinate system attached to the shaft ($r\theta z$). Stress changes induced by shaft sinking are$\Delta\sigma_r$, $\Delta\sigma_\theta$, $\Delta\sigma_z$, $\Delta\tau_{rz}$, $\Delta\tau_{r\theta}$, $\Delta\tau_{\theta z}$. Thus,

$$\sigma_r = S_r + \Delta\sigma_r = 0$$
$$\tau_{rz} = T_{rz} + \Delta\tau_{rz} = 0$$
$$\tau_{r\theta} = T_{r\theta} + \Delta\tau_{r\theta} = 0$$

in consideration of the traction-free shaft wall that is generated through excavation. The remaining three post-shaft wall stresses from Example 3.35 are

$$\sigma_\theta = S_\theta + [(1/2)(S_r + S_\theta) - (3/2)(S_r - S_\theta)]$$
$$\sigma_{zz} = S_z - (2v)(S_r - S_\theta)$$
$$\tau_{\theta z} = T_{\theta z} + T_{\theta z}$$

Alternatively, after collecting terms

$$\begin{aligned} \sigma_\theta &= -S_r + 3S_\theta \\ \sigma_{zz} &= S_z - (2v)(S_r - S_\theta) \\ \tau_{\theta z} &= 2T_{\theta z} \end{aligned} \tag{3.26}$$

The stresses seen in cross-section are the circumferential stress σ_θ and shear stress $\tau_{\theta z}$. These are shown in the sketch. The post-shaft shear stress is given as twice the preshaft shear stress. The preshaft shear stress is

$$T_{\theta z} = -\tau_{ac}\sin(\theta) + \tau_{bc}\cos(\theta) \tag{3.27}$$

where the shear stresses on the right are referred to a rectangular set of axes (*abc*) that are rotated from compass coordinates (*xyz*). The angle θ is measure from the *a*-axis.

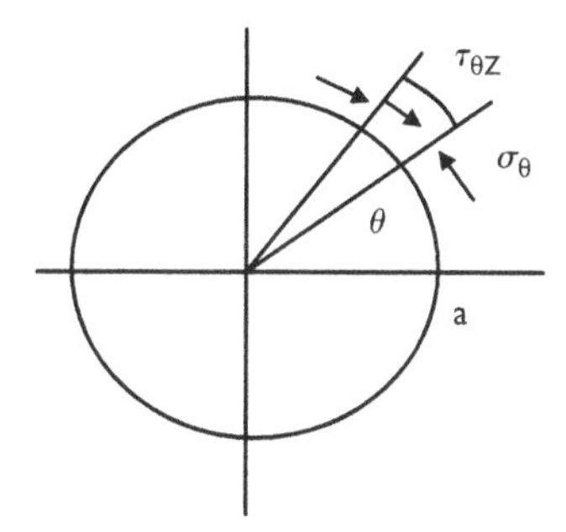

Sketch for Example 3.36 looking down the shaft axis.

To implement the rotation from Example 3.32, the direction cosines are needed. These are:

$$\begin{array}{lll} l_1 = 0, & m_1 = \cos(\delta), & n_1 = -\sin(\delta) \\ l_2 = -1, & m_2 = 0, & n_2 = 0 \\ l_3 = 0, & m_3 = \sin(\delta) & n_3 = \cos(\delta) \end{array}$$

where δ is the *dip of the shaft cross-section* which is 90-shaft axis inclination. For example, when the shaft is vertical, δ=0. The direction cosines are obtained from Table 2.1 in the text with the dip direction α=0.

The shear stresses needed are then

$$\begin{aligned} \tau_{ac} &= [K_o\sin(\delta)\cos(\delta) - \sin(\delta)\cos(\delta)]S_v \\ \tau_{bc} &= 0, \quad \tau_{ab} = 0 \end{aligned} \tag{3.28}$$

When the section is horizontal (shaft) or vertical (tunnel), the shear stresses are nil.

The radial, circumferential, and axial preshaft normal stresses may be computed from the usual equations of transformation (rotation about axis *c*). Thus,

$$\begin{aligned} S_r &= (1/2)(\sigma_{aa} + \sigma_{bb}) + (1/2)(\sigma_{aa} - \sigma_{bb})\cos(2\theta) + \tau_{ab}\sin(2\theta) \\ S_\theta &= (1/2)(\sigma_{aa} + \sigma_{bb}) + (1/2)(\sigma_{aa} - \sigma_{bb})\cos(2\theta) + \tau_{ab}\sin(2\theta) \\ S_{zz} &= \sigma_{cc} \end{aligned} \tag{3.29}$$

Again the preshaft stresses relative to (*abc*) on the right are obtained by rotation of axes from compass coordinates (*xyz*) as in Example 3.2. Thus,

$$\begin{aligned} \sigma_{aa} &= [K_o\cos^2(\delta) + \sin^2(\delta)]S_v = K_1S_v \\ \sigma_{bb} &= K_oS_v \\ \sigma_{cc} &= [K_o\sin^2(\delta) + \cos^2(\delta)]S_v = K_2S_v \end{aligned} \tag{3.30}$$

When the shaft is vertical and $\delta = 0$, $\sigma_{aa} = \sigma_{xx} = K_oS_v$, $\sigma_{bb} = \sigma_{yy} = K_oS_v$, $\sigma_{cc} = \sigma_{zz} = S_v$ which checks. When the shaft is horizontal and $\delta = 90^o$, $\sigma_{aa} = \sigma_{zz} = S_v$, $\sigma_{bb} = \sigma_{yy} = K_oS_v$, $\sigma_{cc} = \sigma_{xx} = K_oS_v$ which also checks.

Equations (3.29) can now be written in terms of the preshaft stresses in compass coordinates. Thus,

$$\begin{aligned}
S_r &= (S_v/2)(K_1 + K_o) + (S_v/2)(K_1 - K_o)\cos(2\theta) \\
S_\theta &= (S_v/2)(K_1 + K_o) - (S_v/2)(K_1 - K_o)\cos(2\theta) \\
S_{zz} &= K_2S_v \\
T_{\theta z} &= K_3S_v
\end{aligned} \tag{3.31}$$

The last equation is added from (3.27) and (3.28) for completeness. A sketch of the shaft coordinates is helpful. From (3.26) and (3.31)

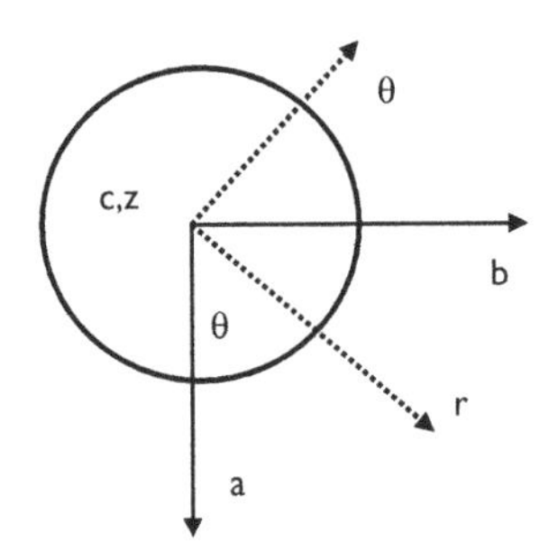

Shaft coordinate sketch looking down the c-axis.

$$\begin{aligned}
\sigma_\theta &= S_r[(K_1 + K_o) - 2(K_1 - K_o)\cos(2\theta)] \\
\sigma_{zz} &= S_v[K_2 - 2v(K_1 - K_o)\cos(2\theta)] \\
\tau_{\theta z} &= 2K_3S_v
\end{aligned} \tag{3.32}$$

The constant K's are

$$\begin{aligned}
K_1 &= [K_o\cos^2(\delta) + \sin^2(\delta)] \\
K_2 &= [K_o\sin^2(\delta) + \cos^2(\delta)] \\
K_3 &= [K_o\sin^2(\delta)\cos(\delta) - \sin(\delta)\cos(\delta)]
\end{aligned}$$

As a check on (3.32), in the vertical case δ=0 and

$$\begin{aligned}
K_1 &= K_o & &\qquad & \sigma_\theta &= 2S_rK_o = 2S_v(S_h)/S_v = 2S_1 \text{ (S}_1\text{ in the view plane)} \\
K_2 &= 1 & &\text{and} & \sigma_{zz} &= S_v \\
K_3 &= 0 & & & \tau_{\theta z} &= 0
\end{aligned}$$

Again, these results are for a gravity only preexcavation stress field.

From (3.32), the stress concentrations are easily obtained. Thus,

$$\begin{aligned}
K_\theta &= \sigma_\theta/S_r = [(K_1 + K_o) - 2(K_1 - K_o)\cos(2\theta)] \\
K_z &= \sigma_z/S_v = [K_2 - 2v(K_1 - K_o)\cos(2\theta)] \\
K_\tau &= \tau_{\theta z}/S_v = 2K_3
\end{aligned}$$

The required maximum and minimum circumferential stress concentration factors are obtained from (3.32). Stationary values of the first derivatives occur at θ=0, $\pi/2$ (measured from the a-axis that points down dip. Thus, with addition of subscript θ for clarity,

$$\begin{aligned} K_\theta(\max) &= 3K_1 - K_o \\ K_\theta(\min) &= -K_1 + 3K_o \end{aligned} \tag{3.33}$$

These expressions have the same structure as those discussed in the text following equation (3.4). Indeed, they reduce to the maximum and minimums given there in case of a vertical shaft excavated in the gravity field.

This example illustrates the great utility of axis rotation. Although lengthy, the derivation is a straightforward application of fundamentals to achieve a useful design goal. The same formulas apply in case of equal but high horizontal stress. In this case, $K_o \geq 1$. A similar procedure can be used to examine stress concentration about shafts of arbitrary orientation excavated in any preshaft stress field. An outline of the solution including displacements as well as stresses and a demonstration of correctness are given by Pariseau (1987).

Comment: There are yet other factors of safety that take into account the biaxial nature of stress at the wall of a naturally supported shaft. These safety factors depend on the failure criterion adopted for design analyses. An example is a factor of safety using the well-known Mohr-Coulomb (MC) failure criterion that is discussed inAppendix B. The MC criterion may be put into a form that indicates the maximum shear stress the material can sustain. Thus,

$$\tau(strength) = \left(\frac{\sigma_1 + \sigma_3}{2}\right)\sin(\phi) + (c)\cos(\phi)$$

where σ_1, σ_3, ϕ, and c are major principal stress, minor principal stress, angle of internal friction, and cohesion, respectively. Compression is considered positive in this expression. The intermediate principal stress σ_2 is assumed to have no effect on failure.

The stress normal to the shaft wall (radial stress) is zero, of course. However, the circumferential stress acting tangent to the shaft wall and the vertical stress are not zero. Moreover, there may be a shear stress present at the shaft wall as well as the two normal stresses. Thus, to apply the MC criterion, one must first determine the principal stress at the shaft wall. The radial stress is a principal stress and the normal (radial) direction to the shaft wall is a principal direction because the shaft wall surface is shear free. Thus, a rotation about a radius will lead to the required principal stresses. If these two principal stresses bracket the zero radial stress, then the radial stress is indeed the intermediate principal stress. This bracketing requires one principal stress to be greater than zero (compression) and one to be less than zero (tension). If both are compression, then the minor principal stress is the zero radial stress. If in the very unlikely event that both are tension, then the intermediate principal stress is the tension of smallest magnitude.

The most likely case is for the radial stress to be the minor principal stress. In the absent of a shear stress at the shaft wall, either the vertical or circumferential stress is the major principal stress. In either case, MC assumes the form

$$\begin{aligned} \tau(strength) &= \left(\frac{\sigma_1}{2}\right)\sin(\phi) + (c)\cos(\phi) \\ \left(\frac{\sigma_1}{2}\right) &= \left(\frac{\sigma_1}{2}\right)\sin(\phi) + (c)\cos(\phi) \end{aligned}$$

When this expression is solved for the major principal stress at failure, the result is simply

$$\sigma_1 = C_o.$$

Thus, the MC criterion leads to a safety factor with respect to compression given in (3.1). There are two such factors of safety, one respect to the circumferential stress and one with respect to the vertical stress. Naturally, the lowest factor of safety guides the design.

The case when the maximum compression is zero and the least principal stress is a tension leads to $\sigma_3 = -T_o$ and a safety factor with respect to tension. As a reminder, only the numerical values of strength and stress are used in factory of safety calculations.

Graphically, the safety factor using the Mohr-Coulomb criterion is simply the ratio of the maximum radius of a Mohr circle representing the state of stress at failure to the actual radius of the Mohr circle representing the given stress state. A subtlety that enters all factor of safety concepts is how the stresses would increase from the current state to a failure state. The process is unknown, so an assumption is needed. One assumption is to suppose that the mean stress remains constant, so the center of the Mohr circle is fixed. The radius of the failure circle is then readily computed from the Mohr criterion. An alternative assumption is to suppose the minor principal stress is fixed, while the major principal stress increases. This situation occurs during conventional triaxial strength testing that fixes confining pressure and increases axial load to failure of a test cylinder.

MC strength is graphically seen in the sketch as the radius of the Mohr circle representing failure **at the given mean normal stress** where the circle is centered. These are the vertical lines extending from the circle centers to the sloping failure line. A maximum radius divided by the radius of the circle representing the actual state of stress is the MC safety factor. Mohr circles representing the actual state of stress have the same radii but different centers and therefore different associated strengths and safety factors.

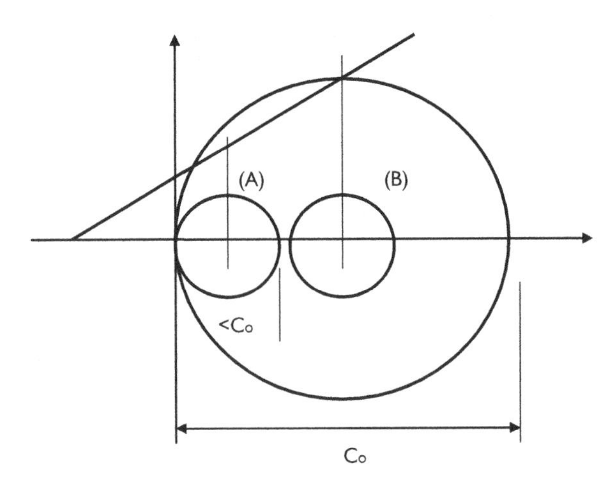

Sketch showing alternative safety factors with respect to Mohr-Coulomb strength.

The circle (A) occurs at a naturally supported excavation wall in consideration of the zero normal stress perpendicular to the wall and in consideration of the remaining two principal stresses at the wall that are assumed compressive. In this common situation, the minor principal stress is zero and the largest of the remaining principal stresses is the major principal stress. In a gravity field, the vertical shaft wall stress increases linearly with depth while the radial stress remains zero. The corresponding Mohr circle increases in diameter as the major principal stress moves to the right while the minor principal stress remains fixed at the origin. At failure $S_v = \sigma_1 = C_o$.

Circle (B) is a stress state where the mean normal stress is just one-half of the unconfined compressive strength. If the mean normal stress is fixed, while the radius increases, then at failure $\sigma_3 = 0$ and $\sigma_1 = C_o$. The safety factor associated with (A) is less than that associated with (B).

3.4 Problems

Single, naturally supported shafts

3.1 A 20 foot diameter circular shaft is planned in a massive rock. Laboratory tests on core from exploration drilling show that

$$C_o = 22{,}000 \text{ psi} \quad T_o = 1{,}200 \text{ psi}$$
$$\gamma = 144 \text{ pcf} \quad E = 5 \times 10^6 \text{ psi}$$
$$G = 2.0 \times 10^6 \text{ psi}$$

Depth is 3,000 ft. Site measurements show that no tectonic stresses are present. Determine the factors of safety with respect to failure in compression and tension.

3.2 If the opening in Problem 3.1 is in an *in situ* stress field such that $\sigma_h = \sigma_v$, what are the safety factors?

3.3 If the opening in Problem 3.1 is in an *in situ* stress field such that $\sigma_h = 0.01\sigma_v$, is there a possibility of failure? Justify your answer.

3.4 A 6 m diameter circular shaft is planned in massive rock. Laboratory tests on core from exploration drilling show that

$$C_o = 152 \text{ MPa} \quad T_o = 8.3 \text{ MPa}$$
$$\gamma = 23 \text{ kN/m}^3 \quad E = 34.5 \text{ GPa}$$
$$G = 13.8 \text{ GPa}$$

Depth is 915 m. Site measurements show that no tectonic stresses are present. Determine the factors of safety with respect to failure in compression and tension.

3.5 If the opening in Problem 3.4 is in an *in situ* stress field such that $\sigma_h = \sigma_v$, what are the safety factors?

3.6 If the opening in Problem 3.4 is in an *in situ* stress field such that $\sigma_h = 0.01\sigma_v$, is there a possibility of failure? Justify your answer.

3.7 Which is preferable from the rock mechanics viewpoint: an elliptical, ovaloidal, or rectangular opening of W_o/H_o of 2 for a stress field $M = 1/3$? $M = 0$? Justify your choices.

3.8 A rectangular shaft 10 ft wide by 20 ft long is sunk vertically in ground where the state of premining stress is

$$\sigma_{xx} = 1{,}141, \quad \sigma_{yy} = 2{,}059, \quad \sigma_{zz} = 1{,}600, \quad \tau_{xy} = 221, \quad \tau_{yz} = 0, \quad \tau_{zx} = 0,$$

with compression positive, x = east, y = north, and z = up, in psi. Rock properties are:

$$E = 4.5 \times 10^6 \text{ psi}, \quad \nu = 0.20, \quad C_o = 15{,}000 \text{ psi}, \quad T_o = 900 \text{ psi}.$$

Find (a) the most favorable orientation of the shaft and (b) the safety factors in this orientation. Indicate by sketch where the peak stresses occur in cross section.

3.9 A rectangular shaft 3 m wide by 6 m long is sunk vertically in ground where the state of premining stress is

$$\sigma_{xx} = 7.9, \quad \sigma_{yy} = 14.2, \quad \sigma_{zz} = 11.0, \quad \tau_{xy} = 1.5, \quad \tau_{yz} = 0, \quad \tau_{zx} = 0,$$

with compression positive, x = east, y = north, and z = up, in MPa. Rock properties are:

$$E = 31.0 \text{ GPa}, \quad \nu = 0.20, \quad C_o = 103 \text{ MPa}, \quad T_o = 6.2 \text{ MPa}.$$

Find (a) the most favorable orientation of the shaft and (b) the safety factors in this orientation. Indicate by sketch where the peak stresses occur in cross section.

3.10 A 12 ft by 24 ft rectangular shaft is sunk to a depth of 3,000 ft in ground where the premining stress field is given by formulas

$$S_v = 1.2h, \quad S_h = 120 + 0.5h, \quad S_H = 3{,}240 + 0.3h$$

where h is depth in feet and the stresses are in psi. Show by sketch, the best orientation of the shaft relative to the directions of h and H. Is this the best orientation at all depths to 3,000 ft?

3.11 A 3.66 m by 7.32 m rectangular shaft is sunk to a depth of 914 m in ground where the premining stress field is given by formulas

$$S_v = 27.2h, \quad S_h = 826 + 11.3h, \quad S_H = 22{,}345 + 6.8h$$

where h is depth in m and the stresses are in kPa. Show by sketch, the best orientation of the shaft relative to the directions of h and H. Is this the best orientation at all depths?

3.12 Thin overburden is scrapped to expose fresh bedrock at the site of a planned shaft. A 0-45-90 strain gauge rosette is bonded to the bedrock after a five inch square portion of the surface is ground smooth. The 0-gauge is oriented N30E. A 6-inch diameter coring bit is then used to relieve the *in situ* strains by overcoring the gauge. Final readings after overcoring are: $\varepsilon(0) = 1{,}000\ \mu$ in./in., $\varepsilon(45) = 500\ \mu$ in./in., $\varepsilon(90) = 3{,}000$ μin./in. where *the strain meter reads tension as positive.*

(a) If Young's modulus is 2.4 million psi and Poisson's ratio is 0.20, what is the state of strain and the state of stress at the gauge site relative to compass coordinates? Note: The *in situ* strains are opposite in sign to the relieved strains.

(b) What is the best orientation of a 10 × 20 ft rectangular shaft at this site?

(c) What unconfined compressive and tensile strengths (C_o, T_o) are needed to ensure safety factors of 2 and 4 in compression and tension, respectively?

3.13 With reference to Problem 3.12, if the state of stress at the surface continues to depth in addition to gravity load, and the specific weight of rock is 28 kN/m^3, what strengths are needed at a depth of 1,234 m for the same safety factors?

3.14 Thin overburden is scrapped to expose fresh bedrock at the site of a planned shaft. A0–45–90 strain gauge rosette is bonded to the bedrock after a 12.7 cm square portion of the surface is ground smooth. The 0-gauge is oriented N30E.A15.2 cm diameter coring bit is then used to relieve the *in situ* strains by overcoring the gauge. Final readings after overcoring are: $\varepsilon(0) = 1{,}000\ \mu$ in./in., $\varepsilon(45) = 500\ \mu$in./in., $\varepsilon(90) = 3{,}000\ \mu$ in./in. where *the strain meter reads tension as positive.*

(a) If Young's modulus is 16.55 GPa and Poisson's ratio is 0.20, what is the state of strain and the state of stress at the gauge site relative to compass coordinates? Note: The *in situ* strains are opposite in sign to the relieved strains.
(b) What is the best orientation of a 3 × 6 m rectangular shaft at this site?
(c) What unconfined compressive and tensile strengths (C_o, T_o) are needed to ensure safety factors of 2 and 4 in compression and tension, respectively?

3.15 With reference to Problem 3.14, if the state of stress at the surface continues to depth in addition to gravity load, and the specific weight of rock is 28 kN/m^3, what strengths are needed at a depth of 1,234 m for the same safety factors?

3.16 A rectangular shaft 10 ft by 20 ft with the long axis parallel to the N-S line exists at a depth of 950 ft. The mining plan calls for deepening the shaft to 1,800 ft. Estimate safety factors for the shaft wall at a depth of 1,750 ft.
Rock properties are:

$$C_o = 23{,}700\,\text{psi},\quad T_o = 1{,}480\ \text{psi},\quad E = 5.29\left(10^6\,\text{psi}\right),\quad \nu = 0.27,$$
$$\gamma = 162\,\text{pcf}$$

The premining stress state relative to compass coordinates is:

$$S_E = 350 + 0.2h,\quad S_N = 420 + 0.35h,\quad S_V = 1.12h$$

where stresses are in psi, h = depth in ft, E, N, V refer to compass coordinates (x = east, y = north, z = up) and compression is positive. Premining shear stresses are nil relative to compass coordinates.

3.17 If the premining stress state in Problem 3.16 were hydrostatic (3D), what shape would be more favorable, an ellipse, rectangle, or ovaloid (with semi-axes ratio of 2)?

3.18 A rectangular shaft 3 × 6 m with the long axis parallel to the N–S line exists at a depth of 290 m. The mining plan calls for deepening the shaft to 549 m. Estimate safety factors for the shaft wall at a depth of 533 m.
Rock properties are:

$$C_o = 164\,\text{MPa},\quad T_o = 10.2\ \text{MPa},\quad \varepsilon = 36.5\,\text{GPa},\quad \nu = 0.27,$$
$$\gamma = 26.6\,\text{kN/m}^3$$

The premining stress state relative to compass coordinates is:

$$S_E = 2{,}414 + 4.5h,\quad S_N = 2{,}897 + 7.9h,\quad S_V = 25.3h$$

where stresses are in kPa, h = depth in m, E, N, V refer to compass coordinates (x = east, y = north, z = up) and compression is positive. Premining shear stresses are nil relative to compass coordinates.

3.19 If the premining stress state in Problem 3.18 where hydrostatic (3D), what shape would be more favorable, an ellipse, rectangle, or ovaloid (with semi-axes ratio of 2)?

3.20 A rectangular shaft 18 ft by 24 ft is planned for a depth of 4,800 ft where the premining stresses relative to compass coordinates (x = east, y = north, z = up) are given by:

$$\sigma_{xx} = 250 + 0.5h, \quad \sigma_{yy} = 800 + 0.2h, \quad \sigma_{zz} = 1.1h,$$

where h is depth and compression is positive. Find the best shaft orientation (explain), then determine if shaft wall support is needed at some depth. Wall rock properties are: E = 6.2 million psi, v = 0.33, c (cohesion) = 5,600 psi, ϕ = 52°, γ = 158 pcf.

3.21 Considering the data and results from Problem 3.20, determine the shaft wall safety factors in tension and compression.

3.22 Given the stresses relative to compass coordinates (x = east, y = north, z = up),

$$\sigma_{xx} = 2{,}155, \quad \sigma_{yy} = 3{,}045, \quad \sigma_{zz} = 4{,}200, \quad \tau_{yx} = -1{,}222,$$

$$\tau_{xz} = 0, \quad \tau_{yz} = 0$$

in psi with compression positive, find the principal stresses σ_1, σ_2, σ_3 (magnitude and orientation). Show with a suitable sketch. Then consider an unlined shaft that will be 13 ft by 26 ft in section, chose the best opening shape (ellipse, ovaloidal, or rectangle) and best orientation from the rock mechanics perspective, assuming the given stress state is critical to the design. Sketch.

3.23 A rectangular shaft 6 × 8 m is planned for a depth of 1,500 m where the premining stresses relative to compass coordinates (x = east, y = north, z = up) are given by:

$$\sigma_{xx} = 1{,}724 + 11.3h, \quad \sigma_{yy} = 5{,}517 + 4.5h, \quad \sigma_{zz} = 2.49h,$$

where h is depth in meters and stress is in kPa; compression is positive. Find the best shaft orientation (explain), then determine if shaft wall support is needed at some depth. Wall rock properties are: E = 42.8 GPa, v = 0.33, c (cohesion) = 38.6 MPa, ϕ = 52°, γ = 25.0 kN/m^3.

3.24 Considering the data and results from Problem 3.23, determine the shaft wall safety factors in tension and compression.

3.25 Given the stresses relative to compass coordinates (x = east, y = north, z = up),

$$\sigma_{xx} = 14.9, \quad \sigma_{yy} = 21.0, \quad \sigma_{zz} = 29.0, \quad \tau_{yx} = -8.4, \quad \tau_{xz} = 0, \quad \tau_{yz} = 0$$

in MPa with compression positive, find the principal stresses σ_1, σ_2, σ_3 (magnitude and orientation). Show with a suitable sketch. Then consider an unlined shaft that will be 4 × 8 m in section, choose the best opening shape (ellipse, ovaloidal, or rectangle) and best orientation from the rock mechanics perspective, assuming the given stress state is critical to the design. Sketch.

3.26 A large vertical shaft is planned for an underground hard rock mine. Laboratory tests on core from exploration drilling show that

$$C_o = 21{,}500 \text{ psi}, \quad T_o = 1{,}530 \text{ psi}, \quad E = 6.25\left(10^6\right) \text{ psi},$$
$$G = 2.5\left(10^6\right) \text{ psi}, \quad \gamma = 144 \text{ pcf}$$

while premining stress measurements can be fit to the formulas

$$S_E = 200 + 0.3h, \quad S_N = 600 + 0.9h, \quad S_V = 1.1h$$

where E, N, V refer to compass coordinates (x = east, y = north, z = up) and compression is positive. Premining shear stresses are nil relative to compass coordinates. Find: (a) the most preferable shape of shaft (elliptical, rectangular, or ovaloidal) when the cross section is 14 × 28 ft and (b) the resulting factors of safety with respect to compression and tension at a depth of 2,800 ft.

3.27 A large vertical shaft is planned for an underground hard rock mine. Laboratory tests on core from exploration drilling show that

$$C_o = 148.3, \quad T_o = 10.6, \quad E = 43.10 \text{ GPa}, \quad G = 17.24 \text{ GPa},$$
$$\gamma = 22.78 \text{ kN/m}^3$$

while premining stress measurements can be fit to the formulas

$$S_E = 1{,}379 + 6.8h, \quad S_N = 4{,}138 + 20.4h, \quad S_V = 24.9h$$

where E, N, V refer to compass coordinates (x = east, y = north, z = up), stress is in kPa, depth h is in m, and compression is positive. Premining shear stresses are nil relative to compass coordinates. Find: (a) the most preferable shape of shaft (elliptical, rectangular, or ovaloidal) when the cross section is 4.5 × 9.0 m and (b) the resulting factors of safety with respect to compression and tension at a depth of 854 m.

3.28 Laboratory tests on core from exploration drilling show that

$$C_o = 21{,}500 \text{ psi}, \quad T_o = 1{,}530, \quad E = 6.25\left(10^6\right) \text{ psi},$$
$$G = 2.5\left(10^6\right) \text{ psi}, \quad \gamma = 144 \text{ pcf}$$

while premining stress measurements can be fit to the formulas

$$S_E = 200 + 0.3h, \quad S_N = 600 + 0.9h, \quad S_V = 1.1h$$

where E, N, V refer to compass coordinates (x = east, y = north, z = up), stress is in psi, depth h is in ft, and compression is positive. Premining shear stresses are nil relative to compass coordinates. Suppose a 12 × 24 ft rectangular shaft is selected and for some reason is oriented with the long axis N30W. Find the resulting factors of safety at 2,150 ft.

3.29 A rectangular shaft 12 ft by 24 ft is planned for deepening from a depth of 3,200 to 4,800 ft in ground where the preshaft stress state relative to compass coordinates (x = east, y = north, z = up) is given by

$$S_{xx} = 2{,}000 + 1.1d, \quad S_{yy} = 50 + 0.9d, \quad S_{zz} = 1.15d, \quad T_{xy} = -350 - 0.3d,$$
$$T_{yz} = 0.0, \quad T_{zx} = 0.0$$

where compression is positive and d is depth below surface.

(a) Find the best orientation of the shaft at a planned 3,850 shaft station and show by sketch.
(b) Also show on the sketch where the peak stress concentrations are likely to occur.
(c) Finally, estimate the peak stress concentrations.

3.30 Laboratory tests on core from exploration drilling show that

$$C_o = 148.3 \text{ MPa}, \quad T_o = 10.55 \text{ MPa}, \quad E = 43.10 \text{ GPa},$$
$$G = 17.24 \text{ GPa}, \quad \gamma = 22.78 \text{ kN/m}^3$$

while premining stress measurements can be fit to the formulas

$$S_E = 1{,}379 + 6.8h, \quad S_N = 4{,}138 + 20.4h, \quad S_V = 24.9h$$

where E, N, V refer to compass coordinates (x = east, y = north, z = up), stress is in kPa, depth h is in m, and compression is positive. Premining shear stresses are nil relative to compass coordinates. Suppose a 3.7 × 7.4 m rectangular shaft is selected and for some reason is oriented with the long axis N30W. Find the resulting factors of safety at 655 m.

3.31 A rectangular shaft 3.7 × 7.4 m is planned for deepening from a depth of 975 to 1,463 m in ground where the preshaft stress state relative to compass coordinates (x = east, y = north, z = up) is given (in kPa) by

$$S_{xx} = 13{,}793 + 24.9d, \quad S_{yy} = 345 + 20.4d, \quad S_{zz} = 26.0d,$$
$$T_{xy} = -2{,}414 - 6.8d, \quad T_{yz} = 0.0, \quad T_{zx} = 0.0$$

where compression is positive and d is depth in m below surface.

(a) Find the best orientation of the shaft at a planned 1,175 m deep shaft station and show by sketch.
(b) Also show on the sketch where the peak stress concentrations are likely to occur.
(c) Finally, estimate the peak stress concentrations.

3.32 A vertical circular shaft will pass through a water bearing stratum at a depth of 2,780 ft where the water pressure is estimated to be 210 psi. Finished shaft diameter must be 18 ft. The preshaft stress state is attributable to gravity alone. Rock properties are: C_o = 6,750 psi, T_o = 675 psi, E = 2.4 million psi, ν = 0.25, γ = 156 pcf. Estimate the unlined shaft wall safety factors.

3.33 A three-compartment rectangular shaft 12 × 24 ft in cross section is planned for a depth of 3,000 ft. The presinking stress field is assumed to be caused by gravity only.

The weakest rock along the proposed shaft route has an unconfined compressive strength of 8,000 psi and a tensile strength of 750 psi. Young's modulus and Poisson's ratio are estimated to be 4.5 million psi and 0.25. Specific weight of rock is 162 pcf. Determine the safety factors, with respect to tension and compression. What is the optimum orientation of the shaft?

3.34 The presinking stress field assumed to be caused by gravity alone (Problem 3.33) turns out to be wrong. The actual stress field has a tectonic component that adds a constant 1,250 psi to the east–west horizontal stress attributable to gravity and twice that amount to the north–south stress attributable to gravity. Orient the shaft in the optimum direction and then determine the unlined shaft wall safety factors.

3.35 A circular shaft 18 ft in diameter is decided upon rather than the proposed rectangular shaft. Determine the unlined shaft wall safety factors. Note: The stress field from Problem 3.34 applies.

3.36 The *in situ* stress field for Problems 3.34 and 3.35 changes between 3,000 and 3,500 ft to one described by the formulas

$$\sigma_{\text{v}} = 1.125h, \quad \sigma_{\text{H}} = 3{,}500 + 0.33h, \quad \sigma_{\text{h}} = 3{,}500 + 0.33h$$

where the stresses are in psi and the depth h is in ft. The stresses σ_{v}, σ_{H}, σ_{h} are principal stresses in the vertical and horizontal directions. Find the safety factor at 4,500 ft for an unlined, circular shaft wall. Assume the same rock properties.

3.37 A three-compartment rectangular shaft 3.7 × 7.4 m in cross section is planned for a depth of 914 m. The presinking stress field is assumed to be caused by gravity only. The weakest rock along the proposed shaft route has an unconfined compressive strength of 55.17 MPa and a tensile strength of 5.17 MPa. Young's modulus and Poisson's ratio are estimated to be 31.0 GPa and 0.25. Specific weight of rock is 25.6 kN/m^3. Determine the safety factors, with respect to tension and compression. What is the optimum orientation of the shaft?

3.38 The presinking stress field assumed to be caused by gravity alone (Problem 3.37) turns out to be wrong. The actual stress field has a tectonic component that adds a constant 8.62 MPa to the east–west horizontal stress attributable to gravity and twice that amount to the north–south stress attributable to gravity. Orient the shaft in the optimum direction and then determine the unlined shaft wall safety factors.

3.39 A circular shaft 5.5 m in diameter is decided upon rather than the proposed rectangular shaft. Determine the unlined shaft wall safety factors. Note: The stress field from Problem 3.38 applies.

3.40 The *in situ* stress field for Problems 3.38 and 3.39 changes between 914 and 1,067 m to one described by the formulas

$$\sigma_{\text{v}} = 25.45h, \quad \sigma_{\text{H}} = 24,138 + 7.47h, \quad \sigma_{\text{h}} = 24,138 + 7.47h$$

where the stresses are in kPa and the depth h is in m. The stresses σ_{v}, σ_{H}, σ_{h} are principal stresses in the vertical and horizontal directions. Find the safety factor at 1,372 m for an unlined, circular shaft wall. Assume the same rock properties.

3.41 A circular shaft liner is sunk to a depth of 3,750 ft (1,143 m). If the premining stress field is caused by gravity alone, what unconfined compressive rock strength in psi (MPa) is needed for a rock factor of safety of 3.0?

Supported shafts, liners, bolts, rings

3.42 A circular vertical shaft is planned to have a finished, inside diameter of 19 ft in an underground hard rock mine. Rock properties are:

$$C_o = 23{,}700 \text{ psi}, \quad T_o = 1{,}480 \text{ psi}, \quad E = 5.29(10^6 \text{ psi}),$$
$$\nu = 0.27, \quad \gamma = 162 \text{ pcf}$$

The premining stress state relative to compass coordinates is:

$$S_E = 350 + 0.2h, \quad S_N = 420 + 0.35h, \quad S_V = 1.12h$$

where stresses are in psi, h = depth in ft, E, N, V refer to compass coordinates (x = east, y = north, z = up) and compression is positive. Premining shear stresses are nil relative to compass coordinates.
Properties of concrete are:
Compressive strength = 5,740 psi, tensile strength = 425 psi, Young's modulus = 4.75 (10^6) psi, Poisson's ratio = 0.25.
Find:

(a) the unlined shaft wall safety factors in tension and compression at a depth of 4,250 ft,
(b) concrete liner thickness in the vicinity of a water-bearing formation where the pressure in the undisturbed ground is 90 psi. Note: The maximum allowable compression in the concrete is 3,500 psi.

3.43 With reference to the previous problem data, if changes in the inside diameter of the lined shaft are monitored for safety, what "reading" in inches would indicate impending failure of the liner?

3.44 A circular vertical shaft is planned to have a finished, inside diameter of 5.8 m in an underground hard rock mine. Rock properties are:

$$C_o = 163.5 \text{ MPa}, \quad T_o = 10.21 \text{ MPa}, \quad E = 36.48 \text{ Gpa},$$
$$\nu = 0.27, \quad \gamma = 25.63 \text{ kN/m}^3$$

The premining stress state relative to compass coordinates is:

$$S_E = 2{,}414 + 4.53h, \quad S_N = 2{,}897 + 7.92h, \quad S_V = 25.34h$$

where stresses are in kPa, h = depth in m, E, N, V refer to compass coordinates (x = east, y = north, z = up) and compression is positive. Premining shear stresses are nil relative to compass coordinates.
Properties of concrete are:
Compressive strength = 39.6 MPa, tensile strength = 2.93 MPa, Young's modulus = 32.76 GPa, Poisson's ratio = 0.25.
Find:

(a) the unlined shaft wall safety factors in tension and compression at a depth of 1,295 m,

(b) concrete liner thickness in the vicinity of a water-bearing formation where the pressure in the undisturbed ground is 0.621 MPa. Note: The maximum allowable compression in the concrete is 24.14 MPa.

3.45 With reference to the previous problem data, if changes in the inside diameter of the lined shaft are monitored for safety, what "reading" in centimeters would indicate impending failure of the liner?

3.46 A circular vertical shaft is planned to have a finished, inside diameter of 26 ft in an underground hard rock mine. Rock properties are:

$$C_o = 27{,}400 \text{ psi}, \quad T_o = 1{,}840 \text{ psi}, \quad E = 6.19\left(10^6 \text{ psi}\right),$$
$$\nu = 0.22, \quad \gamma = 159 \text{ pcf}$$

The premining stress state relative to compass coordinates is:

$$S_E = 600 + 0.3h, \quad S_N = 200 + 0.4h, \quad S_V = 1.12h$$

where E, N, V refer to compass coordinates (x = east, y = north, z = up), h is depth in ft, stress is in psi, and compression is positive. Premining shear stresses are nil relative to compass coordinates. Properties of concrete are: Compressive strength = 5,500 psi, tensile strength = 550 psi, Young's modulus = 5.5(10^6) psi, Poisson's ratio = 0.25. Find: The concrete liner thickness in the vicinity of a water-bearing cavernous dolomite where the pressure in the undisturbed ground is 210 psi. Note: The maximum allowable compression in the concrete is 3,500 psi.

3.47 With reference to the previous problem data, if changes in the *inside* diameter of the lined shaft are monitored for safety, what "reading" in inches would indicate impending failure of the liner?

3.48 A circular vertical shaft is planned to have a finished, inside diameter of 8 m in an underground hard rock mine. Rock properties are:

$$C_o = 189.0 \text{ MPa}, \quad T_o = 12.69 \text{ MPa}, \quad E = 42.69 \text{ GPa},$$
$$\nu = 0.22, \quad \gamma = 25.15 \text{ kN/m}^3$$

The premining stress state relative to compass coordinates is:

$$S_E = 4{,}138 + 6.79h, \quad S_N = 1{,}379 + 9.05h, \quad S_V = 25.34h$$

where E, N, V refer to compass coordinates (x = east, y = north, z = up), h is depth in m, stress is in kPa and compression is positive. Premining shear stresses are nil relative to compass coordinates. Properties of concrete are: compressive strength = 37.93 MPa, tensile strength = 3.79 MPa, Young's modulus = 37.93 GPa, Poisson's ratio = 0.25. Find: the concrete liner thickness in the vicinity of a water-bearing cavernous dolomite where the pressure in the undisturbed ground is 1.45 MPa. Note: the maximum allowable compression in the concrete is 24.14 MPa.

3.49 With reference to the previous problem data, if changes in the *inside* diameter of the lined shaft are monitored for safety, what "reading" in cm would indicate impending failure of the liner?

3.50 A vertical unlined 22 ft diameter circular shaft is fitted with a concrete shaft liner to withstand water pressure of 123 psi in a massive sandstone aquifer at a depth of 2,890 ft. Liner properties are: $E = 3.5$ million psi, $v = 0.30$, $C_o = 4{,}500$ psi, $T_o = 450$ psi, $\gamma = 152$ pcf. A liner safety factor of 3.85 is required in compression. Find the liner thickness (in.) and finished shaft diameter (ft).

3.51 Show why the weight of a concrete shaft liner that is poured to the walls of the unlined shaft is not important to liner stress.

3.52 Alarge vertical finished (inside) shaft diameter of 32 ft is required for hoisting capacity in a planned high volume underground oil shale mine. An aquifer is encountered at a depth of 1,270 ft where the water pressure is 240 psi.

(a) Determine the thickness of a concrete liner needed for a liner safety factor of 2.5 when the concrete compressive strength (C_o) is 3,500 psi, Young's modulus is $5.6(10^6)$ psi, Poisson's ratio is 0.27. Explain, defend your answer.

(b) If changes in the inside diameter of the shaft are monitored for safety, what "reading" in inches would indicate impending failure of the liner?

3.53 Avertical circular shaft will pass through a water-bearing stratum at a depth of 2,780 ft where the water pressure is estimated to be 210 psi. Finished shaft diameter must be 18 ft. The preshaft stress state is attributable to gravity alone. Rock properties are: $C_o = 6{,}750$ psi, $T_o = 675$ psi, $E = 2.4$ million psi, $v = 0.25$, $\gamma = 156$ pcf. What concrete liner thickness is indicated, assuming the maximum allowable concrete liner stress is 3,500 psi?

3.54 Alarge vertical finished (inside) shaft diameter of 9.75m is required for hoisting capacity in a planned high volume underground oil shale mine. An aquifer is encountered at a depth of 387 m where the water pressure is 1.66 MPa.

(a) Determine the thickness of a concrete liner needed for a liner safety factor of 2.5 when the concrete compressive strength (C_o) is 24.2 MPa, Young's modulus is 38.62 GPa, Poisson's ratio is 0.27. Explain, defend your answer.

(b) If changes in the inside diameter of the shaft are monitored for safety, what "reading" in cm would indicate impending failure of the liner?

3.55 A vertical circular shaft will pass through a water-bearing stratum at a depth of 847 m where the water pressure is estimated to be 1.45 MPa. Finished shaft diameter must be 5.5 m. The preshaft stress state is attributable to gravity alone. Rock properties are: $C_o = 46.6$ MPa, $T_o = 4.66$ MPa, $E = 16.6$ GPa, $v = 0.25$, $\gamma = 24.7$ kN/m^3. What concrete liner thickness is indicated, assuming the maximum allowable concrete liner stress is 24.14 MPa?

3.56 An *in situ* stress field between 3,000 ft and 3,500 ft is fit to the formulas

$$\sigma_v = 1.125h, \quad \sigma_H = 3{,}500 + 0.33h, \quad \sigma_h = 3{,}500 + 0.33h$$

where the stresses are in psi and the depth h is in ft. Stresses σ_v, σ_H, σ_h are principal stresses in the vertical and horizontal directions. The weakest rock along the proposed shaft route has an unconfined compressive strength of 8,000 psi and a tensile strength of 750 psi. A circular, concrete liner is planned below 3,500 ft for a shaft with an outside diameter of 18 ft. If the liner is one foot thick and the concrete properties are: $E = 5.0$ million psi, $v = 0.20$, $\gamma = 156$ pcf, $C_o = 4{,}500$ psi and $T_o = 450$ psi, determine what uniform radial stress would just cause the liner to fail, assuming the concrete follows a Mohr–Coulomb criterion.

3.57 With reference to the previous problem, determine the reduction in diameter of the liner when the liner first fails.

3.58 With reference to Problem 3.56, determine the radial displacement of the interface between the liner and shaft wall when the liner first fails.

3.59 With reference to Problem 3.56 data, water pressure of 80 psi is anticipated at 4,500 ft. What liner thickness is indicated, if a liner safety factor of 2.5 is required. Note: Minimum thickness is 1 ft.

3.60 If a steel liner is used in Problem 3.59 instead of concrete and the steel strength is 36,000 psi (compressive and tensile strengths are equal), what is the corresponding steel liner thickness?

3.61 An *in situ* stress field between 914 and 1,067 m is fit to the formulas

$$\sigma_v = 25.45h, \quad \sigma_H = 24{,}138 + 7.47h, \quad \sigma_h = 24{,}138 + 7.47h$$

where the stresses are in kPa and the depth h is in m. Stresses σ_v, σ_H, σ_h are principal stresses in the vertical and horizontal directions. The weakest rock along the proposed shaft route has an unconfined compressive strength of 55.2 MPa and a tensile strength of 5.17 MPa. A circular, concrete liner is planned below 1,067 m for a shaft with an outside diameter of 5.5 m. If the liner is 0.3 m thick and the concrete properties are: E = 34.48 GPa, $\nu = 0.20$, $\gamma = 24.7$ kN/m^3, $C_o = 31.0$ MPa, and $T_o = 3.10$ MPa, determine what uniform radial stress would just cause the liner to fail, assuming the concrete follows a Mohr–Coulomb criterion.

3.62 With reference to the previous problem, determine the reduction in diameter of the liner when the liner first fails.

3.63 With reference to Problem 3.61, determine the radial displacement of the interface between the liner and shaft wall when the liner first fails.

3.64 With reference to Problem 3.61 data, water pressure of 0.552 MPa is anticipated at 1,372 m. What liner thickness is indicated, if a liner safety factor of 2.5 is required. Note: Minimum thickness is 0.3 m.

3.65 If a steel liner is used in Problem 3.64 instead of concrete and the steel strength is 248 MPa (compressive and tensile strengths are equal), what is the corresponding steel liner thickness?

3.66 A circular concrete shaft liner with a Young's modulus of 3.4 million psi, Poisson's ratio of 0.25, unconfined compressive strength 3,500 psi and tensile strength 350 psi is considered for control of water pressure (190 psi) at a depth of 3,750 ft. Inside shaft diameter after lining must be 18 ft. Find the liner thickness necessary to achieve a safety factor of 2.5. What outside shaft diameter is indicated?

3.67 A circular concrete shaft liner with Young's modulus of 3.4 million psi, Poisson's ratio of 0.25, unconfined compressive strength 3,500 psi and tensile strength 350 psi is loaded to the verge of failure in dry ground. If the inside shaft diameter is 22 ft and the liner is one foot thick, what is the contact pressure between rock and liner, assuming it is uniform (psi)?

3.68 With reference to Problem 3.67, measurements are made between points on the inside of the liner on opposite ends of a diametral line. What change in diameter from the no-load condition is indicated at the verge of liner failure (inches)?

3.69 A circular concrete shaft liner with a Young's modulus of 23.45 GPa, Poisson's ratio of 0.25, unconfined compressive strength 24.1 MPa, and tensile strength 2.41 MPa is considered for control of water pressure (1.31 MPa) at a depth of 1,143 m. Inside shaft diameter after lining must be 5.5 m. Find the liner thickness necessary to achieve a safety factor of 2.5. What outside shaft diameter is indicated?

3.70 A circular concrete shaft liner with Young's modulus of 23.45 GPa, Poisson's ratio of 0.25, unconfined compressive strength 24.1 MPa, and tensile strength 2.41 MPa is loaded to the verge of failure in dry ground. If the inside shaft diameter is 6.7 m and the liner is 0.3 m thick, what is the contact pressure between rock and liner, assuming it is uniform (in kPa)?

3.71 With reference to Problem 3.70, measurements are made between points on the inside of the liner on opposite ends of a diametral line. What change in diameter from the no-load condition is indicated at the verge of liner failure (in cm)?

Multiple shafts

3.72 A 12 ft by 24 ft rectangular shaft is sunk to a depth of 3,000 ft in ground where the premining stress field is given by formulas

$$S_v = 1.2h, \quad S_h = 120 + 0.5h, \quad S_H = 3{,}240 + 0.3h$$

where h is depth in feet and the stresses are in psi. Show by sketch, the best orientation of the shaft relative to the directions of h and H. Is this the best orientation at all depths to 3,000 ft?

If mine expansion plans call for two additional shafts of the same size, show by sketch the position and orientation of these additional shafts. Explain.

3.73 A 3.7 × 7.4 m rectangular shaft is sunk to a depth of 914 m in ground where the premining stress field is given by formulas

$$S_v = 27.2h, \quad S_h = 282 + 11.3h, \quad S_H = 22{,}345 + 6.8h$$

where h is depth in m and the stresses are in kPa. Show by sketch, the best orientation of the shaft relative to the directions of h and H. Is this the best orientation at all depths to 914 m?

If mine expansion plans call for two additional shafts of the same size, show by sketch the position and orientation of these additional shafts. Explain.

3.74 A rectangular shaft 10 ft by 20 ft with the long axis parallel to the N–S line exists at a depth of 950 ft. The mining plan calls for deepening the shaft to 1,800 ft. The premining stress state relative to compass coordinates is:

$$S_E = 350 + 0.2h, \quad S_N = 420 + 0.35h, \quad S_V = 1.12h$$

where stresses are in psi, h = depth in ft, E, N, V refer to compass coordinates (x = east, y = north, z = up) and compression is positive. Premining shear stresses are nil relative to compass coordinates. Rock properties are:

$$C_o = 23{,}700 \text{ psi}, \quad T_o = 1{,}480 \text{ psi}, \quad E = 5.29(10^6 \text{ psi}),$$

$$\nu = 0.27, \quad \gamma = 162 \text{ pcf}$$

A second shaft of identical shape is also planned to the new depth of 1,800 ft. Show a location and orientation (plan view) of the two shafts that is most favorable with respect to stress concentration and explain the basis of your choice.

3.75 Given the stresses relative to compass coordinates (x = east, y = north, z = up),

$$\sigma_{xx} = 2{,}155, \quad \sigma_{yy} = 3{,}045, \quad \sigma_{zz} = 4{,}200, \quad \tau_{yx} = -1{,}222,$$
$$\tau_{xz} = 0, \quad \tau_{yx} = 0$$

in psi with compression positive, find the principal stresses $\sigma_1, \sigma_2, \sigma_3$ (magnitude and orientation). Show with a suitable sketch. Then consider an unlined shaft that will be 13 ft by 26 ft in section and (a) choose the best opening shape (ellipse, ovaloidal, or rectangle) and best orientation from the rock mechanics perspective, assuming the given stress state is critical to the design, (b) suppose an identical second shaft is planned near the first. Show the orientation and location of the second shaft relative to the first such that interactions are likely to be acceptable and not unduly high. Sketch. Justify your answer.

3.76 A rectangular shaft 3 m × 6 m with the long axis parallel to the N–S line exists at a depth of 290 m. The mining plan calls for deepening the shaft to 550 m. The premining stress state relative to compass coordinates is:

$$S_E = 3{,}414 + 4.5h, \quad S_N = 2{,}897 + 7.9h, \quad S_V = 25.3h$$

where stresses are in kPa, h = depth in m, E, N, V refer to compass coordinates (x = east, y = north, z = up) and compression is positive. Premining shear stresses are nil relative to compass coordinates. Rock properties are:

$$C_o = 163.5 \text{ MPa}, \quad T_o = 10.2 \text{ MPa}, \quad E = 36.48 \text{ GPa},$$
$$\nu = 0.27, \quad \gamma = 25.6 \text{ kN/m}^3$$

A second shaft of identical shape is also planned to the new depth of 550 m. Show a location and orientation (plan view) of the two shafts that is most favorable with respect to stress concentration and explain the basis of your choice.

3.77 Given the stresses relative to compass coordinates (x = east, y = north, z = up),

$$\sigma_{xx} = 14.86, \quad \sigma_{yy} = 21.00, \quad \sigma_{zz} = 29.00, \quad \tau_{yx} = -8.43, \quad \tau_{xz} = 0, \quad \tau_{yx} = 0$$

in MPa with compression positive, find the principal stresses $\sigma_1, \sigma_2, \sigma_3$ (magnitude and orientation). Show with a suitable sketch. Then consider an unlined shaft that will be 4 m × 8 m in section and (a) choose the best opening shape (ellipse, ovaloidal, or rectangle) and best orientation from the rock mechanics perspective, assuming the given stress state is critical to the design, (b) suppose an identical second shaft is planned near the first. Show the orientation and location of the second shaft relative to the first such that interactions are likely to be acceptable and not unduly high. Sketch. Justify your answer.

Additional problems

3.78 Consider a circular shaft excavated in a stress field caused by gravity only. Inclination of the shaft axis with respect to the vertical varies. Suppose the shaft axis inclination dips δ'degrees from the vertical. Evaluate algebraically the maximum and minimum shaft wall stress concentration factors at the extreme inclinations $\delta' = 0$ and $\delta' = 90$ degrees.

3.79 Consider a circular shaft excavated in a stress field caused by gravity only. Inclination of the shaft axis with respect to the vertical varies. Use the development in Example 3.36 to obtain plots of the maximum and minimum stress concentration factors with respect to the circumferential stress as functions of shaft inclination. Note: dip δ of the shaft cross-section is 90°-shaft inclination. Assume $K_o = 1/3$ and $S_v = 7.5$ MPa for plotting.

3.80 Consider a circular shaft excavated in a stress field caused by gravity only. Inclination of the shaft axis with respect to the vertical varies. Note: dip δ of the shaft cross-section is 90°-shaft inclination. Assume $K_o = 1/3$ and $S_v = 7.5$ MPa and then evaluate the safety of the shaft wall with respect to the axial stress when the dip of the shaft axis is 60°. Assume $C_o = 10$ MPa and $T_o = 1$ MPa. Are the factors of safety greater or less than in the vertical case?

Note: In the vertical case there are three factors of safety, one with respect to circumferential compression (FS_c), one with respect to tensile failure under circumferential stress, (FS_t), when tension is present, and the other with respect to vertical compression (FS_v).

3.81 Consider a circular shaft excavated in a stress field caused by gravity only. Inclination of the shaft axis with respect to the vertical varies. Note: dip δ of the shaft cross-section is 90°-shaft inclination. Assume $K_o = 1/3$ and $S_v = 7.5$ MPa and then evaluate the safety of the shaft wall with respect to the circumferential stress when the dip of the shaft axis is 60°. Assume $C_o = 10$ MPa and $T_o = 1$ MPa. Are the factors of safety greater or less than in the vertical case?

Note: In the vertical case there are three factors of safety, one with respect to circumferential compression (FS_c), one with respect to tensile failure under circumferential stress, (FS_t), when tension is present, and the other with respect to vertical compression (FS_v).

3.82 Consider an element of shaft wall in the inclined case where a circumferential normal stress and shear stress act as shown in the sketch. Inclination of the shaft axis with respect to the vertical varies. Evaluate the safety of the shaft wall at the low point of the shaft section with respect to a Mohr-Coulomb failure criterion when the dip of the shaft axis varies from the vertical to the horizontal and plot. Assume: $K_o = 1/3$, $S_v =$ 7.5 MPa, and $\nu = 0.25$ and further suppose $C_o = 10$ MPa and $T_o = 1$ MPa.

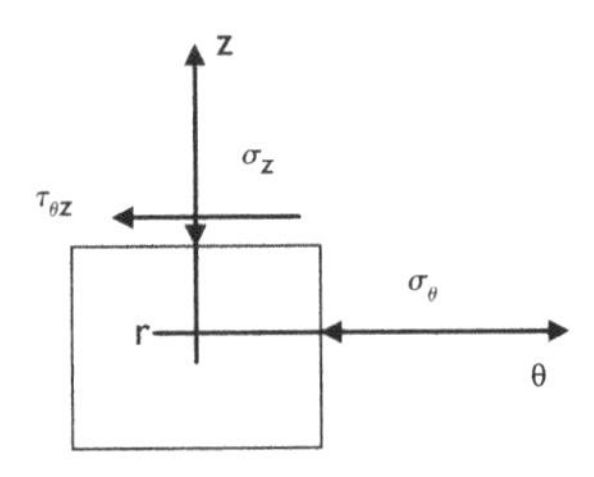

Sketch for Problem 3.82 looking down the radial direction at a shaft wall point.

Chapter 4

Tunnels

Tunnels here include mainline haulage ways in mines and other important, semi-permanent underground passageways as well as conventional tunnels with surface portals at both ends. Like shafts, "tunnels" may be used for transporting personnel, underground supplies, waste rock and for haulage of ore in mines. Like shafts, tunnels also serve as conduits for ventilation air and as pathways for compressed air and water lines. Because of their long service life, tunnels must be carefully designed and constructed with an adequate factor of safety. Tunnel support may be natural with only occasional rock bolting and screening. Support may also be in the form of a permanent, continuous concrete liner, discrete steel sets or both. Use of large bolts placed on an engineered pattern may also be used for robust tunnel support. In the case of naturally supported tunnels, strength failure of the rock mass walls can be designed against using a stress concentration approach, with due consideration of joints. Indeed, in shallow ground tunnels, fall of rock blocks defined by intersecting joints is often the primary threat to safety and stability. In case of parallel or multiple tunnels, pillars between tunnels may be designed using an average stress approach, again with due consideration of joints.

Where additional support or reinforcement is needed beyond that provided by the rock mass proper, difficulties arise that are absent in shaft support analyses. Lack of symmetry and circular sections in tunneling combine to make tunnel design more difficult. Although circular sections are sometimes used in tunneling, the preexcavation principal stresses are not usually equal. Thus, the assumption of a radial "pressure" or support load is not generally justified, although in some special cases, for example, in "squeezing" ground, a radial pressure may be reasonable. In this regard, a reasonable estimate of the support load is a key ingredient to support design.

4.1 Naturally supported tunnels

Estimation of stress concentration for tunnels is similar to that for shafts. Thus, major features of the problem include

1 cross-section shape
2 aspect ratio
3 pretunnel stress field
4 orientation of the tunnel axis.

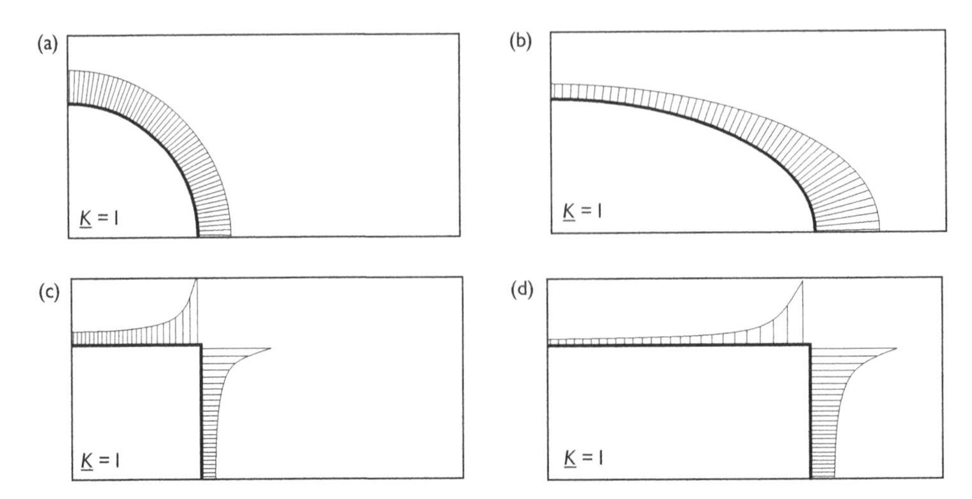

Figure 4.1 Stress distributions about (a) circular section, (b) elliptical section, aspect ratio = 2, (c) square section, and (d) rectangular section, aspect ratio = 2 excavated in a hydrostatic stress field ($S_x = S_y = 1$).

Nearly the entire discussion of stress concentration about single, naturally supported shafts applies to tunnels. The reason is that the mathematics of stress analysis does not distinguish between shafts and tunnels. For example, stress distribution about an elliptical tunnel or an elliptical shaft excavated in a hydrostatic stress field is the same and, as a consequence, so are stress concentration factors. The same is true for circular and rectangular sections. Figure 4.1 shows a comparison of stress distributions about circular, square, elliptical, and rectangular sections excavated in a hydrostatic stress field. The plots in Figure 4.1 indicate magnitude of the tangential stress acting parallel to the tunnel wall by length of a radial line extending from the tunnel outline; tension is plotted inside the tunnel and is considered negative, while compression is plotted outside (positive). No tensile stress appears in any of the distributions in Figure 4.1. The horizontal bar under the note $K = 1$ indicates the length of line for a stress concentration of one. Clearly, a circular section is preferable to a square in an hydrostatic stress field, and an ellipse is preferable to a rectangle of the same aspect ratio ($a/b = 2$) in an hydrostatic stress field when compared on a basis of peak stress (compression).

From a stress analysis view, any distinction between shafts and tunnels is simply one of interpretation. Of course, an elliptical tunnel section would be even more difficult to excavate than an elliptical shaft. Indeed, elliptical shafts are rare and elliptical tunnels are almost unknown. Circular and rectangular tunnel sections are usually more practical. Indeed, tunnel boring machines (TBM's) often cut circular sections. Some boring machines used in mining result in rectangular sections with rounded corners that approach an ovaloid section, while some mechanical excavators (e.g. "continuous miners" used in coal mining) result in rectangular sections with relatively sharp corners. Mechanical excavation is preferred because of low cost associated with a rapid advance rate, often several hundred feet per day in relatively soft ground that requires little temporary support. Because the machine dictates tunnel shape and often aspect ratio, orientation with respect to the preexcavation stress field is the only design parameter available for minimizing stress concentration. When the tunnel route is determined by other overall design considerations, then the orientation is also set.

Single tunnels

Although tunnel shape, size, and orientation may be fixed by overall design and construction considerations and thus there is no opportunity to reduce stress concentration, estimation of stress concentration factors in compression and tension (K_c, K_t) is still important as are determinations of rock mass strengths (C_o, T_o) and preexcavation principal stresses along the tunnel route. Rock mass strength is influenced by joints, of course, and thus depends on the strength of intact rock between joints, joint persistence, and orientation as well as joint strength. Both stress and strength are essential to calculating tunnel wall safety factors (F_c, F_t). Relatively high safety factors, say 4 in compression and 8 in tension, indicate a high degree of stability and little need for an engineered support system. Low safety factors, near 1, indicate the opposite and that some permanent support will be required.

In hardrock, drilling and blasting is necessary for excavation. An advantage in this case is the capability of tailoring the blast design to result in a prescribed tunnel shape. Perhaps the most common shape is a rectangular section with a semi-circular top, as shown in Figure 4.2a. (The bottom in Figure 4.2a is a "semi-square.") The top is also known as the "back" in hardrock tunneling and mining terms (in softrock mining, the "roof"); sidewalls are "ribs," while the bottom is "bottom" (in softrock mining, the "floor"). Drilling and blasting occurs at the "face." The semi-circular top in Figure 4.2a is an arched back and serves the purpose of reducing stress concentration and thus improving safety. This back has a radius equal to one-half the width of the tunnel. However, an arched back need not be a semicircle. A longer radius is possible as are other arch shapes than the circle.

The shape in Figure 4.2a is symmetric about a vertical axis, but there is no symmetry about any horizontal axis as there is for circular, elliptical, and rectangular sections. For this reason, no simple analytical formulas are available for estimating stress concentration factors for an arched back tunnel. Numerical methods serve the purpose instead.

Figures 4.2b,c, and d show distributions of stress about the periphery of an arched back tunnel section with a width (W) to height (H) ratio of one. The radius of the semi-circular back is $W/2$; the height of the rectangular portion is also $W/2$. Figure 4.2b shows the stress distribution that results when the preexcavation stress field is uniaxial and vertical ($S_y = 1$, $S_x = 0$); Figure 4.2c shows the result when the preexcavation stress field is uniaxial and horizontal ($S_y = 0$, $S_x = 1$), and Figure 4.2d shows the hydrostatic case ($S_y = S_x = 1$). In all cases, the tunnel axis is assumed to be parallel to a principal stress direction. These results are based on the assumption of elastic behavior and do not depend on the construction sequence.

These results show that under *vertical* load *tension* nearly equal in magnitude to the applied compression appears at the center of the floor or bottom and at the crown, while high *compressive* stresses appear at the bottom corners and along the shoulders at the top of the ribs (vertical wall sections). In theory, a mathematically sharp, 90° corner, would result in an infinite compressive stress concentration at the corner points. In numerical analysis, the result is simply a high but not infinite stress concentration. A rounded corner done deliberately or left to nature would reduce an extraordinarily high stress concentration. When left to

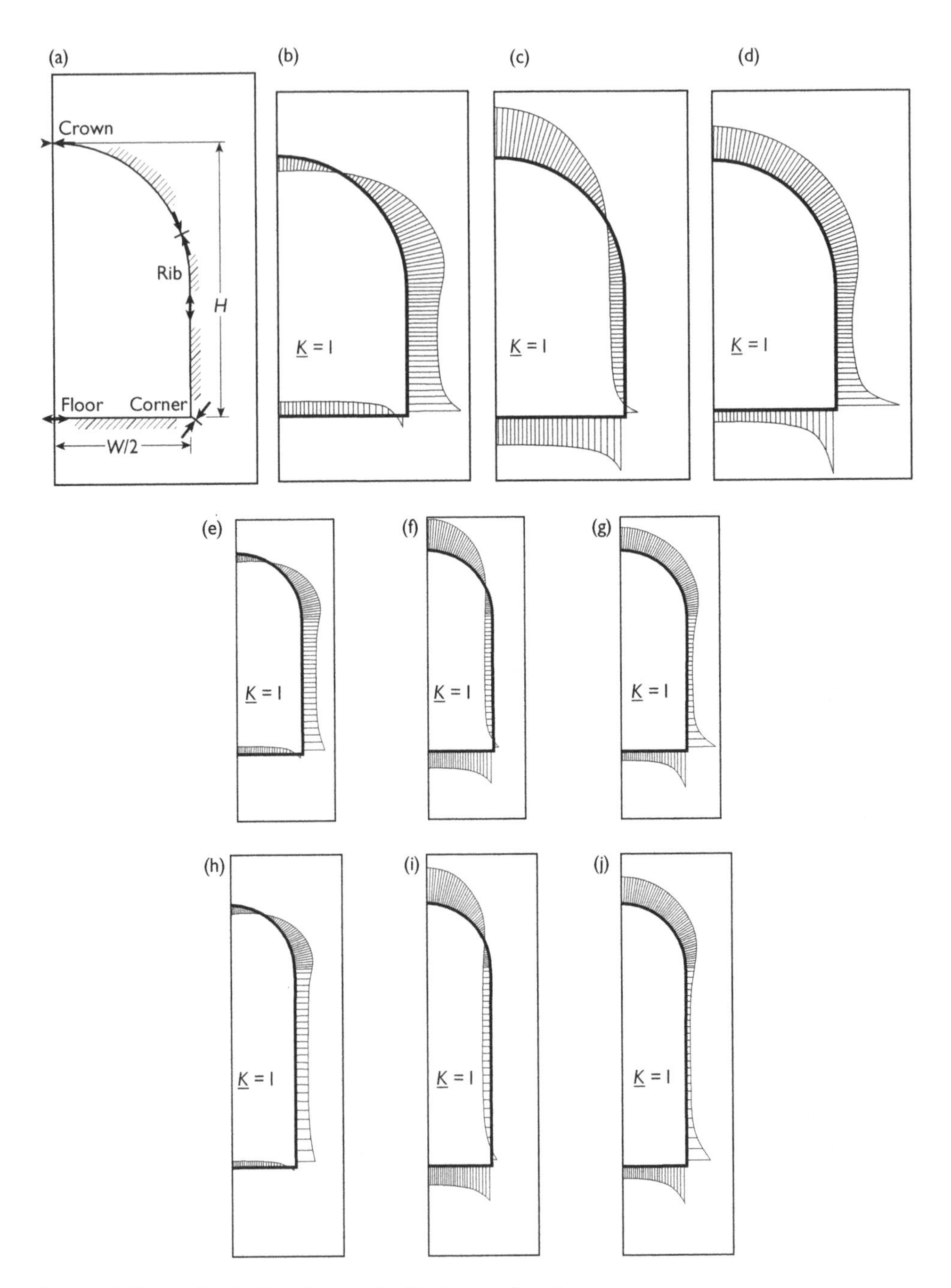

Figure 4.2 Stress distribution about arched back tunnel sections:
(a) section notation, (b) $H/W = 1$, $S_y = 1$, $S_x = 0$,
(c) $H/W = 1$, $S_y = 0$, $S_x = 1$, (d) $H/W = 1$, $S_y = 1$, $S_x = 1$,
(e) $H/W = 1.5$, $S_y = 1$, $S_x = 0$, (f) $H/W = 1.5$, $S_y = 0$, $S_x = 1$,
(g) $H/W = 1.5$, $S_y = 1$, $S_x = 1$, (h) $H/W = 2$, $S_y = 1$, $S_x = 0$,
(i) $H/W = 2$, $S_y = 0$, $S_x = 1$, (j) $H/W = 2$, $S_y = 1$, $S_x = 1$.

nature, reduction in the stress concentration occurs through yielding of the rock mass in the vicinity of the corner point. Yielding at a point elevates stress concentration at neighboring points. Thus, peak stress is lowered, while the average stress is raised in the vicinity of a point stressed beyond the elastic limit. As long as yielding is localized, the threat to tunnel stability is negligible. The distributions shown in Figure 4.2 suggest that this may be the case; the region of high corner compressive stress does not extend far up the ribs. If the stress concentration at the shoulder or top of the rib is tolerable, then the corner stress is also likely to be tolerable.

Under a uniaxial but *horizontal* compression ($S_y = 0, S_x = 1$); the floor and crown are under high *compressive* stress, while the rib is under tension equal in magnitude to the applied compression. When the applied vertical and horizontal stresses are equal (*hydrostatic* pre-excavation stress state), the results show no tension and reduced compressive stress peaks relative to the uniaxial cases.

The shape of an excavation usually changes during construction and consequently so does the distribution of stress. Figures 4.2e,f,g show the same three cases for a higher tunnel that has a height to width ratio of 1.5; Figures 4.2h,i,j show results for an even higher tunnel with a height to width ratio of 2. Under *vertical load* only, the peak tension in the floor and crown changes very little from the first case where the height to width ratio was 1 and the peak tension was nearly equal in magnitude to the applied compression. The high compressive corner stress declines rapidly with increasing tunnel height at fixed width, while the high compressive rib stress declines rather slowly.

These trends are summarized in Figure 4.3 that shows stress concentrations as a function of height to width ratio (H/W). Under *vertical* load, peak tensions in the floor and crown and rib compression are largely unaffected by height to width ratio. Corner compression decreases nonlinearly with height to width ratio (but increases linearly with the reciprocal width to height ratio). Under *horizontal* load, peak tension in the rib remains largely unaffected, while compressive stresses at the floor, corner, and crown increase almost linearly with height to width ratio, as seen in Figure 4.3b. Compressive stress concentration trends with height to width ratio in the *hydrostatic* case, are also linear in height to width ratio, as seen in Figure 4.3c. Tension is absent in all hydrostatic cases.

Trend lines fitted to data in Figure 4.3 show close fits to linearity and for this reason may be extrapolated to somewhat higher height to width ratios than the 2.5 ratio case shown. However, backward extrapolation to ratios less than 1.0 would be ill-advised, and for this reason, the trend lines are not plotted to smaller height to width ratios. In retrospect, numerical values in Figure 4.3 could have been estimated to a degree from consideration of stress concentration about rectangular and ovaloidal openings because of the hybrid shape of the arched back tunnel cases examined.

Superposition of uniaxial cases may be used to obtain stress concentrations associated with other load cases. Care must be exercised to insure that the same location and direction are of stress are being considered.

Example 4.1 An arched back tunnel is planned with a height to width ratio of two. Estimate stress concentration factors in the floor and crown when the preexcavation principal stress ratio $M = 1/4$ and (a) the vertical stress is S_1, (b) the horizontal stress is S_1.

***Solution*:** The stress concentrations are combinations of vertical and horizontal loads and may be estimated from trend lines in Figures 4.3a,b using superposition.

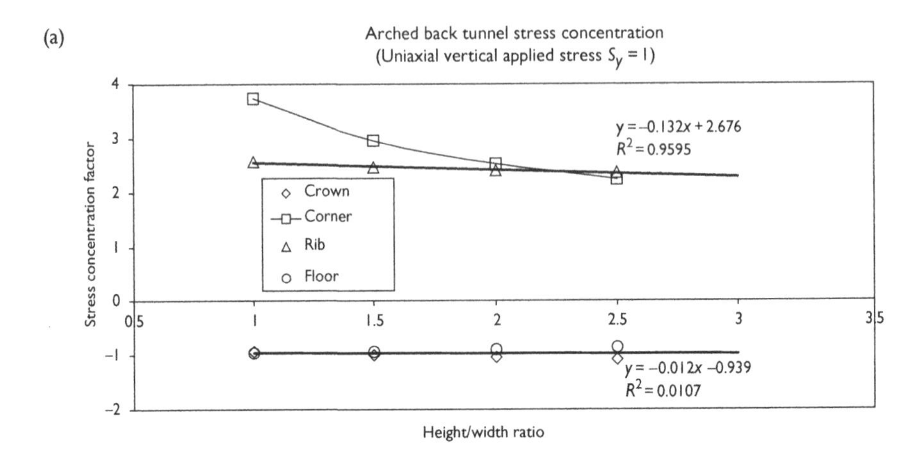

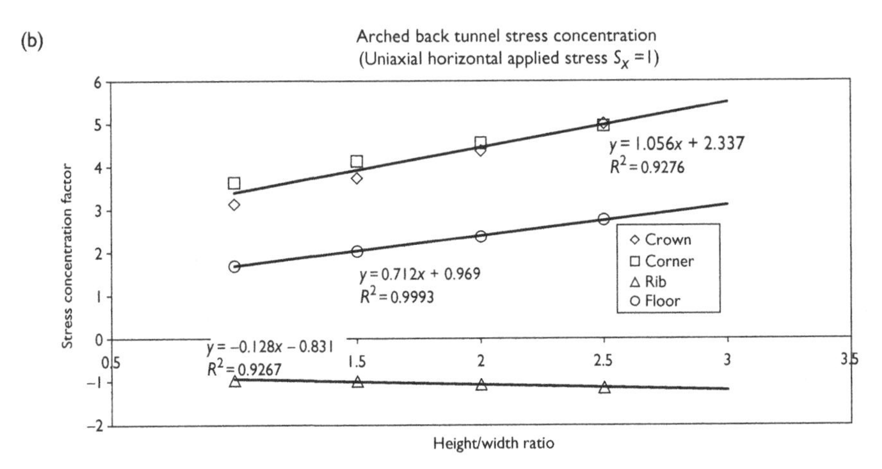

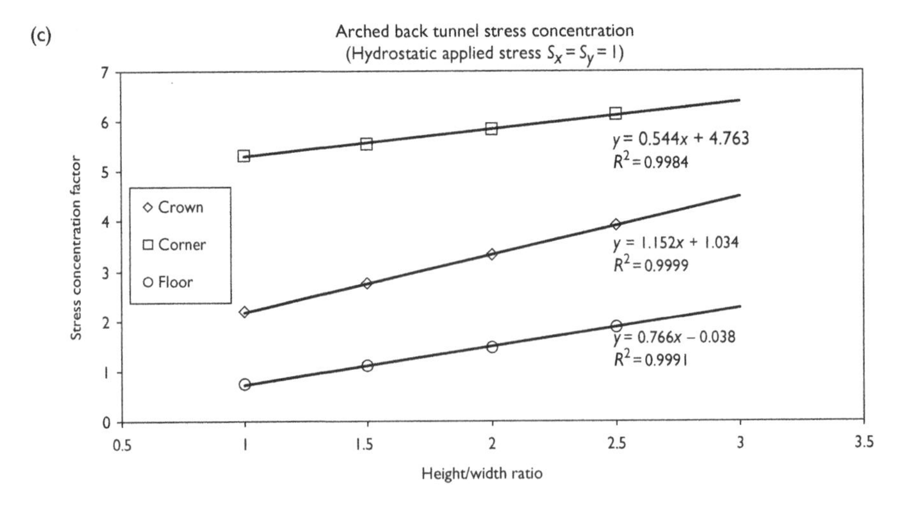

Figure 4.3 Stress concentration trends at key points of single arched back tunnel sections (a) vertical load, (b) horizontal load and (c) hydrostatic load.

In case (a),

K (floor) = (1)[stress concentration from Figure 4.3a] + (1/4)[stress concentration from Figure 4.3b]

$$K_f = (1)[(-0.012)(H/W) - 0.939] + (1/4)[(0.712)(H/W) + 0.969]$$
$$= (1)[-0.963] + (1/4)[2.393]$$
$$K_f = -0.37$$

K (crown) = (1)[stress concentration from Figure 4.3a] + (1/4)[stress concentration from Figure 4.3b]

$$K_{cr} = (1)[(-0.012)(2) - 0.939] + (1/4)[(1.056)(2) + 2.337]$$
$$= (1)[-0.963] + (1/4)[4.449]$$
$$K_{cr} = 0.15$$

In case(b)

$$K_f = (1/4)[-0.963] + (1)[2.393] = 2.15$$
$$K_{cr} = (1/4)[-0.963] + (1)[4.449] = 4.21$$

The orientation effect is considerable. When the major compression is vertical a noticeable tension is induced in the floor, but only a small compression occurs in the crown. When the major compression is horizontal, the floor tension and crown experience substantial compressive stress concentrations.

Example 4.2 An arched back tunnel is planned with a height to width ratio of two. Estimate stress concentration factors in the ribs when the preexcavation principal stress ratio $M = 1/4$ and (a) the vertical stress is S_1, (b) the horizontal stress is S_1. These are same conditions given in Example 4.1.

***Solution*:** In case (a) for the rib,

$$K_r = (1)[(-0.132)(2) + 2.676] + (1/4)[(-0.128)(2) - 0.831]$$
$$= (1)[2.41] + (1/4)[-1.087]$$
$$K_r = 2.14$$

In case (b) for the rib,

$$K_r = (1/4)[2.41] + (1)[-1.087]$$
$$K_r = -0.48$$

Again, there is a strong orientation effect as the rib stress concentration changes from compression to tension with a change in principal stress direction from vertical and parallel to the long axis of the tunnel (height dimension) to horizontal and perpendicular to the long dimension of the tunnel.

Estimation of the corner stress, which is often the peak compressive stress at the tunnel wall, is less easily done because of the change in orientation with applied stresses and height to width ratio. One might suppose that the principal stress at the corner is closely related to the floor and rib stresses very near the corner. These stresses appear as sharp spikes in the stress distributions plotted in the figures. The floor and rib spikes, F_x and R_x under uniaxial horizontal loading ($S_y = 0$, $S_x = 1$) are linear in the height to width ratio (H/W), while under uniaxial vertical loading ($S_y = 1$, $S_x = 0$), the floor and rib spikes, F_y and R_y are linear in the reciprocal ratio (W/H) as shown in Figures 4.4a,b. The trend lines indicate a very good linear fit, so the trend line equations may be used to estimate floor and rib spikes under combined loading. These spikes are generally compressive.

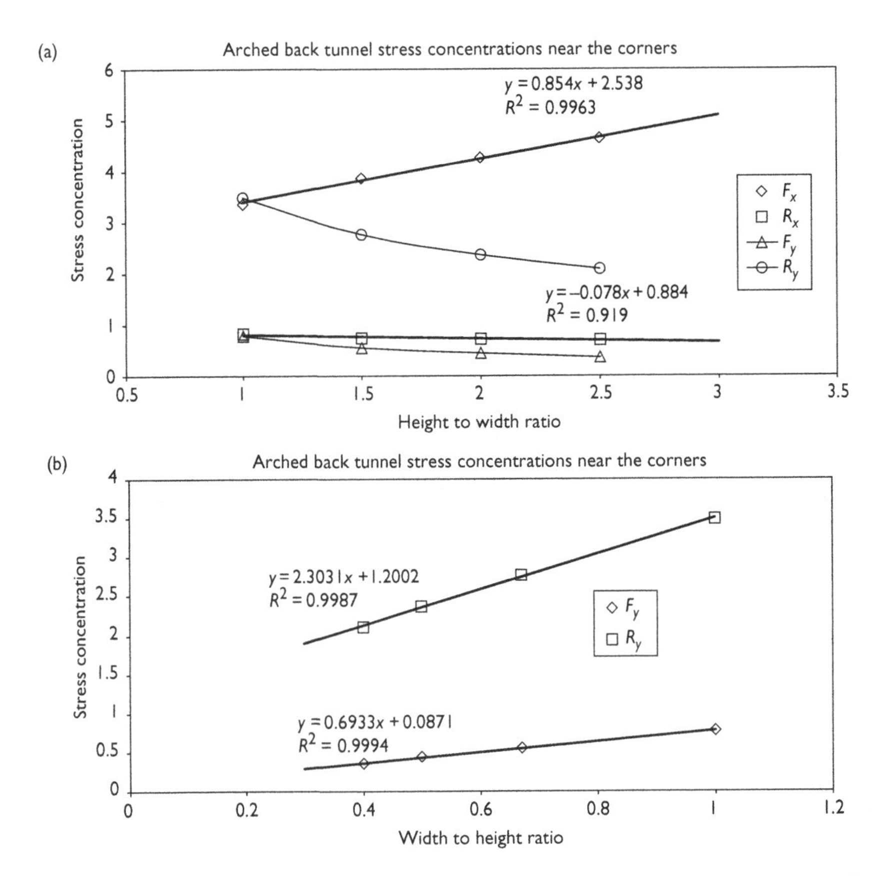

Figure 4.4 Arched back tunnel stress concentration trends near the bottom or floor corners, R = rib, F = floor, x = horizontal, y =vertical (a) horizontal and vertical loading results vs H/W, (b) vertical loading results vs the reciprocal, W/H.

Further examination of floor and rib stress spikes near the corner suggest a simple, empirical procedure for estimating the peak stress at the corner (K). The procedure is to add 10% of the floor and rib spikes to the greater of the two. Clearly, a smaller addition at lower height to width ratios and a somewhat greater percentage addition at higher height to width ratios than the 2 used would improve these estimates. However, the estimates produced by the 10% rule are certainly of sufficient accuracy to allow for a preliminary analysis of safety with respect to strength failure at points of relatively high stress concentration at the corners of an arched back tunnel floor.

Example 4.3 Consider the applied stress: $S_y = 1$ and $S_x = 1/4$ and a height to width ratio of 2, as in Examples 4.1 and 4.2. Estimation of corner stress concentration may be done using data from Figure 4.4.

The floor and rib *corner* stresses induced by the vertical stress are: $R_y = 2.303(1/2) + 1.200$ and $F_y = 0.693(1/2) + 0.087$, that is, $R_y = 2.35$ and $F_y = 0.43$.

Under $S_x = 1$ or full horizontal stress, $R_x = -0.078(2) + 0.884$ and $F_x = 0.854(2) + 2.54$, that is, $R_x = 0.73$ and $F_x = 4.25$.

Superposition of the y and x loading give $R_y + R_x = (1)(2.35) + (1/4)(0.73) = 2.53$, and $F_y + F_x = (1)(0.43) + (1/4)(4.25) = 1.49$.

In consideration of the 10% empirical estimation suggested,

$$K(\text{corner}) = 2.53 + (0.1)(2.53 + 1.49) = 2.93.$$

Example 4.4 Consider the stress concentration factors obtained in Examples 4.1, 4.2, and 4.3 for a tunnel with a height to width ratio of two in a rock mass with unconfined compressive strength and tensile strength of 50 and 5 MPa, respectively. Estimate safety factors with respect to compression and tension with $M = 1/4$ and S_1 is vertical. Assume a tunnel depth of 1,000 m where $S_1 = 20$ MPa. Sketch the results.

***Solution*:** From the previous examples in case (a) – vertical,

$$\begin{aligned} K(\text{floor}) &= -0.37 \\ K(\text{crown}) &= 0.15 \\ K(\text{rib}) &= 2.14 \\ K(\text{corner}) &= 2.93 \end{aligned}$$

Safety factors in compression and tension are

$$FS_c = \frac{50}{(2.93)(20)} = 0.85$$

$$FS_t = \frac{5}{(0.37)(20)} = 0.66$$

which indicate a potential for compressive failure at the floor corners and tensile failure at the floor center for the planned tunnel. The rib has a safety factor of 1.16 which is low and suggests the need to consider reinforcement.

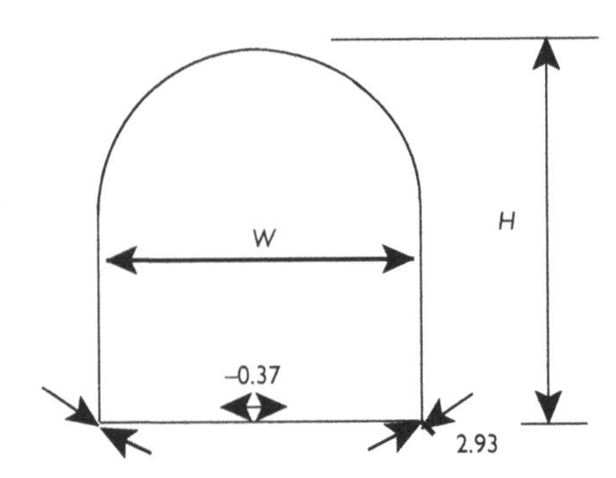

An arched back tunnel is *not* a favorable shape in regions of high horizontal stress, (e.g. where $S_x = 1$ and $S_y = 0$). High horizontal stress is often encountered in relatively deep underground mines. The reasons are evident in Figures 4.2c,f,i that show high compressive stress concentrations over the back and floor and extensive regions of high tensile stress in the ribs. This inference is not surprising in consideration of the fact that the long dimension of the opening is *perpendicular* to the major pretunnel principal stress (compressive is positive here). Recall that the rule of thumb for minimizing stress concentration is to orient the long axis of the cross-section *parallel* to the major principal stress before excavation. This rule also applies to the hybrid shape of the arched back tunnel. Of course, in case of hydrostatic preexcavation stress ($S_x = 1$, $S_y = 1$, Figures 4.2d,g,j), there is no preferred orientation.

Arching the back of a rectangular section does produce a favorable tunnel shape in cases where the preexcavation stress is largely vertical (e.g. where $S_x = 0$, $S_y = 1$). This is often the case in shallow-ground highway and railroad tunnels. Data in Figures 4.2b,e,h show arched back tunnel sections with $H/W \geq 1$ in favorable orientation. Comparisons of stress distributions in backs and floors show peak tensions of similar magnitude at crown points and floor centerlines. Thus, there is no advantage to an arched back relative to peak tension. However, the high floor tension extends nearly the entire width of the section, almost from rib to rib, while the high back tension is contained in a small region adjacent to the crown point. Most of the back is in compression that reaches a high value over the shoulder of the section near the ribs. The ribs are stressed almost uniformly in compression from shoulders to near the floor corners where the compression rises to a peak. One benefit of arching a tunnel back under high vertical stress is therefore in reducing the extent of the tensile zone near the crown. A related benefit follows from the compression induced in the back which may assist in mobilizing resistance to frictional slip on joints and fractures that are usually present. The floor tension is a disadvantage, and may be a threat to stability depending on the associated safety factor. If floor safety is not assured, then an alternative design is indicated, just as unstable ribs under high horizontal stress indicate an alternative design should be considered. An alternative design may retain the original proposed shape but then include additional support and reinforcement.

Single tunnel joints

A zone of compressive stress concentration at the wall of a tunnel may increase or decrease stability of any joints present depending on joint orientation and joint properties. Joint orientation is with respect to the directions parallel and perpendicular to the tunnel wall. The perpendicular or normal to a tunnel wall coincides with a principal stress direction. A common assumption is that the long axis of a tunnel is also a principal direction, so a third principal stress direction is tangential to the tunnel wall as seen in cross section. This orientation need not be the case.

Generally, there is a range of unstable orientations that depends on joint cohesion c_j, friction angle φ_j and tunnel stress state. Cohesive joints that are nearly perpendicular or nearly parallel to a tunnel wall are likely to be safe with respect to slip in shear. However, parallel joints may form slabs through separation under tension. Whether joint slip occurs requires a three-dimensional calculation of normal and shear stresses acting on joints that intersect tunnel walls. The normal joint stress is needed to compute joint strength, while the shear stress is needed to compute a joint safety factor. If the normal joint stress is tension, then one may reasonably suppose separation is likely. If one makes the assumption of Mohr–Coulomb joint strength, then the joint safety factor with respect to slip in shear is

$$\mathrm{FS_j} = \frac{\tau(\text{strength})}{\tau(\text{stress})} = \frac{\sigma_j' \tan(\phi_j) + c_j}{\tau_j}$$

where the subscript j refers to "joint"and the prime denotes effective normal stress. Direction of joint slip is reasonably supposed to occur in the direction of maximum joint shear stress. Because the considered joint intersects the tunnel wall, any joint slip would result in an offset at the tunnel wall. In the common case of cohesionless joints, joint slip may occur when

$$\phi_j < \beta = \tan^{-1}\left(\frac{\tau_j}{\sigma_j}\right)$$

that has the simple geometric interpretation of a "friction cone" shown in Figure 4.5. Even in the simplified case of cohesionless joints and coincidence of the long tunnel axis with a principal stress direction, the calculation of joint normal and shear stresses is lengthy and requires knowledge of the principal stress acting parallel to the tunnel axis as well as wall stress concentration factor and preexcavation stresses.

However, when joint strike is also parallel to the long axis of a tunnel, one may determine algebraically a range of unstable joint normal orientations. Given joint and rock shear

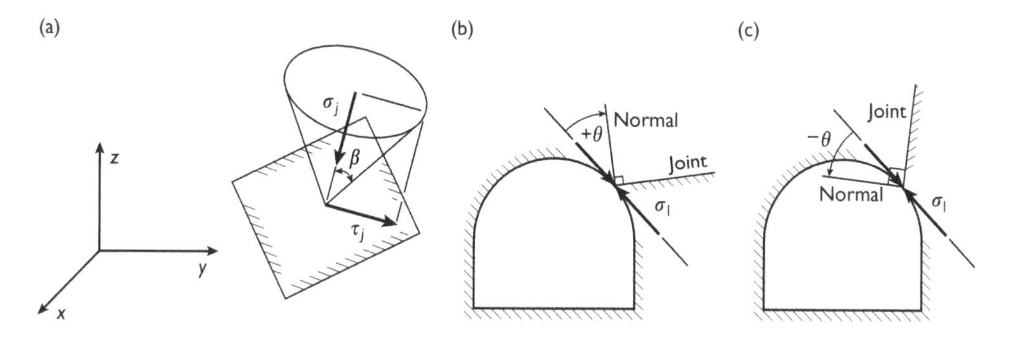

Figure 4.5 Joint plane friction cone and sign convention for orientation angle at tunnel: (a) joint plane friction cone, (b) positive θ, and (c) negative θ.

strength according to Mohr–Coulomb criteria, as shown in Figure 4.6, joint slip is indicated by computed joint safety factors less than one, that is, if

$$FS_j = \frac{[\sigma_m + \tau_m \cos(2\theta)]\tan(\phi_j) + c_j}{\tau_m \sin(2\theta)} < 1$$

then joint slip is possible. The inequality on the right may be solved for the range of the angle θ that defines unstable joint orientations. Thus, if

$$\sin(2\theta - \phi_j) > \sin(\phi_j) + \frac{2c_j \cos(\phi_j)}{\sigma_1}$$

then joint slip is possible. Alternatively, if

$$c_j > \frac{\sigma_1(1 - \sin(\phi_j))}{2\cos(\phi_j)}$$

then joint slip is *not* possible. In this analysis, positive angles in the $\sigma - \tau$ plane are counterclockwise; corresponding angles in the physical plane of a tunnel are opposite, clockwise. This reversal allows one to use the upper half of the $\sigma - \tau$ plane for analysis using the usual mathematical convention of positive angles measured in a counterclockwise direction. Inspection of Figure 4.6 shows that the range of unstable joint orientations is limited by solutions to

$$\sin[(2\theta - \phi_j)] > \sin(\phi_j) + \frac{2c_j \cos(\phi_j)}{\sigma_1}$$

$$\sin[\pi - (2\theta - \phi_j)] < \sin(\phi_j) + \frac{2c_j \cos(\phi_j)}{\sigma_1}$$

In either case, the tunnel wall compression is given by

$$\sigma_1 = KS_1$$

where K and S_1 are stress concentration factor and major principal stress before excavation, respectively. Thus, the limiting range of unstable joint orientations is given by

$$2\theta_1 < 2\theta < 2\theta_2$$

$$2\theta_1 = \phi_j + \sin^{-1}\left[\sin(\phi_j) + \frac{2c_j \cos(\phi_j)}{\sigma_1}\right]$$

$$2\theta_2 = \pi + \phi_j - \sin^{-1}\left[\sin(\phi_j) + \frac{2c_j \cos(\phi_j)}{\sigma_1}\right]$$

The range defined by solutions to these inequalities is ultimately limited by the unconfined compressive strength ($C_o = \sigma_1$) of intact rock between joints.

When the considered joint lacks cohesion, the range of angles between the tangent to the tunnel wall and the normal to the joint is

$$\theta_1 < \phi_j < \pi = \theta_2$$

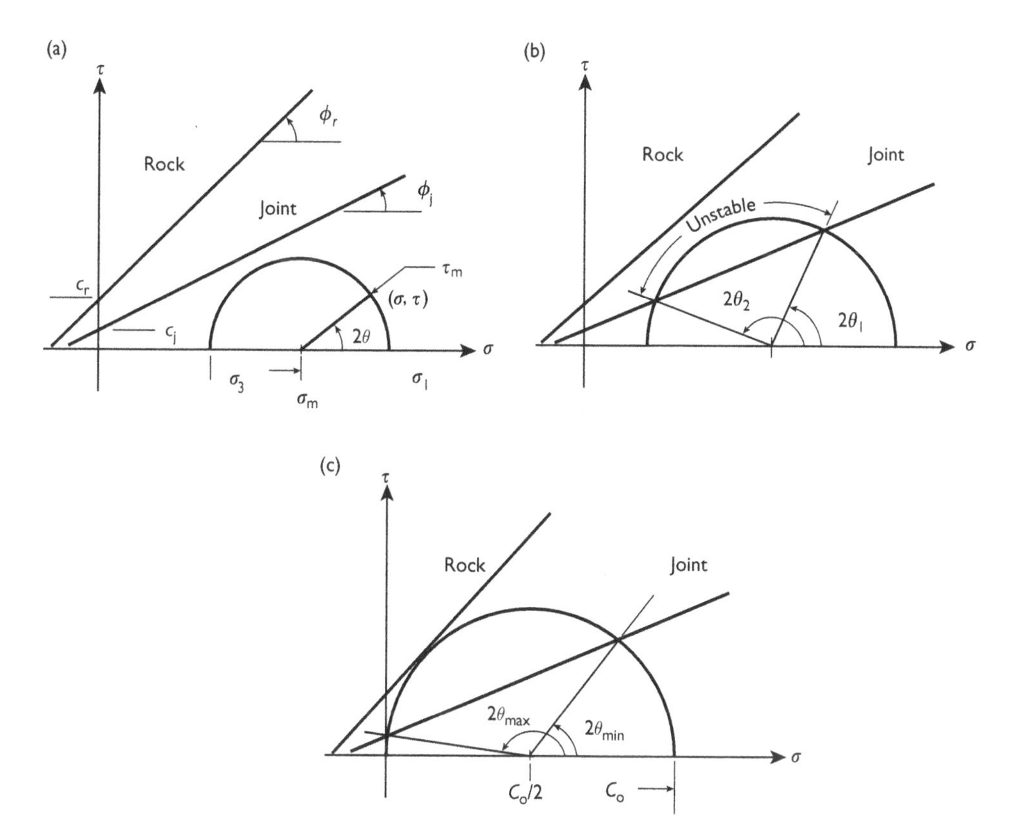

Figure 4.6 Mohr circle construction for a range of joint failure orientations when joints strike parallel: (a) no failure, (b) joint failures $2\theta_1 < 2\theta < 2\theta_2$, and (c) rock and joint failure under uniaxial compression.

which simply states that frictional sliding occurs when the "slope" is greater than the friction angle. A similar analysis may be done in the lower half of the $\sigma - \tau$ plane where shear stresses are negative. A confusion of algebraic signs suggests one handle the situation by inspection rather than analysis. Indeed, a mirror image of the graphics in Figure 4.6 shows analogous results with negative signs for angles θ, reflective of joint orientation symmetry about the direction of σ_1 tangential to the tunnel wall.

Example 4.5 Suppose joint and intact rock properties are determined by laboratory tests on samples acquired at a tunnel site and are: $c_r = 1{,}500$ psi, $c_j = 15$ psi, $\phi_r = 45°$, $\phi_j = 25°$. Analysis indicates that stress concentration in the arch of the tunnel back peaks at 2.87 over the shoulder of the tunnel where the wall slope is inclined 75° to the horizontal. Stress measurements indicate a preexcavation vertical stress of 600 psi and a horizontal stress of 150 psi. Determine if joint slip is possible and, if so, the range of joint normals that allow slip.

***Solution*:** The tunnel wall stress is $\sigma_1 = (2.87)(600) = 1,772$ psi and therefore σ_1 (1 − sin $(\phi_j)/2\cos(\phi_j) = (1,772)[1 - \sin(25)]/[2\cos(25)] = 564$ psi which is greater than $c_j = 15$ psi, so joint slip is possible.

The range of normals may determined from

$$2\theta_1 = 25 + \sin^{-1}\left[\sin(25) + \frac{2(15)\cos(25)}{1{,}772}\right] = 51.0°$$

$$2\theta_1 = 180 + 25 + \sin^{-1}\left[\sin(25) + \frac{2(15)\cos(25)}{1{,}772}\right] = 231.0°$$

The range of normals is therefore (25.5°, 115.5°) measured from σ_1 in either a clockwise or counter-clockwise direction. The joint planes proper prone to slip range between 15.5 and 74.5° on either side of σ_1.

Example 4.6 Consider an arched back tunnel with a height of 6 m and a width of 4 m at a depth of 1,000 m where before excavation the vertical stress is 20 MPa and horizontal stress is 5 MPa as seen is cross-section. A fault dipping 60° intersects the tunnel 3 m above the floor on the right-hand side (looking at the face). Fault friction angle is 32° and fault cohesion is 1 kPa. Determine whether fault slip is possible, and if so, what improvement in cohesion would be required, say by grouting, to give a fault slip safety factor of 1.1.

***Solution*:** The tunnel height to width ratio is 3/2, and the preexcavation principal stress ratio is 1/4th by the given conditions. Intersection of the fault with the tunnel wall is in the rib. According to the trend lines in Figure 4.3, the rib stress concentration is

$$K = (1)[(-0.132(3/2)) + 2.676] + (1/4)[(-0.128(3/2)) - 0.831] = 1.46$$

The rib stress $\sigma_1 = 1.46(20) = 29.2$ MPa which acts vertically at the rib-side. Slip is possible if

$$\frac{\sigma_1[1 - \sin(\phi)]}{2\cos(\phi)} > c_j$$

$$\frac{29.2[1 - \sin(32)]}{2\cos(32)} = 9.09 > 0.1 = \text{MPa}$$

The inequality is satisfied and slip is possible. To prevent slip by increasing cohesion to obtain a safety factor of 1.1 requires

$$\text{FS(fault)} = \frac{[\sigma_1 + \sigma_1\cos(2\theta)]\tan(\phi) + 2c}{\sigma_1\sin(2\theta)} = 1.1$$

$$= \frac{[1 + \cos(2\theta)]\tan(\phi) + 2c/\sigma_1}{\sin(2\theta)} = 1.1$$

$$= \frac{[1 + \cos(180)]\tan(32) + 2c/29.2}{\sin(32)} = 1.1$$

$$\therefore\ c = 8.51\ \text{MPa}$$

When multiple joint sets are present, blocks of rock formed by joint intersections may be liberated during tunnel excavation. Fall of one block may allow other blocks to fall in turn. The threat to tunnel safety in this situation is highly dependent on the presence of a "key block" in allusion to stone arches and a central keystone that is essential to arch stability.

Multiple tunnels

A single tunnel of a given size may not accommodate the transport needs of an operation; additional passageways may be required. Tunnel spacing then becomes an additional design variable. When tunnels are closely spaced, stress concentrations are higher relative to a single, isolated tunnel of the same shape. The effect of tunnel spacing becomes somewhat noticeable when the distance between tunnel walls is about equal to the long dimension of the tunnel section. As spacing is decreased, stress concentration at the tunnel walls increases rapidly, while the *average* stress in pillars between tunnels increases linearly with respect to the ratio of tunnel width (W_o) to pillar width (W_p).

Two tunnels side by side separated by a pillar lead to consideration of tunnels in a row. As the number of tunnels in a row is increased beyond five or so, the addition of one more tunnel has little effect on the tunnels in the center of a row. The row is then effectively one of infinite extent and the situation is relatively easy to analyze because of symmetry with respect to vertical planes through tunnel and pillar centers. There is symmetry in analysis of single, two (twin) tunnels and a row of tunnels, as shown in Figure 4.7, where a vertical plane of symmetry passes through the center of the single opening and a vertical plane of symmetry passes through the center of the pillar between the twin (two) openings. In case of a row of tunnels, vertical planes of symmetry pass through both opening and pillar centers. Comparisons of stress concentration about a single tunnel, two (twin) tunnels and a row of tunnels are shown in Figure 4.8. The analysis results in Figure 4.8 show that for a given tunnel (arched back) and cross-section (*H/W*=1.5), stress concentration induced by excavation is strongly dependent on: (1) the preexcavation stress state, (2) tunnel spacing, $W_o + W_p$, or equivalently, W_o/ W_p ratio, (3) orientation of the section, (4) orientation of the row axis, and (5) number of tunnels excavated. Three preexcavation stress states are considered: uniaxial vertical loading ($S_y = 1$, $S_x = 0$), uniaxial horizontal loading ($S_y = 0$, $S_x = 1$) and hydrostatic loading ($S_y = 1$, $S_x = 1$) with compression positive and plotted outside tunnel walls. Tension is negative and plotted inside.

Under *vertical* loading only, peak compression is about three to four times the preexcavation stress ($K_c \approx 3, 4$) in case of one or two tunnels, but is over five in a row of tunnels. Tension is about 80% to 90% of the preexcavation stress ($K_t \approx -0.8, -0.9$) in all cases. In the row of tunnels case, the average pillar stress is exactly $3S_y$, which is near the relatively uniform stress concentration between corner and shoulder of the section. The average pillar stress S_p in a row of tunnel openings by definition is

$$S_p = \frac{\int_{A_p} \sigma_v dA}{\int_{A_p} dA} = \frac{\int_{A_p} \sigma_v dA}{A_p} = \frac{\int_A S_v dA}{A_p} = \frac{S_v A}{A_p} = \frac{S_v(W_o + W_p)}{W_p}$$

where equilibrium of vertical forces before and after excavation is introduced in the third equation and a unit distance along the tunnel is implied. In this example $W_o = 2W_p$ and $S_v = S_y$, the preexcavation vertical stress, so $\sigma_1 = S_y(2W_p + W_p)/W_p = 3S_y = 3S_1$, that is, $K_c = 3$.

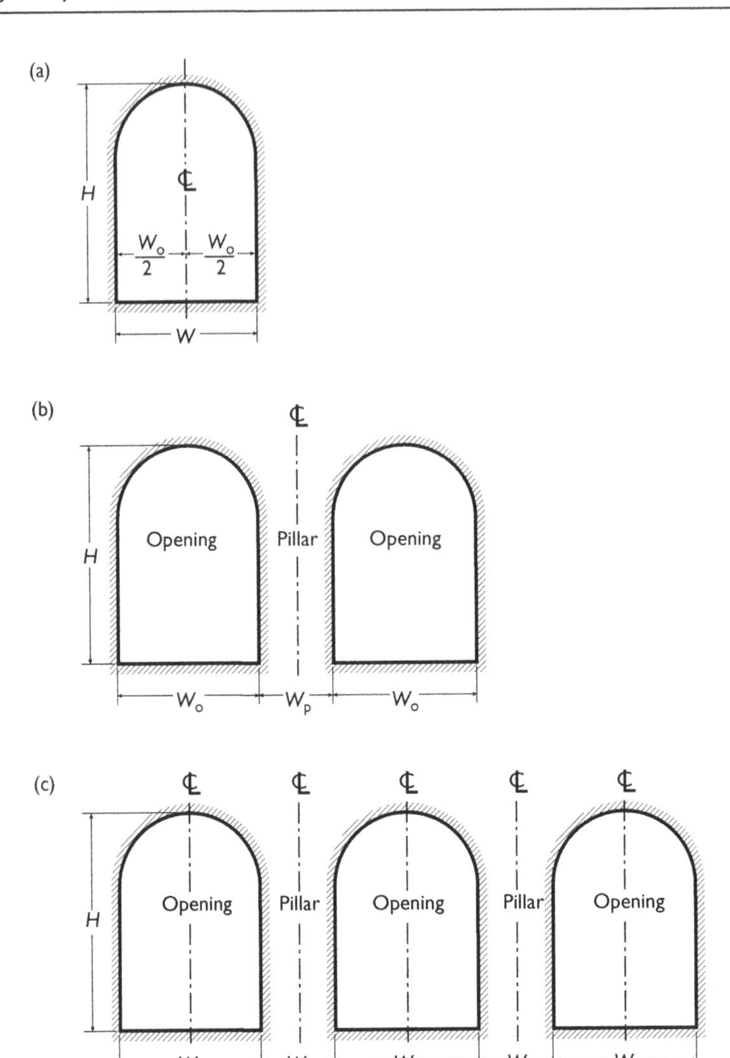

Figure 4.7 Notation and symmetry for (a) single tunnel, $H/W = 1.5$, (b) two tunnels, $H/W = 1.5$, $W_o = 2W_p$, and (c) row of tunnels, $H/W = 1.5$, $W_o = 2W_p$.

In case of *horizontal* loading only, peak compression is greater than three times the preexcavation stress ($K_c > 3$) about one or two tunnels, but is less than two ($K_c < 2$) in a row of tunnels. Tension is again about 80% to 90% of the preexcavation stress ($K_t \approx -0.8, -0.9$) in case of one or two tunnels, but is only about 20% ($K_t \approx -0.2$) in a row of tunnels. In the row of tunnels case, there is considerable reduction in stress concentration in the pillar relative to vertical loading. In fact the average *vertical* stress in the pillar is zero ($3S_y$). The reason is the pillar is in the "shadow" of the adjacent openings ($S_y = 0$, $S_x = 1$). This shadow effect depends on alignment of the row axis with the direction of major principal compression prior to excavation. Generally, the most favorable orientation of a row of openings is with

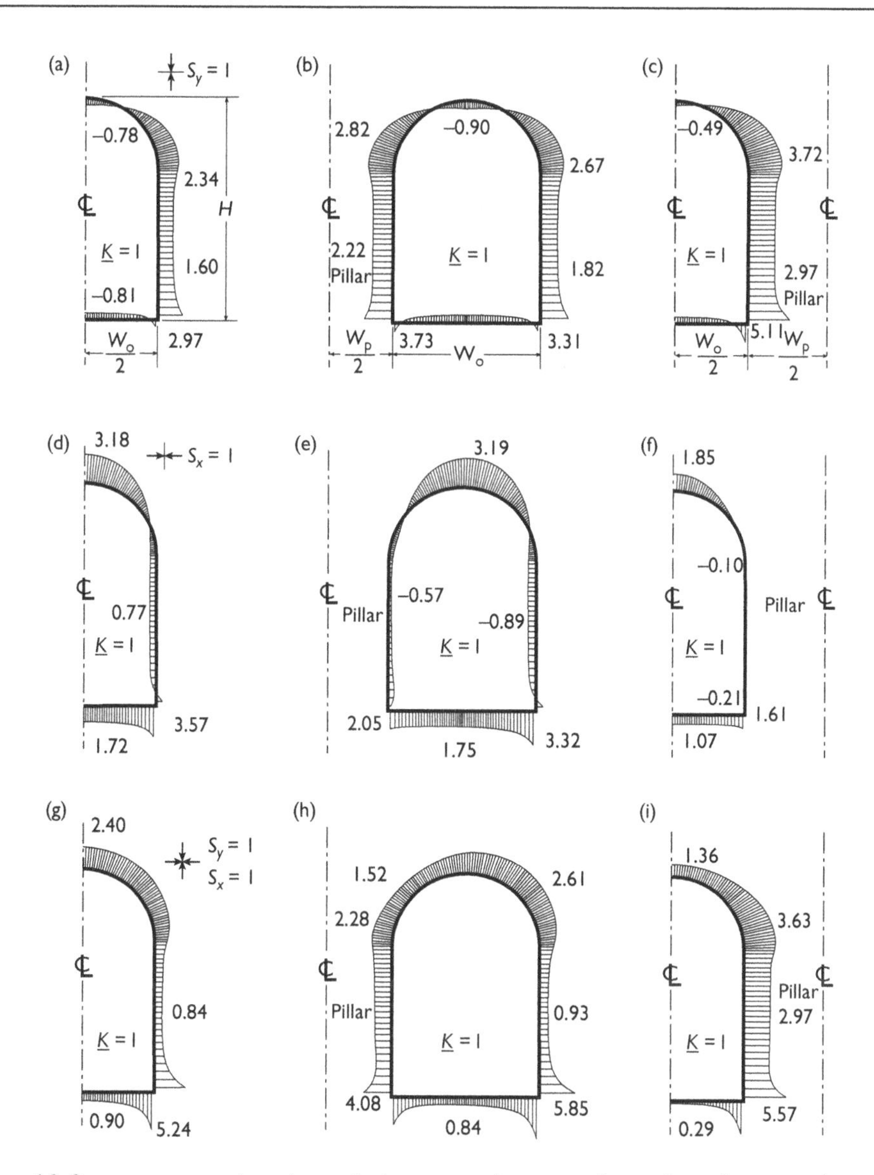

Figure 4.8 Stress concentration about single, two, and a row of tunnels under uniaxial vertical, horizontal, and hydrostatic stress, $H/W = 1.5$, $W_o = 2W_p$.

the row axis parallel to the major preexcavation compression. This shadow effect is present to some degree in the case of two tunnels, where the pillar stresses are less than stresses at the outside tunnel walls.

Under hydrostatic loading, there is no preferred orientation relatively to the preexcavation stress field. In fact, the results in Figure 4.8 may be obtained by superposition of the uniaxial load cases. For example, the floor stress concentration is 0.84 in the two tunnel case and is obtained by adding −0.91 + 1.75. In the row of tunnels under hydrostatic load, the crown stress concentration is 1.36 (= −0.49 + 1.85). Care must be taken to sum stress concentrations at the same point and in the same direction. The direction of corner compression varies, so addition of the peaks is *not* valid at the corners. The average *vertical* stress in the pillar is $3S_y$.

Figure 4.9a shows the distribution of stress about a single tunnel under vertical loading about a typical tunnel of $H/W = 1$ in a row of tunnels as a function of distance from the tunnel centerline. There is a rapid decrease in stress concentration with distance from the wall of a single tunnel. At a distance of one opening width (W_o) into the wall, the vertical stress

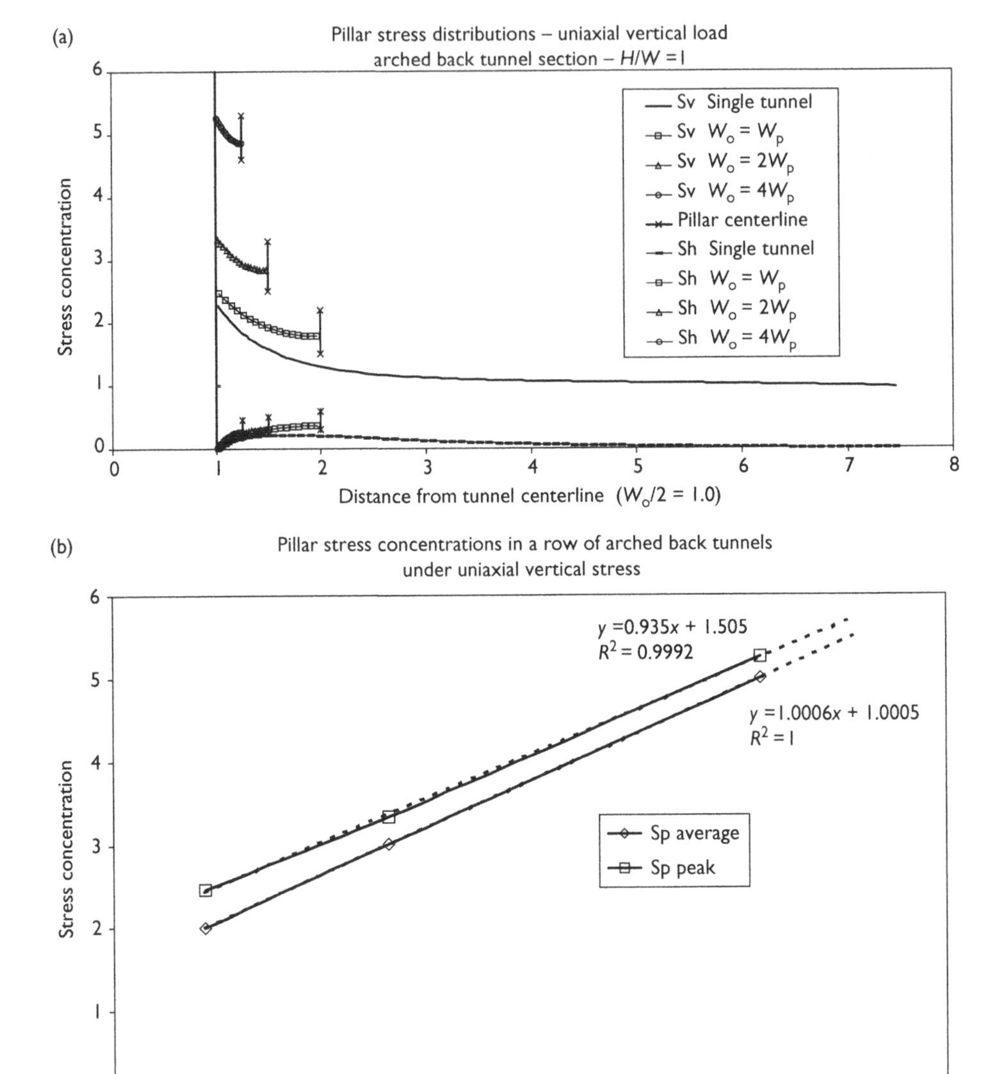

Figure 4.9 Pillar stress distribution and concentration in a row of arched back tunnels ($H/W = 1$) under a uniaxial (vertical) applied stress (a) vertical and horizontal stress distribution, (b) average and peak stress concentrations.

concentration has decreased to nearly the preexcavation value of one, while the horizontal stress is nearly zero. As distance between tunnels decreases, that is, as the pillar width decreases relative to tunnel width (W_p/W_o ratio) to one, a small increase occurs in wall stress concentration, while the average stress in the pillar increases. Interestingly, there is an induced horizontal stress in the pillar between tunnels that increases towards the pillar center. This effect results in a confined pillar core that aids pillar strength. As the pillar width decreases to values less than tunnel width, interaction between adjacent tunnels occurs, as indicated by rapid increases in peak wall stress and average pillar stress. Figure 4.9b shows that the peak tunnel wall stress and average pillar stress are very nearly linear functions of the ratio W_o/W_p over the considered range. Extension of the regression results to ratios greater than four is reasonable. However, backward extension to ratios less than one is not permissible.

Comparison of the single tunnel under vertical and horizontal loading reveals the fact that compressive stress concentration is noticeably less when the long axis of the cross-section is parallel to the direction of the major compression before excavation. There is little difference in tensile stress concentration, although they occur in different locations. From the rock mechanics view, tunnels one below the other would be preferable to tunnels side by side under uniaxial vertical load. Both tunnel axis and row axis are then favorably oriented, parallel to the major preexcavation compression.

Figure 4.10 shows stress concentration about two tunnel sections, one with a long horizontal dimensions (H/W = 0.75) and one with a long vertical dimension (H/W = 1.5). Loading is vertical ($S_y = 1$, $S_x = 0$) in all cases. The results are summarized numerically in Table 4.1.

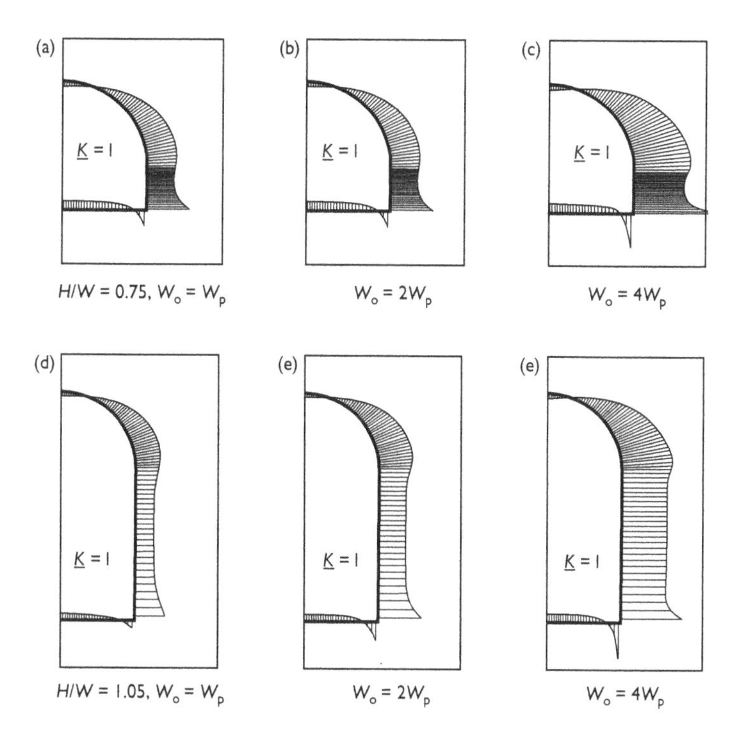

Figure 4.10 Stress concentration about wide tunnels (a, b, c) and tall tunnels (d, e, f).

Table 4.1 Comparison of stress concentrations about arched back tunnels in a row[a]

Location	Tunnel					
	Height/Width = 0.75			*Height/Width=1.50*		
	$W_o = W_p$	$W_o = 2W_p$	$W_o = 4W_p$	$W_o = W_p$	$W_o = 2W_p$	$W_o = 4W_p$
Floor center	−0.81	−0.84	−0.89	−0.71	−0.80	−0.86
Corner	4.54	7.36	11.95	3.85	5.60	8.18
Pillar-Edge / center	2.66/1.66	3.35/2.77	5.25/4.87	2.44/1.82	3.32/2.84	5.22/4.87
Rib/shoulder	2.81	3.66	5.56	2.81	3.72	5.58
Crown	−0.48	−0.49	−0.52	−0.47	−0.49	−0.52

Note
a $S_y = 1$, $S_x = 0$ (uniaxial vertical load).

These results show that there is little difference in floor and crown tensions, but that corner, shoulder and rib compressive stress are noticeably higher when the long dimension of the opening is perpendicular to the load direction. The combination of long tunnel dimension and tunnel row axis oriented perpendicular to the major preexcavation compression (unfavorable opening orientation, unfavorable row axis orientation), leads to high stress concentration, especially when tunnels are closely spaced.

4.2 Tunnel support

Tunnel support may range from none to rock bolts installed as needed (spot bolting) in the opinion of the crew to an engineered design that requires installation of support of a specified size and on a well-defined pattern. Traditional tunnel support in civil projects is by structural steel segments bolted together to form steel sets that conform to section shape. Rock bolts, perhaps with surface support, are more often used in mine operations. Surface support may be in the form of steel mats, welded wire mesh, chain link or similar screen. Rock bolts, wire mesh, and shotcrete may be used in place of steel sets. This combination of support is often referred to as the New Austrian Tunneling Method (NATM), although the method has been in use for many years. Shotcrete is cement that is sprayed onto excavation walls to form a thin layer of support. High strength fibers may be added to shotcrete to increase tensile strength. Various combinations of bolts, steel sets, and shotcrete may be used where ground conditions would lead to large wall displacements without large support load capacity. Even then, large, heavy, relatively stiff support may be inadequate. In such cases, support that is light weight and flexible may provide ground support and excavation safety by accommodating displacements in a controlled manner. Yieldable steel sets and light segment liner with backfill are two support methods that are often used in "squeezing" ground where large time-dependent displacements are anticipated.

Fixed steel sets

Structural analysis of support is relatively simple and has the objective of calculating a *support safety factor* – ratio of structural strength to stress. Strength is a material property and

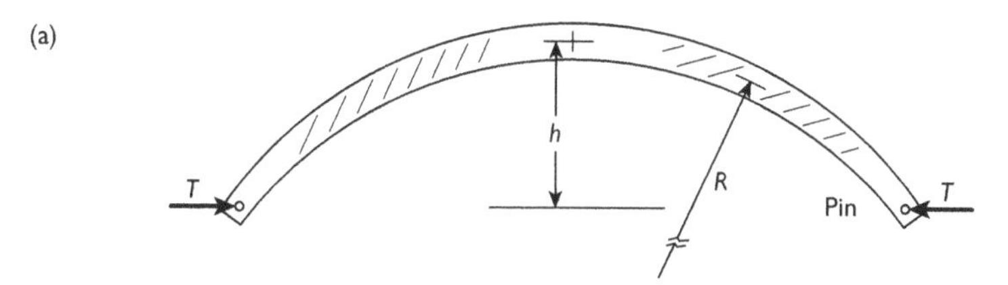

Line of thrust must be on the chord between pin ends.

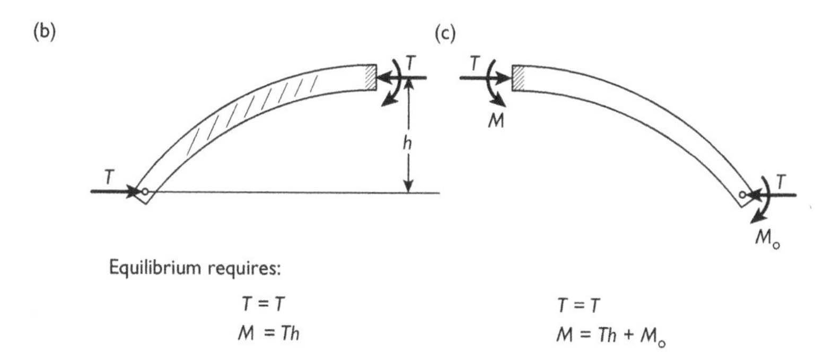

Figure 4.11 Equilibrium of a pin-ended curved beam under thrust *T*.

is considered known. Stress in a steel support "beam" is a combination of an axial stress and bending stress. Thus,

$$\sigma = \sigma_a + \sigma_b = \frac{T}{A} + \frac{Mc}{I} \tag{4.1}$$

where σ_a = axial stress, σ_b = bending stress, T = axial thrust, A = steel cross-sectional area, M = bending moment, c = distance to steel bottom (or top) from the neutral axis, I = area moment of inertia of the steel. The calculation requires knowledge of the steel size. In practice this is assumed on a trial basis and then checked to see if a satisfactory support layout results. Guidelines from past experience are used to make a reasonable first guess.

When the steel beam is circular and simply supported at the ends, no moment is present at the supports. The maximum moment that develops between ends is then related to the thrust T and the rise of the arch h between ends, as shown in Figure 4.11. Under these conditions, $M = Th$. Generally, the ends of a curved beam segment are not pin connected to adjacent beam segments, so some correction is justified. The correction is a reduction in the original moment estimate, that is, M = (constant)Th where the *constant* is some number less than one, perhaps in the range of 0.85–0.67. Regardless, the problem reduces to determining T induced in the steel by motion of the adjacent rock mass.

The standard approach to estimating tunnel steel support load is based on the concept of an arch of rock in the tunnel back that eventually bears on the steel (Proctor and White, 1968). Consider a rectangular tunnel with a semi-circular arched back that is supported by a curved steel beam (a steel "rib") at discrete blocking points as shown in Figure 4.12. The beam ends are supported by columns in the form of straight steel legs of the same size as the beam. Blocking points are wood pieces tightened into place by hammering wedges between block

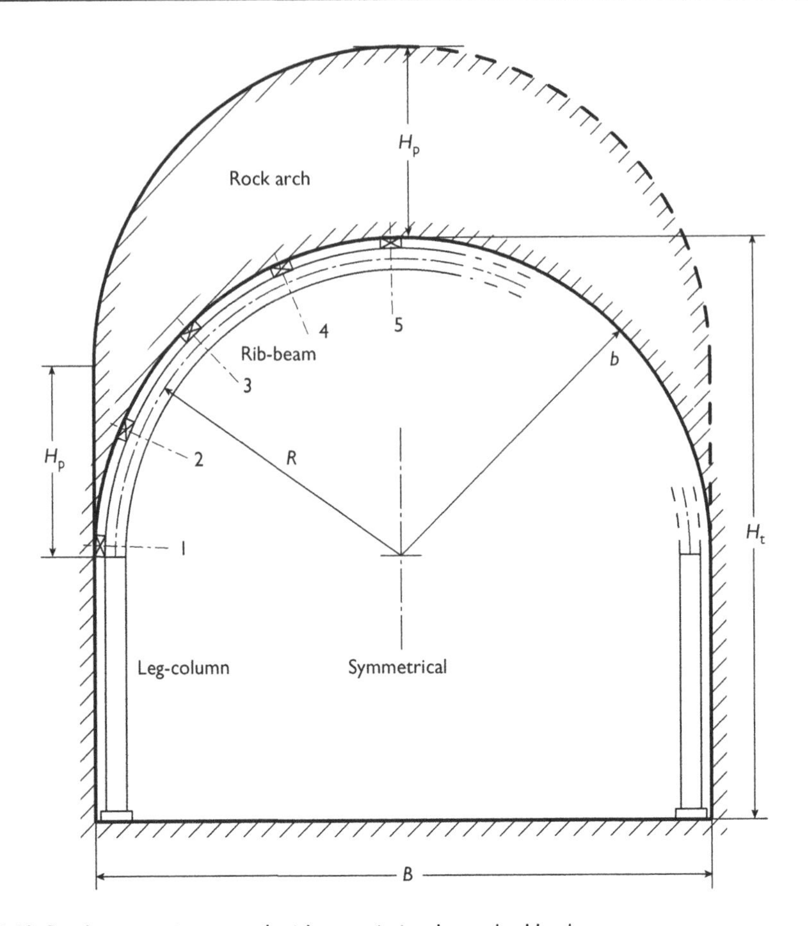

Figure 4.12 Steel support in a tunnel with a semi-circular arched back.

and rock. They serve to hold the set upright and to transmit load from rock to steel. Wedging action induces a small initial load on a set. Any shear loading of the blocking points could easily dislodge a wedge, and for this reason, blocking points effectively transmit normal loads only. Load transmission may be *active* from rock to steel, as expected at the crown point or *passive* as the rock reacts to the steel as the steel springs against the rock near the sides of the tunnel.

Weight of the rock arch in the back must be carried by the internal reaction of the support. The total weight W of the rock arch that has a constant height H_p is γBH_pL where γ is specific weight of rock, B is tunnel breadth, and L is spacing between sets, as shown in Figure 4.12. The "head" of rock bearing on a set is H_p. As a rough rule of thumb, rock "pressure" amounts to about 1 psi per foot of head. Estimation of H_p is an important first step in the analysis and is based on a simple rock classification scheme that is given in Table 4.2. Other rock mass classification schemes for engineering purposes are discussed in Appendix C.

A 10 ft high rock arch is associated with a load of about 10 psi. This seemingly small stress in comparison with rock strength of the order of thousands of psi may appear insignificant.

Table 4.2 Terzaghi rock mass classification for tunnel rock Load H_P

Rock conditions	*Rock load H_P (ft)*[a]	*Remarks*
1 Hard and intact	Nil	Light lining for small spalls
2 Hard stratified, schistose	0.0–0.5B	Light support
3 Massive, some jointing	0.0–0.25B	Light support
4 Moderately blocky, seamy	0.25B–0.35$(B + H_t)$	No side pressure
5 Very blocky and seamy	(0.35–1.10)$(B + H_t)$	Little or no side pressure
6 Completely crushed	1.10$(B + H_t)$	Heavy side pressure
7 Squeezing rock	(1.10–2.10)$(B + H_t)$	High side, bottom pressure
8 Very squeezing rock	(2.10–4.50)$(B + H_t)$	Very high, invert bracing
9 Swelling rock	Up to 250 ft	Full circle ribs needed

Note

a For wet ground. Use one-half table values in dry ground (after Terzaghi in Proctor and White, 1968).

However, rock support and reinforcement seldom provide resistance of more than a few psi and is intended only to control the rock mass in the immediate vicinity of an opening. Support of the entire overburden that generates 1 psi/ft of depth is impractical without refilling the excavation. In fact, temporary mine openings are usually back-filled after ore extraction, although caving methods involve a different type of "back-filling." Main access ways such as shafts and tunnels must remain open for life of the mine and hence require well-engineered support or reinforcement. Reliable control of shaft and tunnel walls for an indefinite period of time is essential to safety.

For each set spacing L, a different steel size results. Obviously, there is no unique support design that meets support objectives. However, in addition to support capacity needed for ground control, the quality of the rock mass also needs to be considered in deciding upon support spacing. Very loose, densely fractured rock that tends to ravel and run will require relatively closely spaced support and wire mesh, plastic net, steel screen, wood lagging, steel lacing, or similar support between sets. In blocky ground, rock bolts may be installed prior to placement of steel sets. In this regard, uncontrollable runs of rock are often prevented by small support forces that maintain the integrity of the rock mass and prevent key blocks from falling.

A free body diagram of the support is shown in Figure 4.13 where R is the radius to the centroid of the steel cross-section. The beam is curved, so a shift of the neutral surface of the beam away from the centroid occurs. However, the effect is small and neglected in analysis; the neutral and centroidal axes are considered coincident. This approximation is certainly acceptable in view of other uncertainties.

The main problem is to estimate the normal loads transmitted from rock to steel via the blocking points. If all forces associated with blocking points are considered acting at points on the neutral axis of the beam, then a diagram of forces that represents equilibrium is possible. Each point on the neutral axis is acted upon by three forces, one normal force and two thrusts from adjacent load points. Equilibrium at each load point is thus represented by a closed triangle of forces.

The analysis treats the continuous beam as a series of pin-connected links between load points. Consequently thrusts between load points are directed along chords between points. *Directions* of all forces at each point are thus known. Figure 4.13 shows the force triangles

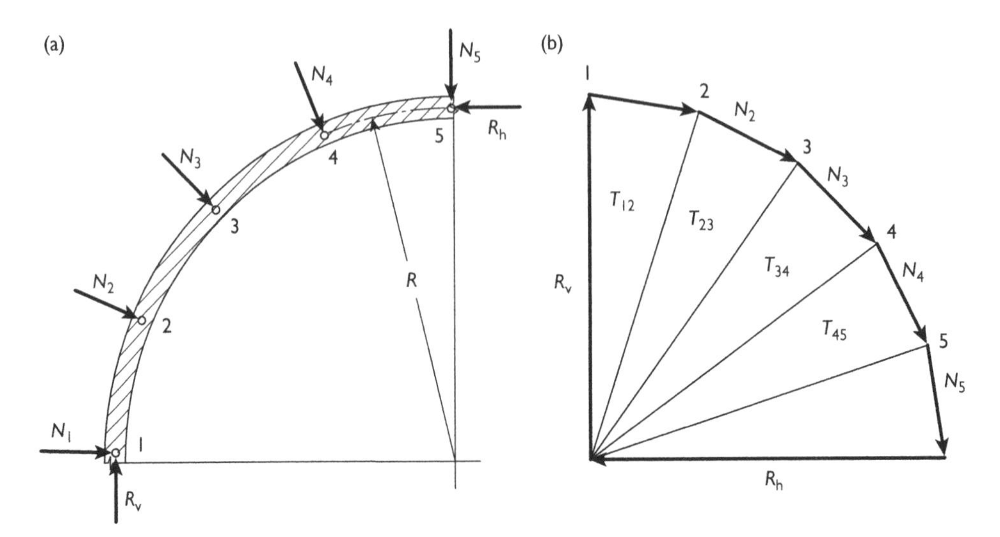

Figure 4.13 Triangles of force equilibrium aligned to form a polygon of forces for overall equilibrium of beam forces (a) free body and (b) force polygon.

associated with the steel arch in Figure 4.12 after arrangement into a polygon of all forces. This polygon is closed and is a graphical representation of overall equilibrium of the set as well as equilibrium at the load points. Although force directions are known, *magnitudes* remain to be determined. In essence the scale or size of the force polygon is unknown. Enlarging or shrinking the force polygon preserves direction of forces while changing magnitudes.

A detailed examination of normal loads is needed to resolve the issue of force magnitudes. The first step requires partitioning the rock arch into blocks by vertical planes half way between blocking points, as shown in Figure 4.14. The next steps are to calculate block weights and to resolve weights into normal and tangential components. These normal forces are *active* loads. Passive loads from rock–steel interaction are also generated and require additional consideration of normal forces as shown in Example 4.7. The force polygon in Figure 4.13 must be large enough to accommodate the largest active normal load. This load is usually at the crown point. After the polygon is properly sized, the largest thrust in the polygon is used to calculate beam stress and safety factor of the trial beam. If the safety factor is too large or small, a different size beam is selected or beam spacing is adjusted.

A subsidiary calculation of arch rise h is needed. Considering the circular arc geometry of radius R and chord length l between blocking points,

$$h = R - [R^2 - (l/2)^2]^{(1/2)} \tag{4.2}$$

If blocking points were placed side by side, the arch rise would be reduced to a small value. The associated moment and bending stress would also be reduced to very small values. Because the bending stress is often a substantial portion of the total beam stress, *blocking point spacing* becomes an important design consideration. While steel costs favor closely

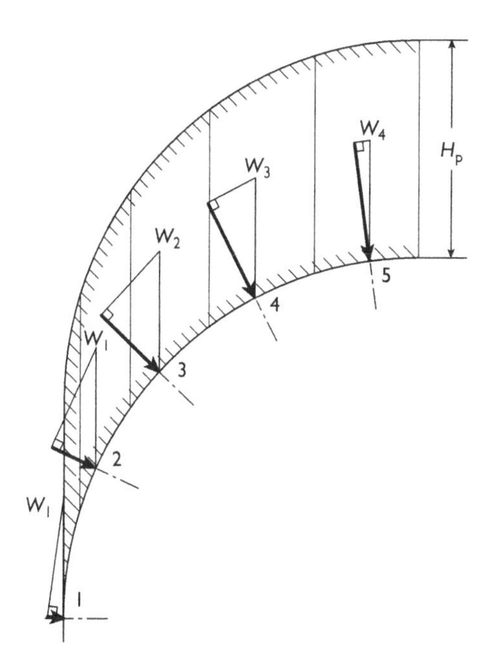

Figure 4.14 Rock arch partitioned into blocks with weights, normal forces, and tangential forces shown.

spaced blocking points (lower stress, smaller beam), installation costs favor widely spaced points (fewer blocks to install). An economic optimum obviously exists in the trade-off. This optimum is addressed by a maximum recommended blocking point spacing that depends on tunnel width. A plot of maximum recommended blocking point spacing as a function of tunnel width is shown in Example 4.7.

Graphic solution of the force polygon size problem requires an accurate rendering of the polygon, of course. The solution could also proceed analytically with repeated application of the sine law, a procedure that is easily programmed for the computer. In case of non-circular arches of rock and steel, either procedure still applies, although the radius of curvature now varies and must be taken into account. A spreadsheet is perhaps the easiest tool to use for this analysis.

Fixed steel sets are intended mainly to resist vertical loads from rock in the back of the opening. The legs are columns that lack effective resistance to side pressure. Although the legs are quite capable of resisting axial load and only need to be sized to prevent buckling, side pressure cannot be accommodated without arching the legs into a horseshoe shape. If side pressure is high, then full-circle support becomes necessary. Even then, fixed steel sets may be inadequate.

Example 4.7 A 26 ft wide by 27 ft high semi-circular arched back tunnel in moderately blocky and seamy ground is proposed where water is present. Estimate the size of steel support needed using blocking point spacing that does not exceed the recommended spacing given in the associated plot. A steel safety factor of 1.5 is required for *A*36 steel.

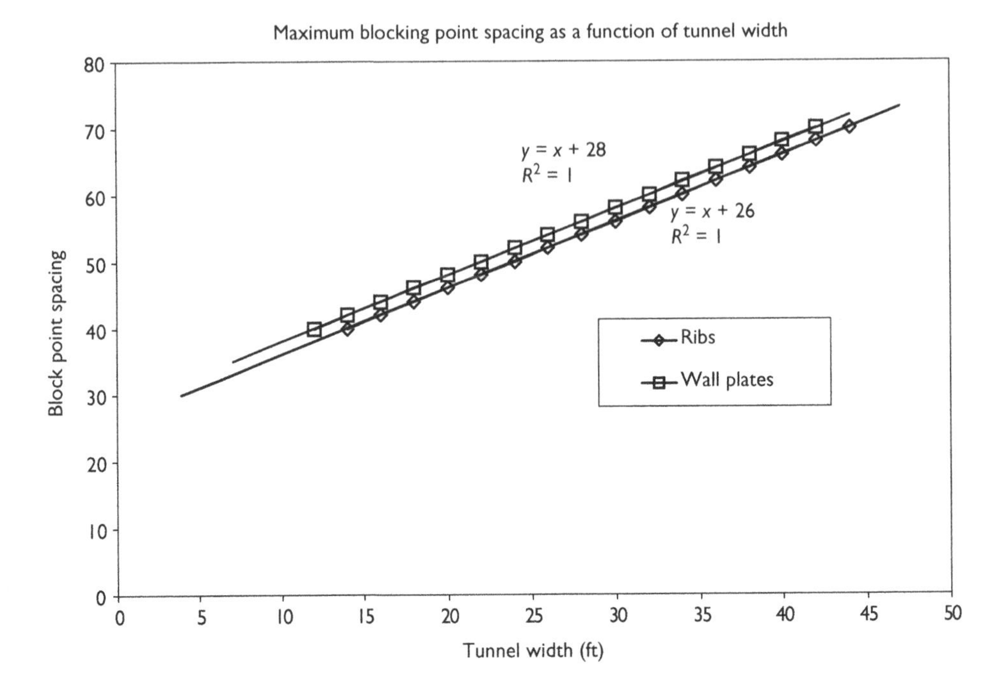

A plot of maximum recommended blocking point spacing (after Proctor and White, 1946, revised, 1968). Tunnel width x = ft. Blocking point spacing y = inches.

***Solution*:** The steel sets will stand inside the tunnel, so some allowance for blocking point thickness and depth of the steel section needs to be made. A reasonable estimate would be to allow 8 in. for blocks and 4 in. for steel half-depth. The neutral surface of the steel is then 1 ft inside the tunnel wall and rises to within 1 ft of the rock crown as shown in the sketch. The radius to the neutral surface is 12 ft.

Maximum recommended spacing is 27 + 26 = 53 in. If one uses equally spaced points, then the number of blocks (points) needed is $[(12)(12)(\pi/2)/53] + 1 = 5.3$ where an additional block is added for the end, that is, blocks are at the crown and springline where the curved steel rib connects to the vertical leg. A fractional block is a mathematical artifact, of course, so after rounding up, the number of blocks is six. The curve of the steel neutral surface is divided into five spaces, so the actual blocking point spacing is $[(12)(12)(\pi/2)/5] = 45.2$ in.

The description of "moderately blocky and seamy" allows for a rock height that ranges from $0.25W$ to $0.35(W+H)$, that is, between 6.5 and 18.6 ft. A mid-range estimate of 12.5 ft is reasonable, so Hp=12.5 ft. Rock specific weight is 158 pcf.

The rock arch and division into slices are shown in the sketch. Blocking points are numbered sequentially beginning at the springline. An origin of coordinates is at the center of the sketch; the x-axis extends horizontally and angles are measured counterclockwise. Only half a slice occurs at the crown. Because the blocking points are equally spaced along the circular arc of the steel, they are also equally spaced angularly. The angular spacing is 90/5=18°. The widths in the table total 13 ft, as they should. The total weight is $25.676(10^3)$ lbf which

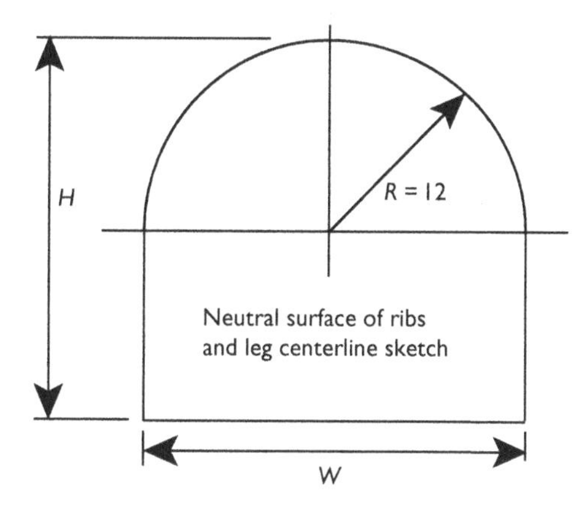

compares with 25.675(10^3) lbf by direct calculation, (γ)(W/2)(Hp). This is just one-half the total weight of the rock arch (per foot of tunnel length), of course.

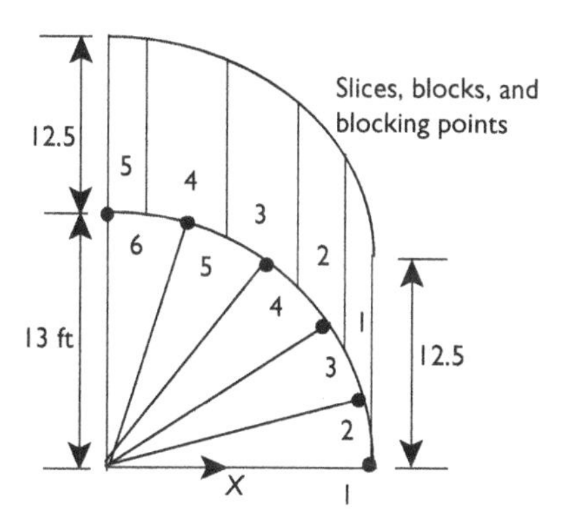

Calculation of block weights requires a trial set spacing S and computation of rock block widths which are defined by mid-points between blocks. In table form:

Point	*Segment*	*Angle*	*X = R cos(θ)*	*Midpts*	*Block*	*Width*	*Wt/ft(10^3lbf)*
1		0	13.00				
	12			12.84	1	1.42	2.805
2		18	12.36				
	23			11.58	2	2.39	4.720
3		36	10.52				
	34			9.19	3	3.29	6.498
4		54	7.64				
	45			5.90	4	3.86	7.624
5		72	4.02				
	56			2.04	5	2.04	4.029
6		90	0.00				

The active normal components of weight may also be tabulated as

Point	Block	Angle	Wt/ft(10^3lbf)	N(10^3)lbf	Slope	N'(10^3)lbf
1		0		0.00		
	1		2.805			
2		18		0.867	72	3.728
	2		4.720			
3		36		2.774	54	4.891
	3		6.498			
4		54		5.250	36	5.991
	4		7.624			
5		72		7.251	18	7.251
	5		4.029			
6		90		4.029	0	4.029

Although the block at the springline has no active normal force from weight because of the flat angle, a reactive component is sure to develop as the points near the crown load the steel and the steel in turn loads the rock.

These normal weight components require additional consideration and modification for blocks near the sides of a tunnel where the steel is steeply inclined to the horizontal. A diagram of such a block shows the forces acting. The combination of a small normal force, a high tangential force, and a small radial force indicate a potentially unstable blocking point because only a slight increase in weight will cause the resultant force to pass outside the rock mass. An increase in radial force supplied by the steel acting against the blocking point will direct the resultant force back into the rock mass and stabilize the situation. Increasing the radial force must also cause an increase in the equal but opposite normal force. An increased normal force must also be equilibrated by an increased radial force. This increase is accomplished by requiring the resultant force to be inclined at 25° to the horizontal, as shown in the sketch.

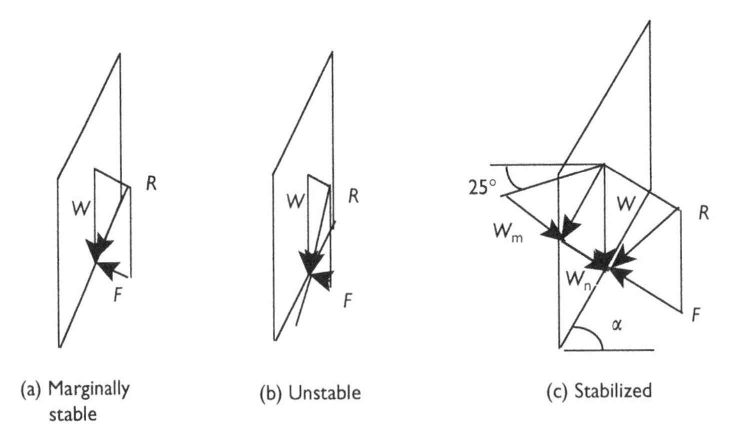

Sketch showing steep block stabilization requirement

The additional normal load is W_m in the sketch and can be computed from W and the sketch geometry. Thus, $W_n = W\cos(\alpha)$, $W_s = W\sin(\alpha)$, and $W_m = (W_s)\tan(\alpha - 25°)$. The total normal force is N' and $N' = W_n + W_m = W[\cos(25)/\cos(\alpha - 25)]$, after some algebra and application of the sine law to the force triangles in the sketch. The augmentation of active normal forces is done only for slices with inclinations greater than 25°. The results for this example are given in the table of normal forces.

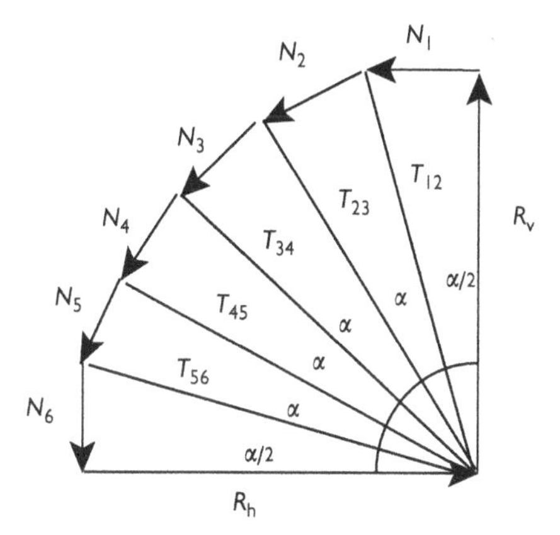

Sketch of polygon of forces

The next step is to impose equilibrium requirements that can be sketched using a polygon of forces. All the directions in the sketch are known in principal. The normal forces are aligned with rays from the origin, while the thrusts are aligned with chords between blocking points. Considering the geometry of equally spaced blocking points with one at the crown and one at the springline leads to the angles shown in the sketch where α is the angular spacing of the blocking points, in this example, 18° A rough estimate of the vertical reaction of the leg is just one-half the weight of the rock arch that loads the set, which was calculated previously as $25.68(10^3)$ lbf of tunnel length. This estimate is likely to be too high as calculation will show.

Equilibrium requires forces to accommodate the largest active normal force. This requirement is examined by a sequential calculation of thrusts and normal forces indicated in the sketch using the sine law beginning with an estimate of R_v. These calculations in tabulated form are

Point	*$N'(10^3)$lbf*	*$T(10^3)$lbf*	*$N''(10^3)$lbf*	*N(new)[a] lbf*	*T(new) lbf*
1	0.00	26.00	4.067	4.029	25.76
2	3.728	26.00	8.135	8.059	25.76
3	4.891	26.00	8.135	8.059	25.76
4	5.991	26.00	8.135	8.059	25.76
5	7.251	26.00	8.135	8.059	25.76
6	4.029	25.68	4.067	4.029	25.44

Note
a Ratio = N'/N'' = 4.029/4.067 = 0.9907.

All the active normal forces are now accommodated by the steel; no normal force N' exceeds the associated estimate N(new). The uniform spacing tends to produce a uniform thrust in the steel. As is often the case, the active force at the crown is the controlling normal load.

In this example, the maximum thrust is $25.76(10^3)$ lbf. If the steel were *A*36 steel with a yield point of 36,000 psi, then a steel area of about 0.72 in.^{2} /ft of tunnel length would be needed even without allowing for a safety factor. At a spacing of 6 ft, about 4.3 in.^{2} would be needed. However, bending stress must also be considered.

The specified safety factor needed is 1.5, so the allowable *A*36 steel stress is 24,000 psi. Thus, at this stage of the analysis,

$$24{,}000 = \frac{25{,}760}{A} + 0.85\frac{(25{,}760)h}{S}$$

where *A* and *S* are steel area and section modulus, respectively, and *h* is the arch rise between blocking points. The coefficient 0.85 in the second term allows for the fact that the segments of steel between blocking points are not actually pin connected, so the maximum moment is somewhat less. The blocking points have an angular spacing of 18° (45.2 in. with respect to steel arc length.), so the arch rise is

$$h = R - R\cos(\alpha/2) = (12)[12 - 12\cos(18/2)] = 1.77\text{in.}$$

Spacing is not unique but ranges from 2 to 10 ft in many cases. A trial spacing of 6 ft leads to

$$24{,}000 = \frac{1.545(10^5)}{A} + \frac{2.325(10^5)}{S}$$

which now requires some trial and error look-up in a handbook of steel beam properties. A *W* 10 × 30 (wide flange) has $A = 8.84$ in.^{2} and $S = 32.4$ in.^{3} which lead to an axial stress 17,478 psi and a bending stress of 7,177 for a total stress of 24,654 psi that does not quite meet the requirements. A somewhat heavier beam perhaps could be found. A *W* 10 × 33 leads to a total stress of 22,554 psi. Because area and section modulus are related, a reduction in section modulus to obtain a lighter beam by allowing for greater bending stress, leads to a smaller area and higher axial stress. By accepting a lower steel safety factor (1.46) the *W* 10 × 30 could be used with a weight savings of 10% (3 lbf per foot of steel). According to requirements, a *W* 10 × 33 *A*36 rib on 6 ft centers with blocking points spaced 45 in. will do.

Obviously larger beams spaced farther apart or smaller beams spaced more closely would also meet the problem specifications. In this regard, other factors remain the same, spacing would be dictated by the surface quality of the tunnel rock and whether there was a tendency to ravel or not. The moderately blocky and seamy description suggests that raveling is not a threat. The details of this lengthy example suggest that a spread sheet approach would indeed be advantageous.

Example 4.8 An arched back tunnel is planned that is similar to the one described in Example 4.7 but with an arch radius *R* of 13.86 ft. Although the steel rib remains circular in this plan, the full rib is no longer semi-circular and joins the leg at the springline at an angle, as shown in the sketch (*O*).

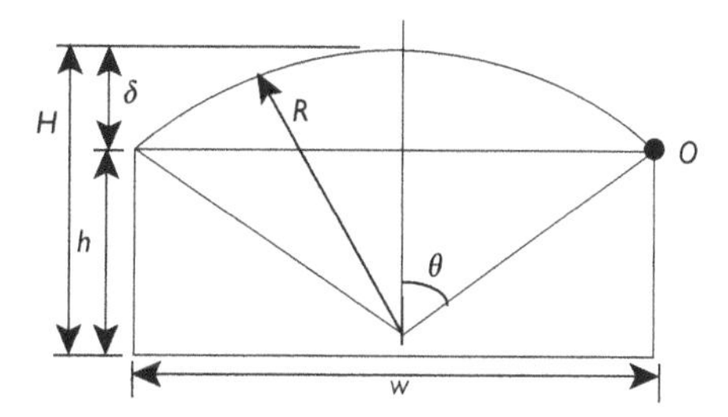

The analysis of the steel rib and determination of steel size is much the same as in Example 4.8 except for the junction at O. Determine the force equilibrium requirements at this junction and show by sketch how the forces contribute to overall equilibrium in the force polygon. Assume six equally spaced blocking points as in Example 4.7.

Solution: The blocking points and rock arch subdivision into slices are shown in the sketch; details of the forces at O are also sketched. Because the 1-block is inclined in this case rather than in a horizontal position, as shown in (b) of the sketch, there is a small rock slice that loads the 1-block. This slice is the 0-slice as shown in the sketch.

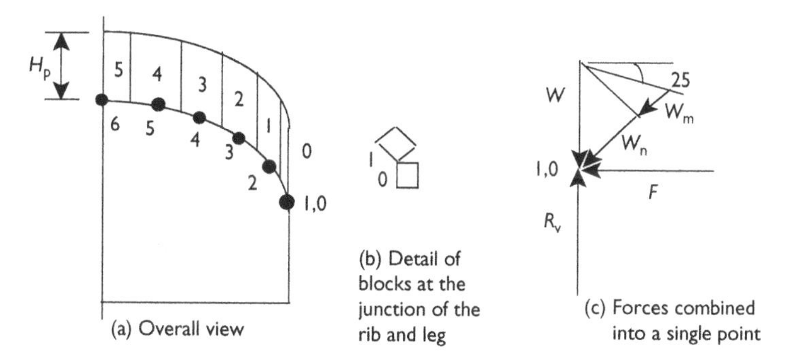

(a) Overall view

(b) Detail of blocks at the junction of the rib and leg

(c) Forces combined into a single point

Treatment of the 0- and 1-blocks as a single point leads to four instead of three forces at the junction as shown in (c) of the sketch. At each of the other blocks only three forces act. A polygon of forces is shown the sketch below.

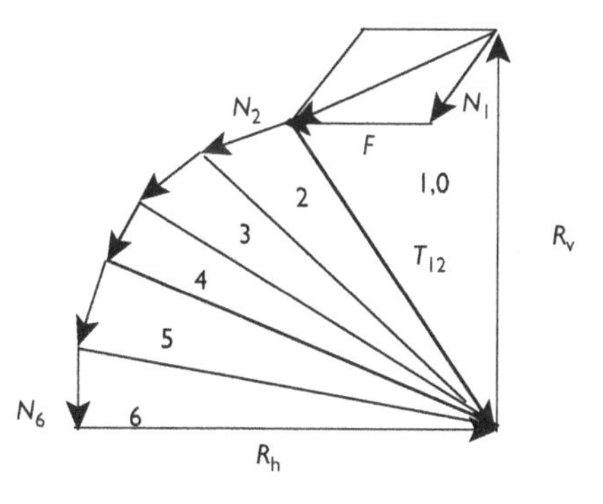

Sketch of force polygon

The vertical leg reaction may be estimated as before, that is, as one-half the rock arch weight. This estimate is likely to be high, so an improved estimate is 0.9 times the one-half the rock arch weight. As in the previous example, all the force *directions* are known including the four forces at the rib–leg junction point 1,0. At this junction R_v is drawn to scale first, then N_1 is drawn in the proper direction and of a scaled length equal to the magnitude calculated from the weight of the 0-block with adjustment for the 25° angle limitation. The horizontal force F is then drawn until it intersects the thrust T_{12}. The remainder of the force polygon is completed sequentially with knowledge of normal force and thrust directions. The order of drawing N_1 and F after plotting R_v is immaterial as the parallelogram of these forces are shown in the sketch. However, if F is drawn before N_1, the process becomes one of trial and error.

The shortened length of the rib relative to a full semi-circular arch would possibly allow for fewer blocking points without violating the maximum recommended blocking point spacing restriction. In this example, the angle θ is 60°[$\sin(\theta) = W/2R = 24/(2)(13.86) = 0.8658$], so the rib length is 14.51 ft [$S = R\theta = (13.86)(60/180)(\pi) = 14.51$]. Instead of six blocking points, in addition to the two blocks at the rib–leg junction, only four more would be needed at a spacing of 43.5 in. which is less than the maximum recommended spacing of 53 in.

Example 4.9 Consider the same tunnel 26 ft wide with an elliptical arch that has a semiaxis ratio *a/b* of two where *a* is the horizontal axis and *b* is the vertical axis, as shown in the sketch. The rectangular bottom section of the tunnel is 13 ft high. If the rock arch is 12.5 ft high and set spacing is 6 ft, what size steel is needed for a safety factor of 1.5?

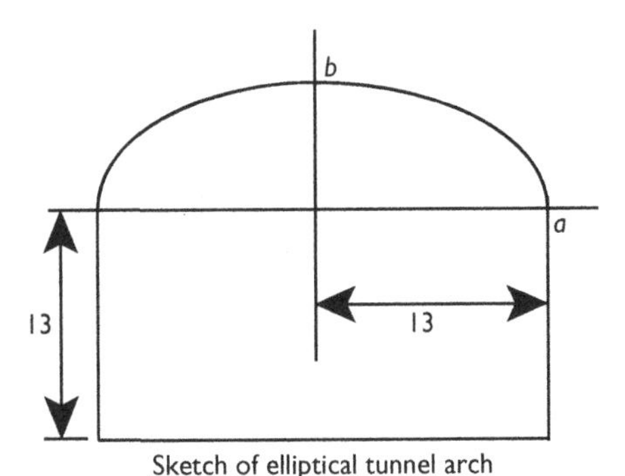

Sketch of elliptical tunnel arch

***Solution*:** Conditions are similar to those in Examples 4.7 and 4.8. Blocking point spacing should not exceed 53 in., as before. The length of the steel rib may be approximated by summation, say, of one degree increments of arc length along the ellipse. A spread sheet calculation is a handy tool for this calculation that results in an arc length of 14.53 ft. The number of equally spaced blocking points is four (rounded up) plus one at the end for a total of five spaced at 43.6 in. which does not exceed the maximum recommended spacing [(4)(43.6)/12 = 14.53, checks]. A sketch shows the blocking points and rock arch subdivided into slices.

The weight of the rock arch is γ (W/2)(Hp)(S) = (158)(13)(12.5) = 25.67(10^3) lbf of tunnel length, as before. Total half-weight is 1.54(10^5) lbf at a 6-ft set spacing. Weights of the slices require estimation of slice widths that, in turn, requires determination of the blocking point positions along the arch and, again, use of spread sheet calculation.

The results are tabulated as

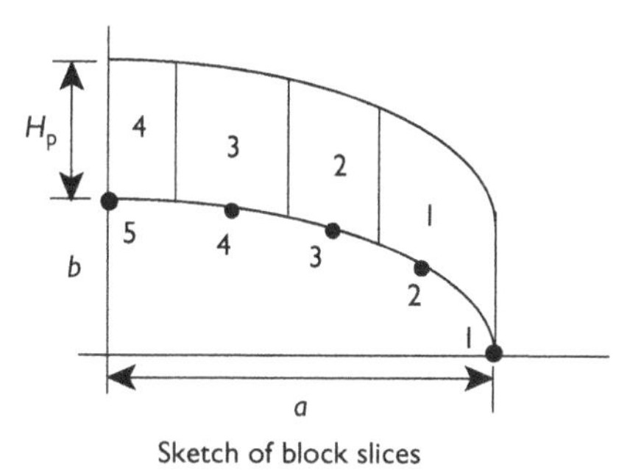

Sketch of block slices

Tabulation:

Point	X	Block	DX	W (ft)	W t (lbf) (10^4)	β/α (deg)[a]	N(lbf) (10^4)	N′(lbf)[b] (10^4)
1	13.0					0/90	0.0	0.0
		1	1.9	3.55	4.21	61		
2	11.1					50/40	3.23	3.95
		2	3.3	3.50	4.15	29		
3	7.8					59/21	3.87	3.87
		3	3.7	3.90	4.62	15		
4	4.1					81/9	4.56	4.56
		4	4.1	2.05	2.43	4		
5	0.0					90/0	2.43	2.43
Sum				13.0	15.4			

Note
a First angle is slope of normal (β), second angle is slope of tangent (α), both measured from the horizontal. Angle between points is inclination of the chord.
b Adjusted for tangent steeper than 25°.

A polygon of forces is

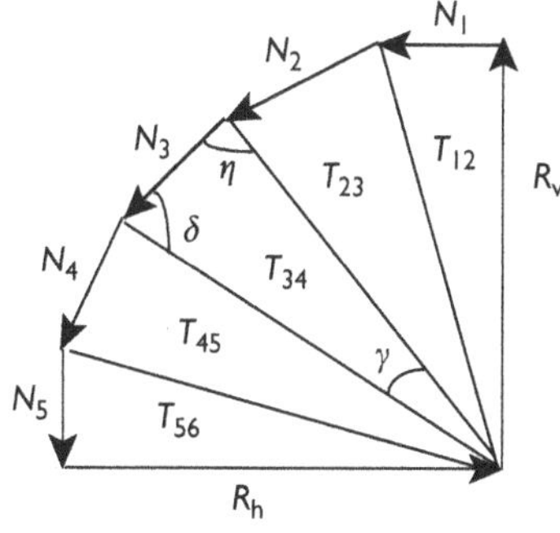

Sketch of polygon of forces

A tabulation of angles and results of application of the sine law to the force polygon:

Point	γ (deg)	δ(deg)	η(deg)	N'^{a}	N''^{a}	T	N(new)[b]	T (new)[b]
1	29	61	90	0.0	8.54	17.6	15.0	30.98
2	32	79	69	4.86	9.50	16.7	16.7	29.40
3	14	74	92	4.04	4.21	17.4	7.41	30.6
4	11	86	82	4.21	3.33	17.3	5.86	30.45
5	4	90	86	2.13	1.21	17.3	2.13	30.45

Note
a Units of force are 10^4 lbf.
b Ratio=1.76. R_v =15.4(10^4) lbf.

Once again, the thrusts tend towards uniformity. The maximum is 310,000 lbf; the maximum allowable stress in the rib, as before, is

$$24{,}000 = \frac{310{,}000}{A} + 0.85\frac{(310{,}000)(h)}{S}$$

The arch rise h is not so easily calculated for the elliptical arch compared with a circular arch, but can be done by first noting that the rise between two blocking points occurs on the ellipse where the tangent is equal to the slope of the chord between blocking points. The rise is then calculated as the distance between this point and the chord. The calculations are readily done in a spreadsheet with the result that the largest rise occurs between points 1 and 2 where the rise is 4.8 in. a rather large value. Arch rises between points 2–3, 3–4, and 4–5 are: 1.9, 0.9, and 0.1 in. respectively.

A trial 12 × 87 W shape has A = 25.6 in.^{2} and S = 118 in.^{3} and leads to a total stress of 22,783 psi. A 12 × 85 W shape has A = 25.0 in.^{2}, and S = 107 in.^{3} and leads to 23,258 psi and meets the requirements. A systematic search could possibly lead to a lighter weight and therefore lower cost rib.

One could decide to reduce set spacing from the given 6-ft spacing. Alternatively, a variable blocking point spacing that would lead to the same arch rise over each segment of the rib could produce a lower stress and lower cost (weight) support. More closely spaced points near the connection of the rib to the leg would reduce rib stress, while more widely spaced points near the crown would have little effect on steel stress.

Pattern bolting – rock reinforcement

Pattern bolting refers to an engineered system of rock reinforcement that uses relatively large bolts. Ordinary rock bolts may still be installed for immediate ground control, as is often the case, but a bolting pattern equivalent to the support obtained from fixed steel sets is also possible. A practical advantage of bolting is the greater use of tunnel cross-sectional area and perhaps lower cost. The concept is based on improving the strength of the rock mass at the tunnel walls by application of confining pressure via the bolts. In this concept, the rock arch formed by the tunnel walls is considered to be a supporting arch capable of sustaining a thrust at the arch ends as shown in Figure 4.15. The analysis follows the concept presented by Bischoff and Smart (1977).

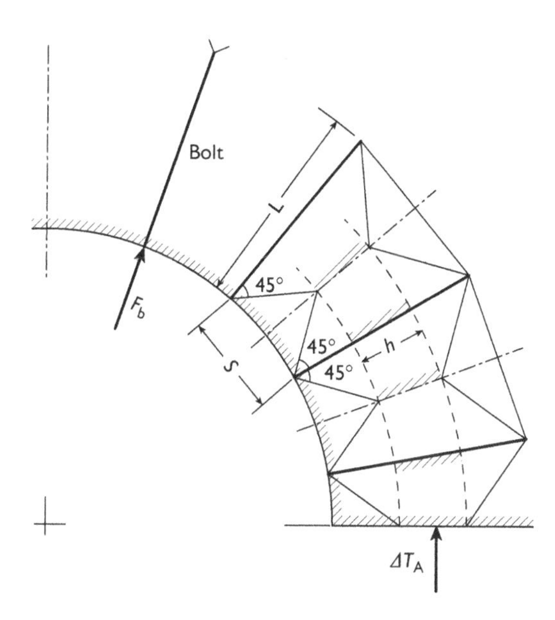

Figure 4.15 Geometry of tunnel bolting and rock arch thickness.

Before bolting, the factor of safety for the supporting rock arch is governed by the unconfined compressive strength of the rock mass. Thus,

$$\mathrm{FS_o} = \frac{C_o}{\sigma_c}$$

After bolting, a confining pressure p_b is applied to the surface of the rock arch. This pressure may be computed from bolt tension F_b and area of influence of the bolt that extends halfway to the next bolt, A_r, "rock area." Thus, $p = F_b / A_r$. According to a Mohr–Coulomb criterion for rock mass strength, compressive strength under confining pressure is

$$C_p = C_o + \left(\frac{C_o}{T_o}\right)p$$

The factor of safety after bolting is then

$$\mathrm{FS} = \frac{C_p}{\sigma_c{}'}$$

where the prime on stress indicates a different stress from the original. In this regard, one expects the stress after bolting to be somewhat less as bolt pressure tends to enlarge the opening and stretch the perimeter. One could also argue that bolting compresses the rock mass between bolt hole collar and anchor and thus expect an increase in rock arch stress. In either case, the effect would be small in consideration of bolting pressures, rock strength, and stress near strength. Assuming that there is no significant change in rock stress after bolting, the factor of safety may be expressed as

$$FS = \frac{C_o + (C_o/T_o)p}{\sigma_c} = FS_o + \Delta FS$$

The improvement in safety factor associated with bolting is therefore

$$\Delta FS = \frac{(C_o/T_o)p}{\sigma_c}$$

while the increase in strength of the supporting rock arch is simply $(C_o/T_o)p$. The increased force capacity of the reinforced rock arch is the product of increased strength by rock arch bearing area, thickness times distance along the tunnel, and is given by

$$\Delta T_A = \left(\frac{C_o}{T_o}\right)phl$$

where h and l are arch thickness and length along the tunnel. If the computation is done per foot of tunnel length, then $l = 1$ ft, and only h remains to be determined.

The thickness of the supporting rock arch depends on bolt length and spacing. With reference to Figure 4.15 that shows the geometry of the bolting pattern and how the point load from the bolt becomes distributed in the rock mass, one concludes: $h = L - S$. This conclusion is based on a 45° "cone angle" to describe the spread of the bolt force with distance into the rock mass.

The increase in thrust capacity of the rock arch may now be equated to the thrust capacity of a steel set for comparison purposes. The thrust T_s in a steel set is the largest thrust obtained from the force polygon. Comparison then amounts to the equality:

$$\Delta T_A = T_S$$

Example 4.10 Suppose the arched back tunnel in Example 4.7 using fixed steel sets with a maximum thrust of $1.545(10^5)$lbf is considered for systematic bolting in a rock mass that has unconfined compressive and tensile strengths of 15,000 and 1,500 psi, respectively. Specify a bolting plan (steel size, grade, spacing, length) that is equivalent to the steel set support plan. Draw a sketch of the proposed bolting plan and label relevant quantities. Use a bolt safety factor of 1.5.

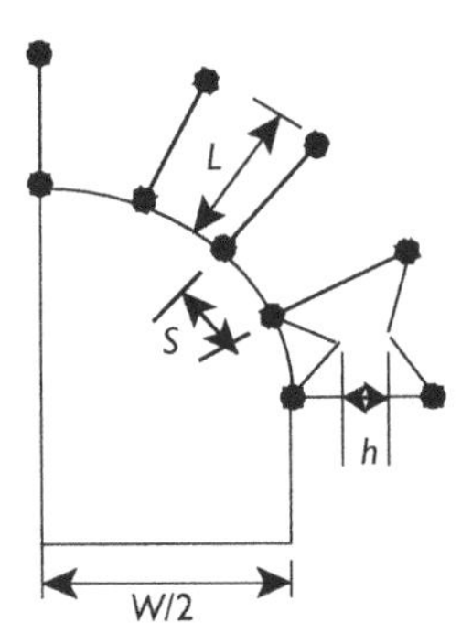

Solution: For equivalency, the increased capacity of the supporting rock arch must equal the thrust capacity of the steel ribs. Thus,

$$\Delta T_A = T_S = \left(\frac{C_o}{T_o}\right) phl$$

$$1.545(10^5)\text{lbf} = \left(\frac{15{,}000}{1{,}500}\right) phl$$

$$\therefore phl = 1.545(10^4)\text{lbf}$$

where p, h, and l are bolting pressure, arch thickness, and set spacing, respectively. As usual, there is no uniqueness to design, so some reasonable combination of variables must be decided upon. A set spacing of 20 ft would be unreasonably high, while a set spacing of 2 ft would likely be too low. A trial set spacing of 5 ft is reasonable. Thickness of the supporting rock arch may be estimated as $L - S$ where L is bolt length and S is in-set spacing. A square pattern is certainly a reasonable trial pattern, so $S = 5$ ft. Bolt length is less easily estimated; one-third tunnel width may do, so $L = 8$ ft and $h = 3$ ft. With these trial values,

$$p = \frac{1.545(10^4)}{(3)(5)} = 1.030 \text{ psf } (7.15 \text{ psi})$$

But also

$$p = \frac{F_b}{A_r} = \frac{\sigma_S A_S}{(5)(5)} = \frac{60{,}000 A_s}{(25)(144)} = 16.67 A_s$$

where 60,000 psi is steel strength, A_s = steel area, A_r = rock area, and F_b = bolt force. Solving for steel area gives $A_s = 0.429$ in.^{2} This result needs to be adjusted for a bolt safety factor of 1.5, that is, the needed $A_s = (1.5)(0.429) = 0.943$ in.^{2} Bolt diameter $d = 0.905$ in. Thus, steel grade = 60,000 psi, bolt diameter = 0.905 in. spacing = 5 × 5 ft, length = 8 ft. As a practical matter, the bolt diameter would be a standard size, say, 1 in.

Example 4.11 Systematic bolting using #10 threaded bars of 75 ksi grade spaced on 5 ft centers within rings and 7 ft centers between rings are planned to support an arched back tunnel 26 ft wide in rock where the cohesion is 3,500 psi and the angle of internal friction is 55°. Estimate the equivalent thrust capacity of fixed steel sets (also spaced on 7 ft centers).

Solution: The equivalency is

$$\Delta T_A = T_s = \left(\frac{C_o}{T_o}\right) phl$$

so estimation of unconfined compressive and tensile strength is required. There is a tacit assumption of Mohr–Coulomb strength in this expression, so

$$\left.\begin{matrix} C_o \\ T_o \end{matrix}\right\} = \frac{2c\cos(\phi)}{1 \mp \sin(\phi)} = \frac{(2)(3{,}500)\cos(55)}{1 \mp \sin(55)} = \left\{\begin{matrix} 22{,}200 \\ 2{,}210 \end{matrix}\right.$$

The bolting pressure is

$$p = \frac{F_b}{A_r} = \frac{95{,}300}{(5)(7)(144)} = 18.9 \text{ psi}$$

where a #10, 75 ksi threaded bar has a yield force of 95,300 and an ultimate load capacity of 127 ksi. A length of 1/3–1/2 tunnel width is reasonable, say, $L = 10$ ft in this example, so $h = L - S = 10 - 5 = 5$ ft. Hence,

$$T_s = \left(\frac{22{,}200}{2{,}210}\right)(18.9)(144)(5)(7) = 95{,}687 \text{ lbf}$$

Combination support

Steel sets and concrete lining may be used in combination. Conventional rock bolting in conjunction with shotcrete, a relatively thin layer of cement that is sprayed onto the tunnel walls, may also be used as a support system and even in combination with steel sets and a concrete liner that encloses the steel. In many tunnels used for civil projects, for example, a highway tunnel, rock bolts and steel sets are considered temporary support. Final support consists of a concrete liner that encloses and protects steel sets, bolts, and any screen that may be used. In soft ground, a thick, precast segmented concrete liner may be used. Wood "squeeze" blocks may then be placed between segments that form a full-circle lining. The analysis here follows that of Kendorski (1977).

A combination of conventional rock bolting for reinforcement and steel sets for support of a jointed rock mass is shown in Figure 4.16. A rock arch in the back loads the steel and supporting rock arch in the ribs. Weight of the rock arch depends on arch height H_p, tunnel dimensions, B and H_t, and on rock quality. For this analysis, the quality of the rock mass is considered "very blocky and seamy," so $H_p = 1.0(B + H_t)$. Weight of the loading rock arch is γBH_p S.

The support loads are T_s and T_m, respectively. Total support is the sum that must equilibrate the weight load with suitable safety factors for supporting rock and steel. Factors of safety for rock and steel may also be used to specify maximum allowable stresses in rock mass and support, $\sigma_m = C_m / FS_r$ and $\sigma_s = C_s / FS_s$, where C_m and C_s are unconfined compressive strengths of rock mass and support, respectively. If bolting is considered to apply a confining pressure to the rock mass, then C_m is compressive strength under confining pressure, perhaps given by a Mohr–Coulomb criterion.

An important consideration in defining rock mass strength is the effect of joints. The Terzaghi model serves the purpose. In this model the shear strength of the rock mass is considered to be a composite of shearing resistance provided by joints and intact rock bridges between joints. Area of shearing A is composed of area occupied by the joints A_j and area occupied by intact rock A_r, as shown in Figure 4.17. Joints and intact rock are assumed to follow a Mohr–Coulomb criterion. Rock mass cohesion and angle of internal friction are given by

$$\begin{aligned} c &= pc_j + (1 - p)c_r \\ \tan(\phi) &= p\tan(\phi_j) + (1 - p)\tan(\phi_r) \end{aligned}$$

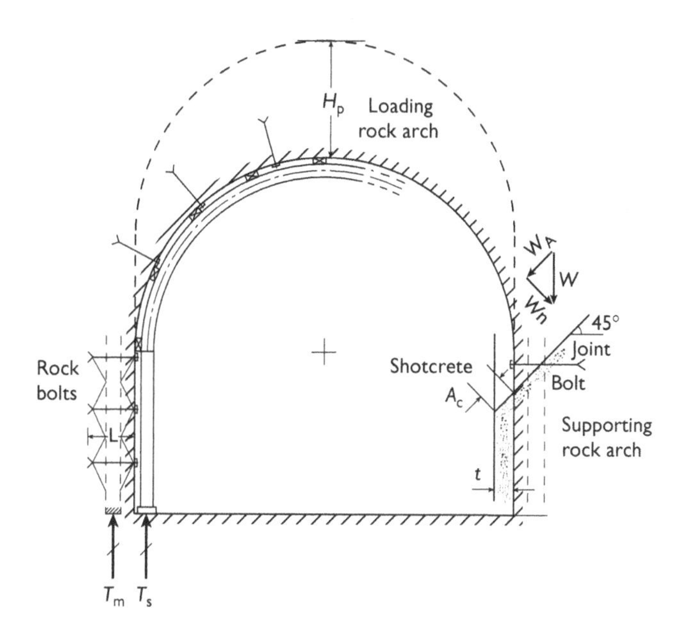

Figure 4.16 Combination support.

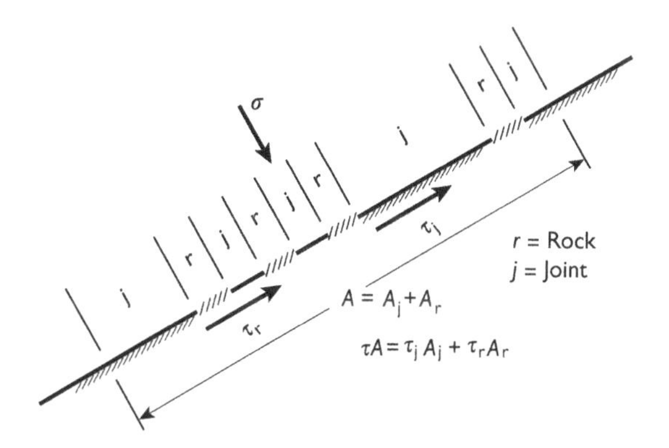

Figure 4.17 Terzaghi joined rock mass shear strength model.

where p is joint persistence, the ratio A_j/A. Rock mass unconfined compressive and tensile strengths are

$$\left.\begin{array}{l} C_o \\ T_o \end{array}\right\} = \frac{2(c)\cos(\phi)}{1 \mp \sin(\phi)}$$

where the notation is now $C_o = C_m$, and

$$\frac{C_o}{T_o} = \left(\frac{1+\sin(\phi)}{1-\sin(\phi)}\right) = \tan^2\left(\frac{\pi}{4}+\frac{\varphi}{2}\right)$$

which follows using double angle formulas and $\mu = \pi/4 - \phi/2$. In case of confining pressure

$$C_p = C_o + \frac{C_o}{T_o}P$$

where c and ϕ are rock mass properties obtained as weighted averages of joint and intact rock and P is confining pressure.

Rock support "size" follows from maximum allowable rock stress σ_m and the relation $T_m = \sigma_m A_m$ where $A_m = hS$, h = supporting rock arch thickness, S = support spacing. Similarly for the support and $T_s = \sigma_s A_s$ where A_s = support cross-sectional area. The allowable support capacity must be equal to or greater than the support load (one-half the "weight"). Thus,

$$T_m + T_s \geq \lambda \gamma B H_p S/2 > 0$$

where a factor λ that ranges between 1 and 2 has been inserted to allow for additional loading induced by nearby mining and excavation. (This feature is explained in more detail in association with chimney forces and caving ground in the chapter on subsidence.) Support, in addition to the support provided by the rock mass, is therefore

$$T_s \geq \lambda \gamma B H_p S/2 - T_m$$

If the capacity of the supporting rock arch is more than adequate to carry the "weight," then no additional structural support is needed to carry the loading arch. This situation is reflected in a negative T_s.

Example 4.12 A 12 ft wide by 12 ft high semi-circular arched back tunnel is planned in a jointed rock mass at a depth of 2,575 ft with the properties given in the table. Joint persistence $p = 0.80$ and specific weight of the rock mass is 158 pcf. Determine whether steel set support is indicated if the rock mass safety factor must be no less than 2.5 and $\lambda = 2$.

Property material	*c (psi)*	*ϕ (deg)*
Rock	3,500	55
Joint	15	25

***Solution*:** Support is indicated if the natural support capacity of the rock mass is inadequate for the given conditions. Two approaches to this issue are possible. One is to consider stress concentration about the tunnel, the second uses the loading rock arch concept that is borrowed from steel set support load estimation. This latter approach considers the weight of the loading rock arch in the crown to be $W = \lambda \gamma B H_p S$ = (load factor for proximity to caving ground)(specific weight)(tunnel breadth, width)(rock arch height)(set spacing along tunnel length). This load must be equilibrated by the supporting rock arch reaction in the rib.

Thus, $W = R = \sigma_m A_m$ where σ_m and A_m are *allowable* rock mass stress in the rib and rock mass support area, respectively. In detail,

$$\lambda\gamma BH_p S = 2\sigma_m A_m$$

$$(2)(158)(12)(12+12)(1) = (2)\left(\frac{C_o}{2.5}\right)(1)(h)$$

$$4.550(10^4) = \left(\frac{C_o}{2.5}\right)(h)$$

Whether this equation is satisfied requires computation of the rock mass compressive strength and an estimate of the supporting rock arch thickness. For the first

$$\left.\begin{array}{l} c \\ \tan(\phi) \end{array}\right\} = \left\{\begin{array}{l} (1-p)c_r + pc_j \\ (1-p)\tan(\phi_r) + p\tan(\phi_j) \end{array}\right\}$$

$$= \left\{\begin{array}{l} (1-0.80)3{,}500 + (0.8)(15) \\ (1-0.8)\tan(55) + (0.8)\tan(25) \end{array}\right\}$$

$$\left.\begin{array}{l} c \\ \tan(\phi) \end{array}\right\} = \left\{\begin{array}{l} 712 \\ 0.6587 \end{array}\right\}$$

$$\therefore c = 712 \text{ psi}$$

$$\phi = 33.4^\circ$$

$$\left.\begin{array}{l} C_o \\ T_o \end{array}\right\} = \frac{2c\cos(\phi)}{1 \mp \sin(\phi)} = \left.\begin{array}{l} 2{,}645 \\ 767 \end{array}\right\} \text{psi}$$

Accordingly, the supporting arch thickness h must be

$$h = \frac{(45{,}500)(2.5)}{2{,}645} = 43.0 \text{ ft}$$

This thickness is more than three times tunnel width and is therefore unreasonable. *Steel support is indicated.* If this thickness were reasonable, then the rock mass about the tunnel would be considered self-supporting and no steel support would be indicated.

Previously, supporting rock arch thickness was estimated as bolt length L less bolt spacing S, and bolt length was estimated to be 1/3–1/2 tunnel width W. In this example, no mention of bolting was made in the problem statement. Estimation of h therefore requires additional analysis. In this regard, the 1/3–1/2 tunnel width estimation of bolt length is made with the expectation that the bolt will be long enough to secure anchorage in relatively undisturbed ground. A one-diameter rule would indicate a distance of tunnel width or height, whichever is greater. Although there are occasions when bolts of such length may be required, a shorter length is practical. The average stress concentrations about a tunnel decrease in proportion to $(a/R)^2$ where a and R are tunnel half-width and distance from the tunnel center, respectively. This observation may also be written as $(W/2\,R)^2$. When R is $W/2 + (1/3)(W/2)$ or $W/2 + (1/2)(W/2)$, then stress concentration is about 1.6 or 1.4. When $R = W$, then the stress concentration is about 1.1. The inference is that steel sets are indicated. However, in consideration of jointing, bolting would almost certainly be done, perhaps, using short bolts (5 ft long) on 5 ft centers or so, depending on joint spacing.

Alternatively, a maximum compressive stress concentration K could be tolerated according to

$$K = \frac{C_o}{FS_c S_1} = \frac{2{,}645}{(1.0)(2{,}575)} = 1.03$$

where S_1 is the preexcavation principal stress estimated at 1 psi ft of depth. The low safety factor indicates a definite possibility of local rock mass failure. The intact rock between joints would not be threatened because of high compressive strength (22,200 psi). Failure would occur in association with joint weakening of the rock mass. For this reason, bolting to prevent blocks formed by joints and fractures from moving into the tunnel opening is indicated.

Although the compressive strength of the rock and any structural support protect an opening against strength failure of the rock mass, there is also a possibility of shear failure along joints. This mode of failure may be considered in relation to a joint factor of safety as a ratio of resisting to driving forces along the joint: $FS_j = R_j/D_j$. The driving force is the tangential or shearing component of rock weight W_s. Resisting forces arise from (1) joint friction and cohesion, (2) rock bolts that cross a joint, and (3) possibly concrete or shotcrete lining the tunnel walls.

In the absence of site-specific data, a generic analysis may be done to illustrate a reasonable procedure for determining a safety factor with respect to joint failure. In this analysis, a single joint dipping 45° into a tunnel wall is assumed, as shown in Figure 4.16. One horizontal rock bolt of length L penetrates the joint. First the driving force is calculated $D = W_s = W\sin(45°) = (1/\sqrt{2})W$

The joint strength resisting force R_j is

$$\tau_j A = \sigma_n A \tan(\phi_j) + c_j A$$

where the subscript j means joint and A is the joint area, $A = LS\sqrt{2}$ which is the length of a rock bolt projected to the inclined joint dipping 45°, and $W_n = \sigma_n A = W\cos(45°)$ is the normal force acting on the considered joint segment.

The bolting resisting shear force is the product of bolt shear strength and cross-sectional area; no allowance is made for dip of the joint through the bolt. Thus, $R_b = \tau_b A_b$. A reasonable estimate of bolt shear strength is one-half bolt tensile (or compressive) strength in consideration of the ductile nature of bolt steel. Nominal bolt diameter is sufficiently accurate in this generic analysis to estimate bolt area.

Resistance to shear down the considered joint by a shotcrete or concrete wall coating using a suitable safety factor, that is, maximum allowable shear force is given by

$$R_c = \tau_c A_c = (2\sqrt{f_c'})tS\sqrt{2}$$

where the terms in parentheses represent the maximum recommended allowable stress of unreinforced concrete in "diagonal tension," that is, pure shear, f_c' is the unconfined compressive strength of shotcrete, t is the horizontal thickness of shotcrete, S is "spacing" (distance along tunnel), and the square root factor accounts for the inclined area of shearing resistance. The units, despite the square root operation, are lbf /in.^{2} or N/m^2.

After consideration of all resisting forces, the joint factor of safety is

$$FS_j = \frac{F_j + F_b + F_c}{D}$$

Example 4.13 Given the rock mass conditions described in Example 4.12, determine the joint safety factor without support. If this number is less than 2.0, determine the joint safety factor under a bolting plan of 3/4-in., high grade (55 ksi), 6-ft bolts, spaced on 4-ft centers. If the result is less than 2.0, estimate the shotcrete thickness needed to obtain a safety factor of 2.0. Shotcrete compressive strength is estimated to be 3,500 psi.

Solution: The joint safety factor without bolting or shotcrete is

$$\begin{aligned} FS_j &= \frac{F_j}{D} \\ &= \frac{N_j \tan(\phi_j) + c_j A_j}{(W/2)(1/\sqrt{2})} \\ &= \frac{(W/2)(1/\sqrt{2})\tan(\phi_j) + c_j(L_j\sqrt{2})(S)}{(W/2)(1/\sqrt{2})} \\ &= \tan(\phi_j) + \frac{2c_j L_j S}{(W/2)} \\ &= \tan(25) + \frac{(15)(144)(6)(\sqrt{2})(4)}{(4.550)(10^4)(1/2)(1/\sqrt{2})(4)} \end{aligned}$$

$$FS_j = 0.4663 + 1.14 = 1.61$$

which does not meet the 2.0 requirement, so bolting must be considered. The bolting force at the yield point is approximately 16,900 lbf. This estimate does not allow for the cut thread and is somewhat optimistic. In shear, this force should be reduced by half. Thus, the bolting addition to the joint safety factor is

$$\begin{aligned} \Delta FS_j &= \frac{F_b}{(W/2)(1/\sqrt{2})(4)} \\ &= \frac{(16{,}900/2)}{4.55(10^4)(1/2)(1/\sqrt{2})(4)} \end{aligned}$$

$$\Delta FS_j = 0.131$$

which is quite small, as expected. The joint safety factor with bolting is 1.74, so shotcrete must be considered. The shotcrete addition to the joint safety factor is

$$\Delta FS_j = \frac{2A_c\sqrt{f'_c}}{(W/2)(1/\sqrt{2})(S)}$$

$$2.0 - 1.74 = \frac{2A_c\sqrt{3{,}500}}{4.550(10^4)(1/2)(1/\sqrt{2})(4)}$$

which can be solved for the shotcrete area. Thus, A_c = 141.4 in.^{2} The thickness required h = 141.4/(12)(4) = 29.5 in. This is a rather large value, so some thought might be given to increasing shotcrete strength, say, by adding fibers or by accounting for wire mesh or by redesign of the bolting pattern. Additional site-specific details would also allow for a more suitable support system design.

Yieldable steel arches

In yielding ground, fixed steel sets or similar stiff, rigid support must be either massive enough to withstand high rock loads, an unlikely possibility, or an alternative ground control approach must be used. Again, even massive support can offer little resistance to rock mass motion that may be generated by thousands of psi of *in situ* stress. The alternative is support that "yields" with the rock mass but still prevents *uncontrolled* flows and falls of rock into the opening. Yieldable steel arches are intended for this purpose. These arches are made of relatively light steel segments that have a "U" or "V" shape. The segments overlap and are tightly clamped with special bolts, "J" or "U," bolts that allow the segments to slip one past the other if the load becomes excessive. Thus, the yielding is not inelastic deformation of the steel that occurs when loaded beyond the elastic limit, but rather frictional sliding.

Resistance to frictional sliding is mobilized by bolt clamping forces. Thus,

$$T = \mu N$$

where T, μ, and N are frictional resistance to slip, coefficient of sliding friction between steel segments and the normal clamping force, as shown in Figure 4.18. Slip should occur just below the elastic limit of the steel to prevent set damage, so T is somewhat less than the axial thrust in the steel at the elastic limit which may be computed as steel strength times section area (σA). In case of the usual $A36$ mild steel used for beams and columns, strength is about 36,000 psi. The corresponding steel safety factor with respect to the elastic limit would be relatively low, say, 1.1.

Thrust in the steel generated by rock "pressure" p is given by the formula $T = pRS$ where R and S are radius and set spacing, respectively. Thus, the structural analysis is simple,

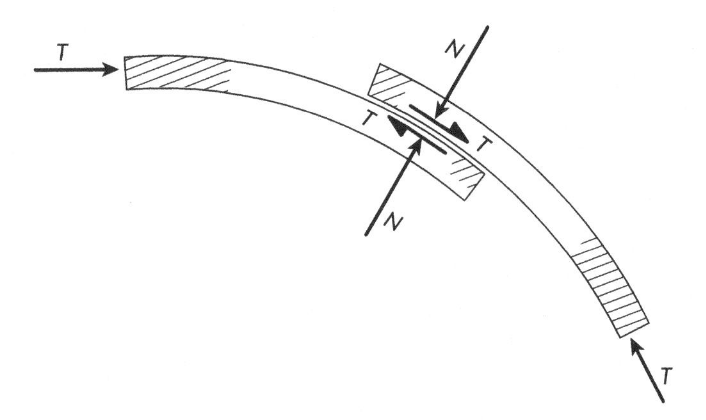

Figure 4.18 Clamping forces, frictional forces, and steel thrust in a schematic of a yieldable steel arch segment.

although there is no unique answer because each spacing tried leads to a different steel size. The main question for support analysis, as before, concerns the load transmitted to the support by the rock mass, p. This load may be estimated for comparison with fixed steel sets, so $p = \gamma H_p$. Spacing of yieldable arches is governed by the same factors that influence spacing of fixed steel sets, mainly, the tendency of the rock mass to ravel or flow between sets.

Example 4.14 Squeezing ground is encountered at a depth of 350 m where yieldable steel arches are being considered for support instead of fixed steel sets for a tunnel 5 m wide by 5 m high. The rock mass is described as "squeezing rock, moderate depth," so $H_p = (1.1–2.1)(B + H_t)$, say, $H_p = 1.5(5 + 5) = 15$ m, a substantial rock load. An arched back section will be used if steel sets are decided upon; a circular section would be used if yieldable arches are used. Determine the cross-sectional area of steel needed for the latter option.

Solution: The formula for thrust $T = pRS = \gamma H_p RS$ where γ, H_p, R, S are specific weight of the rock mass, rock "head," tunnel radius, and set spacing. Thus, for a trial spacing of 1.5 m,

$$T = (25.2)(10^3)(15)(5/2)(1.5) = 1.417(10^6)N$$

$$\therefore A_s = 1.417(10^6)/[(250)(10^6)/(1.1)] = 6.24(10^{-3})m^2 = 62.4\text{cm}^2$$

where the specific weight of rock is assumed to be 25.2 kN/m^3, the yield point of $A36$ steel is 250 MPa and a steel safety factor of 1.1 is used.

Light segment liner

Tunnels and shafts may also be supported by segments of steel panels bolted together to form a continuous liner. Flanges of the curved panel segments may be on the inside or outside of the segment. The outside configuration is used when a smooth inside surface is desirable, for example, when the opening is an ore pass. Relatively lightweight corrugated steel panels may also be used for tunnel and shaft lining. In any case, sand and even concrete is used to fill the void between the panel and rock wall of the opening.

One of the advantages of light-segment liner used with sand backfill is the accommodation of rock wall motion through compaction of the fill, without detriment to the liner. Thickness of the sand fill should be at least twice the maximum rock protuberance. An adequate fill thickness reduces the threat of point loading and buckling of a thin liner. In this regard, corrugated steel, usually galvanized, offers greater resistance to bending than a small, thinwalled steel liner.

Example 4.15 A horseshoe tunnel section is being considered for support by light-segment liner with sand backfill. Local relief of newly blasted tunnel wall is estimated at 25 cm. Sketch the position of the liner with respect to the tunnel wall that allows safety with respect to buckling.

Solution: According to rule of thumb, the liner should stand off the tunnel wall a distance equal to the local relief. In this example, the distance would be 25 cm, as shown in the sketch.

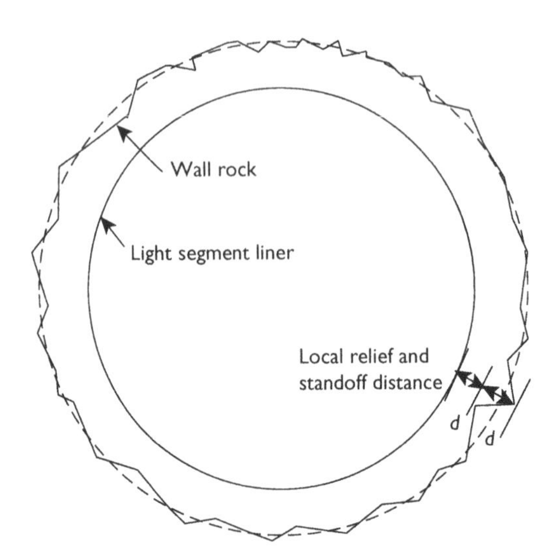

Example 4.16 Consider a semi-circular arched back tunnel at a depth of 300 m in a gravity field where the horizontal preexcavation stress is one-third the vertical stress. The tunnel is 8 m wide by 8 meter high, so the arch radius is 4 m. The tunnel bears due north. Three joint sets are present. Intact rock Young's modulus and Poisson's ratio are 100 GPa and 0.20, respectively. Unconfined compressive and tensile strengths are 100 MPa and 10 MPa, respectively. Young's modulus of the jointed rock mass is estimated at 0.3 times intact rock modulus. Strengths are estimated to be 0.3 times intact values. Determine the tunnel safety factors with respect to tension and compression.

***Solution*:** By definition the safety factors are $FS_c = C_o/\sigma_c$ and $FS_t = T_o/\sigma_t$. The peak stresses at the tunnel periphery are given by stress concentration factors, that is, $\sigma_c = K_c S_1$ and $\sigma_t = K_t S_1$ where S_1 is the preexcavation major principal stress with compression positive. In this example, $S_1 = \gamma H$, the product of unit weight by depth. A reasonable estimate of the vertical gravity stress is 25 kPa/m. Thus, S_1 = (300)(25)or 7.5 MPa. Stress concentration factors for crown, floor, rib, and corner may be obtained from Figures 4.3 and 4.4. Thus,

$$\begin{aligned}
K(crown) &= (1)[(-0.012)(1) - 0.939] + (1/3)[(1.056)(1) + 2.34] \\
K(floor) &= (1)[(-0.012)(1) - 0.939] + (1/3)[(0.712)(1) + 0.969] \\
K(rib) &= (1)[(-0.132)(1) + 2.676] + (1/3)[-(0.128)(1) - 0.83] \\
K(crown) &= 0.181 \\
K(floor) &= -0.391 \\
K(rib) &= 2.22
\end{aligned}$$

The corner stress concentration factor may be estimated by $K\,(corner) = R + 0.1(R + F)$ where R and F are obtained by superposing vertical and horizontal effects. Thus,

$$\begin{aligned}
R_y &= (2.303)(1) + 1.2 \\
F_y &= (0.693)(1) + 0.087
\end{aligned}$$

$$R_x = (-0.078)(1) + 0.884$$
$$F_x = (0.854)(1) + 2.538$$
$$R_y = 3.503$$
$$F_y = 0.780$$
$$R_x = 0.846$$
$$F_x = 3.392$$
$$R = R_y + (1/3)R_x$$
$$R = 3.785$$
$$F = F_y + (1/3)F_x$$
$$F = 1.911$$
$$K(corner) = 3.785 + (0.1)(3.785 + 1.911)$$
$$K(corner) = 4.355$$

The results show the peak compressive stress concentration is at the corner and the peak tensile stress occurs at the floor. Hence,

$$FS_c = C_o/\sigma_c$$
$$= (0.3)(100)/(4.355)(7.5)$$
$$FS_c = 0.92$$

Some yielding at the corner is therefore indicated.

$$FS_t = T_o/\sigma_t$$
$$= (0.3)(10)/(0.319)(7.5)$$
$$FS_t = (1.25)$$

The floor remains safe, but only marginally so in consideration of tensile failure.

Example 4.17 Consider the data in Example 4.16 and the presence of three joint sets with geometric properties shown in the table. Strength properties are given in the second table. Specific weight of rock is 25 kN/m^3. Sketch the tunnel section looking north and joints, then decide if bolting is necessary. Neglect any resistance from the vertical joints.

Joint Geometry

Property Joint Set	Dip (deg)	Dip Direction (deg)	Spacing (m)	Persistence (-)
J1	60	east	3.2	0.8
J2	30	west	2.3	0.5
J3	90	north	1.8	0.9

Joint Strengths

Property Joint Set	Cohesion (kPa)	Friction Angle (deg)
J1	35	35
J2	20	25
J3	10	42

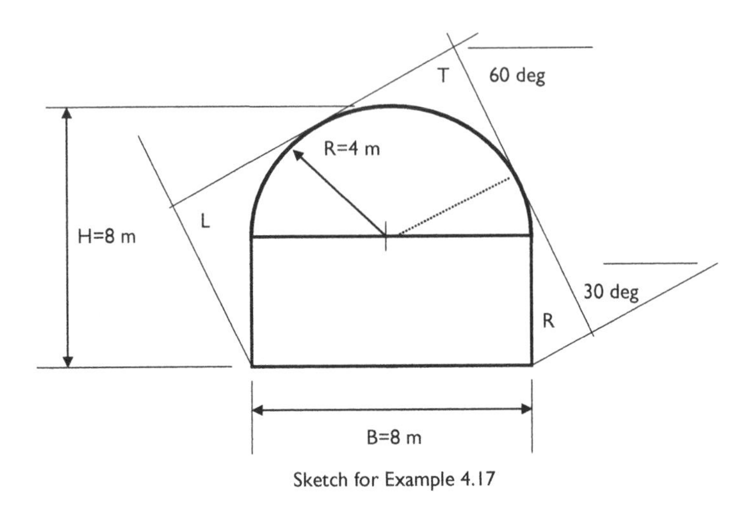

Sketch for Example 4.17

***Solution*:** The section appears in the sketch. Potential rock blocks formed by the joints in the ribs and back are labeled as right (R), left (L), and top (T). Because of the vertical joint parallel to the section, all blocks extend 1.8 m into the page. Failure may occur on the joints, so safety factors with respect to joint block failures are needed. A Mohr-Coulomb failure criterion is reasonable and in accordance with the joint strength properties given in the table. However, strength is stress-dependent and the stress state about the tunnel is complex and, in fact, unknown. An approximate analysis is therefore necessary. A simple approximation is to consider the rib blocks as sliding down an inclined plane.

Rib block R: The plane inclination is 30° while the friction angle is only 25°, so if cohesion were neglected the block would slide as excavation released the preexcavation stress retraining the block. The joint is not 100% persistence and is therefore stronger and has a higher effective friction angle than the given angle. However, at the tunnel wall, the joint persistence is 100%, so strengthening by persistence is negligible for a short distance into the wall. The joint safety factor may be computed as the ratio of driving to resisting forces acting on the block. Thus,

$$FS_j = R/D$$

D = driving force downhill

R = resisting force (strength)

The driving force is the downhill component of block weight, while the resisting force is composed of friction mobilized by the normal component of weight W and the cohesive force of resistance. Hence,

$$FS_j = \frac{\tan(\phi)}{\tan(\delta)} + \frac{cA_j}{W\sin(\delta)}$$

The joint area is the dip length times the spacing of joints along the tunnel (1.8 m); weight of the block is specific weight times volume V. The block volumes may be approximated by triangles and rectangles using a chord length for triangles in the arch. With this approximation, the block weights and areas are:

$$W(R) = (1.8)(5.86)(25) = 264 \text{ kN } (59 \text{ klbf}), A(R) = (1.8)(2.54) = 4.57m^2$$

$$W(L) = (1.8)(14.4)(25) = 648 \text{ kN } (145 \text{ klbf}), A(L) = (1.8)(5.46) = 9.83m^2$$

$$W(T) = (1.8)(3.44)(25) = 155 \text{ kN } (35 \text{ klbf}),$$

The joint safety factors are then

$$FS_j(R) = \frac{\tan(25)}{\tan(30)} + \frac{(20)(4.57)}{(264)\sin(30)} = 1.50$$

$$FS_j(L) = \frac{\tan(35)}{\tan(60)} + \frac{(35)(9.83)}{(648)\sin(60)} = 1.02$$

The safety factors indicate the right rib is stable, but the left rib is near failure with respect to slip on joints. Bolting ribs in this case would be prudent.

The top block is positioned to fall under self-weight. Joint cohesion may be sufficient to support the block in addition to any friction mobilized by joint normal stress. However, the threat of a block fall is evident and support is indicated.

Example 4.18 Consider the data in Example 4.17, and the results that indicate a need for support of ribs and back. Decide on a suitable bolting pattern and specify bolt lengths, spacings, steel size and grade using point-anchored, pretensioned bolts. Show on a sketch of the tunnel cross-section.

***Solution*:** The top block is the most important and must be secured against the possibility of a fall. A reasonable approach is to require the bolts to support the weight of the block as a minimum. Bolts are customarily tensioned to 67% to 75% of the steel yield point. These tensions imply a bolt safety factor of 1.5 to 1.33 (reciprocal of the tension percentage). A bolt safety factor may also be given as a ratio of bolt strength to the weight of a rock block supported by the bolt. Thus, $FS_b = F_b / W_r$. If the bolt is tensioned to 67% of the bolt yield point, then the safety factor is 1.5 and the bolt force should be 1.5 times block weight. The block weight assigned to a bolt is determined by bolt spacing and bolt length less an allowance for anchorage. This allowance is often 0.3 m (1 ft). From the geometry of the top block, the apex of the block measured along a radius is 1.46 m into the solid. This length suggests a bolt length at least 1.76 m, say, 2 m to insure adequate anchorage. Spacing along the tunnel is a

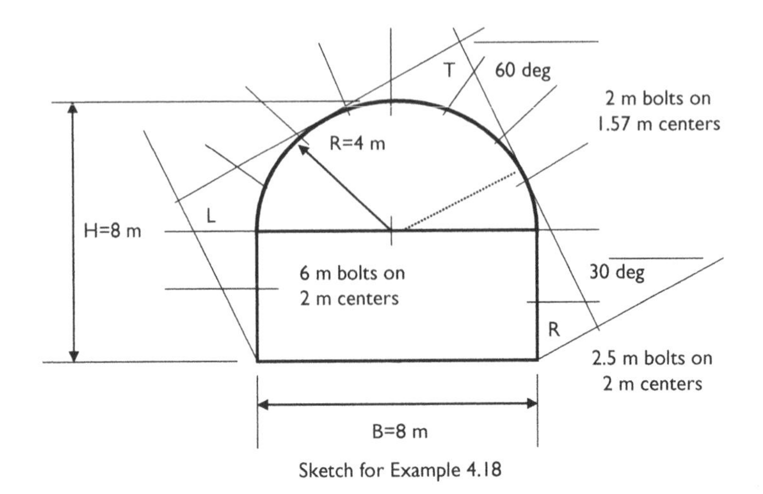

Sketch for Example 4.18

matter of choice and in view of the 1.8 m joint spacing, a spacing of 1.8 m is a reasonable first try.

Square patterns are often the case, but a 1.57 m spacing over the arch of the tunnel leads to a 22.5 degree spacing as shown in the sketch. The center bolt is 1.8 m and has an associated block volume of (1.46)(1.8)(22.5)(4)(π/180)=4.13 and a block weight of 103.2 kN (230 klbf). The bolt load is then (1.5)(25.8)=38.7 kN (86.4 klbf). From Table 3.4, a 7/8 inch, grade 75 bolt will meet the requirement. There are seven such bolts that have a combined capacity of 271 kN (60.5 klbf) that is almost double to total weight of the top block. Again, these bolts are 2 m long.

Bolting the ribs requires bolts to extend beyond the joint plane upon which the block may slip. The larger rib block on the left requires bolts at least 5.65 m long, say, 6 m to allow for anchorage. Bolts in the right rib should be at least 2.2 m long, say, 2.5 m. The rib bolts begin at the base of the arch and are continued down the rib, although typically not continued to the floor. A rib spacing of 2 m would allow for two rib bolts on each side. The lowest bolt would be 2 m above the floor as shown in the sketch. These bolts may also be 7/8 inch diameter, grade 75.

Example 4.19 Consider the data in Example 4.18, and the results that indicate a need for support of ribs and back. Use the bolting specifications given in the example, and then compute the increase in safety factor against rib block sliding obtained by bolting and the resulting safety factors for the right and left rib blocks.

***Solution*:** The increase in safety factor obtained by bolting is caused by mobilization of friction and directly. Thus,

$$\Delta FS_j = \frac{F_b \cos(\delta - \phi)/\cos(\phi)}{W \sin(\delta)}$$

For the block on the right rib, per bolt

$$\Delta FS_j = \frac{(34{,}200)\cos(30 - 25)/\cos(25)}{(59{,}000)\sin(30)}$$

$$\Delta FS_j = 1.27$$

And for the block on the left rib, per bolt

$$\Delta FS_j = \frac{(34{,}200)\cos(60 - 35)/\cos(35)}{(145{,}000)\sin(60)}$$

$$\Delta FS_j = 0.30$$

The improvement for the two rib bolts is

$$\Delta FS_j = 2.54 \text{ (right rib)}$$

$$\Delta FS_j = 0.6 \text{ (left rib)}$$

The resulting safety factors are

$$FS_j = 3.04 \text{ (right rib)}$$

$$FS_j = 1.62 \text{ (left rib)}$$

Example 4.20 Consider the data in Examples 4.16 and 4.17 where there are three joint sets present with geometric properties shown in the table. Strength properties are given in the second table. Specific weight of rock is 25 kN/m^3. FS_j *(R)* = 1.50 and FS_j *(L)* = 1.02. Suppose grouting is used to improve tunnel rib safety by increasing joint cohesion by 1000 kPa (145 psi). Determine the grouted joint safety factors.

***Solution*:** The safety factor of a joint is

$$FS_j = \frac{\tan(\phi)}{\tan(\delta)} + \frac{cA_j}{W\sin(\delta)}$$

And

$$FS_j(R) = \frac{\tan(25)}{\tan(30)} + \frac{(20 + 1000)(4.57)}{(264)\sin(30)}$$

$$FS_j(R) = 36.1$$

$$= \frac{\tan(35)}{\tan(60)} + \frac{(35 + 1000)(9.83)}{(648)\sin(60)}$$

$$FS_j(L) = 18.5$$

These results show that cohesion is important in gravity block slides. Neglecting cohesion would indicate slip of both right and left rib blocks. With the natural cohesion, slip is not indicated and with grouting, slip failure is surely prevented.

4.3 Problems

Naturally supported tunnels

4.1 The stresses relative to compass coordinates (x = east, y = north, z = up) are

$$\sigma_{xx} = 2{,}155,\ \sigma_{yy} = 3{,}045,\ \sigma_{zz} = 4{,}200,\ \tau_{yx} = -1{,}222,$$
$$\tau_{xz} = 0,\ \tau_{yx} = 0$$

in psi with compression positive. Suppose a circular tunnel is driven due east by a tunnel boring machine. Estimate the peak compressive and tensile stress concentrations and show locations in a sketch.

4.2 Consider a horizontal, rectangular opening that is 10 ft high and 20 ft wide and driven due north 5,000 ft. Rock properties are

$$C_o = 23{,}700\text{ psi},\ T_o = 1{,}480\text{ psi},\ E = 5.29(10^6\text{ psi}),$$
$$\nu = 0.27,\ \gamma = 162\text{ pcf}$$

The premining stress state relative to compass coordinates is:

$$S_E = 350 + 0.2\,h,\ S_N = 420 + 0.35\,h,\ S_V = 1.12\,h$$

where stresses are in psi, h =depth in ft, E, N, V refer to compass coordinates (x = east, y = north, z = up) and compression is positive. Premining shear stresses are nil relative to compass coordinates. How wide can this opening be made before failure occurs at a depth of 1,750 ft? Justify.

4.3 A tabular ore body 15 ft thick is mined full-seam height at a depth of 2,300 ft by repeated slices 20 ft wide and 5,000 ft long, so the first drive is simply a 20 ft wide tunnel from the rock mechanics view. Rock properties are

$$E = 5.7 \times 10^6\text{ psi},\ \nu = 0.25,\ C_o = 25{,}300\text{ psi},\ T_o = 2{,}600\text{ psi}.$$

The premining stress field is due to gravity alone. If safety factors of 2.2 in compression and 4.4 in tension are required, how many slices side by side can be safely taken?

4.4 The stresses relative to compass coordinates (x = east, y = north, z = up) are

$$\sigma_{xx} = 14.86,\ \sigma_{yy} = 21.00,\ \sigma_{zz} = 28.97,\ \tau_{yy} = -8.43,\ \tau_{xz} = 0,\ \tau_{yx} = 0$$

in MPa with compression positive. Suppose a circular tunnel is driven due east by a tunnel boring machine. Estimate the peak compressive and tensile stress concentrations and show locations in a sketch.

4.5 Consider a horizontal, rectangular opening that is 3 m high and 6 m wide and driven due north 1,524 m. Rock properties are

$$C_o = 163.5\text{ MPa},\ T_o = 10.2\text{ MPa},\ E = 36.48\text{ GPa},$$
$$\nu = 0.27,\ \gamma = 25.34\text{ kN/m}^3$$

The premining stress state relative to compass coordinates is:

$$S_E = 2{,}414 + 4.53\text{ h},\ S_N = 2{,}897 + 7.92\text{ h},\ S_V = 25.34\text{ h}$$

where stresses are in kPa, h = depth in m, E, N, V refer to compass coordinates (x = east, y = north, z = up) and compression is positive. Premining shear stresses are nil relative to compass coordinates. How wide can this opening be made before failure occurs at a depth of 533 m? Justify.

4.6 A tabular ore body 4.6 m thick is mined full-seam height at a depth of 700 m by repeated slices 6 m wide and 1,520 m long, so the first drive is simply a 6 m ft wide tunnel from the rock mechanics view. Rock properties are

$$E = 39.3\text{ GPa},\ v = 0.25,\ C_o = 174.4\text{ MPa},\ T_o = 17.9\text{ MPa}.$$

The premining stress field is due to gravity alone. If safety factors of 2.2 in compression and 4.4 in tension are required, how many slices side by side can be safely taken?

Supported tunnels

4.7 Consider a pin-connected, two-segment, semi-circular steel rib shown in the sketch.

(a) If the bearing ends are free to rotate, find the thrust and moment in the steel. Data are: H_p = 10 ft, H_t =18 ft, B = 16 ft, R = 7.5 ft, set spacing S = 6 ft and specific weight of rock γ = 156 pcf.

(b) What area of A36 steel is needed to resist thrust?

(c) If the section modulus of the steel beam is 30 in.^{3}, what is the bending stress?

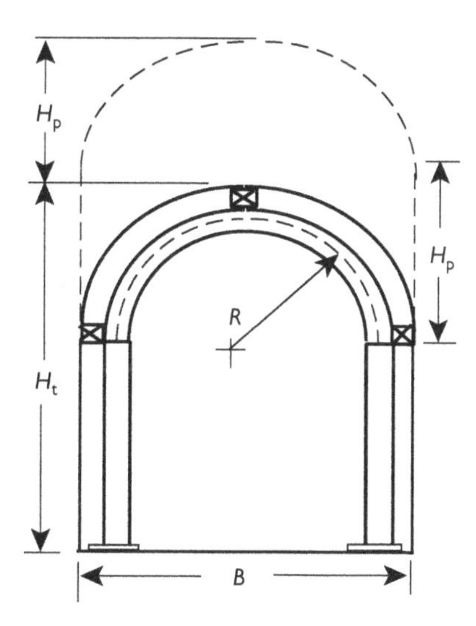

4.8 Two-piece continuous steel sets are to support a tunnel with a semi-circular roof having a radius of 12.67 ft and a straight leg section of 13 ft. The tunnel is therefore 25.33 ft wide by 25.67 ft high. The crown is assumed hinged and the ends of the steel arch are fixed. For preliminary design assume 1 in. of steel rib depth for each 3 ft of tunnel width and a set spacing of 4 ft. Note that the 12 ft radius is to the outside of the steel. Blocks 8 in. thick are placed at the crown and at the base of the arch. In addition, blocks next to these are spaced a maximum of 50 in. Remaining blocks required should not be spaced more than 50 in. A rock load of 10 ft is expected on the basis of rock quality; rock unit weight is 170 pcf.

1 Draw a neat scale diagram of the tunnel and steel set.
2 Show the position of the blocks.
3 Show the rock arch and resolution of forces (to scale) loading the set; indicate magnitudes and directions.
4 Replace the 13 ft steel legs by a vertical reaction R_V at the arch base.
5 Draw the cords between blocking points defining the thrust directions in the steel arch.
6 Construct the force polygon.
7 Calculate the maximum moment in the steel arch.
8 Calculate the maximum stress in the steel.
9 Estimate the safety factor for the steel arch.

4.9 Consider a pin-connected, two-segment, semi-circular steel rib shown in the sketch.

(a) If the bearing ends are free to rotate, find the thrust and moment in the steel. Data are: $H_p = 3$ m, $H_t = 5.5$ m, $B = 5$ m, $R = 2.3$ m, set spacing $S = 2$ m and specific weight of rock $\gamma = 25.0$ kN/m^3.
(b) What area of A36 steel (yield strength 250 MPa) is needed to resist thrust?
(c) If the section modulus of the steel beam is 500 cm^3, what is the bending stress?

4.10 Two-piece continuous steel sets are to support a tunnel with a semi-circular roof having a radius of 3.5 m, and a straight leg section of 4 m. The tunnel is therefore 7 m wide by 7.5 m high. The crown is assumed hinged and the ends of the steel arch are fixed. For preliminary design assume 2.5 cm of steel rib depth for each1 m of tunnel width and a set spacing of 1.2 m. Note that the 3.5 m radius is to the outside of the steel. Blocks 20 cm thick are placed at the crown and at the base of the arch. In addition, blocks next to these are spaced a maximum of 127 cm. Remaining blocks required should not be spaced more than 127 cm. A rock load of 3 m is expected on the basis of rock quality; rock unit weight is 26.9 kN/m^3.

1 Draw a neat scale diagram of the tunnel and steel set.
2 Show the position of the blocks.
3 Show the rock arch and resolution of forces (to scale) loading the set; indicate magnitudes and directions.
4 Replace the 4 m steel legs by a vertical reaction R_V at the arch base.
5 Draw the cords between blocking points defining the thrust directions in the steel arch.
6 Construct the force polygon.
7 Calculate the maximum moment in the steel arch.
8 Calculate the maximum stress in the steel.
9 Estimate the safety factor for the steel arch.

4.11 Consider the analysis of tunnel bolting leading to the equation

$$\Delta T_{\mathrm{A}} = \left(\frac{C_{\mathrm{o}}}{T_{\mathrm{o}}}\right) p_{\mathrm{b}} t l_{\mathrm{r}}$$

If bolting is on a square pattern of spacing S and a 45° cone of influence is assigned to the bolt forces, show that this equation is equivalent to that of Bischoff and Smart (1977), equation 8,

$$\Delta T_{\mathrm{A}} = (q)\left(\frac{\sigma_{\mathrm{b}} A_{\mathrm{b}}}{S^2}\right)(L - S)$$

where

$$q = \tan^2(\pi/4 + \phi/2)$$

provided a Mohr–Coulomb yield condition is assumed for the rock. Note: Equation 8 is per foot of tunnel length.

4.12 Layout an equivalent bolting pattern for the crown portion of the tunnel in Problem 8 using 1 inch diameter rebar tensioned to 60,000 psi. Specify bolt spacing and length. Note: T(max) = 85,000 lbf

4.13 Yieldable steel arches are used to control squeezing ground in a tunnel where the rock pressure corresponds to a rock "head" of 15 ft. Estimate the steel area needed for this situation. What proportion of the area is needed for bending stress?

4.14 Layout an equivalent bolting pattern for the crown portion of the tunnel in Problem 10 using 2.5 cm diameter rebar tensioned to 410 MPa. Specify bolt spacing and length.

4.15 Yieldable steel arches are used to control sqeezing ground in a tunnel where the rock pressure corresponds to a rock "head" of 4.6 m. Estimate the steel area needed for this situation. What proportion of the area is needed for bending stress?

4.16 An arched tunnel is driven in moderately blocky and seamy, wet ground at a depth of 1,970 ft. The tunnel is 14 ft wide by 11 ft high; the back arch is semi-circular.

(a) Specify a suitable steel set size (web depth, flange width, weight per foot), set spacing, and maximum blocking point spacing for these conditions.

(b) Explain how one would handle a relatively high side pressure that may develop crossing a fault zone.

(c) What is the approximate thrust capacity, in lbf, required for equilibrium?

4.17 For Problem 4.16 conditions, find an equivalent bolting pattern (bolt diameter, strength, length, and spacing), that provides the same approximate thrust capacity.

4.18 For the same "rock pressure" and semi-circular arch in Problem 16, determine the cross-sectional area of yieldable steel arches for ground control.

4.19 An arched tunnel is driven in moderately blocky and seamy, wet ground at a depth of 600 m. The tunnel is 4.3 m wide by 3.4 m high; the back arch is semi-circular.

(a) Specify a suitable steel set size (web depth, flange width, weight per foot), set spacing and maximum blocking point spacing for these conditions.

(b) Explain how one would handle a relatively high side pressure that may develop crossing a fault zone.
(c) What is the approximate thrust capacity, in lbf, required for equilibrium?

4.20 For Problem 4.19 conditions, find an equivalent bolting pattern (bolt diameter, strength, length, and spacing), that provides the same approximate thrust capacity.

4.21 For the same "rock pressure" and semi-circular arch in Problem 19, determine the cross-sectional area of yieldable steel arches for ground control.

4.22 An arched tunnel is driven in moderately blocky and seamy, wet ground at a depth of 970 ft. The tunnel is 18 ft wide by 21 ft high; the back arch is semi-circular.

(a) Specify a suitable steel set size (web depth, flange width, weight per foot), set spacing and maximum blocking point spacing for these conditions.
(b) Explain how one would handle a relatively high side pressure that may develop crossing a fault zone.

4.23 For the same "rock pressure" and semi-circular arch in Problem 4.22, determine the cross-sectional area of yieldable steel arches for ground control.

4.24 For Problem 4.23 conditions, find an equivalent bolting pattern (bolt diameter, strength, length, and spacing), that provides the same approximate thrust capacity.

4.25 An arched tunnel is driven in moderately blocky and seamy, wet ground at a depth of 297 m. The tunnel is 5.51 m wide by 61.4 m high; the back arch is semi-circular.

(a) Specify a suitable steel set size (web depth, flange width, weight per foot), set spacing and maximum blocking point spacing for these conditions.
(b) Explain how one would handle a relatively high side pressure that may develop crossing a fault zone.

4.26 For the same "rock pressure" and semi-circular arch in Problem 4.25, determine the cross-sectional area of yieldable steel arches for ground control.

4.27 For Problem 4.26 conditions, find an equivalent bolting pattern (bolt diameter, strength, length, and spacing), that provides the same thrust capacity.

4.28 Semi-circular yieldable steel arches with a nominal radius of 6.75 ft are used in main entries to a coal mine developed from outcrop under plateau overburden. Seam depth at a point of interest is 2,350 ft where ground pressure on the supports is estimated to be 20 psi. Determine cross-sectional area (square inches) of A36 steel needed and a reasonable set spacing for an appropriate steel safety factor.

4.29 With reference to Problem 4.28, an alternative support system in the form of fixed steel sets is considered with entry height 11.75 ft. Select a suitable steel rib for this alternative specifying steel weight and size, set spacing and maximum blocking point spacing.

4.30 With reference to Problem 4.29, develop an approximately equivalent bolting reinforcement system using one-inch diameter, Grade 60 steel (60,000 psi elastic limit) bolts. Rock properties are: E = 4.9 million psi, v = 0.18, C_o = 7,500 psi, T_o = 750 psi, γ = 148 pcf. Specify bolt spacings (in-row, between rows, ft) and bolt lengths (ft).

4.31 Semi-circular yieldable steel arches with a nominal radius of 2 m are used in main entries to a coal mine developed from outcrop under plateau overburden. Seam depth at a point of interest is 716 m where ground pressure on the supports is estimated to be 138 kPa. Determine cross-sectional area (square inches) of A36 steel (250 MPa yield stress) needed and a reasonable set spacing for an appropriate steel safety factor.

4.32 With reference to Problem 4.31, an alternative support system in the form of fixed steel sets is considered with entry height 3.6 m. Select a suitable steel rib for this alternative specifying steel weight and size, set spacing, and maximum blocking point spacing.

4.33 With reference to Problem 4.32, develop an approximately equivalent bolting reinforcement system using one-inch diameter, Grade 60 steel (410 MPa elastic limit) bolts. Rock properties are: $E = 33.8$ GPa, $\nu = 0.18$, $C_o = 51.7$ MPa, $T_o = 5.17$ Mpa, $\gamma = 23.4$ kN/m^3. Specify bolt spacings (in-row, between rows, m) and bolt lengths (m).

4.34 An arched tunnel is driven in moderately blocky and seamy, wet ground at a depth of 2,830 ft where rock properties are:

$$C_o = 14{,}300 \text{ psi}, T_o = 1{,}430 \text{ psi}, E = 4.25\,(10^6)\text{psi},$$
$$G = 1.8(10^6) \text{ psi}, \gamma = 156 \text{ pcf}$$

The tunnel is 14 ft wide by 17 ft high; the back arch is semi-circular. Specify a suitable steel set size (web depth, flange width, weight per foot), set spacing and maximum blocking point spacing for these conditions.

4.35 With reference to Problem 4.34, specify support in the form of rock reinforcement by bolting on a square pattern that has the same support capacity (bolt diameter, spacing, length, steel strength).

4.36 With reference to Problem 4.35, specify support in the form of yieldable steel arches that support the same “rock pressure” (steel area, set spacing, steel strength).

4.37 An arched tunnel is driven in moderately blocky and seamy, wet ground at a depth of 863 m where rock properties are

$$C_o = 98.6 \text{ MPa}, T_o = 9.86 \text{ MPa}, E = 29.3 \text{ GPa}, G = 12.4 \text{ GPa},$$
$$\gamma = 24.7 \text{ kN/m}^3$$

The tunnel is 4.3 m wide by 5.2 m high; the back arch is semi-circular. Specify a suitable steel set size (web depth, flange width, weight per foot), set spacing and maximum blocking point spacing for these conditions.

4.38 With reference to Problem 4.37, specify support in the form of rock reinforcement by bolting on a square pattern that has the same support capacity (bolt diameter, spacing, length, steel strength).

4.39 With reference to Problem 4.37, specify support in the form of yieldable steel arches that support the same “rock pressure” (steel area, set spacing, steel strength).

4.40 An arched (semi-circle) rectangular tunnel 28 ft wide and 28 ft high is driven in dry, moderately blocky, and seamy ground.

(a) Select a steel rib suitable for these conditions; specify flange width, web depth, weight per foot, and set spacing.

(b) Suppose yieldable steel arches were used to support the arched back on the same set spacing under the same rock load ("head"). What thrust in a yieldable set is indicated?

(c) If large, long bolts were used instead, estimate the bolting pressure that is indicated assuming the same spacing and thrust capacity provided by the yieldable arches.

4.41 An arched (semi-circle) rectangular tunnel 8.5 m wide and 8.5 m high is driven in dry, moderately blocky, and seamy ground.

(a) Select a steel rib suitable for these conditions; specify flange width, web depth, weight per foot, and set spacing.

(b) Suppose yieldable steel arches were used to support the arched back on the same set spacing under the same rock load ("head"). What thrust in a yieldable set is indicated?

(c) If large, long bolts were used instead, estimate the bolting pressure that is indicated assuming the same spacing and thrust capacity provided by the yieldable arches.

Rock mass classification schemes, RQD

4.42 Explain the objective of rock mass classification schemes, RMR and Q, why RQD is important to such schemes, what the main components are and what the main differences are. Organize your comparisons and contrasts in itemized lists.

In a run of 5 ft of NQ core (1.875 in. diameter), fractures are observed at 2.4, 5.3, 8.2, 14.2, 17.7, 25.3, 29.3, 34.3, 36.9, 47.5, and 54.8 in. measured along the core from one end to the other. Find the RQD.

4.43 Explain the objective of rock mass classification schemes, RMR and Q, why RQD is important to such schemes, what the main components are and what the main differences are.

In a run of 1.5 m of NQ core (4.75 cm diameter), fractures are observed at 8.4, 18.3, 33.5, 39.9, 66.8, 77.0, 90.7, 95.5, 110.5, 134.1 and 137.7 cm measured along the core from one end to the other. Find the RQD. Explain.

4.44 Explain the objective of rock mass classification schemes, RMR and Q, why RQD is important to such schemes, what the main components are and what the main differences are. Also explain the differences in support strategy concerning fixed steel sets and yieldable arches.

Additional problems

4.45 Consider a semi-circular arched back tunnel at a depth of 984 ft in a gravity field where the horizontal preexcavation stress is one-third the vertical stress. The tunnel is 26 ft wide by 26 ft high, so the arch radius is 13 ft. The tunnel bears due north. Three joint sets are present. Intact rock Young's modulus and Poisson's ratio are 14.5 million psi and 0.20, respectively. Unconfined compressive and tensile strengths are 14,500 psi and 1,450 psi, respectively. Young's modulus of the jointed rock mass is estimated at 0.3 times intact rock modulus. Strengths are estimated to be 0.3 times intact values. Determine the tunnel safety factors with respect to tension and compression.

4.46 Consider the data in Problem 4.45 and the presence of three joint sets with geometric properties shown in the table. Strength properties are given in the second table. Specific weight of rock is 158 pcf. Sketch the tunnel section looking north and joints, then decide if bolting is necessary. Neglect any resistance from the vertical joints.

Joint Geometry

Property Joint Set	Dip (deg)	Dip Direction (deg)	Spacing (ft)	Persistence (-)
J1	60	east	10.5	0.8
J2	30	west	7.5	0.5
J3	90	north	5.9	0.9

Joint Strengths

Property Joint Set	Cohesion (psf)	Friction Angle (deg)
J1	734	35
J2	418	25
J3	209	42

4.47 Consider the data in Problems 4.45 and 4.46, and the results that indicate a need for support of ribs and back. Decide on a suitable bolting pattern and specify bolt lengths, spacings, steel size and grade using point-anchored, pretensioned bolts. Show on a sketch of the tunnel cross-section.

4.48 Consider the data in Problem 4.47, and the results that indicate a need for support of ribs and back. Use the bolting specifications given in Problem 4.47, and then compute the increase in safety factor against rib block sliding obtained by bolting and the resulting safety factors for the right and left rib blocks. Note: Before bolting, $FS_j\ (R) = 1.5$, $FS_j\ (L) = 1.02$. Also, the rib blocks right and left have weights: $W\,(R) = 59$ klbf, $W\,(L) = 145$ klbf.

4.49 Consider the data in Problems 4.45 and 4.46 where there are three joint sets present with geometric properties shown in the table. Strength properties are given in the second table. Specific weight of rock is 158 lbf/ft^3. $FS_j\,(R) = 1.50$ and $FS_j\,(L) = 1.02$. Suppose grouting is used to improve tunnel rib safety by increasing joint cohesion by 145 psi (1000 kPa). Determine the grouted joint safety factors. Note: rib blocks right and left have weights: $W\ (R) = 59$ klbf, $W\ (L) = 145$ klbf and areas $A(R) =$ 49.2ft^2, $A(L) = 105.6$ ft2.

4.50 Three joint sets are present in a rock mass with geometric properties shown in the table above. Joint strength properties in relationship to a Mohr-Coulomb criterion are given in the second table. Intact rock Young's modulus and Poisson's ratio are 14.5 million psi and 0.20, respectively. Unconfined compressive and tensile strengths of intact rock between joints are 14,500 psi and 1,450 psi, respectively. Estimate joint plane strengths (cohesion and friction angles) assuming intact rock is adequately described by a Mohr-Coulomb criterion.

Joint Geometry

Property Joint Set	Dip (deg)	Dip Direction (deg)	Spacing (ft)	Persistence (-)
J1	60	east	10.5	0.8
J2	30	west	7.5	0.5
J3	90	north	5.9	0.9

Joint Strengths

Property Joint Set	Cohesion (psf)	Friction Angle (deg)
J1	734	35
J2	418	25
J3	209	42

4.51 A joint set is present in a rock mass with properties shown in the table. Consider the joint plane as a thin layer of material that is quite different from the adjacent intact rock between joints. Compute the properties of this layer as a function of persistence, then plot ratios of joint layer cohesion to intact rock cohesion and friction angle ratio of intact to joint layer angle on the same plot. Add trend lines to the two graphs.

Properties

Property Material	Cohesion (psi)	Friction Angle (deg)
Rock	1,450	48
Joint	14.5	28

4.52 A joint set is present in a rock mass with properties shown in the table above. Consider the joint plane as a thin layer of material that is quite different from the adjacent intact rock between joints. Compute the properties of this layer as a function of persistence, then plot the ratio of joint layer unconfined compressive strength to intact rock unconfined compressive strength. Do the same for tensile strength. Add trend lines to the two graphs.

Chapter 5

Entries in stratified ground

Entries are "tunnels" of rectangular cross-section. In stratified ground, entries have roof spans that may be determined by beam analysis, provided the immediate roof separates from the overlying strata. This situation is often the case in softrock mines (e.g. coal, salt, trona, potash). Service life is variable, but mainline entries have long service lives, perhaps 20 years or more. Other mining entries may be used for a year or less. In any case, entry roof span is just entry width. If roof strength is inadequate for achievement of an acceptable safety factor, then support or reinforcement is required, usually in the form of roof bolts. The term "roof" bolting implies bolting in stratified ground or softrock mining, for example, coal mining; "rock" bolting is used in hardrock mining and tunneling. There is more than a semantic difference between the two mining environments. If the immediate roof layer is thick relative to entry width or roof span, then bed separation is unlikely. In the absence of bed separation, design analysis requires investigation of stress concentration about the considered entry. This is likely to be the case in hardrock room and pillar mines and excavations for other purposes such as underground manufacturing and storage facilities. In either situation, the usual design criteria are factors of safety in tension and compression,

$$\mathrm{FS_t} = \frac{T_\mathrm{o}}{\sigma_\mathrm{t}}$$
$$\mathrm{FS_c} = \frac{C_\mathrm{o}}{\sigma_\mathrm{c}} \tag{5.1}$$

where T_o and C_o are unconfined tensile and compressive strengths, respectively; σ_t and σ_c and peak tensile and compressive stresses. A complication that may arise is strength anisotropy, for example, tensile strength parallel to stratification planes ("bedding") may be significantly different from tensile strength perpendicular to bedding. Shear strength may also be a factor where shear strengths across and along bedding differ significantly.

In the usual case of multiple entries side by side, pillars are formed by the rock between entries and pillar size becomes a design consideration. Pillar size also raises a question of size effect on pillar strength. Entry mining may occur in several seams, one below the other, either sequentially or concurrently, and thus pose additional design considerations.

5.1 Review of beam analysis

The essential concepts for analysis of beam and slab bending are the usual: equilibrium, geometry of deformation and elasticity. Integration of these fundamental concepts leads to two important formulas: the flexure formula and the Euler–Bernoulli bending formula (see, e.g. Gere, 1997; Hibbeler, 2003; Popov, 1952). Application of these formulas allows for determining bending stress, strain, and displacement. For convenience, tension is considered positive.

Basic beam formulas

Consider an isolated segment of the slab or sheet shown in Figure 5.1a that is deformed into a cylinder about the y-axis and suppose that there is no variation along the y-axis. The "beam" that is isolated has span L and a rectangular cross-section of thickness h and breadth b as shown Figure 5.1b. No loads are applied to the top and bottom surfaces of the beam, so the vertical stress $\sigma_{zz} = 0$. Because no variation occurs along the beam, $\varepsilon_{yy} = 0$. These two observations allow σ_{yy} to be expressed in terms of σ_{xx} through Hooke's law. Thus, $\sigma_{yy} = \nu\sigma_{xx}$ and

$$\varepsilon_{xx} = \left(\frac{1-\nu^2}{E}\right)\sigma_{xx} \tag{5.2}$$

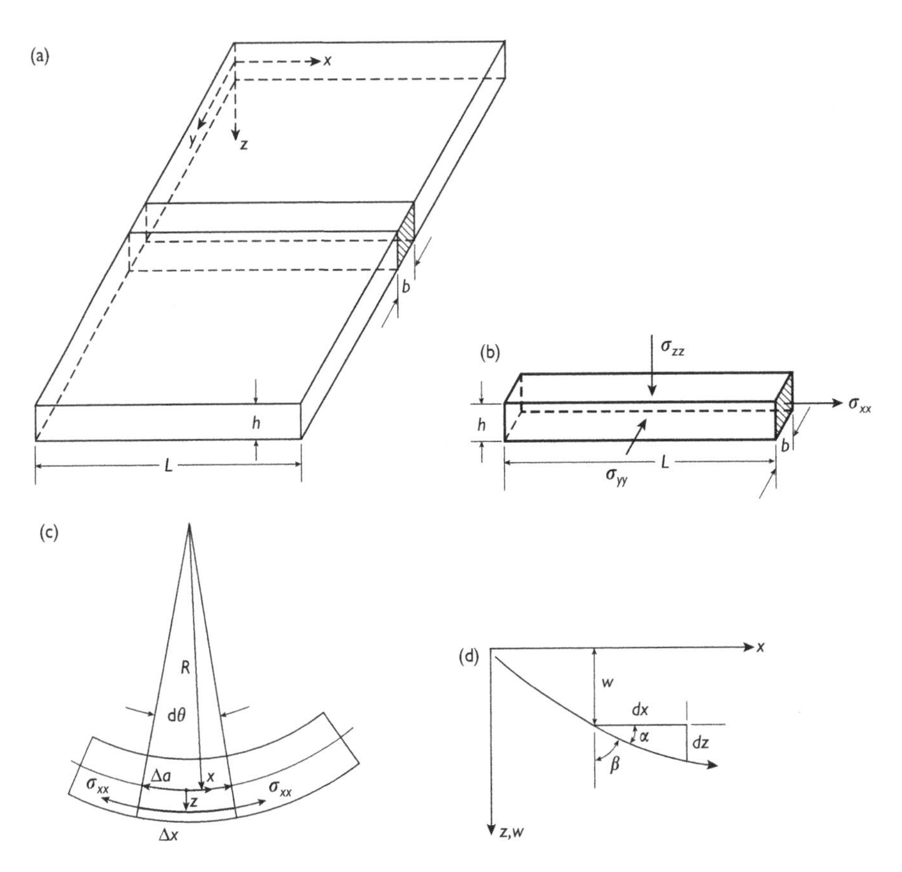

Figure 5.1 (a) Roof sheet geometry, (b) roof beam geometry, (c) details of beam bending and (d) notation.

where x is horizontal and z is positive down. If the slab or sheet were not of indefinite extent in the y-direction and instead were a true beam, then σ_{yy} would be zero and (5.2) would be $\varepsilon_{xx} = \sigma_{xx}/E$. Thus the difference between beam and sheet bending is only about 5% for Poisson's ratio in the range of 0.20–0.25.

The geometry of a beam element in bending is shown in Figure 5.1c. By definition, the strain is a change in length per unit of original length, so

$$\varepsilon_{xx} = \frac{s - s_o}{s_o} = \frac{(R+z)\mathrm{d}\theta - R\mathrm{d}\theta}{R\mathrm{d}\theta} = \frac{z}{R} = \varepsilon \tag{5.3}$$

where R is the radius of curvature of the deformed element and the subscripts have been dropped from the bending strain ε_{xx} $(=\varepsilon)$. Curvature $\kappa = 1/R$ is defined as the rate of change of inclination angle of a curve with respect to arc length measured along the curve as shown in Figure 5.1d. Thus, $\kappa = 1/R = \mathrm{d}\beta/\mathrm{d}s = -\mathrm{d}\alpha/\mathrm{d}s$ where α is the angle of inclination of the considered curve measured from the horizontal. Where α does not change with distance, the curve in fact is straight and the curvature is zero. Thus,

$$\frac{1}{R} = \frac{-(\mathrm{d}^2w/\mathrm{d}x^2)}{[1 + (\mathrm{d}w/\mathrm{d}x)^2]^{3/2}} \tag{5.4}$$

where w is the displacement or deflection of the curve from the original horizontal position; w is "sag" of the beam. When the slope $\mathrm{d}w/\mathrm{d}x$ is small, the curvature is to close approximation

$$\frac{1}{R} = -\left(\frac{\mathrm{d}^2w}{\mathrm{d}x^2}\right) \tag{5.5}$$

Equilibrium requires the internal distribution of stress be equivalent to the external moment. Hence,

$$M = \int_A z\sigma_{xx}\mathrm{d}A \tag{5.6}$$

where $\mathrm{d}A$ is an area element of the beam cross-section, $\sigma_{xx}\mathrm{d}A$ is an element of force and the product $z\sigma_{xx}\,\mathrm{d}A$ is the associated moment that when integrated over the cross-sectional area A is the total internal moment. Examination of Figure 5.1c shows that elements toward the top of the beam are compressed whereas elements toward the bottom are stretched. In between there is a *neutral surface* that is neither stretched nor compressed. A *neutral axis* exists in the neutral surface parallel to the y-axis about which bending occurs.

The geometry of Figure 5.1c suggests that plane sections before deformation remain plane after deformation. If plane sections remain plane during bending, then the bending strain which is zero on the neutral surface is directly proportional to distance above or below the neutral surface. If so, then the same is true for stress because the bending stress is directly proportional to the bending strain according to (5.2), that is, $\sigma_{xx} = (\text{constant})\, z$ where tension is positive. The moment equilibrium requirement (5.6) can then be expressed as

$$\begin{aligned} M &= \int_A Cz^2 \mathrm{d}A \\ &= C\int_A z^2 \mathrm{d}A \\ &= CI \\ M &= \left(\frac{\sigma_{xx}}{z}\right)I \end{aligned} \tag{5.7}$$

where C is a constant and I is the second moment of area about the neutral axis, the area moment of inertia. Equation (5.7) when solved for the bending stress gives the *flexure formula*

$$\sigma = \frac{Mz}{I} \tag{5.8}$$

where the subscripts have been dropped from the bending stress σ_{xx} as they were for the bending strain.

A series of substitutions beginning with (5.2) leads to the Euler–Bernoulli *bending formula*. Thus,

$$\varepsilon = \frac{z}{R} = -z\left(\frac{\mathrm{d}^2w}{\mathrm{d}x^2}\right) = \left(\frac{1-\nu^2}{E}\right)\sigma = \left(\frac{1-\nu^2}{E}\right)\left(\frac{Mz}{I}\right) = \frac{Mz}{E'I} \tag{5.9}$$

where $E' = E/(1-\nu^2)$ for brevity. The variable z can be eliminated from Equation (5.9) that results in

$$\frac{\mathrm{d}^2w}{\mathrm{d}x^2} = -\left(\frac{M}{E'I}\right) \tag{5.10}$$

which is the famous Euler–Bernoulli bending formula. In case of beams with rectangular cross-sections $I = bh^3/12$. Equation (5.10) is an ordinary differential equation that after one integration gives the slope of the deflected beam. Slope and angle of inclination are very nearly equal for small angles α. A second integration gives the deflection or sag. These integrations with respect to x are over the span of the beam; conditions at the beam ends determine the two constants of integration.

When a load acts transverse to a beam, shear forces are generated within the beam that interact with the internal moments. With reference to Figure 5.2 that shows a small segment of a beam under the action of a distributed surface stress ("pressure") p, equilibrium in the vertical z-directions requires $(V + \Delta V) - V + pb\Delta x = 0$. In the limit,

$$\frac{\mathrm{d}V}{\mathrm{d}x} = -pb \tag{5.11}$$

where b is the breadth of the beam. When the surface load is uniformly distributed, moment equilibrium about the centroid of the beam segment requires $(M + \Delta M) - M - (V + \Delta V)(\Delta x/2) - V\Delta x/2 = 0$. In view of (5.11), moment equilibrium requires

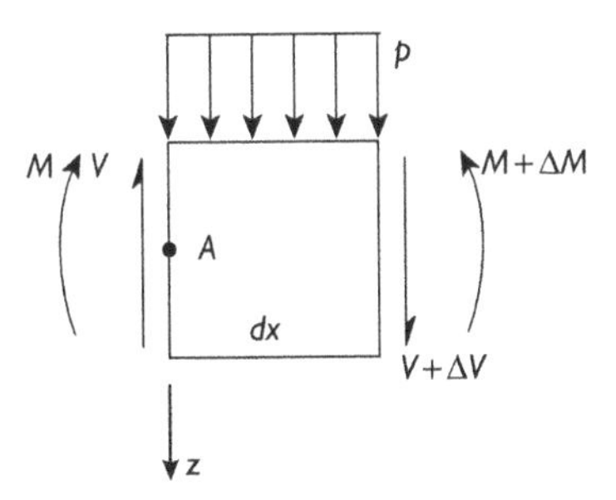

Figure 5.2 A small beam element under a distributed surface load p.

$$\frac{dM}{dx} = V \tag{5.12}$$

A change in moment therefore induces a shear force in a beam. The same result is obtained when the surface load varies along the beam. An average shear stress in the beam can be computed as $\tau_{ave} = V/A$ where A is the cross-sectional area of the beam that is simply bh when the beam is rectangular in section. A graphical interpretation of Equations (5.11) and (5.12) is that the change in shear force over a length x of a beam is the area under the distributed load curve and that the change in moment is the area under the shear force curve. Abrupt changes caused by point loads cause jumps in plots of shear force and bending moment as a functions of x along the beam. Superposition applies, so the result is simply the sum of the effects of distributed and point loads.

The distribution of shear stress in a transversely loaded beam must satisfy the stress equations of equilibrium. Thus,

$$\begin{aligned} \frac{\partial \sigma_{xx}}{\partial x} + \frac{\partial \tau_{zx}}{\partial z} &= 0 \\ \frac{\partial (Mz/I)}{\partial x} + \frac{\partial \tau_{zx}}{\partial z} &= 0 \\ \frac{Vz}{I} + \frac{\partial \tau_{zx}}{\partial z} &= 0 \end{aligned} \tag{5.13}$$

In particular, one must have after integration

$$\tau_{zx} = f(x) - \frac{Vz^2}{2I} \tag{5.14}$$

The function $f(x)$ may be evaluated at the top and bottom of the beam where the shear stress vanishes, that is, at $z = \pm h/2$. The result is

$$\tau = \left(\frac{V}{2I}\right)\left[\left(\frac{h}{2}\right)^2 - z^2\right] \tag{5.15}$$

where the subscript has been dropped for brevity and I is assumed constant along the beam. Inspection of (5.15) shows that the beam shear stress is maximum on the neutral surface where $z=0$, thus $\tau_{max}=Vh^2/8I$. If the distribution (5.15) is averaged over the cross-section of the beam, the result is

$$\tau_{ave}=\frac{Vh^2}{12I}=\frac{V}{A}=\left(\frac{2}{3}\right)\tau_{max} \tag{5.16}$$

The maximum shear stress is therefore 50% greater than the average shear stress.

Important special cases

Two important cases of beam bending involve *distributed* and *point* loads applied to beams of rectangular cross-section. Diagrams of shear force and moment as they vary along the beam are quite helpful in solving these problems. In both cases, beam ends may be simply supported or built-in. A third case involves a beam supported at one end only, a cantilever beam. The main objectives of beam analysis in any case is to determine maximum tension and sag.

Consider a beam under a uniformly distributed load as shown in Figure 5.3a. Equilibrium requires the resultants of the external forces and moments to be zero. In particular, the vertical load $F=pbL$, which acts through the beam center, must be equilibrated by the reactions at the beam ends as shown in Figure 5.3b. Horizontal reactions are omitted from Figure 5.3b. Symmetry of the problem indicates that the end reactions are equal, so $R=R_A=R_B=pbL/2$. Summation of moments about the beam center shows that moment equilibrium is satisfied with $M_A=M_B$. Sectioning of the beam as shown in Figure 5.3c reveals the internal shear force and moment. Summation of forces in the vertical directions shows that

$$V=pb\left(\frac{L}{2}-x\right) \tag{5.17}$$

According to Equation (5.17) the shear force varies linearly along the beam and is zero at the beam center $x=L/2$. Summation of moments about A gives the moment M. Thus,

$$M=M_A+\left(\frac{pb}{2}\right)(L-x)x \tag{5.18}$$

Inspection of (5.18) at $x=0$ and $x=L$ shows that the end conditions are generally satisfied. Integration of (5.10) gives the slope of the beam and a second integration gives the deflection or sag, that is, displacement. The slope is

$$\alpha=-\left(\frac{1}{E'I}\right)\left[M_A x+\left(\frac{pbLx^2}{4}\right)-\left(\frac{pbx^3}{6}\right)\right]+\alpha_o \tag{5.19}$$

where α_o is an integration constant. The deflection is then

$$w=-\left(\frac{1}{E'I}\right)\left[M_A+\left(\frac{x^2}{2}\right)+\left(\frac{pbLx^3}{12}\right)-\left(\frac{pbx^4}{24}\right)\right]+\alpha_o x+w_o \tag{5.20}$$

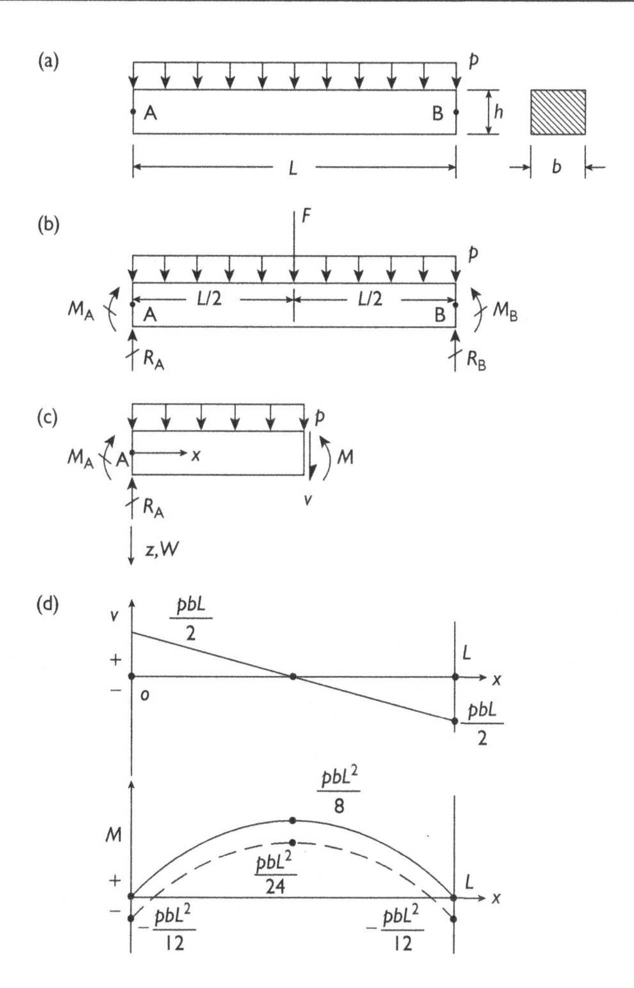

Figure 5.3 Analysis of a beam under a uniformly distributed surface load.

where w_o is another constant of integration. Both may be determined by end conditions as may the end moment. With either simply supported or built-in ends, end sag is zero, so that $w_o = 0$. Also in either case the slope of the beam is zero at mid-span because of symmetry of the problem. Hence,

$$\alpha_o = \left(\frac{1}{E'I}\right)\left[M_A\left(\frac{L}{2}\right) + \left(\frac{pbL^3}{24}\right)\right] \tag{5.21}$$

In the case of simply supported ends, the end moments are zero, that is, $M_A = M_B = 0$; in the case of built-in ends, the ends of the beam are not free to rotate, so the slope is zero. In the latter case, solution of (5.21) shows that

$$M_A = -\left(\frac{pbL^2}{12}\right) \tag{5.22}$$

Inspection of (5.8) shows that at a given point along the beam, the maximum tension and compression occur at the beam top or bottom, depending on the sign of the moment. In both cases, (5.18) has a relative algebraic maximum at mid-span. In the simply supported case this is also the absolute maximum, $pbL^2/8$, which is positive, so the greatest tension occurs at the bottom of the beam. In the built-in end case, the moment at mid-span is $pbL^2/24$ which is less in magnitude than the end moments. Because the end moment is negative in the built-in end case, the greatest tension occurs on top of the beam where $z = -h/2$. Figure 5.3d shows the moment and shear diagrams for the distributed load case with both simply supported and built-in ends. The shear force distribution is the same, while the moment distribution is shifted down in the plots of Figure 5.3d.

The peak tensions computed for the simply supported (SS) and built-in (BI) end cases for a beam of rectangular cross-section under a uniformly distributed load are

$$\begin{aligned} \sigma_t(\text{SS, distributed load}) &= \left(\frac{3}{4}\right)\left(\frac{pL^2}{h^2}\right) \\ \sigma_t(\text{BI, distributed load}) &= \left(\frac{1}{2}\right)\left(\frac{pL^2}{h^2}\right) \end{aligned} \tag{5.23}$$

Inspection of (5.23) shows that the simply supported case leads to a peak tension that is 50 percent greater than the built-in ends case, all other factors being equal. With simply supported ends, the beam ends are free to rotate; no rotation is allowed with built-in ends. However, when the ends are neither free nor fixed but allowed to rotate in proportion to the end moments, the peak stress falls between the two cases in (5.23) (Timoshenko, S. P. and S. Woinowsky-Krieger, 1959).

The maximum deflection or sag occurs at mid-span and is given by

$$\begin{aligned} w_{\max}(\text{SS, distributed load}) &= \left(\frac{5}{32}\right)\left(\frac{pL^4}{E'h^3}\right) \\ w_{\max}(\text{BI, distributed load}) &= \left(\frac{1}{32}\right)\left(\frac{pL^4}{E'h^3}\right) \end{aligned} \tag{5.24}$$

which show that the sag in the simply supported case of a beam deformed under a uniform load is five times that in the built-in end case. Again, in the intermediate case where beam ends are neither free nor fixed but end rotation is proportional to end moment, the sag is between the values given by (5.24).

A simply supported beam under a single point load is shown in Figure 5.4a. The two sections of the beam shown in Figure 5.4b,c are helpful in determining the shear force and moment distributions that are shown in Figure 5.4d. The maximum moment occurs at the load point $x = a$ and the maximum tension for a beam of rectangular section is

$$\sigma_t(\text{SS, } F \text{ at } x = a) = \frac{6Faa'}{Lbh^2} \tag{5.25}$$

If the point load F is equated with an equivalent distributed load such that $F = pbL$ and applied at mid-span the maximum tension according to (5.25) is

$$\sigma_t(\text{SS, } F \text{ equivalent distributed load}) = \left(\frac{3}{2}\right)\left(\frac{pbL^2}{h^2}\right) \tag{5.26}$$

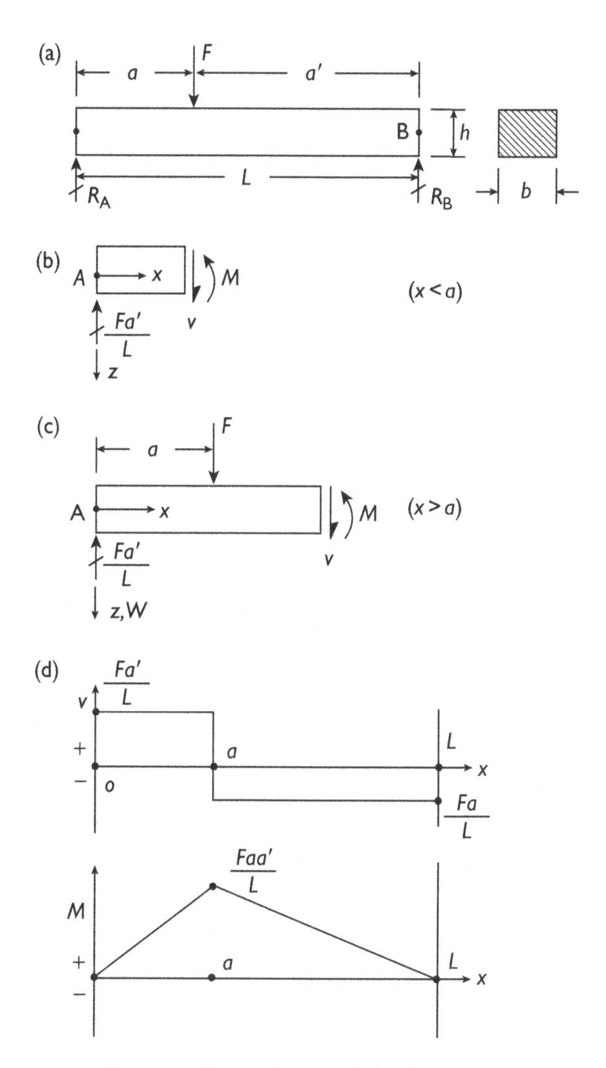

Figure 5.4 A simply supported beam under an offset point load.

which is double the peak tension when the load is actually distributed as seen in comparison with the first of (5.23). For this reason, point loads should be avoided when possible.

Again, two integrations of the Euler–Bernoulli bending formula (10) give the slope and sag of the beam. The maximum sag occurs in the longest segment of the beam and is located at the point x given by

$$x = \left[\frac{L^2 - (a')^2}{3}\right]^{1/2}, \quad (a > a')$$
$$x = \left[\frac{L^2 - a^2}{3}\right]^{1/2}, \quad (a < a') \tag{5.27}$$

where $L = a + a'$ and the load is applied at $x = a$. The value of the maximum sag is

$$w_{max} = \left(\frac{4\sqrt{3}Fa'}{9E'Lbh^3}\right)[L^2 - (a')^2]^{3/2}, \quad (a > a')$$
$$w_{max} = \left(\frac{4\sqrt{3}Fa}{9E'Lbh^3}\right)[L^2 - a^2]^{3/2}, \quad (a < a') \tag{5.28}$$

In case the point load is applied at mid-span, the maximum sag occurs at mid-span and is

$$w_{max}\,(F \text{ at } L/2) = \left(\frac{FL^3}{4E'bh^3}\right), \quad (a = a') \tag{5.29}$$

If the applied load is equated to an imaginary distributed load, so $F = pbL$, then the sag is

$$w_{max}\,(F \text{ equivalent distributed load}) = \left(\frac{8pL^4}{32E'h^3}\right), \quad (a = a') \tag{5.30}$$

Comparison of (5.30) with (5.24) shows that a point load of the same magnitude as a total distributed load causes a sag that is 1.6 times the sag of the same beam under a distributed load with simply supported ends and rectangular cross-section.

Cantilever beams are built-in at one end, a fixed end, but are without support at the opposite or free end. Figure 5.5a shows the case where the distributed load is uniform. The support force reaction R_A is easily computed to be equal to the total load pbL. Sum of the moments for the entire beam gives the moment reaction $M_A = pbL/2$. Sectioning the beam as shown in Figure 5.5b and summing forces in the vertical direction shows that the shear force $V = pb(L - x)$. Summation of moments about A at the fixed end gives the internal moment

$$M = -\left(\frac{pb}{2}\right)(L - x)^2 \tag{5.31}$$

which is greatest in magnitude at the built-in end where $x = 0$. The greatest tension thus occurs on top of the beam at the built-in end where the value is

$$\sigma_t(\text{cantilever, distributed load}) = 3p\left(\frac{L}{h}\right)^2 \tag{5.32}$$

for a beam of rectangular cross-section. Two integrations lead to displacement along the beam. Thus,

$$w = \left(\frac{-pb}{24E'I}\right)(4Lx^3 - x^4 - 6L^2x^2) \tag{5.33}$$

where account is made of zero rotation and zero displacement at the fixed end. The displacement is maximum at the free end of the cantilever. Thus,

$$w_{max}\,(\text{cantilever, distributed load}) = \frac{3pL^4}{2E'h^3} \tag{5.34}$$

where again a rectangular cross-section is assumed.

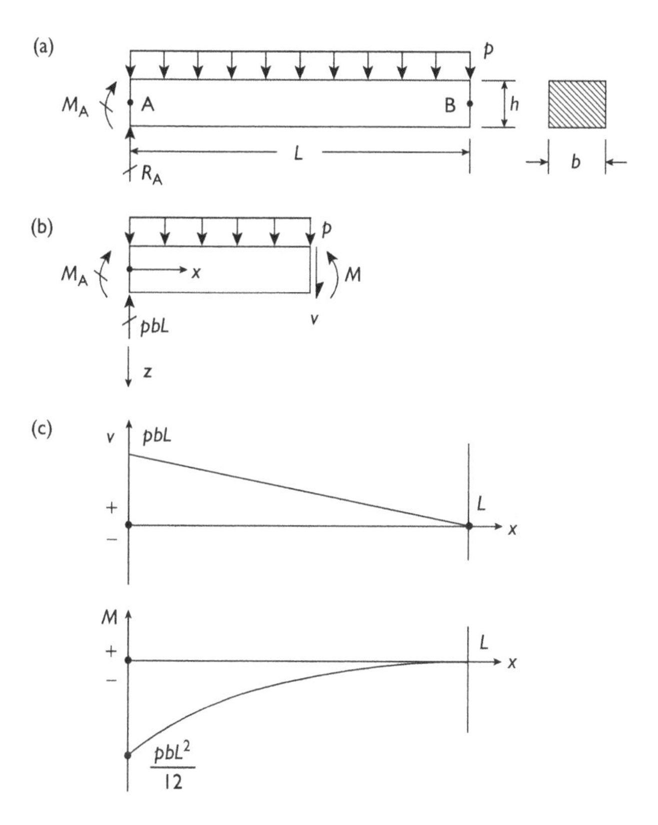

Figure 5.5 Analysis of a uniformly loaded cantilever beam.

A case of a point loaded cantilever is shown in Figure 5.6. The analysis for maximum tension leads to

$$\sigma_t(\text{cantilever}, F \text{ at } x = a) = \frac{6Fa}{bh^2} \tag{5.35}$$

for a rectangular beam. The maximum displacement occurs at the free end and is

$$w_{\max}\,(\text{cantilever}, F \text{ at } x = a) = \left(\frac{2Fa^2}{E'bh^3}\right)(3L - a) \tag{5.36}$$

also for a rectangular beam.

Example 5.1 The linearity of elasticity theory allows for superposition of solutions. This feature simplifies stress analysis of beams under complex loadings. Solution is obtained by simply adding the results of the simpler and separate loadings. Consider a beam of rectangular section of unit breadth and 1.5 ft deep that has a span of 18 ft. Young's modulus is $5(10^6)$ psi. A uniformly distributed load of 20 psi and a point load of 10,000 lbf are applied to the top of the beam. The point load acts at mid-span. Assume simply supported ends, then determine magnitudes and locations of maximum beam tension and sag.

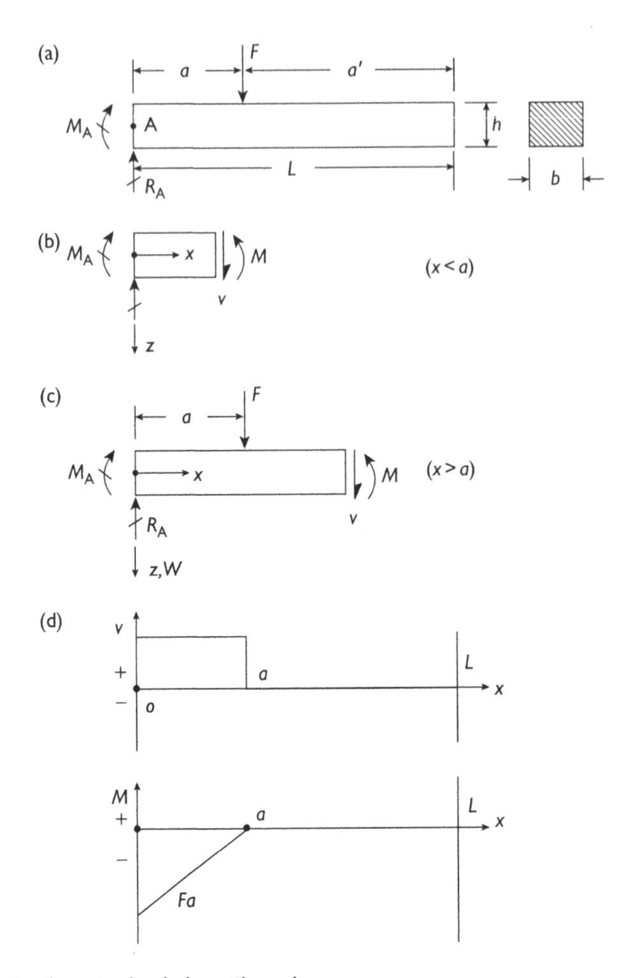

Figure 5.6 Analysis of a point-loaded cantilever beam.

Solution: Under a uniformly distributed load, maximum tension occurs at mid-span on the beam bottom. Maximum sag also occurs at mid-span. Thus, for the distributed load

$$\sigma_t(\max) = \frac{3pbL^2}{4h^2} = \frac{(3)(20)(18)^2}{(4)(1.5)^2} = 1,500 \text{ psi}$$

$$w(\max) = \frac{5pL^4}{32Eh^3} = 0.01944 \text{ ft } (0.233 \text{ in.})$$

and for the point load at mid-span,

$$\sigma_t(\max) = \frac{6Faa'}{Lbh^2} = \frac{(6)(10,000)(18/2)(18/2)}{(18)(1)(1.5)^2}\left(\frac{1}{144}\right) = 833 \text{ psi}$$

$$w(\max) = \frac{FL^3}{4Ebh^3} = \frac{(10,000)(18)^3}{(4)(5)(10^6)(1)(12)(1.5)^3} = 0.072 \text{ in.}$$

where in both cases a beam rather than a sheet computation is done. The total tension and sag are 2,333 psi and 0.305 in., respectively.

Example 5.2 A pattern of multiple, but equal point loads F are applied to the bottom side of a rectangular beam as shown in the sketch where spacing is S and span is L. Develop a formula for superposition of stress for these point loads.

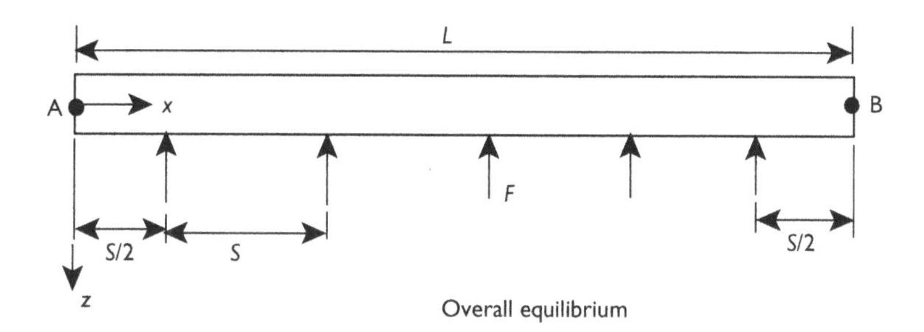

Overall equilibrium

Solution: The point loads, although regularly spaced, induce a bending stress that varies with x and point load position. Overall equilibrium leads to solution for reactions at the ends, while internal equilibrium leads to shear and moment distributions, that in turn, lead to the bending stress. Symmetry of the problem assigns one-half the total load to each end reaction, so at each end $R = nF/2$ where n is the number of point loads. Inspection of the geometry of the loads shows that $n = L/S$, so $R = FL/2S$ where F is the magnitude of the given point load.

For a single point load at a units from the origin A ($a = S/2, 3S/2, \ldots (2n-1)S/2$), internal equilibrium is illustrated in the second sketch

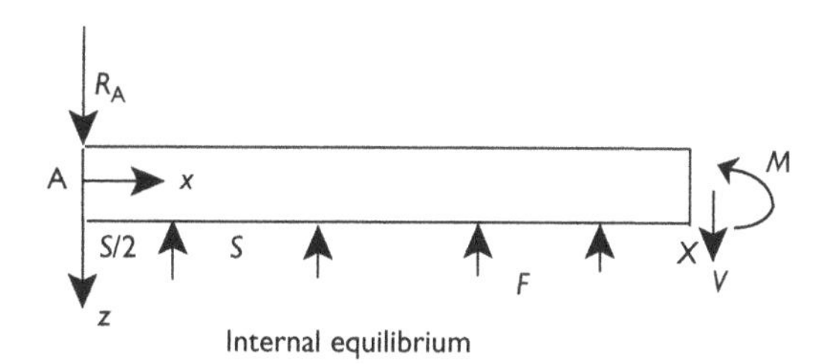

Internal equilibrium

Equilibrium of vertical forces requires $V = -R_A + \sum_1^m F_i$ where m is the number of point forces acting between the origin A and position of the section x. Because the shear force V increases in finite jumps, a plot of shear stress along the beam is an ascending stair case as shown in the plot.

If the applied forces had different magnitudes and locations, then the steps would have different vertical and horizontal lengths, and the end reactions would be different as well. The plot assumes an even number of forces F. In case of an odd number of forces, then the force at mid-span is represented in the shear diagram as one-half a step below and one-half a step above the x-axis. The shear is considered zero at mid-span in both cases.

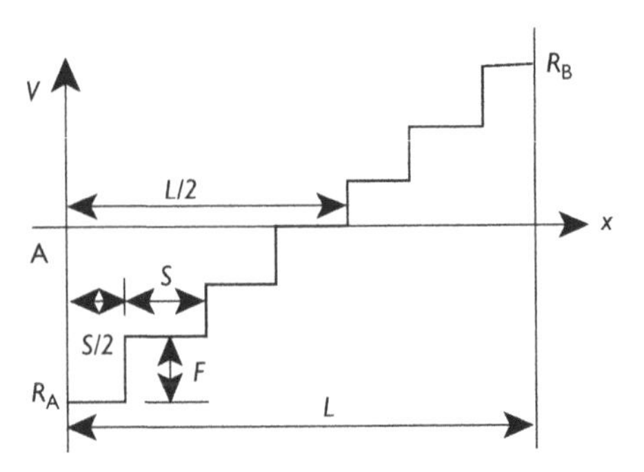

Shear diagram for Example 5.2

Moment equilibrium requires

$$M = \sum_{1}^{m} F_i x_i + Vx = \sum_{1}^{m} F_i x_i + x(R_A - \sum_{1}^{m} F_i)$$

where m is the number of forces to the left of x and n is the total number of applied. This same result is seen in the shear force plot which crosses the x-axis at mid-span. Again the result could be generalized for different forces and locations. The moment diagram is step-like, but the steps are inclined with decreasing slope toward mid-span and then negative and increasingly negative slope past mid-span. They increase from zero at A to a peak at mid-span and then decrease to zero at B. This is known from a graphical interpretation of the relationship $dM = V\,dx$. The numerical greatest moment occurs at mid-span where the shear force is zero. However, the discontinuous nature of the shear force diagram requires some care; substitution of $V = 0$ in the first equation for M should not be done. Use of the second equation is correct. An even simpler method to find the numerically greatest moment is graphically. Thus, the area under the shear force curve to the point on the x-axis where the shear force goes to zero is

$$M = (S/2)(R_A) + S(n/2 - 1)F + S(n/2 - 2)F + \cdots$$
$$M = (S/2)(nF/2) + S(n/2 - 1)F + S(n/2 - 2)F + \cdots$$

where n is even and the count down continues until the shear force meets the x-axis. For example, if $n = 6$ as in the sketch, then $M = (6\,F/2)(S/2) + 2FS + FS = (9/2)SF = (3/4)FL$. When the number of applied forces is odd, then only a half-step occurs at the x-axis crossing. For example, if $n = 3$, then $M = 3(F/2)(S/2) + (FS/2) = (5/4)FS = (5/12)FL$.

Alternatively, one may compute the area to the left, rather than "under" (between the x-axis and the graph) the shear force plot. Thus,

$$M = \sum_{1}^{m} F_i x_i = F(S/2) + F(3S/2) + \cdots + F(2m - 1)(S/2)$$

where $m = n/2$ and n is even. For example, if $n = 6$, then $m = 3$ and $M = 9FS/2 = (3/4)FS$. If $n = 3$, then $m = 3/2$. Summation should be over 2 with the last term a half-step. Thus, summing over 2, $M = (F)[S/2 + (3S/2)(1/2)] = (5/4)FS = (5/12)FL$ where $L = nS$ as in every case. This last case is shown in the sketch.

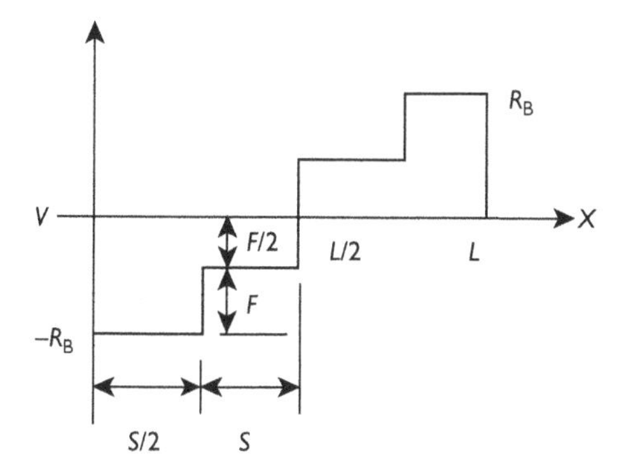

Sketch for the case $n = 3$

The peak tensile occurs on top of the beam, and

$$\sigma_t = \frac{Mc}{I}$$
$$= M\frac{(h/2)}{(bh^3/12)}$$

$$\sigma_t = \frac{6(FL)(\text{number})}{bh^2} = \frac{6(nFS)(\text{number})}{bh^2}$$

where *number* comes from the moment calculation, S = spacing, L = span, and n = number of applied forces (nF = total applied force), b = breadth = 1 unit, h = depth of the rectangular section. Superposition is reflected in the summations that appear in the shear force and moment equations.

5.2 Softrock entries

When bed separation occurs between the immediate roof (the first roof layer above an entry) and the layer above, beam action follows, as indicated in Figure 5.7. Beam action results in equal compressive and tensile stress peaks, so in view of much greater rock compressive strength than tensile strength, the safety factor in tension is usually the governing criterion for safe roof span design analysis. Strengths are material properties, of course, and are considered known. Roof span design then reduces to an analysis of stress in beams that may be naturally self-supporting or require support, usually in the form of bolting. Acceptable safety factors in tension are much higher than for compression because of low tensile rock strength, greater uncertainty, and the potential for fast failure associated with tensile fracture. A tensile safety factor in the range of 4–8 may be desirable (Obert, L. and W. I. Duvall, 1967, chapter 3).

Naturally suppor ted roof

If, under the action of gravity, the immediate roof layer of thickness h and span L separates from the overlying strata, then the magnitude of the maximum tension (and compression) is given by the *flexure* formula:

(a)

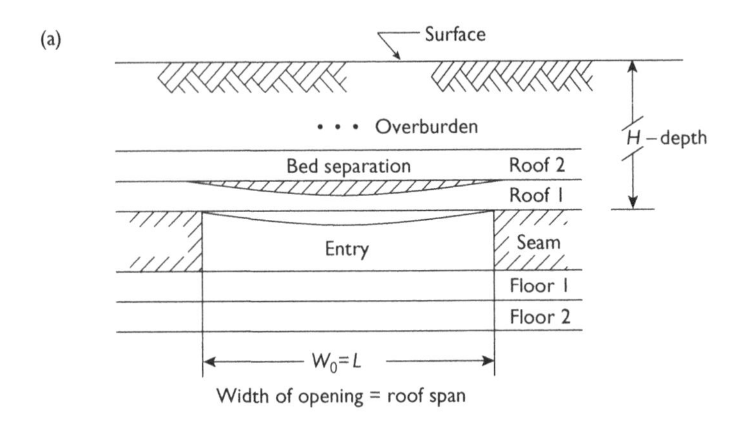

(b)

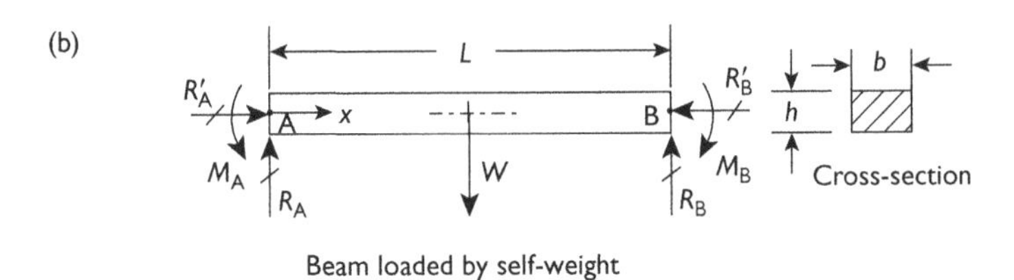

(c)

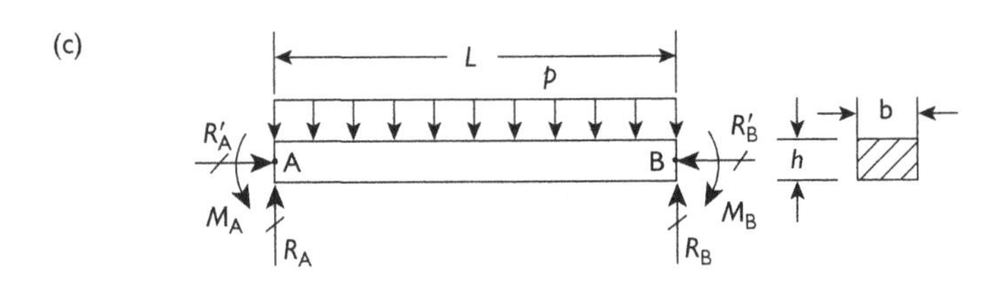

Figure 5.7 Bed separation and beam action of the first roof layer above an entry in stratified ground.

$$\sigma = \frac{Mc}{I} \tag{5.37}$$

where σ is stress, M = maximum bending moment, c = distance to the outermost fiber of the beam ($h/2$), I is the area moment of inertia about the neutral axis ($bh^3/12$), and b is breadth of the beam cross-section. The roof rock beam has a rectangular cross-section of area bh. Often b is assumed to be one foot or meter along the entry.

The bending moment depends upon the nature of the beam action. Although the beam is gravity loaded and deflecting under self-weight, a simpler model is available for analysis. This model assumes that the roof is weightless but deflects under the action of a statically equivalent and uniformly distributed normal stress acting on the top surface of the beam as

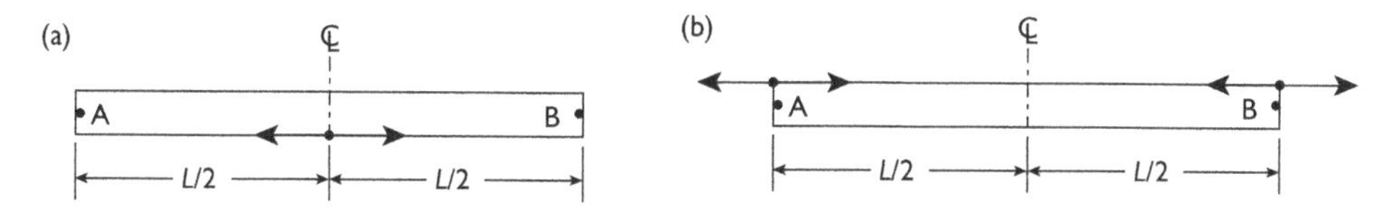

Figure 5.8 Location of maximum roof beam tension: (a) maximum tension in a simply supported (SS) beam and (b) maximum tension in a beam with fixed ends (BI).

shown in Figure 5.7. Static equivalency requires the normal force acting on the top of the weightless beam to be equal to the weight of the original beam. Thus, $F = W = pbL = \gamma bhL$ where p is the normal stress acting on top of the weightless beam. Hence, the equivalent distributed roof beam load is given by $p = \gamma h$.

The maximum moment in a simply supported (SS) beam is $pbL^2/8$ and occurs at beam bottom center. In a beam with built-in ends (BI), the maximum moment occurs on top of the beam at the beam ends and has magnitude $pbL^2/12$. These cases (SS and BI) bound intermediate end behavior that allows some rotation in proportion to bending.

The *maximum tension* occurs where the moment is maximum as indicated in Figure 5.8. Thus,

$$\sigma_t(\text{SS}) = \left(\frac{3}{4}\right)\frac{pL^2}{h^2}$$
$$\sigma_t(\text{BI}) = \left(\frac{1}{2}\right)\frac{pL^2}{h^2} \tag{5.38}$$

Substitution into the tensile factor of safety expression leads to expressions for roof span L,

$$L(\text{SS}) = \sqrt{\left(\frac{4}{3}\right)\left(\frac{T_o}{FS_t}\right)\left(\frac{h^2}{p}\right)} = \sqrt{\left(\frac{4}{3}\right)\left(\frac{T_o}{FS_t}\right)\left(\frac{h}{\gamma_a}\right)}$$
$$L(\text{BI}) = \sqrt{\left(\frac{2}{1}\right)\left(\frac{T_o}{FS_t}\right)\left(\frac{h^2}{p}\right)} = \sqrt{\left(\frac{2}{1}\right)\left(\frac{T_o}{FS_t}\right)\left(\frac{h}{\gamma_a}\right)} \tag{5.39}$$

where $p = \gamma h$ has been modified to $p = \gamma_a h$. The notation γ_a means "apparent" specific weight of roof rock. In the case of the single roof layer considered so far, this is simply the actual specific weight.

The *maximum sag* w of a single roof beam layer, simply supported or with built-in ends, occurs at the beam center and is

$$w(\text{SS}) = \left(\frac{5}{32}\right)\left(\frac{pL^4}{Eh^3}\right)$$
$$w(\text{BI}) = \left(\frac{1}{32}\right)\left(\frac{pL^4}{Eh^3}\right) \tag{5.40}$$

where E is Young's modulus of the roof rock. A somewhat more accurate description would consider the roof to be a long sheet rather than a beam and therefore use $E/(1-\nu^2)$ instead of E. The sag of a sheet is less than a beam, although the difference is less than 10% for Poisson's ratio in the range of 0.20–0.30.

Example 5.3 An entry 6 m wide is driven in a stratum 3 m thick. The immediate roof separates from the overlying strata. Young's modulus and Poisson's ratio of this stratum are 15.17 GPa (2.2×10^6) psi and 0.19, respectively. Stratum thickness and specific weight are 0.5 m and 22.0 kN/m^3 (140 pcf). Determine the maximum roof tension and roof sag that are induced by excavation. Assume simply supported ends.

Solution: According to formula,

$$\sigma_t = \left(\frac{3}{4}\right)(p)\left(\frac{L}{h}\right)^2$$

$$= \left(\frac{3}{4}\right)(\gamma h)\left(\frac{L}{h}\right)^2$$

$$= \left(\frac{3}{4}\right)(22.0 \times 10^3)(0.5)\left(\frac{6}{0.5}\right)^2$$

$$\sigma_t = 1.188 \text{ MPa}(172 \text{ psi})$$

and

$$w(\max) = \left(\frac{5}{32}\right)\left(\frac{pL^4}{Eh^3}\right)$$

$$= \left(\frac{5}{32}\right)\left(\frac{(22 \times 10^3)(0.5)(6)^4}{(15.17 \times 10^9)(0.5)^3}\right)$$

$$w(\max) = 1.175 \times 10^{-3} \text{ m}(1.175 \text{ mm}), \ (0.046 \text{ in.})$$

where the effect of Poisson's ratio is neglected. With consideration of Poisson's ratio, the tension remains the same, but the sag w(max) = (1.175 mm) (15.17 Gpa)/[(15.17 Gpa) / $(1-0.19^2)$] = (1.175 mm) (0.964) = 1.133 mm, about 4% less. The equivalent distributed load of the beam weight applied to the top of the beam is small, that is p = 11.0 kPa (1.6 psi), and the sag slight, as a consequence.

If the beam ends were considered fixed, that is, built-in, then the tension and sag would be less by 2/3 and 1/5 multipliers, respectively. The two cases are extremes. Some rotation would occur at the beam ends that are neither entirely fixed nor free in nature.

Separation may occur between a second and third roof layer. The first and second layers in the near roof then sag the same amount. Indeed, separation may occur in clusters of n-layers, each undergoing the same sag as shown in Figure 5.9. Calculation of a safety factor for any layer in the cluster, say, the ith layer requires an estimate of the load p_i acting on the layer. This load is no longer the product of unit weight and bed thickness ($p_i \neq \gamma_i h_i$). However, once this load is known, the peak tension can be calculated and thence the safety factor.

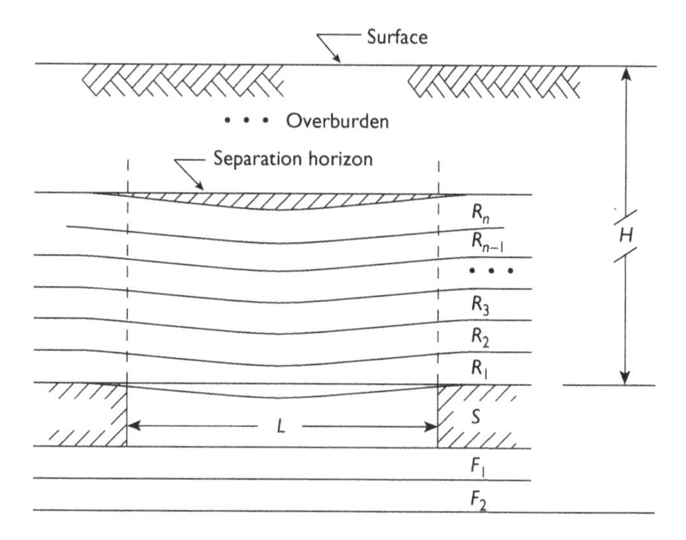

Figure 5.9 A cluster of *n*-layers of roof strata separated from overlying strata. All layers in the cluster sag the same amount.

Because the sag of each layer in the separated cluster is the same, the ratio p/Eh^3, which is directly proportional to maximum sag, is also the same. Thus,

$$\frac{p_1}{E_1h_1^3} = \frac{p_2}{E_2h_2^3} = \cdots = \frac{p_i}{E_ih_i^3}\cdots = \frac{p_n}{E_nh_n^3} \tag{5.41}$$

The total external load P acting on the n-layer cluster is the sum of the individual layer loads, so

$$P = \sum_1^n p_j = \sum_1^n \gamma_j h_j \tag{5.42}$$

where j is the layer number beginning with the immediate roof layer and proceeding upward to the top layer. The jth layer load may be expressed in terms of any other layer load, say, the ith layer. Thus,

$$p_j = \left(\frac{p_i}{E_ih_i^3}\right)E_jh_j^3 \tag{5.43}$$

Hence,

$$P = \sum_1^n p_j = \left(\frac{p_i}{E_ih_i^3}\right)\sum_1^n E_jh_j^3 \tag{5.44}$$

The two expressions for P allow for solution of the load on the ith layer. Thus,

$$p_i(n) = (E_i h_i^3)\left(\frac{\sum \gamma_j h_j}{\sum E_j h_j^3}\right) \quad (i,j = 1,2\ldots,n) \tag{5.45}$$

where $p_i\,(n)$ is the load on the ith layer in a separated cluster of n-layers and summation is over all layers. Again, the E's may be replaced by corresponding values of $E/(1 - \nu^2)$.

If a gas or water pressure is present in the separation gap, then a pressure p_g must be added to the total external load, as shown in Figure 5.10. If the layers or beds dip at an angle δ to the horizontal, then the normal component of specific weight is $\gamma' = \gamma \cos(\delta)$. The normal component of weight deflects roof strata perpendicular to the dip. This "correction" is usually small and is limited to dips, say, less than 15°. With the addition of possible pressure at the bed separation horizon and modification for bed dip, the load on the ith layer is

$$p_i(n) = (E_i h_i^3)\left(\frac{p_g + \sum \gamma'_j h_j}{\sum E_j h_j^3}\right) \quad (i,j = 1,2\ldots,n) \tag{5.46}$$

where $p_i\,(n)$ is the load on the ith layer in a separated cluster of n naturally supported roof layers and summation is over all layers.

A critical question is how to determine if bed separation occurs as entries are advanced. The answer is in the form of a thought experiment. First, the equivalent distributed load

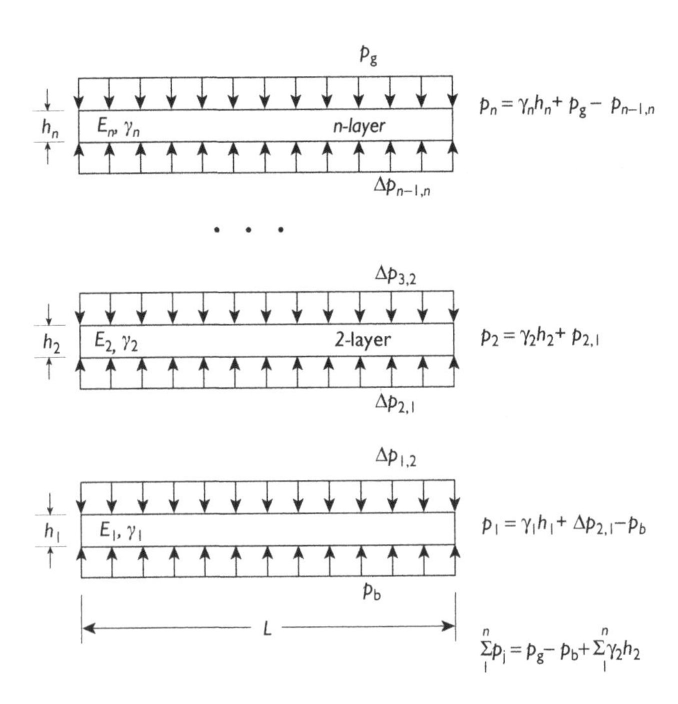

Figure 5.10 Exploded view of an n-layer cluster showing contact stresses ($\Delta p_{3,2}$, etc.), support pressure (p_b), and fluid pressure (p_g) at the separation gap. All layers sag the same amount. Each bed inside the cluster loads the layer below and supports the layer above.

acting on the first roof layer is calculated assuming bed separation between this layer and the layer above. In a gravity loaded beam with no fluid pressure expected in the separation gap, the load on the first layer is $p_1(1) = \gamma_1 h_1$. Next, the layer above is allowed to sag with the first layer; the assumption now is that separation occurs between layer 2 and layer 3. In this case

$$p_1(2) = E_1 h_1^3 \left(\frac{\gamma_1 h_1 + \gamma_2 h_2}{E_1 h_1^3 + E_2 h_2^3} \right) \tag{5.47}$$

If $p_1(2) > p_1(1)$, then the second layer must be loading the first layer. However, if $p_1(2) < p_1(1)$, then the second layer must be supporting the first layer because of the reduction in first layer load. Support action by the second layer requires the contact between layers to sustain a tensile stress. In the absence of tensile strength of the contact or interface between layers, bed separation follows. The process of adding layers to the evolving cluster continues until a load reduction occurs. Thus, if

$$p_1(n+1) < p_1(n) > p_1(n-1) > \ldots > p_1(2) > p_1(1) \tag{5.48}$$

then bed separation occurs between layers $n + 1$ and n. A second cluster of separated layers may form above the first. The analysis is the same provided the first layer in the new cluster is considered to be layer 1, the layer above layer 2, and so on.

As the test for bed separation continues into the roof, a point is reached where beam action is unlikely. Certainly, at a distance of an entry width into the roof, the influence of the entry on the premining stress field is small and beam action is nil.

Example 5.4 Entries 21 ft wide by 15 ft high are driven at a depth of 1,250 ft in flat-lying sedimentary rock. Strata properties are given in the table. Determine bed separation horizons within the given stratigraphic column.

Table of Strata Properties for Examples 5.4–5.9

Property stratum	*Rock type*	*E(10^6) psi*	*ν (–)*	*C_o(10^3) psi*	*T_o(10^2) psi*	*γ pcf*	*h (ft)*
Overburden	—	—	—	—	—	144	1250
Roof 3	Limestone	5.67	0.25	12.3	9.8	158	8.0
Roof 2	Sandstone	3.45	0.27	6.5	6.5	153	1.9
Roof 1	Shale	1.52	0.22	3.9	2.4	142	2.3
Seam	Coal	0.75	0.18	2.5	3.5	105	15.0
Floor	Sandstone	3.45	0.27	6.5	6.5	142	12.0

Notes
E = Young's modulus.
ν = Poisson's ratio.
C_o = unconfined compressive strength.
T_o = tensile strength.
γ = specific weight.
h = stratum thickness.

Solution: Bed separation may be detected by a thought experiment involving a sequential addition of strata on top of the first roof layer until the load on the first layer declines. Thus,

$$p(1,1) = \gamma_1 h_1 = (142)(2.3) = 326.6 \text{ psf}$$

$$p(1,2) = E_1 h_1^3 \left[\frac{\gamma_1 h_1 + \gamma_2 h_2}{E_1 h_1^3 + E_2 h_2^3} \right]$$

$$= \frac{(142)(2.3) + (153)(1.9)}{1 + E_2 h_2^3 / E_1 h_1^3}$$

$$= \frac{(142)(2.3) + (153)(1.9)}{1 + (3.45)(1.9)^3/(1.52)(2.3)^3}$$

$$p(1,2) = 270.8 \text{ psf}$$

When the second roof layer is added to the first, the load on the first decreases. The inference is that bed separation must occur in the absence of inter-bed tensile strength, a reasonable assumption.

Bed separation may also occur above the second roof stratum. The procedure is the same with only a relabeling of the sequence needed. The first stratum is now Roof 2, and

$$p(1,1) = \gamma_2 h_2 = (153)(1.9) = 290.7 \text{ psf}$$

$$p(1,2) = E_2 h_2^3 \left(\frac{\gamma_2 h_2 + \gamma_3 h_3}{E_2 h_2^3 + E_3 h_3^3} \right)$$

$$= \left(\frac{(153)(1.9) + (158)(8)}{1 + [(5.67)/(3.45)](8.0/1.9)^3} \right)$$

$$p(1,2) = 12.57 \text{ psf}$$

where 1 refers to the sandstone and 2 refers to the limestone. Separation is indicated between the sandstone and limestone roof layers.

Example 5.5 Given the data of Example 5.4 and now the presence of gas in the sandstone roof layer between the impermeable shale and limestone strata, determine the effect of 10 psi gas pressure on the bed separation analysis.

Solution: The gas pressure is an additional surface load that increases the gravity load of the layers below. Gas pressure p_g acts to increase $p(1,1)$ because of its presence at the shale–sandstone interface. Thus,

$$p(1,1) = \gamma_1 h_1 + p_g = (142)(2.3) + (10)(144) = 1,767 \text{ psf}$$

$$p(1,2) = E_1 h_1^3 \left[\frac{\gamma_1 h_1 + \gamma_2 h_2}{E_1 h_1^3 + E_2 h_2^3} \right]$$

$$= \frac{(142)(2.3) + (153)(1.9) + p_g}{1 + E_2 h_2^3 / E_1 h_1^3}$$

$$= \frac{(142)(2.3) + (153)(1.9) + (10)(144)}{1 + (3.45)(1.9)^3/(1.52)(2.3)^3}$$

$$p(1,2) = 902.5 \text{ psf}$$

Although $p(1,1)$ and $p(1,2)$ are substantially greater with gas pressure present, separation is still indicated between the shale and sandstone. Gas would be expected to accumulate in the separation cavity and pose a risk to roof stability. In some cases, a drill hole is used to relieve gas pressure. Water pressure would have the same effect.

Repeating the process to detect bed separation above the sandstone,

$$p(1,1) = \gamma_2 h_2 = (153)(1.9) = 290.7 \text{ psf}$$

$$p(1,2) = E_2 h_2^3 \left[\frac{\gamma_2 h_2 + \gamma_3 h_3 - p_g}{E_2 h_2^3 + E_3 h_3^3} \right]$$

$$= \left(\frac{(153)(1.9) + (158)(8) - (10)(144)}{1 + [(5.67)/(3.45)](8.0/1.9)^3} \right)$$

$$p(1,2) = 0.927 \text{ psf}$$

where no gas pressure occurs in $p(1,1)$ because it acts at the bottom and top of the sandstone and therefore imposes no additional load. When the limestone layer is added to the analysis, the gas pressure acts upward on the bottom of the sandstone and is thus negative. In this case, the conclusion is the same as without gas pressure, that is, bed separation also occurs between sandstone and limestone.

Example 5.6 Consider the data in Example 5.4 and further suppose that all strata dip 15°. Determine the effect of this strata dip on bed separation analysis.

***Solution*:** The dip is at the high end of the acceptable range of adjustment using $\gamma' = \gamma \cos(\delta)$ where δ is strata dip. In Example 5.4, only gravity loading was present. In view of the formula

$$p(i,n) = (E_i h_i^3) \left(\frac{\sum_1^n \gamma_i h_i}{\sum_1^n E_i h_i^3} \right) = (E_i h_i^3) \left(\frac{\sum_1^n \gamma'_i h_i}{\sum_1^n E_i h_i^3} \right)$$

all numerical results change by the same factor, $\cos(\delta)$, so no changes in conclusions about bed separations are induced under gravity loading only.

Safety factor $FS_t\,(i, n)$ for the i-th layer in the n-layer cluster is

$$FS_t(i,n) = \frac{T_o(i)}{\sigma_t(i)} \tag{5.49}$$

where the tensile stress is one of

$$\sigma_t(\text{SS}, i, n) = \left(\frac{3}{4}\right)\left(\frac{p_i L^2}{h_1^2}\right) = \left(\frac{3}{4}\right)\left(\frac{\gamma_a(i,n)L^2}{h_i}\right)$$
$$\sigma_t(\text{BI}, i, n) = \left(\frac{1}{2}\right)\left(\frac{p_i L^2}{h_i^2}\right) = \left(\frac{1}{2}\right)\left(\frac{\gamma_a(i,n)L^2}{h_i}\right) \tag{5.50}$$

depending on the assumed beam end conditions, $T_o\,(i)$ is tensile strength of the i-th layer in the separated cluster and where an apparent specific weight may be calculated for any layer in the cluster from the defining relationship

$$\gamma_a(i,n) = \frac{p_i}{h_i} \tag{5.51}$$

Maximum sag of any and all layers in the cluster is given by

$$w(\text{SS}) = \left(\frac{5}{32}\right)\left(\frac{p_i L^4}{E_i h_i^3}\right)$$
$$w(\text{BI}) = \left(\frac{1}{32}\right)\left(\frac{p_i L^4}{E_i h_i^3}\right) \tag{5.52}$$

which are clearly the same for all layers in the separated cluster.

Example 5.7 Consider the results of Example 5.4, then determine safety factors with respect to tension for the immediate roof (shale) and the layer above (sandstone).

***Solution*:** By definition, $FS_t = T_o/\sigma_t$. Under the assumption of simply supported roof beam ends, $\sigma_t = (3/4)(p)(L/h)^2$ and from Example 5.4, p(shale) = 326.6 psf and p(sandstone) = 290.7 psf. Hence

$$\sigma_t(\text{shale}) = \left(\frac{3}{4}\right)\left(\frac{326.6}{144}\right)\left(\frac{21}{2.3}\right)^2 = 141.8 \text{ psi}$$

$$\sigma_t(\text{sandstone}) = \left(\frac{3}{4}\right)\left(\frac{290.7}{144}\right)\left(\frac{21}{1.9}\right)^2 = 185.0 \text{ psi}$$

$$\therefore$$

$$FS_t(\text{shale}) = \frac{240}{141.8} = 1.69$$

$$FS_t(\text{sandstone}) = \frac{650}{185} = 3.51$$

The immediate roof (shale) safety factor is relatively low for a rock mass stressed in tension. The sandstone safety factor is marginal in view of a desired safety factor of 4 or greater in tension.

Example 5.8 Consider the data in Example 5.4 and further suppose that excavation is not full seam height, but rather is 11 ft, measured from the sandstone floor. Determine safety factors with respect to tension for the immediate roof.

***Solution*:** In this case, the immediate roof is 4.0 ft of coal left unmined. A bed separation analysis is needed beginning with this new roof layer as layer 1. Thus,

$$p(1,1) = \gamma_1 h_1 = (105)(4.0) = 420.0 \text{ psf}$$

$$p(1,2) = E_1 h_1^3 \left[\frac{\gamma_1 h_1 + \gamma_2 h_2}{E_1 h_1^3 + E_2 h_2^3} \right]$$

$$= \frac{(105)(4.0) + (142)(2.3)}{1 + E_2 h_2^3 / E_1 h_1^3}$$

$$= \frac{(105)(4.0) + (142)(2.3)}{1 + (1.52)(2.3)^3/(0.75)(4.0)^3}$$

$$p(1,2) = 538.9 \text{ psf}$$

$$p(1,3) = E_1 h_1^3 \left[\frac{\gamma_1 h_1 + \gamma_2 h_2 + \gamma_3 h_3}{E_1 h_1^3 + E_2 h_2^3 + E_3 h_3^3} \right]$$

$$= \frac{(105)(4.0) + (142)(2.3) + (153)(1.9)}{1 + E_2 h_2^3 / E_1 h_1^3 + E_3 h_3^3 / E_1 h_1^3}$$

$$= \frac{(105)(4.0) + (142)(2.3) + (153)(1.9)}{1 + (1.52)(2.3)^3/(0.75)(4.0)^3 + (3.45)(1.9)^3/(0.75)(4.0)^3}$$

$$p(1,3) = 552.3 \text{ psf}$$

$$p(1,4) = E_1 h_1^3 \left[\frac{\gamma_1 h_1 + \gamma_2 h_2 + \gamma_3 h_3 + \gamma_4 h_4}{E_1 h_1^3 + E_2 h_2^3 + E_3 h_3^3 + E_4 h_4^3} \right]$$

$$= \frac{(105)(4.0) + (142)(2.3) + (153)(1.9)}{1 + E_2 h_2^3 / E_1 h_1^3 + E_3 h_3^3 / E_1 h_1^3 + E_4 h_4^3 / E_1 h_1^3}$$

$$= \frac{(105)(4.0) + (142)(2.3) + (153)(1.9) + (158)(8)}{1 + (1.52)(2.3)^3/(0.75)(4.0)^3 + (3.45)(1.9)^3/(0.75)(4)^3 + (5.67)(8.0)^3/(0.75)(4.0)^3}$$

$$p(1,4) = 36.90 \text{ psf}$$

that indicates bed separation occurs between the sandstone and limestone layers. The load on the immediate roof is $p(1,3)$, and the immediate roof bending stress and safety factor with respect to tension are

$$\sigma_t(\text{coal}) = \left(\frac{3}{4}\right)\left(\frac{552.3}{144}\right)\left(\frac{21.0}{4.0}\right)^2 = 79.3 \text{ psi}$$

$$\text{FS}_t = \frac{T_o}{\sigma_t} = \frac{350}{79.3} = 4.41$$

which is an improvement over the case of mining full seam thickness where the immediate roof is the thin, weak shale layer.

Example 5.9 Given the results of Example 5.8, determine safety factors for the shale and sandstone layers.

Solution: Tensile strengths are known from the table of properties in Example 5.4. What is required are the strata loads. These may be obtained from the formula

$$p(i,n) = (E_i h_i^3)\left(\frac{\sum_1^n \gamma_j h_j}{\sum_1^n E_j h_j^3}\right)$$

Alternatively,

$$p(j,n) = p(i,n)\left(\frac{E_j h_j^3}{E_i h_i^3}\right)$$

From Example 5.8, $p(1,3) = 552.3$ psf (coal = 1, coal + shale + sandstone = 3) where separation occurs between sandstone and limestone. Hence,

$$p(\text{shale}) = (552.3)\left[\frac{(1.52)(2.3)^3}{(0.75)(4.0)^3}\right] = 212.7 \text{ psf}$$

$$p(\text{sandstone}) = (552.3)\left[\frac{(3.45)(1.9)^3}{(0.75)(4.0)^3}\right] = 272.3 \text{ psf}$$

$\therefore$

$$\sigma_t(\text{shale}) = \left(\frac{3}{4}\right)\left(\frac{212.7}{144}\right)\left(\frac{21}{2.3}\right)^2 = 92.35 \text{ psi}$$

$$\sigma_t(\text{sandstone}) = \left(\frac{3}{4}\right)\left(\frac{272.3}{144}\right)\left(\frac{21}{1.9}\right)^2 = 173.3 \text{ psi}$$

where simply supported ends are assumed, a conservative assumption relative to fixed ends.

The safety factors are

$$\text{FS}_t(\text{shale}) = \frac{240}{92.35} = 2.60$$

$$\text{FS}_t(\text{sandstone}) = \frac{650}{173.3} = 3.75$$

In this case, the thickness of the coal left in the roof is sufficient to give some support to the overlying shale and sandstone layers. This inference is made in consideration of the previous results when no coal was left in the roof and the shale and sandstone layers separated one from the other and from the limestone above the sandstone (Example 5.7). In that case the safety factors for shale and sandstone were 1.69 and 3.51, respectively.

Bolted roof

Naturally supported roof may not be safe; additional support is then required to achieve an acceptable safety factor. Roof bolting is the method of choice, although additional support may be necessary. Roof bolts are essentially steel rods or tubes; high quality steel is used for manufacturing bolts that are intended to function under tensile loading. There are two main types of roof bolts: (1) mechanical bolts that are point anchored and tensioned during installation, and (2) bolts that have distributed anchorage and are tensioned by roof sag that may occur after installation.

Point anchored, mechanical bolts are the least expensive bolt type. Because of tensioning during installation, point anchored bolts provide immediate support to the roof and are also referred to as *active* support. Tensioning is achieved by tightening a nut against a small plate at the bolt hole collar. This action causes a wedge with small teeth at the top end of the bolt to spread against the walls of the bolt hole and resist the pull induced by tightening the nut. Presence of a strong "anchor" stratum is therefore essential to point anchored mechanical bolting. Bolt diameters of 5/8ths and 3/4ths inches are common. Lengths and spacing vary, but a pattern of 6 ft long bolts on 4 ft centers is perhaps representative. Installation tension ranges from 50% to 75% of the elastic limit of the bolt. Ultimate strength of a bolt is much higher than the initial yield point at the elastic limit, but when a point anchored bolt ruptures all support action is lost. Steel straps and various types of screen, for example, wire mesh, may be used in conjunction with roof bolts. Point anchored bolts have some resistance to shear across a bolt hole, but large shearing displacements may occur because the bolt does not fill the bolt hole.

Bolts with distributed anchorage may be held in the hole by cements of various types or by friction and are sometimes referred to as *passive* support because they provide support only in reaction to roof sag following installation. If the roof does not experience additional sag after installation of these types of bolts, they remain untensioned. Rupture of a distributed anchorage bolt creates two shorter bolts that retain support capacity as long as the anchorage remains effective. This support action is a considerable advantage over point anchored bolts, but is obtained at additional cost. Because distributed anchorage bolts fill the bolt hole, they also have superior resistance to shear. Shear resistance offered by bolts that fill bolt holes is often referred to as *dowel* action.

Point anchored roof bolts act to support the roof by suspension and clamping or beam-building. Distributed anchorage bolts act through suspension and dowel action. However, if distributed anchorage is not installed over the entire hole column, then tensioning during installation is possible. The combination then provides immediate support that includes all mechanisms: suspension, clamping (beam building), and dowel action.

Point anchored roof bolting

The presence of a thick, strong layer in the geologic column above an entry is often associated with bed separation below and allows for the possibility of support by point anchored mechanical bolts. If F_b is bolt tension and bolts are placed on a pattern such that A_r is the area of roof rock normal to the bolt holes, then an equivalent bolting pressure p_b is applied to the first roof layer by the bolt forces acting through the collar plates adjacent to the roof rock. This bolting pressure reduces the natural roof rock load and should be subtracted from the total load acting on the separated cluster. Hence, in this case of bolted roof,

$$p_i = (E_i h_i^3)\left(\frac{p_g - p_b + \sum \gamma_j' h_j}{\sum E_j h_j^3}\right) \quad (i,j = 1,2,\ldots,n) \tag{5.53}$$

gives the distributed load acting on any of layers in the cluster. Inspection of this formula shows that a sufficiently large bolting pressure may reduce layer loads to zero. An implication is that layer sag is also reduced to zero. In this event, the roof becomes fully suspended by the bolts; no bending stress occurs and no natural support is mobilized by roof sag. Reduction of roof sag in this way would close the separation gap at the top of the separated bed cluster. The action would literally "raise the roof." Partial suspension occurs when bolt tension does not completely close the separation gap. Regardless, the anchor stratum must be strong enough to support the suspended load and the anchors must also hold fast and neither slip or creep.

Although bolt forces and rock weights of 10,000 lbf or more may seem high, the associated bolting pressures and layer loads are small relative even to rock tensile strength, perhaps of the order of 100–1,000 psi. For example, the yield point of a high strength 3/4 in. diameter steel bolt is about 18,400 lbf. Distributed over a 4 ft by 4 ft square bolting pattern, this force produces a bolting pressure of about 8 psi which is equivalent to a rock "head" of 8 ft assuming rock specific weight of 144 pcf. Thus, a rock block 4 × 4 × 8 ft could be supported by this bolt. Because the bolt is loaded to the elastic limit, the bolt safety factor is one.

Additional bolt tension may be applied after closure of the bed separation gap by continuing to tighten the bolt nut against the collar plate. This action forcibly clamps together the layers in a separation cluster and thus allows for mobilizing shear resistance to slip between layers. The effect may be seen as beam building. If interlayer slip were entirely prevented, then the cluster would respond as a single, monolithic beam with a thickness equal to the sum of the layer thicknesses in the cluster, as shown in Figure 5.11.

An example calculation serves to scope the magnitude of roof bolt clamping action and beam building by comparing the maximum shear stress in a monolithic beam with shear stresses in a bolted cluster. If the monolithic beam shear stress can be mobilized by bolting, then clamping action and beam building are maximum. Maximum shear stress in a beam under a uniformly distributed normal load occurs on the neutral axis at beam ends. If V is the internal shear force at the ends of a simply supported beam and A is the beam cross-sectional area, then the average shear stress is simply V/A; the maximum shear stress is 3/2 the average shear stress. At the example beam ends $V = pbL/2$, so the maximum shear stress is $(3/4)pL/H$ where H is now the total thickness of the cluster. Interlayer shear resistance with neglect of any cohesion is $\tau = \mu\sigma$ where μ is the interlayer coefficient of friction and σ is the normal stress caused by clamping, that is, $\sigma = p_b$, the bolting pressure. Prevention of interlayer slip then requires

$$\mu \geq \left(\frac{3}{4}\right)\left(\frac{L}{H}\right) \tag{5.54}$$

Thus, if the interlayer coefficient of friction is greater than three-fourths the span to depth ratio of the cluster, then beam building is maximum and the cluster behaves as a monolithic beam. If the span is 18 ft and the cluster thickness is 3 ft, then the coefficient of friction must be greater than 4.5. The corresponding friction angle $\phi = \tan^{-1}\mu$ must be greater than 77°!

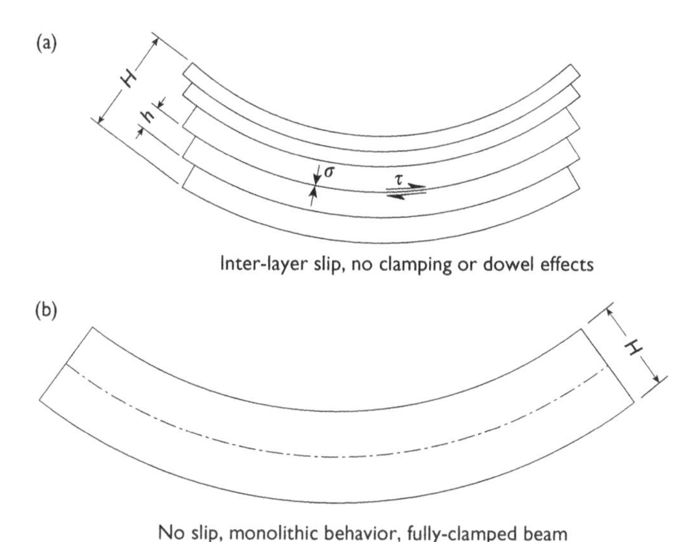

Figure 5.11 Clamping and beam-building action through slip prevention by mobilizing interlayer friction or by dowel action.

The least span to depth ratio needed for beam action is greater than three and, preferably, greater than five. These ratios correspond to friction angles of 66° and 75°, respectively, and are very high friction angles in any situation and therefore seem unlikely to be realized in mine rock, for example, in laminated shale. This analysis also suggests that interlayer shear failure is highly probable near the ends of layered roof rock over entries in stratified ground. The shear force diminishes toward the center of a roof beam, so clamping action would be more effective away from the beam ends. However, while clamping action or beam building is a physically possible roof bolt support mechanism, the effect is probably small relative to suspension.

Full suspension is equivalent to *dead weight* load design that simply equates the weight of rock W between bolts to the bolt load F_b. The approach is conservative in the sense that self-support capacity of the roof rock, that is, rock tensile strength is ignored. If the bolting pattern is square with spacing S, then rock weight is $\gamma S^2 H$ where γ is the average specific weight of the supported block and H is the thickness of the block of suspended strata, usually the distance from roof to a potential anchor stratum. If a thick, strong, and secure anchor stratum is not within reasonable bolt length, then point anchored bolts are unsuitable for controlling the considered roof. Assuming an anchor stratum is present in the geologic column above the entry roof, then bolt length L_b is H plus an anchorage allowance of 0.5 to perhaps 1.0 feet. If the bolting pattern is not square then, $W = \gamma S_r S_b H$ where S_r is the bolt row spacing along an entry and S_b is the spacing between bolts in a row, as shown in Figure 5.12. In this regard, the bolts at the ends of a row are usually placed $S_b/2$ from the entry sides to allow for reduced load bearing capacity of the walls because of local damage incurred during entry advance.

Once the required bolt force is known, bolt cross-sectional area and nominal diameter may be determined for an assumed grade of steel, that is, steel strength. Because point anchored

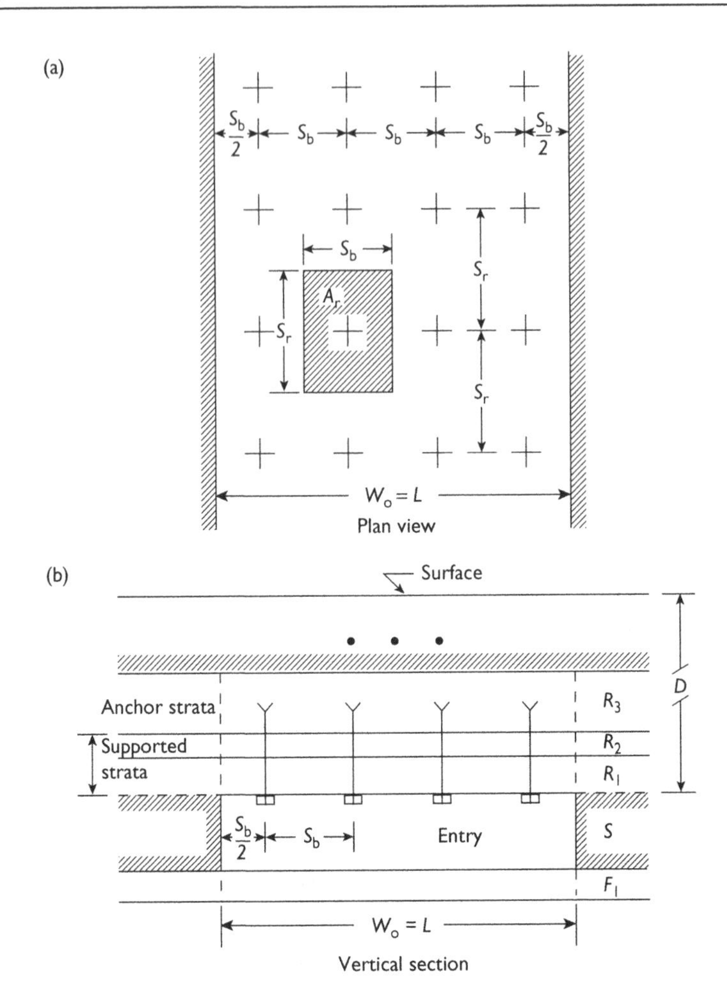

Figure 5.12 Roof bolting pattern in stratified ground.

mechanical bolts are customarily tensioned at the time of installation, the bolt safety factor is simply the reciprocal of the percentage the installation tension is of the bolt yield strength. Specifying bolt installation tension is equivalent to specifying a maximum allowable tensile stress in the bolt (strength/safety factor). This tension is just that required to fully suspend the rock block weight. Thus, $(T_b/FS_b) = F_b = W$ where FS_b is the bolt safety factor with respect to tension. If T is the installation tension fraction of the bolt yield point, then FS_b is $1/T$.

Example 5.10 Entries 21 ft wide by 15 ft high are driven at a depth of 1,250 ft in flat-lying sedimentary rock. Strata properties are given in the table. Bed separation occurs between the shale and sandstone horizons and between sandstone and limestone (Example 5.4). Determine the bolting pressure needed to close the separation gap between shale and sandstone.

Table of Strata Properties for Examples 5.10–5.11

Property stratum	*Rock type*	*E(10⁶) psi*	*ν (–)*	*$C_o(10^3)$ psi*	*$T_o(10^2)$ psi*	*γ pcf*	*h (ft)*
Overburden	—	—	—	—	—	144	1250
Roof 3	Limestone	5.67	0.25	12.3	9.8	158	8.0
Roof 2	Sandstone	3.45	0.27	6.5	6.5	153	1.9
Roof 1	Shale	1.52	0.22	3.9	2.4	142	2.3
Seam	Coal	0.75	0.18	2.5	3.5	105	15.0
Floor	Sandstone	3.45	0.27	6.5	6.5	142	12.0

Notes
E = Young's modulus.
ν = Poisson's ratio.
C_o = unconfined compressive strength.
T_o = tensile strength.
γ = specific weight.
h = stratum thickness.

Solution: The separation gap is the difference in vertical displacement (sag) between the two roof beams. Both sag under self-weight which is treated as an equivalent uniformly distributed load over the top of each roof beam. With a bolting pressure applied to the bottom of the shale layer, the net load $p(\text{shale}) = \gamma h - p_b$. Under the assumption of simply supported ends, the maximum sag of each layer is

$$w(\text{shale}) = \frac{5pL^4}{32E'h^3} = \frac{(5)(142)(2.3)(21)^4}{(32)(1.52)(10^6)(2.3)^3} = 0.5366 \text{ ft}(6.440 \text{ in.})$$

$$w(\text{sandstone}) = \frac{5pL^4}{32E'h^3} = \frac{(5)(153)(1.9)(21)^4}{(32)(3.45)(10^6)(1.9)^3} = 0.3733 \text{ ft}(4.480 \text{ in.})$$

To close the separation gap, shale sag must be reduced to 4.480 in. or (4.480/6.440)100%. The "pressure" required for this amount of sag requires reduction by the same percentage. Thus, $p(\text{new}) = (0.6957)p(\text{old}) = 0.6957(142)(2.3) = 227.2$ psf and $p_b = \gamma h - p(\text{new}) = 99.4$ psf.

As a check,

$$w(\text{shale}) = \frac{5pL^4}{32E'h^3} = \frac{(5)(227.2)(21)^4}{(32)(1.52)(10^6)(2.3)^3} = 0.3733 \text{ ft}(4.480 \text{ in.})$$

which is equal to the sandstone sag, so no gap between shale and sandstone is now present. Application of p_b would be by bolting; p_b is the bolting pressure. The shale layer is now partially suspended.

Example 5.11 Consider the results of Example 5.10, then with neglect of sag of the limestone roof layer, determine the bolting pressure required to fully suspend the shale and sandstone roof layers.

***Solution*:** Full suspension implies reduction of sag to zero. Net loads on the shale and sandstone layers must accordingly be reduced to zero. Hence, p(shale) $= \gamma h - p_b = 0$ and similarly for the sandstone. This reduction has the effect of bringing these beds into contact and forcing them to sag the same amount, ultimately zero. Because the process imposes equal sag, use the formula

$$p_i = (E_i h_i^3)\left(\frac{p_g - p_b + \sum \gamma_j' h_j}{\sum E_j h_j^3}\right) \quad (i,j = 1,2,\ldots,n)$$

The load for full suspension corresponding to zero sag requires the numerator to be zero. Hence, in the absence of gas pressure and zero dip,

$$p_b = \sum_1^2 \gamma_i h_i = (143)(2.3) + (153)(1.9) = 619.6 \text{ psf}$$

With this bolting pressure, the net load on shale and sandstone is zero. Thus, full suspension does indeed correspond to a dead weight load design.

Example 5.12 In view of the results of Example 5.11, specify a bolting plan based on a dead weight design approach for the given entry and stratigraphic column.

***Solution*:** The combined thickness of the shale and sandstone layers is 4.2 ft, soa5 ft long bolt anchored in the relatively thick and stiff limestone layer should do. Bolts spaced on a square pattern S ft × S ft is a reasonable trial design, as shown in the sketch.

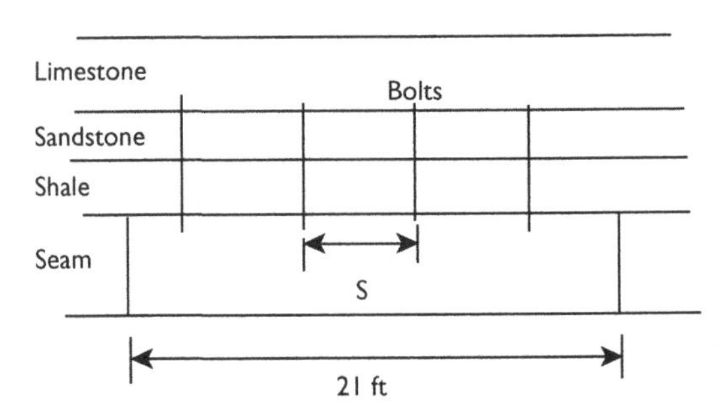

A trial spacing of 5 ft is a reasonable first estimate. The weight W of a block assigned to a bolt is (5)(5)(619.6), that is, W = 15, 490 lbf. This number is also equal to the bolt force F_b, and $F_b = T_b / FS_b$, the ratio of bolt strength to bolt safety factor (allowable bolt tension). A reasonable bolt tension is, say, 2/3rds bolt yield load, T_Y. Hence, $T_Y = (3/2)(15{,}490) = 23{,}235$ lbf. A 3/4 in. bolt, grade 75, has a yield load of 25,100 lbf. Thus, 3/4 in. diameter, grade 75 bolts, 5 ft long and spaced on 5 ft centers will do. Spacing of bolts next to the entry sides should be less than 5 ft, rather than greater than 5 ft. This plan would require 4 bolts per row, spaced 5 ft apart; the two end bolts would be 3 ft from the rib.

No bolting plan is unique. Smaller bolts spaced more closely or larger bolts at greater spacing would also meet requirements. In this regard, spacing is often qualified by any tendency for raveling between bolts, while cost is also a consideration. Fewer setups to drill bolt holes is advantageous. Site-specific data would allow for optimizing a bolting plan with due consideration for geologic variability. A trial design is just a necessary first step, but an essential one that brings out the important features of the problem.

Distributed anchorage roof bolting

Examples of distributed anchorage obtained by cementation of steel in a bolt hole are steel rods cemented along the entire bolt length with a resin compound that is mixed as the bolts are spun into holes. A great advantage to distributed anchorage bolts is that no thick, strong anchor stratum within practical bolt length of the entry roof is needed. Distributed anchorage achieved by frictional contact between bolt and hole wall rather than with cement allows for considerable motion of bolt and rock mass and is a considerable advantage in highly deformable ground.

Collar plates are often used with resin bolts, although the need is not as clear as with point anchored bolts. Reinforcing steel, rebar, that is cemented into a hole using a grout is another example of distributed anchorage roof bolting. The end of rebar steel is often threaded to provide for a nut to retain a collar plate. As with resin bolts, collar plates may serve to hold strapping and screening in place.

Analysis of support action associated with distributed anchorage begins with recognition that one portion of the bolt L_a provides anchorage while the remaining portion provides support L_s, as shown in Figure 5.13 where bolt $L = L_a + L_s$. Equilibrium requires the bolt tension T to be supported by a shear force V transmitted to the rock at the cement–rock contact. Tensile strength of the cement between steel and hole wall is considered negligible. Neither the cement–rock bond nor steel bolt should fail before the other. Thus,

$$T_s = \sigma_Y A_s = \sigma_s \pi \frac{d^2}{4} = \tau_s \pi d L_a \tag{5.55}$$

where σ_Y is the yield point of the steel, A_s is steel area, d is steel diameter, and τ_s is the average shear stress acting over the contact area between cement and steel at failure ("strength"). This equation may be used to estimate the anchorage length needed for a given

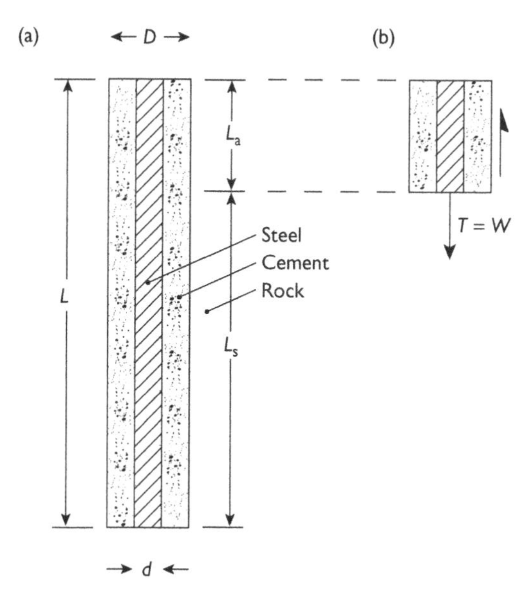

Figure 5.13 Distributed anchorage of a cemented steel bar: (a) geometry and (b) anchorage equilibrium.

bolting plan provided the bond strength between steel and cement is known and the bond between cement and rock does not fail. Anchorage length in terms of steel diameter is

$$\frac{L_a}{d} = \left(\frac{1}{4}\right)\left(\frac{\sigma_s}{\tau_s}\right) \tag{5.56}$$

where τ_s is shear strength of the steel–cement interface that depends more on the steel surface than on the steel grade.

An estimate for steel–cement bond shear strength may be obtained from allowable shear stress in steel reinforced concrete beams which is several hundred psi; strength of the bond may be about 1,000 psi, depending on the safety factor used for allowable bond stress. Steel strength is of the order of 60,000 psi, so the ratio of anchorage length to steel diameter may be roughly 60 / 4 or about 15. A one inch steel rod would require about 15 inches of anchorage to avoid bond failure when the tensile strength was reached. The higher the steel grade (strength), the greater the anchor length required (because of the greater rock load). A very rough rule of thumb then is 1 ft of anchorage per inch of steel diameter.

Bolt tensile strength must be adequate to support rock below. Equilibrium requires bolt tension to be equal to the weight of the supported rock $W = \gamma\, S_r\, S_b\, L_s$. A bolt safety factor in terms of bolt strength and load is

$$FS_b = \frac{T_s}{W} \tag{5.57}$$

Thus, the maximum allowable bolt tension is simply W.

Consideration of equilibrium of the cement annulus shows that the average shear stress acting over the cement–rock interface per unit of anchorage length is less than the aver- age shear stress acting at the steel–cement interface. If the steel does not rupture and the anchorage holds, then failure of the bolt is expected to occur by stripping of cement and rock from the steel. The reason is that both anchorage and steel strengths are greater than rock weight because of the bolt safety factor and the requirement that neither should fail first. The weakest element of the assemblage of steel, cement, and interfaces is then the steel–cement interface. Evidence for this expectation is often seen in falls of bolted roof rock that leave steel protruding from hole remnants above. Of course, bolt breakage is possible when bond strengths and anchorage are high; the steel is then the weakest part of the assemblage.

Thickness of a roof that may be supported by distributed anchorage bolts may be obtained from a dead weight load analysis similar to that used for point anchored bolts of comparable length, that is, by equating bolt tension to rock block weight. As before, there is no unique answer because of the interplay amongst bolt size, steel grade, spacing, and length in equilibrium requirements.

A caution is necessary in considering very long distributed anchorage bolts, for example, friction bolts (steel tubes) 30 ft long with an ultimate strength of 28,000 lbf installed in 1–1/2 in. diameter holes on a 5 × 5 ft square pattern in rock that weighs 150 pcf. Under these conditions, no more than about 7.5 ft of rock can be supported (150×5×5×7.5 = 28,125 lbf) by the bolt. If the elastic limit or yield strength is used, then only 3.75 ft of rock can be supported. Regardless, even if the bolt were a solid one-inch diameter rod, only a few feet of bolt would be necessary for anchorage. The bolt length therefore seems far in excess of what is needed. Alternatively, one may suppose that the anchorage is as strong as the bolt and thus accounts for only a few feet of the bolt. The remainder of the bolt is then the thickness of the supported rock block that weighs far more than the bolt strength, so the bolt would break. The question then is why would long, distributed anchorage bolts be used?

The answer has two parts. The first concerns the load distribution along the bolt; the second relates to the fact that if the bolt breaks, two shorter bolts are created; these bolts continue to support the adjacent rock. Bolts may also be deliberately trimmed or shortened in the course of mining; the portion remaining continues to provide support. Because stress concentration, related strain and associated displacement decay rapidly with distance from a bolt hole collar, most of the bolt load is also concentrated near the collar and diminishes rapidly along the bolt. If the bolt is relatively long, then the bolt is stressed very little over much of the hole length. Alternatively, a tacit assumption in a dead weight load analysis is that the supported rock block is entirely free and unsupported by the adjacent rock mass and therefore the adjacent rock mass has no strength. This assumption is extreme. If slabs and blocks have not actually formed, the actual bolt load may be much less than the assumed dead weight load.

Example 5.13 Entries are driven 18 ft wide at a depth of 987 ft. The stratigraphic column is given in the table. Specify a suitable bolting plan in consideration of laminations in shale that average 3 inches in thickness, while those in sandstone average 6 inches.

Table of Strata Properties for Example 5.13

Property stratum	*Rock type*	*E(10^6) psi*	*ν (–)*	*C_o(10^3) psi*	*T_o(10^2) psi*	*γ pcf*	*h (ft)*
Overburden	—	—	—	—	—	144	987
Roof 3	Sandstone	3.45	0.27	6.5	6.5	158	12.0
Roof 2	Laminated Sandstone	2.45	0.23	6.5	6.5	155	7.5
Roof 1	Laminated Shale	1.32	0.21	3.9	2.4	141	13.8
Seam	coal	0.75	0.18	2.5	3.5	105	8.5
Floor	Sandstone	3.45	0.27	6.5	6.5	158	12.0

Notes
E = Young's modulus.
ν = Poisson's ratio.
C_o = unconfined compressive strength.
T_o = tensile strength.
γ = specific weight.
h = stratum thickness.

Solution: The laminations act as thin roof beams and thus induce a high bending stress, that is,

$$\sigma_t = \frac{3}{4}p\left(\frac{L}{h}\right)^2 = \left(\frac{3}{4}\right)(151)(3/12)\left(\frac{18}{3/12}\right)^2 = 146,772 \text{ psf}(1,019 \text{ psi})$$

which is well above the tensile strengths of 240 and 650 psi for the laminated shale and sandstone roof layers. The thickness of the layers, 13.8 and 7.5 ft, compared with the width of the entry, 18 ft, indicates a conventional point anchored bolt plan would not be feasible because a bolt 22 ft long would be needed to reach the third roof layer that could perhaps serve as an anchor stratum. Such a thickness would require very large bolts. For these reasons, a bolting plan based on distributed anchorage is indicated.

A trial plan using resin bolts spaced 4 ft apart with #6 rebar, grade 60, provides a bolt yield load of $(60{,}000)(\pi/4)(6/8)^2 = 26{,}500$ lbf. (Rebar is reinforcing steel bar. The number is the bar diameter in 1/8 inches) The bolt force is equal to the weight of rock assigned to the bolt, so $26{,}500 = (141)(4)(4)h$ where h is bolt length and thickness of supported roof; $h = 11.7$ ft, which is rather long considering opening width of 18 ft and opening height of just 8 ft. At a spacing of 5 ft, $h = 7.5$ ft. Use of a bolt longer than the height of the opening can be done by bending the bolt while pushing it up into the bolt hole. A mechanized bolting machine would require clearance for the apparatus, so an 8 ft bolt would be impractical unless it were in sections that were coupled during the installation process.

A compromise plan using smaller bolts, say, #5 rebar, same grade 60, 6 ft long on 4 ft centers provide a support force of 18,400 lbf. These bolts are headed and may be used with a collar plate to support wire mesh that would defend against raveling. A rock block 8.2 ft high could be supported by such a bolt. In this plan, the end bolts next to the rib would be installed 3 ft from the rib, although spacing is 4 ft otherwise. In the event experience dictates supplemental support, it would likely be in the form of roof trusses or long, flexible bolts.

Roof trusses

A roof truss is a hybrid support that is a combination of bolts and sets shown in Figure 5.14. The bolts replace the legs or posts of a conventional set; the horizontal tie rod or cable replaces the cap or cross bar. Roof trusses are generally used as supplemental support with conventional bolting, although truss steel bolts and rods are usually much larger than conventional bolts. Spacing between trusses also tends to be greater than spacing of primary bolt support. Trusses are used mainly where point anchored bolts are not practical because of the absence of a suitable anchor stratum, for example, where the roof rock is 20 ft of laminated shale. Such ground would almost certainly require distributed anchorage bolts, with strapping, mesh or screen, as primary support.

A simple equilibrium analysis of the truss in Figure 5.14 shows that the vertical support force F_v provided by the truss acting against the roof through one of the blocking points is

$$F_v = T\sin(\beta) \tag{5.58}$$

where T is bolt and rod tension and β is the bolting angle measured from the horizontal. Bolting angles are often 45°, but may be higher or lower. Bolting pressure p_b associated with roof truss forces is the ratio of force to area. Thus,

$$p_b = \frac{2F_v}{LS} \tag{5.59}$$

where L and S are entry span and truss spacing, respectively. A truss system using one inch diameter steel rods with a tensile strength of 60,000 psi, spaced on 8 ft centers with a 45° bolting angle in an 18 ft wide entry would provide a bolting pressure slightly more than

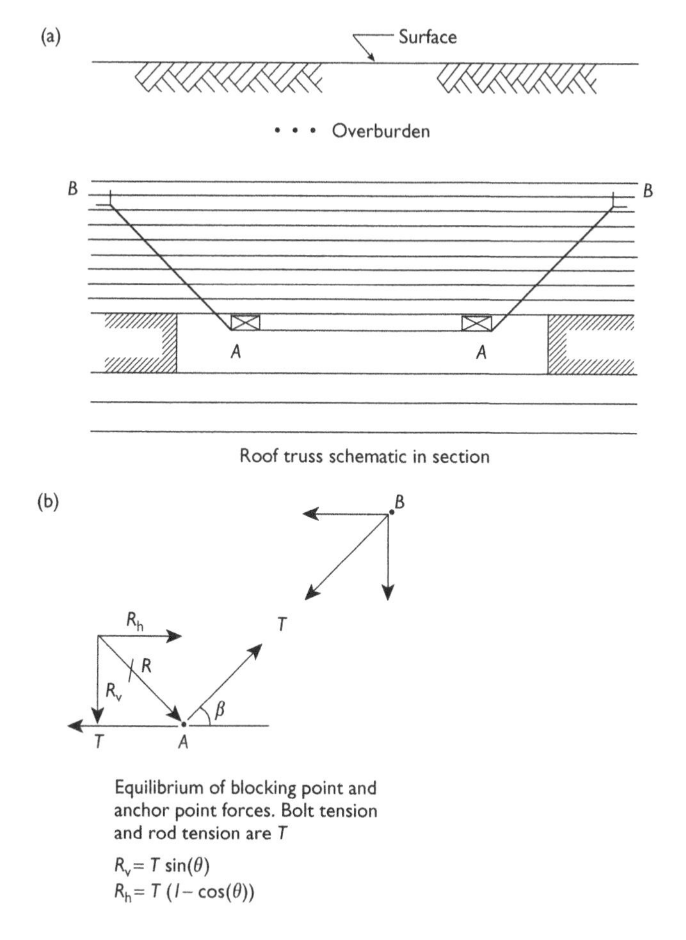

Figure 5.14 Roof truss schematic and equilibrium.

3 psi when tensioned to the elastic limit. Increasing the bolting angle to 60° increases the bolting pressure to about 4 psi. This truss could therefore support a dead weight load of about 3–4 ft of roof rock in an 18 ft wide entry. As is almost always the case, support loads are small relative to intact rock strength and may seem insignificant, but they are essential to ground control and often mean the difference between a safe entry and unstable ground.

If the purpose of a truss is to provide as much roof support as possible, then the bolting angle should be 90°. The question that arises is why not 90°? The reason is for security of the anchorage. Inclined holes that penetrate ground over the shoulder of an entry are in a region of compressive stress that tends to close the bolt hole and thus to grip the bolt anchor more securely. This action is effective whether the bolt is point anchored, is cemented in the hole, or is a friction bolt. The horizontal component of bolt force tends to compress the roof rock between load points and thus reduces tensile stress that may develop from beam action of the roof layers.

Example 5.14 Consider the data and results of Example 5.13 where the immediate roof is 13.8 ft of laminated shale. Evaluate the effect of supplemental roof trusses spaced on 8 ft centers with anchor holes inclined 45° to the horizontal. Roof truss steel is #8 rebar, grade 60; holes are collared 1.5 ft from the ribs.

***Solution*:** The trusses are spaced at twice the bolt spacing of 4 ft. An equilibrium analysis based on the sketch shows that the vertical support pressure amounts to 3.86 psi, the equivalent of about 4 ft of rock. This additional support increases the supported height of roof rock to 10 ft from 6 ft (bolt length in Example 5.14).

Summation of forces horizontally and vertical at a block point shows that

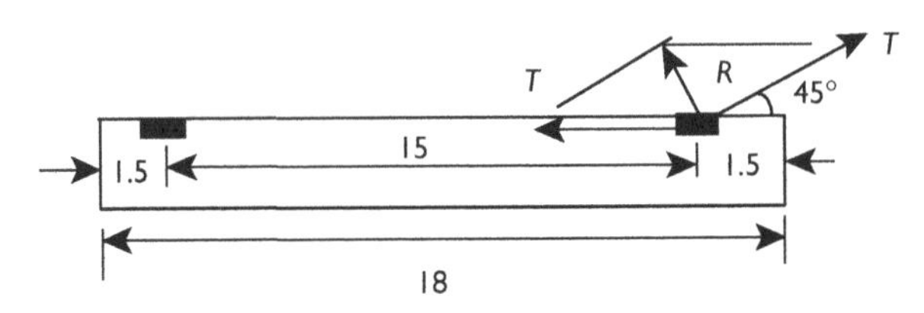

Equilibrium analysis of Example 5.14

$$R_h = T(1 - \cos(45)) = (\pi/4)(1)^2(60,000)(1 - 1/\sqrt{2}) = 13,802 \text{ lbf}$$

$$R_v = T\sin(45) = (\pi/4)(1)^2(60,000)(1/\sqrt{2}) = 33,322 \text{ lbf}$$

The total vertical support load is double the load at one of the two blocking points. This load is distributed over and area equal to (8)(18–3)=120 ft^2. The support pressure is then

$$p_b \frac{(2)(33,322)}{120} = 555.4 \text{ psf} \quad (3.86 \text{ psi})$$

as claimed.

Additional examples

Example 5.15 Consider an immediate roof stratum of span L, thickness h, and breadth b that has a rectangular cross-section. Suppose a uniformly distributed support pressure p' from below acts over a small length L' as shown in the sketch. Let the center of the distribution be a units from the left side and further suppose the ends are simply supported. Derive formulas for the bending moment and bending stress along the beam.

***Solution*:** The sketch is a free body diagram of the beam with equivalent point loads replacing the distributed loads. Overall equilibrium requires resultant forces in the vertical direction and moments equal zero. Thus,

$$\sum F = 0 = F' + R_A + R_B$$

$$\sum M(A) = 0 = (R_B)(L) + (a)(F') \therefore$$

$$R_B = (-a/L)(F'),\ R_A = (-a'/L)(F'),\ F' = p'L',\ L = a + a'$$

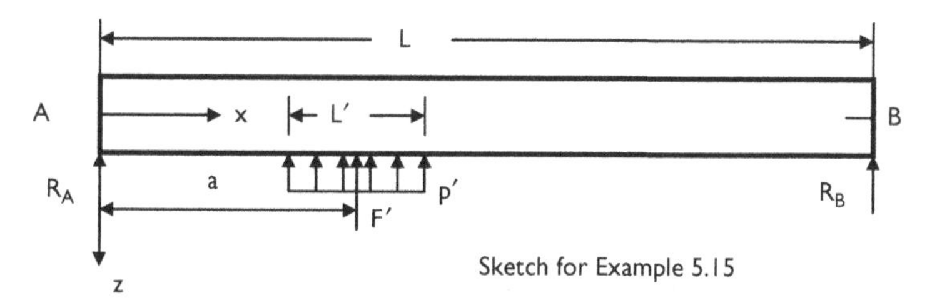

Sketch for Example 5.15

Internal equilibrium for '$x < (a - L'/2) = X_1$ requires a free body diagram.

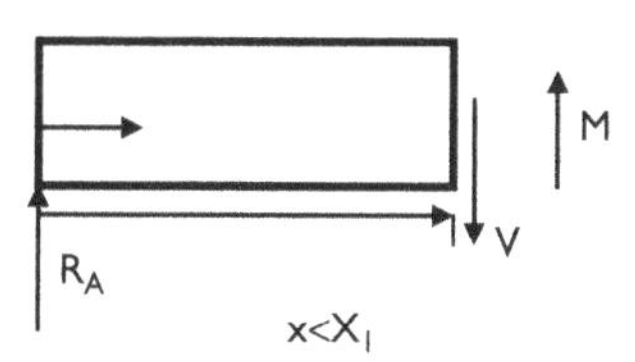

Summation of forces gives

$$\sum(F) = V - R_A, \therefore$$
$$V = R_A$$

Summation of moments gives

$$\sum M(A) = Vx - M, \therefore$$
$$M = R_A x$$

The same result may be obtained by simple integration of $dM/dx = V$.

Internal equilibrium for $X_1 = (a - L'/2) < x < (a + L'/2) = X_2$ requires another free body diagram.

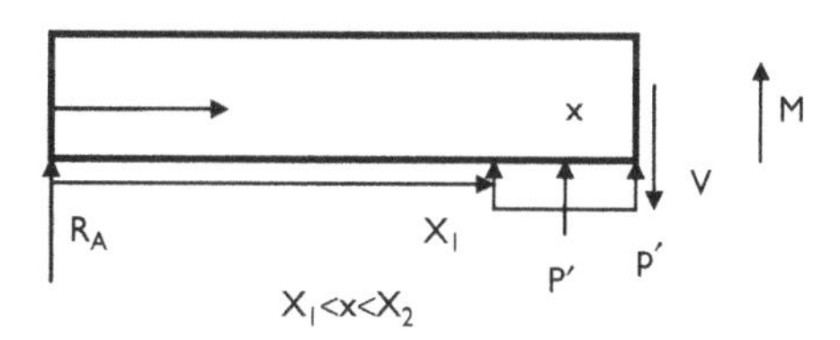

Summation of forces gives

$$\sum(F) = V - R_A - P' \therefore$$
$$V = R_A + p'(x - X_1)$$

Summation of moments gives

$$\sum M(A) = Vx + P'(X_1 + x)/2 - M, \therefore$$
$$M = R_A x + p'(x - X_1)^2/2$$

Internal equilibrium for $X_2 = (a + L'/2) < x$ requires yet another free body diagram.

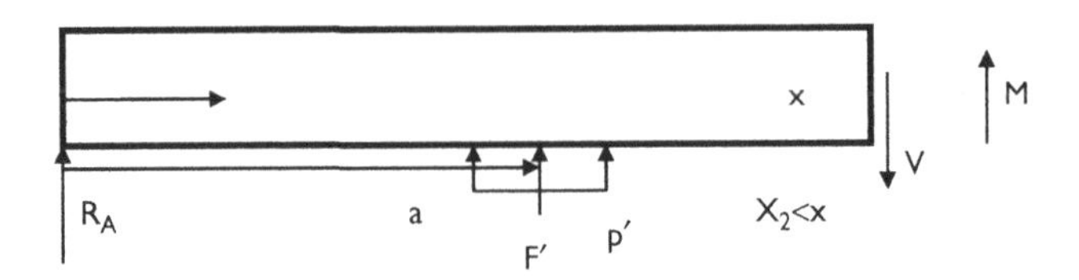

Summation of forces gives

$$\sum(F) = V - R_A - F' \therefore$$
$$V = R_A + F'$$

Summation of moments gives

$$\sum M(A) = Vx - F'a - M, \therefore$$
$$M = R_A x + F'(x - a)$$

Summarizing

$$\begin{aligned} &V = R_A && x < X_1 \\ &V = R_A + p'(x - X_1), && X_1 < x < X_2 \\ &V = R_A + F', && X_2 < x \end{aligned}$$

and

$$\begin{aligned} &M = R_A x && x < X_1 \\ &M = R_A x + p'(x - X_1)^2/2, && X_1 < x < X_2 \\ &M = R_A x + F'(x - a), && X_2 < x \end{aligned}$$

Example 5.16 Use the results from Example 5.15 and plot the shear and moment diagrams for a span of 23 ft, a distributed pressure of 507.5 psi over a length of 1 ft with distribution center 5.75 ft from the left beam end, a beam thickness of 5 ft, a beam breadth of 1 ft (into the page). Also determine the location of the maximum bending stress and magnitude. Consider tension positive.

***Solution*:** Shear and moment diagrams may be plotted from the results in Example 5.15 after making the necessary numerical substitutions. A spreadsheet is an easy path to the plots. Greatest moment magnitude occurs 6 ft from the left side of the stratum where the shear force is zero as seen in the plot and as determined from spreadsheet calculation. Moment

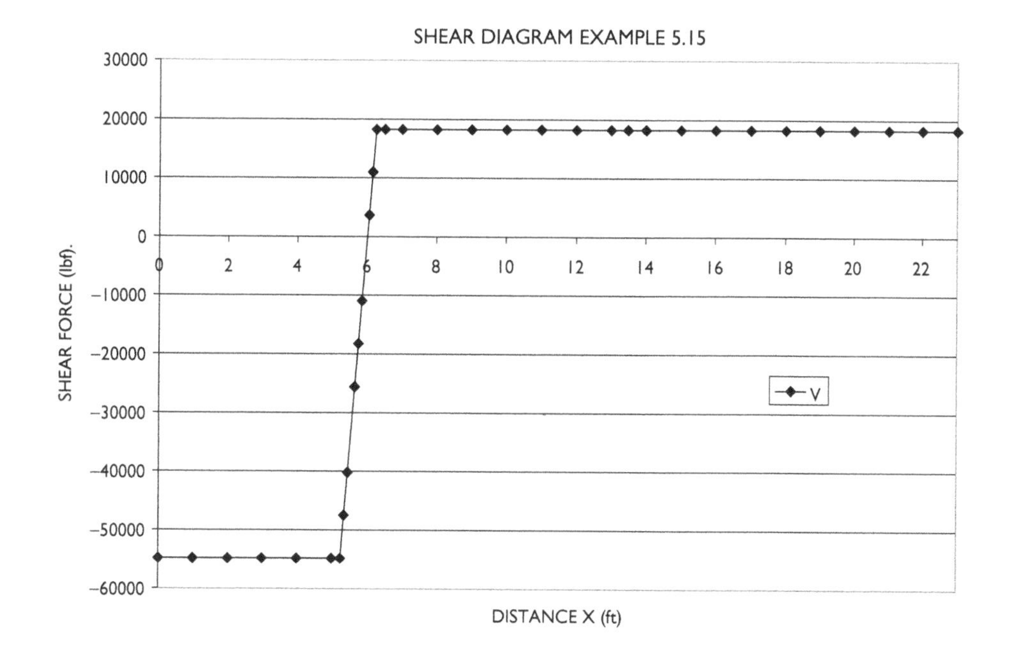
SHEAR DIAGRAM EXAMPLE 5.15
30000
20000
10000
0
-10000
-20000
-30000
-40000
-50000
-60000
0
2
4
6
8
10
12
14
16
18
20
22
SHEAR FORCE (lbf).
DISTANCE X (ft)
V

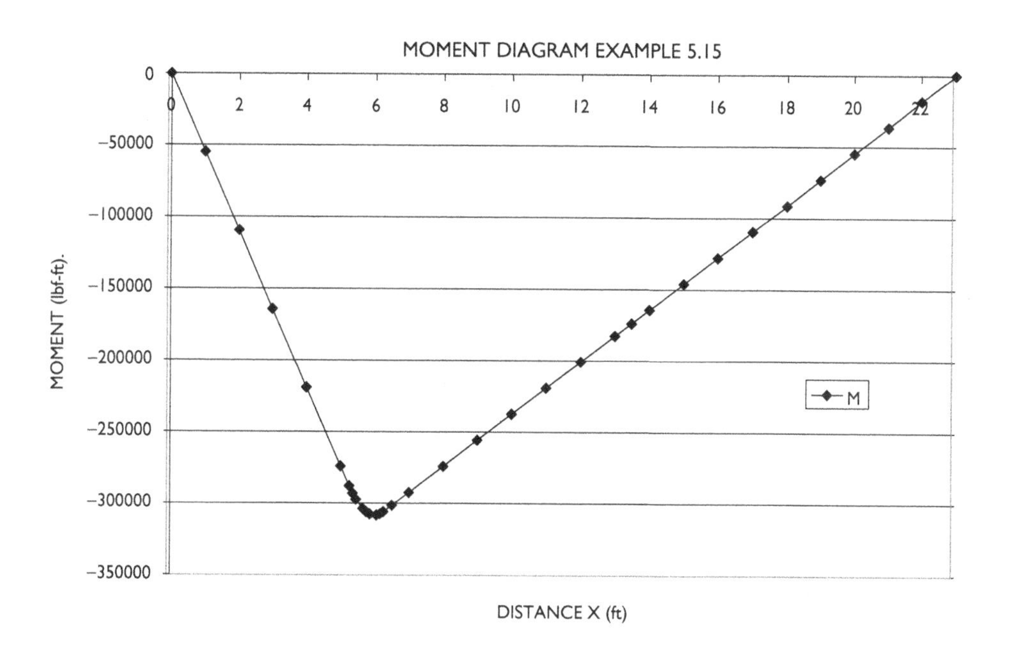
MOMENT DIAGRAM EXAMPLE 5.15
0
-50000
-100000
-150000
-200000
-250000
-300000
-350000
0
2
4
6
8
10
12
14
16
18
20
22
MOMENT (lbf-ft).
DISTANCE X (ft)
M

magnitude is 308,306 lbf-ft. The negative sign indicates peak tension occurs on top of the stratum and

$$\sigma_t = Mc/I$$
$$= (308306)(2.5)/[(1)(5)^3/12]$$
$$\sigma_t = 7.4(10^4)\text{psf} \quad (514 \text{ psi})$$

Example 5.17 The high pressure and small area of distribution in Examples 5.15 and 5.16 are indicative of a bolting force distributed over a collar plate. In Example 5.16 the bolt force is pressure times collar plate area, that is, $F_b = p'A' = (507.5)(1)(1)(144) = 73{,}080$ lbf which is a high bolt force, although not out of reason. The small distribution area suggests approximation of the support pressure by point loading as in Example 5.2 for multiple but equally spaced point loads. Consider the data in Examples 5.15 and 5.16 but develop formulas assuming a point load instead, then plot shear and moment diagrams and determined moment magnitude.

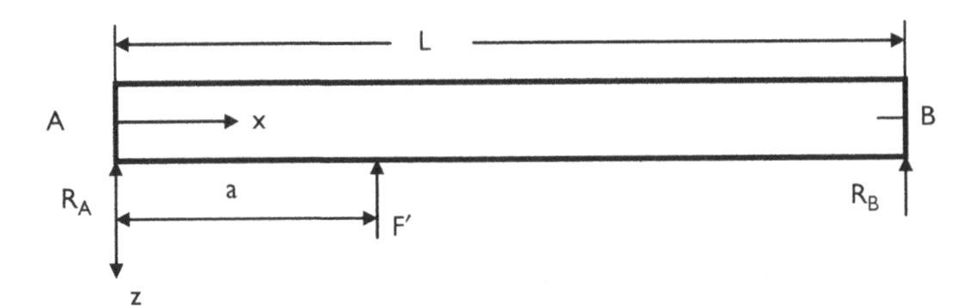

Sketch for EXAMPLE 5.17

Solution: The sketch is a free body diagram of the beam with equivalent point loads replacing the distributed loads. End conditions are not specified, so an assumption of simply supported ends is made. An assumption that $a<L/2$ is also made in view of the sketch. Overall equilibrium requires resultant forces in the vertical direction and moments equal zero. Thus,

$$\sum F = 0 = F' + R_A + R_B$$
$$\sum M(A) = 0 = (R_B)(L) + (a)(F') \therefore$$
$$R_B = (-a/L)(F'),\ R_A = (-a'/L)(F'),\ F' = p'L',\ L = a + a'$$

Internal equilibrium for $x < a$ requires a free body diagram.

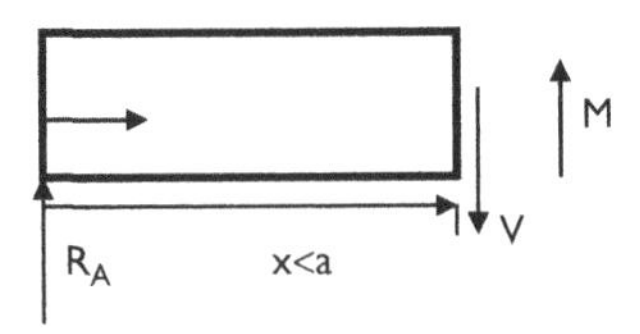

Summation of forces gives

$$\sum(F) = V - R_A, \therefore$$
$$V = R_A$$

Summation of moments gives

$$\sum M(A) = Vx - M, \therefore$$
$$M = R_A x$$

The same result may be obtained by simple integration of $dM/dx = V$. Internal equilibrium for $a < x$ requires another free body diagram.

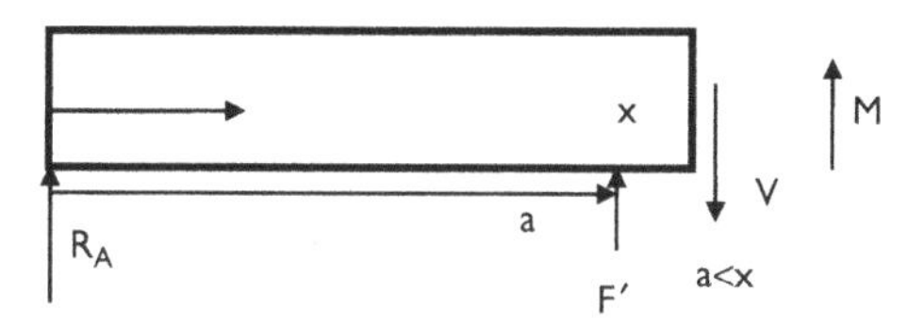

Summation of forces gives

$$\sum(F) = V - R_A - F' \therefore$$
$$V = R_A + F'$$

Summation of moments gives

$$\sum M(A) = Vx - F'a - M, \therefore$$
$$M = R_A x + F'(x - a)$$

Summarizing

$$V = R_A \qquad x < a$$
$$V = R_A + F', \qquad a < x$$
$$M = R_A x, \qquad x < a$$
$$M = R_A x + F'(x - a), \quad a < x$$

Using a span of 23 ft, a 73,080 lbf at 5.75 ft from the left beam end, a beam thickness of 5 ft, a beam breadth of 1 ft (into the page), the shear and moment diagrams plot as: The moment plot and spreadsheet calculation give a maximum moment of 315,158 lbf-ft at 5.75 ft from the left end of the stratum. The sign is negative indicative of tension at the beam top. This magnitude is only slight greater than the one assuming an equivalent but uniformly distributed load.

Example 5.18 Suppose roof bolts are installed on 5.75 ft centers across a 23 ft wide entry. Bolts adjacent to entry ribs are spaced at one half spacing. Bolt rows are also spaced 5.75 ft along the entry. The bolted stratum is 5 ft thick and bolts are 6 ft long as shown in the sketch. The bolt force is a pressure (507.5 psi) times collar plate area (8 inches by 8 inches), so $F = 32,800$ lbf. Consider the bolting forces then as point loads as in Example 5.17. Derive

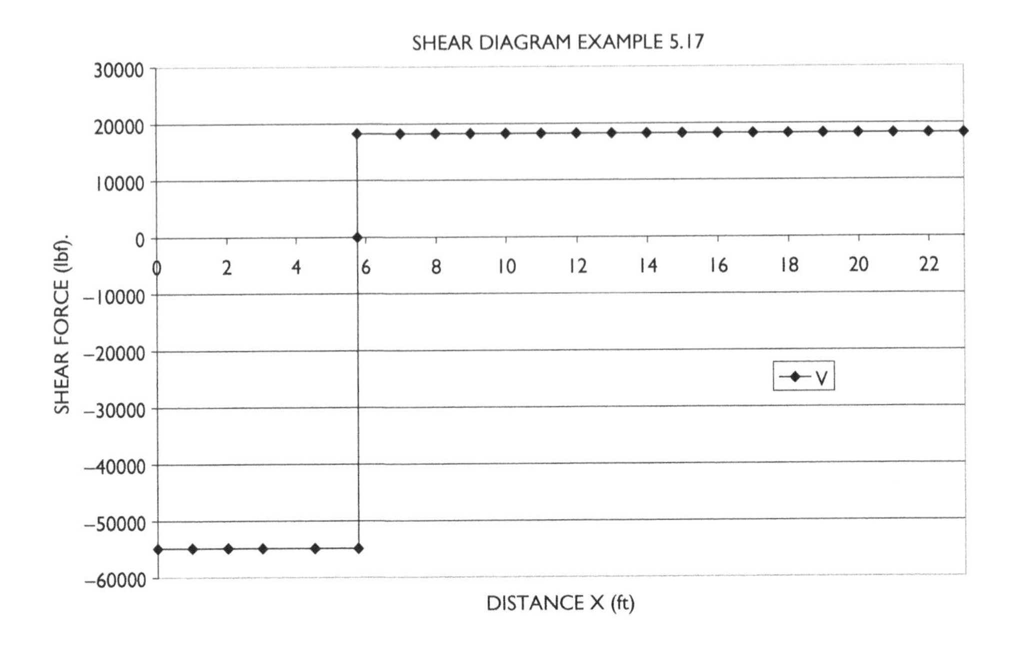

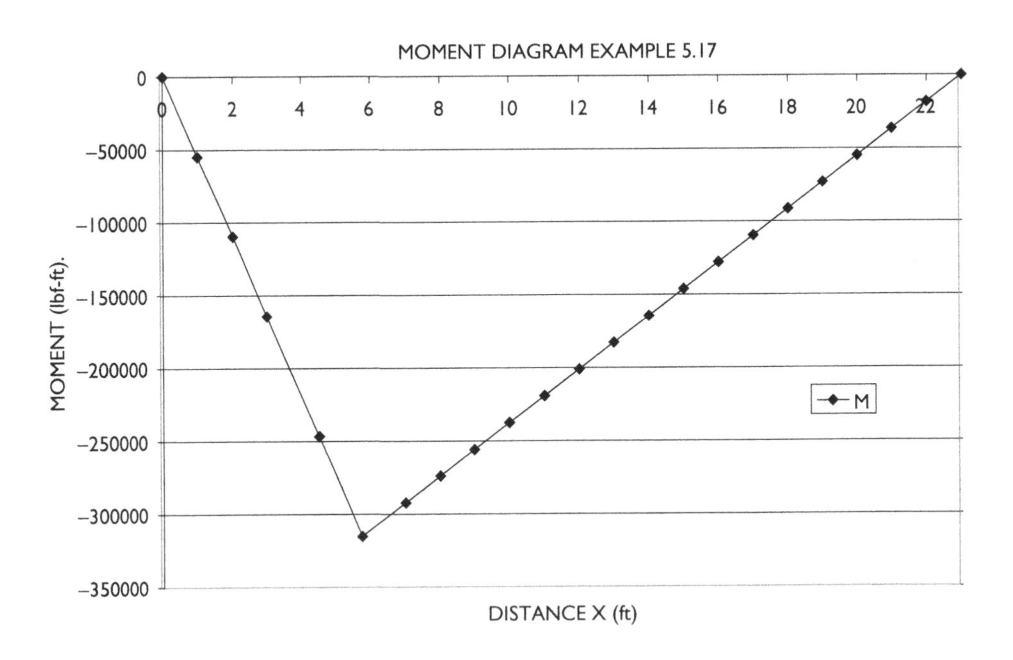

shear and moment formulas; plot these distributions, and find the location and magnitude of the peak moment.

Solution: A free body diagram of the stratum shows the bolt forces F at a spacing S in the sketch. The bolts are given as point loads, but need to be distributed as line loads into the page

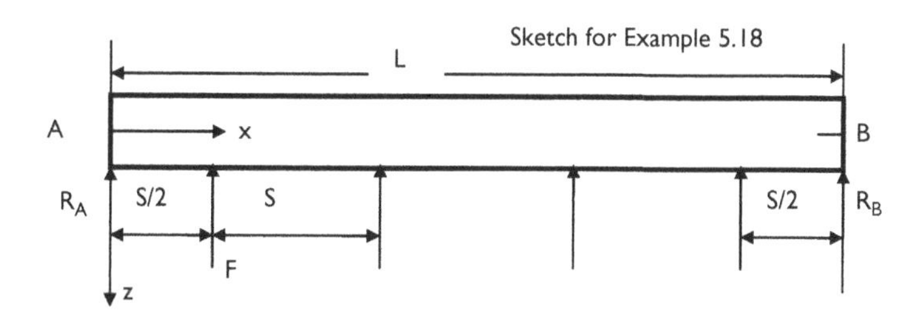

so the analysis can proceed as a beam analysis. Bolt spacing into the page is 5.75 ft, so bolt force per ft into the page is 5,705 lbf. Assume simply supported ends. Overall equilibrium is given by sum of the forces and sum of the moments. Thus,

$$0 = R_A + R_B - nF_b$$
$$0 = LR_B + (F_b/2)(S + 3S + \ldots + (2n - 1)S) = LR_B + (F_bS/2)n^2 \therefore$$
$$R_B = -(nF_b/2), \quad R_A = -(nF_b/2)$$

where n is the number of bolts ($n = L/S$). If n is not an integer, then the spacing must change for a fixed span L. In this example n =4. The reactions may be verified by consideration of symmetry.

Internal equilibrium leads to shear and moment formulas. Thus, with reference to the sketch,

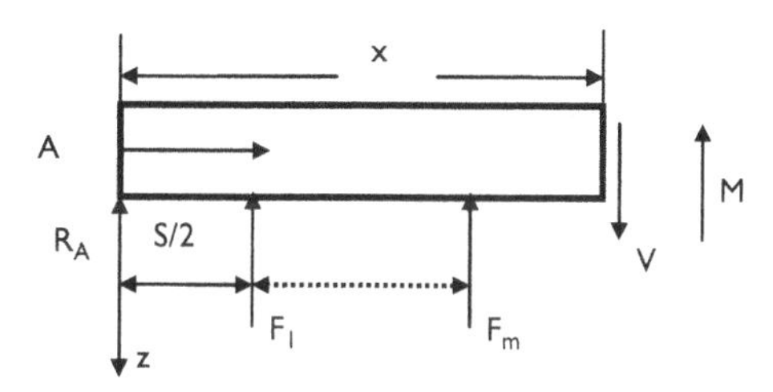

$$0 = V - R_A - mF_b$$
$$V = R_A + mF_b$$

which shows the shear force increases in steps as the count m of bolts (forces) increases to the right. The count can be obtained by rounding down the ratio x/S. For example, if x =12, then m =2; if $x<S$, then m =0.

Internal moment equilibrium is given by

$$0 = Vx - M - (F_b/2)[S + 3S + \ldots + (2m - 1)S]$$
$$M = (R_A + mF_b)x - (m^2SF_b/2)$$

where again m is the count of bolts to the left of x. As a check, $M = 0$ at $x = 0$, and at $x = L$

$$M = (-nF_b/2 + nF_b)L - (n^2SF_b/2)$$
$$= (-nF_b/2 + nF_b)L - n(L/S)SF_b/2$$
$$M = 0$$

which checks.

A plot of shear force shows the expected "staircase" appearance. Because of the interval of zero shear force at the beam mid-span, the moment remains constant and at the peak value over this interval. The moment is also negative in this interval and therefore the peak tension occurs at the top of the stratum. Peak moment magnitude is 65,600 lbf-ft.

This example is similar to Example 5.2 that concerns point loads on a regular spacing.

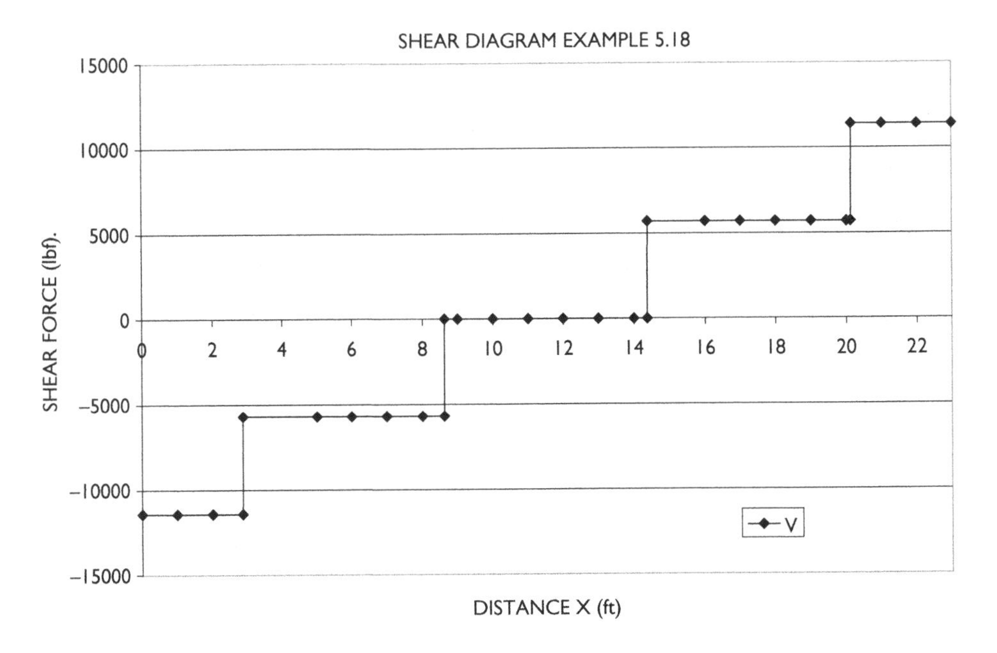

Example 5.19 Consider a roof stratum loaded by a uniformly distributed "pressure" p from above over the entire span L and support by another uniform pressure p' from below over a length L'. Derive formulas for the bending moment and bending stress along the beam (per ft into the page), plot the shear and moment diagrams, and determine the location of the maximum bending stress. Consider tension positive, bolt distance is 5.75 ft from the left end of the stratum, bolting pressure of 200 psi, 1 ft x 1 ft collar plates, bolt row spacing into the page is 6.25 ft, and simply supported ends.

***Solution*:** The sketch is a free body diagram of the beam with an equivalent point load replacing the distributed load. Overall equilibrium requires resultant forces in the vertical direction and moments equal zero. Thus,

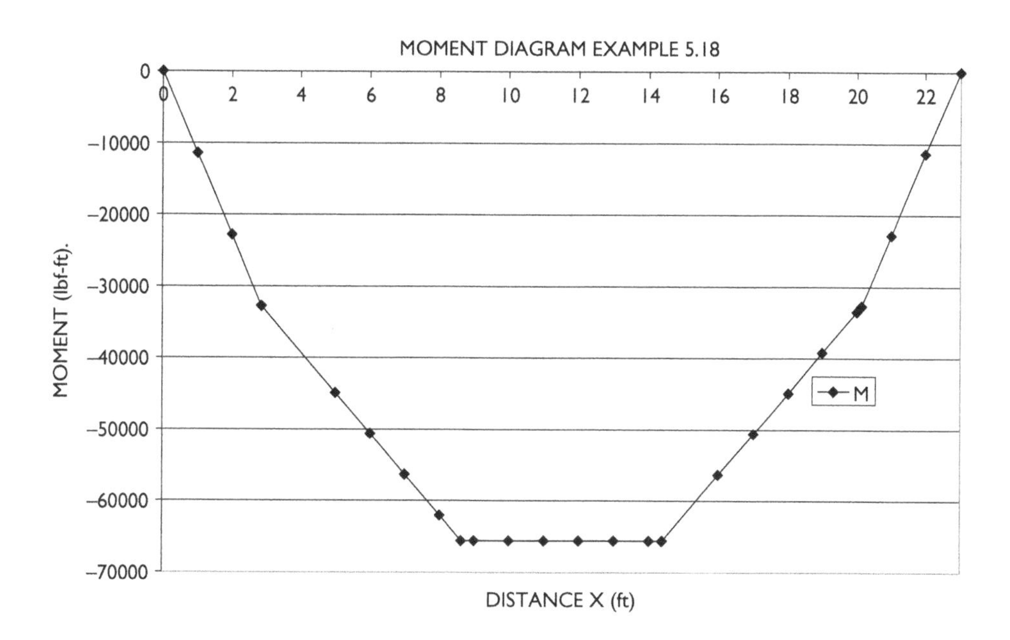

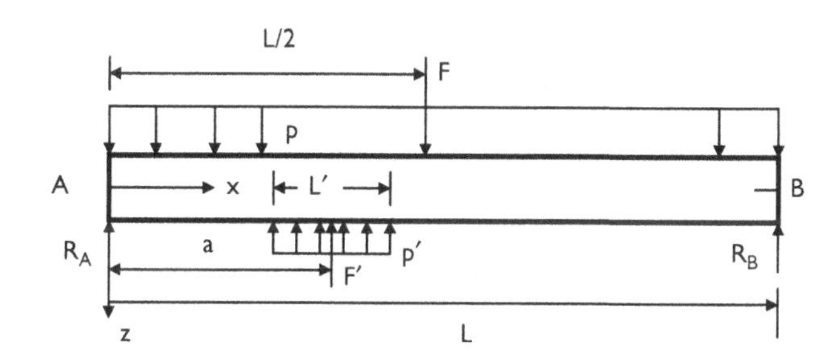

Sketch for EXAMPLE 5.19

$$\sum F = 0 = F - F' - R_A - R_B$$

$$\sum M(A) = 0 = (F)(L/2) - (R_B)(L) - (a)(F')$$

$$\therefore$$

$$R_B = (F/2) - (a/L)(F'),\ R_A = (F/2) - (a'/L)(F'),\ F = pL,\ F' = p'L',\ L = a + a'$$

Internal equilibrium for $x < (a - L'/2) = X_1$ requires a free body diagram. Summation of forces gives

$$\sum(F) = P + V - R_A,\ \therefore$$

$$V = R_A - px$$

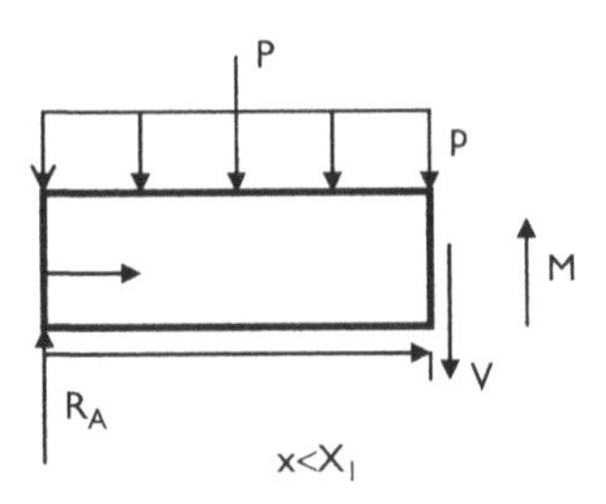

Summation of moments gives

$$\sum M(A) = P(x/2) + Vx - M, \therefore$$
$$M = R_A x - px^2/2$$

The same result may be obtained by integration of $dM/dx = V$.

Internal equilibrium for $X_1 = (a - L'/2) < x < (a + L'/2) = X_2$ requires another free body diagram.

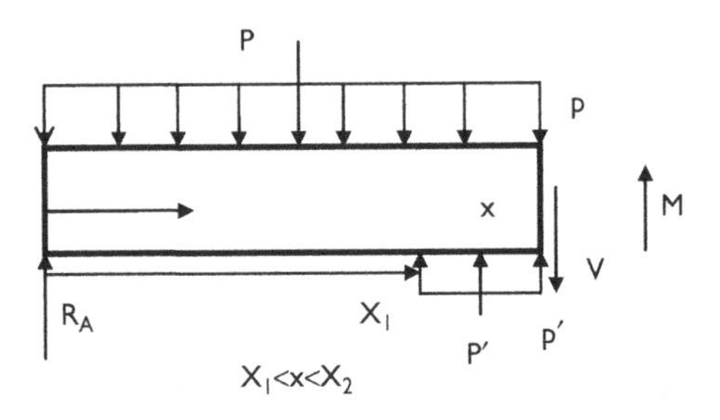

Summation of forces gives

$$\sum(F) = P + V - R_A - P' \therefore$$
$$V = R_A - px + p'(x - X_1)$$

Summation of moments gives

$$\sum M(A) = P(x/2) + Vx - P'(X_1 + x)/2 - M, \therefore$$
$$M = R_A x - px^2/2 + p'(x - X_1)^2/2$$

Internal equilibrium for $X_2 = (a + L'/2) < x$ requires yet another free body diagram. Summation of forces gives

$$\sum(F) = P + V - R_A - F' \therefore$$
$$V = R_A - px + F'$$

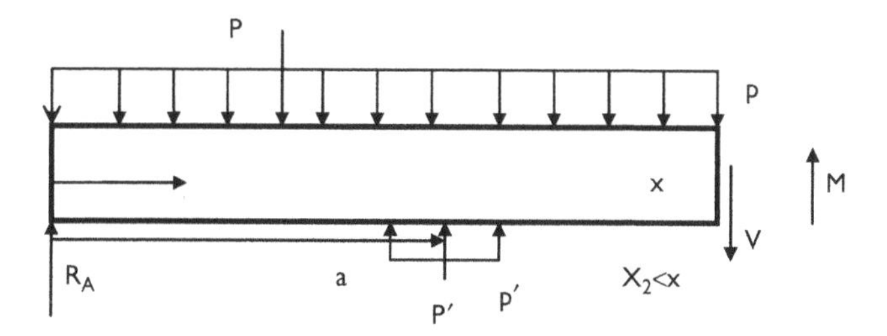

Summation of moments gives

$$\sum M(A) = P(x/2) + Vx - F'a - M, \ \therefore$$
$$M = R_A x - px^2/2 + F'(x - a)$$

Summarizing

$$\begin{aligned} &V = R_A - px, && x < X_1 \\ &V = R_A - px + p'(x - X_1), && X_1 < x < X_2 \\ &V = R_A - px + F', && X_2 < x \end{aligned}$$

and

$$\begin{aligned} &M = R_A x - px^2/2, && x < X_1 \\ &M = R_A x - px^2/2 + p'(x - X_1)^2/2, && X_1 < x < X_2 \\ &M = R_A x - px^2/2 + F'(x - a), && X_2 < x \end{aligned}$$

Shear and moment diagrams may be plotted from the results after making the necessary numerical substitutions. A spreadsheet is a convenient path to the plots after computing forces per ft into the page. The shear force goes to zero at 13.1 ft from the left side of the stratum and is quantified in the spreadsheet used for the calculations.

The maximum moment occurs at 13.1 ft from the left end of the stratum and has a magnitude of 35,284 lbf-ft. The positive sense of the moment indicates peak tension occurs at the bottom of the stratum.

These results could also be obtained by superposition of results from Examples 5.15 and the special case of a beam deformed under a uniformly distributed load illustrated in Figure 5.3.

Example 5.20 Consider a thick, massive stratum between ground surface and mining horizon that behaves as a beam as illustrated in the sketch. Depth to the top of the stratum is H ft (450 ft). The stratum is h ft (50 ft) thick. Support is removed by excavation in a series of openings W_o ft (20 ft) wide that are separated by pillars W_p (60 ft) wide. The series of openings and pillars is continued to a span of L ft. The center to center spacing is $W_p + W_o$ ft (80 ft). Half-pillars are present at the left and right abutments as shown in the sketch. Derive formulas for the shear force and moment distributions along the stratum assuming there are n (5) openings present. Plot the shear and moment distributions and find the location and value of the maximum tensile stress in the stratum. Also determine the location and magnitude of the maximum shear stress.

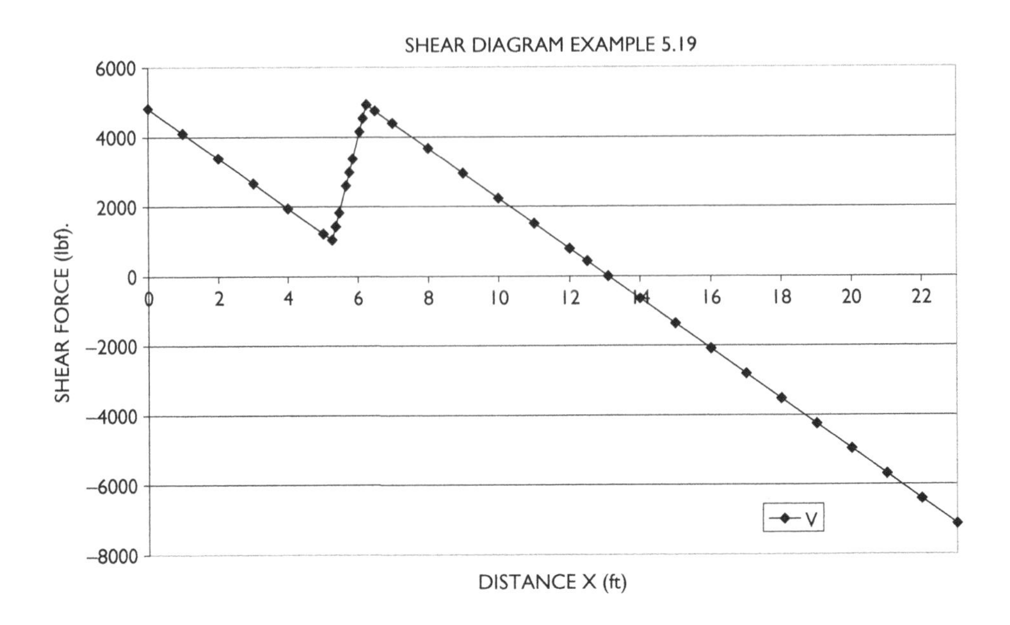

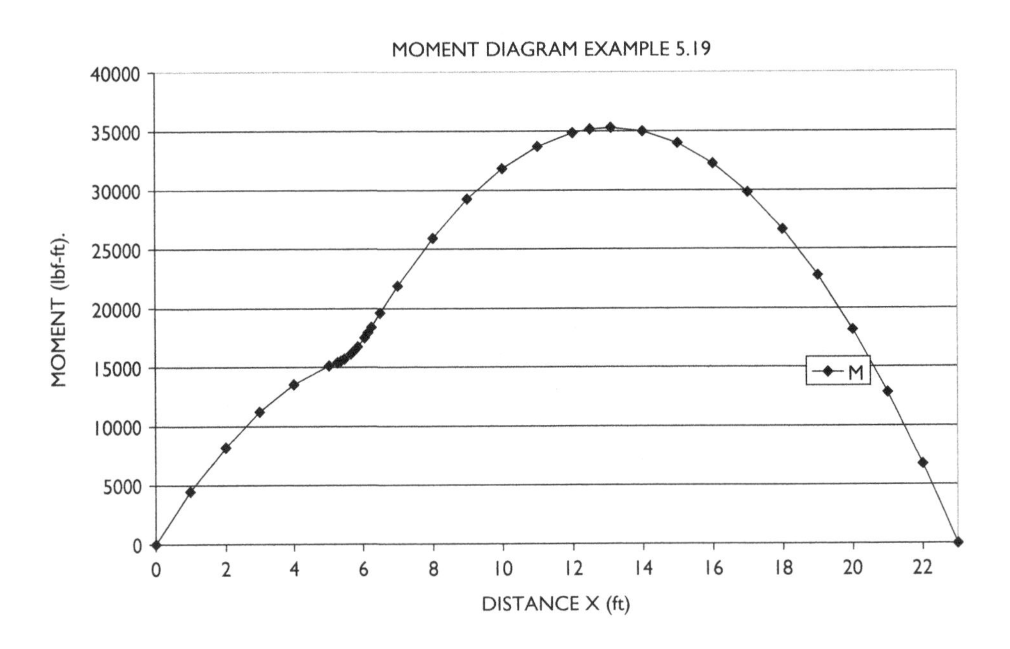

***Solution*:** An algebraic analysis is done first. The formulas for shear force and moment are then evaluated in a spread sheet. Inspection of the sketch shows that the span $L = n(W_p + W_o) = nS$. Force on top of the stratum per ft into the page is $F = pL(1)$; force on the bottom per pillar per ft into the page is $F' = p'W_p\ (1)$. The pressures p and p'

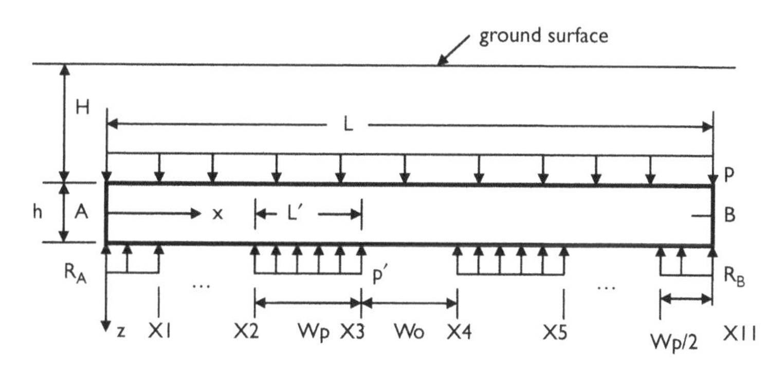

Sketch for EXAMPLE 5.20

are simply overburden stresses: $p = \gamma\ H$ and $p' = \gamma\ (H + h)$ where γ is specific weight of overburden. Any difference between the overburden specific weight and the stratum specific weight may be considered negligible in this example. Because the stratum weight is replaced by an equivalent surface pressure in the analysis, the top and bottom pressures are equal, that is, $p = p' = \gamma\ (H + h)$.

Overall equilibrium is required and leads to expressions for the end reactions. Assuming simply supported ends, gives

$$0 = R_A + R_B + nF' - F$$

$$0 = R_B L - FL/2 + n^2 SF'/2$$

Thus,

$$R_B = F/2 - nF'/2,\ R_A = F/2 - nF'/2,$$

These results could also be obtained by noting the symmetry of the problem and then assigning one-half the net load to each end reaction. At first glance, there seems to be a possibility of the reactions becoming negative in contradiction to common sense. Closer examination of the reaction formulas gives a different view. Thus,

$$\begin{aligned} R_B &= F/2 - nF'/2 = (pL/2) - (npW_p/2) \\ &= np(W_0 + W_p)/2 - npW_p/2 \\ R_B &= npW_0/2 \end{aligned}$$

This result shows that the stratum load is caused by removal of support by excavation of the openings and the abutments carry one-half of the "lost" support each. In this regard, the abutment load, the end reactions, are distributed over the mined stratum abutments, so the stress concentration at the abutments does not grow indefinitely as the abutment force increases with the number of excavated openings. However, abutment stress concentration may reach the elastic limit.

With reference to the second sketch for this example, internal equilibrium leads to expressions for shear and moment distributions. In the second sketch, X5 <x <X6, so the shear force is

$$V = R_A - px + p'X_1 + p'(X_3 - X_2) + p'(X_5 - X_4)$$

and the moment is

$$M = R_A x - px^2/2 + p'X_1(x - X_1/2) + p'(X_3 - X_2)[x - (X_3 + X_2)/2]$$
$$+ p'(X_5 - X_4)[x - (X_5 + X_4)/2]$$

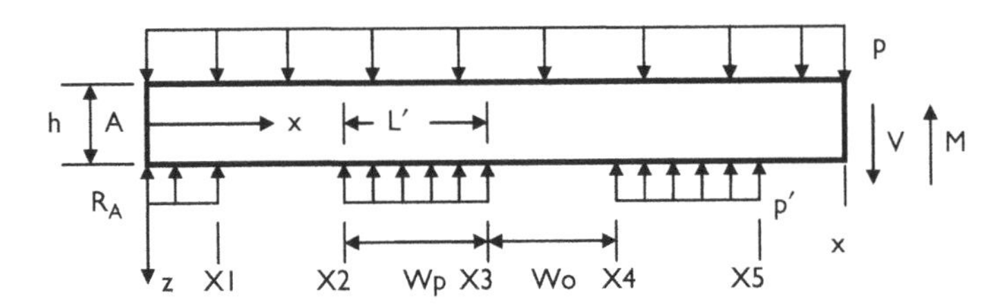

Second Sketch for EXAMPLE 5.20

After a lengthy but straightforward analysis and computation in a spreadsheet, the shear and moment distributions over the entire span are given in the plots for this example. The maximum moment has magnitude $M = 3.708(10^8)$ lbf-ft. This moment leads to a peak tension at the bottom of the stratum at center span.

$$\sigma_t = Mc/I$$
$$= (3.708)(10^8)(50/2)/[(1)(50^3)/12]$$
$$\sigma_t = 6,180 \text{ psi}$$

The maximum shear stress is 1.5 times the average shear stress and occurs at the abutments:

$$\tau(\max) = 1.5\ V/A$$
$$= 1.5(3.6)(10^6)/(1)(50)$$
$$\tau(\max) = 1.08(10^5)\text{psf} \quad (750 \text{ psi})$$

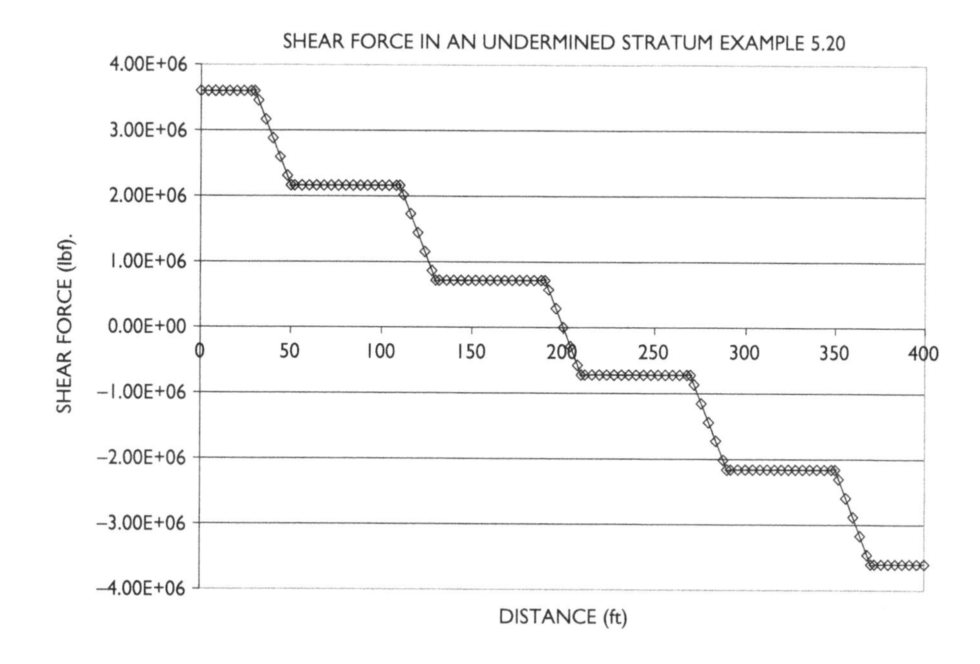
SHEAR FORCE IN AN UNDERMINED STRATUM EXAMPLE 5.20
SHEAR FORCE (lbf).
4.00E+06
3.00E+06
2.00E+06
1.00E+06
0.00E+00
-1.00E+06
-2.00E+06
-3.00E+06
-4.00E+06
0
50
100
150
200
250
300
350
400
DISTANCE (ft)

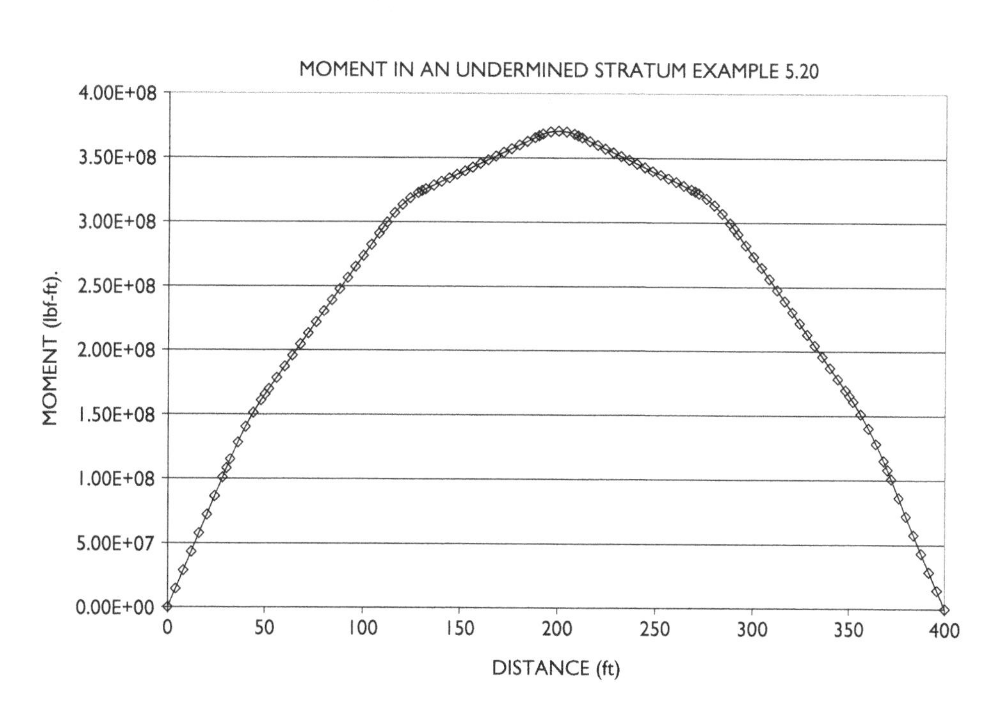
MOMENT IN AN UNDERMINED STRATUM EXAMPLE 5.20
MOMENT (lbf-ft).
4.00E+08
3.50E+08
3.00E+08
2.50E+08
2.00E+08
1.50E+08
1.00E+08
5.00E+07
0.00E+00
0
50
100
150
200
250
300
350
400
DISTANCE (ft)

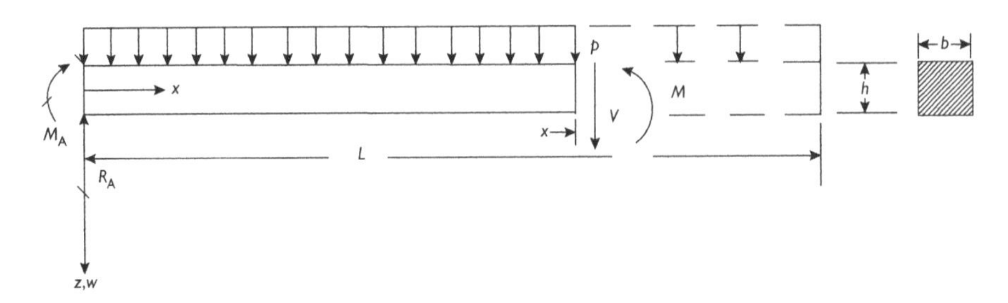

Figure 5.15 Sketch for roof beam analysis.

5.3 Problems

Naturally supported roof

5.1 Derive expressions shear force V, bending moment M and sag w for a "weightless" beam with built-in ends of thickness h, breadth b and span L under a uniform load p. Plot the shear and moment diagrams. Use the sign convention that tension is positive, the x-axis is to the right, the z-axis down, the y-axis is out of the page, a ccw (counterclockwise) moment is positive and a downward shear force is positive as shown in Figure 5.15.

5.2 Show that the maximum tension σ_t in a roof beam with built-in ends occurs at the top of the beam at the beam ends and is given by the expression $(1/2)pL^2/h^2$.

5.3 Show that the maximum sag w_{max} in a roof beam with built-in ends occurs at the center and is given by the expression $(1/32)pL^4/(Eh^3)$.

5.4 Given that the roof beam factor of safety F_t is the ratio of tensile strength T_o to peak tensile stress, find an expression for the maximum possible roof span as a function of the factor of safety in the built-in ends case.

5.5 Consider a rectangular entry 12 ft high and 22 ft wide that is driven at a depth of 950 ft. The immediate roof layer which is 2.25 ft thick separates from the layers above. Layer properties are $E = 3.7 \times 10^6$ psi, $\nu = 0.28$, $\gamma = 134$ pcf, $T_o = 690$ psi. Find the maximum tensile stress σ_t, maximum sag w_{max} and safety factor with respect to tensile failure F_t if (a) the roof ends are simply supported and (b) the ends are built-in.

5.6 A high horizontal stress exists in the vicinity of the excavation given in Problem 5.5. The stress is just sufficient to reduce the tensile bending stress to zero. Find the magnitude of this horizontal stress, assuming the beam ends are simply supported.

5.7 Consider a rectangular entry 3.66 m high and 6.71 m wide that is driven at a depth of 290 m. The immediate roof layer which is 0.69 m thick separates from the layers above. Layer properties are $E = 25.52$ GPa, $\nu = 0.28$, $\gamma = 21.2$ kN/m^3, $T_o = 4.7$ MPa. Find the maximum tensile stress σ_t, maximum sag w_{max} and safety factor with respect to tensile failure F_t if (a) the roof ends are simply supported and (b) the ends are built-in.

Table 5.1 Rock properties for Problems 5.9 and 5.10

Rock type	*Unit*	*γ-pcf (kN/m³)*	*E (10⁶) psi (GPa)*	*ν(−)*	*C₀ (10³) psi (MPa)*	*T₀ (10²) psi (MPa)*	*h-ft (m)*
Sandstone	R5	155 (24.5)	4.7 (32.4)	0.25	11.2 (77.2)	10.3 (7.1)	8.9 (2.71)
Shale	R4	138 (21.8)	2.5 (17.2)	0.20	6.0 (41.4)	4.0 (2.8)	3.6 (1.10)
Coal	R3	90 (14.2)	0.35 (2.4)	0.31	0.35 (2.4)	0.25 (0.2)	0.5 (0.15)
Laminated sandstone	R2	142 (22.5)	3.1 (21.4)	0.18	7.5 (51.7)	6.2 (4.3)	2.5 (0.76)
Shale	R1	138 (21.8)	2.5 (17.2)	0.2	6.0 (41.4)	4.0 (2.8)	1.5 (0.46)
Coal	Seam	90 (14.2)	0.35 (2.4)	0.31	0.35 (2.4)	0.25 (0.2)	16.8 (5.12)
Sandstone	F1	155 (24.5)	4.7 (32.4)	0.25	11.2 (77.2)	10.3 (7.1)	4.7 (1.43)
Claystone	F2	148 (23.4)	6.0 (41.4)	0.21	13.1 (90.3)	15.2 (10.5)	7.9 (2.41)

5.8 A high horizontal stress exists in the vicinity of the excavation given in Problem 7. The stress is just sufficient to reduce the tensile bending stress to zero. Find the magnitude of this horizontal stress, assuming the beam ends are simply supported.

5.9 Consider the geologic column and properties given in Table 5.1 for entries developed on strike at a depth of 1,450 ft (442 m) in strata dipping 8°. Determine the potential for bed separation in the roof rock, that is, find any separation horizons. Mining height is 12 ft (3.7 m).

5.10 With reference to Problem 5.9, determine the maximum possible roof span, given that all roof rock must have a safety factor greater than one.

5.11 Derive the formula for apparent unit weight in bending analysis of roof strata assuming that bed separation has occurred between the nth and $n+1$ layers. Illustrate all terms with sketches. Include the possibilities of strata gas in the separation gap and moderately dipping strata. Note: Apparent unit weight is obtained from $p = \gamma h$.

Bolted roof

5.12 A method of approximating the support action of rock bolts is to equate the bolt tension to a uniformly distributed load acting upward. If the equivalent bolting pressure is p_b, what is the magnitude of the bolting pressure that will reduce the sag in Problem 5.5b or 5.7b to zero?

5.13 Suppose rock bolt load is 16,400 pounds and bolts are spaced on 5 ft centers in a square pattern. Find the bolting "pressure" p_b. What layer thickness is this bolting pressure equivalent to assuming a single layer roof beam sagging under self-weight alone with simply supported ends. Note: Specific weight of the roof layer is approximately 158 pcf.

5.14 Suppose rock bolt load is 73.5 kN and bolts are spaced on 1.5 m centers in a square pattern. Find the bolting "pressure" p_b. What layer thickness is this bolting pressure equivalent to assuming a single layer roof beam sagging under self-weight along with simply supported ends. Note: Specific weight of the roof layer is approximately 25.0 kN/m³.

5.15 With reference to the sketch, a bolting plan consisting of 5/8 in. (1.59 cm) diameter mechanically point anchored bolts spaced on 4 ft centers (1.22 m), square pattern, is proposed. Evaluate this plan from the dead weight view. Show why the plan is adequate or not. Assume high strength. Would 3/4 in. (1.91 cm) diameter bolts on 5 ft (1.52 m) centers be adequate? Which Plan is preferable? Note: Spacing from the rib is one-half bolt spacing.

5.16 Suppose a suspension action is induced by 3/4 in. (1.91 cm) bolts on 5 ft (1.52 m) centers (high strength) installed in the roof shown in the sketch. If the sag is then 40% of the unsupported sag of the suspended strata, what is the reduction in roof tension? Assume that the bolts are tensioned to 75% of the yield point.

5.17 With reference to Problem 5.16, what is the ratio of the supported to unsupported roof rock safety factors?

5.18 If the four shale layers in the sketch are clamped such that they deform as a single beam what is the maximum shear stress, and where does it occur in the clamped roof? In this regard, if the coefficient of friction between shale strata is 0.35, what bolting pressure is required to mobilize sufficient interlayer shear for monolithic roof beam action? Use L algebraically, then set L = 20 ft to evaluate.

Surface

Overburden	E-10^6 psi (GPa)	γ–pcf (kN/m^3)	h –in. (cm)
R5 Shale 5	8.0(55.2)	160(25.3)	59(150)
R4 Shale 4	1.0(6.90)	135(2.14)	7(17.8)
R3 Shale 3	1.0(6.90)	135(2.14)	8(20.3)
R2 Shale 2	1.0(6.90)	135(2.14)	10(25.4)
R1 Shale 1	1.0(6.90)	135(2.14)	12(30.5)
Seam	L		
F1 Sandstone			

Sketch for Problems 5.15–5.18.

5.19 Consider the roof bolt truss shown in the sketch. If the spacing of the trusses along the entry is *S* and the tension in the "bolts" is *T*, what is the equivalent bolting pressure capacity of the truss system?

5.20 With reference to the sketch and Problem 5.19, prescribe *S* and *T* such that the apparent unit weight of the laminated shales is 65% of the unsupported apparent unit weight.

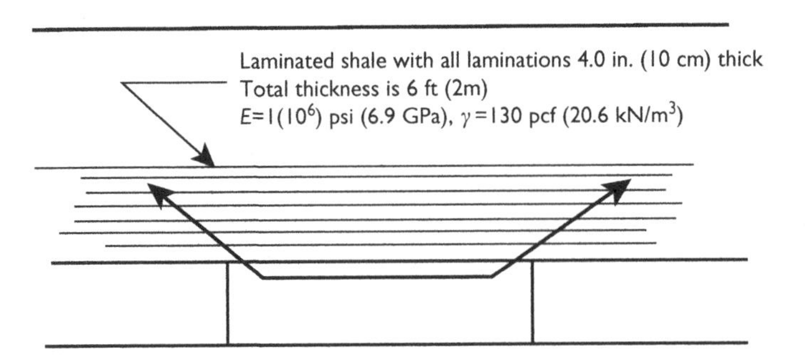

5.21 With reference to the sketch, show in detail the determination of the optimum bolting angle using the given notation.

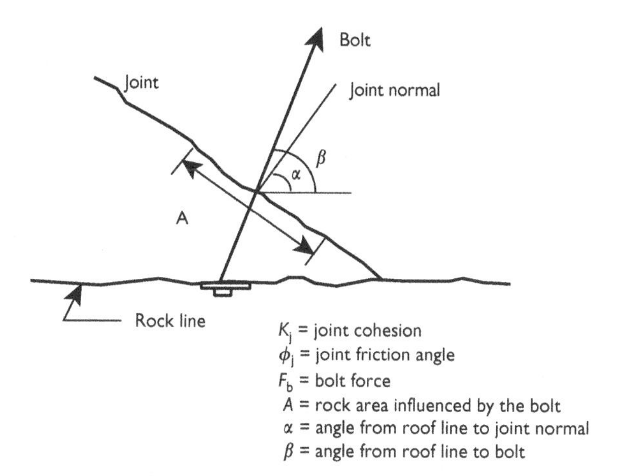

Sketch for joint bolting Problems 5.21, 5.22

5.22 With reference to the sketch, suppose the joint cohesion is zero while the friction angle is 35°. If the force per bolt is 12,000 pounds (53.8 kN)and the area of influence of a bolt is 4 ft^2, find: (a) the unbolted joint safety factor, and (b) the bolted joint safety factor at the optimum angle. Note: The stress state before bolting is a horizontal compression of 648 psi (4.5 MPa). Joint angle is 30°.

5.23 With reference to Problem 5.22, suppose a passive bolt, untensioned and 7/8 in.(2.22 cm) in diameter is grouted in the hole at the optimum angle. Estimate the shear stress in the bolt.

5.24 Coal is mined in two seams that are shown in the sketch. Mining is full seam height in both seams. Entries are 24 ft wide (7.3 m). With reference to the stratigraphic column and rock properties in the sketch, determine for the lower seam:

(a) the safety factor with respect to tension of the immediate roof layer over the entries, and
(b) the second roof layer,
(c) a reasonable bolting plan using the dead weight load approach.

Include (i) bolt length, (ii) nominal diameter, (iii) steel grade, and (iv) spacing.

5.25 Both seams of Problem 5.24 are mined by the longwall method. There are five main entries and sub-main entries used for development in both seams. A three entry system is used for panel development. The sub-main entries are at right angles to the mains, while the panel entries are at right angles to the sub-main entries. Show in vertical section a favorable alignment (a) of the main entries and (b) of the panel entries. A system of steeply dipping lineaments ("joints") in the region strikes due north. (c) Average spacing is 25 ft (7.6 m). Is the mine layout favorably oriented with respect to ground control consequences associated with the joints? Defend your answer.

5.26 A thick seam of coal is mined from the floor to a height of 13 ft at a depth of 1,450 ft. With reference to the stratigraphic column in the sketch and the rock properties given in Table 5.2, assume L = 21 ft, then find:

Rock properties for Problem 5.24 sketch.

Rock type	γ-pcf (kN/m^3)	E-10^6 psi (GPa)	ν	C_o-10^3 psi (MPa)	T_o-10^2 psi (MPa)	h-ft (m)
1 Over-burden	144.0 (22.8)	1.70 (11.7)	0.25	1.00 (6.9)	0.50 (0.34)	1,694 (516)
2 Massive sandstone	148.2 (23.4)	3.00 (20.7)	0.20	19.0 (131)	8.50 (5.86)	1,714 (522)
3 Coal	75.0 (11.9)	0.35 (2.4)	0.30	3.50 (24.1)	1.30 (0.89)	1,719 (524)
4 Layered sandstone	149.5 (23.7)	1.50 (10.3)	0.10	17.5 (120)	7.00 (4.83)	1,741 (531)
5 Massive sandstone	148.2 (23.4)	3.00 (20.7)	0.20	19.0 (131)	8.50 (5.86)	1,744 (532)
6 Sandy shale	170.0 (26.9)	4.50 (31.0)	0.10	15.0 (103)	10.0 (6.90)	1,747 (533)
7 Coal	75.0 (11.9)	0.35 (2.4)	0.30	3.50 (24.1)	1.30 (0.89)	1,754 (535)
8 Massive sandstone	148.2 (23.4)	3.50 (24.1)	0.20	19.5 (134)	10.0 (6.90)	1,778 (542)
9 Layered sandstone	149.5 (23.7)	1.50 (10.3)	0.10	17.5 (120)	7.00 (4.83)	2,250 (686)

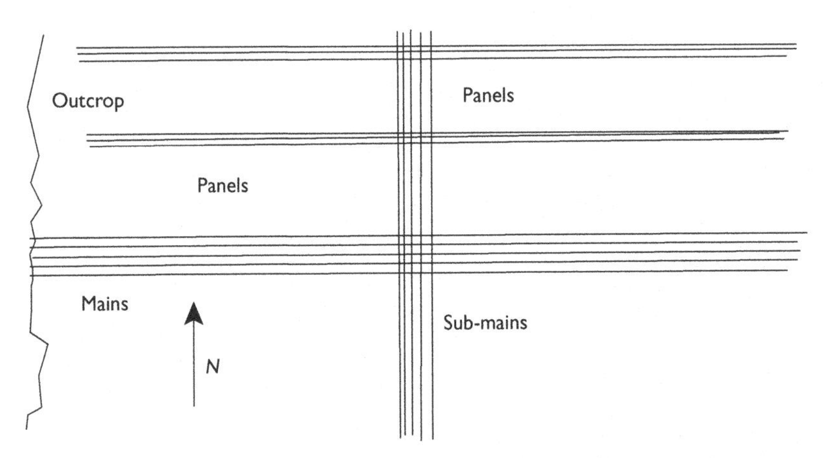

Sketch for joint bolting Problems 5.24, 5.25

(a) the safety factor with respect to tension of the immediate roof layer
(b) the layer above
(c) a reasonable bolting plan using the dead weight load approach.

Include (i) bolt length, (ii) nominal diameter, (iii) steel grade, and (iv) spacing. Note: Entry width is 21 ft and overburden averages 144 pcf.

5.27 With reference to the strata data given in the sketch and Table 5.3, seam depth is 1670 ft (509 m), strata dip is negligible and entries are planned 24 ft (7.32 m) wide. Mining height from the sandstone floor is 12 ft (3.66 m).

1 Determine the *immediate* roof safety factor.
2 Determine bolting pressure needed to reduce sag of all roof layers to zero, assuming bolts are anchored in R3 and the sag of R3 is negligible.

Surface	
(not to scale)	… … overburden
R5 mudstone	
R4 sandstone	
R3 sandstone	
R2 laminated sandstone	
R1 (roof) laminated shale	
S1 (seam – coal)	
F1 (floor) shale	
F2 limestone	

Note
Stratigraphic sequence for Problem 5.26.

Table 5.2 Strata data for Problem 5.26

Rock type	*E(10^6 psi)*	ν–	C_o *(psi)*	T_o *(psi)*	γ *(pcf)*	*h (ft)*
R5	12.6	0.19	12,700	1,270	160	21.5
R4	12.7	0.21	14,200	1,340	158	15.2
R3	12.6	0.19	12,700	1,270	156	6.0
R2	6.3	0.31	8,450	760	146	1.5
R1	5.4	0.23	7,980	610	152	2.0
S1	0.75	0.27	3,750	425	92	15.0
F1	12.3	0.18	11,500	1,050	155	10.3
F2	8.1	0.33	13,700	1,260	160	13.0

Table 5.3 Strata properties

Rock type	*E (10^6) psi (GPa)*	ν–	C_o *(10^3) psi (MPa)*	T_o *(10^3) psi (MPa)*	γ *(pcf) (kN/m^3)*	*h-ft (m)*
R3	4.3 (29.7)	0.25	16.0 (110)	1.2 (8.3)	148 (23.4)	26.0 (7.9)
R2	1.7 (11.7)	0.19	12.5 (86.2)	0.65 (4.5)	142 (22.5)	1.5 (0.46)
R1	4.1 (28.3)	0.20	7.5 (51.7)	0.45 (3.1)	136 (21.5)	2.0 (0.61)
S1	0.35 (2.4)	0.30	3.0 (20.7)	0.25 (1.7)	95 (15.0)	16.0 (4.88)
F1	4.5 (31.0)	0.25	18.0 (124)	1.8 (12.4)	149 (23.6)	18.0 (5.49)

3 Layout a bolting pattern based on dead weight load basis and specify bolt length, diameter, steel grade, spacing, and bolt safety factor using R3 as an anchor stratum.
4 What is the immediate roof safety factor after bolting according to dead weight load design?
5 Explain how gas pressure above the roof shale would affect calculation of the safety factor of R1 before bolting.

5.28 With reference to the strata data given in the sketch and Table 5.4, seam depth is 1,324 ft (404 m), strata dip is negligible and entries are planned 21 ft (6.4 m) wide. Mining height from the sandstone floor is 15 ft (4.6 m).

1 Determine the immediate roof safety factor.

Surface	
(not to scale)	… … overburden
R3 sandstone	
R2 laminated sandstone	
R1 shale	
S1 coal	
F1 sandstone	

Note
Stratigraphic sequence for Problem 5.27.

Table 5.4 Strata properties

Rock type	*E (10^6) psi (GPa)*	*ν–*	*C_o (10^3) psi (MPa)*	*T_o (10^3) psi (MPa)*	*γ (pcf) (kN/m^3)*	*h-ft (m)*
R3	4.5 (31.0)	0.25	16.0 (110)	0.80 (5.52)	148 (23.4)	16.0 (4.9)
R2	2.7 (18.6)	0.19	12.5 (86.2)	0.65 (4.48)	142 (22.5)	2.5 (0.8)
R1	3.1 (21.4)	0.20	8.0 (55.2)	0.40 (2.76)	135 (21.4)	2.0 (0.6)
S1	0.35 (2.4)	0.30	3.0 (20.7)	0.15 (1.03)	95 (15.0)	18.0 (5.5)
F1	3.5 (24.1)	0.25	0.8 (5.5)	0.8 (5.52)	148 (23.4)	14.0 (4.3)

2 Determine bolting pressure needed to reduce sag of all roof layers to zero, assuming bolts are anchored in R3 and the sag of R3 is negligible.
3 Layout a bolting pattern based on dead weight load basis and specify bolt length, diameter, steel grade, spacing, and bolt safety factor using R3 as an anchor stratum.
4 What is the immediate roof safety factor after bolting according to dead weight load design?
5 Explain how gas pressure above the roof shale would affect calculation of the safety factor of R1 before bolting.

Surface	
(not to scale)	… … overburden
R3 sandstone	
R2 laminated sandstone	
R1 shale	
S1 coal	
F1 sandstone	

Note
Stratigraphic sequence for Problem 5.28.

5.29 A room and pillar coal mine is contemplated at a depth of 1,750 ft (533 m) in strata striking due north and dipping 18° east. Entries are driven on strike, crosscuts up and down dip. Mining height is 15.0 ft (4.6 m) measured from the floor; 1.5 ft (0.46 m) of low grade coal is left in the roof.

Table 5.5 Intact rock properties

Property stratum	*γ pcf* (*kN/m*^{3})	*h-ft (m)*	*E 10*^{6} *psi (GPa)*	*G 10*^{6} *psi (GPa)*	C_o *psi (MPa)*	T_o *psi (MPa)*
Limestone (R5)	152 (24.0)	15.9 (4.8)	9.83 (67.8)	4.27 (29.4)	17,690 (122)	1,460 (10.0)
Coal (R4)	93 (14.7)	1.4 (0.4)	0.29 (2.0)	0.12 (0.8)	2,780 (19.2)	390 (2.7)
Mudstone (R3)	153 (24.2)	9.8 (3.0)	7.67 (52.9)	3.07 (21.2)	13,750 (94.8)	1,580 (10.9)
Sandy shale (R2)	142 (22.5)	2.7 (0.8)	4.16 (28.7)	1.79 (12.3)	6,450 (44.5)	720 (5.0)
Shale (R1)	138 (21.8)	1.3 (0.4)	3.62 (25.0)	1.53 (10.6)	7,400 (51.0)	650 (4.5)
Coal (Seam)	90 (14.2)	16.5 (5.0)	0.35 (2.4)	0.14 (1.0)	3,400 (23.4)	310 (2.1)
Sandstone (F1)	147 (23.3)	8.6 (2.6)	8.32 (57.4)	3.35 (23.1)	12,300 (84.8)	1,350 (9.3)

Table 5.6 Joint properties

Property joint set	*c-psi (kPa)*	*φ(deg)*	K_n *10*^{6}*psi/in.* (*GPa/cm*)	K_s *10*^{6}*psi/in.* (*GPa/cm*)	*S-ft (m)*
Set 1	0.0 (0.)	35	1.76 (4.78)	0.56 (1.52)	25.3 (7.71)
Set 2	10.0 (69.0)	30	2.31 (6.27)	3.14 (8.53)	4.7 (1.43)
Set 3	20.0 (138.0)	25	3.29 (8.93)	0.92 (2.49)	6.1 (1.85)

Three joint sets are present. Set 1 is vertical and strikes east-west. Set 2 strikes due north and dips 60° east; Set 3 also strikes due north, but dips 35° west. The average overburden specific weight is 156 pcf (24.7 kN/m^3).

Stress measurements indicate that the premining stresses relative to compass coordinates are: $S_v = 1.05d$, $S_h = 600 + 0.25d$, $S_H = 50 + 0.75d$ ($S_v = 23.8d$, $S_h = 4{,}138 + 5.66d$, $S_H = 345 + 3.45d$,) where v = vertical, h = east, H = north, d = depth in ft and stress units are psi (d = depth in m and stress units are kPa).

Rock properties near the mining horizon are given in Table 5.5. These data were determined from laboratory testing on NX-core at an L/D ratio of two. Joint properties for Mohr–Coulomb failure criteria are given in Table 5.6. Joint normal and shear stiffness (K_n, K_s) that relate normal stress and shear stress to corresponding displacements are also given in Table 5.6 as are the joint spacings (S).

A tensile safety factor of 4.0 is required for the immediate roof and 3.0 for the next two superincumbent strata. Find the maximum safe roof span.

5.30 Given conditions of Problem 5.29, if it were feasible to measure roof sag as entries were mined 15 ft wide, (a) how much sag would be observed at mid-span? (b) What sag measurement would indicate impending roof failure at the same span?

5.31 With reference to Problems 5.29 and 5.30, layout a reasonable bolting plan using a dead weight approach and point anchored mechanical bolts assuming entries and crosscuts are mined 20 ft (6.1 m) wide. Specify bolt length, diameter, spacing, steel grade, and pretension.

5.32 A thin "low" metallurgical coal seam 3 ft thick is mined at a depth of 980 ft. The main entries are mined 4 ft into the R1 roof stratum to give sufficient clearance for track haulage. With reference to the stratigraphic column in the sketch and the rock properties given in Table 5.7, find: (a) the safety factor with respect to tension of the immediate roof layer and (b) the layer above, then (c) specify a reasonable bolting plan using the dead weight load approach. Include (i) bolt length, (ii) nominal diameter, (iii) steel grade, and (iv) spacing. Note: Entry width is 20 ft and overburden averages 144 pcf.

5.33 The immediate roof behind a longwall face often considered to be a cantilever beam as shown in section in the sketch. Beginning with a safety factor design criterion, derive an expression for the distance X the face can advance before R1 failure. Assume bed separation between R1 and R2 occurs, flat strata, no gas pressure and that currently $L = L_o$. Show shear and moment diagrams and location of expected failure.

Surface	
(not to scale)	. . …. overburden
R5 mudstone	
R4 sandstone	
R3 laminated shale	
R2 laminated sandstone	
R1 (roof) sandstone	
S1 (seam – coal)	
F1 (floor) shale	
F2 limestone	

Note
Stratigraphic sequence for Problem 5.32.

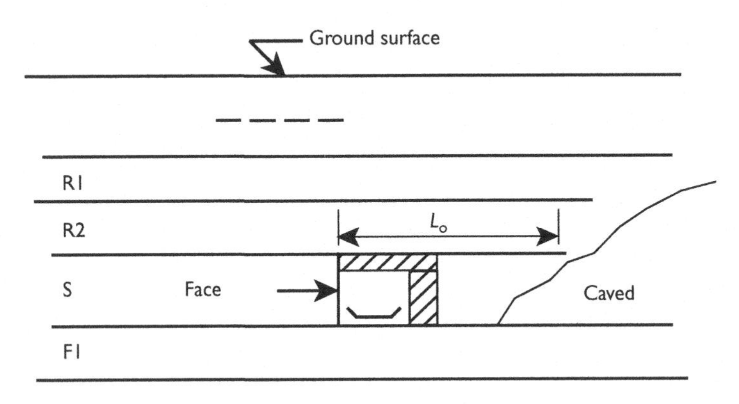

Sketch for Problem 5.33

5.34 Given the geologic column, strata depths and rock properties in Table 5.8, find:

(a) the maximum possible roof span when mining the lower coal seam full height,
(b) the safe roof span, if a safety factor of 4 with respect to tension is required,
(c) the factor of safety of the massive sandstone (M.S.) roof layer above the sandy shale when the layer below is at maximum span.

Table 5.7 Strata data for Problem 5.32

Rock type	$E\ 10^6$ psi (GPa)	v (–)	C_o-psi (MPa)	T_o-psi (MPa)	γ-pcf (kN/m³)	h-ft (m)
R5	12.6 (86.9)	0.19	12,700 (87.6)	1,270 (8.8)	160 (25.3)	21.5 (6.55)
R4	12.7 (87.6)	0.21	14,200 (97.9)	1.34 (9.2)	158 (25.0)	15.2 (4.63)
R3	5.4 (37.2)	0.23	7,980 (55.0)	610 (4.2)	152 (23.1)	1.0 (0.30)
R2	6.3 (43.4)	0.31	8,450 (58.3)	760 (5.2)	146 (24.7)	1.5 (0.46)
R1	12.6 (86.9)	0.19	12,700 (87.6)	1,270 (8.8)	156 (24.7)	6.0 (1.83)
S1	0.75 (5.2)	0.27	3,750 (25.9)	425 (2.9)	92 (14.6)	3.0 (0.91)
F1	12.3 (84.8)	0.18	11,500 (79.3)	1,050 (7.2)	155 (24.5)	10.3 (3.14)
F2	8.1 (55.9)	0.33	13,700 (94.5)	1,260 (8.7)	160 (25.3)	13.0 (3.96)

Table 5.8 Strata data for Problems 5.34 and 5.35

Rock type	γ-pcf (kN/m³)	E-10⁶ psi (Gpa)	v	C_o-psi (Mpa)	T_o-psi (Mpa)	h-ft (m)
1 Over-burden	144.0 (22.8)	1.70 (11.7)	0.25	1,000 (6.9)	50 (0.34)	1,694 (516)
2 Massive Sandstone	148.2 (23.4)	3.00 (20.7)	0.20	19,000 (131.0)	850 (5.86)	1,714 (522)
3 Coal	75.0 (11.9)	0.35 (2.4)	0.30	3,500 (24.1)	130 (0.90)	1,719 (524)
4 Layered sandstone	149.5 (23.6)	1.50 (10.3)	0.10	17,500 (120.7)	700 (4.83)	1,741 (531)
5 Massive sandstone	148.2 (23.4)	3.00 (20.7)	0.20	19,000 (131.0)	850 (5.86)	1,744 (532)
6 Sandy shale	170.0 (26.9)	4.50 (31.0)	0.10	15,000 (103.4)	1,000 (6.90)	1,747 (533)
7 Coal	75.0 (11.9)	0.35 (2.4)	0.30	3,500 (24.1)	1,300 (8.97)	1,754 (535)
8 Massive sandstone	148.2 (23.4)	3.50 (24.1)	0.20	19,500 (134.5)	1,000 (6.90)	1,778 (542)
9 Layered sandstone	149.5 (23.6)	1.50 (10.3)	0.10	17,500 (120.7)	700 (4.83)	2,250 (658)

5.35 Prescribe a bolting plan for entries driven 28 ft (8.5 m) wide in the lower coal seam in the sketch using the dead weight load approach. Include nominal bolt diameter, length, grade of steel, installation tension, and spacing. Support your recommendation with calculations.

Additional problems

5.36 Use the results from Example 5.15 to plot shear and moment diagrams for a span of 7 m, a distributed pressure of 3.5 MPa over a length of 0.3 m with distribution center 1.75 m from the left beam end, a beam thickness of 1.52 m, a beam breadth of 0.3 m (into the page). Also determine the location of the maximum bending stress and magnitude. Consider tension positive.

5.37 The high pressure and small area of distribution in Examples 5.15 and 5.16 are indicative of a bolting force distributed over a collar plate. In Example 5.16 the bolt force is pressure times collar plate area, that is, $F_b = p'A' = (3.5\text{ MPa})(0.3)(0.3) = 3.15$ kN which is a high bolt force, although not out of reason. The small distribution area suggests approximation of the support pressure by point loading as in Example 5.2 for multiple but equally spaced point loads. Consider the data in Problem 5.36 but develop formulas assuming a point load instead, then plot shear and moment diagrams and determined moment magnitude.

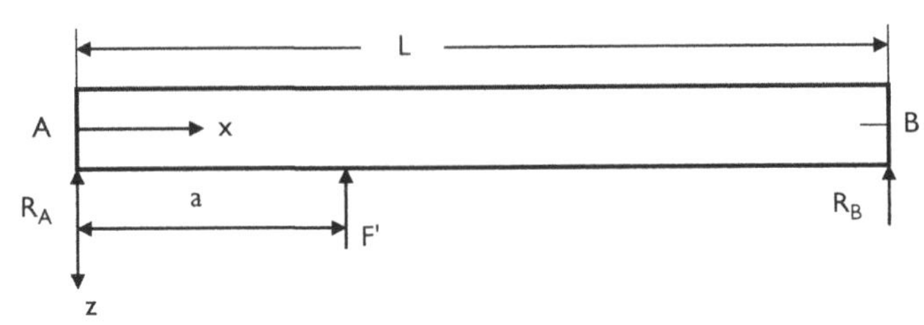

Sketch for Problem 5.37

5.38 Suppose roof bolts are installed on 1.75 m, centers across a 7 m wide entry. Bolts adjacent to entry ribs are spaced at one half spacing. Bolt rows are also spaced 1.75 m along the entry. The bolted stratum is 1.52 m thick and bolts are 1.82 m, long as shown in the sketch. The bolt force is a pressure (3.5 MPa) times collar plate area (0.2 m by 0.2 m), so $F = 140$ kN. Consider the bolting forces then as point loads as in Problem 5.37. Derive shear and moment formulas; plot these distributions, and find the location and magnitude of the peak moment. A free body diagram of the stratum shows the bolt forces F at a spacing S in the sketch.

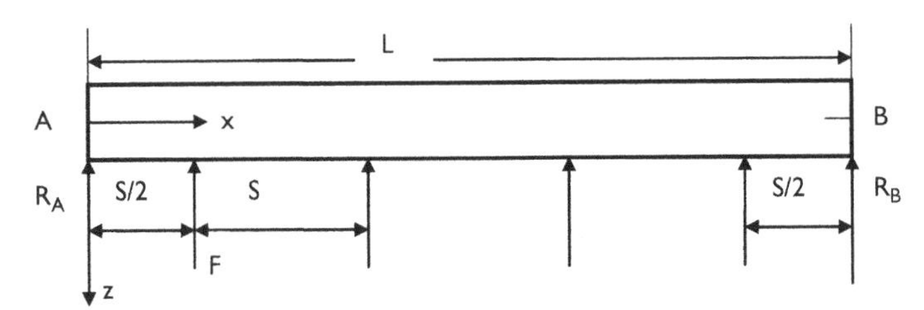

Sketch for Problem 5.38

5.39 Consider a roof stratum loaded by a uniformly distributed "pressure" p from above over the entire span L and support by another uniform pressure p' from below over a length L'. Derive formulas for the bending moment and bending stress along the beam (per 0.3 m into the page), plot the shear and moment diagrams, and determine the location of the maximum bending stress. Consider tension positive, bolt distance is 1.75 m from the left end of the stratum, bolting pressure of 1.38 MPa, 0.3 m x 0.3 m collar plates, bolt row spacing into the page is 1.9 m, and simply supported ends.

5.40 Consider a thick, massive stratum between ground surface and mining horizon that behaves as a beam as illustrated in the sketch. Depth to the top of the stratum is H m (137 m). The stratum is h m (15.2 m) thick. Support is removed by excavation in a series of openings W_o m (6.1 m) wide that are separated by pillars W_p (18.3 m) m wide. The series of openings and pillars is continued to a span of L m. The center to center spacing is $W_p + W_o$ m (24.4). Half-pillars are present at the left and right abutments as shown in the sketch. Derive formulas for the shear force and moment distributions along the stratum assuming there are n (5) openings present. Plot the shear and moment distributions and find the location and value of the maximum tensile stress in the stratum. Also determine the location and magnitude of the maximum shear stress.

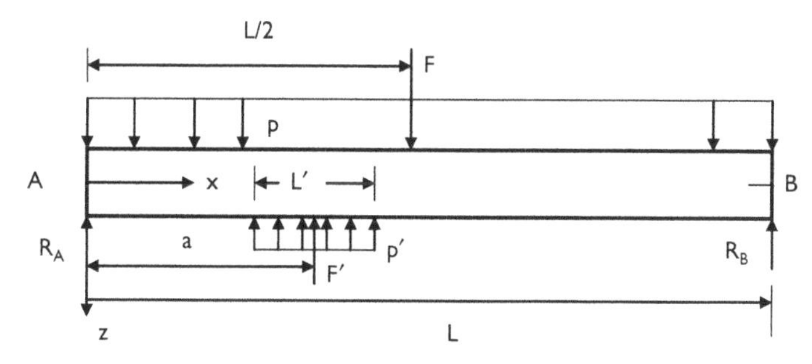

Sketch for Problem 5.39

Notes: Inspection of the sketch shows that the span $L = n(W_p + W_o) = nS$. Force on top of the stratum per 0.3 m into the page is $F = pL(0.3)$; force on the bottom per pillar per 0.3 m into the page is $F' = p'W_p(0.3)$. The pressures p and p' are simply overburden stresses: $p = \gamma H$ and $p' = \gamma (H + h)$ where γ is specific weight of overburden. Any difference between the overburden specific weight and the stratum specific weight may be considered negligible in this example. Because the stratum weight is replaced by an equivalent surface pressure in the analysis, the top and bottom pressures are equal, that is, $p = p' = \gamma (H + h)$.

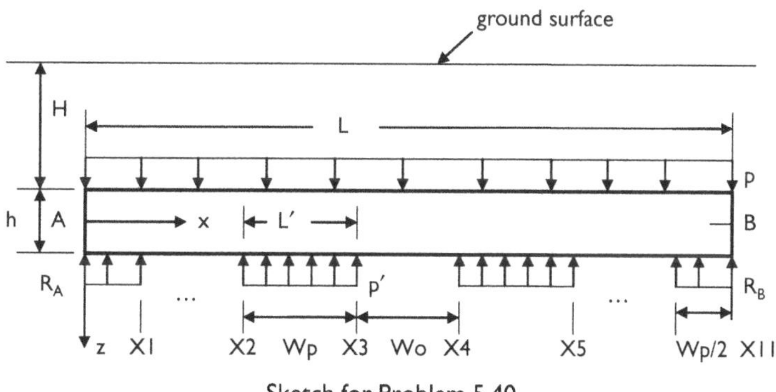

Sketch for Problem 5.40

Chapter 6

Pillars in stratified ground

Pillars in stratified ground are routinely formed between entries and crosscuts during development of main entries, sub-main entries, and panel entries. Main entries that provide primary mine access usually last the life of a mine. Ventilation entries also have relatively long service lives. Sub-mains have shorter design lives. Panel entry service life is typically just a few years at most. Regardless, the purpose of all pillars is to prevent collapse of the adjacent entries.

Pillars formed between entries in a single, flat seam may fail in direct compression. Pillars in dipping seams are loaded in direct compression and in shear as well and require consideration of a biaxial failure criterion. Pillars containing joints may also fail by joint shear or separation. Extraction in several seams raises a question of interaction between seams that is summarized in guidelines for main entry pillars of long life and chain entry pillars of relatively short life. Barrier pillars that defend main entries from effects of adjacent panel excavation that often approaches 100% pose yet additional design challenges.

6.1 Pillars in a single seam

Pillars in single, flat seams are amenable to a force equilibrium analysis that is readily expressed in terms of average stresses. In soft rock, pillars are generally much wider and longer than they are high. In hard stratiform deposits, pillar height is usually much greater than width and length. While short pillars tend to crush slowly, tall pillars may fail catastrophically. In either case, the pillar width to height ratio gives rise to a "size" effect on pillar strength. The effect has several explanations including constraint at pillar top and bottom analogous to end friction in laboratory compression testing.

Tributary area, extraction ratio analysis

Stress concentration occurs at walls of entries whether bed separation occurs in the roof or not. These walls are also pillar walls. Thus, logical extension of the stress concentration approach to design of openings in stratified ground would seek peak compressive stress concentrations in pillar walls. This stress is the requisite stress in application of safety factor criteria to pillar design. In the absence of bed separation, stress concentration in roof and floor would also be sought. In fact, the most popular pillar design approach in stratified ground is based on *average* compressive stress rather than *peak* stress. This approach is based on a *tributary area* concept that requires consideration of a suitable compressive strength for calculating pillar safety factors.

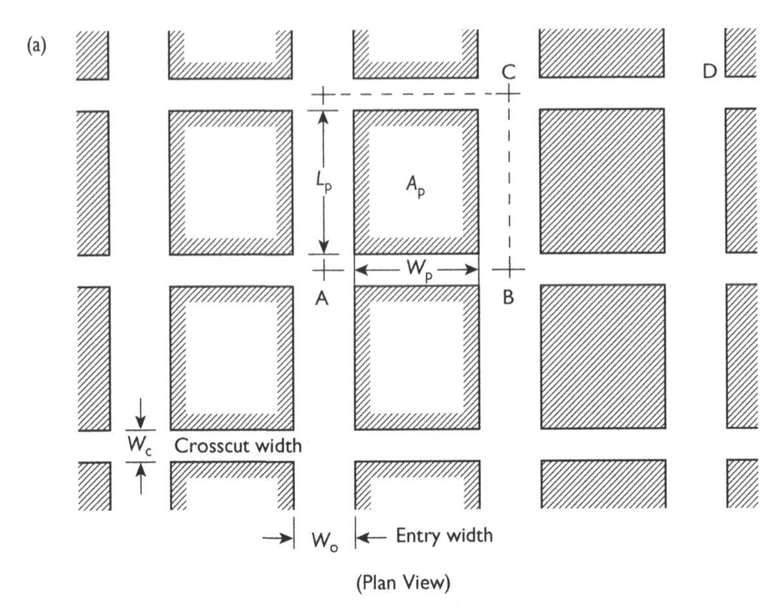

Array of pillars formed by entries and crosscuts. Tributary area $A = \overline{ABCD}$, pillar area $= A_p$, area mined $A_m = A - A_p$

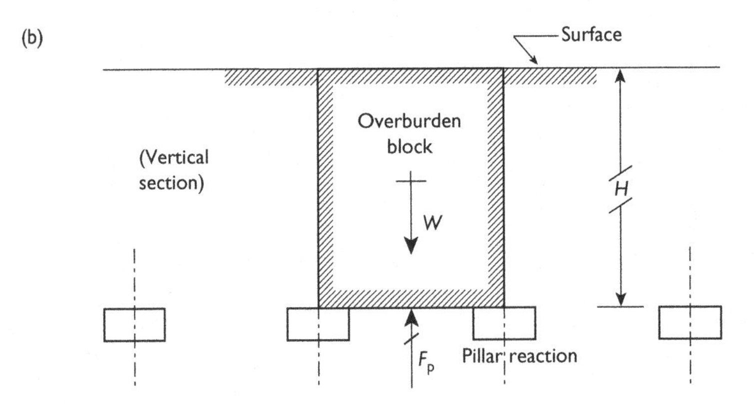

Figure 6.1 Array of entries, pillars, and crosscuts.

The tributary area approach to pillar design is also known as the *extraction ratio* approach and is based on a simple equilibrium force analysis. Consider a large array of pillars formed by a regular pattern of entries and crosscuts shown in Figure 6.1. The periodicity of the array forms vertical planes of symmetry between pillars that extend from seam level to ground surface. Overburden in the prisms formed by these planes rests on the pillars. Equilibrium requires pillar support reaction F_p to be equal to weight W of an overburden block; symmetry precludes vertical shear forces acting on the sides of the block. Thus,

$$W = \gamma HA = S_v A = F_p = S_p A_p \tag{6.1}$$

where γ is average specific weight of overburden, H is height of overburden block (seam depth), S_v is average overburden stress (before excavation), A is cross-sectional area of an overburden block, S_p is average pillar stress, and A_p is pillar cross-sectional area. According to the geometry in Figure 6.1, the *tributary area* $A = A_p + A_m$ where A_m is the excavated area (entries and crosscuts) seen in plan view.

A compressive stress safety factor may be computed as

$$FS_c = \frac{C_p}{S_p} \tag{6.2}$$

where S_p and C_p are *average* pillar stress and corresponding compressive strength, respectively.

The pillar stress S_p may be computed from

$$S_p = \frac{S_v A}{A_p} = \frac{Sv}{A_p/A} = \frac{S_v}{(1 - A_m/A)} = \frac{S_v}{1 - R} \tag{6.3}$$

where $R = A_m / A$ is defined as the *area extraction ratio.* With reference to the notation in Figure 6.1, $A_p = W_p L_p$, $A = (W_o + W_p)(L_p + W_c)$, $A_m = W_o (L_p + W_c) + W_c W_p$, and $R = 1 - L_p W_p / (W_o + W_p)(L_p + W_c)$. A useful form is

$$1 - R = \left[\frac{W_p}{(W_o + W_p)\left(1 + (W_c/L_p)\right)}\right] = \frac{A_p}{A} \tag{6.4}$$

that reduces to a two-dimensional form as the pillar becomes long relative to width.

The extraction ratio R may also be computed from the safety factor. Thus,

$$R = 1 - \frac{(FS_c)(S_v)}{C_p} \tag{6.5}$$

which clearly imposes a physical limit on extraction. In no case may the overburden stress exceed pillar "strength." Most of the coal mines in the US experience an overall extraction of about 50% that suggests an extraction ratio of one-half. In consideration of coal mining depths that exceed 2,500 ft only with great difficulty, these observations suggest coal pillar strength of no more than 5,000 psi, which would be a rather high strength coal.

Example 6.1 A room and pillar operation is planned for a depth of 300 m (984 ft) in a stratum that has a compressive strength estimated at 13.57 MPa (1,968 psi). Entries and crosscuts are planned at 6 m (20 ft). Determine the maximum possible extraction ratio.

Solution: By definition $FS_p = C_p / S_p$ and $S_p = S_v / 1 - R$, so $R = 1 - (FS_p) S_v / C_p$.

An estimate of the overburden stress (premining vertical stress) S_v is 22.6 kPa per meter of depth (1 psi/ft of depth). Hence,

$$R = 1 - \frac{(1)(300)(22.6)\left(10^3\right)}{13.57\left(10^6\right)} = 1 - 0.500 = 0.500$$

Example 6.2 Consider the data and results of Example 6.1 and suppose the pillars are square. Determine the pillar dimensions at R(max).

***Solution*:** By definition $R = A_m / A = 1 - (A_p / A)$ and $A_p = L_p W_p = (W_p)^2$. Thus,

$$1 - R = \frac{(W_p)^2}{(W_p + W_o)^2} = \frac{1}{[1 + (W_o/W_p)]^2}$$

and

$$\left(1 + \frac{W_o}{W_p}\right)^2 = \left(\frac{1}{1 - R}\right) = \left(\frac{1}{1 - 0.500}\right) = 2$$

Hence, $W_p = W_o/(\sqrt{2} - 1) = 6/(\sqrt{2} - 1) = 14.5$ m. Because the pillars are square $W_p = L_p = 14.5$ m which are the pillar dimensions in plan view. Pillar height is equal to the height of the excavation, of course.

Example 6.3 Consider the data of Example 6.1 and suppose a pillar safety factor of 1.5 is required for a room and pillar plan where pillars are square. Determine the *maximum allowable* extraction ratio and pillar dimensions.

***Solution*:** The extraction ratio is given in terms of the factor of safety. Thus,

$$\begin{aligned} R &= 1 - \frac{(\mathrm{FS_p})(S_v)}{C_p} \\ &= 1 - \frac{(1.5)(22.6)(10^3)(300)}{13.57(10^6)} \\ R &= 0.25 \end{aligned}$$

But also, the extraction ratio may be determined by geometry. Thus,

$$\begin{aligned} R &= \frac{A_m}{A} \\ &= 1 - \frac{A_p}{A} \\ R &= 1 - \frac{(W_p)^2}{(W_o + W_p)^2} \end{aligned}$$

Hence, $W_p = 6.464\ W_o = (6.464)(6) = 38.8\ m$. Pillars are square, so pillar length is equal to pillar width; pillar height is equal to opening height.

Example 6.4 Consider the data and results of Example 6.1 and suppose a safety factor of 1.5 is required, as in Example 6.3, but suppose now the pillars are three times as long as they are wide. Determine the maximum allowable extraction ratio and pillar dimensions.

Solution: As in Example 6.3, the maximum allowable extraction ratio is 0.25 from safety factor, strength, and stress considerations. Now from geometric considerations,

$$R = 1 - \frac{W_p L_p}{(W_o + W_p)(W_c + L_p)}$$

$$= 1 - \frac{3W_p W_p}{(W_o + W_p)(W_c + 3W_p)}$$

$$\frac{1}{4} = 1 - \frac{3}{(W_o/W_p + 1)(W_c/W_p + 3)}$$

where substitution for R is made. After some algebra and in recognition that entry and crosscut widths are equal,

$$\left(\frac{W_o}{W_p}\right)^2 + 4\frac{W_o}{W_p} - 1 = 0$$

which is readily solved for W_p. Thus, $W_p = 25.4$ m and $L_p = 76.2$ m. Check:

$$1 - R = \frac{W_p L_p}{(W_o + W_p)(W_c + L_p)}$$

$$1 - 0.25 = \frac{(25.4)(76.2)}{(6.0 + 25.4)(6.0 + 76.2)}$$

$$0.75 = 0.75$$

Size effect on strength

Design of pillars based on average compressive stress ignores

1. details of stress distribution through the pillar;
2. the effect of confining stress in the interior of the pillar on compressive strength;
3. the possibility of progressive pillar failure.

The state of stress in a mine pillar is generally far from uniform and generally varies from point to point. High compressive stresses concentrated at the pillar walls may become high enough to cause local failure, spalling, that leads to an "hour glass" pillar shape often observed even in stable pillars. The interior core of a pillar is confined by the adjacent material and is under horizontal confining stress that increases strength relative to the unconfined outer walls of pillars and assists in pillar stability. Vertical and horizontal stress distributions at pillar midheight are shown in Figure 6.2. Tops and bottoms of pillars are in contact with adjacent strata that may also confine the pillar and reduce any tendency toward lateral expansion under compressive axial load. However, very compliant strata may tend to move laterally more than the pillar and actually tend to split the pillar in tension. This tendency is often the case when the floor beneath a pillar is a soft clay or shale. Such phenomena cannot be directly taken into account by tributary area design of pillars based on average vertical stress.

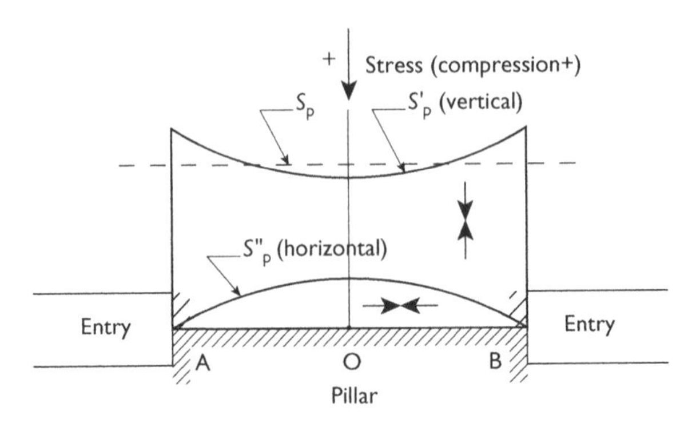

Figure 6.2 Vertical (S'_p) and horizontal (S''_p) stresses in a pillar. S_p is the average vertical stress.

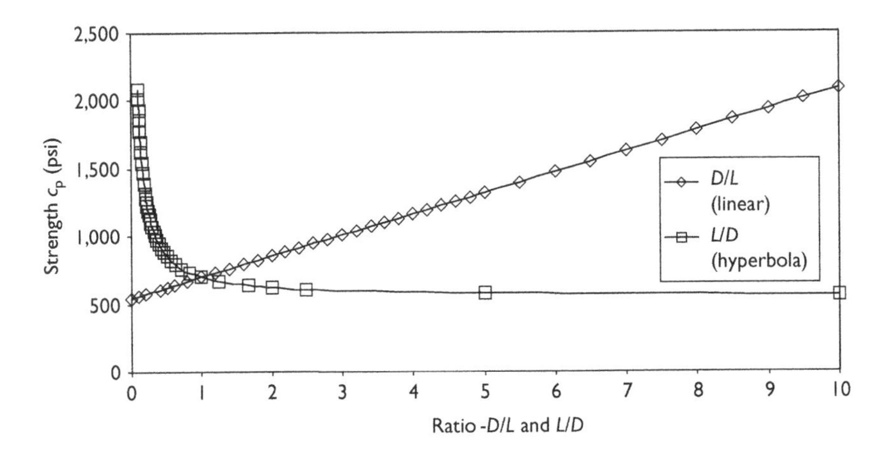

Figure 6.3 Size effect formula for pillar strength as a function of *D/L* (linear) and *L/D* (hyperbola).

Modification of the concept of strength that allows for a *size effect* assists in overcoming the disadvantage of pillar design based on average stress. A size effect is introduced into design procedure by supposing compressive strength is a function of pillar geometry. Full-size mine pillars are not subject to carefully controlled laboratory tests, so size effects formulas are necessarily based on small test specimens, usually cylinders of varying length to diameter ratios. A common size effects formula derived from laboratory data fits is

$$C_p = C_1\left(0.78 + 0.22\frac{D}{L}\right) \tag{6.6}$$

where L is the length of a test cylinder, D is test cylinder diameter, and C_1 is the strength of a cylinder with an L/D ratio of one. Increasing diameter D at fixed height (length L) produces "stubby" pillars of increasing strength. Indeed, strength C_p is a linear function of the ratio D/L and would plot as straight line with D/L on the x-axis and C_p on the y-axis as shown in Figure 6.3. If C_p is plotted as a function of the conventional L/D ratio, the plot is a hyperbola

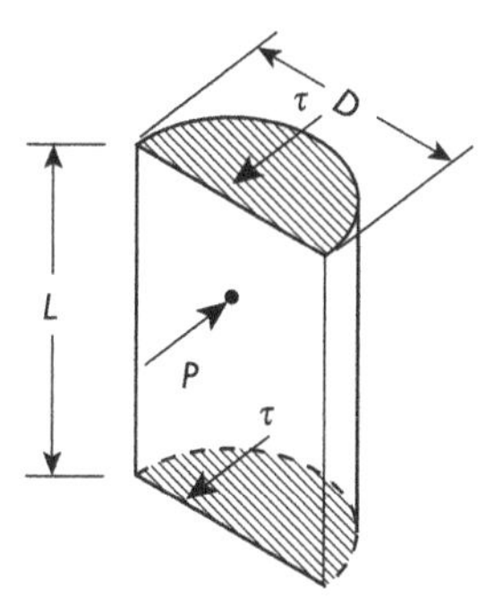

Figure 6.4 End shear stress and pressure reaction of a half-cylinder under uniaxial compression.

that increases without bound along the C_p or y-axis but approaches a horizontal line of $0.78C_1$ with increasing L/D ratio.

The size effects compressive strength C_p is not a material property. A simple explanation of the data fit embodied in the size effects strength formula illustrates the point. During an unconfined compressive strength test, end friction acts to prevent outward motion of the test cylinder. Equilibrium of the half cylinder shown in Figure 6.4 requires an average horizontal stress acting over a diametral section of the cylinder such that

$$pDL = (2)\left[\tau\pi\left(\frac{1}{2}\right)\left(\frac{D^2}{4}\right)\right] \tag{6.7}$$

This simple equilibrium requirement shows that an equivalent confining pressure acts on the nominally unconfined test cylinder with magnitude

$$p = \left(\frac{\tau\pi}{4}\right)\left(\frac{D}{L}\right) \tag{6.8}$$

In view of the Mohr–Coulomb failure criterion $C_p = C_o + (C_o/T_o)p$, test data should plot according to

$$C_p = C_o + \left(\frac{C_o}{T_o}\right)\left(\frac{\tau\pi}{4}\right)\left(\frac{D}{L}\right) \tag{6.9}$$

At a D/L ratio of one, solution of this equation allows for replacement of C_o in terms of C_1 using

$$C_o = \left[\frac{C_1}{1 + (\tau\pi/4T_o)}\right] = C_1 N_1 \tag{6.10}$$

After back-substitution into the Mohr–Coulomb criteria, one obtains

$$C_p = C_1\left[N_1 + (1 - N_1)\left(\frac{D}{L}\right)\right] \tag{6.11}$$

which has the same form as the size effects equation obtained from an empirical fit to laboratory test data. The number N_1 has a value less than one, say, 0.78, $(1 - N_1 = 0.22)$ and thus explains the "size effects" as a simple frictional "end effects" phenomenon.

Other mathematical forms of fits to experimental conditions have been used to obtain size effects formulas for pillar strength. Almost any form can be made to fit the data closely over a limited range of experimental conditions. For example,

$$C_p = 700\left(\frac{D}{L}\right)^{1/2} \tag{6.12}$$

where C_p is in psi and D/L (pillar width to height ratio) is in the interval [0.5, 1.0]. When the two size effects formulas are made to agree at $D/L = 1$, the greatest difference is about 17% as shown in Figure 6.5. This "square root" formula is also referred to as a *shape effects* formula because the test specimens were prisms of square cross section instead of cylinders. Examination of Figure 6.5 shows that the square root formula forecasts zero strength as pillars become tall, while the linear fit gives an intercept and finite compressive strength for tall pillars.

When *cubes* of different sizes are used to examine size effects, end and shape effect explanations do not account for a decrease of strength with size. A statistical, micro-mechanical explanation supposes a laboratory test specimen has numerous grain-scale flaws, microcracks, that locally generate high stress concentrations. Failure of the specimen initiates from these critical flaws. Fast coalescence of micro-cracks propagating from these initial flaws leads to macroscopic failure. The failure is *brittle*, that is, by fracture in tension or shear (compression). Larger test specimens have a greater probability of containing critical flaws and are therefore statistically weaker. Interestingly, some laboratory test data show an increase of strength with size. However, this reverse size effect has not been observed for very large test specimens.

In any case, extrapolation of laboratory test data to full scale mine pillars is risky because of the presence of geologic features (e.g. bedding planes, clay seams, joints) in the mine that are absent in the laboratory. Such features reduce the strength of a mine scale pillar relative to intact laboratory test specimens. Some reduction from laboratory test data

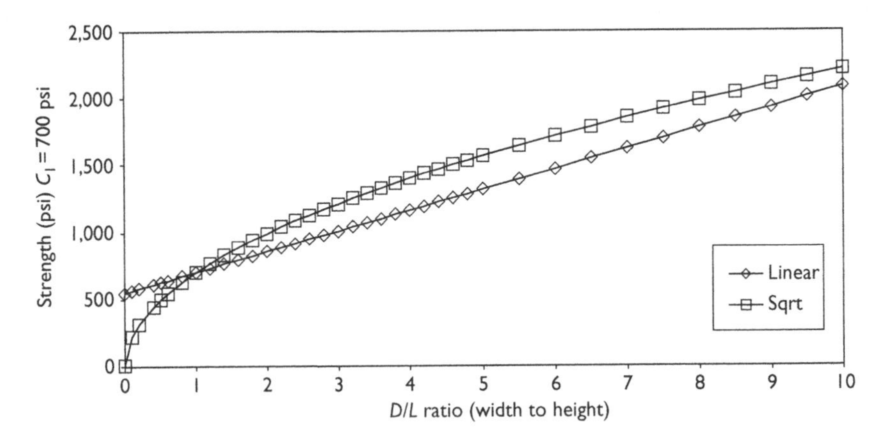

Figure 6.5 Linear and square root fits to laboratory test data for size effects formulas.

would seem needed. However, if joints are recognized as distinct structural features that introduce additional ways a pillar may fail (other than by failure of intact material between joints, bedding planes, and so forth), then no reduction of intact material strength would be indicated. Pillar strength specification for use with average pillar stress and tributary area design is therefore problematic.

One approach to pillar strength that seems reasonable is to invoke the Terzaghi jointed rock mass model that assumes the angles of internal friction of rock and "joint" are the same; joints are cohesionless and only the cohesion of intact rock bridges between joints contribute to rock mass cohesion that is given by the product of intact rock cohesion by the fraction of rock bridges per unit area of potential failure surface. This reduction reduces unconfined compressive strength by the same fraction. Perhaps the best resolution of the question is to do a detailed analysis of pillar stress that allows for local failures and potential pillar collapse. However, in the absence of a detailed stress analysis, specification of pillar strength outright allows for a simple safety factor calculation, given proposed or trial pillar dimensions. Alternatively, specification of a pillar safety factor allows for calculation of one of the pillar dimensions (width or length) in terms of entry and crosscut widths which are usually known from roof span analyses and operating constraints. Use of a size effects relationship complicates this calculation because of the resulting nonlinear expression for one of the pillar dimensions.

Ignoring a size effect for stubby pillars ($L/D < 1$) is usually conservative for two reasons. The first is that any strength enhancing confinement associated with end friction is ignored. If thin or thick clay seams are present at pillar top or bottom, confinement is indeed greatly reduced. In this case, a size effect would not be justified, and some consideration of pillar splitting at top or bottom would be needed. The second reason is related to the geometry of pillar failure. Failure of a stubby pillar is likely to be localized to the pillar walls, while the core of the pillar remains elastic because of interior confinement, as shown in Figure 6.6. Pillar collapse is precluded in this case, although the effect of wall spalls on the roof ends may be an important consideration. Tall pillars ($L/D >> 1$) may be prone to catastrophic failure because spalling quickly reduces the intact load bearing area. In this case, acceptance of a strength reducing size effect is justified, although ignoring the effect and using a higher safety factor may be preferable.

Example 6.5 A room and pillar operation uses entries and crosscuts of widths W_o and W_c, respectively. Pillars are $W_p \times L_p$ in horizontal section and are H_p high. A size effect on strength is given by

$$C_p = C_1\left(0.78 + 0.22\frac{W_p}{H_p}\right)$$

Show that when pillar length is fixed, a quadratic equation for W_p results, but when pillar length is given as k times W_p, a cubic results even at fixed k.

***Solution*:** By definition $FS_p = C_p/S_p$ and $S_p = S_v/1 - R = S_v(A/A_p)$. After combining these two equations and incorporating the size effects formula, the result is

$$(FS_p)(S_v) = C_1\left(0.78 + 0.22\frac{W_p}{H_p}\right)\left[\frac{L_p W_p}{(W_o + W_p)(W_c + L_p)}\right]$$

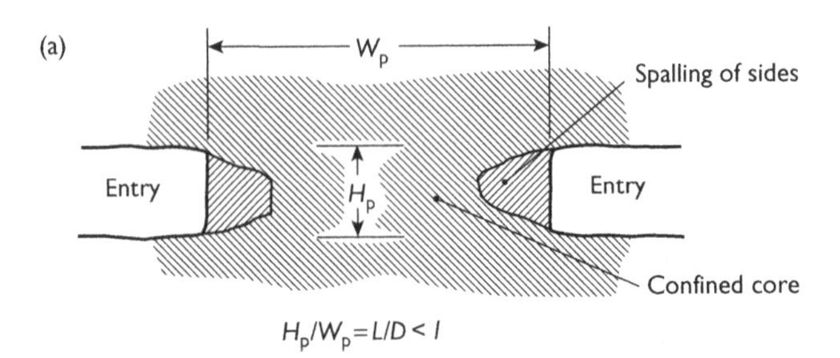

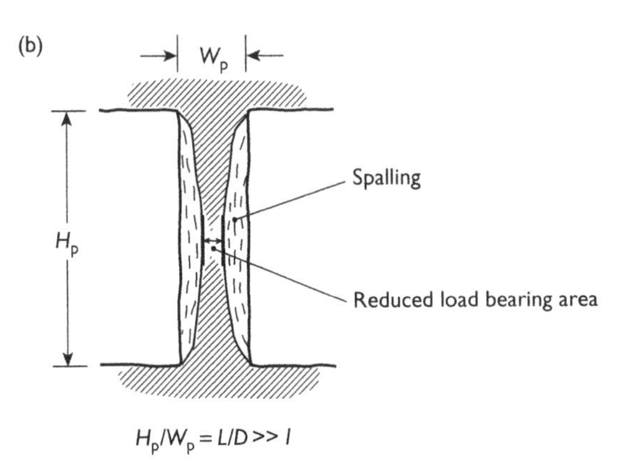

Figure 6.6 Failure modes of (a) stubby, and (b) tall pillars.

that when cleared of fractions results in a quadratic in W_p. Substitution for $L_p = k\ W_p$ gives

$$(\mathrm{FS_p})(S_v) = C_1\left(0.78 + 0.22\frac{W_p}{H_p}\right)\left[\frac{k(W_p)^2}{(W_o + W_p)(W_c + kW_p)}\right]$$

which is a cubic in W_p.

Example 6.6 A room and pillar operation is planned for a depth of 300 m (984 ft) in a stratum that has a compressive strength 13.57 MPa (1,968 psi) determined from laboratory testing on cylinders of *L/D* (length/diameter) of one. Entries and crosscuts are planned at 6 m (20 ft). Square pillars of 4 m high (seam thickness) are anticipated. Determine the pillar dimensions and extraction ratio associated with a pillar safety factor of 1.5.

Solution: From Example 6.5,

$$(\mathrm{FS_p})(S_v) = C_1\left(0.78 + 0.22\frac{W_p}{H_p}\right)\left[\frac{k(W_p)^2}{(W_o + W_p)(W_c + kW_p)}\right]$$

$$(1.5)(22.6)(10)^3(300) = 13.57(10)^6\left(0.78 + 0.22\frac{W_p}{4}\right)\left[\frac{(1)(W_p)^2}{(6.0 + W_p)(6.0 + W_p)}\right]$$

after substitution of known quantities and a reasonable estimate of S_v. Clearing of fractions results in

$$(0.7495)(6.0 + W_p)^2 = 0.78(W_p)^2 + 0.055(W_p)^2$$

that has the form $f(W_p) = g(W_p)$. Plotting these functions with the aid of a spreadsheet defines the solution at the intersection point.

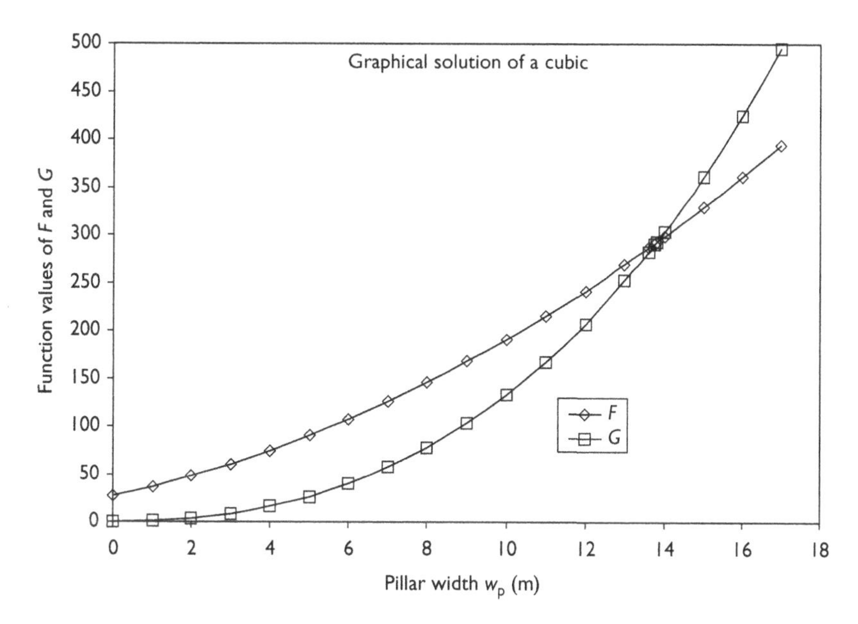

The solution point corresponds to a pillar width of 13.75 m. At this pillar width, the strength is

$$C_p = C_1\left(0.78 + 0.22\frac{13.75}{4.0}\right) = (13.57)(1.536) = 20.84 \text{ MPa}$$

The extraction ratio is then

$$\begin{aligned} R &= 1 - \frac{(FS_p)(S_v)}{C_p} \\ &= 1 - \frac{(1.5)(22.6)(10^3)(300)}{20.84(10)^6} \\ R &= 0.512 \end{aligned}$$

Check:

$$R = 1 - \frac{(13.75)^2}{(6.0 + 13.75)^2} = 0.516$$

which is a valid check in view of roundoff effects in the calculations.

In the absence of a size effect, $R = 0.25$ and $W_p = 38.8$ m, so in this example, size effect on pillar strength is substantial, which could be anticipated in view of the large increase in strength (over 50%).

Example 6.7 A room and pillar operation is planned for a depth of 300 m (984 ft) in a stratum that has a compressive strength 13.57 MPa (1,968 psi) determined from laboratory testing on cylinders of *L/D* (length/diameter) of one. Additional laboratory test data fit a curve for size effects on strength that has the form for application

$$C_p = 13.57\sqrt{\frac{W_p}{H_p}}$$

where C_p is in MPa. Entries and crosscuts are planned at 6 m (20 ft). Square pillars of 4 m high (seam thickness) are anticipated. Determine the pillar dimensions and extraction ratio associated with a pillar safety factor of 1.5.

***Solution*:** From Example 6.5,

$$(\mathrm{FS_p})(S_v) = 13.57(10^6)\left(\frac{W_p}{H_p}\right)^{1/2}\left[\frac{k(W_p)^2}{(W_o + W_p)(W_c + kW_p)}\right]$$

$$(1.5)(22.6)(10)^3(300) = 13.57(10)^6\left(\frac{W_p}{4}\right)\left[\frac{(1)(W_p)^2}{(6.0 + W_p)(6.0 + kW_p)}\right]$$

after substitution of known quantities and a reasonable estimate of S_v. Clearing of fractions results in

$$(0.7495)(6.0 + W_p)^2 = 0.5(W_p)^{2.5}$$

that has the form $f(W_p) = g(W_p)$. Plotting these functions with the aid of a spreadsheet defines the solution at the intersection point. The solution point corresponds to a pillar width of 11.73 m. At this pillar width, the strength is

$$C_p = (13.57)\left(\frac{11.73}{4}\right)^{1/2} = 23.24 \text{ MPa}$$

The extraction ratio is then

$$R = 1 - \frac{(FS_p)(S_v)}{C_p}$$

$$= 1 - \frac{(1.5)(22.6)(10^3)(300)}{23.24(10)^6}$$

$$R = 0.562$$

Check:

$$R = 1 - \frac{(11.73)^2}{(6.0 + 11.73)^2} = 0.562$$

which checks.

In the absence of a size effect, $R = 0.25$ and $W_p = 38.8$ m, so in this example, size effect on pillar strength is substantial, which could be anticipated in view of the large increase in strength (over 50%). This increase is somewhat greater than for the conventional size effects used in Example 6.6 where the extraction ratio was 0.516.

6.2 Pillars in dipping strata

Pillars in dipping strata are loaded in compression and shear and therefore require consideration of a failure criterion that accounts for both. The Mohr–Coulomb criterion serves the purpose quite well. However, use of average stresses leads to an unconventional Mohr circle representation that can be generalized to the case where shear stress loads the pillar top and bottom but is absent at pillar walls. These Mohr circles are not centered on the normal stress axis as is the case when stress at a point is considered and the shear stresses come in pairs. The situation does not arise in flat seam pillars where shear loading is absent. Backfill also affects pillar safety and stability.

Extraction ratio formulas for pillars in dipping seams

The base of a pillar prism in dipping strata is inclined to the horizontal, so even in a preexcavation gravity stress field, pillar top and bottom are loaded in compression perpendicular to the dip and in shear parallel to the dip as shown in Figure 6.7 (compression is positive). A detailed analysis of pillar stress in dipping strata (Pariseau, 1982) shows that to close approximation, gravity forces that an overburden block exerts on a pillar before excavation are the same as the forces after mining, as was the case for pillars in flat seams. Thus,

$$\begin{aligned} S_n A &= S_p A_p \\ T_s A &= T_p A_p \end{aligned} \tag{6.13}$$

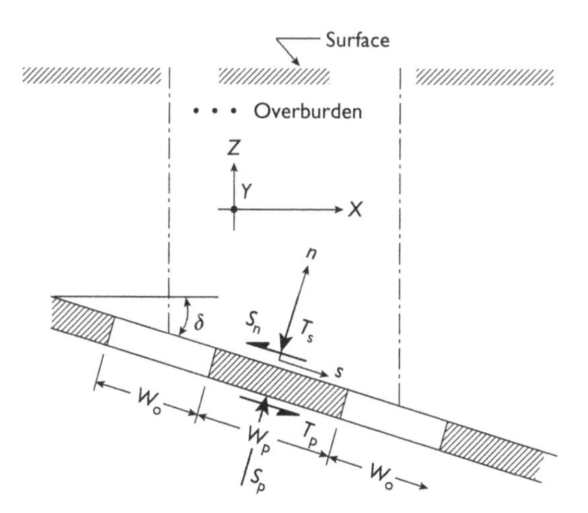

Figure 6.7 Preexcavation pillar normal and shear stresses (S_n, T_s) and postexcavation pillar stresses (S_p, T_p) induced by a tributary overburden block above an inclined bed.

where S_n, S_p, T_s, T_p, A, and A_p are normal stress before mining, average normal stress after mining, shear stress before mining, average shear stress after mining, tributary area, and pillar area, respectively. These stresses act on the pillar top and bottom and are therefore surface tractions. Extraction ratio formulas for pillar stresses are therefore

$$S_p = \frac{S_n}{1-R}$$
$$T_p = \frac{T_s}{1-R} \tag{6.14}$$

where R is the extraction ratio. These formulas reduce to the flat seam case when the dip is zero and shear is not a consideration. The premining stresses (S_n, T_s) relative to seam coordinates (s,n) may be obtained from stresses in compass coordinates (x = east, y = north, z = up) by rotation of the reference axes and the usual equations of transformation.

Because S_p and T_p are average stresses acting over a finite area A_p, they do *not* follow the usual equations of transformation of stress under a rotation of reference axes. The reason is there are no companion shear stresses on the sides of the pillar. Force and moment equilibrium may be satisfied, although the normal stresses are offset, as shown in Figure 6.8(a). If the offset of the normal stress is x, then clearly $x < W_p/2$. Moment equilibrium then requires the pillar width to height ratio satisfy

$$2S_p x = 2T_p \frac{H_p}{2} \tag{6.15}$$

Thus, when an offset of normal stress is required for moment equilibrium, the pillar width to height ratio must satisfy the inequality

$$\frac{W_p}{H_p} < \frac{T_p}{S_p} \tag{6.16}$$

Otherwise, the pillar may overturn and fail by toppling. Of course, if there is no shear load, then no offset is required and toppling is not a concern. An alternative and more conservative toppling analysis that is based on the same resultant normal force but from an assumed triangular distribution of stress, as illustrated in Figure 6.8(b), shows the offset may be no more than one-sixth of the pillar width. If the pillar were considered to be a "tower," then according to this analysis, the resultant force must pass through the middle third of the tower base.

A force analysis shows that the average normal and shear stresses, σ and τ, on an inclined surface within a pillar in a dipping seam shown in Figure 6.8(c) are given by

$$\sigma = \frac{S_p}{2}[1+\cos(2\alpha)] - \frac{T_p}{2}\sin(2\alpha)$$
$$\tau = \frac{S_p}{2}\sin(2\alpha) + \frac{T_p}{2}[1+\cos(2\alpha)] \tag{6.17}$$

where α is the angle between the direction of S_p and the normal to the inclined surface being considered. These appear similar to the usual equations of transformation of stress under a

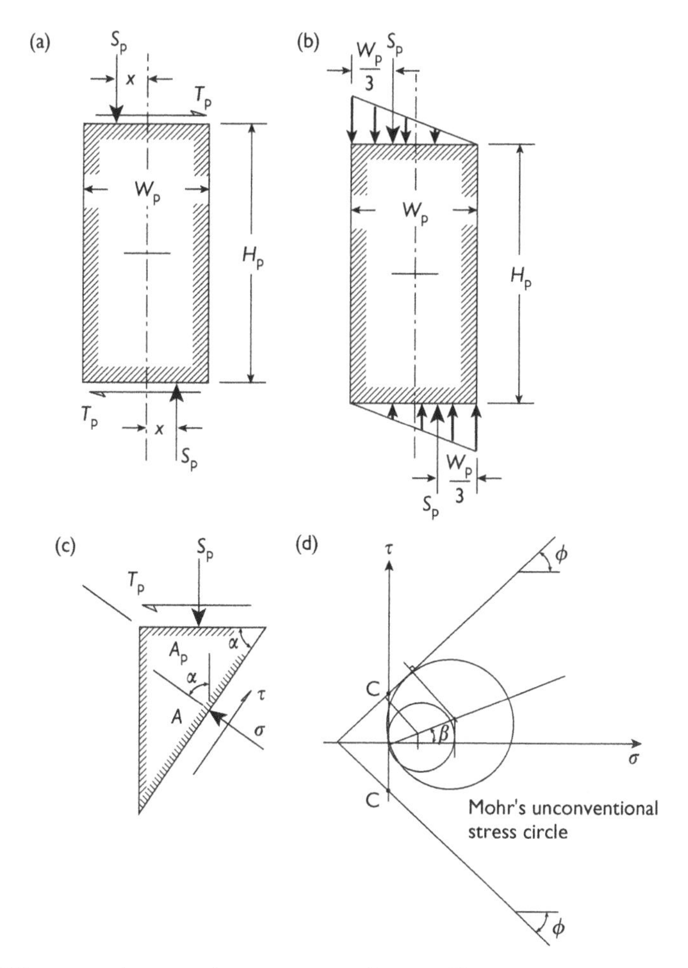

Figure 6.8 (a) Offset normal stresses for moment equilibrium; (b) triangular normal stress distribution; limits offsets to middle-third of pillar width (c) free body diagram for σ, τ on an inclined surface, and (d) unconventional Mohr's circle for stress.

rotation of axes, but differ in the one-half factor multiplying the shear stress term. Again, this result is a consequence of the absence of companion shear forces on the pillar sides and the fact that the pillar stresses are averages acting over surfaces of a body of finite size, not stresses at a point.

The average normal stress has stationary values at rotation angles given by solutions to

$$\tan(2\alpha^*) = \frac{-T_p/2}{S_p/2} \tag{6.18}$$

that may be expressed as

$$\sin(2\alpha^*) = \frac{-T_p/2}{\pm[(S_p/2)^2 + (T_p/2)^2]^{1/2}}$$
$$\cos(2\alpha^*) = \frac{-S_p/2}{\pm[(S_p/2)^2 + (T_p/2)^2]^{1/2}} \quad (6.19)$$

Associated maximum and minimum average normal stresses are:

$$S_1 = \frac{S_p}{2} + \left[\left(\frac{S_p}{2}\right)^2 + \left(\frac{T_p}{2}\right)^2\right]^{1/2}$$
$$S_3 = \frac{S_p}{2} + \left[\left(\frac{S_p}{2}\right)^2 + \left(\frac{T_p}{2}\right)^2\right]^{1/2} \quad (6.20)$$

Interestingly, the minimum average normal stress is a tension. These are not true principal stresses but rather are pseudo-principal stresses because they act on planes that are not free of shear stress. In fact, shear stresses on these planes are

$$\tau_1 = \frac{T_p}{2}$$
$$\tau_3 = \frac{T_p}{2} \quad (6.21)$$

Maximum and minimum average shear stresses are

$$T_1 = \frac{T_p}{2} + [(S_p/2)^2 + (T_p/2)^2]^{1/2}$$
$$T_3 = \frac{T_p}{2} + [(S_p/2)^2 + (T_p/2)^2]^{1/2} \quad (6.22)$$

that occur on planes with normal inclination given by solutions to

$$\tan(2\alpha^{**}) = \frac{S_p}{T_p} \quad (6.23)$$

These directions are at right angles to the directions of maximum and minimum normal stress. Normal stresses on these planes of maximum and minimum shear stress are

$$\sigma_1 = \frac{S_p}{2}$$
$$\sigma_3 = \frac{S_p}{2} \quad (6.24)$$

An unconventional Mohr's circle representation

A useful but unconventional Mohr's circle representation of average normal and shear stress in a pillar loaded in compression and shear may be obtained by observing that

$$
\begin{aligned}
S &= \left(\sigma - \frac{S_p}{2}\right) = \frac{S_p}{2}\cos(2\alpha) - \frac{T_p}{2}\sin(2\alpha) \\
T &= \left(\tau - \frac{T_p}{2}\right) = \frac{S_p}{2}\sin(2\alpha) + \frac{T_p}{2}\cos(2\alpha) \\
&\left(\sigma - \frac{S_p}{2}\right)^2 + \left(\tau - \frac{T_p}{2}\right)^2 = \left[(S_p/2)^2 + (T_p/2)^2\right] \\
&S^2 + T^2 = R^2
\end{aligned}
\tag{6.25}
$$

The last two equations represent a circle of radius R centered at (S_p /2, T_p /2) in a normal stress–shear stress plane shown in Figure 6.8(d). A conventional Mohr's circle would be centered on the normal stress axis. As the pillar stresses are increased proportionally, the radius of the circle R increases; the center of the circle moves outward from the origin in the normal stress–shear stress plane along a line inclined to the normal stress axis at an angle β where $\tan(\beta) = T_p/2/S_p/2$. Figure 6.9 shows details for two such circles, one below failure and

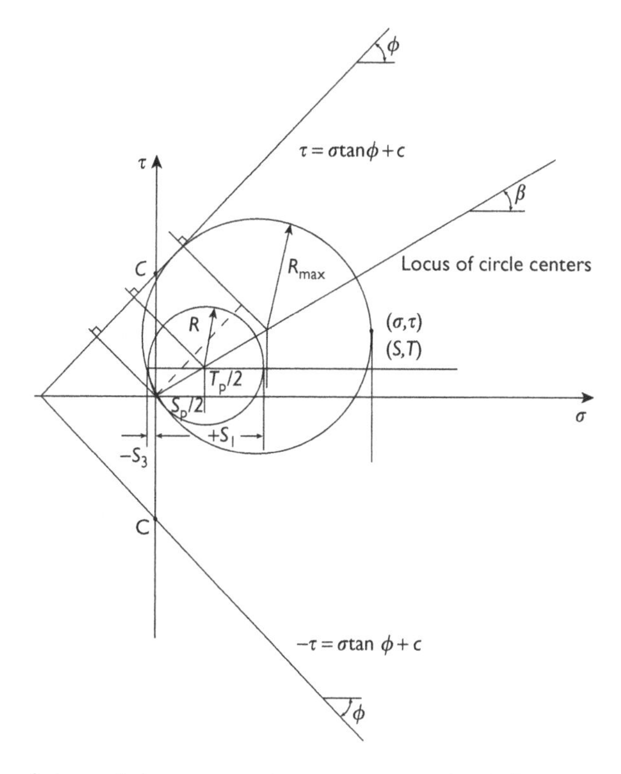

Figure 6.9 Details of stress circles representing average normal and shear stresses on an inclined surface.

one at failure. The circle at failure just contacts the Mohr–Coulomb failure envelope ($\tau = \sigma \tan(\phi) + c$). The radius of this circle is maximum and has the value

$$R(\text{strength}) = \left[\frac{c\cos(\phi)}{1 - \sin(\phi - \beta)}\right] \tag{6.26}$$

where c and ϕ are cohesion and angle of internal friction, respectively. Prior to reaching this maximum, a circle radius is simply

$$R(\text{stress}) = [(S_p/2)^2 + (T_p/2)^2]^{1/2} \tag{6.27}$$

which is a maximum shear stress *relative to the circle center*. A pillar factor of safety with respect to a Mohr–Coulomb criterion is then

$$\text{FS}_p = \frac{R(\text{strength})}{R(\text{stress})} \tag{6.28}$$

Thus,

$$\text{FS}_p = \frac{\{[(2c)\cos(\phi)]/[1 - \sin(\phi - \beta)]\}}{[(S_p)^2 + (T_p)^2]^{(1/2)}}$$

or

$$\text{FS}_p = \frac{(1 - R)\{[(2c)\cos(\phi)]/[1 - \sin(\phi - \beta)]\}}{[(S_n)^2 + (T_s)^2]^{1/2}} \tag{6.29}$$

in the case of dipping seams. Inspection of the first expression shows that under a relatively high shear stress, angle β may be greater than the angle of internal friction ϕ with a consequent reduction in strength. Generally, strength decreases with increasing β.

The last expression highlights the role of extraction. These expressions allow for the additional shear load that appears in dipping seam pillars and the reduction of normal load relative to the vertical. The last expression shows that if the extraction ratio $R = 0$, then the factor of safety applies to just after release of horizontal stress on the pillar walls but before entry widening. The factor $(1 - R)$ is a strength reduction factor in the numerator of this last expression or a stress concentration factor in the denominator.

In the flat seam case ($T_p = 0$), the numerator expressing strength is simply the unconfined compressive strength C_o. In the flat seam case, for comparison,

$$\text{FS}_p = (1 - R)\left(\frac{C_o}{S_v}\right) \tag{6.30}$$

In the extreme dip case of a vertical seam $T_p = 0$ and $S_n = S_h$. The pillar safety factor is then

$$\text{FS}_p = (1 - R)\left(\frac{C_o}{S_h}\right) \tag{6.31}$$

where S_h is the premining horizontal stress.

When roof spans in stratified ground were considered, a correction for dip was suggested that involved using only the normal component of specific weight, although the correction was limited to small dips. A similar approach may be tried for pillars, so instead of using the vertical premining stress $S_v = \gamma H$ one uses $S'_v = \gamma' H$ where $\gamma' = \gamma\cos(\delta)$ and δ is the dip. This approach is obviously in error at a dip of 90°, so a question arises as to how steep a dip is acceptable before the approximation becomes too inaccurate. Detailed analysis shows that this simple approximation is surprisingly accurate, say, within about 10% up to dips of almost 60°. However, there is no pressing need to use this empirical approach to the problem of computing pillar safety factors in dipping seams.

Example 6.8 A coal seam 12 ft thick dips 15° degrees south at a depth of 1,200 ft. Entries are developed on strike due east and are 21 ft wide. Pillars are three times as long as they are wide and are separated by crosscuts on the dip that are 17 ft wide. Seam material compressive and tensile strengths are estimated to be 2,750 and 350 psi, respectively. Neglect any size effect that may be present, then determine the maximum extraction ratio possible. Assume that the horizontal preexcavation stress is one-third the vertical preexcavation stress.

***Solution*:** A pillar safety factor is the ratio of strength to stress, so estimates of both are required. The *average* pillar stresses may be obtained from the extraction ratio formulas

$$S_p = \frac{S_n}{1-R}, \quad T_p = \frac{T_s}{1-R}$$

while the preexcavation stresses S_n and T_s may be obtained by rotation of axis.

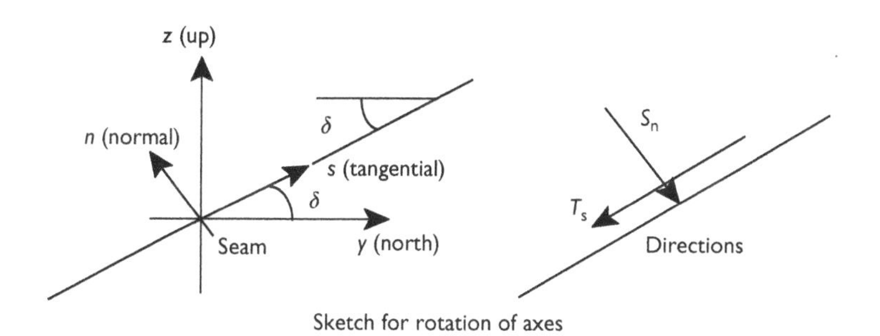

Sketch for rotation of axes

With reference to the sketch, one has

$$S_n = \frac{S_{yy} + S_{zz}}{2} - \frac{S_{yy} - S_{zz}}{2}\cos(2\delta) - T_{yz}\sin(2\delta)$$

$$T_s = -\left(\frac{S_{yy} - S_{zz}}{2}\right)\sin(2\delta) + T_{yz}\cos(2\delta)$$

where $S_{zz} = S_v$, the vertical preexcavation stress, and $S_{yy} = S_h$, the horizontal preexcavation stress. With $S_h = K_o S_v$, and $T_{yz} = 0$, one obtains

$$S_n = \frac{S_v(K_o+1)}{2} - \frac{S_v(K_o-1)}{2}\cos(2\delta)$$
$$= \frac{(1{,}200)(1/3+1)}{2} - \frac{(1{,}200)(1/3-1)}{2}\cos(30)$$
$$S_n = 1{,}146 \text{ psi}$$

$$T_s = -1{,}200\left(\frac{1/3-1}{2}\right)\sin(30)$$
$$T_s = 200 \text{ psi}$$

which are both positive and have the directions shown in the sketch. The average postexcavation stresses may now be obtained from the extraction ratio formulas.

A Mohr–Coulomb strength criterion is reasonable, but because *average* stresses are being considered, an unconventional Mohr's circle is needed to estimate pillar strength. Graphically, when this circle touches the strength envelope, the maximum load combination of S_p and T_p is reached. Figure 6.9 represents the situation and (6.29) gives the associated safety factor. Use of (6.29) requires calculation of cohesion and angle of internal friction from unconfined compressive and tensile strengths for the Mohr–Coulomb criterion. Thus,

$$\sin(\phi) = \frac{C_o - T_o}{C_o + T_o} = \frac{2{,}750 - 350}{2{,}750 + 350} = 0.7742$$
$$\phi = 50.7^\circ$$
$$c = \frac{C_o[1-\sin(\phi)]}{2\cos(\phi)} = \frac{(2{,}750)[1-0.7742]}{2(0.6329)} = 490 \text{ psi}$$

Also required is the angle β that may be computed from the definition. Thus,

$$\tan(\beta) = \frac{T_p/2}{S_p/2}$$
$$= \frac{T_s/2(1-R)}{S_n/2(1-R)}$$
$$= \frac{T_s}{S_n}$$
$$\tan(\beta) = \frac{200}{1,146} = 0.1745$$
$$\beta = 9.9^\circ$$

In (6.29), one has

$$1 = \frac{(1-R)\{[(2)(490)\cos(50.7)]/[1-\sin(50.7-9.9)]\}}{\sqrt{(1{,}146)^2 + (200)^2}} = (1-R)(1.5395)$$

$$1 - R = 0.6495$$

$$R(\max) = 0.350$$

If the seam were flat, then β would be zero, and the extraction ratio would be greater, $R = 0.576$, which shows a considerable effect of the shear load induced by seam dip.

If the seam dip were toward the north instead of south, then the preexcavation shear stress would change sign. The center of the unconventional Mohr's circle would then be located *below* the normal stress axis in Figure 6.9. In this position, angles may be measured positive in the clockwise sense to describe the geometry of the circle. The result is the same expression (6.29) used when the preexcavation shear stress was positive. The maximum extraction ratio possible would be the same, as one would expect.

Example 6.9 Given the data and results from Example 6.8, determine pillar dimensions.

***Solution*:** By definition

$$
\begin{aligned}
1 - R &= \frac{A_p}{A} \\
&= \frac{W_p L_p}{(W_o + W_p)(W_c + L_p)} \\
&= \frac{W_p(3W_p)}{(W_o + W_p)(W_c + 3W_p)} \\
0.6495 &= \frac{3(W_p)^2}{(21 + W_p)(18 + 3W_p)}
\end{aligned}
$$

$\therefore$

$$(1 - 1/0.6495)(W_p)^2 + 27W_p + (21)(6) = 0$$

$\therefore$

$$
\begin{aligned}
W_p &= 54.3 \text{ ft} \\
L_p &= 162.9 \text{ ft} \\
H_p &= 12 \text{ ft}
\end{aligned}
$$

Check:

$$1 - R = \frac{(54.3)(162.9)}{(21 + 54.3)(18.0 + 162.9)} = 0.6494$$

which shows the proper solution to the quadratic was correctly chosen.

Example 6.10 Given the data and results from Examples 6.8 and 6.9, assume that a size effect on pillar strength exists such that pillar compressive strength is given by

$$C_p = C_1\left(0.78 + 0.22\frac{W_p}{H_p}\right)$$

Further assume that "size" does not affect the angle of internal friction φ. In this case, only the cohesion c is influenced by the postulated size effect.

Solution: Cohesion c is directly proportional to compressive strength C_p. If the given compressive strength is considered to be C_1, then a new compressive strength estimate is

$$C_p = (2,750)\left(0.78 + 0.22\frac{54.3}{12.0}\right) = (2,750)(1.776)$$

and a new cohesion estimate is 1.776 times the original cohesion.

In consideration of the safety factor equation, one has

$$1 = (c)(\text{number})(1 - R) \quad \text{and} \quad 1 = c(\text{new})(\text{number})[1 - R(\text{new})]$$

$$\therefore$$

$$1 - R(\text{new}) = (1 - R)\left[\frac{c}{c(\text{new})}\right]$$

$$\therefore$$

$$1 - R(\text{new}) = 0.6495\left[\frac{1}{1.776}\right] = 0.3657$$

Pillar width then is the solution to

$$W_p\left[1 - \frac{1}{1 - R(\text{new})}\right] + 27W_p + 126 = 0$$

The result is $W_p = 19.3$ ft, considerably less than the original 54.3 ft. However, the change in pillar size induces a change in pillar strength that again induces a change in extraction ratio. An iteration procedure suggests itself by repeated substitutions into the strength equation and then into the safety equation and back to the strength equation until convergence is obtained. The extraction ratio formulas serve as a check on the final estimate of pillar width and associated length. An improved estimate that could speed convergence may be to use the average of the preceding two estimates for the next trial pillar width. In this case a new $W_p =$ (54.3 + 19.3)/2 = 36.8 ft. With this estimate, the new cohesion multiplier is [0.78 + 0.22(36.8)/ 12.0] = 1.455, so the new $1 - R$ estimate is (0.6495)(1/1.455) = 0.4464. The new quadratic equation solution for W_p gives an estimate of 25.7 ft. A new average estimate is (36.8 + 25.7)/ 2 = 31.3 ft that leads to a new cohesion multiplier of 1.353. The new $1 - R$ = 0.6495(1/1/353) = 0.4800 and the new W_p = 28.95 ft. A new average is 30.1 ft that leads to a W_p = 29.8 ft which is close to the last estimate of 28.95 ft. A last estimate of $1 - R$ = 0.490 and W_p = 29.8 ft, L_p = 89.4 ft, H_p = 12.0 ft.

Check:$1 - R$ = (29.8)(89.4)/(21 + 29.8)(18 + 89.4) = 0.488 which shows the results accurate to the second decimal place.

Thus, in consideration of a size effect on pillar strength, the extraction ratio may be increased to 0.510 from 0.350, a substantial benefit, provided size effects are realistic.

Generalized Mohr's circle

There are other interesting pillar load circumstances that may also be analyzed using average stresses and an unconventional or generalized Mohr's circle that contains Mohr's circle for stress at a point as a special case. For example, consider a pillar in a dipping seam with

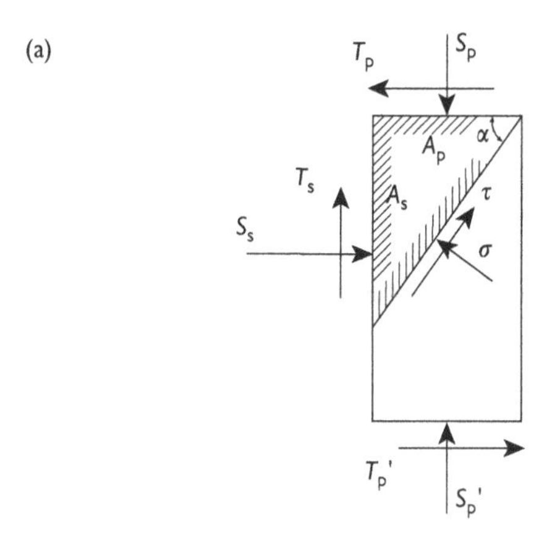

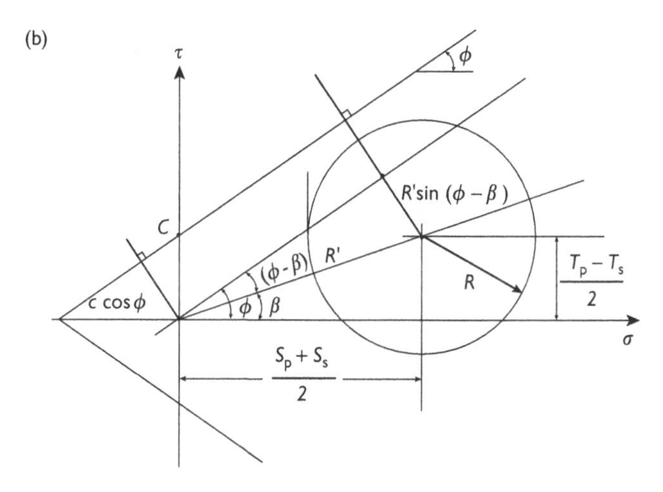

Figure 6.10 Backfill effect on a pillar: (a) pillar with normal load and side shear from backfill on one side and (b) unconventional Mohr's circle.

backfill placed on one side as shown in Figure 6.10(a). Lateral confinement of the pillar by the fill may be expressed as an average horizontal compressive stress acting against the fill side of the pillar. Settlement of the backfill loads the same pillar side in shear. Average normal and shear stresses on an inclined surface in the pillar are

$$\begin{aligned} S &= \left[\sigma - \left(\frac{S_p + S_s}{2}\right)\right] = \left(\frac{S_p + S_s}{2}\right)\cos(2\alpha) - \left(\frac{T_p + T_s}{2}\right)\sin(2\alpha) \\ T &= \left[\tau - \left(\frac{T_p - T_s}{2}\right)\right] = \left(\frac{S_p - S_s}{2}\right)\sin(2\alpha) + \left(\frac{T_p + T_s}{2}\right)\cos(2\alpha) \end{aligned} \tag{6.32}$$

where α is the inclination of the considered surface. Again, the shear stress averages are not paired and equal as they are in case of stress at a point. However, if $T_p = T_s$, then these equations of transformation have the same form as those for stress at a point. Pseudo-principal stresses, directions and associated shears, principal shears, and so forth may still be computed, all of which may also be represented graphically by a generalized Mohr's circle. This circle has a radius $R = [S^2 + T^2]^{1/2}$ and center at $[(1/2)(S_p + S_s), (1/2)(T_p - T_s)]$. The locus of circle centers is a line inclined to the normal stress axis at an angle β where $\tan(\beta) = [(1/2)(T_p - T_s)/(1/2)(S_p + S_s)]$, as shown in Figure 6.10(b).

Transformation of plot plane coordinates from (σ, τ) to (σ', τ') with a shift in origin to the stress circle center leads to expressions for normal and shear stress

$$\begin{aligned} \sigma' &= R\cos(2\alpha') = \left(\frac{S'_1 - S'_3}{2}\right)\cos(2\alpha') \\ \tau' &= R\sin(2\alpha') = \left(\frac{S'_1 - S'_3}{2}\right)\sin(2\alpha') \end{aligned} \tag{6.33}$$

where R is the circle radius as before, $2\alpha'$ is the angle from the σ' - axis to the point on the circle with coordinates (σ', τ'), and S'_1 and S'_3 are maximum and minimum normal stresses in the shifted and rotated coordinate plane (the "prime" plane). In this plane, they are true principal stresses. Inspection of Figure 6.10b shows that $S'_1 = +R$, $S'_3 = -R$ and the mean normal stress $(1/2)(S'_1 + S'_3) = 0$ which explains why this term is missing from these equations of transformation of stress. The principal shears are $T'_1 = +R$, $T'_3 = -R$; planes of the principal shears bisect the directions of major and minor (maximum and minimum) principal stresses. These shear planes are free of normal stress because the mean normal stress is zero in the prime plane. In fact, the 1- and 3-directions are parallel to the prime axes.

The Mohr–Coulomb failure criterion in the prime plane is simply $\tau' = \sigma' \tan(\phi') + c'$ where $\phi' = \phi - \beta$ and c' is given by

$$c' = \frac{(c)\cos(\phi)}{\cos(\phi - \beta)} + R_c\tan(\phi - \beta) \tag{6.34}$$

where $2R_c = [(S_p + S_s)^2 + (T_p - T_s)^2]^{1/2}$, that is, R_c is the slope distance to the circle center that touches the Mohr–Coulomb failure envelope. Inclinations of potential failure surfaces in the prime plane are at angles of $\pm(\pi/4 - \phi'/2)$ to the direction of the major (maximum) principal stress.

A factor of safety with respect to a Mohr–Coulomb criterion in the prime plane has the same representation as in a conventional plot. Thus,

$$\text{FS} = \frac{\tau'_m(\text{strength})}{\tau'_m(\text{stress})} = \frac{\sigma'_m\sin(\phi') + c'\cos(\phi')}{\tau'_m} = \frac{R(\text{strength})}{R(\text{stress})} \tag{6.35}$$

In the prime system, the mean normal stress is zero, so only the cohesive term adds to the numerator.

Backfill effects on pillar safety factors

The pillar safety factor in this case of additional loading by backfill is

$$\text{FS} = \frac{\text{"strength"}}{\text{"stress"}} = \frac{R(\text{strength})}{R(\text{stress})} = \frac{[2R_c\sin(\phi - \beta) + (2c)\cos(\phi)]}{[(S_p - S_s)^2 + (T_p - T_s)]^{1/2}} \tag{6.36}$$

where $2R_c = [(S_p + S_s)^2 + (T_p - T_s)^2]^{1/2}$ and is twice the slope distance from origin to circle center. This safety factor formula reduces to the previous case of a pillar in a dipping seam in the absence of backfill ($S_s = T_s = 0$). Examination of this expression shows prefill shear loading decreases strength by increasing β and further decreases safety by increasing stress. Postfill shear loading has the opposite effect as can be seen by the opposite algebraic sign. Postfill "confining pressure" S_s decreases "stress" while increasing "strength" even though application is to only one side of the considered pillar.

In practice, fill may be placed sequentially along the other pillar sides as mining and filling proceed. In each case, an equilibrium analysis beginning with a free body diagram and proceeding through force and moment summations is possible. Usually, offsets of normal forces needed for moment equilibrium can only be bounded by pillar geometry and the fact that forces must act within pillar dimensions for equilibrium. However, safety factor analysis can still proceed using unconventional Mohr's circles as these analyses show.

While extraction ratio formulas lead to values for S_p and T_p, estimates of S_s and T_s remain. One approach is to use the famous Janssen formulas (Janssen, 1986, chapter 8) that were originally derived within a context of bulk materials stored in silos. Concern for estimates of stresses transmitted to silo walls by bulk materials within lead to the Janssen formulas that are also used to estimate stresses between caving ground in vertical chimneys and surrounding rock walls. The Janssen formulas are

$$\begin{aligned} S_v &= \left(\frac{\gamma}{C_1}\right)[1 - \exp(-C_1 z)] \\ S_s &= kS_v \\ T_s &= \mu S_s \\ C_1 &= \frac{\mu k P}{A} \end{aligned} \tag{6.37}$$

where S_v, S_s, T_s, k, μ, P, A, γ, and z are vertical fill stress, horizontal fill stress at the pillar wall, shear stress at the pillar wall, a material constant, coefficient of friction between fill and pillar rock, perimeter of the fill column, cross-sectional area of the fill column, fill specific weight, and depth below the fill top. The top of the fill is assumed to be stress-free. When the fill top is not stress-free or if the column is inclined, adjustments can be computed. Adjustments for water or other fluid saturation can also be made.

Example 6.11 A lane and pillar excavation plan is executed in a flay-lying, thick, strong limestone formation. Depth is 500 ft, lanes are 28 ft wide, pillars are 135 ft high, unconfined compressive strength is 15,000 psi, and the ratio of unconfined compressive to tensile strength is 15. No size effect is present. If a safety factor of 6.0 is required, determine pillar width and then estimate the improvement in the pillar safety factor by sand fill placed along one side of the pillars. Fill properties are: $\gamma = 100$ pcf, $\mu = 1$, and $k = 1/3$.

Solution: Geometry of excavation is shown in the sketch. By definition

$$\begin{aligned} FS_p &= \frac{C_o}{S_p} = \frac{C_o(1-R)}{S_v} \\ 6.0 &= \frac{(15,000)(1-R)}{500} \end{aligned}$$

$$\therefore$$

$$1 - R = 0.2$$
$$R = 0.8$$

where an estimate of the preexcavation vertical stress of 1 psi per foot of depth is used.

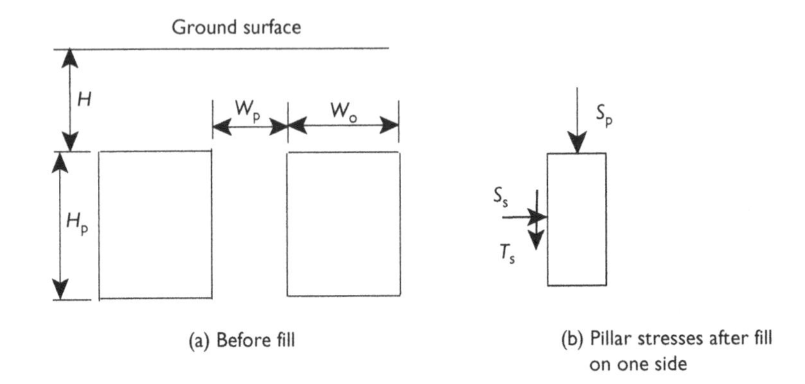

(a) Before fill

(b) Pillar stresses after fill on one side

The extraction ratio formula for this essentially two-dimensional geometry gives

$$1 - R = \frac{W_p}{W_o + W_p}$$

$$0.2 = \frac{W_p}{28 + W_p}$$

$$\therefore$$

$$W_p = 7.0 \text{ ft}$$

The pillars are indeed tall relative to pillar width and would likely fail catastrophically. For this reason a rather high safety factor is needed despite high strength and shallow depth.

Pillar stress is

$$S_p = \frac{S_v}{1 - R} = \frac{500}{0.2} = 2,500 \text{ psi}$$

No horizontal and no shear stress are present because of the horizontal attitude of the limestone unit.

However, after placement of the sand fill, horizontal and shear stresses appear at the fill–pillar contact as shown in the sketch. These are *average* stresses that may be computed from distributions given by application of Janssen-type formulas obtained by adaptation of the three-dimensional formulas to this two-dimensional geometry. The adaptation consists of observing that the area A is now taken to be entry width times one foot along the entry, while the perimeter P is simply 2, one for each foot of lateral area on a side. Thus,

$$\sigma_v = \left(\frac{\gamma W_o}{2\mu k}\right)\left[1 - \exp\left(-\frac{2\mu k}{W_o}z\right)\right]$$

$$= \left(\frac{(100)(28)}{2(1)(1/3)}\right)\left[1 - \exp\left(-\frac{2(1)(1/3)}{28}z\right)\right]$$

$\therefore$

$$\sigma_v(z = H_p) = 4,031 \text{ psf}(28.0 \text{ psi})$$

$$\sigma_h(z = H_p) = k\sigma_v = 1,334 \text{ psf}(9.33 \text{ psi})$$

$$\tau(z = H_p) = \mu\sigma_h = 1,334 \text{ psf}(9.33 \text{ psi})$$

which are maximum values that occur at the bottom of the fill.
Average values may be obtained by integration. Thus,

$$\bar{\tau} = \left(\frac{\mu k}{H_p}\right)\int_{H_p}\sigma_v dz = \left(\frac{\mu k}{H_p}\right)\int_{H_p}\left(\frac{\gamma W_o}{2\mu k}\right)\left[1 - \exp\left(-\frac{2\mu k}{W_o}z\right)\right]dz$$

$$= \left(\frac{\mu k}{H_p}\right)\left(\frac{\gamma W_o}{2\mu k}\right)\left[z - \left(\frac{W_o}{2\mu k}\right)\exp\left(-\frac{2\mu k}{W_o}z\right)\right]\Big|_0^{H_p}$$

$$= \left(\frac{(1)(1/3)}{135}\right)\left[\frac{(100)(28)}{(2)(1)(1/3)}\right]$$

$$\left[135 - \left(\frac{(28)}{2(1)(1/3)}\right)\exp\left(-\frac{2(10)(1/3)}{28}135\right) - \left(\frac{(28)}{2(1)(1/3)}\right)\right]$$

$$\bar{\tau} = 982 \text{ psf}(6.82 \text{ psi})$$

A check may be obtained from vertical force equilibrium. Thus,

$$W = F(\text{bottom}) + F(\text{side}) = \sigma_v(z = H_p)A(\text{bottom}) + 2\bar{\tau}A(\text{side})$$

$$W = (100)(135)(28)(1)$$

$$F(\text{bottom}) = (4,031)(28)(1)$$

$$2\bar{\tau}A(\text{side}) = 2\bar{\tau}(135)(1) = 2.651(10)^5 \text{ lbf}$$

$\therefore$

$$\bar{\tau} = 982 \text{ psf}$$

which checks.

The average horizontal stress is then

$$\bar{\sigma}_h = (1/\mu)(\bar{\tau}) = 982 \text{ psf } (6.82 \text{ psi })$$

If there were fill on both sides of the pillar, then this stress could be considered a confining pressure. Under a Mohr–Coulomb strength criterion, the effect would be to increase the compressive strength according to

$$C_{\mathrm{p}} = C_{\mathrm{o}} + \frac{C_{\mathrm{o}}}{T_{\mathrm{o}}}p = 15,000 + 15(6.82) = 15,102 \text{ psi}$$

which would result in a slight increase in the pillar safety factor ($FS_p = 6.04$). An even smaller increase is anticipated for fill placed on only one side of the pillar. However, there is an additional effect of shear stress that should be taken into account.

One may use the generalized Mohr's circle development for considering the effect of fill on only one side of the pillar after determining the horizontal shear stress induced by the fill side stresses. An equilibrium sketch showing the average stresses is helpful.

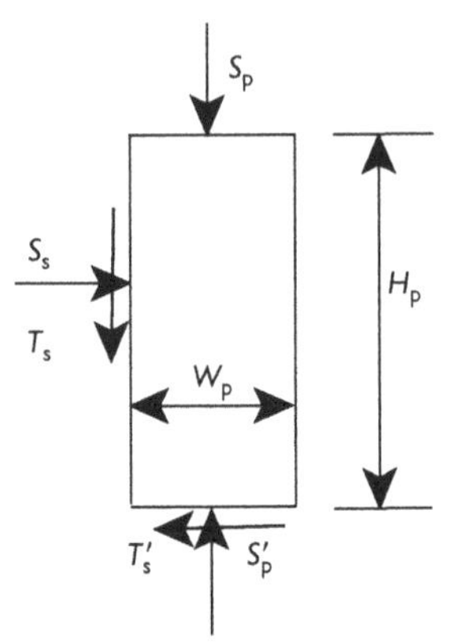

Sketch for stress equilibrium.

The stress system in the sketch allows for satisfaction of force and moment equilibrium. For example, the normal stress at the bottom of the pillar will be slightly greater than at the top because of the vertical shear on the fill side of the pillar. Thus,

$$\Sigma F_{\mathrm{v}} = 0 = S_{\mathrm{p}}A_{\mathrm{p}} + T_{\mathrm{s}}A_{\mathrm{s}} - S'_{\mathrm{p}}A_{\mathrm{p}}$$

$$0 = (144)[(2,500)(7)(1) + (6.82)(135)(1) - S'_{\mathrm{p}}(7)(1)]$$

$$\therefore$$

$$S'_{\mathrm{p}} = 2,632 \text{ psi}$$

which is only sightly greater than the initial 2,500 psi. The effect is to reduce the original safety factor from 6.0 to 5.7 in the vicinity of the pillar bottom without any consideration of other stress effects induced by the fill such as strength increase with confinement and shear loading at the base of the pillar.

The horizontal shear force on the pillar bottom is equilibrated by the horizontal normal fill side wall force. Thus,

$$\Sigma F_{\mathrm{h}} = 0 = S_{\mathrm{s}}A_{\mathrm{s}} - T'_{\mathrm{p}}A_{\mathrm{p}} = (144)[(6.82)(135)(1) - T'_{\mathrm{p}}(7)(1)]$$

$$\therefore$$

$$T'_{\mathrm{p}} = 132 \text{ psi}$$

Moment equilibrium results in shifting of normal force points of application.

The pillar safety factor with fill is given by (6.36) that may be brought into the form

$$\mathrm{FS} = \frac{\{[(2c)\cos(\phi)]/[1-\sin(\phi-\beta)\sin(\eta)/\sin(\beta)]\}}{\sqrt{(S_\mathrm{p}-S_\mathrm{s})^2+(T_\mathrm{p}-T_\mathrm{s})^2}}$$

which reduces to (6.29) in the absence of pillar side wall stresses. Here

$$\tan(\beta) = \frac{(T_\mathrm{p}-T_\mathrm{s})/2}{(S_\mathrm{p}+S_\mathrm{s})/2} = \frac{(132-6.82)}{(2,632+6.82)} = 0.04744, \quad \beta = 2.72^\circ$$

$$\tan(\eta) = \frac{(T_\mathrm{p}-T_\mathrm{s})/2}{(S_\mathrm{p}-S_\mathrm{s})/2} = \frac{(132-6.82)}{(2,632-6.82)} = 0.04768, \quad \eta = 2.73^\circ$$

These angles are very nearly equal with a ratio of sines of 1.004. The numerator has the value given by

$$N = \frac{(2)(1,936)\cos(61.05)}{1-\sin(61.05-2.72)(1.004)} = 12,880 \text{ psi}$$

while the denominator is

$$D = \sqrt{(2,632-6.82)^2+(132-6.82)^2} = 2,628 \text{ psi}$$

Hence, FS = 12, 880/2, 628 = 4.90 and the one-sided fill is seen to decrease the pillar safety factor. The reason for the decrease is in the additional loading.

Introduction of the extraction ratio R leads to a relationship between the pillar safety FS and R, a quadratic in $(1-R)$. Thus,

$$a\left(\frac{1}{1-R}\right)^2 + b\left(\frac{1}{1-R}\right) + c = 0$$

$$a = [(S_n)^2 + (T_n)^2]$$

$$b = [(2)(S_n S_s + T_n T_s)]$$

$$c = -\left[\left(\frac{N}{\mathrm{FS}}\right)^2 - (S_s)^2 - (T_s)^2\right]$$

which may be used when both the pillar safety factor and extraction ratio are not specified.

Example 6.12 Consider the data and results in Example 6.11, then estimate the pillar factor of safety when fill is placed on both sides of the pillar.

***Solution*:** In this case the horizontal shear at the pillar bottom vanishes because of the symmetry of loading. At the same time, the vertical compression in the pillar increases twice the previous amount because of doubling of the vertical shear on the pillar walls caused by settlement of the fill. If one considers the pillar base where compression is greatest, but also where confinement is greatest, then pillar strength at the pillar wall is estimated from

$$C_\mathrm{p} = C_\mathrm{o} + \frac{C_\mathrm{o}}{T_\mathrm{o}} S_s = 15,000 + (15)(6.82) = 15,102 \text{ psi}$$

while one-half of the pillar stress is 2,500 + 2(135) = 2,770 psi. The safety factor at the bottom of the pillar is then 5.46; at the top, the safety factor is 6.04, and on average the pillar safety factor is 5.75, which may be compared with the one-sided fill case safety factor of 4.90 and the no-fill case of 6.0. Evidently in these examples the additional loading of the pillar by the fill exceeds the benefit of fill confinement.

If the pillar were to yield and crush, a large lateral expansion would impend with a consequent large increase in confinement caused by horizontal compression of the fill. The fill reaction would then mitigate against catastrophic failure and allow for a controlled yielding even as the pillar safety factor was reduced to one with the onset of failure. Such action in the post-elastic range would be a significant benefit of the fill.

6.3 Pillars with joints

Geological discontinuities such as faults, bedding plane contacts, fractures – "joints" for brevity – that transect pillars may fail even though the pillar proper does not. Joint failure mechanisms as well as strength failure of a pillar therefore need to be examined for pillar design. An appropriate safety factor for joints is

$$FS_j = \frac{\tau_j(\text{strength})}{|\tau_j(\text{stress})|} \tag{6.38}$$

where shear stress and shear strength relate to the joint. A Mohr–Coulomb criterion for joint strength is reasonable, so shear strength is given by

$$\tau_j = \sigma_j \tan\left(\phi_j\right) + c_j$$

where the subscript j refers to the joint. Joint properties are considered known, but stress analysis is necessary to determine the normal stress acting across the joint and the shear stress acting along the joint.

Flat seam pillars with joints

A simple force equilibrium analysis suffices for the determination of joint stresses that, in fact, are average stresses. With reference to Figure 6.11, equilibrium in the *flat seam* case requires

$$\begin{aligned} \sigma_j A_j &= S_p A_p \cos(\alpha) \\ \tau_j A_j &= -S_p A_p \sin(\alpha) \end{aligned} \tag{6.39}$$

where the stresses are indeed averages over the respective areas acted upon. In view of the relationships $A_p = A_j \cos(\alpha)$,

$$\begin{aligned} \sigma_j &= S_p \cos^2(\alpha) = S_p\left(\frac{1+\cos(2\alpha)}{2}\right) \\ \tau_j &= -S_p \sin(\alpha)\cos(\alpha) = -\left(\frac{S_p}{2}\right)\sin(2\alpha) \end{aligned} \tag{6.40}$$

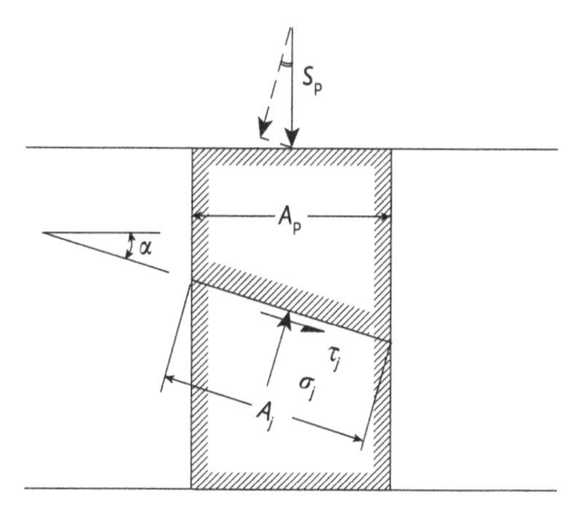

Figure 6.11 Pillar in a flat seam with a joint.

where the negative sign on the shear stress relates to the directions shown in Figure 6.11. The absolute value is used in the safety factor calculation.

A variation on the question of pillar safety when a joint is present is a question concerning dangerous joint dips. Is there a range of joint dips that are safe? If slip is impending, then

$$\tau_j(\text{stress}) = \left(\frac{S_p}{2}\right)\sin(2\alpha) > \tau_j(\text{strength}) = \left(\frac{S_p}{2}\right)[1+\cos(2\alpha)]\tan\left(\phi_j\right) + c_j \tag{6.41}$$

where absolute shear stress value is used. After rearrangement, this criterion is

$$\left(\frac{S_p}{2}\right)\sin\left(2\alpha - \phi_j\right) > \left(\frac{S_p}{2}\right)\sin\left(\phi_j\right) + \left(c_j\right)\cos\left(\phi_j\right) \tag{6.42}$$

A graphical interpretation of this criterion is shown in Figure 6.12 that contains Mohr–Coulomb failure criteria for pillar and joint and the Mohr circle that represents the stress state in the pillar. Figure 6.12 shows that in the range (α_A, α_B) joint slip is possible. This range increases with pillar stress and is maximum when the pillar stress equals pillar unconfined compressive strength, as shown in Figure 6.12 where the Mohr circle just touches the pillar strength line. Formal solution requires finding the inverse sine of the function containing α in the slip condition. There are actually four solutions because there is symmetry to the problem. This symmetry is graphically represented in the lower half of Mohr's circle where shear stresses and (strengths) are negative. Physically, there is symmetry of dangerous and safe dips about the vertical load axis. Near vertical and near horizontal joints will be safe as one would intuitively suppose.

Example 6.13 A room and pillar excavation plan is executed in a flat-lying, thick, strong limestone formation. Depth is 500 ft, entries and crosscuts are 28 ft wide, pillars are 135 ft high, unconfined compressive strength is 15,000 psi, and the ratio of unconfined compressive to tensile strength is 15. No size effect is present. However, a joint transects the pillar at a dip of 55°. Joint friction angle is 32°. Determine the pillar safety factor without a joint when the

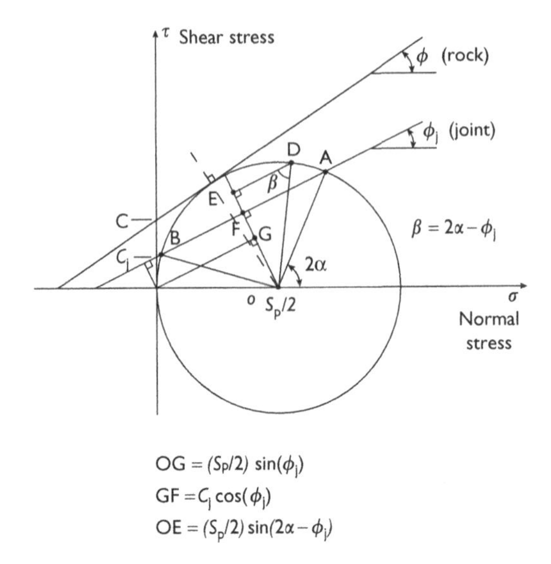

Figure 6.12 Mohr circle geometry for Mohr–Coulomb joint and pillar failure.

fextraction ratio is 0.80. Also determine the joint cohesion required for a pillar safety factor with respect to joint slip of 1.5 at the same extraction ratio.

Solution: By definition,

$$\begin{aligned}
\mathrm{FS} &= \frac{C_o}{S_p} \\
&= \frac{C_o(1-R)}{S_v} \\
&= \frac{(15{,}000)(1-0.8)}{500}
\end{aligned}$$

$$\mathrm{FS} = 6.0$$

which agrees with Example 6.11. The pillar safety factor without a joint is 6.0.

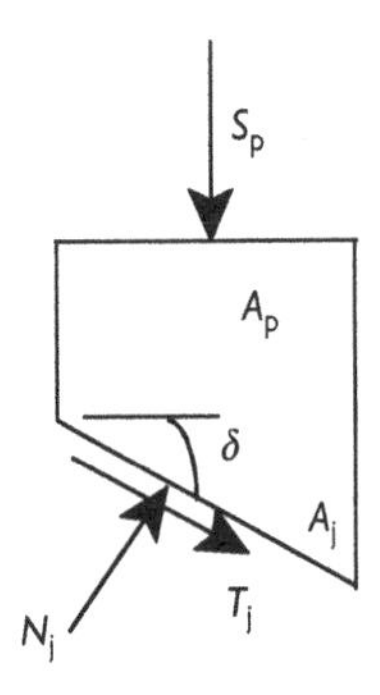

The pillar stress may be obtained from the extraction ratio formula, so after assuming a Mohr–Coulomb joint strength criterion expressed in terms of joint forces, one has

$$
\begin{aligned}
\mathrm{FS_j} &= \frac{N_j \tan(\phi_j) + c_j A_j}{T_j} \\
&= \frac{S_v A_p \cos(\delta)\tan(\phi_j)/(1-R) + c_j A_j}{S_v A_p \sin(\delta)/(1-R)} \\
&= \frac{\tan(\phi_j)}{\tan(\delta)} + \frac{c_j A_j (1-R)/A_p}{S_v \sin(\delta)} \\
1.5 &= \frac{\tan(32)}{\tan(55)} + \frac{(0.2)[1/\cos(55)](c_j)}{500\sin(55)} \\
\therefore & \\
c_j &= 1,248 \text{ psi}
\end{aligned}
$$

Here, an estimate of the preexcavation vertical stress of 1 psi/ft is used. This analysis shows the separate contributions of joint friction and cohesion to the joint safety factor and how the extraction ratio enters the safety factor calculation. Without cohesion joint slip impends whenever joint dip exceeds the friction angle. In this case, the required joint cohesion is substantial.

Example 6.14 A room and pillar excavation plan is executed in a flat-lying, thick, strong limestone formation. Depth is 500 ft, entries and crosscuts are 28 ft wide, pillars are 135 ft high, unconfined compressive strength is 15,000 psi, and the ratio of unconfined compressive to tensile strength is 15. No size effect is present. Joint friction angle is 32° and joint cohesion is 97.5 psi. Determine the range of joint dips that may be possible when the extraction ratio is 25%.

***Solution*:** The joint safety factor formula in terms of stresses is

$$
\begin{aligned}
\mathrm{FS_j} &= \frac{\sigma_j \tan(\phi_j) + c_j}{\tau_j} \\
&= \frac{(S_p/2)[1+\cos(2\delta)]\tan(\phi_j) + c_j}{(S_p/2)\sin(2\delta)} \\
\mathrm{FS_j} &= \frac{(S_v/2)[1+\cos(2\delta)]\tan(\phi_j) + c_j(1-R)}{(S_v/2)\sin(2\delta)} < 1
\end{aligned}
$$

where the inequality defines a range of joint dips that slip. After rearrangement of the inequality, one has

$$
\begin{aligned}
\sin(2\delta - \phi_j) &> \sin(\phi_j) + \frac{2c_j(1-R)\cos(\phi_j)}{S_v} \\
&> \sin(32) + \frac{2(97.5)(1-0.25)\cos(32)}{500}
\end{aligned}
$$

$$\sin\left(2\delta - \phi_j\right) > 0.7797$$

$\therefore$

$$\left(2\delta - \phi_j\right) > 51.075 \quad \text{and} \quad \left(2\delta - \phi_j\right) < 180 - 51.075$$

$\therefore$

$$41.5^\circ < \delta < 80.5^\circ$$

Joint dips in this range with the specified cohesion would slip and thus threaten pillar stability at the given extraction ratio.

Dipping seam pillars with joints

Pillars in dipping seams that contain joints may also be analyzed through force equilibrium requirements. With reference to Figure 6.13 that shows two cases, one with seam and joint dipping in opposite directions, the other with seam and joint dipping in the same direction, equilibrium in the first case (opposite dips) requires

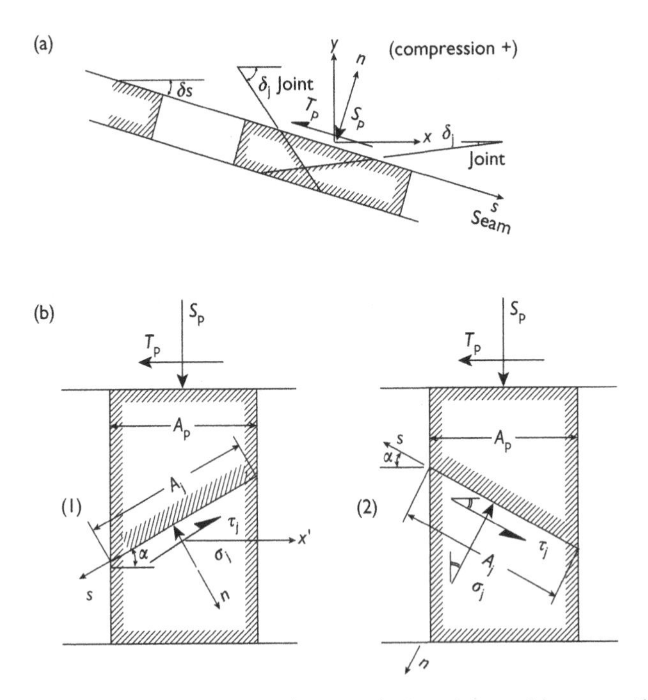

Figure 6.13 Pillars with a joint in dipping seams (seam and joints shown with same strike direction). (a) Pillar in a dipping seam and (b) dipping pillars with a joint.

$$\sigma_j = \left(\frac{S_p}{2}\right)[1 + \cos(2\alpha)] - \left(\frac{T_p}{2}\right)\sin(2\alpha)$$
$$\tau_j = \left(\frac{S_p}{2}\right)\sin(2\alpha) + \left(\frac{T_p}{2}\right)[1 + \cos(2\alpha)] \tag{6.43}$$

and in the second case

$$\sigma_j = \left(\frac{S_p}{2}\right)[1 + \cos(2\alpha)] - \left(\frac{T_p}{2}\right)\sin(2\alpha)$$
$$\tau_j = \left(-\frac{S_p}{2}\right)\sin(2\alpha) + \left(\frac{T_p}{2}\right)[1 + \cos(2\alpha)] \tag{6.44}$$

These expressions are *not* the equations of transformation of stress under a rotation of reference axis and do not have a Mohr's circle representation. They are formulas for average stresses. Alternatives are certainly possible, depending on the equilibrium diagram. In any case, a consistent sign convention is essential. Once the normal and shear stresses acting on a joint are obtained, a safety factor calculation is possible that is, again, the ratio of shear strength to shear stress. Assumption of Mohr–Coulomb failure criteria is reasonable, but other failure criteria may be used. The analysis of stress (averages) is the same.

A local analysis of joint slip in terms of stress and strength *at a point* is certainly feasible. In this case, the usual Mohr's circle representation is available. Figure 6.14 shows a Mohr's circle representing a state of stress characterized by principal stresses (σ_1, σ_3) with compression positive. Also shown are Mohr–Coulomb strength criteria for joint and host rock which is considered stronger than the joint. Joint failure impends if

$$\tau_m \sin\left(2\alpha - \phi_j\right) > \sigma_m \sin\left(\phi_j\right) + \left(c_j\right)\cos\left(\phi_j\right) \tag{6.45}$$

where the term on the left is joint shear stress and the term on the right is joint shear strength. The range of joint dips related to the angle α for which slip impends under the given stress

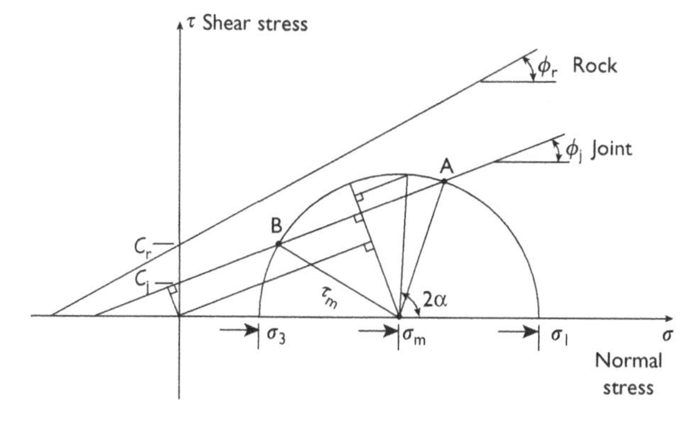

Figure 6.14 Joint slip analysis at a point.

state is (α_A, α_B), as shown in Figure 6.14. If α is outside this range slip does not impend. Again there is symmetry about the direction of the major principal stress α_1 and a second pair of angles that also bound a range of dips for which slip impends.

Example 6.15 A room and pillar operation is planned at a depth of 964 ft in a seam dipping 13°. An extraction percentage of 50% is anticipated. The ratio of preexcavation horizontal to vertical stress is estimated to be one-fourth. A joint set is present that dips 60° in the same direction as the seam. Rock and joint properties are given in the table. Determine if there is a threat to pillar stability.

Property material	*Cohesion (psi)*	*Friction angle (°)*
Rock	2,300	35.0
Joint	23.0	25.0

Solution: A threat to pillar stability exists if either the pillar safety factor or joint safety factor is less than one. Pillar strength and joint properties are given, while pillar stress may be determined from extraction ratio formulas. Joint stresses may be determined from equilibrium requirements. With reference to the sketch, the preexcavation normal and shear stresses are

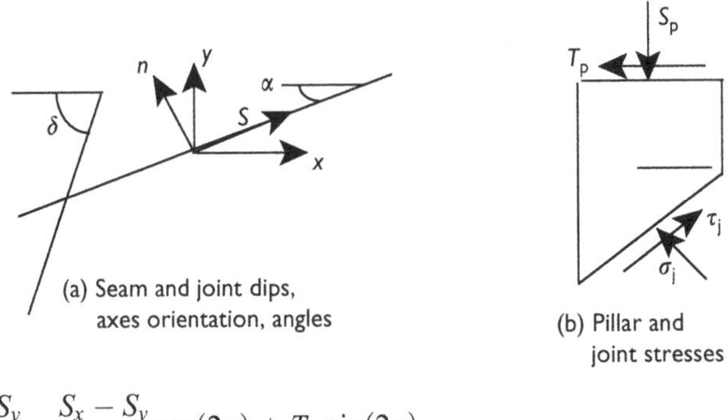

(a) Seam and joint dips, axes orientation, angles

(b) Pillar and joint stresses

$$S_n = \frac{S_x + S_y}{2} - \frac{S_x - S_y}{2}\cos(2\alpha) + T_{xy}\sin(2\alpha)$$

$$= \frac{S_v(K_o + 1)}{2} - \frac{S_v(K_o - 1)}{2}\cos(26) + (0)\sin(26)$$

$$= \frac{964(1/4 + 1)}{2} - \frac{964(1/4 - 1)}{2}\cos(26) + (0)\sin(26)$$

$$S_n = 927 \text{ psi}$$

$$T_s = -\left(\frac{S_x - S_y}{2}\right)\sin(2\alpha) + T_{xy}\cos(2\alpha)$$

$$= -(964)\left(\frac{1/4 - 1}{2}\right)\sin(26)$$

$$T_s = 158 \text{ psi}$$

The pillar stresses are

$$S_{\mathrm{p}} = \frac{S_n}{1-R} = \frac{927}{1-1/2} = 1,854 \text{ psi}$$

$$T_{\mathrm{p}} = \frac{T_s}{1-R} = \frac{158}{1-1/2} = 316 \text{ psi}$$

The pillar factor of safety is given by

$$\mathrm{FS_p} = \frac{[(2c)\cos(\phi)]/[1-\sin(\phi-\beta)]}{\sqrt{(S_{\mathrm{p}})^2 + (T_{\mathrm{p}})^2}}$$

$$= \frac{[(2)(2,300)\cos(35)]/[1-\sin(35-9.7)]}{\sqrt{(1,854)^2 + (316)^2}}$$

$$\mathrm{FS_p} = 3.50$$

Thus, the pillar safety factor indicates no threat of failure.

The joint stresses are

$$\sigma_{\mathrm{j}} = \left(\frac{S_{\mathrm{p}}}{2}\right)\{1+\cos[2(\delta-\alpha)]\} - \left(\frac{T_{\mathrm{p}}}{2}\right)\sin[2(\delta-\alpha)]$$

$$= \left(\frac{1,854}{2}\right)[1+\cos(94)] - \left(\frac{316}{2}\right)\sin(94)$$

$$\sigma_{\mathrm{j}} = 705 \text{ psi}$$

$$\tau_{\mathrm{j}} = \left(\frac{S_{\mathrm{p}}}{2}\right)\sin[2(\delta-\alpha)] + \left(\frac{T_{\mathrm{p}}}{2}\right)\{1+\cos[2(\delta-\alpha)]\}$$

$$= \left(\frac{1,854}{2}\right)\sin(94) + \left(\frac{316}{2}\right)[1+\cos(94)]$$

$$\tau_{\mathrm{j}} = 1,072 \text{ psi}$$

and so the joint safety factor is

$$\mathrm{FS_j} = \frac{\sigma_{\mathrm{j}}\tan\left(\phi_{\mathrm{j}}\right) + c_{\mathrm{j}}}{\tau_{\mathrm{j}}}$$

$$= \frac{(705)\tan(25) + 23.0}{1,072}$$

$$\mathrm{FS_j} = 0.328$$

Thus, joint failure poses a threat to pillar stability in this excavation plan.

Example 6.16 Given the data and results from Example 6.15, determine the range of joint dips that would slip were the excavation plan realized.

Solution: The range of dips is governed by the range of joint safety factors that exceed one. with reference to the sketch that illustrates various angles, joint dips may be extracted from the angles α_A and α_B. In this example $\alpha = \delta_j - \delta_s$ that is the difference between joint and seam dip.

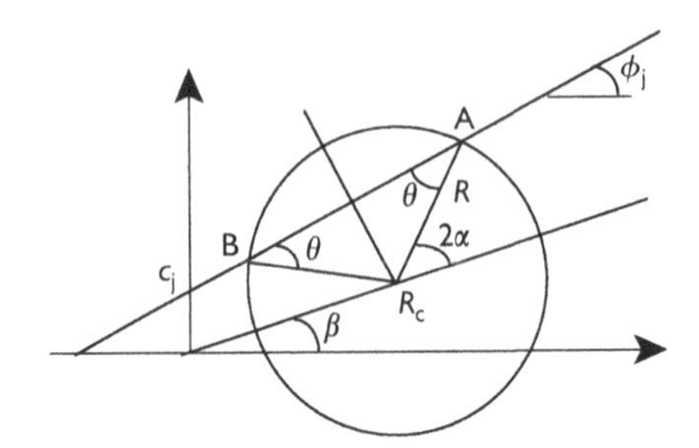

Points on the stress circle above the joint failure line are in the range of joint dips that slip. Thus, if

$$R\sin(\theta) > c\cos\left(\phi_j\right) + R_c\sin\left(\phi_j - \beta\right)$$

slip is possible. Here R and R_c are the stress circle radius and the slope distance to the circle center measured along the line that has slope $\tan(\beta)$. The center of the circle is located at $(S_p/2, T_p/2)$ and from Equations (6.43) and (6.44),

$$R = R_c = \sqrt{(S_p/2)^2 + (T_p/2)^2}$$

$$= \sqrt{(1,854/2)^2 + (316/2)^2}$$

$$R = R_c = 940 \text{ psi}$$

After solving for $\sin(\theta)$

$$\sin(\theta) = \sin\left(\phi_j - \beta\right) + \frac{c_j\cos\left(\phi_j\right)}{R}$$

$$= \sin(25 - 9.7) + \frac{(23)\cos(25)}{940}$$

$$\sin(\theta) = 0.2861$$

$$\therefore$$

$$\theta = 16.6^\circ$$

From the geometry of the stress circle $\theta = 2\alpha_A - (\phi_j - \beta)$, so $2\alpha = 16.6 + (25-9.7) = 31.9^\circ$. But $\alpha_A = \delta_j - \delta_s$, so $\delta_j = \alpha_A + \delta_s = 31.9/2 + 13.0 = 29.0^\circ$. Joints with a lesser dip will not slip. A similar analysis leads to $\alpha_B = 57.4^\circ$ and a $\delta_j = 70.4^\circ$. Hence the range of joint dips that will slip is $29.0^\circ < \delta_j < 70.4^\circ$.

6.4 Pillars in several seams

Multiseam mining involves simultaneous or sequential excavation in two or more seams. Large sedimentary basins typically contain a repeating stratigraphic sequence (e.g. sandstone, coal, shale sandstone, coal, etc., or limestone, shale, coal, limestone, etc.). Consequently, these basins may contain a number of horizons suitable for mining. Distance between horizons may range from hundreds of feet to nil when two seams merge in the stratigraphic column. Although interaction between seams may be insignificant at first when extraction is small, that is, when dimensions of the area mined are less than the distance between seams, as mining progresses some interaction is likely. In the Eastern US soft (bituminous) coal region, seam depths range above and below 900 ft, which is comparable with topographic relief in the region. Lateral dimensions of extraction range to thousands of feet, so seam interaction and surface subsidence are inevitable. In the Western US, depths and topographic relief are greater, but with extraction dimensions in excess of 10,000 ft, subsidence and seam interactions are again inevitable. Multilevel mining of bedded salt, potash, and trona also pose questions of how seam interaction may affect safe roof span calculations and pillar safety factors.

Conventional wisdom provides qualitative guidance for entries and pillars in multiseam mining and simply states that one should: "*columnize the mains and stagger the chains.*" Parametric study of a large number of combinations of entry widths, pillar widths, seam thicknesses, mining dimensions, and importantly, seam separation, coupled with simple stratigraphy supports this guidance (Pariseau, 1983, contains a short bibliography on multiseam mining).

Columnized main entry pillars

Main entries provide the principal avenues of ingress and egress from soft rock mines and often serve for the life of a mine. Stability of main entries and pillars between is thus essential to mine safety. Examination of conventional wisdom is therefore in order. Figure 6.15 shows main entries in a vertical section through two seams. Entry width Wo is typically 20 ft. Pillar width W_p is larger and may be as much as five times entry width. Mining height H_p (also entry or opening height H_o) may be more or less than 10 ft. The pillars are generally short because $W_p > H_p$. In Figure 6.15(a), entries and pillars are shown during development when extraction is typically less than 50%. In Figure 6.15(b), a mining "panel" is shown between pillars. The panel may be formed by removing pillars (room and pillar mining) or by the longwall mining method.

Figure 6.16(a) shows a three-seam mining configuration; Figures 6.16(b) and (c) show entries columnized and staggered, respectively. In the columnized case, pillars between entries have the same dimensions and experience about the same average stress, $S_p = S_v (W_o + W_p)/W_p$, when seam separation is small compared with seam depth. Thus, there is no great difference in pillar safety factors compared with single seam tributary pillar analysis. However, the entries in Figure 6.16(b) shadow one from the other between seams and thus experience a reduction in stress concentration. From another viewpoint, these entries have a row axis (vertical) that is favorable because it is parallel to the direction of major premining compression (vertical). With respect to stress concentration, "columnizing the mains" is advantageous, although, if bed separation occurs in the immediate roof, then there is a lessened advantage.

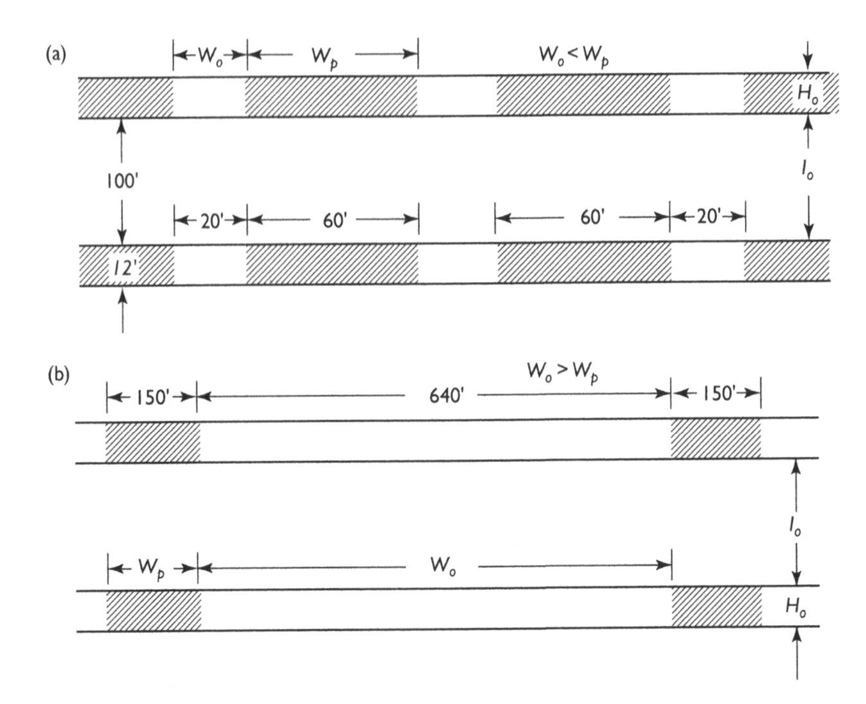

Figure 6.15 Multi-seam room–pillar geometry: (a) rooms and pillars on development; (b) mined section and pillars.

In the staggered case, the lower entries are offset beneath pillars between the entries above. These lower entries may undermine the pillars between the entries above. In the extreme, when seam separation is zero, pillar width W_p would be reduced by entry width W_o to $W_p' = W_p - W_o$. The average pillar stress S_p would increase in this two-dimensional view to $S_p' = S_v(W_o + W_p)/W_p'$ from $S_p = S_v(W_o + w_p)/W_p$. The ratio $S_p'/S_p = W_p/W_p' = /1(1 - W_o/W_p)$. If W_o were a 20 ft and W_p were 60 ft, then pillar stress would increase by a substantial 50%. This is an extreme example of possible interaction.

Details of stress distribution about entry roofs, floors, and walls, which are also pillar walls, are shown in Figure 6.17 for an entry in the top seam in a two-seam mining configuration with columnized entries. The distribution of stress about the bottom seam entries is similar. Examination of hundreds of such results shows that (1) peak tension always occurs at the center of roof and floor; (2) peak compression in roof and floor is always less than the peak compression in the pillar, and (3) tension is absent in the pillar rib when the premining stress is from gravity alone and the long axis of the opening is normal to the gravity axis. These results are for development mining when extraction is less than 50% and entry width is greater than entry height. These are the usual conditions in soft rock mines.

Two-seam analysis results are summarized in highly condensed form in Figures 6.18(a) and (b) for top and bottom entries when columnized, and in Figures 6.19(a) and (b) when entries are staggered. These figures show stress concentrations in roof, pillar, and floor (KR, KP, and KF) as functions of seam separation with extraction as a parameter. In all cases, there

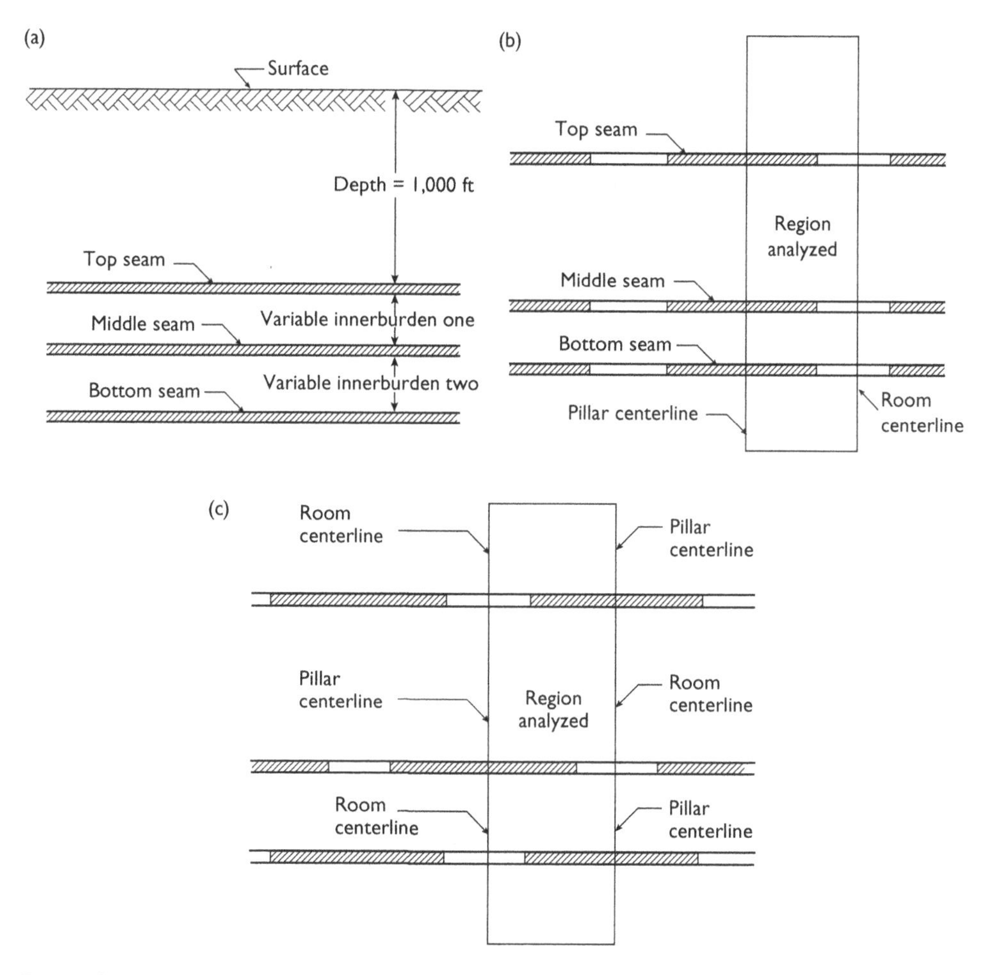

Figure 6.16 (a) Three-seam geometry, (b) three-seam mesh region – columnized case, and (c) three-seam mesh region – staggered case.

is little interaction until seam separation (innerburden) is reduced to less than 40 ft, which is two times entry width (20 ft). In the columnized case (Figure 6.18), the interaction is favorable with reduction in top seam entry floor stress concentration and bottom seam roof stress concentration (shadow effect). These results are almost symmetric and would be except for a small increase of gravity loading from top to bottom seams. By contrast, in the staggered case (Figure 6.19), the same stress concentrations remain almost constant as seam separation decreases below 40 ft or two times entry width. However, pillar stress concentration, top seam floor stress concentration and bottom seam roof stress concentration increase considerably as innerburden is decreased below 40 ft.

When seams are widely separated, interaction is negligible. Widely separated entry width may be taken to be Wo after invoking the one-diameter rule of thumb for reducing interaction

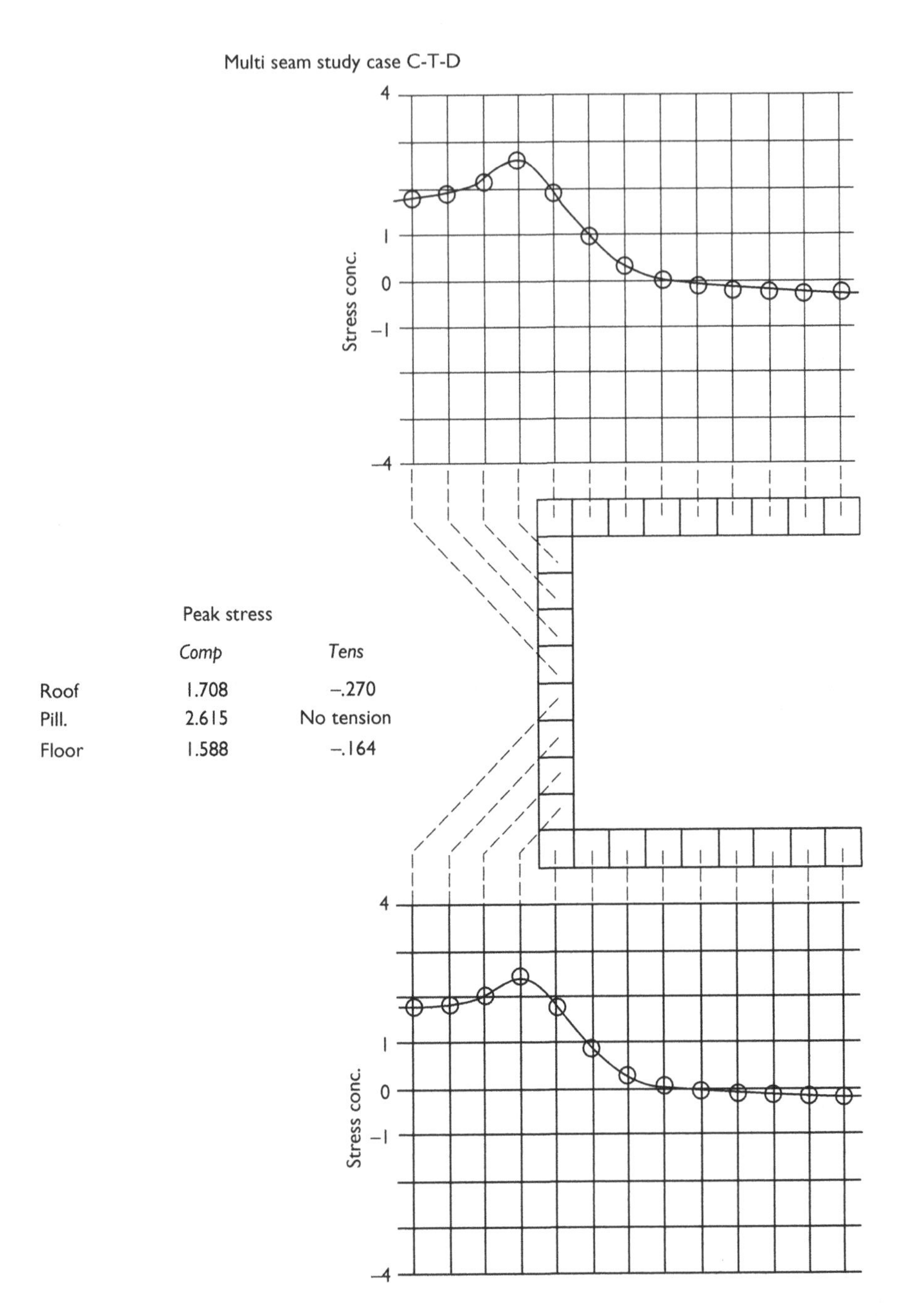

Figure 6.17 Detail plot example. Top seam, two-seam case.

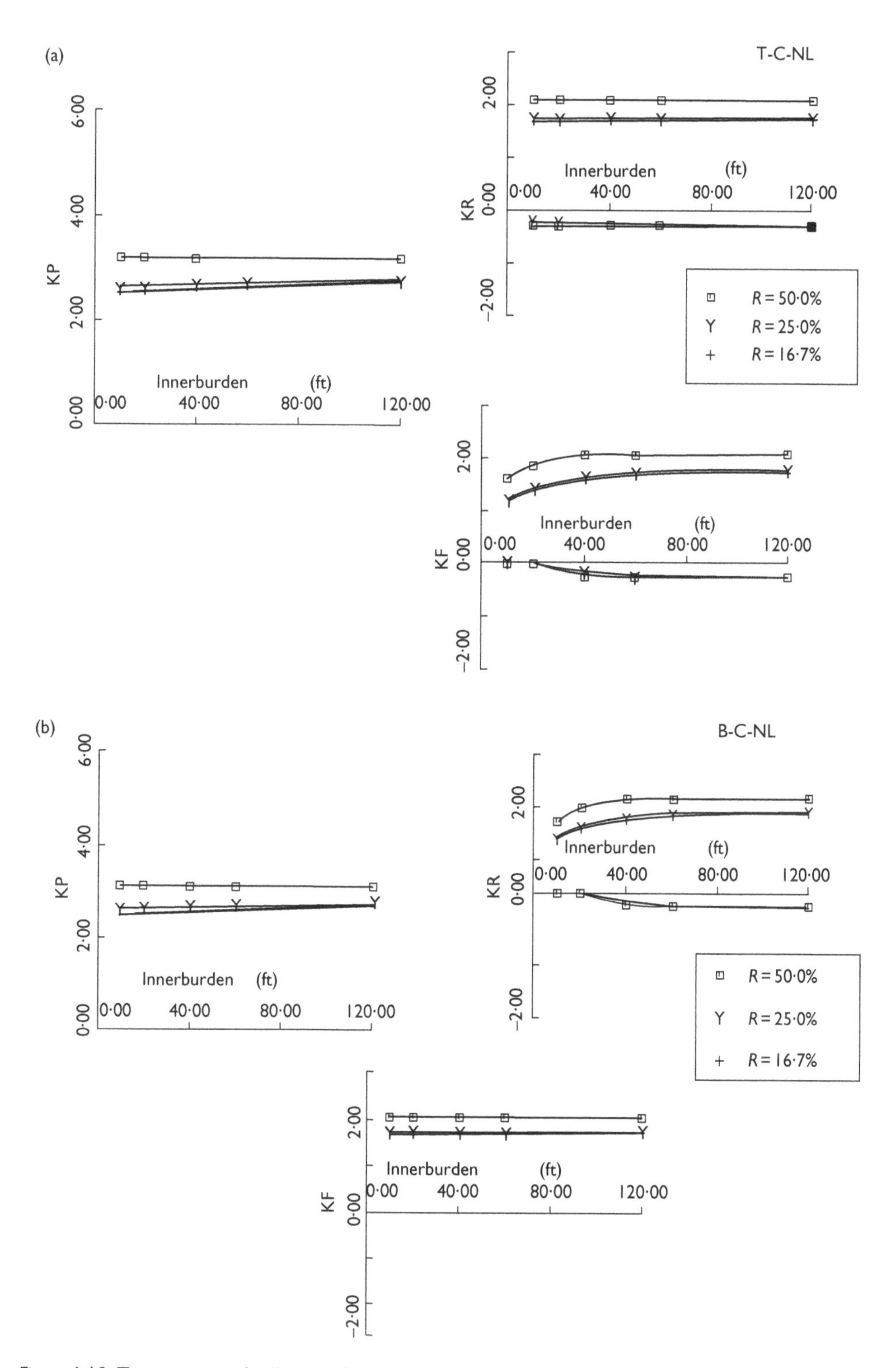

Figure 6.18 Two-seam roof, pillar, and floor stress concentration factors: (a) top seam, columnized and (b) bottom seam, columnized.

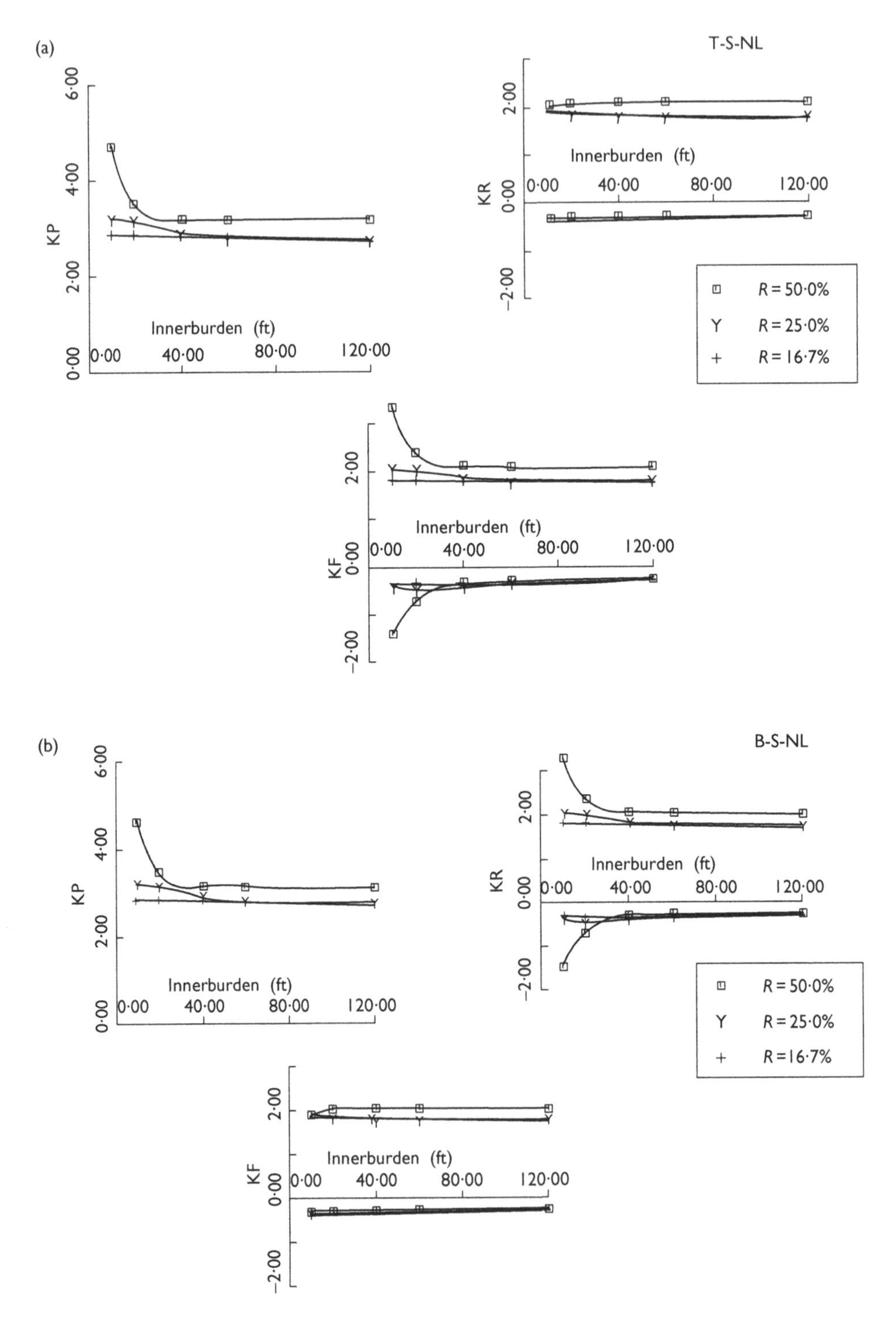

Figure 6.19 Two-seam roof, pillar, and floor stress concentration factors: (a) top seam, staggered and (b) bottom seam, staggered.

between adjacent entries to an acceptable level, that is, to the point when stress concentrations are no more than 10% of those experienced by a single, isolated entry. Because of the importance of main entry stability, more than one-diameter or entry width may be invoked. Indeed, one might suppose that pillar width Wp should be used, not only to isolate entries, but to reduce the effect of pillar undermining by staggered entries. In the example, seams separated by more than 60 ft experience negligible interaction even when entries are staggered. Of course, the better practice would be to take care to "columnize the mains." In both cases, the data indicate that interaction is negligible when seam separation is greater than two times entry width. In view of the limited amount of data from just one parametric study, a more conservative approach may be advisable and that would allow for uncertainties in stratigraphy. As a rule of thumb to decide whether seam interactions are negligible is to use the greater of two times entry width or pillar width in comparison with actual seam separation.

Staggered chain entry pillars

Chain entries and pillars may be considered temporary because of a relatively short life that ranges from months to perhaps a year or so. They provide access to the area where mining occurs, that is, to "panels" that eventually become large tabular excavations with complete removal of the coal, potash, trona, or other commodity of interest. Chain entries and pillars are also known as panel entries and pillars. These tabular excavations are of the order of hundreds of feet wide and thousands of feet long. Caving of overlying strata into panels naturally occurs as mining progresses. Figure 6.20 shows a vertical section through mining panels, chain entries, and chain pillars in two seams. In Figure 6.20, the chain pillars are staggered by placement of the lower chain pillars beneath the center of the panel above. Conventional wisdom states that the staggered chain pillar configuration is preferable because the overlying panel shadows the chain pillars and entries below and thus reduces chain pillar stress and stress concentration about the entries. The argument is plausible at first glance, but in consideration of caving action that allows the overburden weight to bear on the strata below after mining, any advantage to staggering the chains would be transitory. Published parametric study of chain pillar location and interactions in multiseam mining operations appears to be lacking. In any event, mining the top seam first is preferable. Otherwise, mining is likely to be through caved ground.

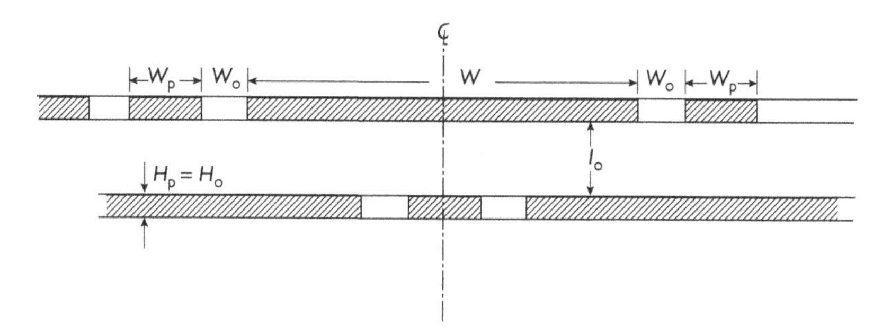

Figure 6.20 Two-seam layout with a double entry system and staggered chain pillars.

Entries adjacent to a previous mined panel in the same seam are lost and chain pillars are crushed as mining progresses past each in turn ("tailgate" entries and pillars). Temporary additional support in tailgate entries is often required in the form of cribs of various types. Entries and pillars adjacent to the next panel scheduled for mining ("headgate" entries and pillars) experience large increases in stress with approach of mining and may also require additional support beyond that installed during development. Panels usually end near main or sub-main entries. Protection of these semipermanent entries requires consideration of a new type class of pillars – barrier pillars.

Example 6.17 A two-level room and pillar operation is conducted in hardrock at a depth of 550 m. Unconfined compressive strength of the rock mass is 85.0 MPa. The area extraction ratio is 0.64. Geometry of excavation is shown in the sketch. Height and width of openings are the same on each level as are crosscut widths, pillar widths, and pillar lengths. A plan is proposed to remove pillars and sill between levels as shown in the sketch. Determine if the plan is feasible with respect to pillar safety. Ignore any size effect on pillar strength. Crosscut width is equal to opening width and pillar length is three times pillar width.

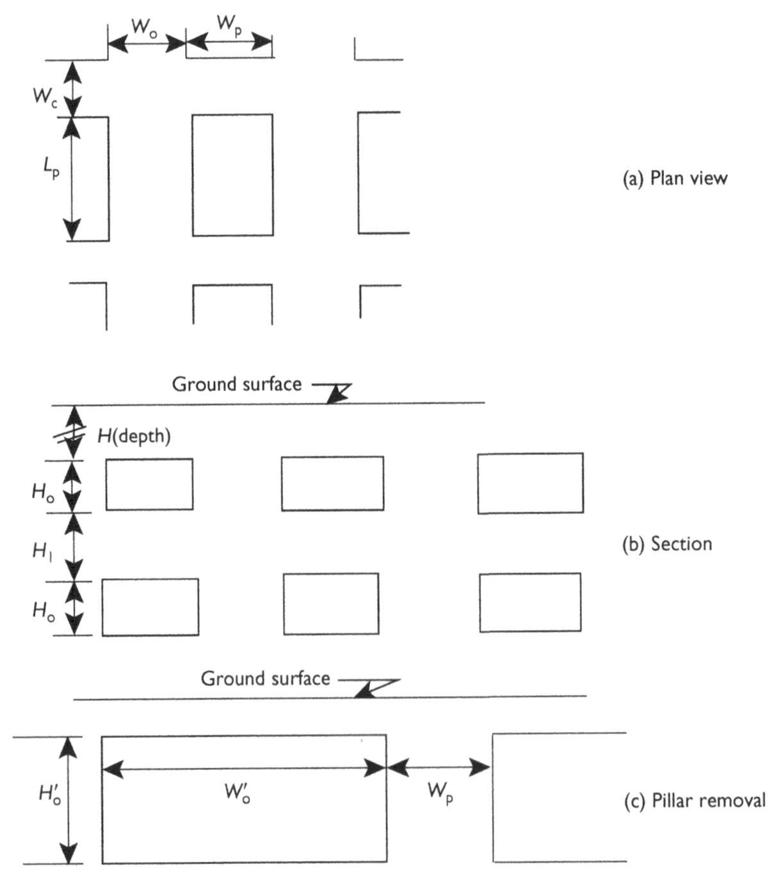

Sketch of excavation and pillar removal geometry

Solution: There is no size effect, so the pillar safety factor before pillar extraction is simply

$$FS_p = \frac{C_o}{S_p} = \frac{C_o(1-R)}{S_v}$$

$$= \frac{(85)(10)^6(1-0.64)}{(25)(10^3)(550)}$$

$$FS_p = 2.23$$

where an estimate of the preexcavation vertical stress is 25 kPa per meter of depth. The safety factor after pillar removal is given by the same formula, so that

$$FS_p{}' = (1-R')\left(\frac{FS_p}{1-R}\right) = (1-R')(6.194)$$

But also

$$1-R = \frac{W_p L_p}{(W_o + W_p)(W_c + L_p)}$$

$$= \frac{3W_p W_p}{(W_o + W_p)(W_c + 3W_p)}$$

$$0.36 = \frac{3}{(W_o/W_p + 1)(W_o/W_p + 3)}$$

$$\therefore$$

$$\frac{W_o}{W_p} = 1.055$$

This ratio may be used in computing the new extraction ratio. Thus,

$$1-R' = \frac{3W_p W_p}{(2W_o + W_p)(W_c + 3W_p)}$$

$$= \frac{3}{(2W_o/W_p + 1)(W_o/W_p + 3)}$$

$$= \frac{3}{[(2)(1.055)+1](1.055+3)}$$

$$1-R' = 0.2379$$

The safety factor of pillars remaining after removal of other pillars is then

$$FS_p{}' = (0.2379)(6.182) = 1.47$$

which is a substantial reduction in pillar safety factor but indicates the remaining pillars would be stable. The result depends only on the ratio of dimensions that characterize change and is therefore independent of the actual excavation dimensions.

Example 6.18 Consider the data and results in Example 6.17 but suppose the pillars are square. Derive a formula for the number of pillars that can be extracted without threatening the remaining pillars.

***Solution*:** The remaining pillars must have a safety factor no less than one. Hence,

$$\mathrm{FS}' = (1 - R')\left(\frac{\mathrm{FS}}{1 - R}\right) > 1$$

$$\therefore$$

$$(1 - R') > (1 - R)/\mathrm{FS} = (0.36)/(2.23) = 0.1614$$

But also

$$1 - R' = \frac{(W_\mathrm{p})^2}{[nW_\mathrm{o} + (n-1)W_\mathrm{p}](W_\mathrm{o} + W_\mathrm{p})}$$

that may be solved for n provided the ratio of opening to pillar width is known. Here n is the number of openings that are incorporated into the new opening that results from sill and pillar removal. There are $n - 1$ original pillars removed. The required ratio may be computed as in Example 6.17 with the result $W_\mathrm{o}/W_\mathrm{p} = r = 2/3$. Thus,

$$n = \left[\left(\frac{1}{1 - R'}\right)\left(\frac{1}{1 + r}\right) + 1\right]\left(\frac{1}{1 + r}\right)$$

$$= \left[\left(\frac{1}{0.1614}\right)\left(\frac{1}{1 + 2/3}\right) + 1\right]\left(\frac{1}{1 + 2/3}\right)$$

$$n = 2.83$$

and the final value of the integer n is 2. Two original openings and one pillar may be removed safely.

6.5 Barrier pillars

An example identification of barrier pillars is shown in Figure 6.21 for a two-seam mining configuration. The barrier pillars are columnized in Figure 6.21. Also shown in Figure 6.21 is the local area extraction ratio (ratio of area mined to original area in plan view). Under gravity loading in soft rock mining, compression stress concentration at the panel ends (faces) increases linearly with the ratio of panel width to height. Consequently, an enormous stress concentration could develop on the panel side of a barrier pillar, if yielding at the elastic limit does not relieve this tendency. Yielding on the panel side of a barrier pillar reduces the peak stress concentration, while safety on the main entry side of a barrier pillar prohibits yielding.

Vertical stress distributions over barrier pillars and main entries are shown in Figure 6.22 as functions of barrier pillar size (W_b) and innerburden distance (seam separation) in a twoseam mining configuration with columnized main entries and pillars. Figure 6.22(a) relates to the top seam and Figure 6. 22(b) to the bottom seam. Common to Figures 6.22(a) and (b) is the elevated stress in the innerburden between barrier pillars. Another feature is the rapid decrease in stress concentration into the barrier pillars. However, this decrease is relative because the stress level remains high well into the barrier pillars.

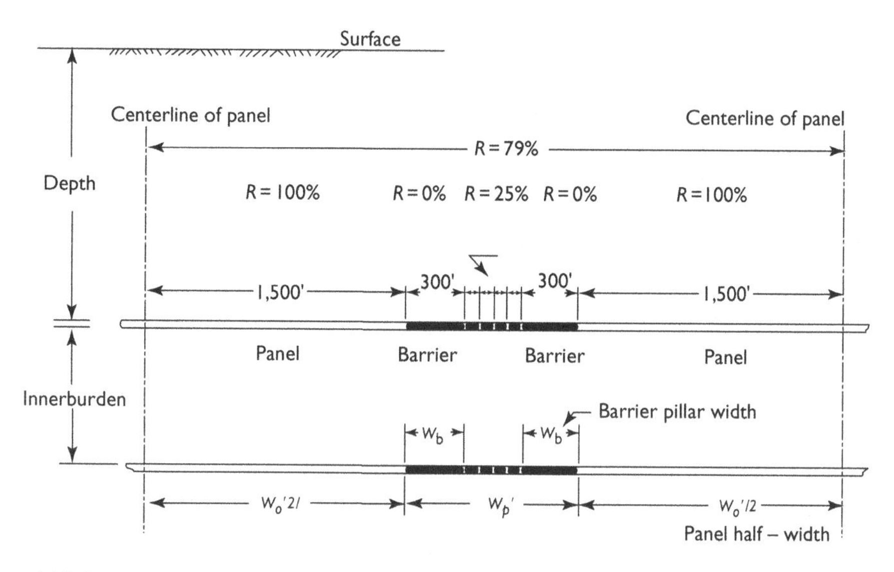

Figure 6.21 Panel extraction notation and example dimensions – two-seam columnized case.

If the region of high stress and yield does not extend beyond the barrier pillars to the main entries and pillars then the main entries should be safe. Should the barrier pillars yield entirely, then the pillar safety factor is one, but if only portions of the barrier pillars yield, then the pillar safety factor is greater than one. Thus, if the barrier pillar safety factor FS_b is given as C_p / S_b, then the barrier pillars are of sufficient size to prevent collapse of the main entries provided $FS_b > 1$. The barrier pillar stress Sb may be estimated by an extraction ratio formula applied to panels and barrier pillars illustrated in Figure 6.21, while strength may be estimated from the popular Mohr–Coulomb criterion.

Before panel excavation but after entry excavation with neglect of crosscuts

$$\begin{aligned} S_p A_p &= S_v A \\ S_p W_p &= S_v (W_p + W_e) \end{aligned} \tag{6.46}$$

where W_p and W_e are total pillar and main entry widths. After panel excavation,

$$S_p W_p/2 + S_b W_b = S_v (W_p/2 + W_e/2 + W_b + W_o/2) \tag{6.47}$$

where $W_o/2$ is panel half-length, Wb is barrier pillar width, and symmetry at the center of the main entries and the panel is assumed. In view of (6.46), the average vertical stress in the barrier pillar is

$$S_b = S_v \left(1 + \frac{W_o}{2W_b}\right) \tag{6.48}$$

Panels are measured in thousands of feet while barrier pillars may be a few hundred feet wide. Thus, the barrier pillar stress may be more than 10 times overburden stress S_v.

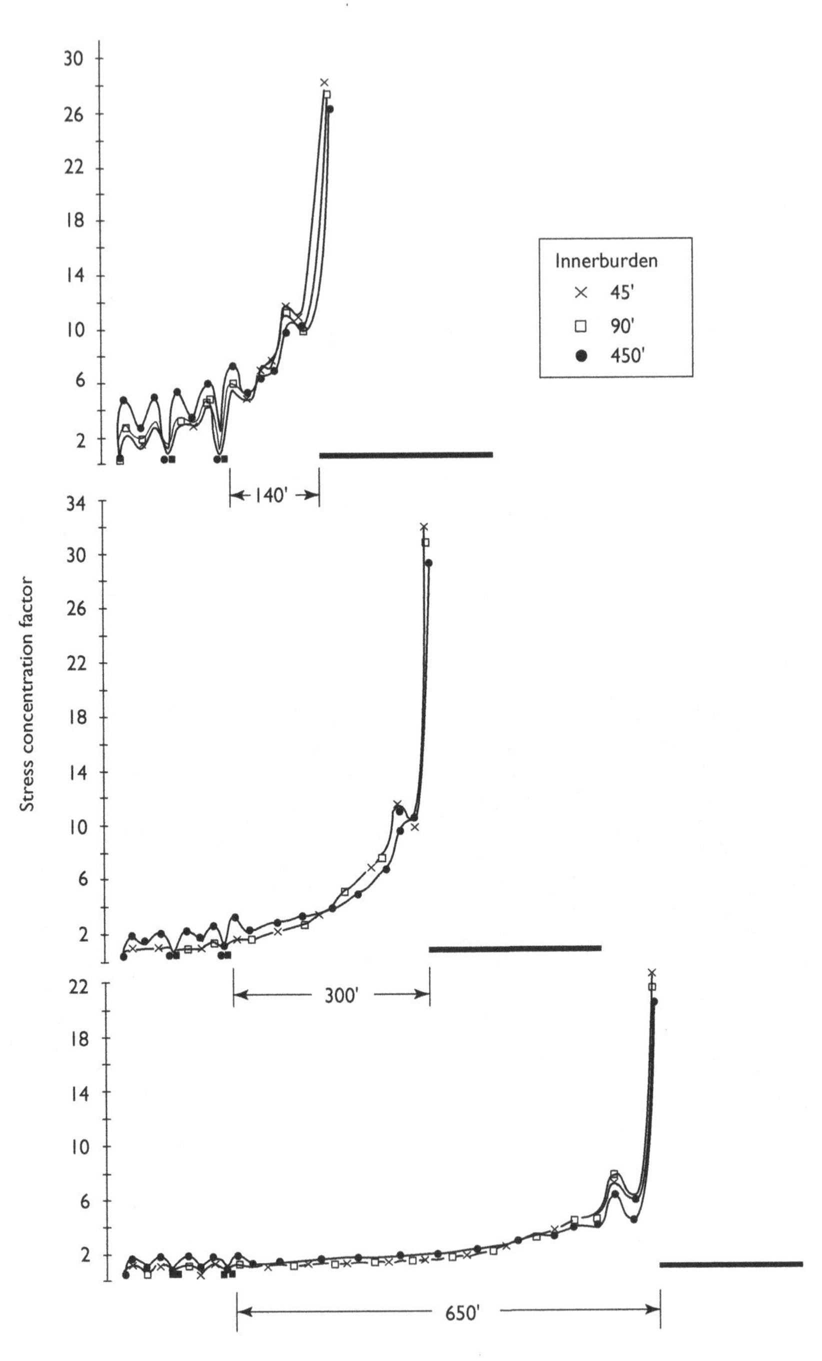

Figure 6.22a Vertical stress distribution as a function of barrier pillar size, panel length, and innerburden. Top seam.

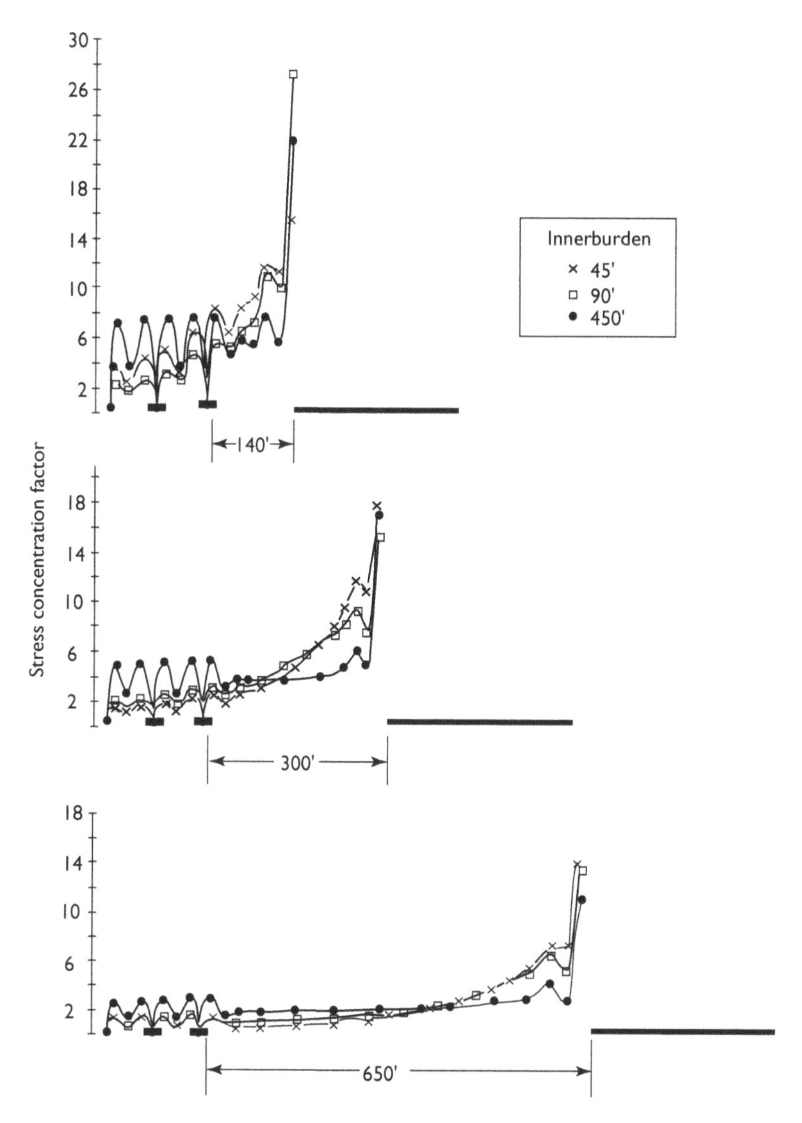

Figure 6.22b Vertical stress distribution as a function of barrier pillar size, panel length, and innerburden. Bottom seam.

Pillar strength C_p is linked to pillar width that should be sufficient to allow the horizontal stress to return to the premining value. Pillars that are much wider than they are high, that is, when $W_b >> H_p$, satisfy the condition. Assuming a Mohr–Coulomb criterion,

$$C_p = C_o + \frac{C_o}{T_o}p = C_o + \frac{C_o}{T_o}S_h = C_o + (C_o/T_o)K_o\gamma H \tag{6.49}$$

where S_h is K_o times overburden stress (vertical premining stress) which is estimated as unit average unit weight of overburden γ times seam depth H. Barrier pillar stress should satisfy a safety factor criterion $FS_b = C_p / S_b > 1$, hence

$$W_b > \left[\frac{W_o/2}{C_o/S_v + (C_o/T_o)K_o - 1} \right] \tag{6.50}$$

that leads to a rather wide barrier pillar.

Example 6.19 Suppose seam depth is 1,000 ft, overburden specific weight is 144 pcf, panel length is 10,000 ft, coal compressive strength is 2,500 psi, tensile strength is 250 psi, and the horizontal premining stress is one-fourth the vertical stress, then for $F_p > 1$. Estimate barrier pillar width.

Solution: According to (6.50)

$$W_b > \frac{(10,000/2)}{[2,500/1,000 + (10)(1/4) - 1]} = 1,250 \text{ ft}$$

so a barrier pillar at least 1,250 ft is required.

This estimate is certainly on the high side because of the barrier pillar load estimate. Caving above a panel would reduce this load and subsequent compaction and reloading of caved material by overlying strata would further reduce the barrier pillar load estimate substantially, especially for very long panels. The result would be to shorten the excavated panel half-width used in the derivation leading to (6.50). Instead of $W_o/2$, one might use $W_o/6$, for example, that leads to a barrier width of 413 ft.

Example 6.20 Another allowance for caving consists of assigning an overburden block defined by vertical plane extending to the surface from the entry side of a barrier pillar and a second inclined line extending from the panel side of the barrier pillar to the surface. If the angle from the vertical to this second line is δ, then force equilibrium requires the pillar reactions to support the weight of the overburden block, as shown in the sketch. Develop a formula for barrier pillar width based on this loading concept then evaluate for conditions given in Example 6.19 with an angle of influence of 35°.

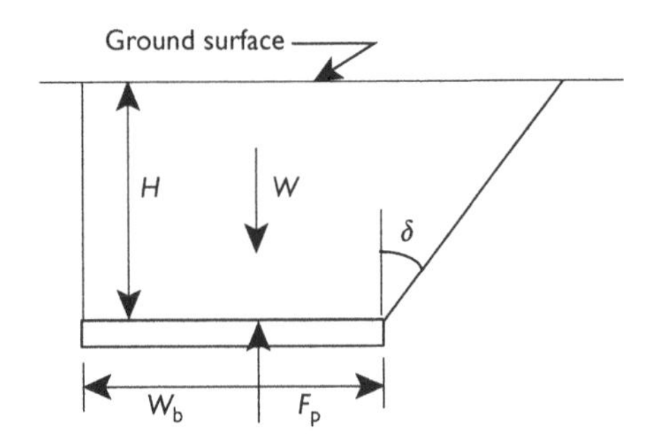

Solution: Equilibrium requires $W = F_p$, so with neglect of crosscuts

$$S_p W_b = \gamma [H\ W_b + (H^2/2) \tan\ \delta]$$

In view of the barrier pillar safety factor requirement

$$W_b > \frac{H \tan\delta}{2[C_o/\gamma H + (C_o/T_o)K_o - 1]}$$

In this case, with an "influence" angle of 35°, the barrier pillar width is

$$W_b > \frac{(1,000)\tan(35)}{2[2,500/1,000 + (10)(1/4) - 1]} = 87.5 \text{ ft.}$$

If confinement were neglected and the unconfined compressive strength were used, then $W_b >$ 233 ft.

These two examples indicate a wide difference in barrier pillar width required for safety under the same given conditions. Barrier pillars are generally much wider than they are high, so use of strength under confining pressure is reasonable, although the ratio of unconfined compressive to tensile strength may be uncertain. This ratio in the Mohr–Coulomb criterion is also the ratio $[1 + \sin(\phi)]/[1 - \sin(\phi)]$. A ratio of 10 implies an angle of internal friction of about 55° and may be an overestimate in the purely compressive region. A reduced ratio would lead to even greater barrier pillar widths. The uncertainty in barrier pillar width is therefore in the estimate of barrier pillar stress, and not so much in strength.

6.6 Problems

6.1 A regular room and pillar method is used in a hardrock lead–zinc mine. The ore horizon is at a depth of 1,180 ft where the overburden averages 159 pcf. Laboratory tests on core samples ($L/D = 2$) show that the ore has an unconfined compressive strength of 13,700 psi and a tensile strength of 1,250 psi. Find an algebraic expression for the maximum safe extraction ratio, if no size effect exists and a safety factor of 1.75 is required, then evaluate the expression for the given data.

6.2 With reference to Problem 6.1, if pillars are square and rooms (entries and crosscuts) are 45 ft wide, what size of pillars is indicated?

6.3 With reference to Problems 6.1 and 6.2, suppose the pillars are 45 × 45 ft and that a size effect exists such that $C_p = C_1\,(0.78 + 0.22W_p/H_p)$. If the mining height changes from 30 to 120 ft, as a sill between two ore horizons is removed, what is the resulting pillar safety factor, all other parameters remaining the same? What is the pillar safety factor if the size effect is neglected?

6.4 Suppose that entries and crosscuts are 45 ft wide and that pillars are square and 45 ft wide, as in Problem 6.3. Laboratory properties are the same as in Problem 6.1; depth and specific weight of overburden are also the same as in Problem 6.1. In addition, a joint set is present in the mine. Joint set dip is 68°; dip direction is 54° clockwise from North.

(a) Find the joint shear strength necessary to just prevent joint slip within a typical pillar.

(b) If the joint is cohesionless and obeys a Mohr–Coulomb slip criterion, what joint friction angle is needed to just prevent slip?

(c) If the joint is frictionless, what cohesion is necessary?

6.5 A regular room and pillar method is used in a hardrock lead–zinc mine. The ore horizon is at a depth of 360 m where the overburden averages 25.2 kN/m^3. Laboratory tests on core samples ($L/D = 2$) show that the ore has an unconfined compressive strength of 94.5 MPa and a tensile strength of 8.63 MPa. Find an algebraic expression for the maximum safe extraction ratio, if no size effect exists and a safety factor of 1.75 is required, then evaluate the expression for the given data.

6.6 With reference to Problem 6.5, if pillars are square and rooms (entries and crosscuts) are 13.7 m wide, what size of pillars is indicated?

6.7 With reference to Problems 6.5 and 6.6, suppose the pillars are 13.7 × 13.7 m and that a size effect exists such that $C_p = C_1\,(0.78 + 0.22W_p/H_p)$. If the mining height changes from 9.14 to 36.6 m, as a sill between two ore horizons is removed, what is the resulting pillar safety factor, all other parameters remaining the same? What is the pillar safety factor if the size effect is neglected?

6.8 Suppose that entries and crosscuts are 13.7 m wide and that pillars are square and 13.7 m wide, as in Problem 6.7. Laboratory properties are the same as in Problem 6.5; depth and specific weight of overburden are also the same as in Problem 6.5. In addition, a joint set is present in the mine. Joint set dip is 68°; dip direction is 54° clockwise from North.

(a) Find the joint shear strength necessary to just prevent joint slip within a typical pillar.
(b) If the joint is cohesionless and obeys a Mohr–Coulomb slip criterion, what joint friction angle is needed to just prevent slip?
(c) If the joint is frictionless, what cohesion is necessary?

6.9 Suppose that the average vertical pillar stress is 26.7 MPa in a room and pillar mine and that a joint set is present with an average bearing of N45W. If the joints obey a Mohr–Coulomb slip criterion and have a friction angle of 28° and a cohesion of 0.90 MPa, what range of joint dips is unsafe with respect to joint slip?

6.10 Coal mine entries are driven 20 ft (6.1 m) wide on strike at a depth of 1200 ft (366 m) in a coal seam that dips 15°. Crosscuts are driven up and down dip and are also 20 ft (6.1 m) wide. The production pillars that are formed are twice as long in the strike direction as they are in the dip direction. Compressive strength of coal is estimated to be 2,000 psi (13.8 MPa); tensile strength is estimated at 350 psi (2.41 MPa). The premining stress state is considered to be attributable to gravity only, so the vertical stress is unit weight times depth and the horizontal stress is one-fourth the vertical stress. Assume that a Mohr–Coulomb failure criterion applies and that a safety factor with respect to compressive strength of 1.5 is required. Find the extraction ratio possible under these conditions, and then estimate the pillar safety factor with respect to shear strength. Neglect any size effect that may exist.

6.11 A room and pillar trona mine at a depth of 1,560 ft (475 m) develops pillars 30 ft (9.14 m) wide and 60 ft (18.3 m) long in a flat seam 12 ft (3.66 m) thick. Entries and crosscuts are 24 ft (7.32 m) wide. Laboratory tests of trona core 2-1/8 in. (5.4 cm) in diameter and 4-1/4 in. (10.8 cm) long give an average unconfined compressive strength of 6,740 psi (46.5 MPa) and a coefficient of variation of 38%. A size effect is indicated. If a pillar safety factor with respect to compressive failure of 1.67 is required, estimate the maximum allowable extraction ratio?

6.12 A narrow, 20 ft (6.10 m) thick vein dips 60°. Stoping occurs at a depth of 5,100 ft (1,555 m) in rock that averages 165 pcf (26.1 kN/m^3). The vertical stress before mining is equal to the unit weight of overburden times depth; the horizontal premining stress is twice the vertical stress. Unconfined compressive strength is 25,000 psi (172 MPa); tensile strength is 2,850 psi (19.7 MPa). Find the maximum possible extraction ratio assuming no size effect.

6.13 If mining in Problem 6.12 is by conventional overhand stoping, and round length (measured on the dip) is 15 ft (4.57 m), is failure likely to be stable or not as mining proceeds past the maximum extraction ratio previously calculated? Explain your answer.

6.14 Consider a large array of similar pillars on a regular grid that have a safety factor with respect to compressive failure of *Fc*. Suppose a single pillar fails and the load is then shared equally by the nearest neighboring pillars.

(a) Find the safety factor *Fc* needed to avoid failure of the nearest neighbor pillars and thus a domino effect and catastrophic collapse of the section.
(b) If the load is shared equally by all adjacent pillars, including the corner pillars, what is the safety factor necessary to avoid a cascade of failing pillars?

6.15 A regular room and pillar method is used in a hardrock lead–zinc mine. The ore horizon is at a depth of 980 ft where the overburden averages 156 pcf. Laboratory tests on core samples ($L/D = 2$) show that the ore has an unconfined compressive strength of 12,400 psi and a tensile strength of 1,050 psi. Find an algebraic expression for the maximum extraction ratio possible, if no size effect exists, then evaluate the expression for the given data.

6.16 With reference to Problem 6.15, if pillars are square and rooms (entries and crosscuts) are 30 ft wide, what size of pillars are indicated when the pillar safety factor is 2?

6.17 With reference to Problems 6.15 and 6.16, suppose the pillars are 30 × 30 ft and that a size effect exists such that $C_p = C_1\,(0.78 + 0.22W_p\,/H_p)$. If the mining height changes from 25 to 90 ft as a sill between two ore horizons is removed, what is the resulting pillar safety factor, all other parameters remaining the same? What is the pillar safety factor if the size effect is neglected?

6.18 Suppose that entries and crosscuts are 30 ft wide and that pillars are square and 30 ft wide, as in Problem 6.17. Laboratory properties are the same as in Problem 6.15; depth and specific weight of overburden are also the same as in Problem 6.15. In addition, a joint set is present in the mine. Joint set dip is 60°; dip direction is 45° clockwise from North.

(a) Find the joint shear strength necessary to just prevent joint slip within a typical pillar.
(b) If the joint is cohesionless and obeys a Mohr–Coulomb slip criterion, what joint friction angle is needed to just prevent slip?
(c) If the joint is frictionless, what cohesion is necessary?

6.19 A regular room and pillar method is used in a hardrock lead–zinc mine. The ore horizon is at a depth of 300 m where the overburden averages 25 kN/m3. Laboratory tests on core samples ($L/D = 2$) show that the ore has an unconfined compressive strength of 86 MPa and a tensile strength of 7 MPa. Find an algebraic expression for the maximum extraction ratio possible, if no size effect exists, then evaluate the expression for the given data.

6.20 With reference to Problem 6.19, if pillars are square and rooms (entries and crosscuts) are 9 m wide, what size of pillars are indicated when the pillar safety factor is 2?

6.21 With reference to Problems 6.19 and 6.20, suppose the pillars are 9 × 9 m and that a size effect exists such that $C_p = C_1 (0.78 + 0.22W_p / H_p)$. If the mining height changes from 8 to 27 m as a sill between two ore horizons is removed, what is the resulting pillar safety factor, all other parameters remaining the same? What is the pillar safety factor if the size effect is neglected?

6.22 Suppose that entries and crosscuts are 9 m wide and that pillars are square and 9 m wide, as in Problem 6.21. Laboratory properties are the same as in Problem 6.19; depth and specific weight of overburden are also the same as in Problem 6.19. In addition, a joint set is present in the mine. Joint set dip is 60°; dip direction is 45° clockwise from North.

(a) Find the joint shear strength necessary to just prevent joint slip within a typical pillar.
(b) If the joint is cohesionless and obeys a Mohr–Coulomb slip criterion, what joint friction angle is needed to just prevent slip?
(c) If the joint is frictionless, what cohesion is necessary?

6.23 Suppose that the pillar stress is 4,247 psi in a room and pillar mine and that a joint set is present with an average bearing of N45 W. If the joints obey a Mohr–Coulomb slip criterion and have a friction angle of 35° and a cohesion of 600 psi, what range of joint dips is unsafe with respect to joint slip?

6.24 Suppose that the pillar stress is 29 MPa in a room and pillar mine and that a joint set is present with an average bearing of N45W. If the joints obey a Mohr–Coulomb slip criterion and have a friction angle of 35° and a cohesion of 4 MPa, what range of joint dips is unsafe with respect to joint slip?

6.25 Coal is mined in two seams that are shown in the sketch. Mining is full seam height in both seams. Entries are 24 ft (7.3 m) wide. Crosscuts associated with driving main

Rock type	γ-pcf (kN/m³)	E(10⁶)psi (GPa)	ν	C_o(psi) (MPa)	T_o(psi) (MPa)	h(ft) (m)
1 Over burden	144.0 (22.8)	1.70 (11.7)	0.25	1,000 (6.9)	50 (0.34)	1,694 (516)
2 Massive sandstone	148.2 (23.4)	3.00 (20.7)	0.20	19,000 (131.9)	850 (5.86)	1,714 (522)
3 Coal	75.0 (11.9)	0.35 (2.4)	0.30	3,500 (24.1)	130 (0.90)	1,719 (524)
4 Layered sandstone	149.5 (23.6)	1.50 (10.3)	0.10	17,500 (120.7)	700 (4.83)	1,741 (531)
5 Massive sandstone	148.2 (23.4)	3.00 (20.7)	0.20	19,000 (131.9)	850 (5.86)	1,744 (532)
6 Sandy shale	170.0 (26.9)	4.50 (31.0)	0.10	15,000 (103.4)	1,000 (6.90)	1,747 (533)
7 Coal	75.0 (11.9)	0.35 (2.4)	0.30	3,500 (24.1)	1,300 (8.97)	1,754 (535)
8 Massive sandstone	148.2 (23.4)	3.50 (24.1)	0.20	19,500 (134.5)	1,000 (6.90)	1,778 (542)
9 Layered sandstone	149.5 (23.6)	1.50 (10.3)	0.10	17,500 (120.7)	700 (4.83)	2,250 (658)

Sketch and geologic column, strata depths and rock properties

Table 6.1 Properties for Problem 6.26

Property material	*C_o psi (MPa)*	*T_o psi (MPa)*	*c psi (MPa)*	Φ(°)
Rock	15,000 (103)	1,500 (10.3)	2,372 (16.4)	54.9
Joint	66.6 (0.459)	24.0 (0.166)	20.0 (0.138)	28.0

entries are 18 ft (5.5 m) wide and driven on 120 ft (36.6 m) centers. Associated pillar length is three times pillar width. With reference to the stratigraphic column in the sketch and the rock properties tabulated there, determine the minimum pillar dimensions possible. Show solution in a carefully labeled plan view sketch. Note: No size effects are present.

6.26 A pillar in a hardrock, strata-bound, room, and pillar mine is transected by a small fault that dips 75° due east and strikes due north. The ore horizon dips due west 15° and also strikes due north. Depth to the pillar of interest is 1,230 ft (375 m). Overburden specific weight averages 158 pcf (25 kN/m^3). Rock and joint properties are given in Table 6.1. The preexcavation stress is caused by gravity alone; the horizontal stress is estimated to be one-fourth the vertical stress before mining. Determine whether the pillar is stable in consideration of a planned extraction ratio of 75%.

6.27 A thick seam of coal is mined from the floor to a height of 13 ft at a depth of 1450 ft. Entry width is 21 ft and overburden averages 144 pcf. Crosscuts are 18 ft wide and driven on 100 ft centers. Pillar length is three times pillar width. With reference to the stratigraphic column in the sketch and the rock properties given in Table 6.2, determine

(a) the maximum safe extraction ratio, and
(b) pillar dimensions for a pillar safety factor of 1.5. Show solution in a carefully labeled plan view sketch.

Surface

(Not to scale)	*Overburden*
R5 mudstone	
R4 sandstone	
R3 sandstone	
R2 laminated sandstone	
R1 (roof) laminated shale	
S1 (seam - coal)	
F1 (floor) shale	
F2 limestone	

Stratigraphic sequence.

6.28 A pillar in a hardrock, strata-bound, room, and pillar mine is transected by a small fault that dips 75° due east and strikes due north. The ore horizon dips due west 15° and also strikes due north. Depth to the pillar of interest is 1,230 ft (375 m). Overburden specific weight averages 158 pcf (25.0 kN/m^3). Rock and joint properties are given in Table 6.3. The preexcavation stress is caused by gravity alone; the horizontal stress is estimated to be one-fourth the vertical stress before mining.

Table 6.2 Strata data for Problem 6.27

Rock type	*E 10^6 psi (GPa)*	*ν (–)*	*C_o-psi (MPa)*	*T_o psi (MPa)*	*γ -pcf (kN/m^3)*	*h ft (m)*
R5	12.6 (86.9)	0.19	12,700 (87.6)	1,270(8.8)	160 (25.3)	21.5 (6.55)
R4	12.7 (87.6)	0.21	14,200 (97.9)	1,34 (9.2)	158 (25.0)	15.2 (4.63)
R3	5.4 (37.2)	0.23	7,980 (55.0)	610 (4.2)	152 (23.1)	1.0 (0.30)
R2	6.3 (43.4)	0.31	8,450 (58.3)	760 (5.2)	146 (24.7)	1.5 (0.46)
R1	12.6 (86.9)	0.19	12,700 (87.6)	1,270 (8.8)	156 (24.7)	6.0 (1.83)
S1	0.75 (5.2)	0.27	3,750 (25.9)	425 (2.9)	92 (14.6)	3.0 (0.91)
F1	12.3 (84.8)	0.18	11,500 (79.3)	1,050 (7.2)	155 (24.5)	10.3 (3.14)
F2	8.1 (55.9)	0.33	13,700 (94.5)	1,260 (8.7)	160 (25.3)	13.0 (3.96)

Table 6.3 Properties for Problem 6.28.

Property material	*C_o psi (MPa)*	*T_o psi (MPa)*	*c psi (MPa)*	*Φ(°)*
Rock	15,000(103)	1,500(10.3)		
Joint			20 (0.14)	28

Determine:

(a) the premining stress acting normal to the ore horizon and the shear stress acting parallel to the ore horizon at the depth of interest;
(b) the post mining pillar normal and shear stresses acting on the pillar;
(c) the maximum safe extraction ratio consistent with a pillar safety factor of 1.5 (ignore any size effect and neglect the joint);
(d) joint shear strength needed for a safety factor of 1.5.

6.29 With reference to the strata data given in the sketch and Table 6.4, seam depth is 1,670 ft, strata dip is negligible, and entries are planned 24 ft wide. Mining height from the sandstone floor is 12 ft. If crosscuts are as wide as entries and pillar length is three times pillar width:

1 Determine the extraction ratio possible with a pillar safety factor of 1.4 and no size effects.
2 Determine the pillar dimensions, entry spacing (center to center), and crosscut spacing (c-c) at the FS and R determined previously.
3 If strata dip were 25°, estimate pillar stresses at a depth of 1,670 ft, assuming an extraction ratio of 20%.
4 A joint set is present, strata are *flat-lying* and extraction is 20%; determine the range of joint dips that may lead to joint slip, assuming Mohr–Coulomb criteria for strata and joints with joint friction angle at 28° and negligible joint cohesion.

Surface

(Not to scale)	*Overburden*
R3 sandstone	
R2 laminated sandstone	
R1 shale	
S1 coal	
F1 sandstone	

Table 6.4 Strata properties for Problem 6.29.

Rock type	*E* 10^6 *psi (GPa)*	v	C_o 10^3 *psi (MPa)*	T_o 10^3 *psi (MPa)*	γ *pcf* (*kN/m*^{3})	*h ft (m)*
R3	4.3 (29.7)	0.25	16.0 (110)	1.2 (8.3)	148 (23.4)	26.0 (7.9)
R2	1.7 (11.7)	0.19	12.5 (86.2)	0.65 (4.5)	142 (22.5)	1.5 (0.46)
R1	4.1 (28.3)	0.20	7.5 (51.7)	0.45 (3.1)	136 (21.5)	2.0 (0.61)
S1	0.35 (2.4)	0.30	3.0 (20.7)	0.25 (1.7)	95 (15.0)	16.0 (4.88)
F1	4.5 (31.0)	0.25	18.0 (124)	1.8 (12.4)	149 (23.6)	18 (5.49)

6.30 Several proposals have been presented to increase the extraction ratio in a room and pillar system to 50% from 33%. The pillars are long compared with width, so the problem is essentially two-dimensional.

Proposal 1. Increase the ratio of opening width to pillar width from 0.5 to 1.0 by increasing entry width to 30 ft (9 m) from 20 ft (6 m) while decreasing pillar width. Spacing is held constant.
Proposal 2. Increase the ratio of opening width to pillar width from 0.5 to 1.0 by keeping pillar width constant while increasing entry width to 40 ft (12 m).
Proposal 3. Increase the ratio of opening width to pillar width from 0.5 to 1.0 by keeping entry width constant while decreasing pillar width to 20 ft (6 m).
Which proposal is preferable from the rock mechanics view and why?

6.31 With reference to the strata data given in the sketch and Table 6.5, mining height is 15 ft, seam depth is 1,324 ft, entries are 21 ft wide, and strata dip is negligible. If crosscuts are as wide as entries and pillar length is three times pillar width:

1. Determine the extraction ratio possible with a pillar safety factor of 1.8 and no size effects.
2. Determine the pillar dimensions, entry spacing (center to center), and crosscut spacing (c-c) at the FS and R determined previously.
3. If strata dip were 30°, estimate pillar stresses at a depth of 1,324 ft, assuming an extraction ratio of 20%.
4. A joint set is present, strata are flat-lying and extraction is 20%; determine the range of joint dips that may lead to joint slip, assuming Mohr–Coulomb criteria for strata and joints with joint friction angle at 35° and negligible joint cohesion.

Surface

(Not to scale)	*Overburden*
R3 sandstone	
R2 laminated sandstone	
R1 shale	
S1 coal	
F1 sandstone	

Stratigraphic sequence.

Table 6.5 Strata properties for Problem 6.31

Rock type	*E* 10^6*psi (GPa)*	v	C_o 10^3*psi (MPa)*	T_o 10^3*psi (MPa)*	γ *pcf* (kN/m^3)	*h ft (m)*
R3	4.5 (31.0)	0.25	16.0 (110)	0.80 (5.52)	148 (23.4)	16.0 (4.9)
R2	2.7 (18.6)	0.19	12.5 (86.2)	0.65 (4.48)	142 (22.5)	2.5 (0.8)
R1	3.1 (21.4)	0.20	8.0 (55.2)	0.40 (2.76)	135 (21.4)	2.0 (0.6)
S1	0.35 (2.4)	0.30	3.0 (20.7)	0.15 (1.03)	95 (15.0)	18.0 (5.5)
F1	3.5 (24.1)	0.25	0.8 (5.5)	0.8 (5.52)	148 (23.4)	14.0 (4.3)

6.32 A multilevel room and pillar metal mine is under consideration in flat strata striking N60E. Entries are planned on strike, crosscuts up, and down dip. Table 6.6 shows depths and material properties associated with the geologic column. Rock properties data were determined from laboratory testing on NX-core at an *L/D* ratio of two. Stress measurements indicate that the premining stresses relative to compass coordinates are: $S_v = 1.1d$, $S_h = 300 + 0.3d$, $S_H = 500 + 0.5d$, where v = vertical, h = azimuth is 150°, H = azimuth is 60°, d = depth in ft, and stress units are psi. Three joint sets are present. Set 1 is vertical and strikes parallel to the strata. Set 2 strikes due north and dips 30° east; Set 3 also strikes due north, but dips 65° west.

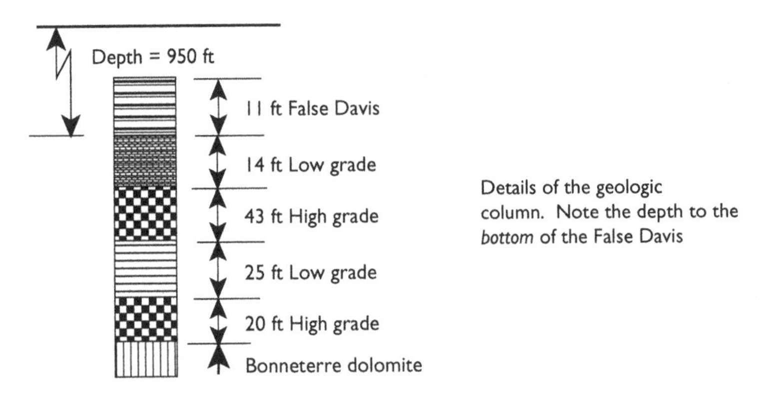

Details of the geologic column. Note the depth to the *bottom* of the False Davis

Joint properties for Mohr–Coulomb failure criteria are given in Table 6.7. Joint normal and shear stiffness (K_n, K_s) that relate normal stress and shear stress to corresponding displacements are also given in Table 6.7 as are the joint spacings (S).

One mining plan calls for mining a lower B level 102 thick in high and low grade ore at a depth of 950 ft and an upper A level of high grade ore 18 ft thick at a depth of 921 ft.

(a) Assume there is no size effect on compressive strength, then determine the maximum possible extraction ratio based on ore strength (ignore joints).

(b) Assuming square pillars, show in plan and vertical section, the relationship between rooms and pillars in the two levels that would be dictated by rock mechanics and explain your recommendation.

(c) What safety factor would you recommend for pillars in the upper and lower mining horizons and why?

Table 6.6 Rock properties, strata depths, and thickness

	γ pcf	$E(10^6)$ psi	ν	$Co(10^3)$ psi	$To(10^2)$ psi	$C(10^2)$ psi	ϕ (°)	Depth ft (thickness)
1	130.0	0.003	0.30	0.10	0.05	—	—	0 (60)
2	163.2	6.92	0.34	11.91	11.74	25.17	49.0	60 (50)
3	167.2	10.65	0.31	16.12	7.34	27.33	53.1	110 (195)
4	170.8	12.09	0.26	26.87	12.28	32.67	59.2	305 (350)
5	162.5	7.58	0.32	23.76	12.85	34.30	51.5	655 (110)
6	161.3	5.35	0.21	19.61	12.05	37.67	49.3	765 (150)
7	166.3	6.75	0.31	29.26	9.35	43.67	49.0	915 (6)
Ore	218.7	8.75	0.30	18.27	10.07	18.50	60.4	921 (18)
9	152.0	3.79	0.22	5.21	7.30	6.80	36.0	939 (11)
Ore	218.7	8.75	0.30	18.27	10.07	18.50	60.4	950 (102)
11	166.3	6.75	0.31	29.26	9.35	43.67	49.0	1,052 (163)
12	146.2	3.94	0.41	11.24	7.90	31.00	42.3	1,215 (330)
13	167.2	10.65	0.31	16.12	7.34	27.33	53.1	1,545 (100)

Source: 1 = overburden, 2 = Gasconade Dolomite, 3 = Eminence Dolomite, 4 = Potosi Dolomite, 5 = DerbyDoerum Dolomite, 6 = Davis Shale, 7 = Bonneterre Dolomite, 8 = Ore, 9 = False Davis, 10 = Ore, 11 = Bonneterre Dolomite, 12 = Lamotte Sandstone, 13 = Precambrian Felsites.

Table 6.7 Joint properties

Property joint set	*c psi*	$\Phi°$	K_n 10^6*psi/in.*	K_s 10^6*psi/in.*	*S ft*
Set 1	0.0	35	1.76	0.56	25.3
Set 2	10.0	30	2.31	3.14	4.7
Set 3	20.0	25	3.29	0.92	6.1

6.33 With reference to Problem 6.32 data, roof rock above the upper A Level is Bonneterre dolomite that is overlain by Davis shale.

(a) Determine the maximum opening width that is physically possible in the upper A Level ore horizon.
(b) With reference to Problem 6.32, assume square pillars, then determine minimum pillar width with respect to ore strength, given the results of part a.
(c) What safety factor for the A Level roof would you recommend and why?

6.34 With reference to Problem 6.33, concerning opening width in the A Level, determine a bolting plan that would allow for 90 ft wide rooms.

(a) Specify, bolt length, safety factor, diameter, steel grade, and spacing.
(b) Also specify the associated pillar size and pillar safety factor with respect to ore strength for the mining plan.

6.35 An alternative mining plan is to develop two levels, below the upper A Level, and mine only the high grade. The B Level would be in the 43 ft of high grade, while the C level would be in the lower 20 ft of high grade in the 102 ft column shown in the detailed geologic column. Determine the maximum possible roof span for the B Level.

6.36 With reference to Problem 6.32, if pillars are sized according to the maximum extraction ratio and pillar compressive strength, determine if the pillars are safe with respect to failure of Joint Set 2.

6.37 A multilevel room and pillar metal mine is under consideration in flat strata striking N60E. Entries are planned on strike, crosscuts up, and down dip. Table 6.8 shows depths and material properties associated with the geologic column. Rock properties data were determined from laboratory testing on NX-core at an L/D ratio of two. Stress measurements indicate that the premining stresses relative to compass coordinates are: $S_v = 25d$, $S_h = 2{,}069 + 6.8d$, $S_H = 3, 448 + 11.4d$, where v = vertical, h = azimuth is 150°, H = azimuth is 60°, d = depth in m, and stress units are kPa. Three joint sets are present. Set 1 is vertical and strikes parallel to the strata. Set 2 strikes due north and dips 30° east; Set 3 also strikes due north, but dips 65° west.

Joint properties for Mohr–Coulomb failure criteria are given in Table 6.9. Joint normal and shear stiffness (K_n, K_s) that relate normal stress and shear stress to corresponding displacements are also given in Table 6.9 as are the joint spacings (S).

One mining plan calls for mining a lower B level 31.1 m in high and low grade ore at a depth of 290 and an upper A level of high grade ore 5.5 m thick at a depth of 281 m.

(a) Assume there is no size effect on compressive strength, then determine the maximum possible extraction ratio based on ore strength (ignore joints).
(b) Assuming square pillars, show in plan and vertical section, the relationship between rooms and pillars in the two levels that would be dictated by rock mechanics and explain your recommendation.

Table 6.8 Rock properties, strata depths, and thicknesses

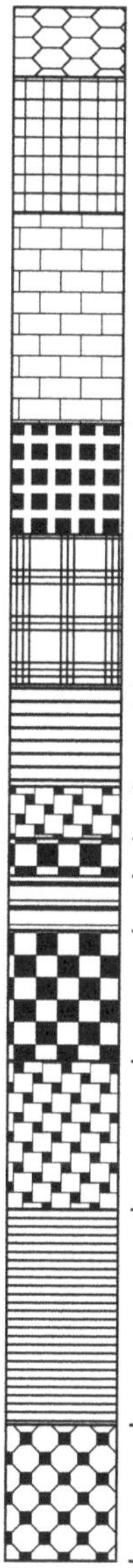

	γ kN/m^3	*E* GPa	ν	C_o MPa	T_o MPa	*c* psi	ϕ (°)	Depth ft (thickness)
1	130.0	0.003	0.30	0.10	0.05	—	—	0 (18.3)
2	163.2	6.92	0.34	11.91	11.74	25.17	49.0	181.3 (15.2)
3	167.2	10.65	0.31	16.12	7.34	27.33	53.1	33.5 (59.4)
4	170.8	12.09	0.26	26.87	12.28	32.67	59.2	92.9 (106.7)
5	162.5	7.58	0.32	23.76	12.85	34.30	51.5	199.6 (33.5)
6	161.3	5.35	0.21	19.61	12.05	37.67	49.3	233.1 (45.7)
7	166.3	6.75	0.31	29.26	9.35	43.67	49.0	278.8 (1.83)
Ore	218.7	8.75	0.30	18.27	10.07	18.50	60.4	280.6 (5.5)
9	152.0	3.79	0.22	5.21	7.30	6.80	36.0	286.1 (3.35)
Ore	218.7	8.75	0.30	18.27	10.07	18.50	60.4	289.5 (31.1)
11	166.3	6.75	0.31	29.26	9.35	43.67	49.0	320.6 (49.7)
12	146.2	3.94	0.41	11.24	7.90	31.00	42.3	370.3 (100.5)
13	167.2	10.65	0.31	16.12	7.34	27.33	53.1	470.8 (30.5)

Source: 1 = overburden, 2 = Gasconade dolomite, 3 = Eminence dolomite, 4 = Potosi dolomite, 5 = DerbyDoerun dolomite, 6 = Davis Shale, 7 = Bonneterre dolomite, 8 = ore, 9 = False Davis, 10 = ore, 11 = Bonneterre dolomite, 12 = Lamotte sandstone, 13 = Precambrian felsites.

Table 6.9 Joint properties

Property joint set	*c MPa*	$\Phi(°)$	K_n *GPa/cm*	K_s *GPa/cm*	*S m*
Set 1	0.0	35	4.77	1.52	7.71
Set 2	0.069	30	6.27	8.53	1.43
Set 3	0.138	25	8.43	2.49	1.86

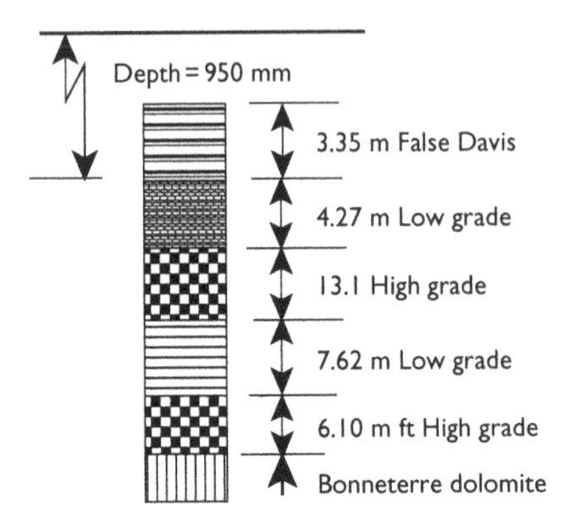

Details of the geologic column. Note the depth to the *bottom* of the False Davis

(c) What safety factor would you recommend for pillars in the upper and lower mining horizons and why?

6.38 With reference to Problem 6.37 data, roof rock above the upper A Level is Bonneterre dolomite that is overlain by Davis shale.

(a) Determine the maximum opening width that is physically possible in the upper A Level ore horizon.
(b) With reference to Problem 6.37, assume square pillars, then determine minimum pillar width with respect to ore strength, given the results of part a.
(c) What safety factor for the A Level roof would you recommend and why?

6.39 With reference to Problem 6.38, concerning opening width in the A Level, determine a bolting plan that would allow for 27.4 m wide rooms.

(a) Specify, bolt length, safety factor, diameter, steel grade, and spacing.
(b) Also specify the associated pillar size and pillar safety factor with respect to ore strength for the mining plan.

6.40 An alternative mining plan is to develop two levels, below the upper A Level and mine only the high grade. The B Level would be in the 13.1 m of high grade, while the C level would be in the lower 6.10 m of high grade in the 31.1 m column shown in the detailed geologic column. Determine the maximum possible roof span for the B Level.

6.41 With reference to Problem 6.37, if pillars are sized according to the maximum extraction ratio and pillar compressive strength, determine if the pillars are safe with respect to failure of Joint Set 2.

Table 6.10 Intact rock properties

Property stratum	*γ pcf (kN/m³)*	*h ft (m)*	*E 10⁶psi (GPa)*	*G 10⁶psi (GPa)*	*C₀ psi (MPa)*	*T₀ psi (MPa)*
Limestone (R5)	152(24.0)	15.9 (4.8)	9.83 (67.8)	4.27 (29.4)	17,690 (122)	1,460 (10.0)
Coal (R4)	93 (14.7)	1.4 (0.4)	0.29 (2.0)	0.12 (0.8)	2,780 (19.2)	390 (2.7)
Mudstone (R3)	153 (24.2)	9.8 (3.0)	7.67 (52.9)	3.07 (21.2)	13,750 (94.8)	1,580 (10.9)
Sandy shale (R2)	142 (22.5)	2.7 (0.8)	4.16 (28.7)	1.79 (12.3)	6,450 (44.5)	720 (5.0)
Shale (R1)	138 (21.8)	1.3 (0.4)	3.62 (25.0)	1.53 (10.6)	7,400 (51.0)	650 (4.5)
Coal (Seam)	90 (14.2)	16.5 (5.0)	0.35 (2.4)	0.14 (1.0)	3,400 (23.4)	310 (2.1)
Sandstone (F1)	147 (23.3)	8.6 (2.6)	8.32 (57.4)	3.35 (23.1)	12,300 (84.8)	1,350 (9.3)

Table 6.11 Joint properties

Property joint set	*c psi (kPa)*	*Φ (°)*	*K_n 10⁶psi/in. (GPa/cm)*	*K_s 10⁶psi/in. (GPa/cm)*	*S ft (m)*
Set 1	0.0 (0.)	35	1.76 (4.78)	0.56 (1.52)	25.3 (7.71)
Set 2	10.0 (69.0)	30	2.31 (6.27)	3.14 (8.53)	4.7 (1.43)
Set 3	20.0 (138.0)	25	3.29 (8.93)	0.92 (2.49)	6.1 (1.85)

6.42 A room and pillar coal mine is contemplated at a depth of 533 m in strata striking due north and dipping 18° east. Entries are driven on strike, crosscuts up, and down dip. Mining height is 4.6 m measured from the floor; 0.5 m of low grade coal is left in the roof.

Three joint sets are present. Set 1 is vertical and strikes east–west. Set 2 strikes due north and dips 60° east; Set 3 also strikes due north, but dips 35° west. The average overburden specific weight is 24.7 kN/m³.

Stress measurements indicate that the premining stresses relative to compass coordinates are: S_v = 23.8d, S_h = 4, 138 + 5.66*d*, S_H = 345 + 17.0*d*, where *v* = vertical, *h* = east, *H* = north, *d* = depth in m, and stress units are kPa.

Rock properties near the mining horizon are given in Table 6.10. These data were determined from laboratory testing on NX-core (5.4 cm diameter) at an L/D ratio of two. Joint properties for Mohr–Coulomb failure criteria are given in Table 6.11. Joint normal and shear stiffness (K_n, K_s) that relate normal stress and shear stress to corresponding displacements are also given in Table 6.11 as are the joint spacings (S). What extraction ratio is possible with no size effect on pillar strength and no joints present when a safety factor with respect to maximum shear stress of 1.75 is required? What is the maximum possible extraction ratio? Comment.

6.43 With reference to the data given in Problem 6.42, if a joint safety factor of 1.75 with respect to shear is required, what extraction ratio is possible considering only joint Set 3 as important? What is the maximum possible extraction ratio allowed by joint Set 3? Comment.

6.44 With reference to the data in Table 6.10, occasionally, caving to the coal "rider" seam just below the limestone stratum happens at intersections of the 20 ft wide entries and crosscuts. Specify a cable bolt plan that will prevent these caves assuming the steel–

Table 6.12 Strata data for Problem 6.46

Rock type	E 10^6 psi (GPa)	v	C_o psi (MPa)	T_o psi (MPa)	γ pcf (kN/m^3)	h ft (m)
R5	12.6 (86.9)	0.19	12,700 (87.6)	1,270 (8.8)	160 (25.3)	21.5 (6.55)
R4	12.7 (87.6)	0.21	14,200 (97.9)	1,34 (9.2)	158 (25.0)	15.2 (4.63)
R3	5.4 (37.2)	0.23	7,980 (55.0)	610 (4.2)	152 (23.1)	1.0 (0.30)
R2	6.3 (43.4)	0.31	8,450 (58.3)	760 (5.2)	146 (24.7)	1.5 (0.46)
R1	12.6 (86.9)	0.19	12,700 (87.6)	1,270 (8.8)	156 (24.7)	6.0 (1.83)
S1	0.75 (5.2)	0.27	3,750 (25.9)	425 (2.9)	92 (14.6)	3.0 (0.91)
F1	12.3 (84.8)	0.18	11,500 (79.3)	1,050 (7.2)	155 (24.5)	10.3 (3.14)
F2	8.1 (55.9)	0.33	13,700 (94.5)	1,260 (8.7)	160 (25.3)	13.0 (3.96)

grout interface bond strength is 750 psi. This support is in addition to the earlier roof bolting plan.

6.45 After grouting, a cemented fill, that has completely filled the old workings, has a Young's modulus of 100 ksi (690 MPa), and an unconfined compressive strength of 3,400 psi (23.4 MPa), as determined from unconfined compressive strength tests. The original extraction ratio in this area is estimated to be about 40%. Assume that the old pillars deteriorate to the point where they have no support capability. Estimate the maximum subsidence that might occur in this event. Explain why this is a maximum or upper bound estimate and why the actual "settlement" would probably be less. Or is it?

6.46 A thin "low" metallurgical coal seam 3 ft (1 m) thick is mined at a depth of 980 ft (300 m). The main entries are mined 4 ft into the R1 roof stratum to give sufficient clearance for track haulage. With reference to the stratigraphic column in the sketch and the rock properties given in Table 6.12, assume rectangular pillars that are twice as long as they are wide and crosscuts are the same width as the entries. Find:

1 the maximum possible extraction ratio;
2 pillar width (at a safety factor of 1.5 with respect to compression), and
3 the crosscut spacing (at $F_c = 1.5$). Show by sketch.

Surface

(Not to scale)	Overburden
R5 mudstone	
R4 sandstone	
R3 laminated shale	
R2 laminated sandstone	
R1 (roof) sandstone	
S1 (seam – coal)	
F1 (floor) shale	
F2 limestone	

Sketch of the stratigraphic sequence

6.47 Main entries are driven in sets of seven at a depth of 1,530 ft in flat-lying strata. Entries are 20 ft wide (W_o); pillars are 80 ft wide (W_p). In vertical section showing entries and pillar widths, the two-dimensional extraction ratio $R' = W_o/(W_o + W_p)$.

Three proposals are presented for increasing this extraction ratio to 33% from the existing 20%. (1) Decrease pillar width to 40 ft while keeping entry width the same; (2) increase entry width while keeping pillar width the same, and (3) decrease pillar width while increasing entry width to 33-1/3 ft, while keeping the center to center spacing of pillars and entries the same. Which plan is preferable from the rock mechanics viewpoint? Explain.

6.48 Main entries are driven in sets of seven at a depth of 1,466 mt in flat-lying strata. Entries are 6 m wide (W_o); pillars are 24 m wide (W_p). In vertical section showing entries and pillar widths, the two-dimensional extraction ratio $R' = W_o/(W_o + W_p)$. Three proposals are presented for increasing this extraction ratio to 33% from the existing 20%. (1) Decrease pillar width to 12 m while keeping entry width the same; (2) increase entry width while keeping pillar width the same, and (3) decrease pillar width while increasing entry width to 10 m, while keeping the center to center spacing of pillars and entries the same. Which plan is preferable from the rock mechanics viewpoint? Explain.

6.49 Development entries 16 ft (4.9 m) wide are driven in a steeply dipping anthracite coal seam at a depth of 760 ft (232 m). Crosscuts are not a factor with distance between raises on the dip at 300 ft (91 m). (The extraction ratio in cross section is thus R′.) Strata dip is 55°; seam thickness is 28 ft (8.5 m). The premining stress field is considered to be attributable to gravity alone, although no measurements are available. Laboratory tests for unconfined compressive strength show a size effect given by $C_p = C_1\,[0.78 + 0.22(D/L)]$. Specifically, $C_1 = 5,250$ psi (36.2 MPa). Triaxial test show data that fit a Mohr–Coulomb criterion reasonably well with $\phi = 35^\circ$ and $c = 1,210$ psi (8.34 MPa). A safety factor with respect to pillar compression of 2.75 is desired. Estimate the maximum safe extraction ratio and corresponding pillar width?

6.50 Consider an underground limestone mine in flat strata where pillars are 95 ft high and and depth is 845 ft. Rock properties are: $E = 11.3$ million psi, $v = 0.20$, $C_o = 21,500$ psi, $T_o = 1,680$ psi, unit weight = 156 pcf. A joint set of variable dip pervades the mine. Mohr–Coulomb properties are: $c = 630$ psi, $\phi = 40°$. Pillars must have a safety factor of 2.5 with respect to any potential failure mechanism. Is this possible? Show why or why not.

6.51 Consider an underground limestone mine in flat strata where pillars are 29 m high and depth is 258 m. Rock properties are: $E = 77.9$ GPa, $v = 0.20$, $C_o = 148$ MPa, $T_o = 11.6$ MPa, unit weight = 24.7 kN/m^3. A joint set of variable dip pervades the mine. Mohr–Coulomb properties are: $c = 4.34$ MPa, $\phi = 40°$. Pillars must have a safety factor of 2.5 with respect to any potential failure mechanism. Is this possible? Show why or why not.

6.52 With reference to the sketch, Table 6.13, and the lower coal seam, what is the maximum possible extraction ratio?

6.53 With reference to the sketch, Table 6.13 and the lower coal seam, if a factor of safety of 1.5 with respect to compression is required when entries and crosscuts are actually 18 ft (5.5 m) wide and pillars are twice as long as they are wide, find (with neglect of any size effect on strength):

1 width of pillars and length of pillars;
2 entry and crosscut spacing (center to center).

Table 6.13 Strata properties for Problems 6.52, 6.53

Rock type	γ-pcf (kN/m^3)	$E(10^6)$psi (GPa)	ν	Co(psi) (MPa)	To(psi) (MPa)	h(ft) (m)
1 Over burden	144.0 (22.8)	1.70 (11.7)	0.25	1,000 (6.9)	50 (0.34)	1,694 (516)
2 Massive sandstone	148.2 (23.4)	3.00 (20.7)	0.20	19,000 (131.9)	850 (5.86)	1,714 (522)
3 Coal	75.0 (11.9)	0.35 (2.4)	0.30	3,500 (24.1)	130 (0.90)	1,719 (524)
4 Layered sandstone	149.5 (23.6)	1.50 (10.3)	0.10	17,500 (120.7)	700 (4.83)	1,741 (531)
5 Massive sandstone	148.2 (23.4)	3.00 (20.7)	0.20	19,000 (131.0)	850 (5.86)	1,744 (532)
6 Sandy shale	170.0 (26.9)	4.50 (31.0)	0.10	15,000 (103.4)	1,000 (6.90)	1,747 (533)
7 Coal	75.0 (11.9)	0.35 (2.4)	0.30	3,500 (24.1)	1,300 (8.97)	1,754 (535)
8 Massive sandstone	148.2 (23.4)	3.50 (24.1)	0.20	19,500 (134.5)	1,000 (6.90)	1,778 (542)
9 Layered sandstone	149.5 (23.6)	1.50 (10.3)	0.10	17,500 (120.7)	700 (4.83)	2,250 (658)

Sketch and geologic column

6.54 A tabular ore body is 32 ft (9.75 m) thick, strikes due north, and dips 22° east. The premining stress field is due to gravity alone. Stopes are excavated at a depth of 2,590 ft (179 m) in a regular room and pillar configuration. A lane and pillar method is used, so the system appears two-dimensional in section. Lanes (rooms) are 52 ft (16.9 m) wide. Rock properties are:

$$C_o = 18,500 \text{ psi}(128 \text{ MPa}), \quad T_o = 1,650 \text{ psi}(11.5 \text{ MPa}),$$
$$E = 11 \times 10^6 \text{ psi}(75.9 \text{ GPa}), \quad \nu = 0.19$$

A safety factor of 1.5 with respect to the maximum shear stress is required for pillar design. Find the extraction ratio and pillar dimensions assuming no size effect on strength.

6.55 With reference to the data given in Problem 6.54, suppose that a joint system is present. If the joints also strike due north and dip to the 45° east, find the joint cohesion needed for a safety factor or 1.5 with respect to joint shear. Note: the joints are clay-filled and have a friction angle of 18°.

Chapter 7

Three-dimensional excavations

Three-dimensional excavations include caverns such as hydroelectric machine halls for civil engineering purposes and stopes in hardrock mines. Stopes are regions in mines where ore is excavated. In hardrock mines, stopes range from relatively large brick-shaped excavations to thin, tabular zones that may be flat lying, gently to steeply dipping, or vertical. Stopes differ from shafts, tunnels, and entries. The latter excavations are long relative to cross-sectional dimensions and are amenable to two-dimensional analysis (plane strain, plane stress, axial symmetry). Cavern and stope design usually requires a three-dimensional analysis of stress. The purpose of an analysis is the same in any case: determination of magnitudes and locations of peak tensile and compressive stresses. From these data, stress concentration factors may be obtained for use in design safety factor calculations:

$$\begin{aligned} F_c &= \frac{C_o}{\sigma_c} = \frac{C_o}{K_c S_1} \\ F_t &= \frac{T_o}{\sigma_t} = \frac{T_o}{K_t S_1} \end{aligned} \tag{7.1}$$

where C_o, σ_c, K_c, T_o, σ_t, K_t, and S_1 are unconfined compressive strength, peak compressive stress, compressive stress concentration factor, tensile strength, peak tension, tensile stress concentration factor, and major principal stress before excavation. When compressive stress is considered positive, then a positive stress concentration factor implies a peak compressive stress; a negative stress concentration factor implies tension. A safety factor greater than one at a point of peak stress, assures that all other points at the excavation wall are also safe with respect to compression or tension or both.

The safety factors defined in (7.1) are not the only safety factors possible because definitions depend on the measures of strength and stress. An alternative to the Mohr–Coulomb definition in (7.1) is a safety factor based on the well-known Drucker–Prager (DP) yield condition. This yield condition is based on all three principal stresses. If one considers a root-mean-square measure of the principal shear stresses, that is,

$$\mathrm{RMS} = \left\{\left(\frac{1}{3}\right)\left[\left(\frac{\sigma_2-\sigma_3}{2}\right)^2 + \left(\frac{\sigma_3-\sigma_1}{2}\right)^2 + \left(\frac{\sigma_1-\sigma_2}{2}\right)^2\right]\right\}^{1/2} = (1/\sqrt{2})J_2^{1/2}$$

where J_2 is a symbol for what is known as the second invariant of deviatoric stress and supposes that J_2 is a function of the mean normal stress $\sigma_m = (\frac{1}{3})(\sigma_1+\sigma_2+\sigma_3) = (\frac{1}{3})I_1$ where I_1 is own as the first invariant of stress, then $J_2^{1/2} = AI_1 + B$ which is the Drucker–Prager yield

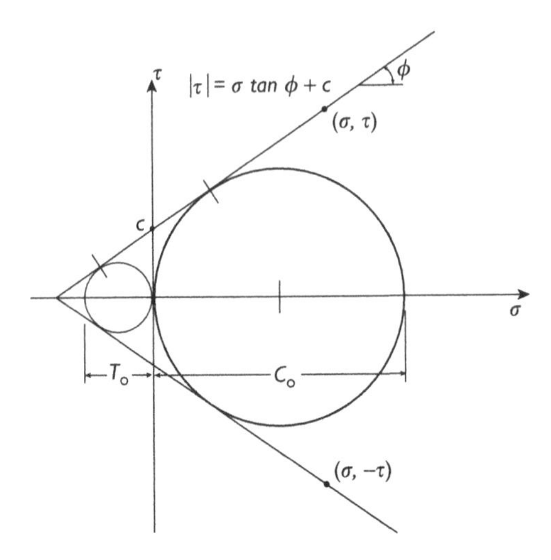

Figure 7.1 Mohr–Coulomb shear strength criterion with C_o and T_o strength circles shown.

condition and compression is considered positive. The left side is the greatest possible value of $J_2^{1/2}$ and represents the strength of the material; A and B are strength properties.

A factor of safety at a point using DP, is then $\text{FS} = J_2^{1/2}(\max)/J_2^{1/2}(\text{actual})$. Under strictly uniaxial conditions, compression or tension, this safety factor reduces to those defined in (7.1). Additional discussion of strength and yield criteria including the popular Hoek–Brown (HB) criterion and n-type criteria that provide technically sound fits to test data are presented in Appendix B.

Failures on geologic structures such as joints that intersect excavation walls may be anticipated in a similar manner using joint strength properties and joint stresses in analogous safety factor calculations. Thus,

$$\text{FS}_j = \frac{\tau_j(\text{strength})}{\tau_j(\text{stress})} \tag{7.2}$$

that includes uniaxial tensile and compressive failures as special cases, as illustrated in Figure 7.1 where Mohr–Coulomb shear strength is used.

7.1 Naturally supported caverns and stopes

There are far fewer analytic solutions available for stress concentration factors for even simple three-dimensional shapes such as a cube than for two-dimensional openings. Ellipsoids with three semi-axes (a, b, c) parallel to x, y, z are an exception. Solutions exist for stresses about spheres $(a = b = c)$, oblate spheroids $(a = b > c)$, prolate spheroids $(a = b < c)$, and triaxial ellipsoids $(a > b > c)$. However, considerable numerical work is still required for evaluation. Direct numerical solution is needed for other excavation shapes. In all cases, application here is restricted to homogenous, isotropic, linearly elastic rock masses. Although, spheroids and ellipsoids are not often practical excavations shapes, they offer insight into stress concentration in three dimensions and allow comparisons to be made with associated two-dimensional openings that may qualitatively be extended to other shapes.

Despite restrictions of the elastic model of rock mass behavior, useful design guidance may still be obtained by identifying locations and magnitudes of peak stresses at excavation walls. Such guidance is especially helpful in consideration of extreme preexcavation stress states including the hydrostatic case ($S_1 = S_2 = S_3$) and the uniaxial case ($S_1 \neq 0, S_2 = S_3 = 0$) where S_1, S_2, and S_3 are the major, intermediate, and minor principal stresses before excavation.

Spheroidal excavations

A *spherical* excavation is shown in Figure 7.2; because of symmetry, only the first octant is depicted in a rectangular coordinate system. Compressive stresses S_x, S_y, and S_z are pre-excavation stresses in the direction of the *x*-, *y*-, and *z*-axes, respectively. Stress concentration factors for unit applied loads are given in Table 7.1. Multiplication of the actual applied stress by the stress concentration factor for the direction (*x*, *y*, or *z*) and point (*A*, *B*, or *C*) of interest gives the stress component at the considered point. Superposition allows for arbitrary loads applied parallel to the coordinate axes.

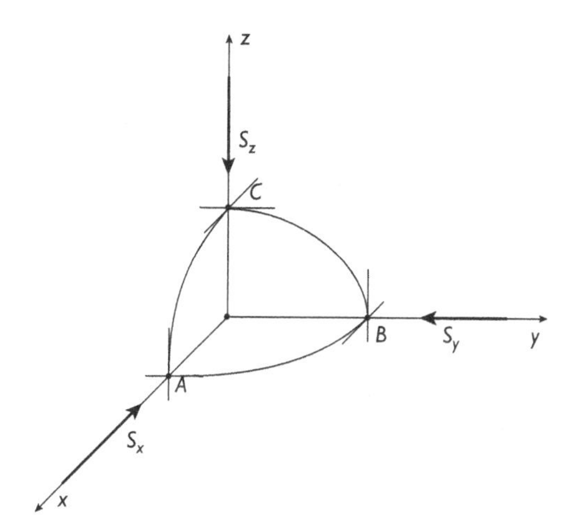

Figure 7.2 A sphere under load.

Table 7.1 Stress concentration factors for a spherical hole under uniaxial load

Factor	Load[a]	$S_x = 1.0$ $S_y = S_z = 0.0$			$S_y = 1.0$ $S_z = S_x = 0.0$			$S_z = 1.0$ $S_y = S_x = 0.0$		
	Point	A	B	C	A	B	C	A	B	C
$K_x =$		0	2.00	2.00	0	−0.50	0.00	0	0.00	−0.50
$K_y =$		−0.50	0	0.00	2.00	0	2.00	0.00	0	−0.50
$K_z =$		−0.50	0.00	0	0.00	−0.50	0	2.00	2.00	0

Notes

a *A* = on *x*-axis at hole wall, *B* = on *y*-axis at hole wall, *C* = on *z*-axis at hole wall. Poisson's ratio = 0.20 (derived from Coates, 1970).

Example 7.1 As a simple example, under hydrostatic loading ($S_x = S_y = S_z = 1.0$), the stresses σ_{xx}, σ_{yy}, and σ_{zz} at B are 1.5 (2.00 − 0.50 + 0.00), 0.0 (0.00 + 0.00 + 0.00), and 1.5 (0.0 − 0.50 + 2.00), respectively, with the assumption of Poisson's ratio equal to 0.20. In fact, uniform or hydrostatic loading results in a normal stress acting tangential to the hole wall of 1.5. By comparison, hydrostatic loading of a circular opening gives a peak compressive stress of 2.0 that is 33% higher in comparison.

Example 7.2 If $S_x = (1/3)S_z$ and $S_y = (1/3)S_z$ and $S_z = 1.0$, find the vertical and horizontal stresses at A, B, and C, assuming Poisson's ratio is 0.20.

Solution: From the data in Table 7.1,

$$K_v\,(A, B) = [-0.50(1/3) + 0.0(1/3) + 2.00(1.0)] = 1.83$$
$$K_h\,(A, B) = [-0.50(1/3) + 2.00(1/3) + 0.00(1.0)] = 0.500$$
$$K_v\,(C) = [0.00(1/3) + 0.00(1/3) + 0.00(1.0)] = 0.00$$
$$K_h\,(C) = [2.00(1/3) + 0.00(1/3) + (-0.50)/(1.0)] = 0.167$$

Under a unit hydrostatic load, simple formulas exist for the radial and tangential normal stresses (see, for example, Love, 1944). Thus,

$$\sigma_{rr} = 1 - \left(\frac{a}{r}\right)^3, \qquad \sigma_{tt} = \sigma_{pp} = 1 + 0.5\left(\frac{a}{r}\right)^3 \tag{7.3}$$

where a is the sphere radius and r is the radial coordinate. Clearly, at the hole wall the radial stress is zero, as it must be, while the tangential stresses are indeed concentrated by a factor of 1.5. It is also clear that peak (tangential stress) decreases with distance from the hole wall in proportion to r^{-3}, which is even more rapid than the r^{-2} decay in the two-dimensional, circular hole case. At the same time, the radial stress rapidly increases to the preexcavation value as shown in Figure 7.3. Thus the zone of influence of three-dimensional openings tends to be less than analogous two-dimensional openings and the interaction between adjacent cavities also tends to be less.

The hydrostatic case is quite exceptional because of the absence of elastic properties in the stress concentration formulas. Generally, Poisson's ratio must be considered in three-dimensional analyses, unlike analyses in two dimensions. In Table 7.1, Poisson's ratio is assumed to be 0.20, a realistic value for many materials, but by no means always a good estimate. Table 7.2 gives stress concentration factors for a sphere loaded uniaxially in the z-direction as functions of Poisson's ratio, ν. When $\nu = 0.2$, the results are the same as in Table 7.1 for $S_z = 1.0$ and $S_x = S_y = 0.0$. The expressions in Table 7.2 may be used to obtain stress concentration factors for loading along the x- and y-axes with appropriate reorientation. The tension at the top of the sphere according to Table 7.2 is sensitive to Poisson's ratio.

Example 7.3 If $S_x = (1/3)S_z$ and $S_y = (1/3)S_z$ and $S_z = 1.0$, find the vertical and horizontal stresses at A, B and C, assuming Poisson's ratio is 0.0.

***Solution*:** From the data in Table 7.1,

$K_v (A, B) = [-(3/14)(1/3) + (-3/14)(1/3) + (27/14)(1.0)] = (25/14) = 1.79$
$K_h (A, B) = [-(3/14)(1/3) + (27/14)(1/3) + (-3/14)(1.0)] = (5/14) = 0.357$
$K_v (C) = [0.00(1/3) + 0.00(1/3) + 0.00(1.0)] = 0.00$
$K_h (C) = [(27/14)(1/3) + (-3/14)(1/3) + (-3/14)(1.0)] = (5/14) = 0.357$

A summary for the Cartesian stress concentration factors is given in the Example Tablewhere *A*, *B*, and *C* are the *x*, *y*, and *z* axis intercepts.

Table for Example 7.3[a]

Concentration	*Point*		
	A	*B*	*C*
K_x	0	5/14	5/14
K_y	5/14	0	5/14
K_z	25/14	25/14	0

Notes
a $\nu = 0.0$.

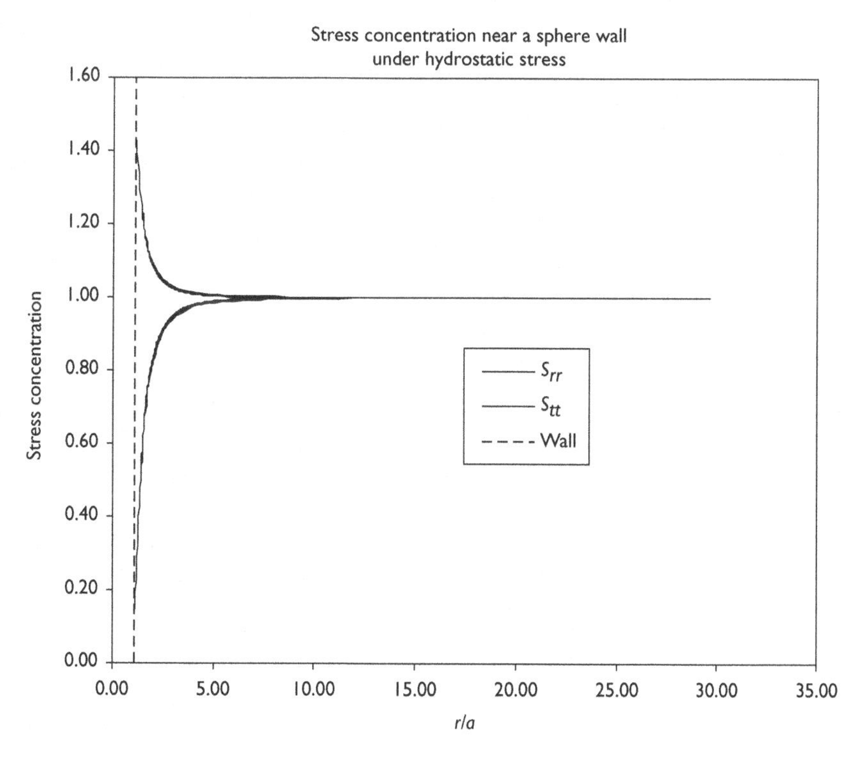

Figure 7.3 Stress distribution about a spherical cavity under hydrostatic stress obtained from finite element analysis. Peak compressive stress concentration is 1.5. Tension is absent. The wall is at $r/a = 1.0$.

Table 7.2 Stress concentration factors for a sphere as functions of Poisson's ratio

Factor	*Load*	$S_z = 1.0 \quad S_x = S_y = 0.0$		
	Point	A	B	C
$K_x =$		0	$\frac{30v}{14-10v} - \frac{3(1+5v)}{14-10v}$	$-\frac{3(1+5v)}{14-10v}$
$K_y =$		$\frac{30v}{14-10v} - \frac{3(1+5v)}{14-10v}$	0	$-\frac{3(1+5v)}{14-10v}$
$K_z =$		$\frac{30v}{14-10v} - \frac{3(1+5v)}{14-10v}$	$\frac{30v}{14-10v} - \frac{3(1+5v)}{14-10v}$	0

Source: Developed from Terzaghi and Richart, 1952.

Table 7.3 Peak tension and compression at sphere walls under uniaxial load $S_z = 1.0$

Poisson's ratio v	*Compression at sides A and B*	*Tension at top C*
0.0	1.929	−0.214
0.1	1.962	−0.346
0.2	2.00	−0.500
0.3	2.046	−0.682
0.4	2.100	−0.900
0.5	2.167	−1.167

Compression at the sides of the sphere is relatively insensitive as the data in Table 7.3 show.

Figure 7.4 shows stress distribution about a sphere under uniaxial load ($S_z = 1.0$, $S_x = S_y = 0.0$) with Poisson's ratio of 0.2. Peak compressive stress concentration is 2.0 and is vertical at the equator (*A*, *B*) of the sphere. Peak tensile stress concentration is −0.5 and is horizontal at the pole (*C*).

An *oblate spheroid* is shown in Figure 7.5 with stresses applied along the *x*, *y*, and *z* axes. The plane $z = 0$ is a circle bounded by the equator of the spheroid, while the end of the *z*-axis *c* is the pole of this figure. Vertical sections containing the *z*-axis are identical ellipses; each has a major semi-axis $a(= b)$ and minor the semi-axes *c*. There is rotational symmetry about the *z*-axis. Under uniaxial vertical loading ($K_o = 0$) and hydrostatic loading ($K_o = 1$) there is also a symmetry of stress about the *z*-axis.

Figure 7.6 shows peak stress concentrations for oblate spheroids (thickened at the equator with major and minor semi-axes *a* and *c*, respectively) as a function of the aspect ratio *a/c*. The prehole stress state has little influence on the peak compressive stress concentration that occurs at the ends of the major axis of the opening. In both load cases, this peak rises rapidly with increasing aspect ratio. However, the stress concentration at the opening top and bottom (ends of minor axis) changes from compression to tension as the load changes from hydrostatic to uniaxial. The tensile stress concentration does not exceed −1. The same phenomenon is present in two dimensions, although the peak compressive stress concentrations are much greater in two dimensions. For example, under uniaxial loading, a two-dimensional elliptical opening with an aspect ratio of 5, has a compressive stress concentration of 11 (compared

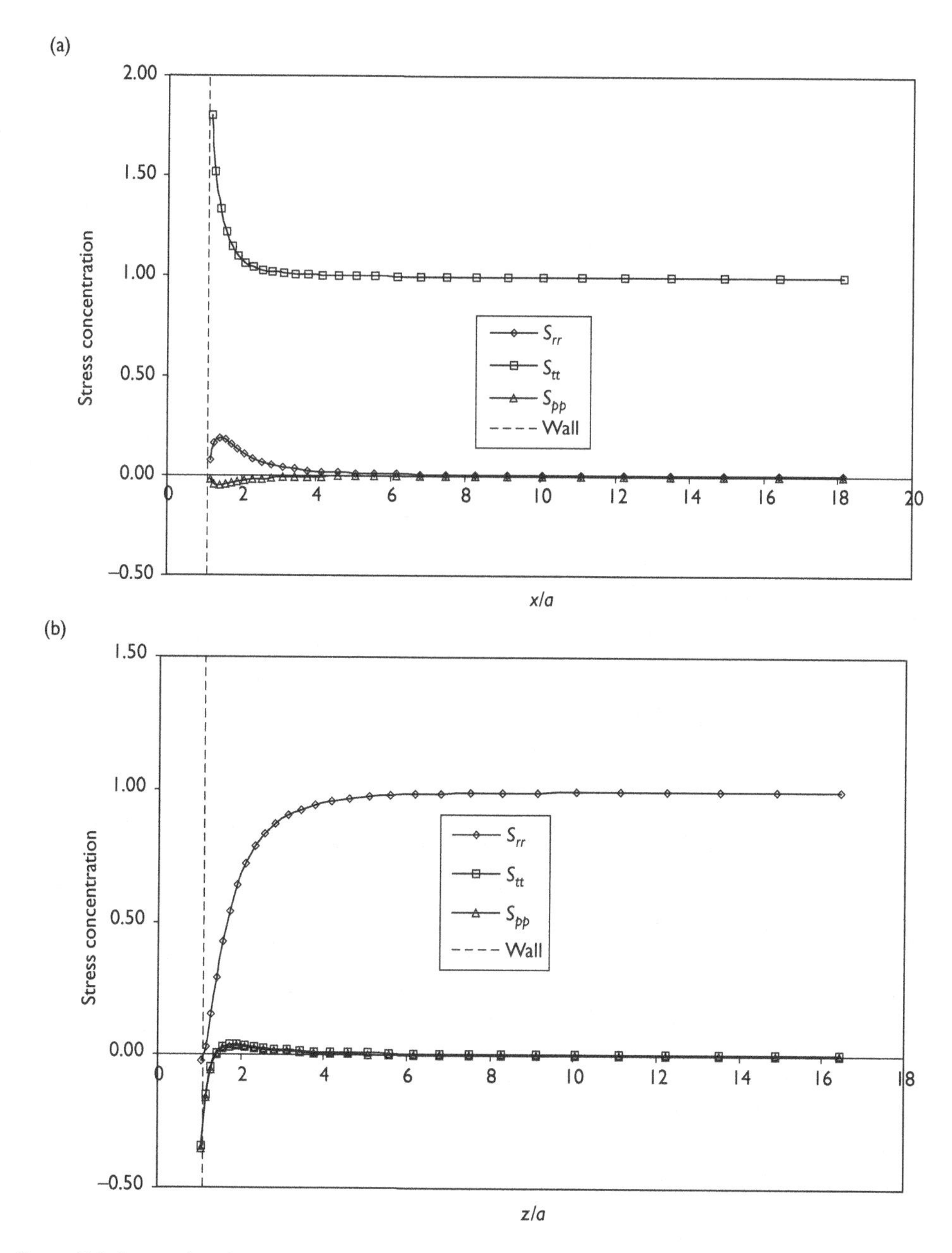

Figure 7.4 Stress distribution about a spherical cavity under vertical load (S_z) only: (a) along the horizontal *x*-axis. $S_{tt} = S_{zz}$, $S_{rr} = S_{xx}$, $S_{pp} = S_{yy}$ and (b) along the vertical z-axis. $S_{rr} = S_{zz}$, $S_{tt} = S_{yy}$, $S_{pp} = S_{xx}$.

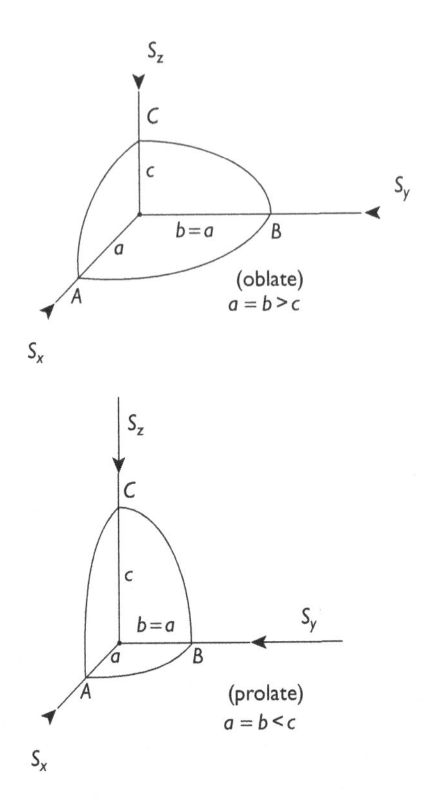

Figure 7.5 Spheroids under normal stress loadings.

with 7 in three dimensions, Figure 7.6). An aspect ratio of 10 gives a stress concentration of 21, but only 13.4 in three dimensions, while an aspect ratio of 50 in two dimensions gives a stress concentration of 101, Figure 7.6. An interesting feature of the peak compressive stress concentration along the equator of the spheroid is the linear dependency on aspect ratio. This feature is similar to that for a two-dimensional ellipse where $K = 2(a/c) + (1 - K_o)$ that has the linear form $y = ax + b$.

Peak stress concentration, although extremely high at high aspect ratios, decreases rapidly with distance from the excavation surface as shown in Figure 7.7.

Figure 7.8 shows a triaxial ellipsoid and the stress concentration along the x-axis in the case of $c = 1$, $b = 3$, and $a = 9$ and Poisson's ratio = 0.3. At the end of the x-axis the peak stress concentration is vertical and is 6.0. By comparison, in two dimensions, stress concentration at the same location but about an ellipse with a semi-axis ratio of 9 is 19, under the same vertical load. Again, the three-dimensional stress concentration is much less than an associated two-dimensional stress concentration.

The decrease of stress concentration with distance from the ellipsoid end (point A) is rapid. At a distance of $x/a = 1.1$, the effect of stress concentration is slight. This point is at a distance of $0.1x/a = 0.3b/a = 0.9x/c$ into the solid along the x-axis. If the minor semi-axis c is indicative of a "radius," so $2c = D$ (diameter), then the zone of negligible

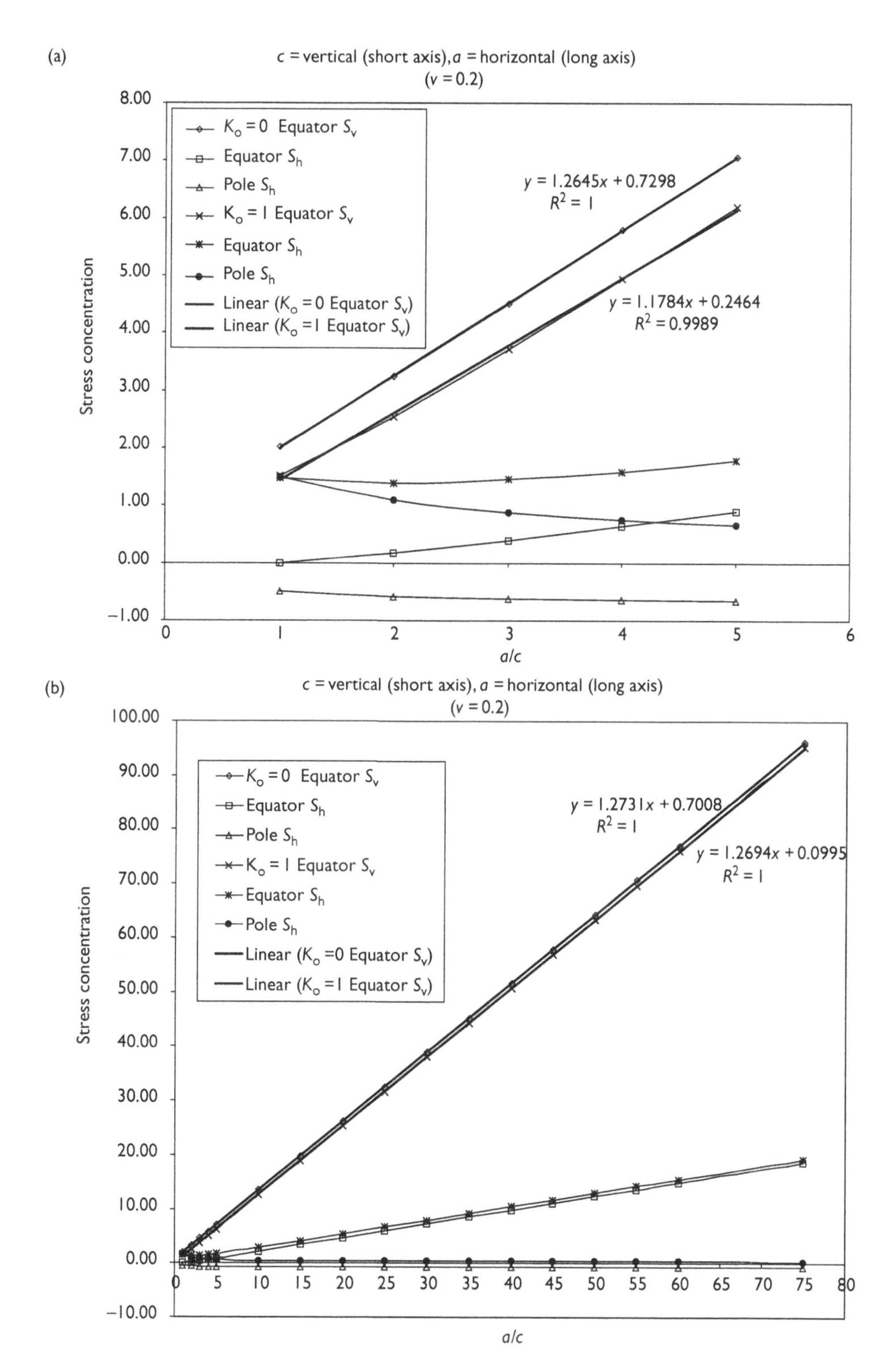

Figure 7.6 Stress concentration data about oblate spheroids under uniaxial vertical ($K_o = 0$, open symbols) and hydrostatic ($K_o = 1$, solid symbols) loading: (a) low aspect ratio and (b) high aspect ratio (after Terzaghi and Richart, 1952).

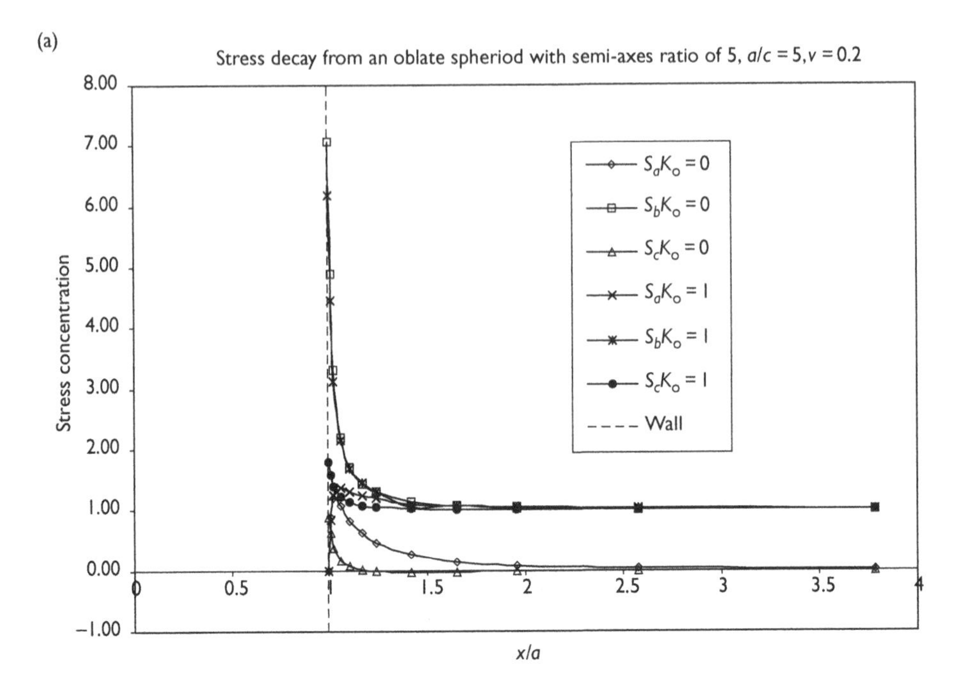

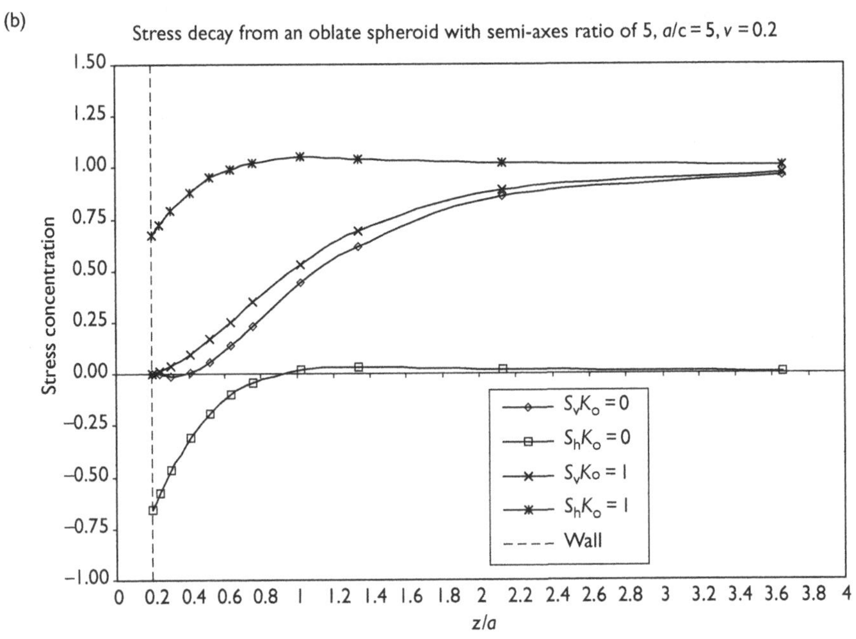

Figure 7.7 Rapid decrease of stress away from the surface of an oblate spheroid with $c/a = 5$: (a) stress decay along the horizontal x-axis and (b) stress decay along the vertical z-axis (after Terzaghi and Richart, 1952).

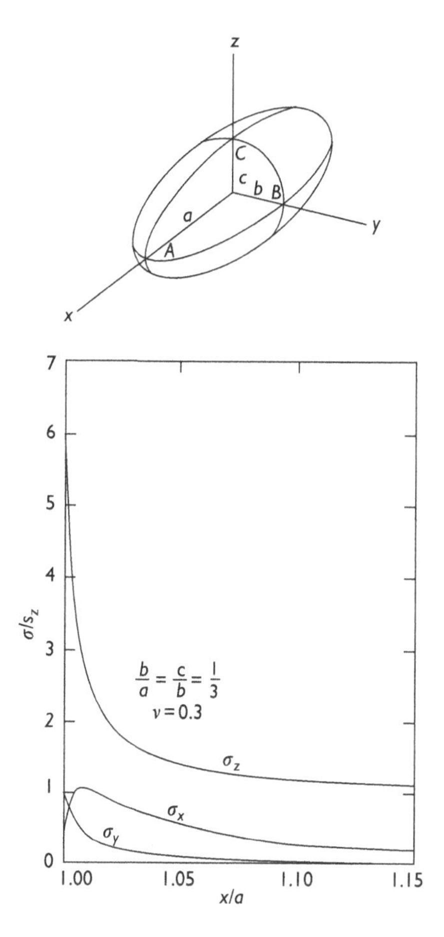

Figure 7.8 Stress concentration about a triaxial ellipsoid with $a = 3b = 9c$ under uniaxial (vertical) loading with Poisson's ratio of 0.3 (after Sadowsky and Sternberg, 1949).

stress concentration is only $0.45D$ away from the hole wall. This result indicates that the zone of influence in three dimensions is smaller than in two dimensions and is proportional to the shortest dimension of the opening rather than the long dimension (as for tunnels and shafts).

Consideration also needs to be given to stress concentration and the rate of decrease of stress along the y and z axes as well as the x axis. As the major semi-axis a is increased indefinitely and the ellipsoid becomes very long in the x direction, the shape approaches a tunnel-like two-dimensional opening with an elliptical cross section in the yz plane that has dimensions b and c with an aspect ratio of 3. Peak compressive and tensile stress concentrations about an ellipse with an aspect ratio of three under uniaxial compression normal to the long axis are 7.0 and −1.0, respectively. These concentrations occur at points corresponding to B and C in Figure 7.8. Both are greater than corresponding three-dimensional stress concentrations. In two dimensions, the long dimension of the opening is used in the "one diameter" rule of thumb that defines the zone of hole influence. Stress concentration at the ends of the opening along the x axis still decay very rapidly, but less so in the y and z direction. These limited observations suggest that the *intermediate dimension of a three-dimensional opening be used in the "one-diameter" rule of thumb for defining a zone of influence.*

Analysis of stress about mathematically simple shapes such as spheres, spheroids, and ellipsoids and plots of some of the results allows one to observe the nature of stress concentration in three dimensions and to compare with analogous results in two dimensions. The main results of comparisons of three-dimensional with two-dimensional stress analyses are:

1 stress concentrations are lower about three-dimensional openings than two-dimensional openings of similar cross section;
2 the zone of influence of a three-dimensional opening may be approximated by a "one diameter" rule of thumb using the intermediate dimension of the opening;
3 when the long axis of a three-dimensional opening is more than twice the intermediate axis length, the peak stress concentrations approach those of a two-dimensional section.

Cubical and brick-shaped excavations

A step toward more realistic shapes is the calculation of stress about a cube that is necessarily done numerically. Brick-shaped stopes are, technically, rectangular parallelopipeds with semi-axes (a, b, c) parallel to Cartesian coordinate axes (x, y, z) that may also be considered compass coordinates (x = east, y = north, z = up). Beginning with a cube shown in Figure 7.9, as the vertical c-axis increases at fixed a- and b-axes, a shaft-like opening is generated. Fixing b and c, while *increasing a* creates a tunnel-like opening. Fixing a and c, while increasing b also creates a similar tunnel-like opening. Tabular openings have a large in-plane extent relative to thickness. Fixing b and c, while *decreasing a* creates a tabular opening. Fixing a and c, while decreasing b creates a similar tabular opening. A tabular opening is also generated by fixing a and b, while decreasing c, as shown in Figure 7.9.

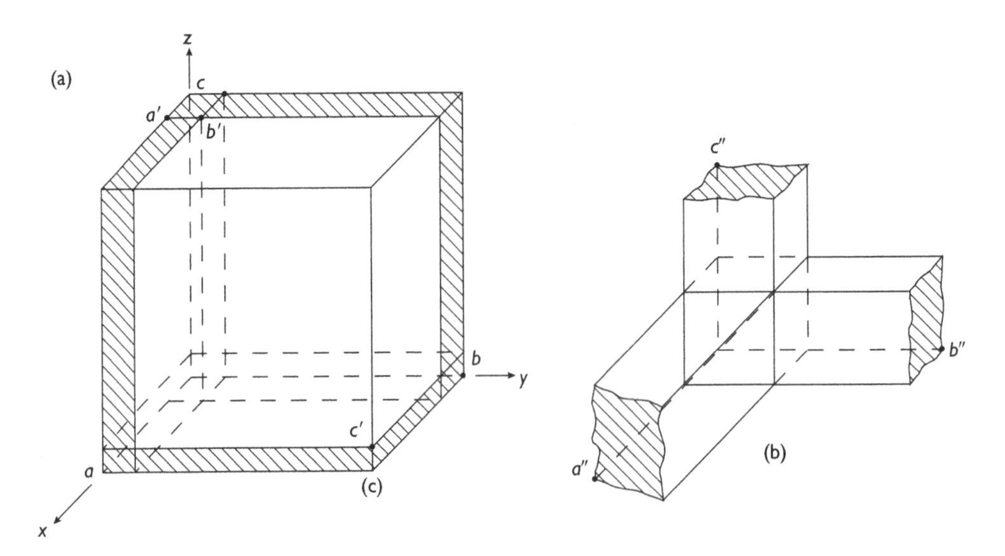

Figure 7.9 (a) Cube, (b) shaft and tunnel-like openings, and (c) tabular openings.

The distinction between the shaft and tunnel-like openings is one of orientation. The same is true of the three tabular openings generated from the basic cube, only orientation changes. In this regard, orientation is important with respect to the preexcavation stress field that may be characterized by magnitudes of premining principal stresses (compression positive) S_1, S_2, S_3, and direction angles θ_x, θ_y, θ_z of the major principal stress (S_1). Directions of the other two principal stresses (S_2, S_3) are fixed by orthogonality and right-handedness of the principal axes. The most convenient orientation occurs when the geometric axes of the excavation are parallel to the preexcavation principal stress directions. In these cases, the excavation of interest is not loaded in shear. Otherwise shear stresses also contribute to stress concentration at the excavation walls. Six stress concentration factors are needed in this general loading case. Each factor is a function of opening geometry.

The many combinations of stress and shape in three dimensions militates against any simple codification. There is also an associated question of when a three-dimensional opening may be approximated with a two-dimensional analysis. Consider a cubical excavation initially and then suppose further excavation occurs along the x-axis. The question is at what distance along the x-axis does the brick-shaped excavation appear essentially two-dimensional. Computational results in Figure 7.10 using a Drucker–Prager strength criterion shows the evolution of safety factor contours in long section as an originally cubical excavation becomes a brick-shaped excavation and approaches tunnel-like conditions. The long section is a vertical plane passing through the center of the excavation; coordinate planes are planes of symmetry, so only one octant is shown. The preexcavation stress state is a uniaxial vertical stress for illustrative purposes; the excavation cross section remains square as advance is made along the x-axis. The black regions in Figure 7.10 are elements that have reached the elastic limit and have yielded with a local safety factor of one. A lobe of relative high stress appears at the edge of the opening within the 1.5 safety factor contour. Interestingly, there appears a "bubble" of relative safety above the failed region that is especially noticeable in (a), the cubical excavation. However, as the excavation is advanced along the x-axis, this region diminishes somewhat, as seen in (b) and (c) as the semi-axes ratio a/c increases to about 2 and then 3. With further advance, this region "pinches off" altogether in (e) and (f) when the semi-axis ratio reaches five and then six. Formation of a safety bubble under the load point (under the applied stress S_z), so to speak, is not unusual.

Another phenomenon is the growth in the height of the yield zone (black region), as excavation progresses from the cube (a) to a tunnel-like excavation (f). This progression is from a three-dimensional opening to one that in cross section could be analyzed as a two-dimensional excavation. A two-dimensional view (yz coordinate plane) would show higher stress and greater extent of yielding than the section through the cube (a). The reason why openings that are long, compared with cross-sectional dimensions (tunnels, shafts), experience higher stress concentrations is the restriction of load redistribution to the sides only. In three dimensions, load is redistributed to sides and ends thus reducing stress concentration and extent of yielding. The presence of yield zones suggests reduction of stress by excavating a more favorable *shape*, for example, by arching the top of an excavation when yielding is indicated there by stress analysis.

Evolution of the three-dimensional cubical excavation toward a tunnel-like two-dimensional excavation is also shown in Figure 7.11. The vertical sections in Figure 7.11 are cross sections through the center of the excavation. The presence of a safety bubble is clear in the cubical excavation case (a) as is the diminution of the bubble with advance of the excavation along the x-axis into the page. Yielding reaches highest extent when the semi-axis

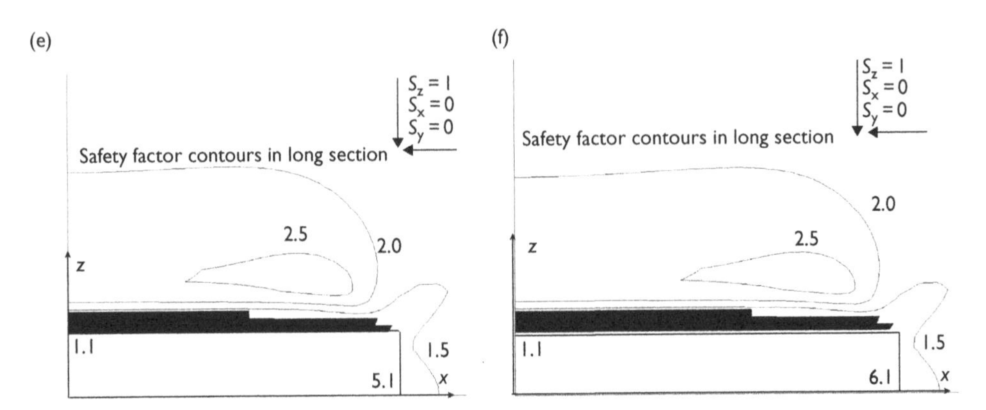

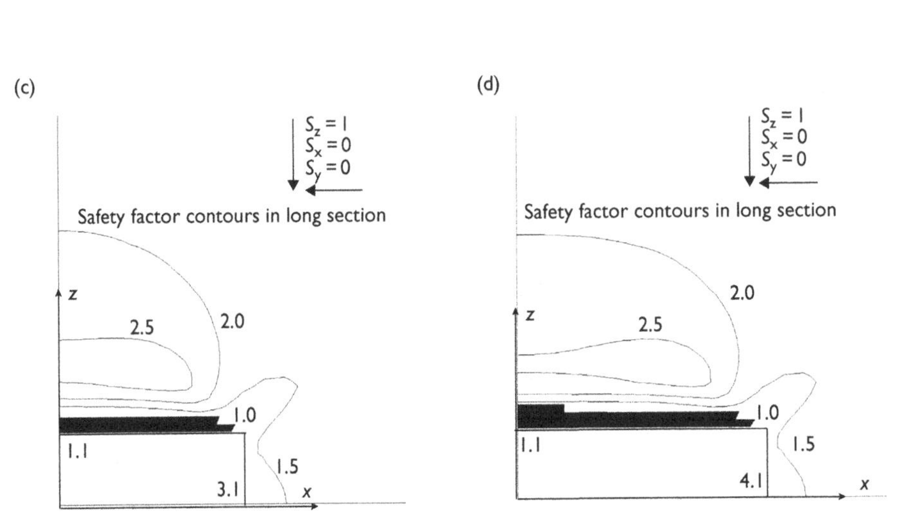

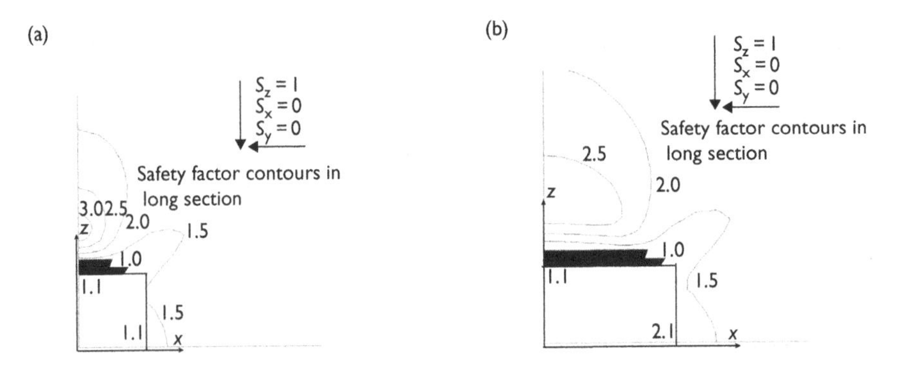

Figure 7.10 Evolution of a cubical excavation to a tunnel-like excavation of square section as seen in contours of safety factor using the Drucker-Prager strength criterion: (a) cube, $a/c = 1$, $b/c = 1$, (b) brick, $a/c \approx 2$, $b/c = 1$, (c) brick, $a/c \approx 3$, $b/c = 1$, (d) brick, $a/c \approx 4$, $b/c = 1$, (e) brick, $a/c \approx 5$, $b/c = 1$, and (f) brick, $a/c \approx 6$, $b/c = 1$.

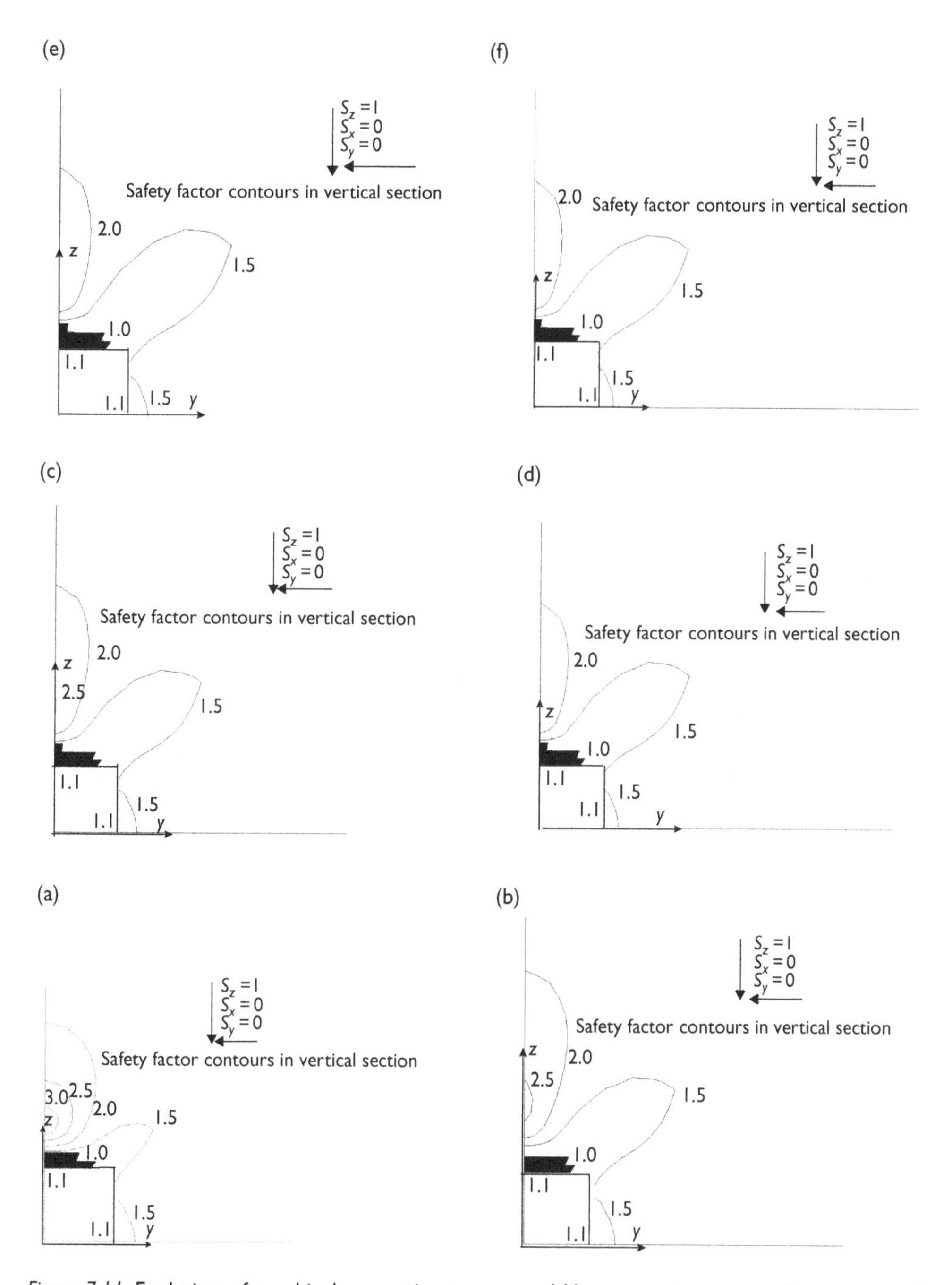

Figure 7.11 Evolution of a cubical excavation to a tunnel-like excavation as seen in contours of safety factor about vertical cross sections using Drucker–Prager strength. (a) Cube, $a/c = 1$, $b/c = 1$; (b) brick, $a/c \approx 2$, $b/c = 1$; (c) brick, $a/c \approx 3$, $b/c = 1$; (d) brick, $a/c \approx 4$, $b/c = 1$; (e) brick, $a/c \approx 5$, $b/c = 1$, and (f) brick, $a/c \approx 6$, $b/c = 1$.

ratio $a/c \approx 3$, although the safety bubble diminishes further as $a/c \approx 4$. Beyond $a/c \approx 4$ there is little difference in cross section safety factor distributions. These results suggest that when excavation length exceeds three times cross section dimensions, tunnel-like conditions are approached. Additional advance then no longer causes further decrease in excavation wall safety.

The best orientation of a three-dimensional excavation of brick shape in a given stress field may also be studied numerically. Under uniaxial load S with other principal preexcavation stresses zero there are three orientations possible when the excavation dimensions are unequal. If a, b, c are the excavation semi-axes and $a > b > c$, the three orientations are S parallel to a, b, or c. One may hypothesize that the most favorable orientation is with S parallel to the long axis a. This orientation minimizes the area exposed perpendicular to the major preexcavation principal stress S. The least favorable would be with S parallel to the short axis c resulting in maximum exposure.

Figure 7.12 shows safety factor contours about a brick-shaped excavation with semi-axes ratios of about $a/c \approx 3$ and $b/c \approx 1.5$. The most favorable orientation is clearly with the long axis of the excavation parallel to the load direction. Extent of the yield zone (black region) is much less in Figures 7.12(a) and (b) than in Figures 7.12(c) and (d) which is the least favorable orientation with the load axis parallel to the short axis of the excavation. In both orientations, a safety bubble occurs under the load above the zones of yielding.

A more challenging situation occurs when the preexcavation stress state is triaxial rather than uniaxial. If the major principal stresses before excavation are S_1, S_2, and S_3 such that $S_1 > S_2 > S_3$ and the excavation dimensions are $L_1 = 2a$, $L_2 = 2b$, and $L_3 = 2c$ with $L_1 > L_2 > L_3$, then there are six distinct orientations of the excavation with respect to the preexcavation stress field (Pariseau, 2005). These six combinations are shown in Table 7.4. One might suppose upon inspection of Table 7.4 that combination 5 would be least favorable because of the switch between the 1 and 3 directions. However, when combinations 5 and 6 are compared as if they were tunnel-like openings, then the greatest stress difference in cross section occurs in combination 6.

Generally speaking, stress concentration in three dimensions can be expected to be lower than in corresponding two-dimensional openings for the physical reason that stress can be redistributed from side to side and back to front in three dimensions but only side to side in two dimensions. As the distance between back and front increases relative to cross-sectional dimensions, stress concentration away from the front and back ends in the midsection of an opening tends toward the higher two-dimensional values. *The most favorable orientation of a three-dimensional excavation is with the long dimension parallel to the greatest preexcavation principal stress and with the short dimension parallel to the least preexcavation principal stress.* The intermediate excavation dimension is then necessarily parallel to the intermediate excavation dimension. This rule places the excavation wall area defined by the product of intermediate by least dimension normal to the greatest preexcavation principal stress. As a consequence the long axis of the excavation is parallel to the greatest preexcavation principal stress. If the excavation wall area is *exposure*, then a rule of thumb for greatest safety is to minimize exposure.

Some choice may be present in design of excavations for civil purposes. If so, then the orientation that favors safety should be specified. In mining, excavation geometry and orientation are constrained by the size and shape of the ore body. Excavation must follow the ore, so there may be little choice for orientating stopes in the most favorable way. Yet another design consideration relates to geologic structure. Shallow excavations may be prone to joint

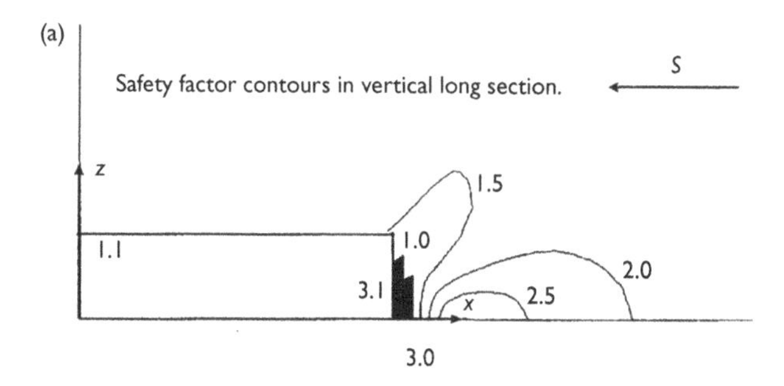

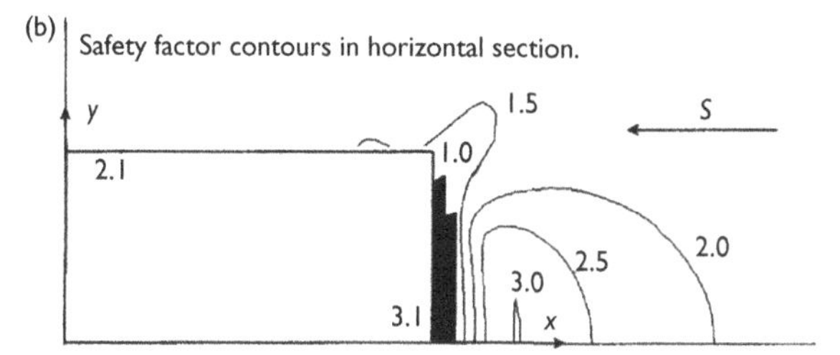

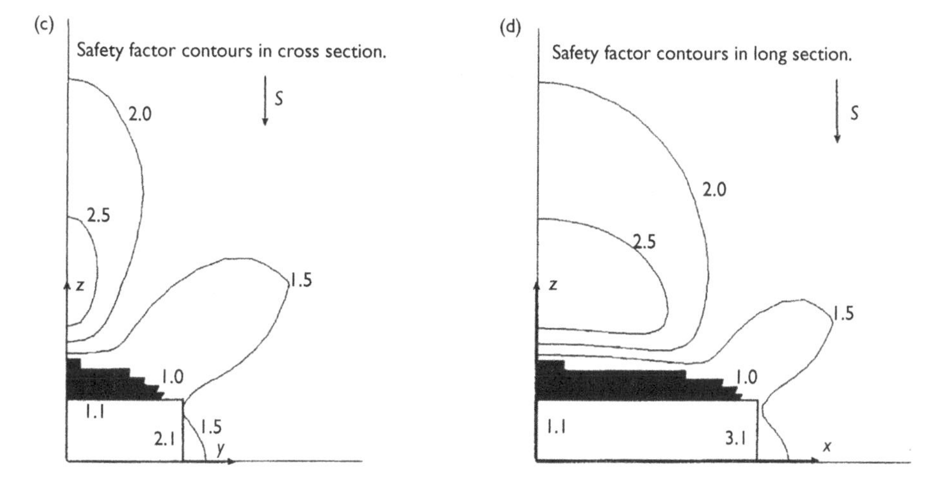

Figure 7.12 Safety factor contours about a brick-shaped excavation with semi-axes $a/c \approx 3$, $b/c \approx 1.5$ under uniaxial load S parallel to the long axis of the excavation. Most favorable orientation shown in (a), (b). Least favorable orientation shown in (c), (d).

Table 7.4 Orientation combinations of preexcavation stress and excavation dimensions from most favorable to least favorable

Stress	*Dimension*		
	S_1	S_2	S_3
1 (most)	L_1	L_2	L_3
2	L_1	L_3	L_2
3	L_2	L_3	L_1
4	L_2	L_1	L_3
5	L_3	L_2	L_1
6 (least)	L_3	L_1	L_2

failure rather than failure under high stress. If so, then orientation with respect to geologic structure is a major design consideration.

Example 7.4 Consider a large underground cavern 10 × 20 × 40 m that is being planned for a depth of 300 m in a stress field such that the vertical preexcavation stress is simply unit weight times depth. The north–south stress is 1.2 times the vertical stress, while the east–west stress is 0.60 times the vertical stress. Determine the most favorable orientation of the proposed excavation.

***Solution*:** According to rule of thumb, the long axis of an excavation should be parallel to the greatest preexcavation principal stress, and the short axis should be parallel to the least preexcavation principal stress. Thus, the 40-m length should be north-south, and the 10-m length should be east–west as shown in the sketch. The 20-m length is vertical.

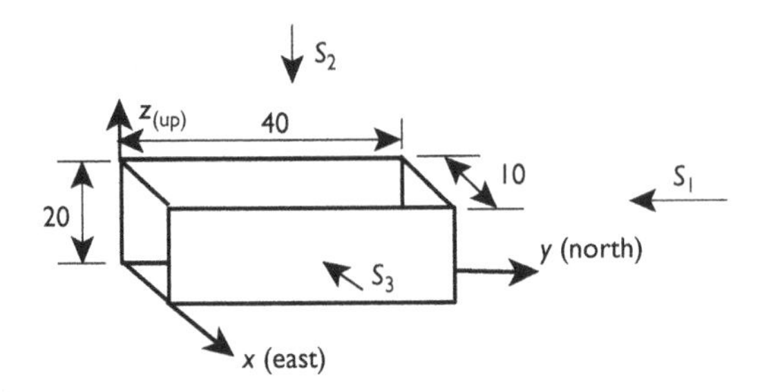

Should a stress analysis show high stress concentration and therefore low safety factors near the excavation top, bottom, and sides, then arching the top may be considered as may be rock mass reinforcement and support.

Example 7.5 Consider the evolving tabular excavation in the sketch. A "drift" is a mine excavation along the strike of the ore body and is a tunnel-like excavation advanced horizontally to the right in the sketch a distance L. After completion, backfill is introduced before

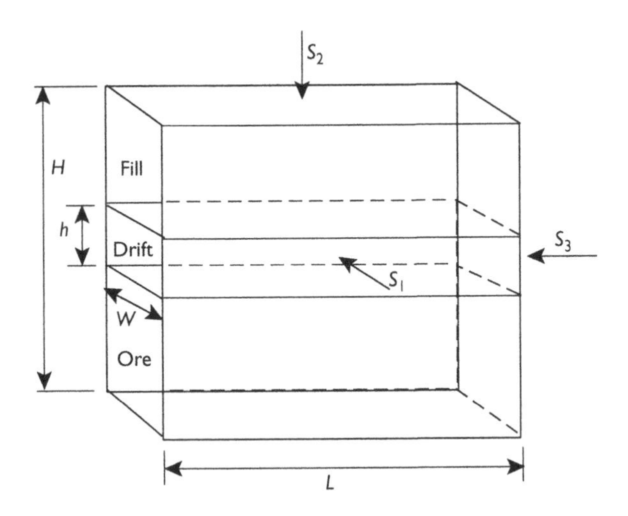

starting another drift below. Drift width W is equal to width of the vein; drift height is h. Progress downward continues along a vertical distance H in this example. Stress measurements show that the greatest premining stress is a compression across the vein; the least premining stress is parallel to the strike of the vein, and the intermediate premining stress is vertical, as shown in the sketch. Determine whether this excavation sequence is favorable to safety and whether the drift dimensions can be adjusted to improve safety about the drift.

***Solution*:** The final configuration results in maximum exposure to the greatest premining stress and, according to the rule of thumb for favorable orientation of three-dimensional excavations, is unfavorable. However, consideration should be given to the evolving geometry as mining progresses and to the influence of fill.

The first drift at the top of the future tabular excavation evolves into a tunnel-like excavation of width W and height h. If $h < W$, then S_1 is parallel to the long dimension of the drift cross section which is favorable to safety. For example, if $W = 15$ and $h = 10$, safety is favored, but if $W = 5$ and $h = 10$, the section has a disadvantage with respect to stress concentration and safety, as illustrated in the second sketch.

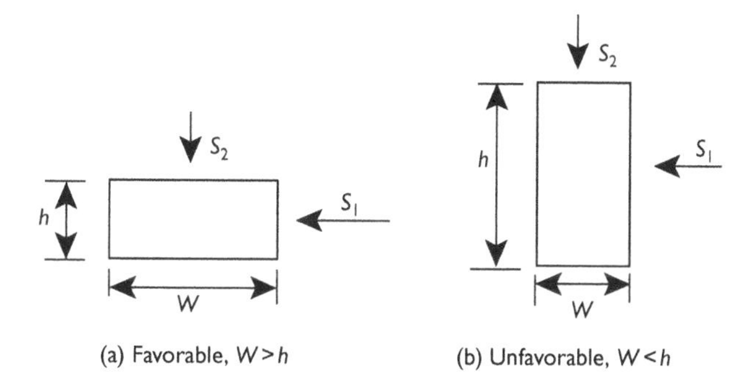

(a) Favorable, $W > h$ (b) Unfavorable, $W < h$

Mine fill is usually much more compliant than the rock mass and has little effect on the deformation and stress concentration that are induced by excavation. For this reason, fill may be neglected. However, this is not to say that fill is unimportant. Fill prevents detached rock slabs and blocks from falling into an otherwise empty stope and is thus essential to safety and stability in the overall scheme of mining.

As the number of drifts (n) increases, so does the depth of excavation. When $(n)(h) > W$, maximum exposure occurs under S_1, which is definitely unfavorable to safety and stability.

An alternative mining sequence would be to create vertical shaft-like openings of height H and then advance to the right in the sketch by excavating successive "shafts" until the stope is completed. This sequence would only be feasible if fill were not needed to prevent falls of rock into the mined region to the left of the next "shaft." There would be no advantage in safety from the rock mechanics view because the exposure is the same. However, there would be a considerable cost advantage.

7.2 Joints in cavern and stope walls

Slip and separation of joints at cavern and stope walls constitute additional failure mechanisms that require design consideration beyond local safety factor analysis based on an elastic analysis for stress and rock mass strengths or strength of intact rock between joints. Although the presence of joints affects stress distribution relative to the joint-free case, elastic analysis of stress based on rock mass properties is a reasonable basis for an initial consideration of joint safety or stability. If results indicate a potential for extensive joint failures, then a more detailed design analysis may be considered.

Analysis for wall joint safety requires estimation of two numbers, joint shear strength, and joint shear stress. Shear strength relates to joint properties and may be considered known for design analysis. If one adopts the well-known Mohr–Coulomb criterion, then joint shear strength is characterized by cohesion c and angle of friction ϕ. Thus, $\tau_j(\text{strength}) = \sigma_j \tan(\phi) + c$. Estimation of joint shear stress is more involved and requires several distinct steps. The first and most difficult step is analysis of stress. Short of a detailed stress analysis, stress concentration factors from known results may be used, if available and suitable for the problem at hand.

Joint plane orientation (dip direction and dip) with respect to compass coordinates ($x = E$, $y = N$, $z = \text{Up}$) is known, so joint normal and shear stresses (σ_j, τ_j) may be obtained by rotation of stress axes from compass coordinates to joint plane coordinates (a = down dip, b = on strike, c = normal direction). Symbolically, this rotation is $[\sigma\,(abc)] = [R][\sigma\,(xyz)][R]^t$ where $[R]$ is a 3×3 matrix of direction cosines between compass and joint coordinates, the superscript "t" means transpose and $[\sigma\,(.)]$ is a 3×3 matrix of stresses. Joint normal stress is then $\sigma_j = \sigma_{cc}$ and joint shear stress is $\tau_j = [(\tau_{ca}{}^2) + (\tau_{cb})^2]^{1/2}$. These calculations are readily done with the aid of standard personal computer software and allow for estimation of joint plane safety factors.

The difficulty is that the requisite stress analysis results $[\sigma\,(xyz)]$ must be available to start the calculation. In two-dimensional analyses, wall stress concentration data were used in lieu of a detailed stress analysis and required consideration only of the stress acting tangential to the wall. In three dimensions, there are two such tangential stresses to consider, but generally only the highest-stress is reported as stress concentration data. In this case, only when the strike of a joint plane is parallel to the intermediate principal wall stress, may the peak three-dimensional stress concentration data be used for a joint plane safety factor calculations. The joint plane normal is then in a plane defined by the major and minor principal stresses at the excavation wall.

An additional complication in three-dimensional analysis is the fact that the stress state varies along the trace of joint intersection with excavation walls. Thus while slip may be indicated at some points along the wall joint trace, safety may be indicated at other points. Whether excavation safety is threatened is then problematical, although prudence would dictate consideration of reinforcement or support.

7.3 Tabular excavations

Tabular excavations such as hardrock stopes and softrock panels may be flat, dipping, or vertical and pose a special challenge even for direct numerical calculations. Reasonably accurate determination of stress distribution along the least dimension of a tabular excavation requires five to ten subdivisions (elements). In a vein 10 ft thick a spacing of about 1 ft would be required. If the stope dimensions are $10 \times 200 \times 200$, then 400,000 elements would be required just to fill the planned excavation. A longwall panel with dimensions $10 \times 750 \times 7{,}500$, would require over 56 million elements of the same size just to fill the panel. Total elements required would be easily 100 to 1,000 times excavation elements. Even one million elements is an enormous number and that requires a large, fast computer. Use of variable size elements improves computational efficiency, but the problem of stress analysis about

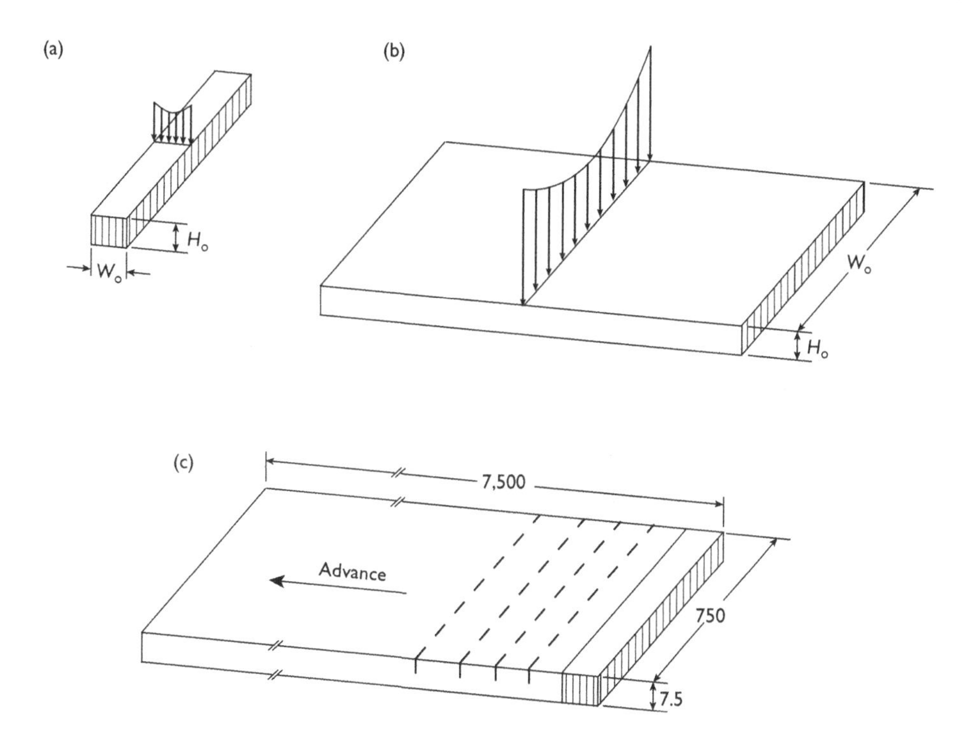

Figure 7.13 Hypothetical vertical stress distributions about an evolving tabular excavation: (a) initial tunnel-like opening, W_o/H_o ratio and location of peak compression, (b) panel extended, tunnel-like opening new W_o/H_o ratio and peak compression, and (c) a single tabular excavation: face advance changes stress concentration.

tabular excavations in three dimensions is still numerically intensive. Programs that operate only on the boundary surface of an excavation rather than through the volume enclosed have a distinct advantage but also have restrictions and are numerically intensive as well. A closely related problem of stress analysis about tabular excavations is mine subsidence and the calculation of surface displacement in response to mining. While analysis of progressive mining in tabular ore bodies is possible, they are not routine and results from parametric studies for compiling stress concentration factors are not available.

Thought experiments in view of the three-dimensional results that are available provide some guidance to stress concentration about tabular excavations. For simplicity, consider a single, flat-lying tabular excavation similar to a longwall panel. At the start, the excavation is a tunnel-like opening of rectangular section as shown in Figure 7.13. Under gravity loading only, tension is absent and peak compressive stress concentration appears along the face until within about one diameter of the ends. In this situation, D is the width of the rectangular cross section. As the face advances, the width to height ratio of the cross section increases and consequently so does the stress concentration at the face. In two dimensions, the increase in stress concentration is directly proportional to the width to height ratio. At an advance of just 75 ft, this ratio is 10 (mining height is 7.5 ft); the corresponding stress concentration factor is also about 10, which is relatively insensitive to the premining stress state. If mining depth in this example is 1,000 ft the stress at the face is 10,000 psi. At an advance of 150 ft, the stress at the face is about 20,000 psi which is likely to be near laboratory compressive strength of hardrock and well above unconfined compressive strength of most softrock. With continued face advance the tunnel-like geometry wanes as face advance becomes equal to face length (750 ft in this example). At this stage, peak compression occurs at the midsides of the panel. Further advance to 1,500 ft, twice face length begins to produce tunnel-like geometry in the direction of the advance. The aspect ratio is now quite high, 100 (750/7.5), so stress concentration at the panel sides is enormous. Continued face advance increases the zone of high stress concentration as the tunnel-like geometry is extended, but the magnitude changes little. The very high stress concentrations associated with tabular excavations undoubtedly results in stress that exceeds strength and causes local yielding near the excavation walls. Excavation of adjacent panels creates still higher stresses which are relieved beyond the elastic limit by inelastic fracture and flow.

7.4 Cavern and stope support

Any tendency for stress to rise above the elastic limit is necessarily relieved by inelastic deformation. A support strategy that takes into account rock mass failure mechanisms is therefore one that accounts for yielding, fracture, and flow. There are numerous mechanisms of yielding in rock masses: local crushing, fracture and slab formation, spalling, slip on joints in blocky ground, shear along foliation, kinking and buckling of thin laminations and spalls, squeezing and flow in soft ground, and so forth. A common feature of these mechanisms is loss of cohesion from fracturing of intact rock between joints, shear along joints, and shear of intact rock bridges between joints. The result is partial detachment of the inelastic zone from the parent rock mass that may no longer be self-supporting. Sometimes detachment is complete. Without adequate support, rock falls ensue. An important question in cavern and stope stability thus concerns the extent of any zone of inelasticity that may evolve into a fall of rock. Although wall stress concentration and safety factor calculations based on elasticity theory do not directly indicate the extent of potentially unsafe, unstable regions of rock, a computed local safety factor slightly less than one suggests a relatively small zone of inelasticity in the vicinity of the considered point. This zone may be controlled with local support such as spot rock bolting; very low safety factors suggest a larger yield zone and therefore a need for an engineered support system.

A support strategy based on dead weight load would size support to prevent a potential rock fall. The direction of the fall whether from the back or ribs is important. Bolting is the usual form of support and method of defending against rock falls when the excavation method requires personnel within the cavern or stope. Conventional rock bolts and cable bolts are used extensively for stope support in some hardrock mines and in caverns for civil works. Timber support is used rather sparingly in mines because of high cost and hardly at all in caverns for civil works.

A companion support strategy is to maintain the integrity of cavern and stope walls by prevention of raveling or the fall of small blocks that would allow additional blocks to fall and lead to caving, piping, rat-holing, chimney formation, or similar phenomena. In mines, this strategy is commonly implemented using some form of stope fill. Fill may be placed concurrently with mining or after extraction of ore from stopes. There are many varieties of fill used in mining. In civil works, bolts in conjunction with screening of some type, are commonly used for temporary support. Shotcrete or flexible coatings may also be used before permanent support is installed in the form of a concrete lining.

Hardrock mine fill

Pillars are the primary defense against opening collapse, but local support and reinforcement are also required for safety. Bolting most often is used for local support and strapping or mesh provides containment of loose rock at excavation walls. Timber sets and cribs also provide local support. Densely fractured ground may be grouted. If the mining method does not involve caving, but pillars are extracted, then almost certainly some form of back fill is used. Fill provides for regional ground control at the scale of multiple stopes, panels, and levels.

Fills used in mining range from waste rock generated during advance of development openings to engineered fills where careful attention is paid to particle size, cement added, water content, and mineral chemistry. Fill may be used concurrently with mining or may be placed after mining. Eventually, almost all mine openings are filled. Exceptions include mining systems associated with caving methods and those where pillars are left unmined. Fills may be placed dry, hydraulically in slurry form or in paste form. Dry distribution systems are generally combinations of fill passes and mine cars or conveyor belts in some very large mines. Pipelines are used to distribute hydraulic and paste fills.

Hydraulic or slurry fills are used extensively in many deeper underground hardrock mines. A typical hydraulic fill consists of water and sand-size particles obtained from mill tailings by removal of fines (clay-size particles) and is about 65% solids by weight. Hydraulic fill is delivered to stopes by pipeline. A minimum velocity of flow is required in pipes to avoid settling of solids. The main rock mechanics purpose of fill is to provide surface support. Fill also inhibits closure of openings. Desirable properties of hydraulic fill include high strength, high elastic modulus, and high permeability (hydraulic conductivity). High hydraulic conductivity allows for relatively fast drainage of the fill and therefore rapid increase in effective stress and strength. Fines (small particles) are removed in fill preparation plants to improve permeability of hydraulic fills.

In this regard, soil classification schemes for engineering purposes generally distinguish between sands, silts, and clays on the basis of particle size. Particle size conveys much qualitative information about expected mechanical behavior of soils and mine fills. For example, dry sand is free-running and cohesionless. Typically, gravel-size particles are greater than 2 mm in diameter. Sand-size particles range between 2 and 0.05 mm. Silt- and clay- size particles, that is, "fines" are smaller than 0.05 mm. A popular alternative classification scheme specifies fines as particles that pass through a 200-mesh sieve. These particles are less than 0.074 mm in size.

Cement is a common additive to hydraulic fill and serves to increase modulus and strength. Cement addition is usually in the range of one-fifteenth to one-seventh by weight and is often introduced at the top of a fill to improve bearing capacity of the fill top that becomes an equipment floor for the next mining advance. Strength of the fill must provide adequate bearing capacity in a timely manner to avoid delays in the mining cycle. Modulus of cemented fills range from less than 10 K (10,000 psi) to over 100 K in the extreme. Unconfined compressive strengths of cemented sand fills generally range from 100 to 1,000 psi after 28 days of cure time. Strain to failure is about 1%. In this regard, lower strength, lower modulus fills may deform beyond an elastic limit by plastic flow, while high strength fills may actually fail by

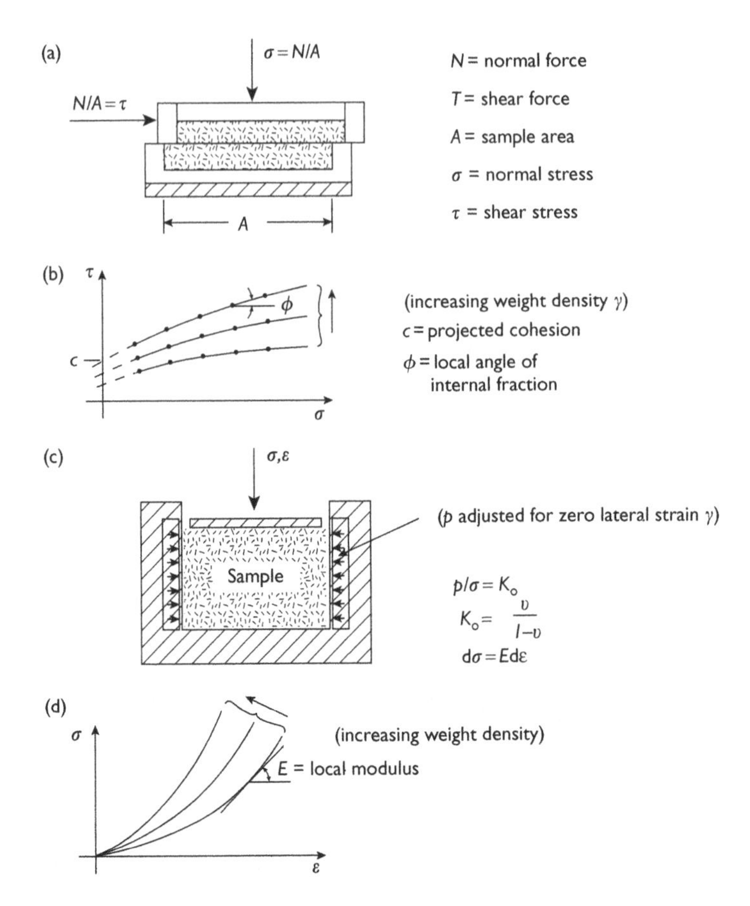

Figure 7.14 Schematic apparatus for fill strength and modulus determinations.

fracture and even form slabs and spalls when exposed by subsequent mining. This brittle-like behavior is undesirable and wasteful of cement. In time, oxidation of sulphide minerals (especially pyrrhotite) in hydraulic fills may assist in cementation, although heat and acid mine water are also produced. Smelter slag that involves a pozzolanic action (a natural cementing action that occurs in some materials such as slag, ash, and some manufactured materials) may be used to partially replace more costly cement addition to fill. Simple compaction from inward motion of stope walls also tends to increase fill strength and modulus.

Direct shear testing apparatus is often used to determine strength of fill as illustrated in Figure 7.14. Determinations of cohesion c and angle of internal friction ϕ follow. Determination of a fill modulus is often done using a K_0-chamber that is essentially a hollow, steel cylinder that allows for uniaxial compression under confining pressure that prevents radial strain. Strength and modulus generally depend on the initial weight density (specific weight) of the fill and tend to increase with increasing density. Mohr–Coulomb strength criteria and elastic behavior (Hooke's law) are usually assumed, although the stress–strain response is generally nonlinear as particles are brought into closer contact during settlement and subsequent compaction by inward motion of stope walls.

Hydraulic conductivity of two to four inches per hour (100 mm/h) usually provides a satisfactory seepage rate through sand fill. Drainage may occur by decantation of water that forms a pond on top of a settled fill and by wells placed in the fill. Decantation is often the method used in narrow vein mines where a sand wall of timber and burlap is constructed at a raise that provides access to the stope. Wells may be used in wide ore bodies mined by cut and fill methods. In time, cement tends to lower hydraulic conductivity and often leads to stratification of the fill.

Darcy's law governs seepage and relates seepage velocity to pressure drop along a flow path. Flow is slow and considered laminar for application of Darcy's law. Thus,

$$v = kh \tag{7.4}$$

where v is a nominal seepage velocity, k is hydraulic conductivity, and h is hydraulic gradient. Alternatively,

$$Q = vA = khA \tag{7.5}$$

where Q is volume flow rate and A is total cross-sectional area perpendicular to the flow. The true seepage velocity V is greater than the nominal velocity or Darcy velocity because the flow occurs in the void space of a porous material. The area of flow is composed of void area and solid area, so $A = A_v + A_s$ and $Q = vA = VA_v$. Hence $V = v/n$ where $n = A_v/A$ is porosity of the fill and is a number less than one (assuming that the area porosity is equal to the volume porosity).

Permeameters are used to determine hydraulic conductivity. Figure 7.15 shows a falling head permeameter that allows for a relatively quick test even for fills of low hydraulic

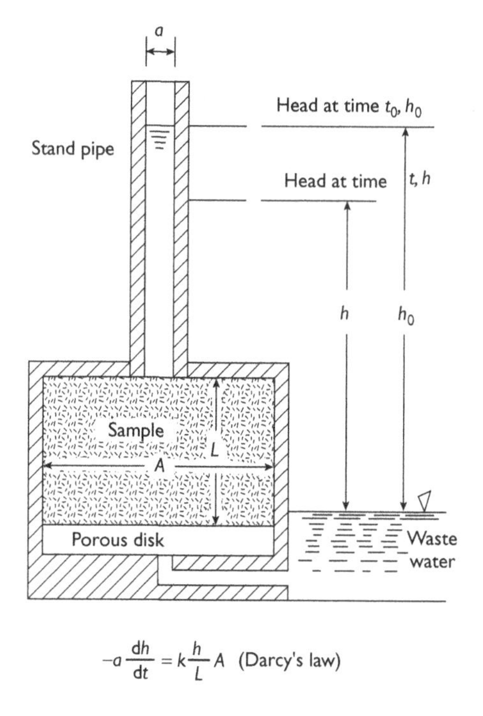

Figure 7.15 Falling head permeameter.

conductivity. A constant head permeameter would involve water addition at the top of the apparatus. Strictly speaking, hydraulic conductivity, k with units of ft/s or m/s, is not a material property because it depends on fluid viscosity and density, which are usually temperature dependent. However, permeability is a material property. The relationship between hydraulic conductivity and permeability, K with units of ft^2 or m^2, is

$$K = k\mu/\gamma \tag{7.6}$$

where γ and μ are specific weight of fluid (lbf/ft^3, N/m^3) and fluid viscosity (absolute, lbf-s/ft^2, N-s/m^2), respectively. Representative values of γ and μ for water are 62.4 (lbf/ft^3) and 2.1×10^{-5} (lbf-s/ft^2), although viscosity varies with temperature.

Materials may have a large porosity because of relatively high percentage of voids, but if the voids are very small, then permeability will be low and seepage velocity small. A high porosity does not necessarily mean high permeability because of the role of void size. Clays and shales are often very porous but also have very low permeabilities because of the small particle and therefore small void size. In soil mechanics, consolidation refers to soil settlement in response to exudation of the pore fluid, usually water. The rate of consolidation depends on permeability. Porosity is more indicative of the amount of fluid stored in the pores. Clays may require decades to become fully consolidated in response to loads imposed by buildings and other structures that are completed in just a few years. Surface settlement also occurs in response to removal of fluids such as oil and water from strata below the surface.

Because the modulus of hydraulic fill (10 K) is quite low relative to the elastic modulus, say, Young's modulus of rock (1,000 K), there may be little difference in computed stope wall closure (relative displacement across a vein) with and without fill. For this reason, some engineers hold the opinion that fill is not effective for ground control and provides nothing more than a platform from which to mine. Providing a mining platform in a timely manner is quite important to operations, of course. However, this opinion ignores one of the basic tenets of ground control that relies on small forces applied to opening walls, not to prevent wall motion, but rather to prevent small falls of rock that may lead to larger rock falls, caving, and collapse. Detailed numerical simulations of cut and fill vein mining indicate that a fill to rock modulus ratio as small as 1/100 is effective in controlling wall displacement (Pariseau *et al.*, 1976).

For comparison, Young's modulus, E, for water is about 0.30×10^6 psi (300 K) which is greater than even high modulus fills but much less than Young's modulus for rock. For this reason, water is usually considered incompressible in soil and mine fill mechanics. An incompressibility assumption in analysis of saturated, porous, fractured rock is not justified.

Example 7.6 Consider the evolving tabular excavation in the sketch. A "drift" is a mine excavation along the strike of the ore body and is a tunnel-like excavation advanced horizontally a distance L to the right in the sketch. After completion, backfill is introduced before starting another drift below. Drift width W is equal to width of the vein; drift height is h. Progress downward continues a vertical distance H in this example. Stress measurements show that the greatest premining stress is a compression across the vein; the least premining stress is parallel to the strike of the vein, and the intermediate premining stress is vertical, as shown in the sketch. Determine whether this excavation sequence is favorable to safety and whether the drift dimensions can be adjusted to improve safety about the drift.

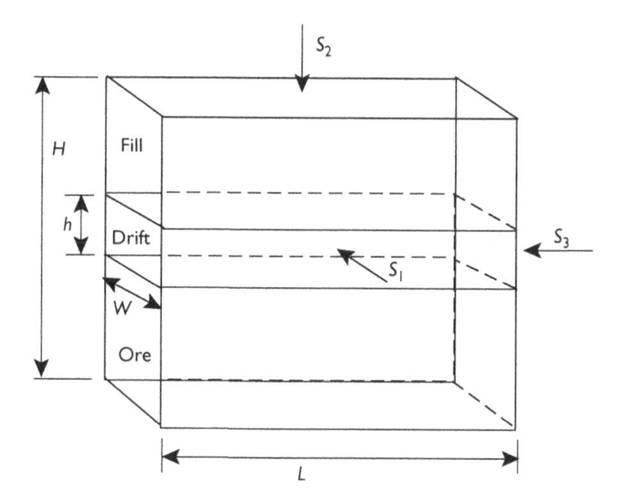

***Solution*:** The final configuration results in maximum exposure to the greatest premining stress and, according to the rule of thumb for favorable orientation of three-dimensional excavations, is unfavorable. However, consideration should be given to the evolving geometry as mining progresses and to the influence of fill.

The first drift at the top of the future tabular excavation evolves into a tunnel-like excavation of width W and height h. If $h < W$, then S_1 is parallel to the long dimension of the drift cross section which is favorable to safety. For example, if $W = 15$ and $h = 10$, safety is favored, but if $W = 5$ and $h = 10$, the section has a disadvantage with respect to stress concentration and safety, as illustrated in the second sketch. As the number of drifts (n) increases, so does the depth of excavation. When $(n)(h) > W$, maximum exposure occurs under S_1, which is definitely unfavorable to safety and stability.

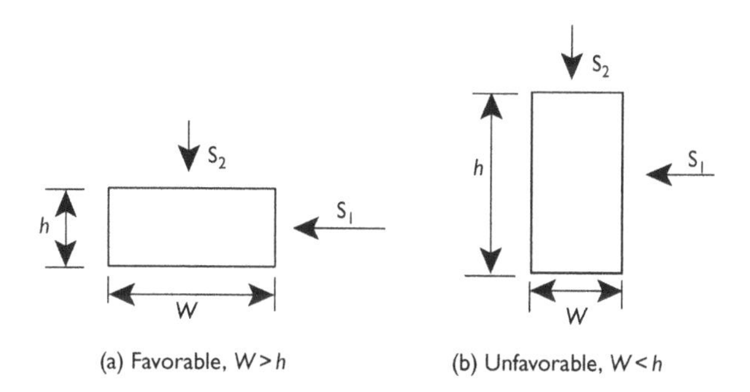

(a) Favorable, $W > h$ (b) Unfavorable, $W < h$

Mine fill is usually much more compliant than the rock mass and has little effect on the deformation and stress concentration that are induced by excavation. For this reason, fill may be neglected. However, this is not to say that fill is unimportant. Fill prevents detached rock slabs and blocks from falling into an otherwise empty stope and is thus essential to safety and stability in the overall scheme of mining.

An alternative mining sequence would be to create vertical shaft-like openings of height H and then advance to the right in the sketch by excavating successive "shafts" until the stope is completed. This sequence would only be feasible if fill were not needed to prevent falls of rock into the mined region to the left of the next "shaft." There would be no advantage in safety from the rock mechanics view because the exposure is the same. However, there would be a considerable cost advantage.

Example 7.7 Direct shear tests are conducted on a sand fill under different normal stresses. Results are given in the Table. The first sand fill has no cement added, while the second fill is 10% cement by weight. Estimate the Mohr–Coulomb strength properties of the two sand fills after plotting the data in the normal stress shear stress plane.

Table for Example 7.7[a]

Without cement		*With cement*	
Normal force (lbf)	*Shear force (lbf)*	*Normal force (lbf)*	*Shear force (lbf)*
60	42	60	70
120	80	120	135
180	128	180	186
240	170	240	242
300	203	300	308
360	247	360	361

Notes
a Sample area 6 sqin.

***Solution*:** Plots of both data sets are shown in the figure. Trend lines fitted to each set show an excellent linear fit in consideration of the R values. Slopes of the trend lines are equal to the tangent of the angle of internal friction, ϕ, one of the Mohr–Coulomb strength properties. Thus, ϕ(no cement) = $\tan^{-1}$ (0.6838) = 34.4°, and ϕ(cement) = $\tan^{-1}$ (0.9667) = 44.4°. The shear axis intercepts are cohesions. Thus, c(no cement) = 0.23 psi, an almost negligi- ble amount, and c(cement) = 2.33 psi or 336 psf. Corresponding tensile and compressive strengths may be computed from cohesion and angle of internal friction. Thus,

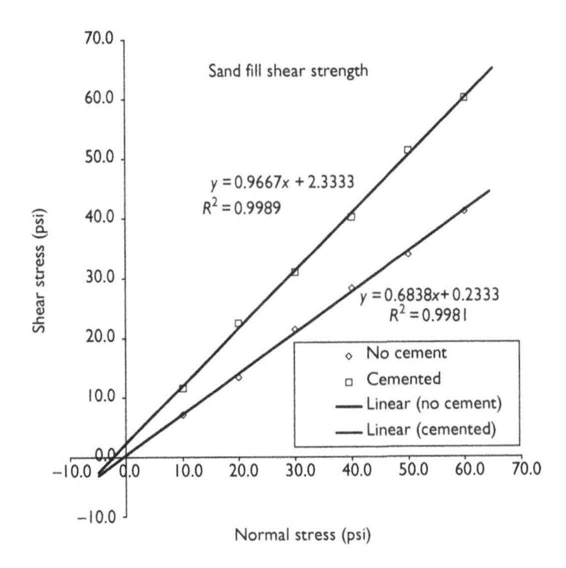

$$\left.\begin{matrix} C_o \\ T_o \end{matrix}\right\} = \frac{2(c)\cos(\phi)}{1 \pm \sin(\phi)}$$

$$\left.\begin{matrix} C_o \\ T_o \end{matrix}\right\} = \frac{2(0.233)\cos(34.4)}{1 \pm \sin(34.4)} = \left.\begin{matrix} 0.88 \\ 0.25 \end{matrix}\right\} \text{psi(no cement)}$$

$$\left.\begin{matrix} C_o \\ T_o \end{matrix}\right\} = \frac{2(2.333)\cos(44)}{1 \pm \sin(44)} = \left.\begin{matrix} 11.0 \\ 1.98 \end{matrix}\right\} \text{psi(cement)}$$

The sand fill without cement is essentially cohesionless, but that is not the case with addition of cement. These results are summarized in the results table.

Property fill	*Cohesion (psi)*	*Friction angle (°)*	*Compressive strength C_o (psi)*	*Tensile strength T_o (psi)*
No cement	0.23	34.4	0.88	0.25
Cemented	2.33	44.0	11.0	1.98

Example 7.8 A sand fill is tested in uniaxial compression using a K_o chamber. The data with and without the addition of cement are fitted to a curve such that

$$(\sigma + \sigma_o) = a(\varepsilon + \varepsilon_o)^2$$

which is a parabola that has the form shown in the figure. Determine Young's modulus E at an expected stress of 100 psi for both fills. Also determine the initial Young's modulus for both fills (at the origin).

Note:

a(no cement) = 0.5(10^6) psi, a(cemented) = 1.0(10^6) psi

$\varepsilon_o = 0.001$,

σ_o (no cement) = 0.5 psi, σ (cemented) = 0.1 psi

***Solution*:** The sand fill is considered a nonlinear elastic material with changes in stress and strain derived from the linear form of Hooke's law. Thus,

$$E d\varepsilon_{zz} = d\sigma_{zz} - v d\sigma_{xx} - v d\sigma_{yy}$$
$$E d\varepsilon_{xx} = d\sigma_{xx} - v d\sigma_{yy} - v d\sigma_{zz}$$
$$E d\varepsilon_{yy} = d\sigma_{yy} - v d\sigma_{zz} - v d\sigma_{xx}$$

where v is Poisson's ratio. Because lateral strain is prevented, the horizontal stress x- and y-stresses are equal and may be expressed as functions of the vertical or axial applied stress,

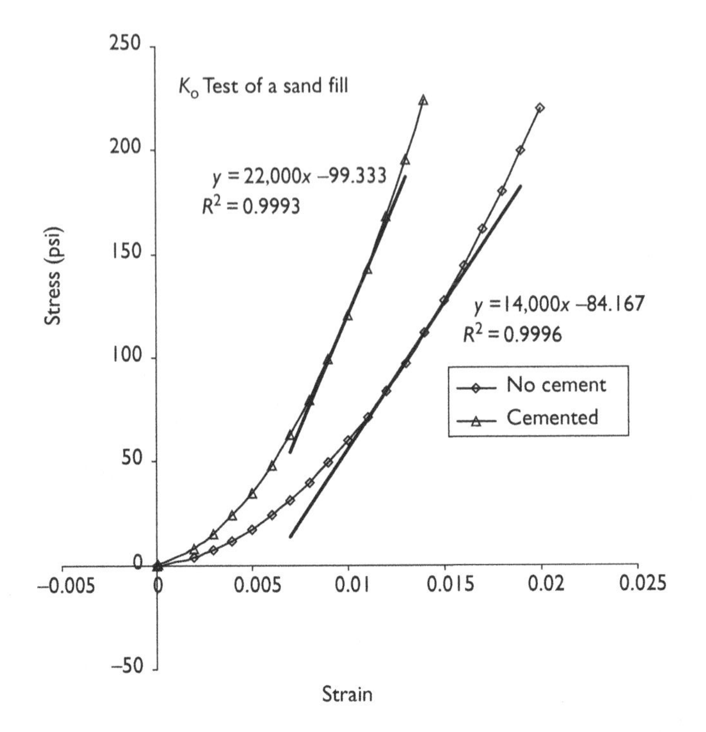

that is,

$$E d\varepsilon_{zz} = \left[1 - 2\nu\left(\frac{\nu}{1-\nu}\right)\right] d\sigma_{zz}$$

When this result is substituted into the expression for the vertical (axial) strain, one has

$$d\sigma_{xx} = d\sigma_{yy} = \left(\frac{\nu}{1-\nu}\right) d\sigma_{zz}$$

After dropping the subscripts and solving for stress as a function of strain,

$$d\sigma = K_o d\varepsilon$$

$$K_o = \left(\frac{1-\nu}{1-\nu-2\nu^2}\right) E$$

The slope of the assumed parabola is also K_o, so with a knowledge of Poisson's ratio, one can determine E from the test data at any point along the experimental plots. With the assumption of 0.2 for Poisson's ratio, one has $K_o = E/0.9$. Thus, E is about 10% greater than the slope of a K_o data plot.

There are two ways of determining K_o : from the constants of the fitted parabolas or graphically from tangent lines at points of interest. Because trend lines are available near the required 100 psi, slopes of uncemented and cemented fills are 14,000 and 22,000 psi, respectively. Assuming Poisson's ratio is 0.2, the Young's moduli are 15.6 and 24.4 ksi, respectively.

At the origin, the slopes of the parabolas are $2a\varepsilon_o$, so K_o (no cement) = 1,000 psi and K_o (cemented) = 2,000 psi. Hence, the initial Young's moduli are 1.1 and 2.2 ksi for the uncemented and cemented fills, respectively. In table form the results are:

Modulus fill	*E (ksi) (initial)*	*E (ksi) (at 100 psi)*
No cement	1.1	15.6
Cemented	2.2	24.4

Example 7.9 Data obtained from a falling head permeameter test on sand fill without cement and a cemented fill are given in the table. Estimate the hydraulic conductivity of both fills.

Note: Standpipe diameter $a = 0.5$ in., sample height $L = 6.0$ in., sample diameter = 6.0 in. The water in the standpipe is initially 36.0 inches above the datum.

Time (sec)	*No cement*	*Cemented*
	h (in.)	*h (in.)*
0	36.0	36.0
10	29.2	31.9
20	24.1	28.4
30	20.0	25.0
40	16.9	22.5
50	12.9	20.3
60	11.5	17.7
70	9.8	15.5
80	7.2	14.6
90	5.6	12.8
100	5.0	10.7
110	4.2	10.0
120	3.1	8.1

Solution: The data table may be expanded to allow for computation of hydraulic conductivity according to the integration of Darcy's law when applied to the apparatus. Thus,

$$k = \left(\frac{aL}{At}\right)\ln\left(\frac{h_o}{h}\right)$$

where A is the sample cross-sectional area. These results are given in the next table.

Time (s)	*No cement h (in.)*	*k (in./h)*	*Cemented h (in.)*	*k (in./h)*
0	36.0		36.0	
10	29.2	3.14	31.9	1.81
20	24.1	3.01	28.4	1.78
30	20.0	2.94	25.0	1.82
40	16.9	2.84	22.5	1.76
50	12.9	3.08	20.3	1.72
60	11.5	2.85	17.7	1.77
70	9.8	2.79	15.5	1.81
80	7.2	3.02	14.6	1.69
90	5.6	3.10	12.8	1.72
100	5.0	2.96	10.7	1.82
110	4.2	2.93	10.0	1.75
120	3.1	3.07	8.1	1.86
	AVE	2.98		1.78
	STD	0.11		0.05
	CV%	3.76		2.86

A plot of hydraulic conductivity at each data point is shown in the figure.

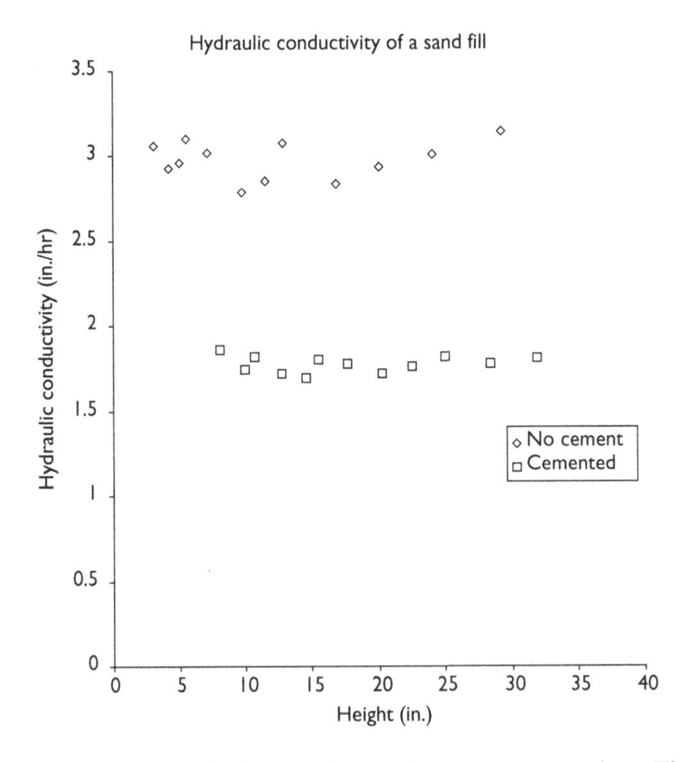

While there is some scatter in the data, the variation appears random. Thus, hydraulic conductivity for the uncemented sand fill is 2.98 in./h and for the cemented fill 1.78 in./h.

The uncemented fill would be acceptable because the criterion is 2.0 to 4.0 in./h. The cemented fill is not acceptable by this criterion.

Paste fill does not have fines removed before placement and may be up to 85% solids by weight. In fact, about 15% (by weight) clay-size fines that are smaller than 20 microns (0.20 mm) are needed to form a paste (Landriault, 1996). These colloidal size particles retain water and are responsible for the "paste." Solid settlement does not occur in pipes, so there is no critical velocity that must be achieved for flow. An important operational benefit is reduced surface area for disposal of mine waste. Paste fill without cement is subject to liquefaction and conversion to a mud slurry, so cement of some kind must be added to the paste. In this context, the cement is also called a "binder" which may be ordinary Portland cement, fly ash, or smelter slag, as is the case with cemented hydraulic fills. Paste fill strength and modulus are similar to those of cemented hydraulic fills and depend on the amount of cement added. Unlike hydraulic fills, drainage is not a consideration. Indeed, the water retention property of paste fill assists in reducing acid mine water and leaching of heavy metals from fill. Paste fill distribution is more difficult and costly than hydraulic fill systems, but additional benefits often outweigh the added cost and, as experience is gained with paste systems, costs may be expected to be reduced.

Cemented rock fill or fillcrete, as the name suggests, is almost the opposite of paste fill because of the removal of fines and the addition of coarse particles. Cemented rock fill is similar to concrete because of the presence of coarse aggregate that must be produced from surface quarries. Mill tailings generally lack coarse- or gravel-size particles. The advantages of cemented rock fill are high strength and high modulus; the disadvantage is high cost relative to other fills. The high cost is related to production of the aggregate and placement. Cement in slurry form is often added to rock fill underground just prior to delivery.

With reference to Figure 7.16, a simple mechanical analysis of equilibrium that sets the forces acting across a narrow vein before mining equal to the forces after mining shows that

$$S_n A = S_p A_p + S_f A_f \tag{7.8}$$

where S_n, S_p, S_f, A, A_p, A_f are average normal stress before mining, average pillar stress, average fill stress, total area, pillar area, and fill area, respectively, and $A = A_p + A_f$. The area extraction ratio $R = A_f/A$, so

$$S_n = (1 - R)S_p + RS_f \tag{7.9}$$

According to this analysis, as overhand mining proceeds up dip in the usual manner, R increases. When extraction is complete, the fill supports the entire normal load. Pillar stress is given by a modified extraction ratio formula. Thus,

$$S_p = \frac{S_n}{(1 - R)} - \frac{RS_f}{(1 - R)} \tag{7.10}$$

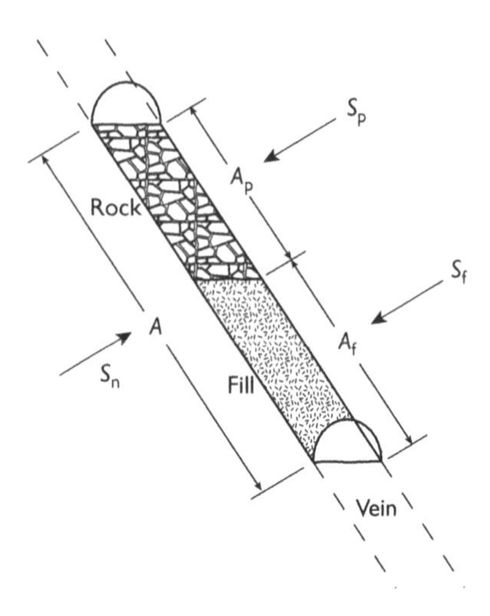

Figure 7.16 Narrow vein cut and fill geometry for equilibrium analysis.

which shows that without fill, the pillar stress is given by the usual extraction ratio formula and that fill reduces pillar stress from this value. Equation (7.10) is an equilibrium model. However, prior to complete extraction, the partitioning between fill and pillar stress is unknown. Pillar stress may actually be many times fill stress. Suppose fill stress is some fraction f of pillar stress, so $S_f = f S_p$, then

$$S_p = \left[\frac{S_n}{1 - R(1 - f)}\right] \tag{7.10a}$$

An alternative analysis based on an assumed uniform displacement across the vein and linearly elastic behavior shows that

$$S_f = \left(\frac{E_f}{E_p}\right) S_p \tag{7.11}$$

where E_f and E_p are fill and pillar modulus, respectively. Hence,

$$S_p = \frac{S_n}{[1 - R(1 - E_f/E_p)]} \tag{7.12}$$

which indicates a role of fill modulus in reducing pillar stress. Equation (7.12) is a "compatible" model that also satisfies equilibrium. In comparison with (7.10(a)), one has $f = E_f/E_p$ as an estimate. In either case, force equilibrium is satisfied.

At a relatively high fill to modulus ratio of 0.1 and low extraction, say, 0.1, pillar stress without fill is 1.11 times S_n. With fill, pillar stress is almost the same at 1.10 times S_n.

However, at 90% extraction, pillar stress without fill would be 10.0 times S_n, but with fill would be about 5.3 times S_n, a considerable reduction from the unfilled stope case. When $R = 0$ before excavation, pillar stress is just preexcavation stress, and when $R = 1$ and excavation is complete, fill stress is equal to the preexcavation stress. Thus, according to this model analysis, fill becomes progressively more effective in reducing stope pillar stress as extraction proceeds and pillar stress increases. This stage of stoping is just when fill is most needed for ground control. While this analysis is highly idealized and based on averages, the interaction between fill and pillar properties and geometry is at least qualitatively correct.

Example 7.10 Consider a sand fill without cement and with cement addition, then determine the influence of cement on pillar stress in a tabular, narrow vein excavation as extraction increases.

***Solution*:** A plot of pillar stress versus extraction ratio with fill stress fraction of pillar stress as a parameter addresses the question in an efficient way. An equilibrium model that equates pillar force and fill force to the preexcavation force, all expressed in terms of stress averages and areas, leads to

$$\frac{S_p}{S_n} = \left[\frac{1}{1 - R(1 - f)}\right]$$

where $f = S_f / S_p$. An estimate of f is the ratio of fill to rock mass Young's modulus, that is, $f = E_f / E_p$. A range of this ratio may be obtained from a range of fill modulus of 10 to 100 ksi while rock mass modulus is perhaps 1,000 ksi. A plot of pillar stress ratio versus extraction ratio is shown in the first figure. This figure shows the two fill plots as almost coincident. A different perspective is given in the second figure for this example.

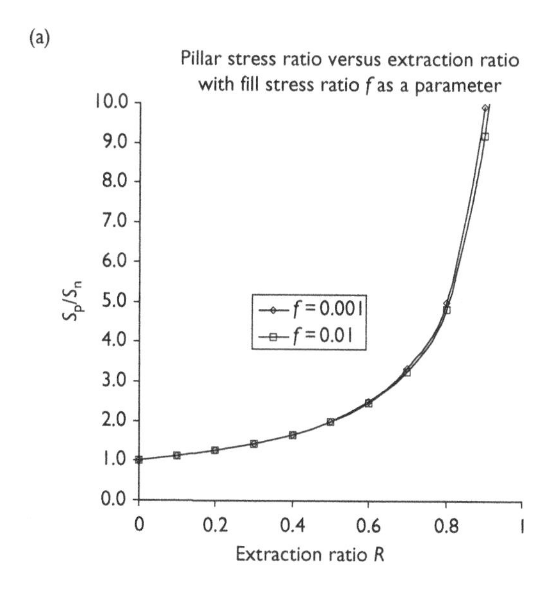

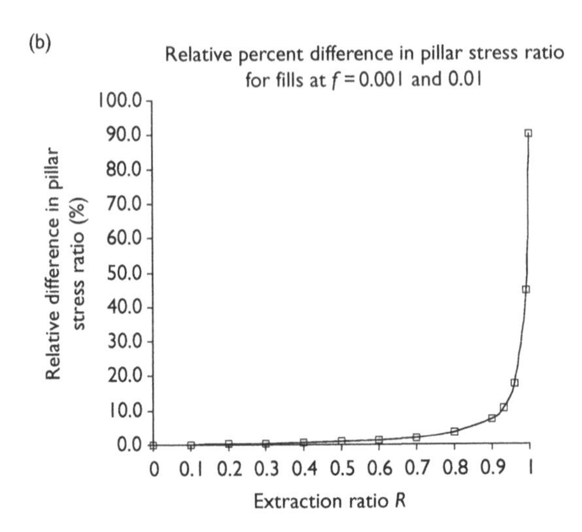

In the second figure the relative percentage difference given by

$$y = (100)\left[\frac{(S_p/S_n)_{0.001} - (S_p/S_n)_{0.01}}{(S_p/S_n)_{0.001}}\right]$$

where the subscripts 0.001 and 0.01 refer to the ratio of fill modulus to rock mass (pillar) modulus which is an estimate of fill stress to pillar stress. Data in the second figure show that there is little difference in pillar stress up to about 90% extraction beyond which the difference that the two fills induce in pillar stress ratio becomes considerable. Thus, one infers that the direct effect of fill modulus on pillar stress only becomes significant at high extraction.

Stopes in large ore bodies in wide vein mines are usually extracted in a checker board pattern of primary or first mined stopes, secondary, and then tertiary stopes as shown in Figure 7.17. Such stopes may be several hundred feet or more in height and one hundred feet or so on edge in plan view. Primary stope walls are rock, of course, but secondary stopes have a mix of rock and fill walls; tertiary stopes are pillars formed by the mining sequence that become enclosed in fill. The pattern leads to 25% extraction on first mining and 50% on mining secondary stopes. Mining the pillars increases the extraction to 100%. Fill–pillar interaction is strongest during the latter part of this final stage of mining. Another pattern in medium width vein mines is to simply alternate stopes and pillars. Stopes are mined and filled; the pillars between are then mined. First mining extracts 50% of the ore; pillar mining extracts the remaining 50%.

Completely filling large, open stopes so the fill makes intimate contact with the back is difficult. Usually a gap is left between stope backs and fill tops. This gap prevents direct support of the back by the fill. Consequently, pillars provide all the support. If pillars shorten sufficiently as mining proceeds and extraction increases, then contact with fill occurs and the fill begins to share the support load. However, shortening of the pillars is unlikely to be sufficient in the elastic range. Inelastic pillar compression and yielding may be necessary before substantial load is carried by the fill. Lateral confinement of pillars by fill increases

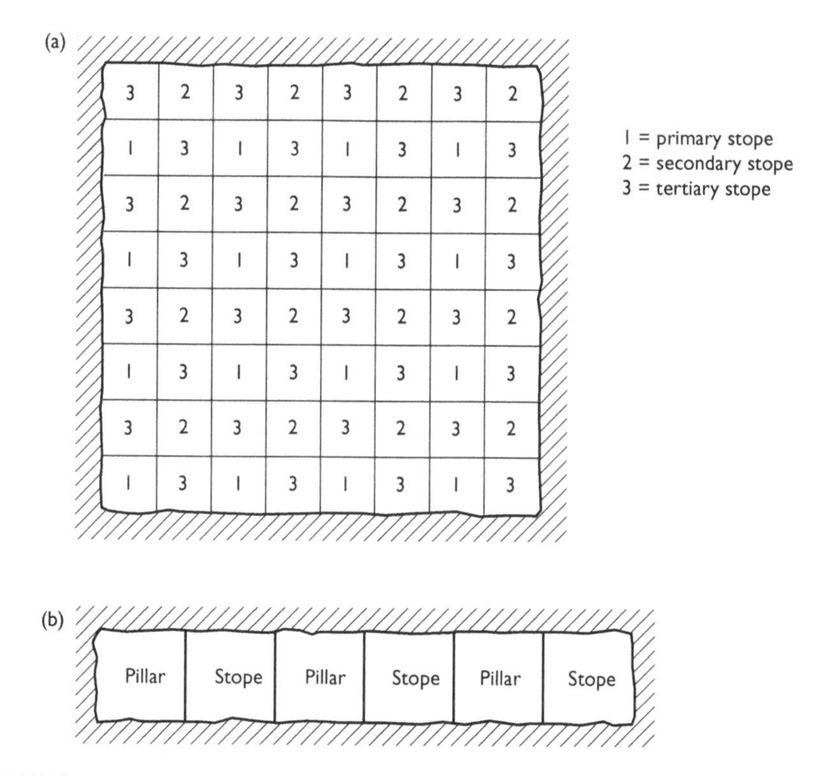

Figure 7.17 Planview: mining sequence in (a) wide ore bodies and (b) medium width ore bodies.

pillar strength and inhibits pillar compression in the elastic range and, importantly, prevents fast failure and collapse beyond the elastic limit.

Stopes that expose fill walls are at some risk of fill failure as well as rock wall failure. Both must be considered for safe mining. Fill wall failure is not only a threat to safety but may also be a serious source of dilution. In this regard, as long as fill is not in contact with stope backs, the main source of fill load in high rise stopes is the weight of fill itself. A representative specific weight of fill is 100 lbf/ft^3 (15.8 kN/m^3) leads to about 2/3 psi per foot (15 kPa/m) of fill height. A three hundred foot high fill wall would develop a 200 psi (1.38 Mpa) wall stress at the fill bottom. Although the top of a fill column may be stress-free, horizontal stresses may be developed through interaction with stope and pillar walls. Friction between fill and walls inhibits upward vertical motion of the fill and thus in reaction develops additional vertical stress. Fill strength must be adequate to support such loads.

Cemented fill is also used for subsidence control over abandoned room and pillar mines in many cases. Pea-size gravel is placed underground via boreholes and then grouted to provide a stiff, strong fill. The fill provides lateral pillar confinement and thus increases existing roof support and, if brought into close contact with the roof, provides additional resistance to the overburden tendency to subside. Fly ash may also be used to fill abandoned mine workings where pillars tend to deteriorate with time.

Example 7.11 A vein of ore 5 m wide bearing due north and dipping 60° is mined full width starting at a depth of 1,700 m. Advance is up dip for a distance of 50 m by a sequence of horizontal drifts 100 m long. Each drift is followed by sand fill. Rock mass and fill properties are given in the table. Stress measurements indicate the premining stress state as a function of depth is given by

$$S_v = 25h$$
$$S_{NS} = 1{,}000 + 40h$$
$$S_{EW} = 500 + 15h$$

where compass coordinates are used, depth h is in meters, and stress is in kPa. Estimate pillar stress and fill stress in MPa at 50% and 95% extraction. Also estimate average vein wall closure (relative displacement across the vein) at 50% and 95% extraction.

Table of Properties

Material property	*Rock*	*Fill*
E (Young's modulus)	13.8 GPa/2(10^6) psi	0.138 GPa/20 ksi
Poisson's ratio	0.20	0.20
C_o (compressive strength)	138 MPa/20,000 psi	2.76 MPa/400 psi
T_o (tensile strength)	10.3 MPa/1,500 psi	0.21 MPa/30 psi
k (hydraulic conductivity)	7.621(10^{-4}) mm/h/3(10^{-5}) in./h	76.2 mm/h/3.0 in./h
γ (specific weight)	24.5 kN/m³/156 pcf	14.1 kN/m³/90 pcf

Solution: The first step is to develop an approximate free body diagram of the excavation at 50% extraction, as shown in the sketch.

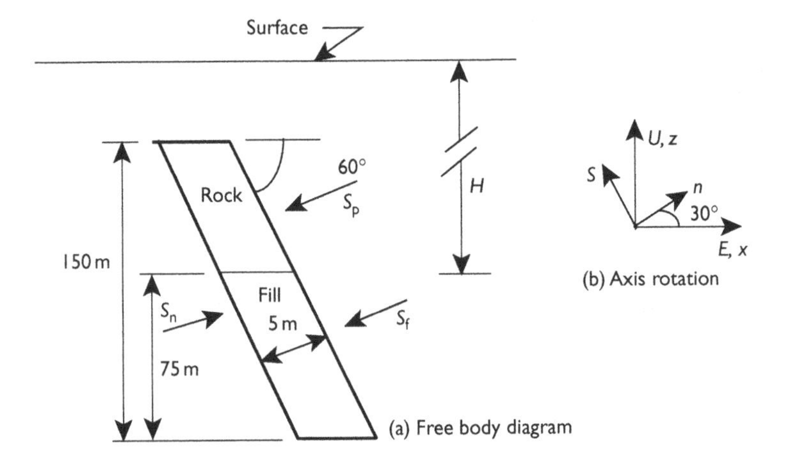

(a) Free body diagram
(b) Axis rotation

Next, the preexcavation stress acting normal to the vein is computed through the usual equations of transformation. For this calculation, the given stress formulas may be used.

A depth to the midheight of the excavation is reasonable, that is $h = 1,700 - 150/2 = 1,625$ m. At this depth $S_v = (25)(1,625) = 40.63$ MPa, $S_{NS} = 1,000 + (40)(1,625) = 66.00$ MPa, $S_{EW} = 500 + (15)(1,625) = 24.88$ MPa.

Rotation about the NS axis is then

$$S_n = \frac{S_{EW} + S_v}{2} + \left(\frac{S_{EW} - S_v}{2}\right)\cos(2.30)$$

$$= \frac{24.88 + 40.63}{2} + \left(\frac{24.88 - 40.63}{2}\right)\cos(60)$$

$$S_n = 28.86 \text{ MPa}$$

The equilibrium model is

$$S_p = \left[\frac{S_n}{1 - R(1 - f)}\right]$$

$$= \left[\frac{28.86}{1 - (0.5)(1 - 0.01)}\right]$$

$$S_p = 57.1 \text{ MPa}$$

where the ratio of fill to rock modulus is used as an estimate of f.

Hence, at 50% extraction, the average pillar stress is estimated to be 57.1 MPa which is less than the pillar unconfined compressive strength of 138 MPa, so no yielding is indicated. The fill stress is 0.571 MPa which is below the fill compressive strength so fill yielding is not indicated.

As a check: $(57.1)(0.5) + (0.571)(0.5) = 28.86$ which is the preexcavation normal stress, as it should be.

At 95% extraction, the pillar and fill stresses are 485 and 4.85 MPa, respectively. Both are well above compressive strength and indicate yielding would occur before reaching 95% extraction, if unconfined compressive strength is a reliable indicator for such. However, there is confinement normal to the cross section shown in the figure. Under plane strain conditions, the confining "pressure" according to Hooke's law (before yielding) is $\nu/(1 - \nu)$ times the normal stress. The confining pressure then is about one-fourth the normal stress. According to a Mohr–Coulomb strength criterion, pillar strength under confining pressure is

$$C_p = C_o + \frac{C_o}{T_o}p$$

$$= 138 + \frac{138}{10.3}(0.25)(485)$$

$$C_p = 1{,}763 \text{ MPa}$$

Fill strength by the same criterion is 18.7 MPa. Both strengths under confining pressure are above the estimated stresses, so yielding is unlikely.

Relative displacement U across the vein may be roughly approximated from vein width W and a strain. Thus,

$$\varepsilon = \frac{\Delta L}{L_o} = \frac{U}{W} = \frac{S_p}{E_p} = \frac{S_f}{E_f}$$

$\therefore$

$$U = (W)\left(\frac{S_p}{E_p}\right)$$

$$U(50\%) = (5)\frac{57.1}{13.8(10^3)} = 0.021 \text{ m}$$

$$U(95\%) = (5)\frac{485}{13.8(10^3)} = 0.18 \text{ m}$$

The strains across the vein at 50% and 95% extraction are 0.42% and 3.5%, respectively. At 95% extraction and a strain of 3.5% again raises a concern about yielding. A more detailed analysis of stress is needed to resolve such a questions.

Example 7.12 Consider a vertical ore body of medium width where extraction occurs full ore body width *w,* W_o in the horizontal direction, and over vertical height h with pillars of width W_p left between excavations (stopes). Fill is introduced after each stope is mined. This pattern is followed along the vein for a great distance. Creating the pattern shown in the sketch.

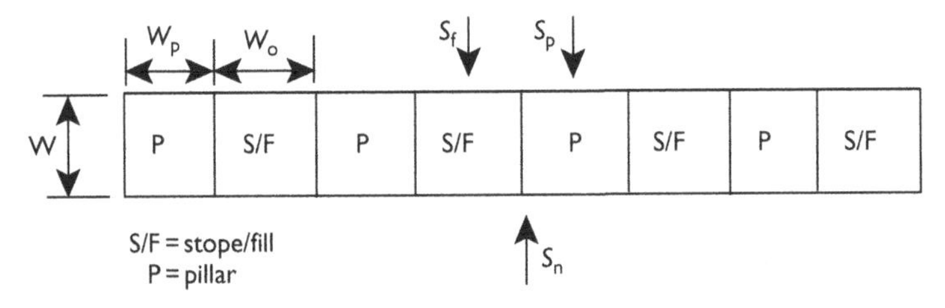

Stope, fill, pillar sequence in plan view

The preexcavation stress state is caused by gravity alone at a depth H. Laboratory testing shows that the ratio of fill to rock modulus is 0.10. After filling the last stope in the sequence, pillars are mined creating an empty void between columns of fill.

Suppose the initial extraction ratio is 50% ($W_o = W_p$), derive an expression for pillar stress after fill is placed but before pillar mining begins, and after every other pillar is extracted.

Also determine whether fill failure would be expected from compression across the vein or by vertical compression under self-weight after pillar mining begins. Neglect any sidewall friction between fill and rock and assume excavation height is 225 ft.

***Solution*:** A free body diagram may be inferred from the sketch such that for equilibrium (before pillar mining) in the horizontal direction across the vein per foot of vertical depth.

The extraction ratio R for the given geometry is 50%. Thus, from the definition,

$$R = \frac{W_o}{W_o + W_p} = \frac{1}{2}$$

$$\therefore$$

$$W_o = W_p = W_f$$

A formula for pillar stress after filling but before pillar mining may be obtained from force equilibrium across the vein. Thus,

$$F_n = F_p + F_f$$

$$S_n A = S_p A_p + S_f A_f$$

$$\therefore$$

$$S_p = 2(S_p + S_f)$$

$$\therefore$$

$$\frac{S_p}{S_n} = \frac{2}{1 + f}$$

where f is the ratio of fill to pillar stress and the equality of pillar and fill areas is used ($A_p = A_f$ and $A = A_p + A_f$).

Once pillar mining begins the geometry changes, as shown in the second sketch where P = pillar, F = fill, and the open space is where a pillar was mined.

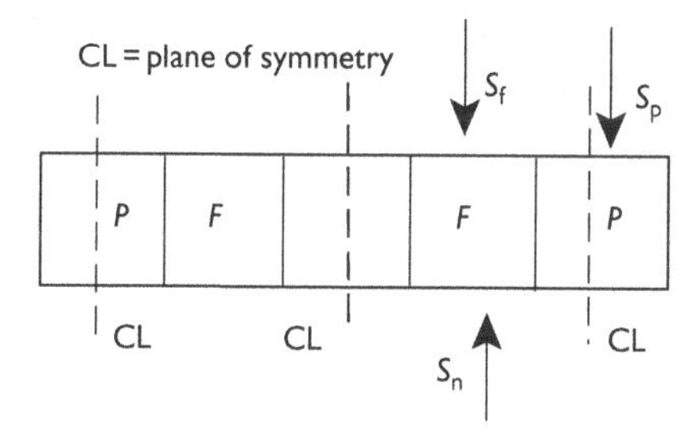

Equilibrium after mining the first set of pillars requires

$$S_n(W_o/2 + W_f + W_p/2) = S_p W_p/2 + S_f W_f$$

$$\therefore$$

$$\frac{S_p}{S_n} = \frac{4}{1 + 2f}$$

in view of the equal widths and where f is the ratio of fill to pillar stress. An estimate of f is the ratio of fill modulus to rock modulus which is given as 0.10. Thus, $S_p = (10/3)S_n$ that may be compared with pillar stress after first mining and filling, but before the start of pillar extraction, $S_p = (20/11)S_n$. During the first phase of pillar extraction, pillar stress increases to 3.3 S_n from 1.8 S_n, not quite double. In each stage, fill stress is 0.1 times pillar stress.

The preexcavation state of stress is caused by gravity alone. Thus, $S_v = \gamma H$, $S_h = K_o S_v$ where a reasonable estimate of K_o is 0.25. Hence, S_n is known at any given depth, $S_n = S_h$, so that pillar and fill stresses are computable.

After pillar mining, $S_f = 0.1\ S_p = (4/12)\ S_n = (4/12)(0.25)S_v$ which acts across the vein. With neglect of side friction, the vertical stress at the bottom of a fill column is $S_f' = \gamma h$. If $S_f' > S_f$, then the fill column would be expected to fail under self-weight. Thus if,

$$S_f' = \gamma(\text{fill})h > S_f = (1/12)\gamma(\text{rock})H$$

Estimates of rock and fill specific weights are about 156 and 90 pcf, so if

$$H < (1/0.14)h$$

then the fill failure would be because of fill weight. Stope height is 250 ft, so below 1,731 ft, gravity failure of fill might be expected. However, side wall friction would reduce the vertical fill stress and so the critical depth would be even less.

Cable bolts support

Cable bolts commonly used in hardrock mines are usually 5/8 in. in diameter, have seven strands and have ultimate strengths ranging from 55,000 to 60,000 lbf and an elastic modulus of about 30×10^6 psi (ASTM, 1998). They are grouted in holes commonly 2-1/4 in. in diameter and thus have distributed anchorage. Figure 7.18 shows a typical cable bolt assemblage ready for grouting in a hole. Attention is therefore given to strength of the steel–grout interface and the grout–rock interface as well as to the cable bolt. When rock falls occur in cable bolted backs, the rock and grout are often observed to be stripped from the cable bolt, although cable bolts may

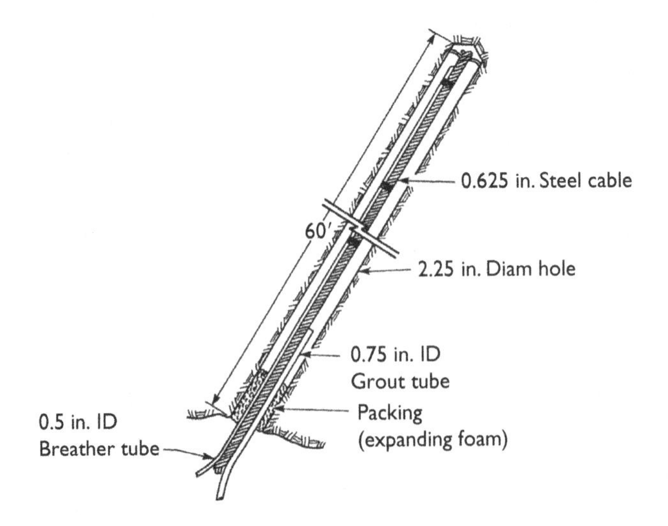

Figure 7.18 A typical cable assemblage ready for grouting (after Schmuck, 1979).

occasionally rupture. More than one cable bolt may be installed in a hole. A great advantage of cable bolts is flexibility that allows installation of long bolts, 60 ft or so, from a small drift or crosscut. Length itself is an advantage. Much practical information on cable bolting in underground excavations is presented by Hutchinson and Diedrichs (1996).

Additional transmission of load to the cable is warranted when failure of an interface occurs because neither interface or steel should fail before the other in an optimum installation. Steel rings threaded onto cable bolts provide for additional load transmission from supported rock to cable. Various methods of attaching such rings, buttons, collars, and so forth are used, but at added cost. Most cable bolt installations do not use rings. Another method of increasing steel utilization is to separate the wires of the bolt into a "bulb" or "birdcage." Grout readily flows into the space created and increases anchorage and pullout resistance. Bolt hole size needs to be large enough to allow easy push of cable into the holes, so the birdcage cannot be too large without increasing hole size.

A roller device that pushes cable into up-holes is an important feature of safe cable bolting. Pushing cables into up-holes manually incurs a risk of a cable sliding out of the hole and injuring nearby personnel. In this regard, bending strands at a cable bolt end into "fishhooks" provides temporary anchorage of bolts prior to grouting. Special fixtures are also available for temporary cable bolt anchorage.

Laboratory pull tests indicate a linear relationship between anchor length and pullout load, although factors such as water to cement ratio used in the grout, surface condition of the cable, and confinement are important to secure anchorage and effective cable bolting. Confinement comes into consideration because of the tendency of the cable to expand the adjacent grout during pullout. Lay of the cable creates a spiral mold about the cable, so when pulled the tendency is for the cable to twist and thus expand against the grout. Confinement prevents such expansion and thus improves anchorage. High modulus grouts and rock masses provide superior confinement. Application of the previously suggested rule of thumb for distributed anchorage bolting of one foot of hole per inch of steel diameter gives an anchorage length of about 6 inches for a 5/8-in.cable bolt.

Cable bolts may be tensioned during the installation process to obtain immediate support action. However, this "pretensioning" process requires two installation steps because tensioning can only be done after anchor grout has developed sufficient strength. Pretensioning cable bolts is not usually done, although special circumstance may justify the effort. Cable bolts may also be fitted with point anchors and tensioned immediately at installation time. In this case, the cable acts much like a point anchored mechanical bolt. However, the higher strength of cable, the greater length and greater flexibility give cable bolts superior support performance. In some cases, a portion of the hole may be left ungrouted to improve deformability of the bolt system and thus accommodate larger wall displacements. This provision may be important in burst-prone mines that experience transitory wall movements leading to wall disintegration. In this case, a yielding bolt (and mesh) system provides for containment of loose rock.

A dead weight load approach to cable bolting is an equilibrium approach that relates the main variables influencing bolt safety factor. Some of the geometries are shown in Figure 7.19. Each requires a separate analysis and some decision as to the slab thickness expected or else one may compute a slab thickness that is supportable with a given bolting plan. In this regard, a square pattern is often used, but in-row and between-row spacings may certainly differ. Cable bolt spacing in practice ranges from a close 4 ft to perhaps 15 ft; 8 to 10 ft spacing is common.

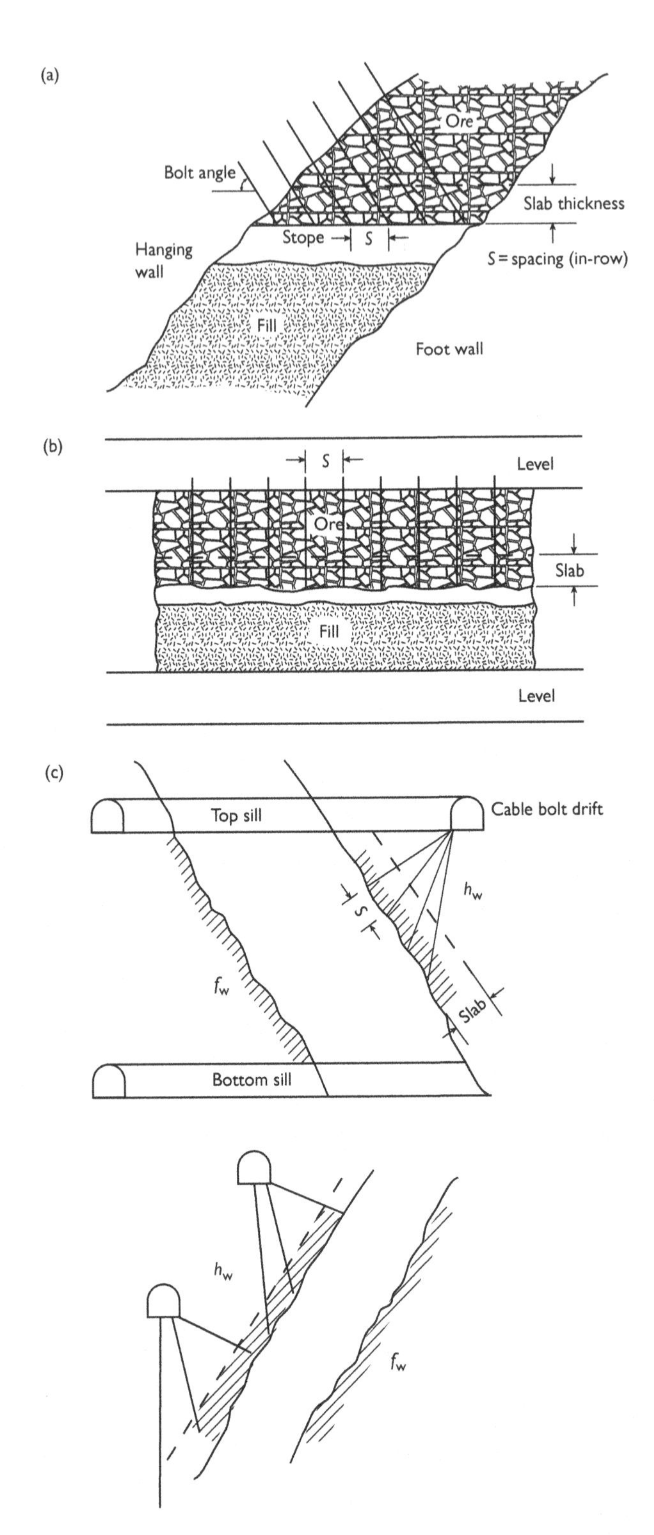

Figure 7.19 Cable bolt geometries: (a) medium width vein – cross section, (b) narrow vein – long section, wide vein – single row in long section, and (c) use of a cable drifts.

Example 7.13 A dipping ore body is mined across the vein with fill below and rock above the excavation as shown in the sketch. Cable bolts are used for support. Develop an expression that relates the main features of the support system using a dead weight load approach.

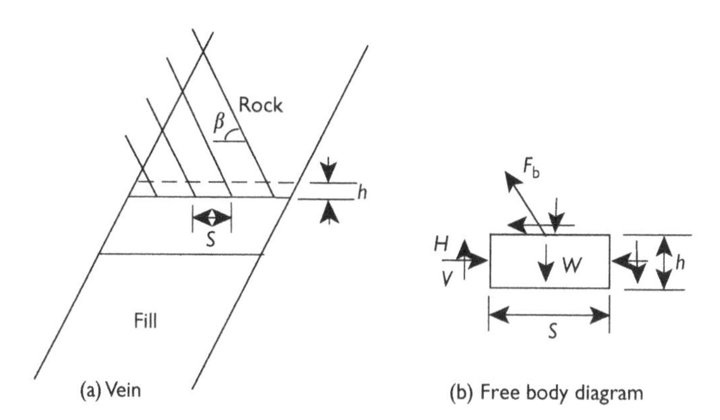

(a) Vein (b) Free body diagram

***Solution*:** A free body diagram is shown in the sketch, but under a dead weight load approach internal horizontal and vertical forces are neglected. Only the weight of the supported block and the support load provided by the cable bolt are considered. With these assumptions, equilibrium requires

$$F_b \sin(\beta) = W$$

$$\frac{F_b(\max)}{\text{FS}} = W$$

$$\therefore \text{FS(bolt)} = \frac{F_b(\max)}{W}$$

$$\text{FS(bolt)} = \frac{n f_b \sin(\beta)}{\gamma S^2 h}$$

where FS = cable bolt safety factor (breaking load/actual load), β = bolting angle, n = number of cable bolts per hole, f_b = strength of one cable bolt, γ = specific weight of rock, S = bolt spacing, and h = slab thickness. If the in-row and between-row spacing differ, then S^2 is replaced by the product of these spacings.

Example 7.14 Cable bolts are used to support the hanging wall of a steeply dipping stope as shown in the sketch. Use a dead weight load approach to develop an expression for the safety factor of the indicated bolt.

***Solution*:** A dead weight load free body diagram and force diagram are shown in the sketch. The potential rock block may move into the excavation in a direction perpendicular to the hanging wall. The normal component of bolting force must therefore be at least equal to the normal component of block weight. By definition, bolt safety factor is bolt strength

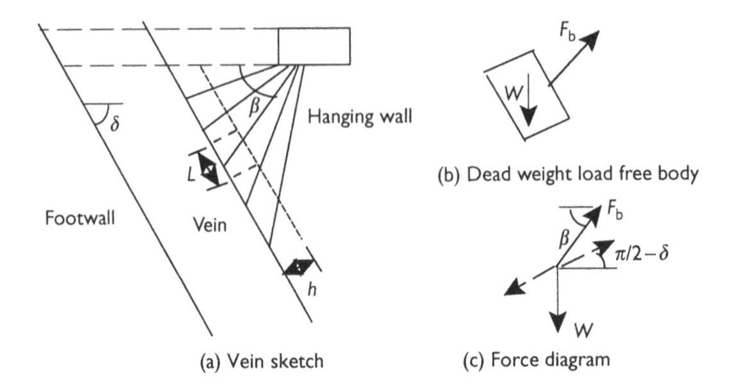

(a) Vein sketch

(c) Force diagram

(maximum load possible) divided by actual bolt force. Thus,

$$\begin{aligned}
\text{FS} &= \frac{F_b(\max)}{F_b(\text{actual})} \\
&= \frac{F_b(\max)}{N_b/\cos[\beta - (\pi/2 - \delta)]} \\
&= \frac{nf_b(\max)}{W\sin(\pi/2 - \delta)/\cos[\beta - (\pi/2 - \delta)]} \\
\text{FS} &= \frac{nf_b(\max)\sin[\beta + \delta)]}{\gamma hLS\cos(\delta)}
\end{aligned}$$

where n = number of cables per bolt hole, f_b = bolt strength, β = bolting angle (dip), δ = vein dip, γ = rock specific weight, h = slab thickness, L = in-fan spacing, and S = spacing between fans (into the page).

Additional examples

Example 7.15 Consider the sphere shown in Figure 7.2 and suppose a unit compressive load $S_x = 1.0$ is applied parallel to the x-axis. This load is the preexcavation stress state. Use the formulas in Table 7.2 to construct a table of stress concentration factors for x-axis loading.

Solution: First, Table 7.2 is rewritten in simple form. Thus,

Symbols for Sz=1.0 loading

Point	*Factor*		
	A	B	C
Kx east	xA	xB	xC
Ky north	yA	yB	yC
Kz vertical	zA	zB	zC

Here, for example, xA=0, and xB is given by

$$\mathrm{xB} = \frac{30v}{14-10v} - \frac{3(1+5v)}{14-10v}$$

For clarity, use compass coordinates: x=east (horizontal), y=north (horizontal), z=up (vertical) Because of symmetry, one only needs to do a thought experiment that interchanges the x- and z-axes and the associated point labels. Thus,

Example 7.15 Symbols for Sx=1.0 loading

Point	*Factor*		
	A	*B*	*C*
Kx east	zC	zB	zA
Ky north	yC	yB	yA
Kz vertical	xC	xB	xA

Hence,

Example 7.15 Sx=1.0, Sy=0.0, Sz=0.0

Point	*Factor*		
	A	*B*	*C*
Kx east	0	$\frac{30}{14-10v} - \frac{3(1+5v)}{14-10v}$	$\frac{30}{14-10v} - \frac{3(1+5v)}{14-10v}$
Ky north	$-\frac{3(1+5v)}{14-10v}$	0	$\frac{30v}{14-10v} - \frac{3(1+5v)}{14-10v}$
Kz vertical	$-\frac{3(1+5v)}{14-10v}$	$\frac{30v}{14-10v} - \frac{3(1+5v)}{14-10v}$	0

Example 7.16 Consider the sphere discussed in Example 7.15 under a unit compressive load $S_x = 1.0$ is applied parallel to the x-axis. Use the example results to construct a table of peak tensile (negative) and compressive stress concentration factors for values of Poisson's ratio on the interval [0.0, 0.5] analogous to Table 7.3. Indicate in the table the location of the peak stress concentrations found.

***Solution*:** A spread sheet calculation gives the results:

Poisson's ratio	Peak Compression (B and C)	Peak Tension (A)
0.0	1.929	−0.214
0.1	1.962	−0.346
0.2	2.000	−0.500
0.3	2.045	−0.682
0.4	2.100	−0.900
0.5	2.167	−1.167

The numerical results are the same as in Table 7.3. Peak tension always occurs under the load point and acts horizontally. Peak compression occurs at points away from the load point. In this case, the points are B and C. The peak compression acts parallel to the direction of the applied load, i.e., parallel to the x-direction in this example. As so often is the case, peak tension (for mid-range values of Poisson's ratio) seldom exceeds the applied stress in magnitude.

Example 7.17 Consider a spherical cavern of radius a (65 m) at a depth of 1,000 m under gravity loading only where the vertical stress is estimated at 24 kPa/m and the horizontal stresses are one-third the vertical preexcavation stress. Poisson's ratio is 0.30; unconfined compressive and tensile strengths are: $C_o = 100$ MPa, $T_o = 10$ MPa. Determine the cavern wall safety factors.

Solution: By definition, the factors of safety are

$$FS_c = C_o/\sigma_c \text{ and } FS_t = T_o/\sigma_t$$

and the peak stresses are given by stress concentrations

$$\sigma_C = K_c S_1 \text{ and } \sigma_t K_t S_1$$

where S_1 is the reference stress (the major preexcavation principal stress). Tables of stress concentration factors for z- and z-direction unit loads are available (Table 7.2 and the table in Example 7.16). In case of y-direction loading, following the process in Example 7.16, one obtains

Example 7.16 Symbols for Sy=1.0 loading

Point	*Factor*		
	A	*B*	*C*
Kx east	xA	xC	xB
Ky north	zA	zC	zB
Kz vertical	yA	yC	yB

Hence,

Example 7.16 Sx = 0.0, Sy = 1.0, Sz = 0.0

Point	*Factor*		
	A	*B*	*C*
Kx East	0	$-\frac{3(1+5v)}{14-10v}$	$\frac{30v}{14-10v}-\frac{3(1+5v)}{14-10v}$
Ky North	$\frac{30}{14-10v}-\frac{3(1+5v)}{14-10v}$	0	$\frac{30v}{14-10v}-\frac{3(1+5v)}{14-10v}$
Kz vertical	$\frac{30v}{14-10v}-\frac{3(1+5v)}{14-10v}$	$-\frac{3(1+5v)}{14-10v}$	0

There are four distinct values in all the tables. In Table 7.2, one has

Example 7.16 Symbols for Sz=1.0, loading, v=0.30

Point	*Factor*		
	A	*B*	*C*
Kx east	xA=0	xB=0.136	xC= −0.682
Ky north	yA=0.136	yB=0	yC= −.682
Kz vertical	zA=2.046	zB=2.046	zC=0

After computing all table entries, one obtains symbolically

Example 7.16 Sx=1/3, Sy=1/3, Sz=1, v=0.30

Point	*Factor*		
	A	*B*	*C*
Kx east	0	(1/3)zB+(1/3)xC+(1)xB	(1/3)zA+(1/3)xB+(1)xC
Ky north	(1/3)yC+(1/3)zA+(1)yA	0	(1/3)yA+(1/3)zB+(1)yC
Kz vertical	(1/3)xC+(1/3)yA+(1)zA	(1/3)xB+(1/3)yC+(1)zB	0

Example 7.16 Sx=1/3, Sy=1/3, Sz=1, v=0.30

Point	*Factor*		
	A	*B*	*C*
Kx east	0	(1/3)(2.046) + (1/3) (−0.682)+(1)(0.136)	(1/3)(2.046) + (1/3)(0.136) +(1) (−0.682)
Ky north	(1/3)(−0.682) +(1/3) (2.046)+(1)(0.136)	0	(1/3)(0.136) + (1/3)(2.046) (1/3) (2.046)+(1)(−0.682)
Kz vertical	(1/3)(−0.682) +(1/3) (0.136)+(1)(2.046)	(1/3)(0.136) + (1/3) (−0.682)+(1)(2.046)	0

Example 7.16 Sx=1/3, Sy=1/3, Sz=1, v=0.30

Point	*Factor*		
	A	*B*	*C*
Kx east	0	0.591	0.453
Ky north	0.591	0	0.453
Kz vertical	1.864	1.864	0

The vertical loads at A and B are equal as they should be by symmetry. The east and west loads are equal at C as they should be. The east load at B is equal to the north load at A which is also as it should be in consideration of the problem symmetry. Peak compression occurs at the equator of the sphere and is vertical.

There is no tensile stress concentration, so the safety factor with respect to tension is not a consideration. The compressive stress safety factor is

$$
\begin{aligned}
FS_c &= C_o/\sigma_c = C_o/K_c S_1 \\
&= (100)(10^6)/(1.864)(24)(10^3)(1000) \\
FS_c &= 2.24
\end{aligned}
$$

Example 7.18 A storage cavern is dissolved into an oblate spheroid at a depth of 300 m (cavern center) in a thick, massive salt formation where the stress state is hydrostatic before excavation. Cavern height is 30 m; width is 90 m. Estimate salt formation unconfined compressive and tensile strengths needed for safety factors of 3 and 6 in compression and tension, respectively. Also locate the position of peak tension and compression.

Solution: The semi-axis ratio a/c is 3 and $K_0 = 1$ from the given data. For estimation, a reasonable value of Poisson's ratio is 0.20 which also allows use of plots in the text. A vertical stress gradient of 24 kPa/m is reasonable under gravity loading. At 300 m, the vertical stress is 7.2 MPa. By definition, $Fs_c = C_o/\sigma_c$ and $FS_t = T_o/\sigma_t$. The peak stresses are given by stress concentration formula $\sigma_c = K_c S_1$ and $\sigma_t = K_t S_1$. From Figure 7.6(a), tension is absent, so no restriction is imposed on T_0. The compressive stress concentration factor may be obtained from the regression line equation. Thus, $K_c = (1.1784)/(3) + (0.2464) = 3.782$. Hence

$$
\begin{aligned}
C_o &= (FS_c)(K_C)(S_1) \\
&= (3)(3.782)(7.2) \\
C_o &= 81.7 \text{ MPa}
\end{aligned}
$$

The peak compression occurs along the equator and acts in the vertical direction, as one might suppose in consideration of the cavern geometry.

Example 7.19 Consider the data given in Example 7.18, but suppose the long dimension (90 m) of the cavern is vertical, so the cavern has the shape of a prolate spheroid. Estimate the required tensile and unconfined compressive strengths to achieve the specified factors of safety.

Solution: Because the stress state is hydrostatic, the problem geometry can be viewed from the horizontal direction as an oblate spheroid which is the problem in Example 7.18. However, the equator is now a vertical circle rather than a horizontal circle. Tension is still absent and the peak compression occurs at the equator but acts horizontally. The required unconfined compressive strength is the same, $C_o = 81.7$ MPa.

Example 7.20 A tunnel of rectangular cross-section 7 m wide by 4 m high is advanced at a depth of 340 m where the vertical preexcavation stress is 24 kPa/m; the north-south stress is 1.4 times the vertical stress and the east-west stress is 0.7 times the vertical stress. Determine the most favorable direction of advance and then estimate the stress concentration factors for tension and compression when the tunnel has advanced 100 m.

***Solution*:** The most favorable orientation of a three-dimensional opening is one with dimensions in proportion to the preexcavation stresses. In particular the long axis should be parallel to the greatest preexcavation principal stress. In this case, the tunnel axis should be parallel to the greatest horizontal stress. Also, one should have $L_2/L_3 = S_2/S_3$. However, according to the given conditions $L_2/L_3 = 7/4$ and $S_2/S_3 = S_v/S_h = 1/0.7$, so a most favorable orientation of the section is not possible. In this orientation, the two-dimensional principal stress ratio is $M = S_3/S_1 = 0.7/1.0$ with the major principal stress acting perpendicular to the opening width of 7 m. Stress concentration factors for a rectangular opening are given in Table 3.2a. In this case the semi-axes ratio is 4/7. At M=0.75 and $k' = 2$, tension is absent and the compressive stress concentration factor is 5.92. The actual stress ratio is less as is the semi-axes ratio. The first tends to reduce the stress concentration factor, while the latter tends to increase the stress concentration factor. The value 5.92 is therefore a reasonable estimate.

Example 7.21 Consider the data in Example 7.20 and further suppose the tunnel is advanced in an east-west direction rather than along the given north-south route. Estimate the stress concentration factors in this orientation.

***Solution*:** Again, the most favorable orientation of a three-dimensional opening is one with dimensions in proportion to the preexcavation stresses. In particular the long axis should be parallel to the greatest preexcavation principal stress. However, in this case the tunnel is parallel to the minimum principal stress, so a most favorable orientation of the section is not possible. In this orientation, the two-dimensional principal stress ratio is $M = S_3 /S_1 = 1.0/1.4 = 0.71$ with the major principal stress acting parallel to the opening width of 7 m. Stress concentration factors for a rectangular opening are given in Table 3.2a. The semi-axes ratio is 4/7. At $M = 0.75$ and $k = 0.5$, tension is absent and the compressive stress concentration factor is 5.48. The actual stress ratio is less as is the semi-axes ratio. The first tends to reduce the stress concentration factor, while the latter tends to increase the stress concentration factor. The value 5.48 is therefore a reasonable estimate. Comparison with results in Example 7.20 indicates there is little to choose from with respect to tunnel direction under the given conditions.

Example 7.22 A tunnel of rectangular cross-section 7m wide by 4m high is advanced along a north-south route at a depth of 340m where the vertical preexcavation stress is 24 kPa/m; the north-south stress is 1.4 times the vertical stress and the east-west stress is 0.7 times the vertical stress. A second excavation pass widens the tunnel to 14 m. Determine the unconfined compressive and tensile strengths needed for safety factors of 2 and 4, respectively.

***Solution*:** A cross-section has semi-axes ratio of 4/14 and a $k' = 14/4 = 3.5$ because the major principal stress in cross-section is acting parallel to b or y axis. Recall a is the semi-axis measured along the x axis. The principal stress ratio $M = 0.7/1.0 = 0.7$ according to the given conditions. From Table 3.2a, for $M = 0.75$ and $k' = 4$, $K_c = 7.99$ and $K_t = 0.00$. Inspection of the table indicates slightly lower stress concentration factors, so a slight but negligible tension may be present. No limitation is therefore placed on tensile strength and the factor of safety with respect to tension is not applicable.

The unconfined compressive strength needed is

$$
\begin{aligned}
C_o &= (FS_c)(K_C)(S_1) \\
&= (2)(7.99)(24)(10^3)(340) \\
C_o &= 130 \text{ MPa}
\end{aligned}
$$

Example 7.23 A tunnel of rectangular cross-section 7m wide by 4m high is advanced along a north-south route at a depth of 340m where the vertical preexcavation stress is 24 kPa/m; the north-south stress is 1.4 times the vertical stress and the east-west stress is 0.7 times the vertical stress. A second excavation pass widens the tunnel to 14 m. Unconfined compressive and tensile strengths are 150 MPa and 15 MPa, respectively. Determine the number of additional passes of the same 7 m width that can be made before failure occurs.

***Solution*:** This problem asks for the aspect ratio at failure. Stress concentration depends on aspect ratio and inconsideration of the long dimension of the cumulative cross-section being normal to the major principal stress in cross-section, one has $FS_c = C_o/K_c\,(aspect\ ratio)S_1$ that requires inversion for the sought aspect ratio (width to height ratio). Hence,

$$
\begin{aligned}
K_c &= C_o/FS_cS_1 \\
&= (150)(10^6)/(1)(24)(10^3)(340) \\
K_c &= 18.38
\end{aligned}
$$

Inspection of Figure 3.9a shows a regression lines close but slightly above the $M = 1/3$ line. According to this line

$$
\begin{aligned}
K_{\text{max}} &= a_1 + a_2k + a_3/k + a_aM \\
18.38 &= 0.69 + 1.27k + 0.68/k + 3.83(1/3) \\
&\therefore \\
k &= 12.9
\end{aligned}
$$

Because M is greater than 1/3 used in the regression equation, k should be somewhat lower, say, 12. At this ratio, the tunnel width is (12)(4) = 48 m. At 7 m per pass, six passes could be made.

Example 7.24 A spherical storage cavern with center at a depth of 800 ft and a diameter of 90 ft in hydrostatic stress field begins to spall. Spalling spreads uniformly into the cavern walls and ceases when the diameter reaches 120 ft and the caved material fills the cavern. The cavern is filled with a liquid that has a specific gravity of 1. Estimate the void volume of fluid in the cavern after spalling.

***Solution*:** Volume of the cavern at a diameter of 120 ft is composed of solid fragments and voids. Hence, $V = V_s + V_v$. Volume of the solids is the difference between the new and original cavern volumes, so $V_s = V - V' = (4/3)\pi\,(60^3 - 45^3) = 5.231(10^5)$ ft^3. Void volume is therefore $V_v = V - V_s = (4/3/)\pi\,(60)^3 - 5.231(10^5) = 3.817(10^5)$ ft^3. Note that the volume of the voids is simply the volume of the original cavern. The ratio of caved material volume to the original solid volume is $V/V_s = 5.231/3.817 = 1.37$ that indicates a swell of 37%, a reasonable amount for rubble piles.

Example 7.25 A spherical storage cavern with center at a depth of 800 ft is planned with a diameter of 90 ft in hydrostatic stress field ($\sigma = 800$ psi). The walls of the cavern may spall, so a cable bolt pattern of support is installed *before* excavation from a spiral ramp about the future cavern. Suppose the cable bolts stretch in a compatible mode with the inward displacement of the cavern. Estimate relative displacement of 30 ft long, 5/8 inch diameter, 7-strand, bolts with a breaking load of 25 tons. Bolts extend from cavern walls into the solid and are spaced on a square pattern, 10 ft × 10 ft. Note: Young's modulus is $5(10^6)$ psi and Poisson's ratio is 0.35. Young's modulus of the cable bolt is $30(10^6)$ psi.

Solution: A formula for the displacement induced about a spherical cavern excavated in a hydrostatic stress field is needed. Expressions for stresses are given in Example 7.2. Use of Hooke's law yields an expression for the radial displacement derivative after substitution for the radial strain. From Example 7.2 that gives stress concentration for a unit load one has $\sigma_r/\sigma = 1 - (a/r)^3$, $\sigma_\theta/\sigma = \sigma_\phi/\sigma = 1 + (1/2)(a/r)^3$. These formulas are composed of two parts, an original part and an induced change. The original part may be obtained by a thought experiment that allows the radius to become very large and thus locates a point remote from the cavern wall where the original stresses act. This thought experiment shows that the stress changes are simply $\Delta\sigma_r/\sigma = -(a/r)^3$, $\Delta\sigma_\theta/\sigma = \Delta\sigma_\phi/\sigma = +(1/2)(a/r)^3$. The associated radial strain change from Hooke's law is then $E\Delta\varepsilon_r = (\Delta\sigma_r - v\Delta\sigma_\theta - v\sigma_\phi)$ where E and v are Young's modulus and Poisson's ratio, respectively. Hence,

$$
\begin{aligned}
E\partial u/\partial r &= \Delta\sigma_r - v\Delta\sigma_\theta - v\sigma_\phi \\
&= [-(a/r)^3 - 2v(1/2)(a/r)^3]\sigma \\
E\partial u/\partial r &= -(1+v)(a/r)^3 \\
&\therefore \\
u &= [(1+v)/2E](a^3)(r^{-2})(\sigma)
\end{aligned}
$$

where u is the radial displacement induced by excavation and inward displacement is positive. The relative displacement between cable bolt ends, one of which is at the cavern wall while the other is at r is simply the difference $\Delta u = [(1+v)/2E](a^3)(1/a^2 - 1/r^2)(\sigma)$. Thus, bolt elongation is given by

$$
\begin{aligned}
\Delta u &= [(1+0.35)/2(5)(10^6)](45^3)(1/45^2 - 1/75^2)(800)(12) \\
\Delta u &= 3.73(10^{-2}) \text{ inches}
\end{aligned}
$$

Cable bolt strain is small, that is,

$$
\begin{aligned}
\varepsilon &= 3.73(10)^{-2}/(30)(12) \\
\varepsilon &= 1.04(10^{-4}) \text{ in./in.}
\end{aligned}
$$

Bolt tension is also small,

$$\sigma = E\varepsilon = 30(10^6)(1.04)(10^{-4})$$
$$\sigma = 3{,}120 \text{ psi}$$

Comments: The force corresponding to this stress is an obviously small 957 lbf relative to breaking load of 50,000 lbf and raises a question whether bolts are needed. The actual cavern wall displacement is 0.0583 inches, about 56% greater than the relative displacement between bolt ends. A longer bolt would therefore experience greater elongation, higher strain, and higher stress. Thus, longer bolts do not necessarily provide better support. Indeed, there may be a case for shorter bolts or for no cable bolts at all. Much depends on geologic structure and the presence of joints. Should blocks become detached from cavern walls, cable bolt stress would be entirely different depending on the dead weight load of blocks supported by the cable bolts. This problem also illustrates the application of fundamentals (physical laws, kinematics, and material laws) to engineering design.

Example 7.26 A spherical storage cavern with center at a depth of 800 ft is planned with a diameter of 90 ft in hydrostatic stress field ($\sigma = 800$ psi). The walls of the cavern may spall, so a cable bolt pattern of support is installed *before* excavation from a spiral ramp about the future cavern. Suppose the cable bolts stretch in a compatible mode with the inward displacement of the cavern. Estimate relative displacement of 30 ft long, 5/8 inch diameter, 7-strand bolts with a breaking load of 25 tons when supporting a dead weight load equal to the mass of rock supported by each cable. Bolts extend from cavern walls into the solid and are spaced on a square pattern, 10 ft × 10 ft. Note: Young's modulus is $5(10^6)$ psi and Poisson's ratio is 0.35. Young's modulus of the cable bolt is $30(10^6)$ psi and elongation at the breaking load is 3.5%.

***Solution*:** The volume of rock supported by each cable is approximately (10)(10)(30) = 3,000 cubic feet. A key block at the top of the cavern is likely to require maximum support. Weight of this volume is estimated as (144)(3,000)/2000 = 216 tons which exceeds bolt capacity of 25 tons. Elongation at the breaking point is (30)(0.035) = 1.05 ft.

Example 7.27 A spherical storage cavern with center at a depth of 800 ft is planned with a diameter of 90 ft in hydrostatic stress field ($\sigma = 800$ psi). The walls of the cavern may spall, so a cable bolt pattern of support is installed *before* excavation from a spiral ramp about the future cavern. Suppose the cable bolts stretch in a compatible mode with the inward displacement of the cavern. Determine the number of cable bolts per hole needed to support a dead weight load of rock. Bolts are 30 ft long, 5/8 inch diameter, 7-strand with a breaking load of 25 tons when supporting a dead weight load equal to the mass of rock supported by each cable. Bolts extend from cavern walls into the solid and are spaced on a square pattern, 10 ft x 10 ft. Note: Young's modulus is $5(10^6)$ psi and Poisson's ratio is 0.35. Young's modulus of the cable bolt is $30(10^6)$ psi and elongation at the breaking load is 3.5%.

***Solution*:** From Example 7.26, the weight of a rock block at the top of the cavern is 216 tons. At 25 tons per cable, at least 9 cables per hole are required.

Comment: This is a large number for underground bolting, so consideration should be given to a smaller spacing and also to a more careful estimate of what is a most probable rock block size. In this case, the bolts would restrain small spalls and thus provide support to intact rock behind the spalls, so quite possibly only a few feet into the cavern wall would need bolt support. At even half the bolt length, block size would be reduced by half as would the number of cables required. A spacing of 8 x 8 ft with a block length of 15 ft leads to a dead weight load 69 tons and 3 strands per hole. This example points out the risk of excessive conservatism and the need for accurate estimation of bolt loads for design.

Example 7.28 Given the data in Example 7.27 and the rock block size 10×10×30 ft. Determine the block stiffness under axial load and the additional stiffness provided by a single cable bolt.

***Solution*:** Stiffness is defined by a force-displacement equation. In this uniaxial loading case, $F = KU = \sigma A = E\varepsilon A = (EA/L)(U)$where $F, K, U, \sigma, \varepsilon, E, A$, and L are axial force, stiffness, displacement, axial stress, axial strain, Young's modulus, cross-section area, and block length, respectively. One has $K = EA/L$. Hence for the rock block,

$$K(rock) = 5(10^6)(144)(10)(10)/30 = 2.4(10^9)\ \text{lbf/ft}\ [2.0(10^8)\ \text{lbf/in.}].$$

For the cable bolt

$$K(bolt) = 30(10^6)(\pi/4)(5/8)^2/(30)(12) = 25.6(10^3)\ \text{lbf/in.}$$

Comment: Cable bolt stiffness is orders of magnitude smaller than rock stiffness and for this reason has little effect on cavern wall displacement. This is frequently the case with reinforcement and is often seen in displacements computed via detailed numerical analyses.

Another similar computation is to compute the effect of a roof bolt on displacement. For example, consider a 6 ft long, 3/4 inch diameter roof bolt with a Young's modulus of $30(10^6)$ psi. If bolts are spaced on 4 ft centers, then block stiffness is

$$K(rock) = 5(10^6)(144)(4)(4)/6 = 1.92(10^9)\ \text{lbf/ft}\ [1.6(10^8)\ \text{lbf/in.}]$$

Bolt stiffness is

$$K(bolt) = 30(10^6)(\pi/4)(3/4)^2/(6)(12) = 1.8410^5)\ \text{lbf/in.}$$

Again, bolt stiffness is orders of magnitude smaller than rock stiffness and would have little effect on displacement as a consequence. However, in stratified roof where bed separation occurs, the displacement is beam-like rather than massive. Roof bolts may have a substantial effect on roof beam sag and, indeed, even reverse sag to the point of eliminating bed separation and actually induce a clamping action between strata.

Example 7.29 A tabular excavation is oriented vertically and extends in a north-south direction. Excavation width is 20 ft, height is 100 ft, and length is 500 ft north-south. A hydraulic sand backfill is place in the excavation. The fill Young's modulus is 20 ksi. Determine the fill stiffness with respect to horizontal force per ft of length and the force required to compress the fill by 1 inch. Also determine the equivalent horizontal stress for this force.

***Solution*:** By definition, stiffness $K = EA/L$. Here A = (100)(1) square feet and L = 20 ft. Hence $K = (20)(1000)(144)(100)(1)/20 = 14.4(10^6)$ lbf /ft [$1.2(10^6)$ lbf /in.]

A force of $1.2(10^6)$ lbf is required to compress the fill. This force acts over area A, so the equivalent stress is

$$\sigma = F/A$$
$$= 1.2(10^6)/(100)(1)(144)$$
$$\sigma = 83.3 \text{ psi}$$

Example 7.30 Given the data and results from Example 7.29 and the presence of two joint sets, with dip direction east-west. The first set dips 30° east; the second set dips 60° west. Both joint sets have friction angles of 25° and zero cohesion. Joint spacing is 8 ft in both sets and are fully persistent. These data are cautious and represent a worst case situation where blocks form in the cavern walls. Consider the east wall and the largest possible block that may form per ft of length. Determine whether the block will slide. If so, compute the horizontal force that would develop without fill and suppose this force actually compresses the fill across the opening. Estimate the unbalanced horizontal force and associated fill displacement.

Solution: A sketch illustrates the situation in cross section.

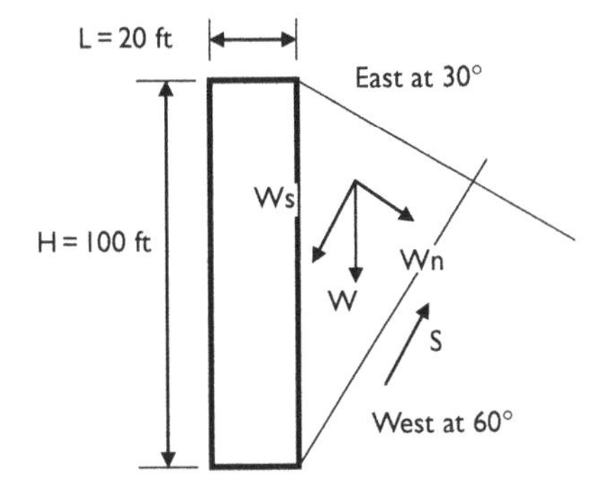

Sketch for Example 7.30

Block weight is $W = \gamma V$ and $V = H \sin(60) H \cos(60)/2$ per foot into the page. A reasonable specific weight is 156 pcf. Hence, $W = (156)(100)\sin(60)(100)\cos(60)/2 = 3.38(10^5)$ lbf. The downhill component of weight is $Ws = W\sin(60) = 3.38(10^5)\sin(60) = 2.93(10^5)$ lbf and the normal component is $Wn = W\cos(60) = 3.38(10^5)\cos(60) = 1.69(10^5)$ lbf. The maximum uphill resisting force mobilized by the normal force is $S = Wn \tan(\phi) = 1.69(10^5)\, tan(25) = 0.79(10^5)$. The unbalanced downhill force is therefore $F = (2.93 - 0.79)(10^5) = 2.14(10^5)$ lbf and the block will slide. This result is also evident because the friction angle is less than the slope angle.

The horizontal component of this force is

$$Fh = 2.14(10^5)\cos(60) = 1.07(10^5) \text{ lbf}.$$

From example 7.29, fill stiffness is $K = 1.2(10^6)$ lbf /in. Hence, the fill will compress 0.146/1.2, that is, *0.12 inches*.

Comment: Although fill stiffness is generally much smaller than rock stiffness, fill is quite effective at restraining blocks that would otherwise cave into an opening. The stabilizing effect of block retention, in this case a 169 ton block, is considerable. The acceleration of the block in the absence of fill is given by $a = Fg/W = 2.92g/3.38 = 0.864g$ where g is the acceleration of gravity. This acceleration is a large fraction of free fall acceleration. Because the block also tends to move downwards, the fill is also sheared at the fill-block contact and complicates the situation. However, the general conclusion that fill is highly effective in controlling wall block movement still holds.

7.5 Problems

3D Caverns

7.1 An oblate spheroid is solution mined in a salt bed with $a = 75$ ft and $b = 150$ ft. Center of the cavity is 650 ft deep. Estimate peak stresses and show locations on a cavity sketch.

7.2 An oblate spheroid is solution mined in a salt bed with $a = 23$ m and $b = 46$ m. Center of the cavity is 200 m deep. Estimate peak stresses and show locations on a cavity sketch.

7.3 An underground storage cavern is solution mined in a thick salt bed where

$$C_o = 13{,}200 \text{ psi} \qquad T_o = 1{,}230 \text{ psi}$$
$$\gamma = 144 \text{ pcf} \qquad E = 5.0 \times 10^6 \text{ psi}$$
$$G = 2.0 \times 10^6 \text{ psi}$$

and depth to the cavern center is 1,450 ft. The cavern has a spherical shape and is 150 ft in diameter. The premining stress is hydrostatic. Estimate the wall safety factor.

7.4 An underground storage cavern is solution mined in a thick salt bed where

$$C_o = 91.0 \text{ MPa} \qquad T_o = 8.5 \text{ MPa}$$
$$\gamma = 22.8 \text{ kN/m}^3 \qquad E = 34.5 \text{ GPa}$$
$$G = 13.8 \text{ GPa}$$

and depth to the cavern center is 442 m. The cavern has a spherical shape and is 49 m in diameter. The premining stress is hydrostatic. Estimate the wall safety factor.

7.5 An underground storage cavity is excavated in a massive salt formation by solution mining. Depth to the center of the cavity is 1,340 ft (408 m). Borehole surveying shows the cavity has the shape of an oblate spheroid 100 ft (30.5 m) in diameter and 50 ft (15.25 m) high. Estimate the salt strength necessary for stability. Show the stresses in vertical section at the top and sides of the opening.

7.6 Mining over the years produces a tabular excavation that extends about 2,000 ft (610 m) down dip from the surface. Dip of the ore zone is 65°, width is 40 ft (12.2 m). Strike length is over 6,000 ft (1,839 m). Estimate the peak stresses and indicate their location on a sketch of the mine.

7.7 Stopes mined along plunging folds of Precambrian rock range between depths of 869 m (2,850 ft) and 1,174 m (3,850 ft) below surface. The stopes are up to 30.5 m (100 ft) wide. Fold plunge is 12°. Fold dip varies, but stope walls are vertical. A vertical, rectangular shaft 4.6 × 6.4 m (15 × 21 ft), provides access. How close to the shaft should stoping be allowed? Explain.

Back fill

7.8 Consider a cut and fill stope in a 15 ft wide vein where raise and level intervals are 175 ft. A hydraulic fill is placed along the entire stope length to a height of 12 ft. If the specific weight of fill is 100 pcf and the porosity is 35%, how many tons of solids are in the fill and how many gallons of water?

7.9 Consider a cut and fill stope in a 4.6 m wide vein where raise and level intervals are 53.3 m. A hydraulic fill is placed along the entire stope length to a height of 3.66 m. If the specific weight of fill is 15.8 kN/m^3 and the porosity is 35%, how many tons of solids are in the fill and how many liters of water?

7.10 Void ratio e is defined as the ratio of void volume to volume of solids in porous material. Show that void ratio may be computed from porosity by the formula $n/(1 - n)$ where n is porosity.

7.11 Consider a narrow vein overhand stope using cut and fill, as shown in the sketch. Cemented, hydraulic sand fill of modulus E_f is placed in the stope as mining proceeds up dip in a rock mass of modulus E_r. (a) Develop a formula for crown pillar safety factor assuming uniform closure across the vein, whether across fill, open area, or unmined ore. (b) If the vein is vertical, vein width is 15 ft, level and raise intervals are 175 ft, stope strike length is 175 ft, $E_f = 25$ ksi, $E_r = 1{,}000$ ksi, rock compressive strength is given by a size effects formula $C_p = C_1\ (0.78 + 0.22H_p/W_p)$ and $C_1 = 28{,}000$ psi; fill cohesion and angle of internal friction are 300 psi and 35°, respectively, then determine pillar width measured down dip when failure impends. Note: the premining vertical stress in psi is $1.1d$ where d is depth in ft; the horizontal premining stresses are equal and equal to twice the vertical premining stress.

7.12 Consider a narrow vein overhand stope using cut and fill, as shown in the sketch. Cemented, hydraulic sand fill of modulus E_f is placed in the stope as mining proceeds up dip in a rock mass of modulus E_r. (a) Develop a formula for crown pillar safety factor assuming uniform closure across the vein, whether across fill, open area, or unmined ore. (b) If the vein is vertical, vein width is 4.6 m, level and raise intervals are 53.3 m, stope strike length is 53.3 m, $E_f = 172$ MPa, $E_r = 689$ MPa, rock compressive strength is given by a size effects formula $C_p = C_1\ (0.78 + 0.22H_p/W_p)$ and $C_1 = 193$ MPa; fill cohesion and angle of internal friction are 2.07 MPa and 35°, respectively, then determine pillar width measured down dip when failure impends. Note: the premining vertical stress in kPa $24.4d$ where d is depth in meters; the horizontal premining stresses are equal and equal to twice the vertical premining stress.

7.13 Laboratory test data from 4-in. (10.2 cm) diameter samples, 4 in. (10.2 cm) long (not cemented) are given in Table 7.5. Estimate the hydraulic conductivity of the fill (in./h).

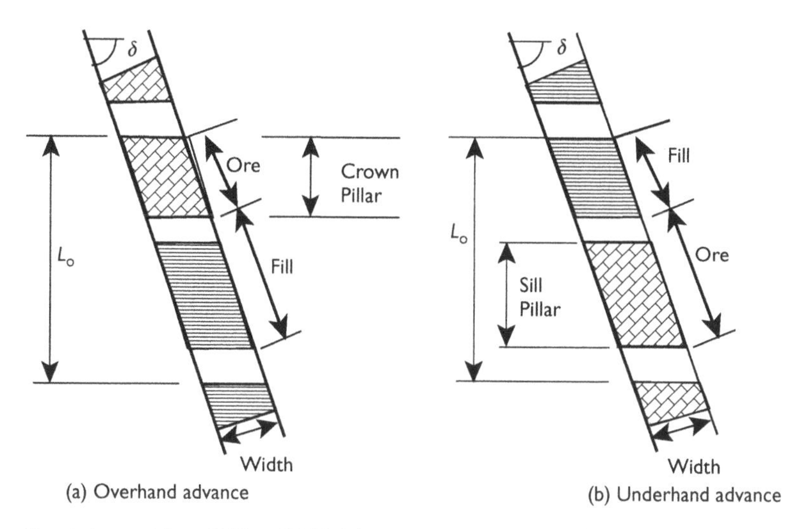

Sketch for problems 7.11 and 7.12. Narrow vein overhard and underhard cut and fill sto ping.

Table 7.5 Fill test data

Time (min)	0.0	1.90	4.22	7.22	11.4
Head (in.)	36.0	30.0	24.0	18.0	12.0
(cm)	91.4	76.2	61.0	45.7	30.5

Note
The standpipe inside diameter is 0.5 in. (1.27 cm).

Table 7.6 Data for Problem 7.14

Head (in.)	35.8	29.9	16.2	10.8	7.3	4.9
(cm)	91.0	76.0	41.1	27.4	18.5	12.4
Time (s)	0	45	90	135	180	225

7.14 A hydraulic sand fill is tested using a falling head permeameter that has a tube diameter of 1.27 cm (0.50 in.), sample diameter of 10.2 cm (4.0 in.), sample length of 20.4 cm (8.0 in.), and a starting head of 91 cm (35.8) in. The data collected are heads vs times and are given in Table 7.6.

Is this fill likely to be satisfactory? Justify your answer.

7.15 Old room and pillar workings extend under Hometown where the strata roll to the horizontal (depth = 1,750 ft, 533 m) and where additional protection in the form of river-run sand fill is placed hydraulically through boreholes. The extraction ratio in this area is estimated to be about 40%. Once in place, the fill is grouted for additional strength. Laboratory test data from 4-in. (10.2 cm) diameter samples, 4 in. (10.2 cm) long (not grouted) are given in Table 7.7. Estimate the hydraulic conductivity of the fill (in./h).

Table 7.7 Fill test data for Problem 7.15

Time (min)	0.0	1.90	4.22	7.22	11.4
Head (in.)	36.0	30.0	24.0	18.0	12.0
(cm)	91.4	76.2	61.0	45.3	30.5

Note
The standpipe inside diameter is 0.5 in. (1.27 cm).

Cable bolting

7.16 A large mechanized cut-and-fill stope in hard rock is cable bolted on a square pattern as shown in the sketch. A single 5/8 in. (1.59 cm) diameter cable bolt is installed per hole; cables have an ultimate strength of 56,000 lbf (251 kN). Bond strength between grout and cable steel is 700 psi (4.83 MPa). Shear failure is expected to occur at the steel–grout interface rather than at the grout–rock interface.

(a) How many inches of hole length are required to provide anchorage capacity in shear equal to the bolt tensile strength?
(b) Develop a formula that relates the thickness h of a slab that could be supported in the stope back to the spacing S of the bolts assuming that there are n bolts per hole.
(c) Plot the formula with h on the x axis and S on the y axis for slab thickness ranging from 0 to 10 ft (3 m); use the number of bolts per hole as a parameter and plot curves for $n = 1, 2, 3$, and 4.
(d) If the spacing is 10 ft (3 m), what slab thickness could be supported with two bolts per hole?

7.17 Consider the cable bolting array shown in the sketch. Develop an expression based on a requirement for equilibrium in the direction normal to the vein dip that brings into association vein dip δ, bolting angle (from the horizontal) β, bolt spacing S (assume a square pattern), bolt tension T, and so forth. Note: specific weight of rock is γ and two bolts per hole are used.

7.18 Consider a vein of medium width (50 ft, 15.2 m) that dips 65° and is mined by vertical crater retreat (VCR), as shown in the sketch for the previous problem. Level interval is 150 ft (45.7 m). Layout a hanging wall cable bolt pattern using two 5/8-in. (1.6 cm) diameter, 25 ton (224 kN) capacity bolts per hole with toe spacing a maximum of 10 ft (3 m). Specify hole angles and lengths. Justify in detail your bolting plan including spacing of bolt fans on strike.

7.19 Consider a medium width vein (50 ft, 15.2 m) dipping 65° and mined by overhand mechanized cut and fill (MCF) with a level interval of 200 ft (61 m) as shown in the sketch. (a) Derive a bolting formula algebraically for bolt safety factor assuming n bolts per hole using cable of capacity T_b (lbf, kN) force in ore (rock) having specific weight γ (pcf, kN/m^3) while bolting on a square pattern with spacing S (ft, m). (b) Layout a cable bolt pattern specifying bolt length, spacing, and angle using one 5/8-in., 25 ton (1.6 cm, 224 kN) bolt per hole.

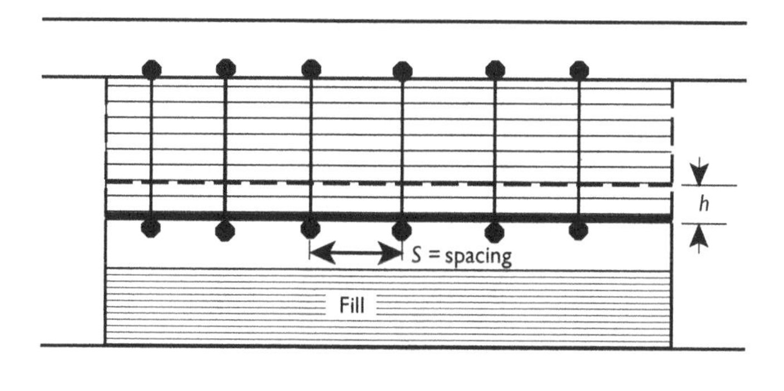

Sketch for Problem 7.16.

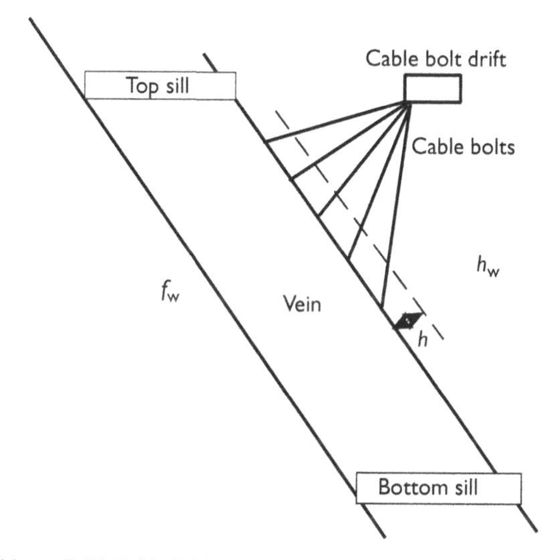

Sketch for Problems 7.17, 7.18, 7.19, vertical section across the strike of the ore body.

7.20 With reference to the sketch, verify the cable bolting formula below. If the formula is unsatisfactory, then derive a substitute formula based on a dead weight load analysis.

$$S^2 = \frac{NUV}{T(\text{SF})\sin^3\alpha}$$

Here S = spacing (ft, m) assuming a square pattern, N = number of cables per hole, U = ultimate tensile strength (tons, kN), V = specific volume of the rock mass (ft^3 /ton, m^3 /kN), T = thickness of suspended rock mass (ft, m), α = bolting angle from the horizontal, SF = bolt safety factor. Note: this formula appeared in the technical literature some years ago.

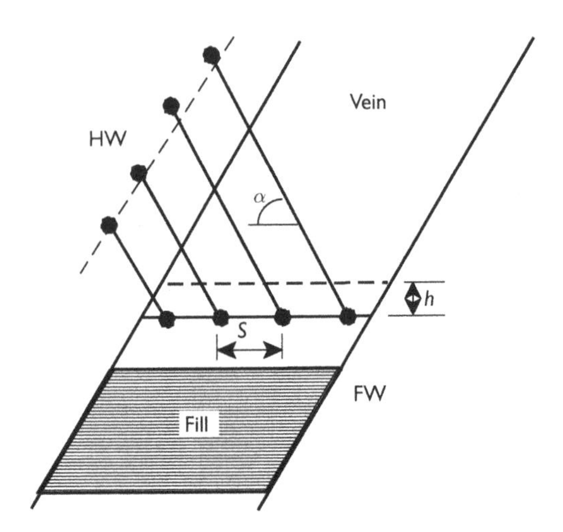

Additional problems

7.21 Consider the sphere shown in Figure 7.2 and suppose a unit compressive load $S_x = 1.0$ is applied parallel to the x-axis. This load is the preexcavation stress state. Use the formulas in Table 7.2 to construct a table of stress concentration factors for x-axis loading, then evaluate for Poisson's ratio of 0.0 and 0.5.

7.22 Consider the sphere discussed in Example 7.15 under a unit compressive load $S_x = 1.0$ is applied parallel to the x-axis. Use the example results to construct a table of peak tensile (negative) and compressive stress concentration factors for values of Poisson's ratio on the interval [0.0, 0.5] analogous to Table 7.3. Indicate in the table the location of the peak stress concentrations found.

7.23 Consider a spherical cavern of radius a (213 ft) at a depth of 3,281 ft under gravity loading only where the vertical stress is estimated at 1.06 psi/ft and the horizontal stresses are one-third the vertical preexcavation stress. Poisson's ratio is 0.30; unconfined compressive and tensile strengths are: $C_o = 14,500$ psi, $T_o = 1,450$ psi. Determine the cavern wall safety factors.

7.24 A storage cavern is dissolved into an oblate spheroid at a depth of 984 ft (cavern center) in a thick, massive salt formation where the stress state is hydrostatic before excavation. Cavern height is 98.4 ft; width is 295.3 ft. Estimate salt formation unconfined compressive and tensile strengths needed for safety factors of 3 and 6 in compression and tension, respectively. Also locate the position of peak tension and compression.

7.25 Consider the data given in Example 7.18, but suppose the long dimension (90 m) of the cavern is vertical, so the cavern has the shape of a prolate spheroid. Estimate the required tensile and unconfined compressive strengths to achieve the specified factors of safety.

7.26 A tunnel of rectangular cross-section 23 ft wide by 13.1 ft high is advanced at a depth of 340 m where the vertical preexcavation stress is 1.06 psi/ft; the north-south stress is 1.4 times the vertical stress and the east-west stress is 0.7 times the vertical stress. Determine the most favorable direction of advance and then estimate the stress concentration factors for tension and compression when the tunnel has advanced 328 ft.

7.27 Consider the data in Example 7.20 and further suppose the tunnel is advanced in an east-west direction rather than along the given north-south route. Estimate the stress concentration factors in this orientation.

7.28 A tunnel of rectangular cross-section 23 ft wide by 13.1 f high is advanced along a north-south route at a depth of 1,116 ft where the vertical preexcavation stress is 1.0 psi/ft; the north-south stress is 1.4 times the vertical stress and the east-west stress is 0.7 times the vertical stress. A second excavation pass widens the tunnel to 46 ft. Determine the unconfined compressive and tensile strengths needed for safety factors of 2 and 4, respectively.

7.29 A tunnel of rectangular cross-section 23 ft wide by 13.1 ft high is advanced along a north-south route at a depth of 1,116 ft where the vertical preexcavation stress is 1.06 psi/ft; the north-south stress is 1.4 times the vertical stress and the east-west stress is 0.7 times the vertical stress. A second excavation pass widens the tunnel to 46 ft. Unconfined compressive and tensile strengths are 21,750 psi and 2,175 psi, respectively. Determine the number of additional passes of the same 23 ft width that can be made before failure occurs.

7.30 A spherical storage cavern with center at a depth of 244 m and a diameter of 27.4 m in hydrostatic stress field begins to spall. Spalling spreads uniformly into the cavern walls and ceases when the diameter reaches 36.6 m and the caved material fills the cavern. The cavern is filled with a liquid that has a specific gravity of 1. Estimate the void volume of fluid in the cavern after spalling.

7.31 A spherical storage cavern with center at a depth of 244 m is planned with a diameter of 27.4 m in hydrostatic stress field ($\sigma = 5.52$ MPa). The walls of the cavern may spall, so a cable bolt pattern of support is installed *before* excavation from a spiral ramp about the future cavern. Suppose the cable bolts stretch in a compatible mode with the inward displacement of the cavern. Estimate relative displacement of 9.14 m long, 1.59 cm diameter, 7-strand, bolts with a breaking load of 224 kN. Bolts extend from cavern walls into the solid and are spaced on a square pattern, 3 × 3 m. Note: Young's modulus is 34.5 GPa and Poisson's ratio is 0.35. Young's modulus of the cable bolt is 208 GPa.

7.32 A spherical storage cavern with center at a depth of 244 m is planned with a diameter of 27.4 m in hydrostatic stress field ($\sigma = 5.52$ MPa). The walls of the cavern may spall, so a cable bolt pattern of support is installed *before* excavation from a spiral ramp about the future cavern. Suppose the cable bolts stretch in a compatible mode with the inward displacement of the cavern. Estimate relative displacement of 9.1 m long, 1.59 cm diameter, 7-strand bolts with a breaking load of 224 kN when supporting a dead weight load equal to the mass of rock supported by each cable. Bolts extend from cavern walls into the solid and are spaced on a square pattern, 3 m x 3 m. Note: Young's modulus is 34.5 GPa and Poisson's ratio is 0.35. Young's modulus of the cable bolt is 208 GPa and elongation at the breaking load is 3.5%.

7.33 A spherical storage cavern with center at a depth of 244 m is planned with a diameter of 27.4 m in hydrostatic stress field ($\sigma = 5.52$ MPa). The walls of the cavern may spall, so a cable bolt pattern of support is installed *before* excavation from a spiral ramp about the future cavern. Suppose the cable bolts stretch in a compatible mode with the inward displacement of the cavern. Determine the number of cable bolts per hole needed to support a dead weight load of rock. Bolts are 9.1 m long, 1.59 cm diameter, 7-strand with a breaking load of 224 kN when supporting a dead weight load equal to the mass of rock supported by each cable. Bolts extend from cavern walls into the solid and are spaced on a square pattern, 3 m × 3 m. Note: Young's modulus is 35.4 GPa and Poisson's ratio is 0.35. Young's modulus of the cable bolt is 208 GPa and elongation at the breaking load is 3.5%.

7.34 Given the data in Example 7.27 and the rock block size 3×3×9.1 m. Determine the block stiffness under axial load and the additional stiffness provided by a single cable bolt.

7.35 A tabular excavation is oriented vertically and extends in a north-south direction. Excavation width is 6.1 m, height is 30 m, and length is 152 m north-south. A hydraulic sand backfill is place in the excavation. The fill Young's modulus is 138 MPa. Determine the fill stiffness with respect to horizontal force per m of length and the force required to compress the fill by 1 cm. Also determine the equivalent horizontal stress for this force.

7.36 Given the data and results from Example 7.29 and the presence of two joint sets, with dip direction east-west. The first set dips 30o east; the second set dips 60o west. Both joint sets have friction angles of 25° and zero cohesion. Joint spacing is 2.44 m in both sets and are fully persistent. These data are cautious and represent a worst case situation where blocks form in the cavern walls. Consider the east wall and the largest possible block that may form per m of length. Determine whether the block will slide. If so, compute the horizontal force that would develop without fill and suppose this force actually compresses the fill across the opening. Estimate the unbalanced horizontal force and associated fill displacement.

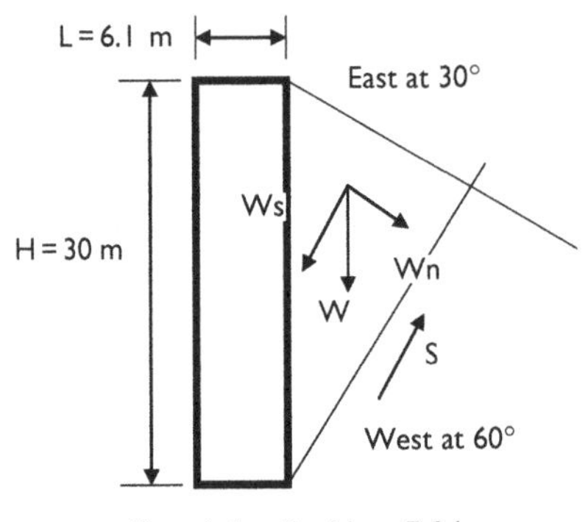

Sketch for Problem 7.36

Chapter 8

Subsidence

Subsidence usually refers to downward motion of a ground surface, "settlement," but may also refer to below ground motion or internal subsidence. The most common causes of subsidence are: (1) withdrawal of fluids from pores and fractures and (2) disturbance of the ground by excavation processes. Water well pumping and consequent drawdown of the water table is a very common cause of surface subsidence. Withdrawal of oil, and to a lesser extent, withdrawal of gas from wells is also a significant source of subsidence in some areas. Superincumbent loads on saturated, porous ground may also cause settlement of the ground surface by expulsion of pore fluids, a consolidation process. Surface and underground excavation cause subsidence through induced changes in stress, strain, and displacement fields in adjacent soil and rock masses. Sometimes excavation is a natural process, for example, formation of solution cavities by circulating ground water in limestone formations. If the induced displacements from underground excavation extend to the surface, then surface subsidence as well as internal subsidence result. In cases where damage to surface structures or underground utilities occurs, there is considerable importance attached to the determination of the cause of damage in the presence of subsidence and the actual subsidence mechanism.

8.1 Chimneys

Caving is an often observed phenomenon associated with underground mining and begins with collapse of the back, roof, or hanging wall. Caving may occur unexpectedly or be deliberately induced in conjunction with the mining method (block caving, sub-level caving, longwall mining, room and pillar mining with pillar extraction). Caving begins because locally stress tends to exceed strength, loading progresses beyond the elastic limit, fracture and failure occur, and broken rock is liberated from the parent rock mass and falls to the floor. Failure of the rock mass is greatly assisted by the presence of joints, of course, especially if the intact rock, between joints is strong, but both mechanisms of rock mass failure (slip and separation of joints, fracture of intact rock, and extension of such fractures) contribute to caving. Cessation of caving may occur because an equilibrium shape forms as the cave evolves (natural "arching"), a strong rock formation is encountered during the upward progress of the cave or the void becomes filled with broken rock (swell).

Flow of broken rock in ore passes is closely related to the motion of broken rock, "muck," in caving ground. Design of storage bins, bunkers, and silos for handling crushed rock and other bulk materials also has features in common with the flow of muck in caving ground and raises important questions about drawpoint size and spacing and support of nearby access ways. In this regard, a redistribution of stress occurs in the vicinity of a caved zone, at the walls of a chimney cave, near ore pass walls and so forth, that has implications for ground control in nearby openings, drifts, entries, crosscuts, and raises. This redistribution often leads to heavy support requirements in the form of combinations of bolts, shotcrete, steel sets, and continuous concrete liners. Understanding stress associated with muck flow in silos and manufactured storage and handling facilities is also essential to structural design.

Chimney cave geometry

Chimney caves are characterized by vertical walls that are generated directly over an initial opening. The cave forms when the back falls into the initial void. Because caved rock in bulk occupies greater volume than in place ("bank" measure), bulking may eventually cause caved rock to fill the void. Bulking porosity B characterizes this property of the rock mass. By definition

$$B = \frac{V_v}{V} = \frac{V_v}{V_s + V_v} \tag{8.1}$$

where V_v is void volume, V_s is solid volume, and V is total volume.

With reference to Figure 8.1 that illustrates a chimney cave, the height H of the caved portion of a chimney that forms over an opening initially h ft high with cross-sectional A is

$$H = h\left(\frac{1}{B} - 1\right) \tag{8.2}$$

provided H is less than the depth measured from ground surface to the top of the initial void.

If caving reaches the surface, the ground surface subsides a distance S as shown in Figure 8.2. The chimney height is now less than H which becomes depth to the top of the initial void. This depth is given by

$$H = (h - S)\left(\frac{1}{B} - 1\right) \tag{8.3}$$

which may be solved for the surface subsidence. Thus,

$$S = h - H\left(\frac{B}{1 - B}\right) \tag{8.4}$$

that reduces to the formula for subsurface chimney cave height when $S = 0$.

These simple formulas based on the concept of bulking mask an interesting phenomenon that is revealed when observations of chimney caves are used to estimate an associated bulking porosity. One might suppose that a reasonable estimate of bulking porosity B could be obtained by using swell factors for rock rubble produced by blasting in open pit mines, for example. A ratio of bank specific weight (before blasting) to bulk specific weight (after blasting) of 1.3 to 1.4 would be reasonable, say 1.35. The associated porosity $B = 0.35$. Chimney height above an initial opening height of 15 ft would then be 27.9 ft, not a very high cave. In fact, chimney caves are observed to be much higher, so the bulking porosity must be much less than that associated with rock rubble. The implication is that caving may appear at first glance to produce a random pile of broken rock but then not nearly as random as

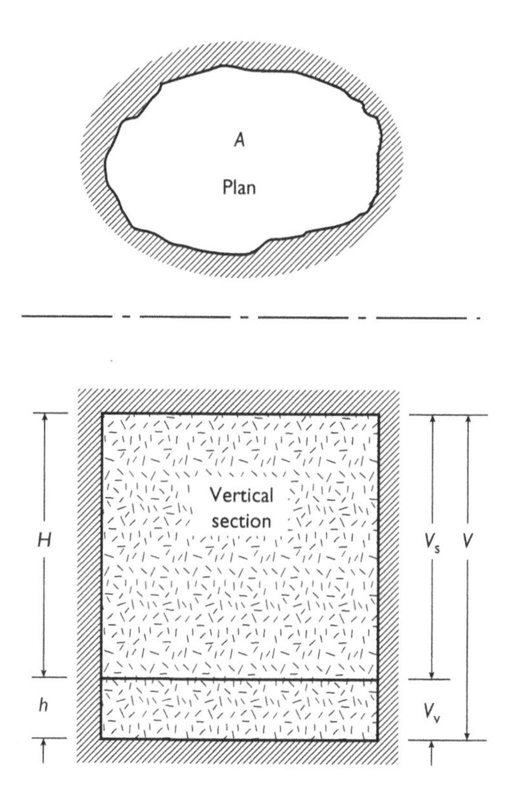

Figure 8.1 Sketch of a chimney cave below ground surface.

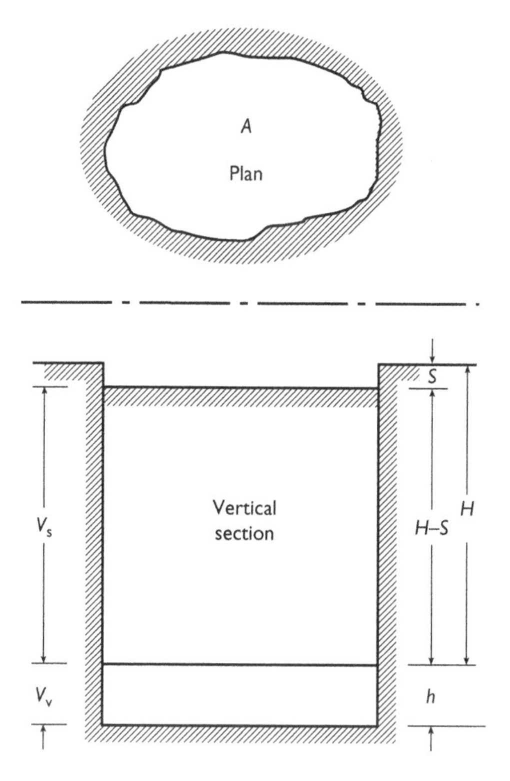

Figure 8.2 Sketch of a chimney cave resulting in surface subsidence.

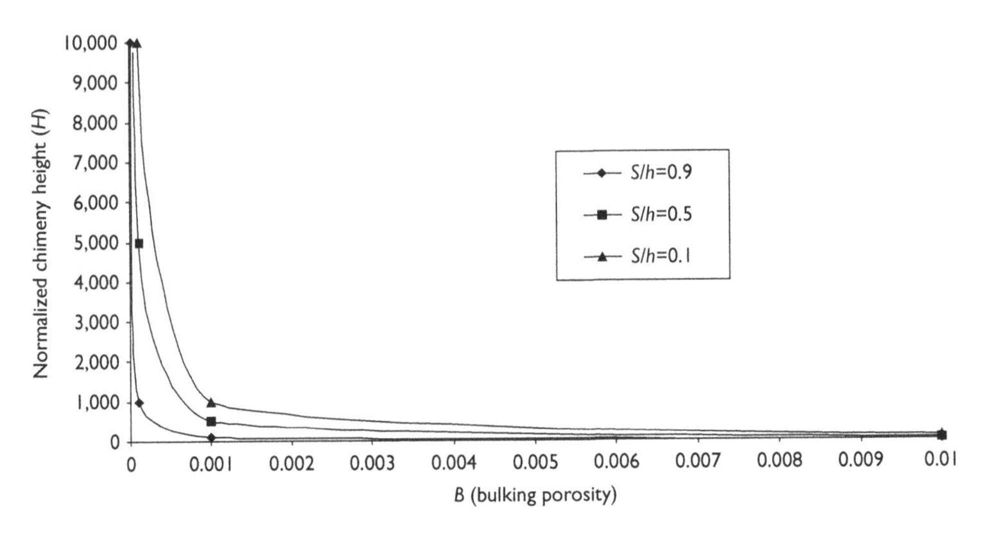

Figure 8.3 Chimney cave height as a function of bulking porosity and surface subsidence.

supposed. Caving over large vertical heights must occur with little bulking; the rock fragments somehow must maintain fits with adjacent fragments and blocks, so the chimney appears to subside almost as if it were a solid plug. For example, suppose a stope 100 ft high at a depth of 2,400 ft caves and the cave just reaches the surface. The associated bulking porosity is just 0.04, an order of magnitude less than a bulking porosity that might be estimated from well-blasted rock.

Figure 8.3 shows a plot of normalized chimney height as a function of bulking porosity with surface subsidence normalized by initial void height as a parameter. Clearly, substantial chimney cave heights must be associated with very small bulking porosities.

A sequential cave may occur with a fall of rock into an opening that is subsequently removed only to be followed by another fall of rock that continues to form a chimney. In this case the void volume of the bulk material in the chimney is the initial void of the opening plus the void formed by the first rock fall. The volume of solid in the chimney is the volume of rock that caved after the first fall of rock.

Example 8.1 A chimney cave forms over a square excavation of side length L and height h. The caved zone above the chimney is a cylinder of height H and forms an inscribed circular cross section. If the bulking porosity is B, how high will the caved zone extend above the excavation? If $B = 0.01$ and $h = 5$ m, determine H.

Solution: With reference to the sketch and the definition of bulking porosity,

$$
\begin{aligned}
B &= \frac{V_v}{V} \\
&= \frac{hA}{HA' + hA} \\
&= \frac{hL^2}{H(\pi/4)D^2 + hL^2}
\end{aligned}
$$

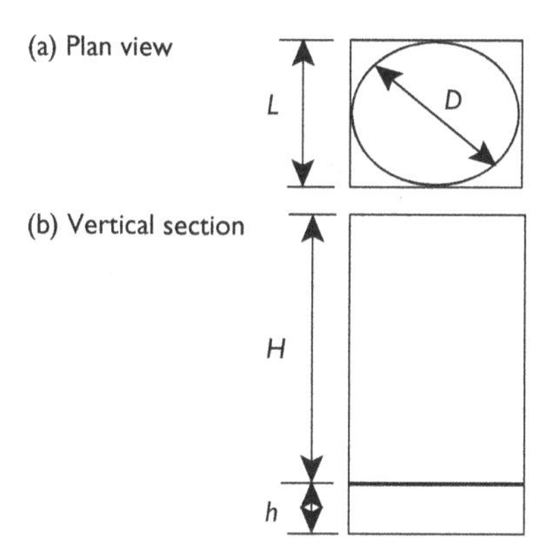

$$B = \frac{h}{(\pi/4)H + h}$$

that may be solved for H. Thus, caving height is

$$H = \left(\frac{4h}{\pi}\right)\left(\frac{1}{B} - 1\right)$$

when $B = 0.01$, $h = 5$ m, $H = 630.3$ m

Example 8.2 A rectangular excavation of height h, width w, and length l induces a vertical cave of elliptical cross section with semi-axes $w/2$ and $l/2$ (inscribed ellipse). Develop a cave height formula in terms of bulking porosity and caving geometry. If excavation is at a depth of 1,000 m, bulking porosity is 0.01, excavation height is 5 m, width is 12 m, and length is twice width, is there a danger of surface subsidence?

***Solution*:** From the definition of bulking porosity,

$$B = \frac{V_v}{V_v + V_s}$$

$$= \frac{hwl}{hwl + \pi ab}$$

$$= \frac{hwl}{hwl + \pi(w/2)(l/2)H}$$

$$B = \left[\frac{1}{1 + (\pi/4)H/h}\right]$$

$$\therefore$$

$$H = (4h/\pi)\left(\frac{1}{B} - 1\right)$$

$$H = (4h/\pi)\left(\frac{1}{0.01} - 1\right) = 630.3 \text{ m}$$

where the geometry is shown in the sketch. Thus, there is no danger of surface subsidence.

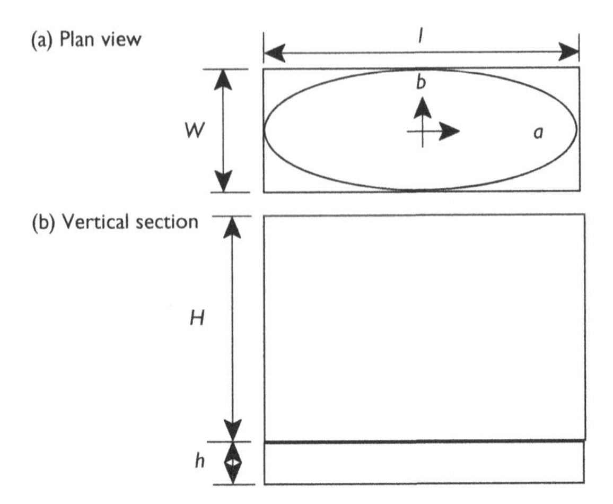

Note: This result is a somewhat more generalized result than that in Example 8.1, which is seen as a special case of the elliptical caving cylinder postulated in this example.

Example 8.3 Suppose an initial excavation is a cylinder 15 ft high with a 64 ft diameter. A hemispherical fall of rock occurs. Muck is removed and chimney formation begins. The hemispherical shape of back is maintained during caving that proceeds for 482 ft (back to top of chimney). No surface subsidence occurs. What bulking porosity is indicated?

Solution: A sketch and application of the bulking porosity definition leads to a solution.

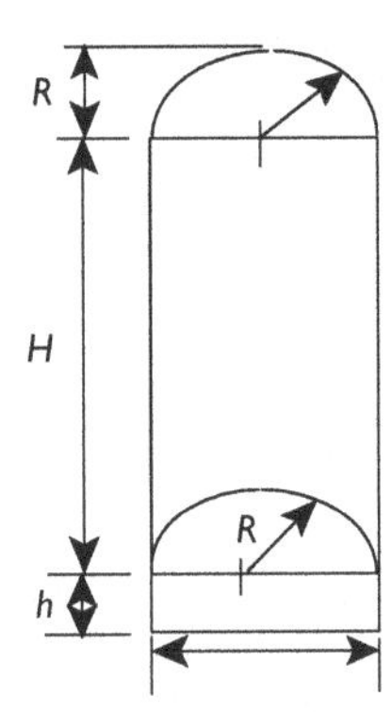

Thus, using expressions for cylinder and sphere volumes, one has for bulking porosity,

$$B = \frac{V_v}{V} = \frac{V_v}{V_s + V_v}$$

$$= \frac{\pi R^2 h + (2/3)\pi R^3}{\pi R^2 H + \pi R^2 h + (2/3)\pi R^3}$$

$$B = \frac{h + (2/3)R}{H + h + (2/3)R}$$

$\therefore$

$$B = \frac{15 + (2/3)(32)}{(482 - 32) + 15 + (2/3)(32)} = 0.0747$$

Thus, bulking porosity indicated by the given data is 0.075. This same problem is illustrated in Figure 8.4.

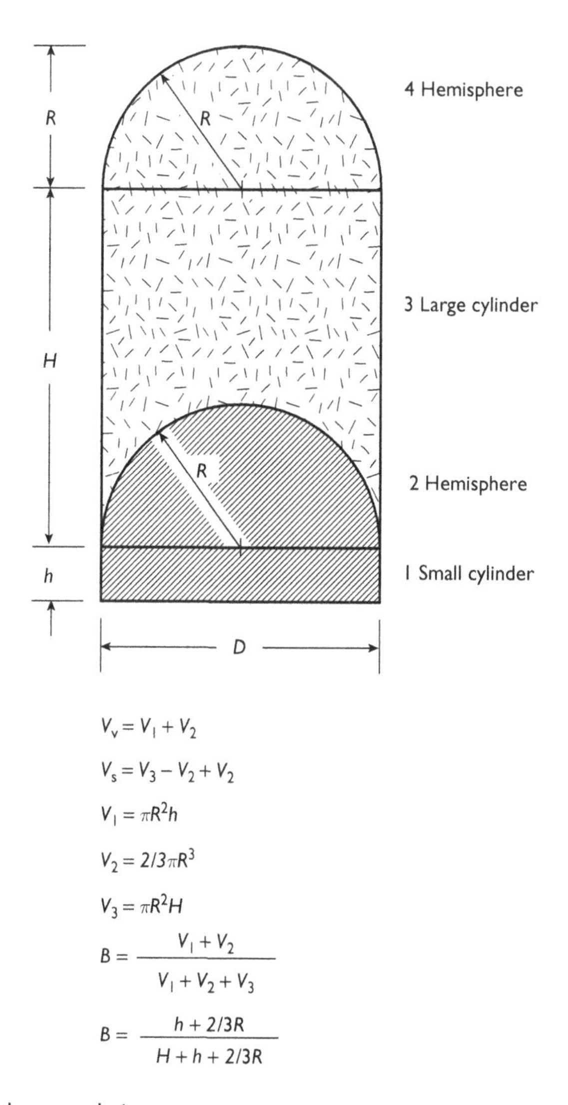

Figure 8.4 Sequential cave analysis.

Example 8.4 Consider a prismatic excavation of initial height h at depth H that caves to the surface and causes subsidence S. Develop an expression for the normalized depth ratio H/h in terms of bulking porosity B, h, and S.

Solution: With reference to sketch, $V = A(H - S + h)$ and $V_s = HA$, so $V_v = A(h - S)$. From the definition of bulking porosity,

$$B = \frac{h - S}{H + h - S}$$

which may be solved for H/h. Thus,

$$\frac{H}{h} = \left(1 - \frac{S}{h}\right)\left(\frac{1}{B} - 1\right)$$

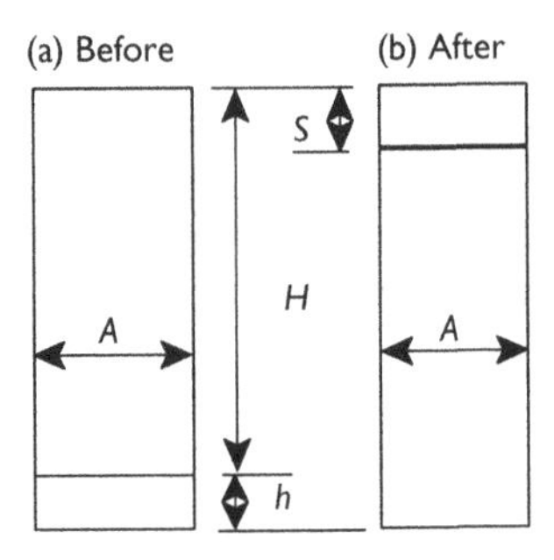

Example 8.5 Surface subsidence is observed over an area of 2.25(10^4) sqft above an excavation 1,500 ft deep. Original excavation height is 15 ft and surface subsidence occurs eventually reaching 30 ft. There are 12 ft^3/ton of material. Estimate the bulking porosity and the tonnage of material removed after the original opening is excavated.

Solution: From Example 8.4, when caving has just reached the surface, $B = h/(H + h) = 15/(1{,}500 + 15) = 0.099$. Also from Example 8.4

$$B = \frac{h' - S}{H + h' - S}$$

where h' is a fictitious opening height associated with the given surface subsidence, that is $h'A$ is void volume created by the initial excavation and the additional removal of solid. In this case

$$h' = S + H\left(\frac{B}{1 - B}\right)$$
$$= 30.0 + 1,500\left(\frac{0.0099}{1 - 0.0099}\right)$$
$$h' = 45.00 \text{ ft}$$

The tonnage of solid material removed is

$$\begin{aligned} \text{tons} &= V(\text{solids})/12 \\ &= (h' - h)(A)/12 \\ &= (45.00 - 15)(2.25)\left(10^4\right)/12 \\ \text{tons} &= 5.625\left(10^4\right) \end{aligned}$$

In block caving, continual withdrawal of muck from a caving panel creates high, steep slopes at the surface. These slopes eventually fail to form a "glory hole" at the surface that has an upside down bell shape. Caving over softrock above a coal mine longwall panel seldom reaches the surface and instead causes flexing of strata that form a trough-like surface depression rather than a glory hole typical of hardrock chimney caves. Caves are also referred to as "ratholes" and "pipes" and the caving process as "ratholing," "piping," or "chimneying." Caving is also a widespread natural phenomenon associated with "sinkhole" formation from dissolution of limestone and subsequent collapse of the roof into the void below formed by circulating groundwater.

Caving rock flow

Cave geometry associated with ore extraction may expand laterally and "bell out," as shown in Figure 8.5, with a resulting ore flow of muck on muck. The diameter of the chimney is difficult to predict with confidence. However, observations of sand models and continuum models of flow indicate the presence of a fast flow zone at the core near the drawpoint. This zone is bounded by thin, intense shear zones that are associated with discontinuities in flow velocity. Avelocity discontinuity is a rapid change in magnitude and direction of the flow. The outer boundaries of the chimney are also intense shear zones and velocity discontinuities. In principle, these discontinuities may be calculated, but the computation is numerically

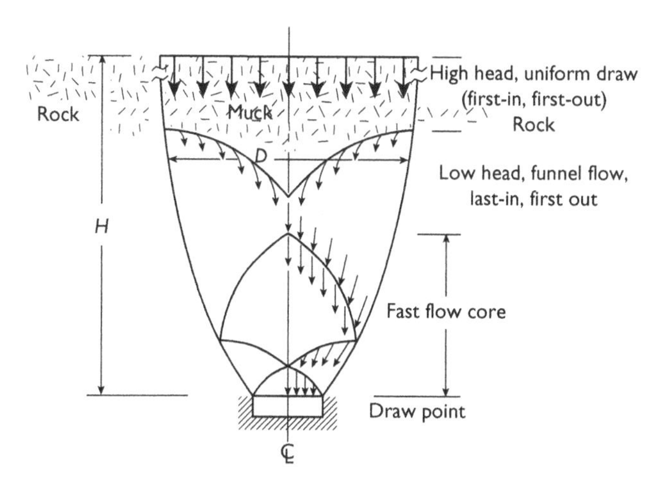

Figure 8.5 Single draw point and chimney of muck flowing on muck.

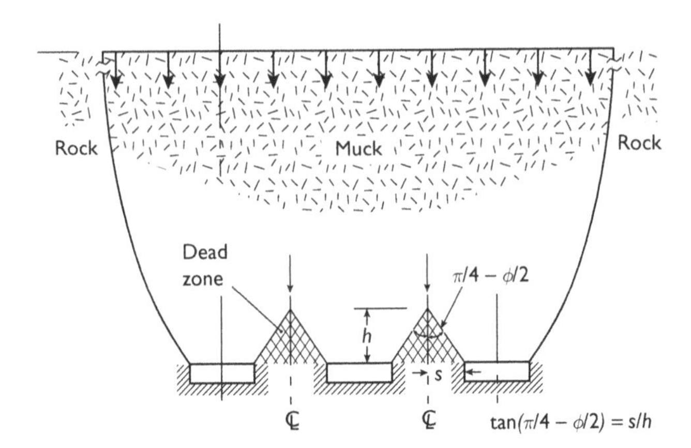

Figure 8.6 Multiple draw points and dead zones.

intensive and not at all routine. As the top of the muck is pulled close to the drawpoint, the flow changes from a uniform first-in, first-out draw to a *funnel flow* pattern where the top material moves into the fast flowing core while the material near the side and below moves slowly and passes through the drawpoint only during the final stage that empties the chimney.

An important engineering question relates to the spacing of drawpoints that allows for a uniform drawdown of ore, as shown in Figure 8.6, and a minimum of ore trapped in dead zone between drawpoints. Unfortunately, general rules for drawpoint spacing remain elusive. One relationship between spacing s and dead zone height h is $s/h = \tan(\pi/4 - \phi/2)$ where ϕ is the angle of internal friction of the muck. A prudent estimate of acceptable dead zone height, then leads to s. Width of the pillar between drawpoints is then $2s$. Drawpoint width should be three to five times the maximum block size in the ore drawn to prevent hangups in a draw chute.

Example 8.6 Caving is assisted by several joint sets that have average spacings ranging from 3 to 12 ft. Estimate the maximum block size that would be reasonably expected and the minimum size of drawpoint for efficient operations.

***Solution*:** The maximum block size generated by caving action can only be roughly estimated from the limited information available. If blocks are well defined because the joints are continuous or highly persistent then linear dimensions of blocks may range from 3 to 12 ft. Some breakage would be expected, especially of long blocks, so perhaps linear dimensions of large blocks reaching drawpoints is about 3 ft. A few larger blocks and many smaller blocks would be expected.

A rule of thumb for drawpoints states that drawpoint width should be three to five times maximum block size. In this case, the drawpoint width should be at least 9 ft to avoid frequent blockages caused by interlocking of large blocks at drawpoints.

Example 8.7 Caving rock is estimated to have zero cohesion and a friction angle of 35°. If drawpoints are spaced on 20 m centers and are 3.5 m wide, estimate the height of the dead zone between drawpoints.

***Solution*:** If the dead zone is just at yield, but remains immobile because of lateral confinement, then the height of the dead zone is given by

$$\frac{s}{h} = \tan(\mu)$$

$$\therefore h = (S/2 - W/2)\cot(\pi/4 - \phi/2)$$

$$= (20 - 3.5)(1/2)\cot(\pi/4 - 35/2)$$

$$h = 15.8\,\text{m}$$

The geometry is shown in the sketch. This is a relatively large height and suggests that closer spacing of drawpoints be considered to reduce the loss of ore in the dead zones.

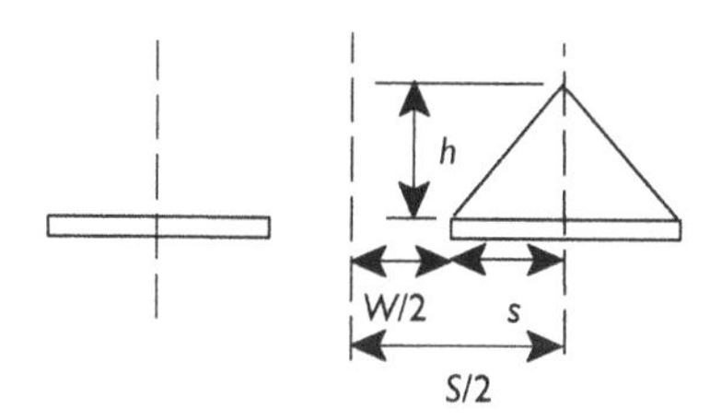

Chimney cave forces

An equilibrium analysis reveals important features of stress redistribution associated with chimney formation caving. Figure 8.7 shows a typical slab of caved material within a cave zone. The equilibrium requirement in the vertical direction in terms of forces is

$$\Delta F_v - W + F_s = 0 \tag{8.5}$$

where ΔF_v is the difference in vertical forces acting on the top and bottom surfaces of the "slab" of material in the chimney, W is slab weight, and F_s is the total vertical shear force acting over the lateral surface of the slab. In terms of average stresses and related areas, $\Delta F_v = A\Delta\sigma_v$, $W = \gamma A\Delta z$, and $F_s = \tau C\Delta z$ where σ_v and τ are vertical normal stress and shear stress, respectively, Δz is slab thickness. Hence,

$$A\Delta\sigma_v - \gamma A\Delta z + \tau C\Delta z = 0 \tag{8.6}$$

where A and C are cross-sectional area and circumference, respectively, shown in Figure 8.7.

Because the broken rock, the "muck", in the chimney is sliding along rock walls, a frictional relationship between muck and rock is present. Thus,

$$\tau = \mu\sigma_h = \sigma_h\tan(\phi') \tag{8.7}$$

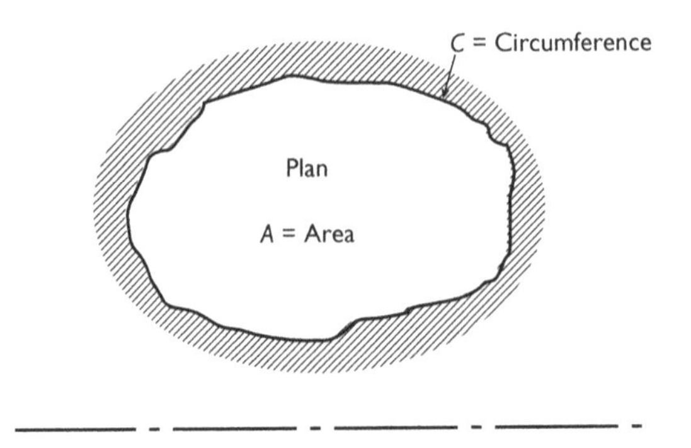

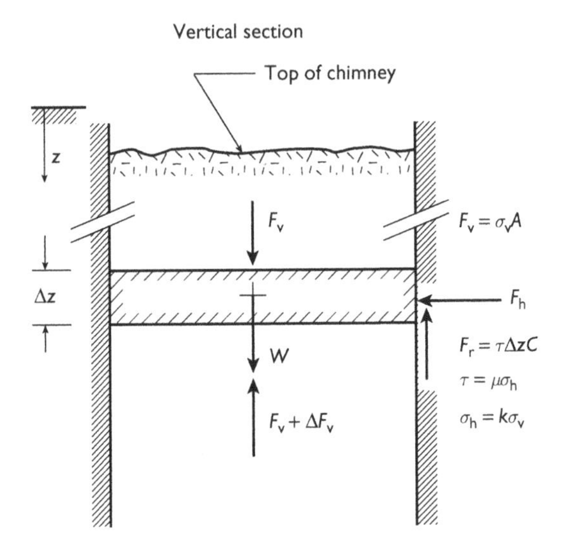

Figure 8.7 Chimney cave forces.

where σ_h is the average horizontal stress acting over the lateral surface area of the slab, ϕ' is the angle of sliding friction of muck on rock, and μ is the coefficient of sliding friction. This frictional condition tacitly implies no adhesion between muck and rock. If the muck is "sticky," this assumption may need to be modified.After some rearrangement, the equilibrium requirement is,

$$\left(\frac{\Delta\sigma_v}{\Delta z}\right) + \left(\frac{\mu C}{A}\right)\sigma_h = \gamma \tag{8.8}$$

which suggests a differential equation for the normal stresses.

However, there are two unknown stresses and only one equation. This deficit of information is made up by assuming a relationship between the vertical and horizontal stresses in the form of a simple ratio: $k = \sigma_h/\sigma_v$. Justification for such a relationship and an estimation of k is obtained by assuming the muck motion is inelastic and follows a Mohr–Coulomb yield condition. Cohesion is considered negligible because of muck flow. The ratio of normal stresses cannot be less than the ratio of principal stresses in any case, and under the assumption of cohesionless Mohr–Coulomb flow, $k > [1 - \sin(\phi)]/[1 + \sin(\phi)]$ where ϕ is the angle of internal friction of the muck. If ϕ is 30°, then $k = 0.3$ is a minimum estimate. The resulting equilibrium requirement is now

$$\frac{d\sigma_v}{dz} + C_1\sigma_v = \gamma \tag{8.9}$$

where the constant $C_1 = kC\mu/A$. In this regard, the ratio A/C is sometimes called the "hydraulic" radius, but such a label is better used in fluid mechanics. At the top of the muck where $z = 0$, $\sigma_v = 0$, so after integration

$$\sigma_v = \frac{\gamma}{C_1}[1 - \exp(-C_1 z)] \tag{8.10}$$

Figure 8.8 shows a plot of stress versus depth in a cylindrical column of broken rock that is stress-free at the column top, $z = 0$. In this example, $C_1 = 0.01$. The asymptotic limit is 67 psi in this example. At a depth of just $2D$ the vertical stress is already 54 psi or slightly greater than 80% of this limit.

When a surcharge σ_o acts at the top of the muck column, the formula for vertical stress is

$$\sigma_v = \left(\frac{\gamma}{C_1}\right)[1 - \exp(-C_1 z)] + \sigma_o \exp(-C_1 z) \tag{8.11}$$

The asymptotic limit to the vertical stress is the same as before. For example, a surcharge or additional load on a muck column may be caused by impact of additional muck falling into an

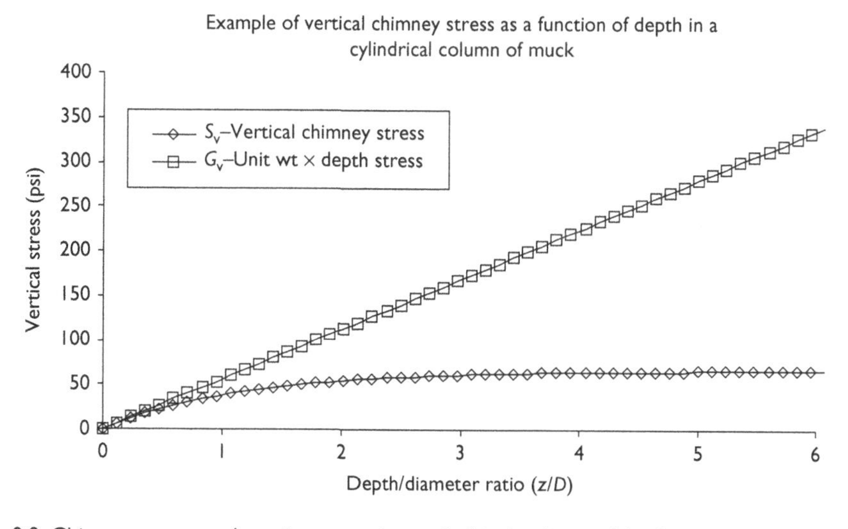

Figure 8.8 Chimney stress and gravity stress in a cylindrical column of broken rock versus depth in terms of diameter.

underground mine ore pass. The muck in the ore pass is essentially a chimney of broken rock. Impact loading on chutes from falling rock may damage the chutes or load-out facilities if not protected by a muck cushion near the bottom of the ore pass. The height of the muck cushion may be found by setting the additional vertical stress to some fraction, say, 0.20 or 20% of its peak value. A depth equal to about two diameters is necessary to meet this requirement. A three-diameter cushion reduces the additional bottom stress to 8% of the surcharge. In a 10 ft diameter ore pass, this cushion would be 30 ft high and much less than level interval (usually 100 ft or more), so the ore pass could be drawn down considerably between rock and ore dumps.

Example 8.8 Compare the forces generated by flow of muck in circular and square ore passes on the basis of equality of diameter and diagonal, other factors being equal.

***Solution*:** The sketch illustrates the geometry and meaning of terms in the formulas for stresses that follow.

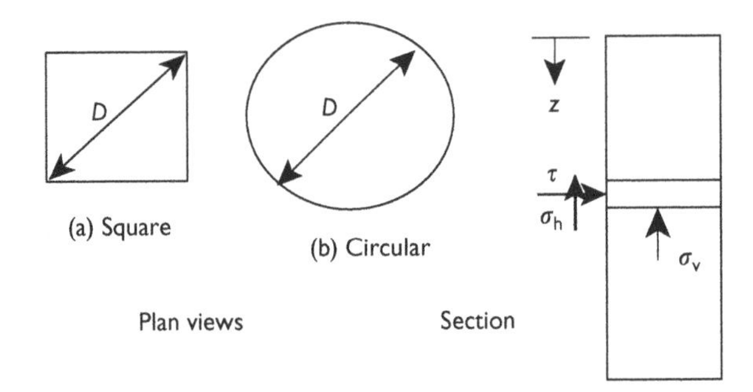

From the text,

$$\sigma_v = \left(\frac{\gamma A}{\mu k C}\right)\left[1 - \exp\left(-\frac{\mu k C}{A} z\right)\right]$$
$$\sigma_h = k\sigma_v$$
$$\tau = \mu\sigma_h$$

Properties of the muck (k, μ) are the same, so the difference between the two sections (square, circular) can only be in circumference and area. In fact, any difference can only be in the ratio of area to circumference. In the circular case, $A/C = D/4$; in the square case, $A/C = D/4$, and therefore the stress distributions are the same.

However, there are practical advantages to the circular shape including a larger cross-sectional area and therefore greater storage capacity, relative ease of excavation and perhaps less likelihood of stoppages from arching, that is, formation of stable arches caused by interlocking of larger blocks in the muck.

Example 8.9 Suppose caving occurs above a relatively long excavation such that the situation appears two-dimensional, as a "slot" illustrated in the sketch. Derive expressions for average stresses in this "plane strain" condition.

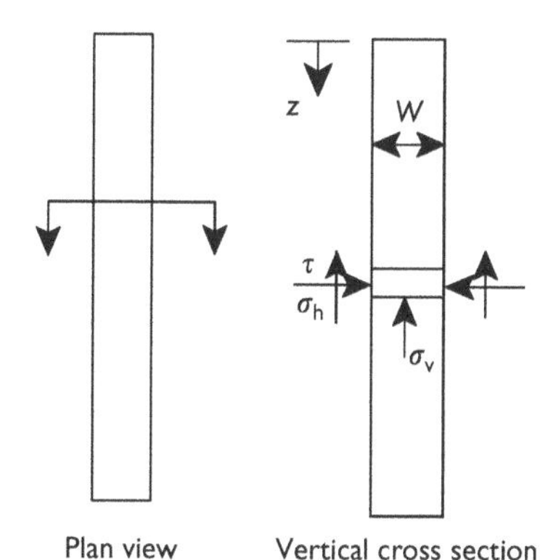

***Solution*:** Equilibrium of vertical forces per unit length of excavation (into the page) acting on the element of caving rock shown in cross section requires

$$(1)(W)(\Delta\sigma_v) + 2\tau\Delta z - \gamma(1)(W)\Delta z = 0$$

where Δz is the thickness of the element considered. This expression leads to the ordinary differential equation

$$\frac{d\sigma_v}{dz} + \left(\frac{2\mu k}{W}\right)\sigma_v = \gamma$$

where the substitutions $\tau = \mu\sigma_h$ and $\sigma_h = k\sigma_v$ have been made. This result is similar to the corresponding three-dimensional equation but with the ratio $2/W$ replacing C/A. In three dimensions the shear stress acts about circumference C; in this two-dimensional case, the shear stress acts per foot of length on the two sides of the cave. Thus,

$$\sigma_v = \left(\frac{\gamma W}{2\mu k}\right)\left[1 - \exp\left(-\frac{2\mu k}{W}z\right)\right]$$
$$\sigma_h = k\sigma_v$$
$$\tau = \mu\sigma_h$$

In comparison with a circular or square cave, $W/2$ would be replaced by $D/4$. At a given depth, the two-dimensional stresses are more than twice the three-dimensional stresses for $W = D$.

Example 8.10 With reference to Example 8.9, at a given depth of caved material, one would expect that with increasing width, the vertical stress would approach the usual expression for a gravity load, unit weight times depth. Show that this expectation is indeed the case.

***Solution*:** Examination of the results in Example 8.9 shows that

$$\sigma_v \rightarrow (\infty)[0]$$

as W becomes indefinitely large. After rearrangement of the original expression, one has

$$\sigma_{\mathrm{v}} = \left(\frac{1}{2\mu k/\gamma W}\right)\left[1 - \exp\left(-\frac{2\mu k}{W}z\right)\right]$$

which then shows

$$\sigma_{\mathrm{v}} \rightarrow (1/0)[0]$$

Application of l'Hopital's rule for indeterminant forms gives the desired result. Thus,

$$\sigma_{\mathrm{v}} = \left[\frac{1}{(2\mu k/\gamma)(-1/W^2)}\right]\left(\frac{2\mu kz}{-W^2}\right) = \gamma z$$

This result reflects the physical expectation that frictional shear stress retardation of the caving material becomes less effective with increasing width or diameter of the caving zone or ore pass. In three dimensions, the same result follows from allowing the ratio of area to circumference (A/C) to increase indefinitely.

Example 8.11 Caving is induced in a rock mass 50 m by 50 m by 180 m high. The angle of internal friction of the caving material ϕ is estimated to be 28°. Friction angle of muck on rock is estimated at 32°. Because of the motion of the caving mass, cohesion is considered zero. Estimate the average vertical stress at the bottom of the cave.

***Solution*:** The stresses associated with caving are

$$\sigma_{\mathrm{v}} = \left(\frac{\gamma A}{\mu kC}\right)\left[1 - \exp\left(-\frac{\mu kC}{A}z\right)\right]$$
$$\sigma_{\mathrm{h}} = k\sigma_{\mathrm{v}}$$
$$\tau = \mu\sigma_{\mathrm{h}}$$

The first equation gives the vertical stress. A reasonable estimate for specific weight is 15.8 $\mathrm{kN/m^3}$, $\mu = \tan(32) = 0.625$, k is approximately [1−sin(28)]/[1+sin(28)] = 0.361, so

$$C_1 = \frac{\mu kC}{A} = \frac{(0.625)(0.361)(4)(150)}{(150)(150)} = 6.017\left(10^{-3}\right)(1/\mathrm{m})$$

and thus,

$$\sigma_{\mathrm{v}} = \frac{15.8\left(10^3\right)}{6.017\left(10^{-3}\right)}\left\{1 - \exp\left[-6.017\left(10^{-3}\right)150\right]\right\} = 1.561\mathrm{MPa}(226\mathrm{psi})$$

that may be compared with a vertical gravity stress of (15.8)(150) = 2.37 MPa (344 psi).

Example 8.12 With reference to Example 8.11, determine the vertical force over the cave bottom and the shear force over the cave sides.

***Solution*:** The vertical force at the cave bottom is simply average vertical stress times bottom area, while the side shear force is the difference between weight of the caved material and the bottom force. Thus,

$$\begin{aligned}
F_v &= \sigma_v A \\
&= (1.561)(50)(50) \\
F_v &= 3.90GN(0.871 \times 10^9 \text{1bf}) \\
S_v &= W - F_v \\
&= (15.8)(10^3)(50)(50)(150) - 3.90(10^9) \\
S_v &= 2.03GN(0.452 \times 10^9 \text{1bf})
\end{aligned}$$

The vertical shear force may be obtained in another way as a check. From the formulas for stress

$$\begin{aligned}
\tau &= \mu k \sigma_v \\
S_v &= \int_A \tau \mathrm{d}A \\
&= \int_A \mu k \sigma_v \mathrm{d}A \\
&= \int_H \mu k \left(\frac{\gamma A}{\mu k C}\right)\left[1 - \exp\left(-\frac{\mu k C}{A} z\right)\right] C\mathrm{d}z \\
&= \gamma A z + \gamma A \left[\exp\left(-\frac{\mu k C}{A} z\right)\right]\left(\frac{A}{\mu k C}\right)\Big|_0^H \\
&= W - A\left\{\gamma\left(\frac{A}{\mu k C}\right)\left[1 - \exp\left(-\frac{\mu k C}{A} H\right)\right]\right\} \\
&= W - A\sigma_v \\
S_v &= W - F_v
\end{aligned}$$

which was the starting point for the original calculation of side wall shear force. In the integration the area element $\mathrm{d}A$ is associated with the shear stress about the circumference of the chimney and is therefore $C\mathrm{d}z$.

The difference between the gravity stress of unit weight times depth and the average vertical stress in a column of broken rock is a consequence of the vertical shear stress at the column walls. This difference is an important structural consideration in ore storage bins and silos as well as for drifts and similar access openings in caving operations. The total vertical shear is just the difference between the weight of the muck column and the bottom force. Thus,

$$F_s = W - \sigma_v A \tag{8.12}$$

where the average vertical stress is given by the formula developed in the previous section and A is the cross-sectional area at the considered depth. In the example muck column used to plot Figure 8.8, at a depth of three times diameter (3 × 84 ft), the vertical stress is about 61 psi, W is about 134 million pounds, but the bottom load is only about 49 million pounds force. The difference of about 85 million pounds is transmitted through frictional shear to the wall of the chimney.

If the "chimney" is actually a steel storage silo, say, one inch thick, then the vertical compression induced in the silo wall at bottom is almost 27,000 psi. Wall stress near a chimney of broken rock would also be elevated, although distributed into the rock mass. This additional stress may again be concentrated about openings near the cave zone. A "hoop" stress (normal stress acting in the circumferential direction) induced in the silo wall is $\sigma_t = pR/h$ where h is thickness, in this case, one inch, and p is the horizontal pressure acting against the silo wall, that is, $p = \sigma_h = k\sigma_v = 18.3$ psi in the example calculation. The hoop stress is about 9,220 psi, which indicates the vertical compression is the critical stress for design. These stresses are illustrated in Figure 8.9.

Tall, thin-walled structures such as steel silos also need to be designed for stability against buckling. When such structures buckle, they often have the appearance of an aluminum can

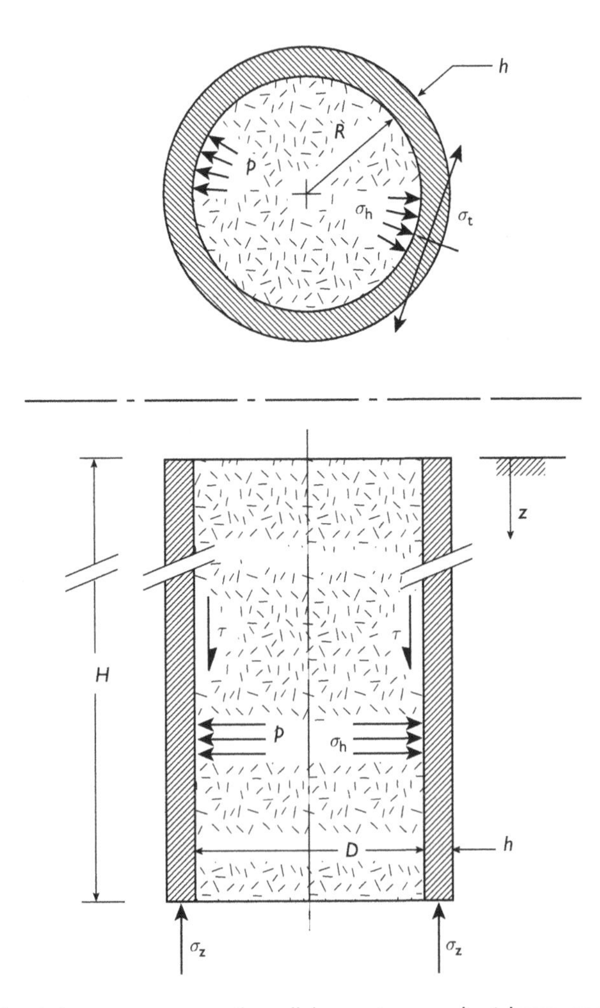

Figure 8.9 Normal and shear stresses at a silo wall, hoop stress, and axial compression in wall.

with the side pressed inward and the ends folded together. Because buckling is associated with eccentric axial loading, silos should be emptied and filled at the center of the cross section. Side draws should be avoided.

The formula for vertical stress is the famous *Janssen formula* that was published near the turn of the century in response to silo design needs. This formula is the basis of most silo design codes around the world. Prior to the acceptance of the Janssen formulas, bulk materials in silos were considered to behave like fluids. The horizontal pressure against the silo wall was then calculated as specific weight times depth. In the example calculation, this pressure is about 167 psi which leads to a hoop stress of over 84,000 psi. Some wood silo walls for grain storage had thicknesses measured in feet!

When a silo chute is opened, support is removed from material at the gate and a switch in the direction of the major compression occurs near the chute. The switch is a change from vertical to horizontal and during draw may cause high hoop stress that must be designed against. The point of this switch and therefore of peak horizontal stress against the silo wall is usually where the flow first occurs over the full cross section of the silo or at the transition between conical bottom hopper and cylindrical silo above. Reinforcement may be needed at this elevation, especially if the silo is a cement block silo that has low tensile strength and prone to crack under hoop stress.

Example 8.13 Acylindrical concrete silo is used to store bulk coal that has a specific weight of 80 pcf. In motion, coal cohesion is zero, the angle of internal friction is 28°, and the friction angle of coal on concrete is 23°. The silo is 30 ft in diameter (inside diameter) and 120 ft high. Estimate the peak vertical, horizontal, and shear stresses acting on the silo.

***Solution*:** The formulas for the required stresses are

$$\sigma_{\rm v} = \left(\frac{\gamma}{C_1}\right)[1 - \exp(-C_1 z)]$$

$$\sigma_{\rm h} = k\sigma_{\rm v}$$

$$\tau = \mu\sigma_h$$

$$C_1 = \frac{\mu k C}{A}$$

$$= \frac{\tan(23)\Big(1 - \sin(28)/1 + \sin(28)\Big)(\pi 30)}{(\pi/4)\left(30^2\right)}$$

$$= (0.425)(0.361)(4/30)$$

$$C_1 = 0.0204$$

Thus, after evaluating the constant C_1,

$$\sigma_v = 24.85\ \text{psi}$$
$$\sigma_{\rm h} = 8.47$$
$$\tau = 3.81$$

Example 8.14 Consider the data from Example 8.13 and suppose the concrete has a compressive strength $C_o = 4{,}500$ psi and a tensile strength $T_o = 350$ psi. Estimate the thickness of the silo wall needed for a concrete safety factor of 2.2.

***Solution*:** The stresses are greatest at the silo bottom when the silo is full. Compression in the silo walls is generated by shear load at the coal concrete interface. Equilibrium requires

$$\sigma_c A_c = S_v$$

$$\therefore \sigma_c = \frac{W - F_v}{(\pi/4)\left(D_o^2 - D_i^2\right)}$$

$$W = \gamma(\pi/4)D_i^2 H = (80)(\pi/4)(30)^2(120) = 6.79\left(10^6\right)\text{lbf}$$

$$F_v = \sigma_v A = (24.85)(144)(\pi/4)\left(30^2\right) = 2.53\left(10^6\right)\text{lbf}$$

$$S_v = 6.79\left(10^6\right) - 2.53\left(10^6\right) = 4.27\left(10^6\right)\text{lbf}$$

as illustrated in the sketch.

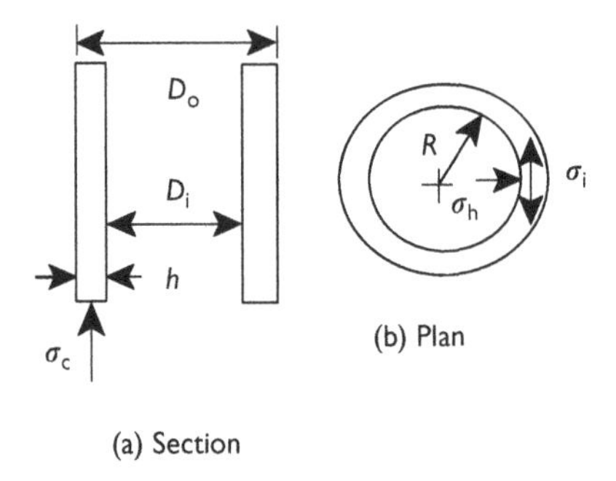

(a) Section

Substituting into the expression for the wall compression shown section gives

$$\left(\frac{C_o}{\text{FS}}\right)\left(\frac{\pi}{4}\right)\left(D_o^2 - D_i^2\right) = S_v$$

that may be solved for the outside diameter and hence thickness h. Thus,

$$D_o^2 = D_i^2 + \frac{(\text{FS})(S_v)(4/\pi)}{C_o} = (30)^2 + \frac{(2.2)(4.27)\left(10^6\right)(4/\pi)}{(4,500)(144)}$$

$$D_o^2 = 918.5$$

$$D_o = 30.306$$

$$h = 0.306\ \text{ft}(3.67\ \text{in.})$$

The required thickness of concrete to meet the compressive stress safety factor requirement is 3.67 in.

The tensile "hoop" stress is

$$\sigma_t = \sigma_{\mathrm{h}}\left(\frac{R}{h}\right)$$
$$\therefore h = \sigma_h(R)\left(\frac{\mathrm{FS}}{T_{\mathrm{o}}}\right)$$
$$= (8.47)\frac{(30/2)(12)(2.2)}{350}$$
$$h = 9.58\text{ in.}$$

and requires a wall thickness of 9.58 in. The required wall thickness to meet the safety factor requirement in tension and compression is 9.58 in.

Chimney cave water forces

When water infiltrates a chimney cave and saturates the muck below the water table, frictional resistance to muck flow is reduced while bottom stress is increased. Total stresses are still required for equilibrium as illustrated in Figure 8.7, but shearing resistance at the chimney walls requires consideration of *effective* stress. Effective stress is also referred to as *intergranular* stress and is the force per unit area transmitted through point contacts among particles and between particles and chimney wall. Effective vertical and horizontal normal stresses are

$$\begin{aligned} \sigma'_{\mathrm{v}} &= \sigma_v - p_{\mathrm{f}} \\ \sigma'_h &= \sigma_h - p_{\mathrm{f}} \end{aligned} \tag{8.13}$$

where p_{f} is pore fluid pressure, that is, water pressure given as specific weight of water γ_{w} times depth below the water table z_{w}. The ratio k that relates principal stresses at failure now pertains to the effective stresses. Thus,

$$k = \frac{\sigma'_{\mathrm{h}}}{\sigma'_{\mathrm{v}}} \tag{8.14}$$

which reduces to the case when pore fluid pressure is negligible and the total and effective stresses are the same.

The equilibrium equation is now

$$\frac{\mathrm{d}\sigma_{\mathrm{v}}}{\mathrm{dz}} - \gamma_{\mathrm{sat}} + C_1(\sigma_{\mathrm{v}} - \gamma_{\mathrm{w}} z_w) = 0 \tag{8.15}$$

where the specific weight of the muck in the column below the water table is labeled as the saturated specific weight, γ_{sat}, that is indicative of the total weight per unit volume of material with void space completely filled with water.

If the water table is at the top of the muck column, then

$$\sigma_{\mathrm{v}} = \left(\frac{\gamma_{\mathrm{sat}} - \gamma_{\mathrm{w}}}{C_1}\right)[1 - \exp(-C_1 z)] + \gamma_{\mathrm{w}} z \tag{8.16}$$

The difference between the saturated specific weight of the muck and the specific weight of water is the *buoyant* specific weight which is also known as the *submerged* specific weight. If wall friction was absent, the vertical stress transmitted through particle contacts would be just the buoyant specific weight times depth. If water is absent, then this formula reduces to the previous formula for muck not submerged. When the muck column is submerged, water pressure adds to the vertical load in amount of unit weight of water times depth, as seen in the formula.

When the muck column is saturated but *not* submerged, then water clings to the muck through surface tension but does not form a water column that generates the water pressure p_f. In this case the water terms vanish and the previous formula for vertical stress applies.

Once submerged, surface meniscii between particles are destroyed; the water no longer clings to the solid particles and hydrostatic pressure develops in the connected water-filled voids of the muck.

When the water table is below the top of the muck column, say, at $z = z_o$, then

$$\sigma_v = \left(\frac{\gamma_{sat} - \gamma_w}{C_1}\right)[1 - \exp(-C_1 z_w)] + \gamma_w z_w + \left(\frac{\gamma}{C_1}\right)[1 - \exp(-C_1 z_o)]\exp(-C_1 z_w) \tag{8.17}$$

where the last term on the right is a contribution of load by the muck above the water table with specific weight γ that includes moisture present in the muck as well as solid particles. This contribution decreases exponentially with depth below the water table. If $z_o = 0$, so $z_w = z$, then the previous case is recovered where the water table is at the top of the muck column. If the muck is saturated, but not submerged, then the original Janssen formula is recovered. In all cases, the effective normal stresses (vertical and horizontal) are obtained by subtracting the hydrostatic water pressure from total normal stress.

Example 8.15 Consider the silo data in Examples 8.13 and 8.14 and suppose water fills the voids in the bulk coal because of heavy rain and a leaky roof. Estimate the silo stresses σ_v, σ_h, τ. Note: The specific weight of saturated bulk coal is 25% greater than dry bulk coal.

Solution: Formulas for the stresses when water is present to the top of a filled silo are

$$\sigma_v = \left(\frac{\gamma_{sat} - \gamma_w}{C_1}\right)[1 - \exp(-C_1 z)] + \gamma_w z$$

$$\sigma_v{}' = \sigma_v - \gamma_w z$$

$$\sigma_h{}' = k\sigma_v{}', \quad k = 0.361$$

$$\sigma_h{}' = \sigma_h - \gamma_w z$$

$$\tau = \mu\sigma_h{}', \quad \mu = 0.425$$

$$C_1 = 0.0204$$

$$\gamma_w = 62.4\,\text{pcf}$$

$$\gamma_{sat} = 1.25\gamma_{dry} = (1.25)(80) = 100\,\text{pcf}$$

where data from Examples 8.13 and 8.14 are used. The total vertical stress and effective vertical stress are therefore

$$\sigma_v = \left(\frac{100 - 62.4}{0.0204}\right)\{1 - \exp[-(0.0204)(120)]\} + (62.4)(120)$$
$$= 1,684 + 7,488$$
$$\sigma_v = 9,172\,\text{psf}\;\;(63.7\,\text{psi})$$
$$\sigma_v{}' = 1,684\,\text{psf}\;\;(11.7\;\text{psi})$$

The water load more than doubles the dry load vertical stress as seen in comparison with Example 8.14 results.

The horizontal stresses are

$$\sigma_h{}' = (0.361)(1,684) = 608\,\text{psf}(4.22\,\text{psi})$$
$$\sigma_h = \sigma_h{}' + \gamma_w z = 608 + (62.4)(120) = 8,096\,\text{psf}(56.2\,\text{psi})$$

Although the effective horizontal wall stress at the bottom of the silo is only about half the same stress in the dry case, the total horizontal stress is more than five times the horizontal stress in the dry condition. Because the horizontal stress induces a hoop stress tension in the silo walls, this increase would likely threaten to crack the walls. Keeping bulk materials dry is an important consideration in many storage schemes and is obviously important to structural safety of silos.

The shear stress is

$$\tau = \mu\sigma_h{}' = (0.425)(608) = 258\,\text{psf}(1.79\,\text{psi})$$

which is less than one half the same stress in the dry case, showing the reduction in shear that has the consequence of increasing bottom load.

Example 8.16 An underground mine has a stope filled with well-blasted ore. The stope (shrinkage stope) is 20 ft wide, 250 ft long, and 150 ft high. Broken ore is estimated to have an angle of internal friction of 41° and a friction angle of muck on rock of 33°. Dry specific weight of broken ore is 95 pcf. Wet specific weight is 115 pcf. A cross section of the stope is shown is the sketch. Suppose water becomes trapped in the stope and rises to the top. Estimate the changes in vertical, horizontal, and shear stress at the stope bottom that are induced by the water.

***Solution*:** The changes are just the differences between the wet and dry states. In either case, a plane assumption is reasonable. From Example 8.9 in the dry case

$$\sigma_v = \left(\frac{\gamma W}{2\mu k}\right)\left[1 - \exp\left(-\frac{2\mu k}{W}z\right)\right]$$
$$\sigma_h = k\sigma_v$$
$$\tau = \mu\sigma_h$$

where W is stope width shown in the sketch and $W/2$ plays the role of area / circumference.

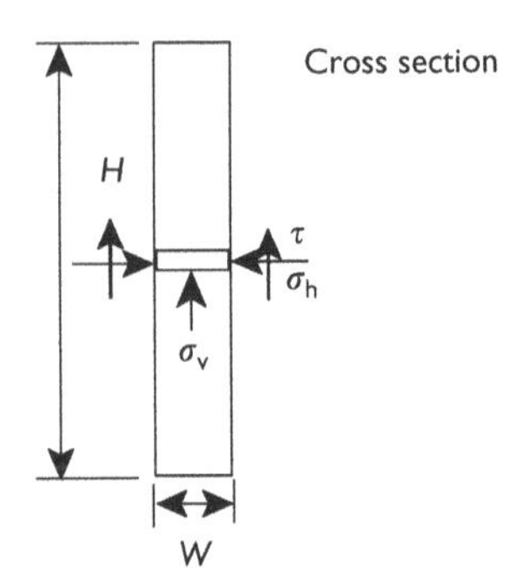

The constants required for stress computation are

$$\mu = \tan(33) = 0.6494$$

$$k = \frac{1 - \sin(\phi)}{1 + \sin(\phi)} = \frac{1 - \sin(41)}{1 + \sin(41)} = 0.277$$

$$\gamma_{\text{dry}} = 95\,\text{pcf}$$

$$W = 20\,\text{ft}$$

Thus, in the dry case the stresses are

$$\sigma_{\text{v}} = \left(\frac{(95)(20)}{(2)(0.6494)(0.277)}\right)\{1 - \exp[-(2/20)(0.6494)(0.277)(1500)]\}$$

$$= \left(\frac{95}{0.01799}\right)\{1 - \exp[-0.01799(150)]\}$$

$$\sigma_{\text{v}} = 4,925\,\text{psf}(34.2\,\text{psi})$$

$$\sigma_{\text{h}} = k\sigma_{\text{v}} = (0.277)(4,925)$$

$$\sigma_{\text{h}} = 1,364\,\text{psf}(9.47\,\text{psi})$$

$$\tau = \mu\sigma_{\text{h}} = (0.6494)(1364)$$

$$\tau = 886\,\text{psf}(6.15\,\text{psi})$$

In the wet case

$$\sigma_{\text{v}} = \left(\frac{\gamma_{\text{sat}} - \gamma_{\text{dry}}}{2\mu k/W}\right)\left[1 - \exp\left(-\frac{2\mu k}{W}z\right)\right] + \gamma_{\text{w}}z$$

$$\sigma_{\text{v}}{}' = \sigma_{\text{v}} - \gamma_{\text{w}}z$$

$$\sigma_{\text{h}}{}' = k\sigma_{\text{v}}{}'$$

$$\tau = \mu\sigma_{\text{h}}{}'$$

All data needed for evaluation are available, so

$$\sigma_v = \left(\frac{115 - 95}{0.01799}\right)\{1 - \exp[-(0.01799)(150)]\} + (62.4)(150)$$

$$= 1,037 + 9,360$$

$$\sigma_v = 10,397\text{psf } (72.2\text{ psi})$$

$$\sigma_v{}' = 1,037\text{psf } (7.2\text{psi})$$

$$\sigma_h{}' = k\sigma_v{}'$$

$$= (0.277)(1,037)$$

$$\sigma_h{}' = 287\text{psf } (1.99\text{ psi})$$

$$\sigma_h = \sigma_h{}' + \gamma_w z$$

$$= 287 + (62.4)(150)$$

$$\sigma_h = 9,647\text{psf } (67.0\text{ psi})$$

$$\tau = \mu\sigma_h{}'$$

$$= (0.6494)(287)$$

$$\tau = 186\text{ psf } (1.29\text{ psi})$$

Changes in stress induced by the trapped water are

$$\Delta\sigma_v = \sigma_v(\text{wet}) - \sigma_v(\text{dry}) = 10,397 - 4,925$$
$$\Delta\sigma_v = 5,472\text{ psf } (38\text{ psi})$$
$$\Delta\sigma_h = 9,647 - 1,364 = 8,283\text{ psf } (57.5\text{ psi})$$
$$\Delta\tau = 186 - 886 = -700\text{ psf } (4.86\text{ psi})$$

The increase in vertical stress is over a 100%. The increase in horizontal stress is even greater, by a factor of seven over the dry case. Wall shear stress decreases to less than one-third the dry case value. Practical consequences are a large increase in bottom load where material handling facilities are located and a serious threat of muck rushes when ore is "pulled from the chutes" at the bottom of the stope. The effect of water in reducing shear stress is sometimes used to advantage by introducing a flow from a water hose to free arches that may have formed in the stope muck.

Support near caving ground

The transfer of load from a chimney cave to the walls of the cave zone leads to an elevated state of stress that increases the need for support in nearby access drifts, crosscuts, and raises. The effect is not easy to quantify and poses a difficult challenge to stress analysis because of the complications of three-dimensional mining geometry, the mechanics of caving and the role of geologic discontinuities, all of which come into play in block caving mines. A practical compromise is to adapt existing "tunnel" support approaches to the block caving environment (Kendorski, 1977, chapter 4).

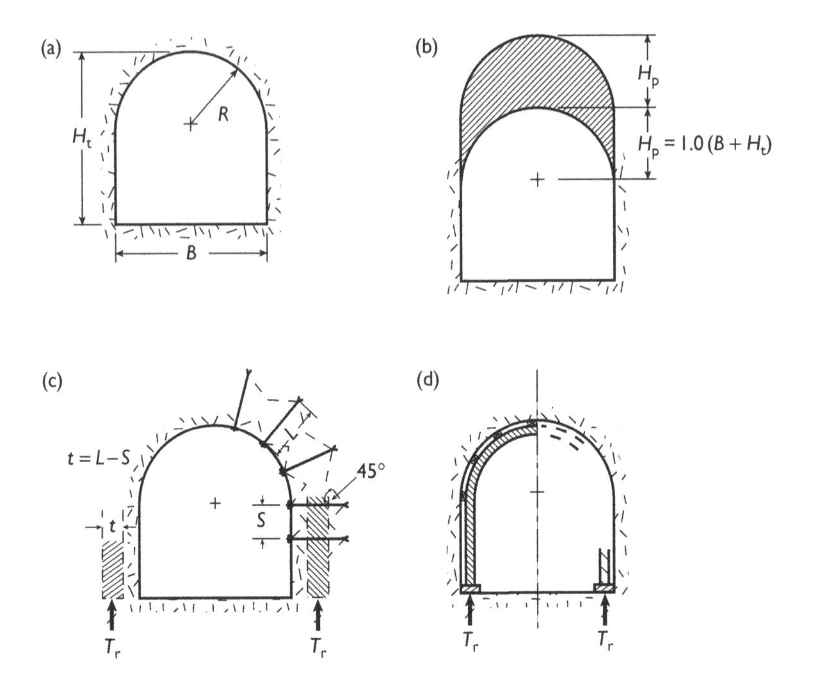

Figure 8.10 Combination drift: (a) geometry, (b) loading rock arch, (c) supporting rock arch and bolting geometry, and (d) steel set support.

Existing approaches include steel set support, rock bolt reinforcement and concrete or "shotcrete" lining. Shotcrete is a sand–cement mixture that is sprayed onto tunnel walls in thicknesses of several inches or so and is often used with wire mesh or chain link fence that is secured to the wall with rock bolts. Fibers may be added to shotcrete to improve strength. Figure 8.10 shows a rectangular drift with a circular arched back supported by steel sets, rock bolts, and shotcrete.

Weight W of an imaginary rock arch in the back loads natural support provided by the rock mass in the walls of the drift and steel sets, if present. Dimensions of the drift, breadth B, and total height H_t, dictate the height of the rock arch, that is, $H_p = 1.0(H_t + B)$. No attempt is made to fine-tune classification of the rock mass as in the case of sizing steel sets for tunnel support according to the traditional Terzaghi classification scheme. In essence, a judgment is made at this stage of design that the rock mass is toward the "very blocky and seamy" condition, a reasonable judgment in view of the caving method being used and the expected ground conditions in jointed ore bodies amenable to caving. This load may be increased, say, up to twice the initial value because of the elevated stress state about the considered opening. Drift or "tunnel" load may thus be as high as $2W$ and

$$W = \gamma B H_p S \tag{8.18}$$

where γ is rock mass specific weight and S is "spacing," perhaps bolt spacing or simply one foot along the drift, so weight is then per foot of drift length.

Unconfined compressive strength of the rock mass multiplied by the *supporting rock* arch cross-sectional area is the primary and natural support available in the walls of the drift. Rock bolts reinforce the rock mass and increase the natural support capacity of the supporting rock arch in the ribs of the drift. This *increase* in strength of the rock mass may be estimated from bolting pressure considerations or ignored without serious detriment to the analysis because the effect is likely to be rather small. Bolts are still essential to prevention of rock block falls and to maintaining integrity of drift walls, of course. In any case, the thickness t of the supporting arch shown in Figure 8.10 must be estimated before computing the natural support capacity of the walls (T_r). Several estimates of t are possible: (1) bolt length L; (2) length less spacing, $L - S$; (3) S, if $S < L$ (usually); (4) $1D$ where D is drift width; (5) $B/2$, and so forth. The first seems reasonable; the second and third choices are more conservative, but still depend on bolting geometry. The last two choices relate to stress concentration. Stress concentration decreases rapidly with distance from an opening wall, usually in proportion to r^{-2}. The average stress taken over a distance into the wall decreases more slowly, in proportion to r^{-1}, so the fifth choice may be appropriate. However, the one diameter rule is optimistic because of the greater support area provided for a given rock arch weight. Bolts are almost always used, so bolt length is a reasonable choice unless additional site-specific information is at hand. The capacity of the natural supporting rock arch is

$$T_r = C_o S t \tag{8.19}$$

where bolt length L may be substituted for arch width t and S is "spacing," as used to calculate W.

A safety factor for the naturally supported drift may be defined as the ratio of resisting force R to driving force D. Thus,

$$\text{FS}_o = \frac{R}{D} = \frac{T_r}{\lambda W/2} \tag{8.20}$$

where λ is multiplier that allows for increasing weight up to twice the initial weight load. An allowable stress for the rock mass may be introduced through the usual definition,

$$\sigma_{al} = \frac{C_o}{\text{FS}_o} \tag{8.21}$$

in place of compressive strength for calculating the minimum support capacity required of the natural rock T_r (min), which could also be obtained directly as T_r/FS_o.

Failure in a jointed rock mass involves slip on joints and shear through intact rock bridges between joints, as shown in Figure 8.11. Shear strength is the total of rock bridge and joint shear resistances. Thus,

$$\tau A = \tau_r A_r + \tau_j A_j \tag{8.22}$$

With the assumption of Mohr–Coulomb failure criteria for joints and rock, and introduction of a normalized joint area $A' = A_j/A$, rock mass shear strength becomes

$$\tau = \sigma[A' \tan(\phi_j) + (1 - A')\tan(\phi_r)] + [A' c_j + (1 - A')c_r] \tag{8.23}$$

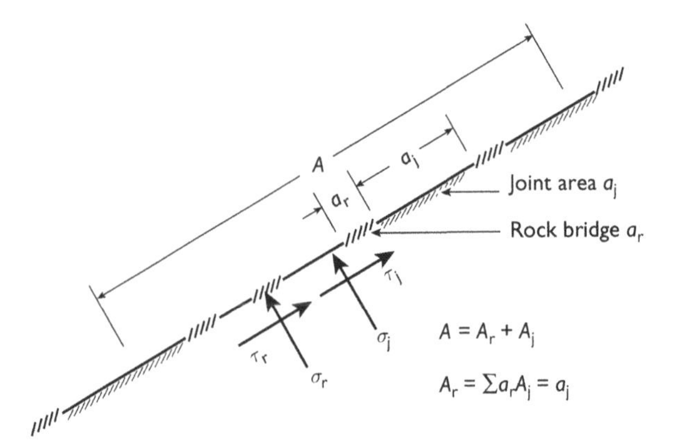

Figure 8.11 Shear in a jointed rock mass.

where equality of normal stress acting through intact rock bridges and across joints is implied. The parameter A' is a measure of *joint persistence* in the considered rock mass. Each joint set would be expected to have a different persistence. The form of rock mass shear strength is also Mohr–Coulomb; the equivalent angle of internal friction and cohesion of the rock mass with respect to the considered joint set are given by

$$\begin{aligned} \tan(\phi) &= A'\tan\left(\phi_j\right) + (1 - A')\tan(\phi_r) \\ c &= A'c_j + (1 - A')c_r \end{aligned} \tag{8.24}$$

These relations could be applied to each joint set in the rock mass. Equivalent unconfined compressive and tensile strengths can be obtained from cohesion and angle of internal friction by the usual formula. Thus,

$$\begin{aligned} C_o &= \left[\frac{2(c)\cos(\phi)}{1 - \sin(\phi)}\right] \\ T_o &= \left[\frac{2(c)\cos(\phi)}{1 + \sin(\phi)}\right] \end{aligned} \tag{8.25}$$

The rock mass compressive strength is the appropriate strength for use in computing capacity of the natural support arch.

If the minimum natural support capacity is not available, then additional support T_s is required. The amount of additional support required may be obtained from an *equilibrium* condition,

$$\lambda \frac{W}{2} = T_r + T_s \tag{8.26}$$

where T_r and T_s may be minimum acceptable amounts. In this regard, a minimum acceptable amount of additional support may be obtained from another safety factor FS_s, that is,

$$T_s(\min) = \frac{T_s}{FS_s} \tag{8.27}$$

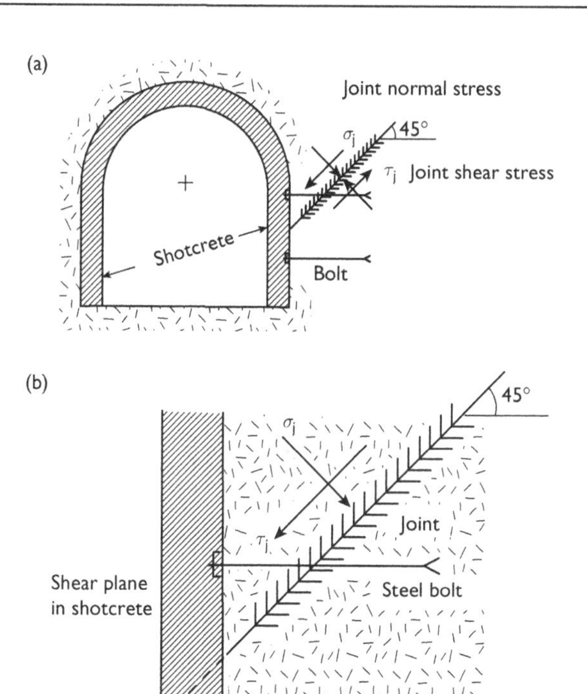

Figure 8.12 Shotcrete resistance to shear failure on joints with bolts: (a) shotcrete lining and (b) shotcrete shear plane.

Joints in the rock mass are controlled by bolts that cross joints and by shotcrete which is strong in shear and thus aids in the prevention of rock blocks slipping along joints. The numerous combinations of joint orientation, properties, and persistence precludes development of detailed design formulas. Instead, a generic analysis for joint shear failure is done to demonstrate the procedure and to develop a method of determining shotcrete thickness to defend against rock blocks shearing along joints into the opening. Dip of the generic joint is assumed (45°), as shown in Figure 8.12. Joint slip is driven by the component of rock arch weight acting "downhill," that is, parallel to the joint in the direction of the opening. This driving force is

$$D_j = \left(\frac{\lambda W}{2}\right)\sin(\delta_j) \tag{8.28}$$

where δ_j is joint dip and λ is a weight multiplier that accounts for additional load transmitted to the rock mass from caving chimneys nearby. The normal component of weight acting across the joint is

$$N_j = \left(\frac{\lambda W}{2}\right)\cos(\delta_j) \tag{8.29}$$

Joint resistance to shear is therefore

$$R_j = N_j \tan\left(\phi_j\right) + c_j A_j \tag{8.30}$$

where A_j is area of the considered joint. Rock bridges do not affect joint failure in this calculation. If bolt length is L and spacing is S, then a reasonable estimate of $A_j = LS/\cos(\delta_j)$.

Resistance is also provided by bolts. If one bolt is assumed to intersect the considered joint area, as shown in Figure 8.12, then bolt resistance is estimated to be

$$R_b = S_b A_b \tag{8.31}$$

where S_b is bolt shear strength, which may be estimated at one-half bolt tensile strength, and A_b is the cross-sectional area of the bolt. No allowance is made for the joint intersecting the bolt at an angle.

A factor of safety for joint shear failure is the ratio of resisting to driving forces. Thus

$$\mathrm{FS}_j = \frac{R}{D} = \frac{R_j + R_b}{D} \tag{8.32}$$

If the joint safety factor is insufficient, then additional resistance is needed and may be added in the form of shotcrete.

Shotcrete resistance to shear is given by the product of shear strength and shotcrete area. According to the Mohr–Coulomb criterion, strength under pure shear S_o is $(c)\cos(\phi)$, that is

$$S_o = (c)\cos(\phi) = C_o\left[\frac{1}{C_o/T_o + 1}\right] \tag{8.33}$$

Another estimate of an *allowable* shear stress for concrete loaded in pure shear is $2\sqrt{f'_c}$ in the notation commonly used for concrete ($f'_c = C_o$ is unconfined compressive strength). Unconfined compressive strength of concrete and shotcrete generally ranges between 3,000 and 5,000 psi. In the notation used here

$$2\sqrt{f'_c} = \frac{S_o}{\mathrm{FS_s}} = \frac{C_o[1/(C_o/T_o + 1)]}{\mathrm{FS_s}} \tag{8.34}$$

A safety factor with respect to shear of about 3.2 is implied in the allowable stress recommendation when using an unconfined compressive strength of 4,000 psi and a compressive to tensile strength ratio of nine. Use of shear strength from the Mohr–Coulomb criterion and a specified safety factor in shear is a simple method of prescribing allowable stress.

The specified allowable stress in shotcrete multiplied by shotcrete area is shotcrete resistance to shear. Because of joint dip, shotcrete shear area appears to vary with dip, if taken parallel to joint dip. This approach would be impractical at very steep joint dips and suggests a flaw in the approach. A simple and more practical view is to consider shear across the shotcrete thickness measured perpendicular to the opening wall. In this case, shotcrete area is simple hS and shotcrete resistance is

$$R_c = \left(\frac{S_o}{\mathrm{FS_s}}\right) hS \tag{8.35}$$

where shear strength may be determined from unconfined compressive strength and a reasonable assumption of compressive to tensile strength ratio, say, nine.

Solution of the joint safety factor equation

$$\mathrm{FS_j} = \frac{R_j + R_b + R_c}{D} \tag{8.36}$$

for R_c allows for determination of shotcrete thickness h required to achieve the specified joint safety factor under the given allowable shear stress.

Example 8.17 Aconcrete lining is being considered for support of drifts near caving blocks of ore. Drifts are rectangular in section with a semicircular arched back. Width is 12 ft and height is 15 ft. Several joint sets are present at various spacings and persistences with different properties. A simplified view considers joint persistence to be 0.87, joint cohesion $c_j = 5$ psi, and joint angle of friction $\phi_j = 28°$. Intact rock between joints has unconfined compressive and tensile strengths $C_o = 18{,}500$ psi and $T_o = 1{,}350$ psi, respectively. Pattern bolting is used to support drifts during development. Bolts are 6 ft long, 3/4 in. in diameter, and of high strength steel. They are spaced on 4 ft centers in a square pattern. The concrete liner would have compressive and tensile strengths of 4,500 and 350 psi, respectively. Safety factors for the rock mass, concrete liner, and joint slip are all 1.75. Determine whether a concrete liner is needed and, if so, the required thickness.

***Solution*:** If the rock mass and bolts provide adequate support, then no concrete is needed.

$$\mathrm{FS} = \frac{R}{D} = \frac{C_m S h}{\lambda W/2}$$

The bolts are considered as reinforcement against joint slip which is a separate failure mechanism from rock mass failure. A safety factor for the rock mass at the drift wall is where C_m, S, h, λ, W are unconfined compressive strength of the rock mass, spacing of bolts along the drift, supporting rock arch thickness, caving multiplier, and weight of the loading arch, respectively. Each factor needs to be calculated in turn.

Rock mass compressive strength may be estimated from rock mass cohesion c and angle of internal friction ϕ. In consideration of joint persistence,

$$c = (1-p)c_r + pc_j, \quad \tan(\phi) = (1-p)\tan(\phi_r) + p\tan(\phi_j)$$

Joint cohesion and friction angle are known, but intact rock and angle of internal friction need to be determined. With the common assumption of Mohr–Coulomb strength, one has

$$\begin{aligned}
\sin(\phi_r) &= \frac{C_o - T_o}{C_o + T_o} \\
&= \frac{18{,}500 - 1{,}350}{18{,}500 + 1{,}350} \\
\sin(\phi_r) &= 0.86398 \\
\phi_r &= 59.8°
\end{aligned}$$

$$c_r = \frac{C_o[1 - \sin(\phi_r)]}{2\cos(\phi_r)}$$
$$= \frac{(18{,}500)[1 - \sin(59.8)]}{2\cos(59.8)}$$
$$c_r = 2,496 \text{ psi}$$

Rock mass cohesion and angle of internal friction are then

$$c = (1 - p)c_r + pc_j, \qquad \tan(\phi) = (1 - p)\tan(\phi_r) + p\tan\left(\phi_j\right)$$
$$= (1 - 0.87)(2,496) + (0.87)(5) \qquad = (1 - 0.87)\tan(59.8) + (0.87)(\tan 28)$$
$$c = 329 \text{ psi} \qquad \tan(\phi) = 0.68595$$
$$\phi = 34.4°$$

and rock mass compressive and tensile strengths are

$$\left.\begin{matrix} C_o \\ T_o \end{matrix}\right\} = \frac{2(c)\cos(\phi)}{1 \mp \sin(\phi)}$$
$$= \frac{(2)(329)\cos(34.4)}{1 \mp \sin(34.4)}$$
$$\left.\begin{matrix} C_o \\ T_o \end{matrix}\right\} = \left.\begin{matrix} 1,248 \\ 347 \end{matrix}\right\} \text{psi}$$

Bolt spacing S is given, and an estimate of supporting arch thickness h is simply bolt length. Weight of the loading arch is given by

$$W = \gamma BSH_p = (158)(12)(4)(12 + 15) = 2.05\left(10^5\right) \text{lbf}$$

where H_p is height of the loading arch estimated about caving ground as the sum of the drift width and height, $(B + H_t)$ and an estimate of 158 pcf is made for rock mass specific weight.

The rock mass safety factor is

$$\text{FS} = \frac{R}{D} = \frac{C_m Sh}{\lambda W/2}$$
$$= \frac{(1,248)(144)(4)(6)}{2.05\left(10^5\right)}$$
$$\text{FS} = 21.0$$

where a caving multiplier of 2 is used. Rock mass safety factor is more than adequate to meet the 1.75 requirement.

However, there is the possibility of joint slip that needs to be examined. A joint safety factor is

$$\text{FS}_j = \frac{R_j + R_b}{D}$$

that is the ratio of joint resisting plus bolt resisting forces to driving force. Joint resistance is calculated from the geometry of the drift and assumed 45° dipping joint and a Mohr–Coulomb strength criterion. Bolt resistance to shear is bolt shear strength, estimated at one-half tensile strength, and bolt cross-sectional area. Thus,

$$
\begin{aligned}
R_j &= N\tan\left(\phi_j\right) + c_j A_j \\
&= (2.05)\left(10^5\right)\cos(45)\tan(28) + (5)(144)[6/\cos(45)](4) \\
&= 7.708\left(10^4\right) + 2.444\left(10^4\right) \\
R_j &= 1.015\left(10^5\right)\text{lbf} \\
R_b &= (55,000/2)(\pi/4)(3/4)^2 \\
R_b &= 1.215\left(10^4\right)\text{lbf} \\
D &= W\sin(45) = 2.05\left(10^5\right)\sin(45) \\
D &= 1.450\left(10^5\right)
\end{aligned}
$$

Hence,

$$
\text{FS}_j = \frac{1.015 + 0.122}{1.450} = 0.78
$$

that indicates the bolting pattern may not be adequate to hold against the joint slip before a concrete liner can be installed. Certainly, a concrete liner will be needed to prevent joint slip at the drift walls.

A revised joint slip safety factor is now

$$
\begin{aligned}
\text{FS}'_j &= \frac{R_j + R_b + R_c}{D} = \text{FS}_j + \frac{R_c}{D} \\
R_c &= S_c h S \\
S_c &= C_o\left(\frac{1}{C_o/T_o + 1}\right) = (4,500)\left(\frac{1}{4,500/350 + 1}\right) = 325\text{ psi} \\
R_c &= (325)(h)(4)(12) = 15,600h \\
&\therefore \\
2.5 &= 0.78 + \frac{15,600h}{145,000} \\
&\therefore \\
h &= 16.0\text{ in.}
\end{aligned}
$$

Thus, the required concrete liner thickness is 16.0 inches that would not be unusual in block caving drifts.

Example 8.18 With reference to the data and computations in Example 8.17 where there was an indication that the proposed bolting system would not be adequate to prevent joint slip, define an alternative but still square pattern that would result in a safety factor of 1.1 before any consideration of concrete lining.

Solution: The required safety factor is given by

$$FS_j = \frac{R_j + R_b}{D}$$

$$\therefore$$

$$(1.1)(1.45)(10^5) = 1.015(10^5) + F_b$$

$$\therefore$$

$$F_b = 49,000 \text{ lbf}$$

so each bolt must resist 49,000 lbf when installed on the original 4 ft centers. The force is one of shear and as shear strength is just one-half tensile strength, bolt tensile strength would indicate a force of 98,000 lbf. Bolts of 60,000 psi steel strength, 1 inch in diameter have a yield load of only 47,000 lbf. As a practical matter then, the pattern must be changed as well as bolt capacity.

Changing bolt spacing does not change the quotient of R_j/D but will change the ratio R_b/D. The required ratio is

$$\frac{R_b}{D} = 1.1 - \frac{1.015}{1.45} = 0.400$$

where $D = \gamma BH_pS \sin(45) = 1.45(10^5)(1/4)S$.

Thus, $R_b = 0.400(1.45)(10^5)(1/4)S = 14,500$ lbf/ft is the bolt force per foot of drift needed in shear. In tension, the bolt force needed is 29,000 lbf per foot of drift. One-inch diameter bolts of 75,000 psi steel spaced on 2-ft centers would meet the requirement.

The spacing is quite close and suggests consideration of alternatives. One alternative would be to use shotcrete concurrently with bolting to maintain safety during drift development. Steel sets might also be considered instead of the concrete liner of Example 8.17.

8.2 Troughs

Trough subsidence is closely linked to excavation in stratified ground where extraction is 100% as in mining a longwall panel or during pillar mining on retreat from a room and pillar panel. Some caving occurs in the immediate roof, but remote strata flex in bending and create a "trough" at the surface. Of course there are other causes of land subsidence, for example, drawdown of the water table. In any case, a major objective of subsidence analysis is the description of the trough.

If an origin of coordinates $Oxyz$ is fixed at the center of a subsidence trough with the z-axis vertical, then a subsidence trough is symbolically a function $s(x,y)$ where s is the downward vertical displacement of a surface point at (x,y). Displacement of the considered point may have horizontal components, but traditionally only the vertical component is used to define a subsidence trough. Methods of determining the function $s(x,y)$ include numerical analysis, curve fitting to data, and an empirical approach based on surveys of surface subsidence

(Voight and Pariseau, 1970). This last approach is embodied in a handbook published by the National Coal Board (NCB, 1975) of the UK and is the approach described here. The reason for this emphasis is the relatively large database that underpins the NCB approach. As with all empirical methods, some modification is expected when applied to conditions outside the UK where the NCB handbook data were obtained.

In the NCB approach, the problem of estimating a subsidence trough is reduced to the description of a subsidence profile $s(x)$ where the x-axis is parallel to the face of a long-wall panel edge and extends from panel center laterally to the limit of subsidence. This profile is a rib-side profile. Plotting the same function parallel to the direction of mining and across the face of a panel produces a face-side profile. A face-side profile could also be plotted over the rear abutment of a panel. Face- and rib-side profiles are considered to have the same shape. The function $s(x)$ is not actually determined in the NCB approach, but rather judiciously spaced points along the subsidence profile are determined from a look-up table. A smooth curve drawn through these points then gives the subsidence profile; implementation of the NCB approach is essentially graphical. Two important characteristics of this approach are the maximum possible subsidence and the lateral extent of subsidence.

Damage to surface structures may occur because of subsidence and is linked to differential settlement (vertical displacement) that occurs at the surface. Severity of damage is also linked to strain in the horizontal direction. Consequently, a second important objective of subsidence analysis is estimation of horizontal surface strains. Horizontal surface strains are described, in principle, by some function $e(x,y)$. In the NCB approach, a strain profile $e(x)$ is determined at discrete points along the subsidence profiles, face-side, and rib-side. The ground surface in a subsidence trough is both stretched and compressed causing tensile and compressive strains. Important characteristics of surface strain profiles are thus peak tensions and peak compressions. A secondary compression maximum may also occur.

Limit of subsidence

An *angle of draw* defines the practical extent of surface subsidence, $s(x) = 0$. In elasticity theory, any excavation regardless of depth and extent causes some motion of the ground surface above. However, only displacements that are measurable by practical methods, say, by surveying, are important. This angle is measured from the *vertical* to a line drawn from a panel edge to the surface where subsidence is nil. Figure 8.13 illustrates the angle of draw concept which is a purely geometric construction that has no implications for strata mechanics. In particular, there is no implication that strata fracture along the angle of draw. In the UK, the angle of draw is typically 35°. However, observations in other coal basins of the world indicate different angles of draw (Peng, 1978). A reason for differences between coal basins is in the stratigraphic column. In particular, the presence of much sandstone is a likely cause for reduction in the angle of draw. In the US, 28° degrees is a reasonable estimate for the angle of draw. However, there are also differences between coal mining districts within the US. Western coal fields, for example, tend to have thick, massive sandstone in the overburden and consequently smaller angles of draw.

The same concept applies to dipping strata, but the subsidence trough is shifted according to the "rule of the normal". This rule simply projects the subsidence trough center to be at the surface where a line drawn normal from panel center intersects the surface, as shown in Figure 8.13. The result is differing lateral extent up and down dip. This empirical rule is

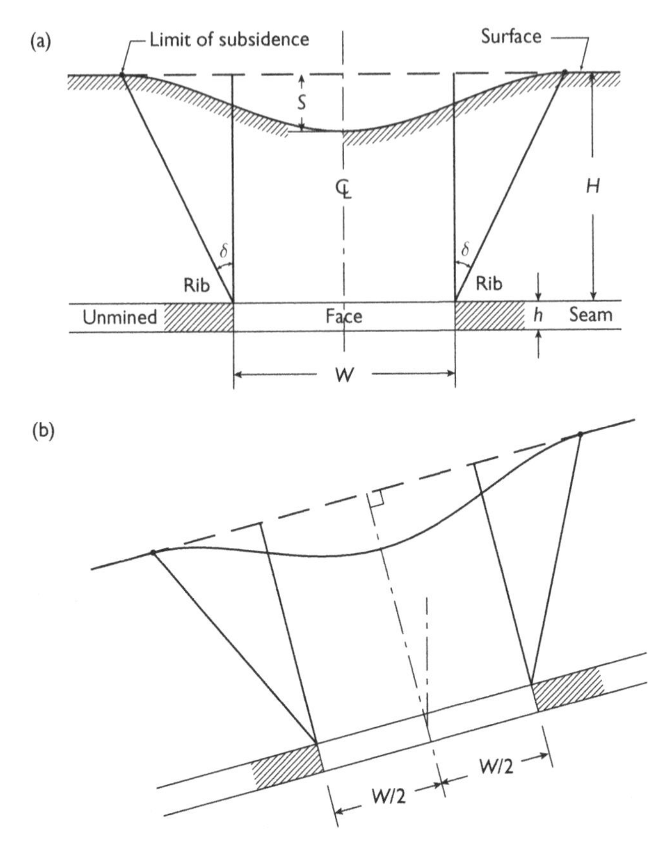

Figure 8.13 Trough subsidence notation, angle of draw, and rule of the normal: (a) rib-side subsidence profiles, angle of draw δ, mining height h, mining depth H, maximum subsidence S, and (b) rule of normal for dipping seams.

limited, as consideration of a vertical seam shows, and becomes progressively less reliable with increasing dip. As a practical matter, dips much above 10° (grades above 15%) are not amenable to mechanized coal mining. Hence, a limit of 20° to the rule of the normal is reasonable.

Time and topography also affect surface subsidence, but the effects are difficult to quantify. Observations in the UK indicate that subsidence is mostly instantaneous and occurs concurrently with face advance. Only up to about 10% of total surface subsidence is expected to occur after cessation of mining. In the Eastern bituminous coal fields, topographic relief and mining depths are roughly equal, about 800 ft. In Western coal fields, coal mines are usually developed from outcrop in plateau topography where not only rapid increase in overburden occurs, but also rapid decreases occur across canyon walls. Mining depths and topographic relief over 2,000 ft are encountered. However, the effect of topography is not well known and is ignored in the NCB approach to subsidence estimation.

Maximum subsidence

A *subsidence factor* S_f is defined as the ratio of maximum possible subsidence S_{max} to mining thickness h. Unless excavation reaches a minimum size, no surface subsidence occurs. This minimum size is expected to depend on depth. In flat strata, *maximum* subsidence is expected to occur in the center of a trough directly above the center of the excavated panel. As mining proceeds *laterally* in all directions beyond the minimum size, subsidence increases until a *critical area* A_c is reached beyond which no additional surface subsidence is possible. Further excavation then does not cause additional subsidence near the trough center. This maximum (S_{max}) is the greatest subsidence that is physically possible at the given excavation height h.

In the UK, the subsidence factor is 0.9, that is, the maximum possible subsidence is 0.9 times seam thickness when mining full seam height. As with the angle of draw, the subsidence factor is different in coal basins throughout the world. Again, the reason is in the nature of the overburden which is not taken into account by the empirical NCB method of subsidence analysis. In the US, a subsidence factor of 0.65 is reasonable.

Critical width

A *critical width* W_c is associated with the critical area A_c and is defined using the angle of draw, as shown in Figure 8.14. Critical area A_c is then a square with side length W_c. For

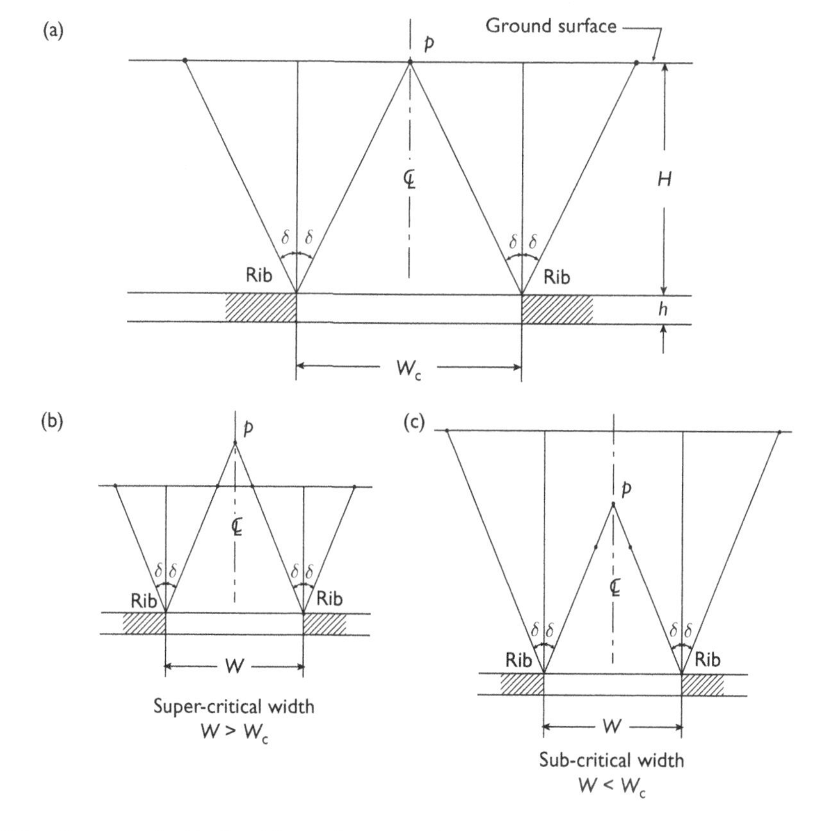

Figure 8.14 Concept of critical width in relation to rib-side subsidence profiles.

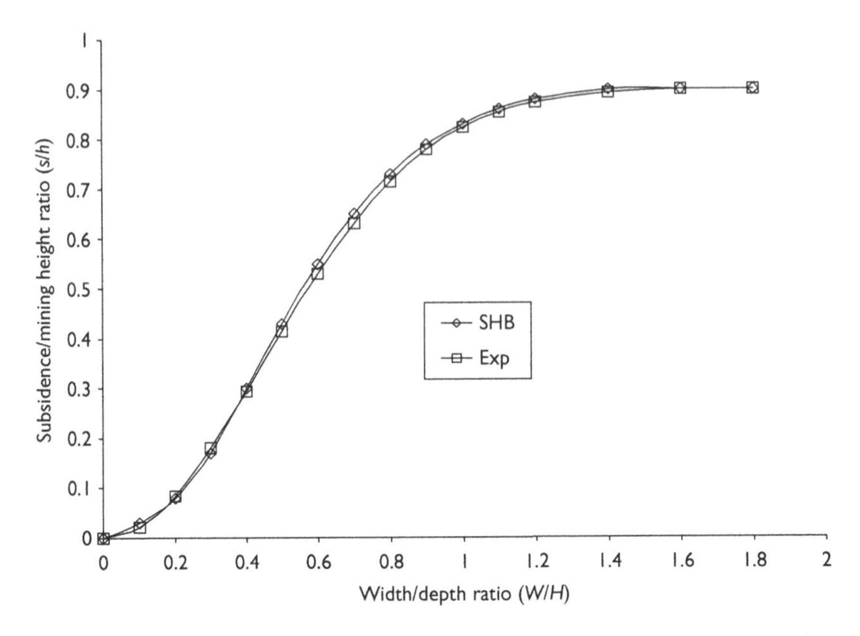

Figure 8.15 Ratio of maximum subsidence to mining height as a function of panel width to depth ratio for UK conditions. SHB = *Subsidence Engineer's Handbook* (1966), EXP = empirical fitted curve.

maximum subsidence to occur, panel width *and* face advance must be equal to or greater than the critical width. If the angle of draw is 35°, then the critical width is 1.4 times seam depth. Panel widths less than critical do not result in maximum possible subsidence regardless of panel length. A supercritical panel width leads to maximum possible subsidence only after face advance occurs that is equal to critical width. In any case, the subsidence trough width is panel width W (face length) plus 1.4 times seam depth.

If the panel width (face length) is subcritical, then $W/h < W_c/h$ where h is mining height and the actual maximum of subsidence S is less than the maximum possible S_{max}. This observation indicates that S depends on the width to depth ratio of the considered panel. Indeed, according to NCB data, there is an almost universal curve of S/h versus W/H where H is seam depth. Figure 8.15 shows this curve. Also shown in Figure 8.15 is a curve that has the form

$$\frac{S}{h} = a\left\{1 - \exp\left[-\frac{1}{2}\left(\frac{W/H}{a/2}\right)^2\right]\right\}$$

where the constant a is the *subsidence factor*, that is, $a = (S/h)_{\text{max}} = S_f$, which is 0.9 in the UK but generally depends on the coal basin stratigraphic sequence.

Once S/h is determined in consideration of excavation width and depth, maximum subsidence S is simply the product $h(S/h)$. The parameter S is important because all subsequent subsidence amounts s along the subsidence profile are initially given in terms of S, that is, as

Table 8.1 Subsidence profile points at various width to depth ratios (from SHB, 1975)

s/S W/H	*0.00*	*0.05*	*0.10*	*0.20*	*0.30*	*0.40*	*0.50 X/H*	*0.60*	*0.70*	*0.80*	*0.90*	*0.95*	*1.00*
0.20	0.80	0.64	0.57	0.48	0.41	0.37	0.32	0.28	0.23	0.19	0.13	0.08	0.00
0.24	0.82	0.62	0.53	0.43	0.36	0.32	0.28	0.24	0.20	0.16	0.11	0.07	0.00
0.28	0.84	0.61	0.51	0.39	0.33	0.28	0.24	0.21	0.18	0.14	0.09	0.07	0.00
0.30	0.85	0.61	0.50	0.38	0.32	0.27	0.23	0.20	0.17	0.13	0.09	0.06	0.00
0.34	0.87	0.60	0.49	0.36	0.30	0.25	0.22	0.19	0.16	0.12	0.08	0.06	0.00
0.38	0.89	0.60	0.48	0.35	0.29	0.24	0.21	0.18	0.15	0.12	0.08	0.06	0.00
0.40	0.90	0.59	0.47	0.34	0.28	0.24	0.21	0.18	0.15	0.12	0.08	0.06	0.00
0.44	0.92	0.59	0.47	0.33	0.28	0.23	0.20	0.17	0.15	0.12	0.08	0.06	0.00
0.48	0.94	0.59	0.47	0.33	0.28	0.23	0.20	0.17	0.15	0.12	0.08	0.06	0.00
0.50	0.95	0.59	0.47	0.34	0.28	0.24	0.21	0.17	0.15	0.12	0.08	0.06	0.00
0.54	0.97	0.59	0.47	0.34	0.29	0.25	0.21	0.18	0.15	0.12	0.08	0.06	0.00
0.58	0.99	0.59	0.47	0.35	0.30	0.25	0.22	0.18	0.16	0.13	0.09	0.06	0.00
0.60	1.00	0.59	0.47	0.36	0.30	0.26	0.22	0.19	0.16	0.13	0.09	0.06	0.00
0.64	1.02	0.59	0.48	0.37	0.31	0.27	0.23	0.20	0.17	0.13	0.09	0.06	0.00
0.68	1.04	0.60	0.49	0.38	0.32	0.28	0.24	0.21	0.17	0.14	0.10	0.07	0.00
0.70	1.05	0.60	0.49	0.39	0.33	0.29	0.25	0.21	0.18	0.14	0.10	0.07	0.00
0.74	1.07	0.61	0.50	0.40	0.34	0.30	0.26	0.23	0.19	0.15	0.10	0.07	0.00
0.78	1.09	0.63	0.52	0.42	0.36	0.32	0.28	0.24	0.20	0.16	0.11	0.08	0.00
0.80	1.10	0.63	0.52	0.42	0.36	0.32	0.28	0.25	0.21	0.17	0.11	0.08	0.00
0.84	1.12	0.65	0.54	0.44	0.38	0.34	0.30	0.26	0.22	0.18	0.12	0.09	0.00
0.88	1.14	0.67	0.56	0.45	0.40	0.36	0.32	0.28	0.24	0.20	0.13	0.10	0.00
0.90	1.15	0.68	0.57	0.46	0.40	0.36	0.32	0.29	0.25	0.20	0.14	0.10	0.00
0.94	1.17	0.69	0.58	0.48	0.42	0.38	0.34	0.31	0.26	0.22	0.16	0.11	0.00
0.98	1.19	0.71	0.60	0.50	0.44	0.40	0.36	0.33	0.28	0.24	0.17	0.12	0.00
1.00	1.20	0.72	0.61	0.51	0.45	0.41	0.37	0.33	0.29	0.24	0.18	0.13	0.00
1.10	1.25	0.77	0.65	0.55	0.50	0.45	0.42	0.38	0.34	0.29	0.21	0.16	0.00
1.20	1.30	0.81	0.70	0.60	0.54	0.50	0.46	0.42	0.38	0.33	0.25	0.19	0.00
1.30	1.35	0.86	0.75	0.65	0.59	0.55	0.51	0.47	0.43	0.38	0.30	0.23	0.00
1.40	1.40	0.91	0.80	0.70	0.64	0.60	0.56	0.52	0.48	0.43	0.35	0.27	0.00
1.60	1.50	1.01	0.90	0.80	0.74	0.70	0.66	0.62	0.58	0.53	0.45	0.37	0.05
1.80	1.60	1.11	1.00	0.90	0.84	0.80	0.76	0.72	0.68	0.63	0.55	0.47	0.10
2.00	1.70	1.21	1.09	0.99	0.94	0.90	0.86	0.82	0.78	0.73	0.65	0.57	0.16
2.20	1.80	1.31	1.19	1.09	1.04	1.00	0.96	0.92	0.88	0.83	0.75	0.67	0.23
2.40	1.90	1.41	1.29	1.19	1.14	1.10	1.06	1.02	0.98	0.93	0.85	0.77	0.31
2.60	2.00	1.51	1.39	1.29	1.24	1.19	1.16	1.12	1.08	1.03	0.95	0.87	0.41

s/S in Table 8.1. Of course, S does not appear immediately at the start of excavation, so a correction for face advance may be expected.

NCB subsidence profile

The relative subsidence s/S along a rib-side profile occurs in relation to the center of the face, that is, as a function of the ratio x/H where x is measured from the origin on the surface and H is seam depth. Once the relative subsidence s/S and distance x/H are known, the actual subsidence and distance, s and x, are given by the products $S(s/S)$ and $H(x/H)$, respectively. The relationship between s/S and x/H depends parametrically on the panel width to depth ratio W/H. These data are presented in Table 8.1.

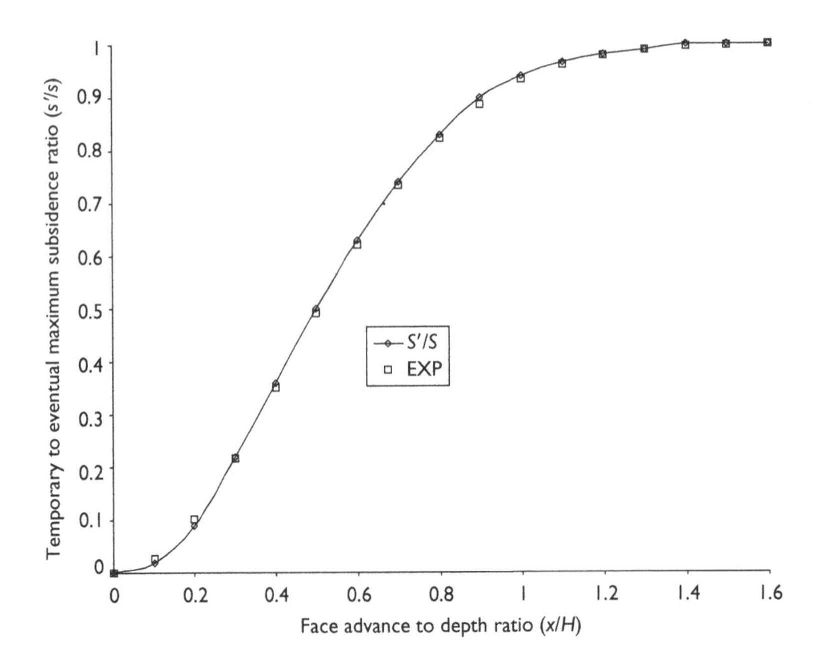

Figure 8.16 Ratio of temporary subsidence maximum to eventual subsidence maximum (S′/S) as a function of face advance to depth ratio (x/H).

Although panel dimensions may eventually lead to a maximum of subsidence S, which may be less than the maximum possible S_{max}, until face advance reaches the critical width W_c, a temporary maximum S' occurs that is less than S. A correction to S for face advance is therefore needed to plot a subsidence profile at limited face advance. If this temporary subsidence maximum is S' , then $S' = S(S'/S)$ where the ratio S'/S is obtained from Figure 8.16.

For example, if the ratio of face advance to seam depth is 0.6, then $S' = 3(0.622)$ where seam depth is 200 m and S is estimated to be 3 m. Figure 8.17 shows how a face-side profile develops with face advance and how the maximum subsidence increases to the value allowed by panel width and face advance beyond the critical distance (width).

Steps to constructing a subsidence profile are:

1. Determine the subsidence factor S/h from Figure 8.15 or use the equation

$$\frac{S}{h} = 0.9\left\{1 - \exp[-2.47(W/H)^2]\right\}$$

which is the fitted curve shown in Figure 8.15 and where S = maximum subsidence with W/H = excavation width to depth ratio and h = excavation height.

2. Correct S for face advance, if desired, from Figure 8.16 or use the equation

$$\frac{S'}{S} = \left\{1 - \exp[-2.70(x/H)^2]\right\}$$

which is the fitted curve shown in Figure 8.16 and where S' is the maximum subsidence at face advance x and depth H.

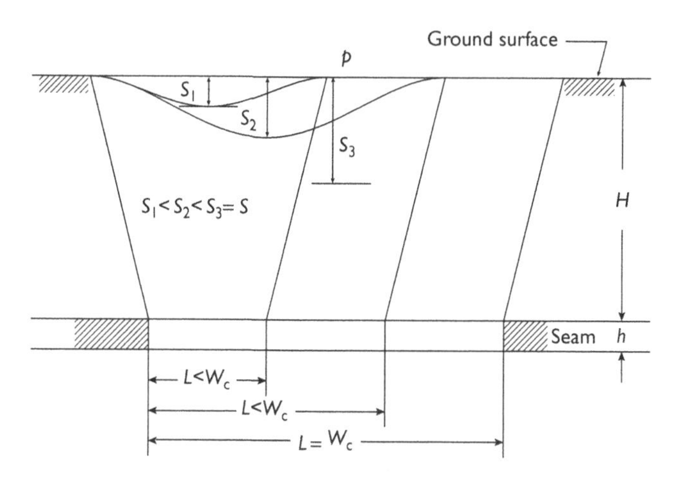

Figure 8.17 Face-side subsidence profile and increasing subsidence maximum with face advance to critical width.

3 Determine s/S at selected points x/H along the profile from Table 8.1.
4 Calculate actual values of s and x, plot, and draw a smooth curve through the plotted points. An exaggerated vertical scale for subsidence s is needed relative to distance scale for x to present the data in useful form.

Example 8.19 An underground coal mine uses the longwall mining method in a seam 3 m thick at a depth of 300 m. Panels are planned to be 240 m wide (face length) and 2,000 m long. Determine the maximum subsidence that will occur when the panel is completely mined and the maximum subsidence that is expected after the face has advanced 100 m. Assume UK conditions.

***Solution*:** Under UK conditions, the data in Figure 8.15 may be used to estimate the maximum subsidence that occurs with complete excavation of a panel. The width to depth ratio is $W/H = 240/300 = 0.8$. From the equation associated with Figure 8.15,

$$\frac{S}{h} = 0.9\left\{1 - \exp[-2.47(0.8)^2]\right\} = 0.715$$

Hence, the maximum subsidence after complete extraction is $S = 0.715(3) = 2.14$ m.

When the face has advanced 100 m, a correction factor to S may be obtained from Figure 8.16 or the associated equation. Thus,

$$\frac{S\prime}{S} = \left\{1 - \exp[-2.7(100/300)^2]\right\} = 0.259$$

Hence, at 100 m of face advance, the maximum subsidence is $S' = (2.14)/(0.259) = 0.56$ m.

Example 8.20 Consider the data given in Example 8.19 and assume UK conditions. Estimate the width of the subsidence trough as the panel nears completion.

***Solution*:** Width of the subsidence trough is the horizontal distance measured at the surface between points of negligible surface movement. The sketch shows the situation.

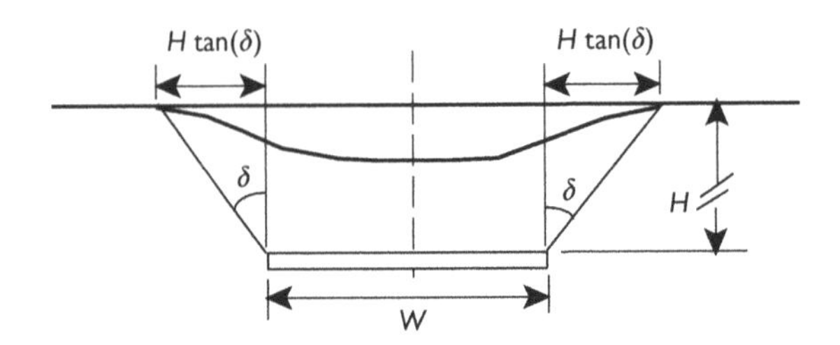

The trough width $w = W + 2H\tan(\delta)$, that is, $w = (240) + (2)(300)\tan(35) = 660$ m. Note that under UK conditions, an angle of draw of 35° is used.

Example 8.21 Consider the data given in Example 8.19 and suppose that three panels are mined side by side. Estimate the maximum subsidence expected upon completion of the third panel What is the maximum subsidence upon completion of the second panel?

***Solution*:** After mining three adjacent panels, the excavated area is (3)(240) × 2,000 in plan view. The width of the excavation is 720 m, so the width to depth ratio is 720/300 = 2.40. This number is off the graph in Figure 8.15, and indicates that the ratio of maximum subsidence to excavation height $S/h = 0.9$. Hence, $S = (0.9)(3) = 2.7$ m. This is the physically maximum possible subsidence.

Completion of the second panel leads to an excavated width of 480 m and width to depth ratio of 1.6. According to the data in Figure 8.15, the ratio $S/h = 0.9$, and so the maximum subsidence is 2.7 m. The same result is obtained using the associated equation. Evidently, excavation of the third panel does not cause additional subsidence.

Example 8.22 Develop a rib side subsidence profile for conditions described in Example 8.19 (3 m high excavation, depth of 300 m, panels 240 m wide (face length) and 2,000 m long, UK conditions).

***Solution*:** The width to depth ratio is 240/300 = 0.8. At this ratio, the subsidence points

$$s/S = 0.00\ 0.05\ 0.10\ 0.20\ 0.30\ 0.40\ 0.50\ 0.60\ 0.70\ 0.80\ 0.90\ 0.95\ 1.00$$

are located at x/H ratios from the trough center

$$0.80 \quad 1.10\ 0.63\ 0.52\ 0.42\ 0.36\ 0.32\ 0.28\ 0.25\ 0.21\ 0.17\ 0.11\ 0.08\ 0.00$$

These data transform using $S = 2.14$ m from Example 8.19 and $H = 300$ from the given data. Thus,

$$s(m) = 0.00\ \ 0.11\ \ 0.21\ \ 0.42\ \ 0.64\ \ 0.86\ \ 1.07\ \ 1.28\ \ 1.50\ \ 1.71\ \ 1.93\ \ 2.03\ \ 2.14$$
$$x(m) = 330\ \ 189\ \ 156\ \ 126\ \ 108\ \ 96\ \ 84\ \ 75\ \ 63\ \ 51\ \ 33\ \ 24\ \ 0$$

Aplot of these data may be obtained after importing to a spreadsheet and then exporting to this example. As seen in the plot, the trough is symmetric about the panel centerline because of the flat seam. The vertical displacement scale is greatly exaggerated relative to the horizontal distance scale in the plot. A dipping seam would require application of the rule of the normal.

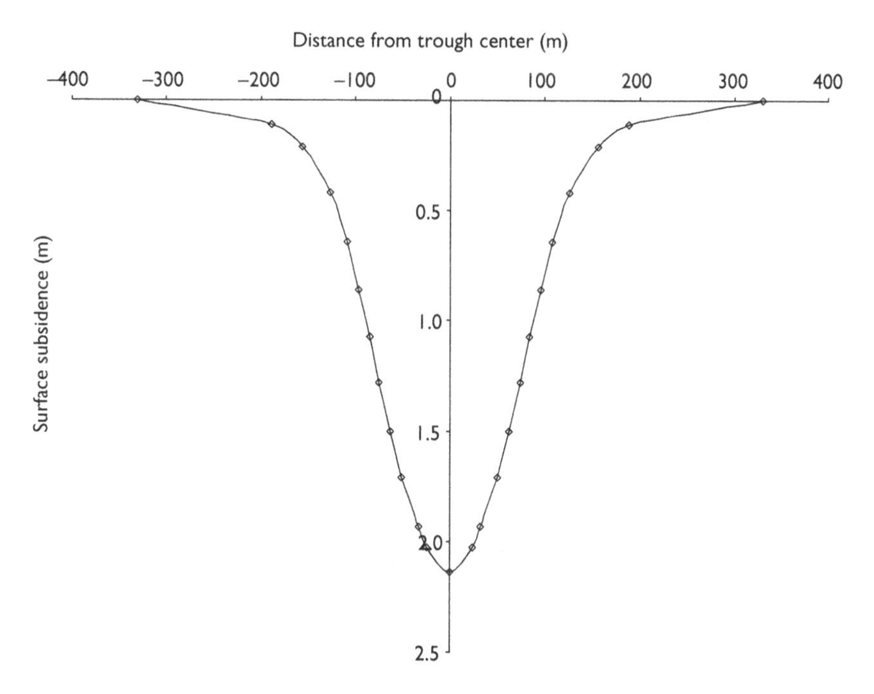

Example 8.23 Consider the data from Example 8.22 and further suppose the seam dips 12° and that depth H of 300 m is depth to the center of the panel. The long dimension of the panel is in the direction of the coal seam, while the panel width is measured parallel to the dip of the seam. Develop a rib side subsidence profile for these conditions (3 m high excavation, depth of 300 m, panels 240 m wide (face length), and 2,000 m long, UK conditions, 12° dip).

***Solution*:** According to the "rule of the normal," the subsidence trough is developed as for a flat seam at the average depth of the panel, but is shifted in the dip direction to a point where a normal line from the panel center intersects the surface. The situation is shown in the sketch.

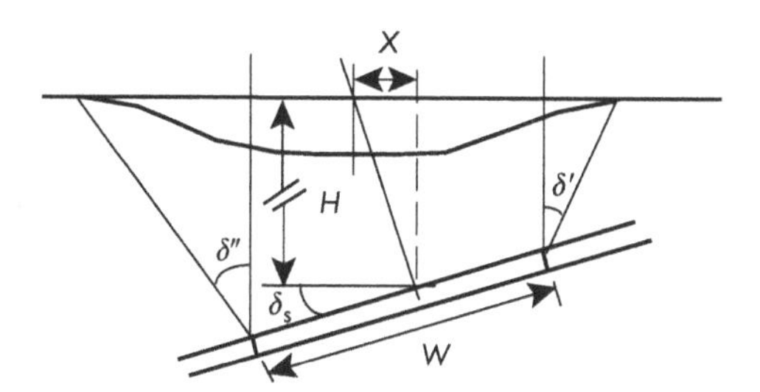

The shift towards the down dip direction is indicated by X in the sketch and may be simply calculated as $H\tan(\delta_s) = (240)\tan(12) = 51$ m where δ_s is seam dip.

Shift of the trough position induces angles of draw that differ from the flat-seam value of 35°. The up-dip angle diminishes, while the down-dip angle increases.

Angle of draw and subsidence factor adjustments

One approach to adjusting SHB subsidence troughs that are based on a limiting angle of draw of 35° and a subsidence factor of 0.9 is to scale the results by a new angle of draw and subsidence factor. If x is the horizontal distance from a panel center to a point where the relative subsidence is s/S and x' is a scaled distance that locates a point on an adjusted trough with the same relative subsidence, then a scaling formula is

$$x\prime = x\left(\frac{W\prime}{W_o}\right) = x\left[\frac{W + 2H\tan(\delta\prime)}{W + 2H\tan(\delta)}\right] \tag{8.37}$$

where W' is the trough width for δ', the new angle of draw; W_o and δ are the old trough width, and angle of draw (35°). At x' the relative subsidence s'/S' is the same as is the actual relative subsidence s/S.

Figure 8.18 shows a SHB subsidence trough over a panel 500 m deep and 250 m wide in a seam 4 m thick. This trough is based on the 35° angle of draw implied in the SHB and has a maximum subsidence of 0.45 times mining height, that is, $S = 1.8$ m. Adjustments of this trough to 28° and 21° are also shown in Figure 8.18. Without further adjustment, these two new troughs have the same maximum subsidence, while the limit to subsidence is shifted toward the panel center ($x = 0$). The inward shift steepens the trough and thus increases slope.

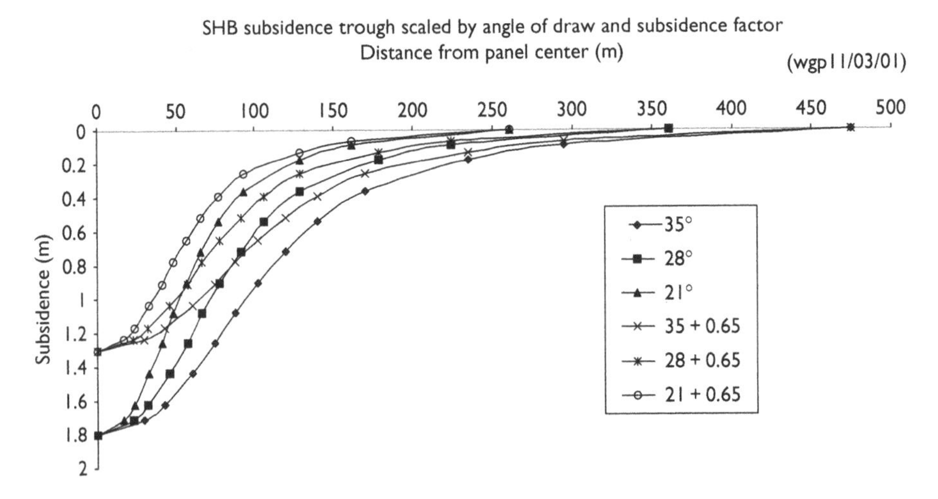

Figure 8.18 An example SHB subsidence trough adjusted for angle of draw and subsidence factor.

Adjustment of a subsidence trough for a limiting subsidence factor that differs from the 0.9 value used in the SHB is obtained by

$$s\prime = s\left(\frac{S_f\prime}{S_f}\right) \tag{8.38}$$

where s' is relative to the new limiting subsidence factor $S_f{}'$, s is the subsidence from the SHB and S_f is the limiting subsidence factor S_{max}/h from the SHB that has the value 0.9. Figure 8.18 shows three troughs at angles of draw of 35°, 28°, and 21° after adjustment to a limiting subsidence factor of 0.65. The maximum subsidence S' for the example data used in Figure 8.18, after adjustment to a subsidence factor of 0.65, is 1.3 m [1.3 = (1.8) (0.65/0.9)].

This reduction in subsidence reduces the slope of the trough and tends to counteract steeping that occurs in response to a lesser angle of draw.

Example 8.24 A mine in a flat seam at a depth of 520 m excavates panels with a face width of 260 m and a length of 2,500 m. Mining is full seam height at 4.2 m. The angle of draw and subsidence factor are estimated to be 27° and 0.67, respectively. Determine the maximum subsidence and trough width for a single panel, then sketch the subsidence profile and locate the one-half subsidence points from panel center.

***Solution*:** The width to depth ratio $W/H = 260/520 = 0.5$, so from the equation associated with Figure 8.15,

$$\frac{S}{h} = 0.9\left\{1.0 - \exp\left[-\frac{1}{2}\left(\frac{0.50}{0.45}\right)^2\right]\right\} = 0.415$$

$$\therefore$$

$$S = (4.2)(0.415) = 1.74\text{m}$$

that would be the maximum subsidence over a single panel excavated under UK conditions. Under the given conditions

$$S\prime = S\left(\frac{S_f\prime}{S_f}\right) = (1.74)\left(\frac{0.67}{0.90}\right)$$

$$\therefore$$

$$S\prime = 1.30\text{m}$$

From Table 8.1

s/S	=	0	0.05	0.1	0.2	0.3	0.4	0.5	0.6	0.7	0.8	0.9	0.95	1
x/H	=	0.95	0.59	0.47	0.34	0.28	0.24	0.21	0.17	0.15	0.12	0.08	0.06	0

Thus,

S(m)	0	0.065	0.13	0.26	0.39	0.52	0.65	0.78	0.91	1.04	1.17	1.235	1.3
x(m)	394.9	245.2	195.4	141.3	116.4	99.8	87.3	70.7	62.3	49.9	33.3	24.9	0.0

after adjustments for the angle of draw and subsidence factor, Equations (8.37) and (8.38), are made.

The half-subsidence point occurs where $s/S = 0.5$, that is, where $S = 0.65$ m and $x = 87.3$ m from the panel center.

Trough width is

$$W\prime = W + 2H\tan(\delta)$$

$$= 260 + (2)(520)\tan(27)$$

$$W\prime = 790\text{m}$$

which agrees with the data in the table where zero subsidence occurs at 395 m from the trough center, so the trough width is 790 m.

A plot of the data for the subsidence profile is shown in the sketch.

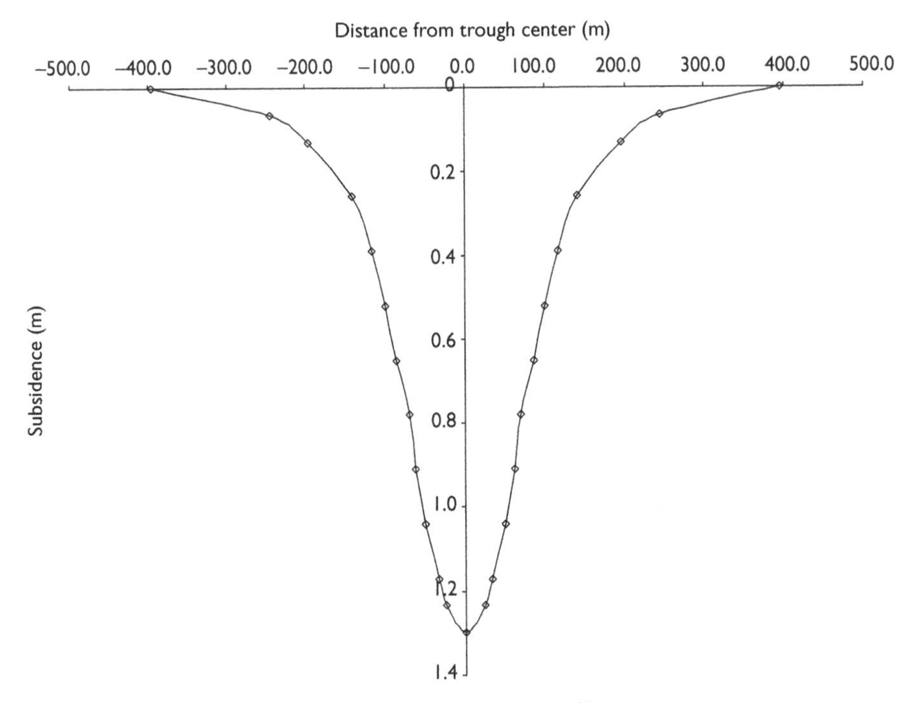

Sketch of the subsidence profile

Example 8.25 Consider the data in Example 8.24 and suppose the seam dips 8° and panels are developed on strike so the face runs up and down dip. Show the location of the trough in relation to the panel using the rule of the normal. Assume the panel is at a depth of 520 m.

***Solution*:** According to the rule of the normal, the trough is shifted towards the down dip side of the panel by a horizontal amount of H tan(δ_s) or (520) tan(8) = 73.1 m. The shifted trough relative to the dipping panel is shown in the sketch.

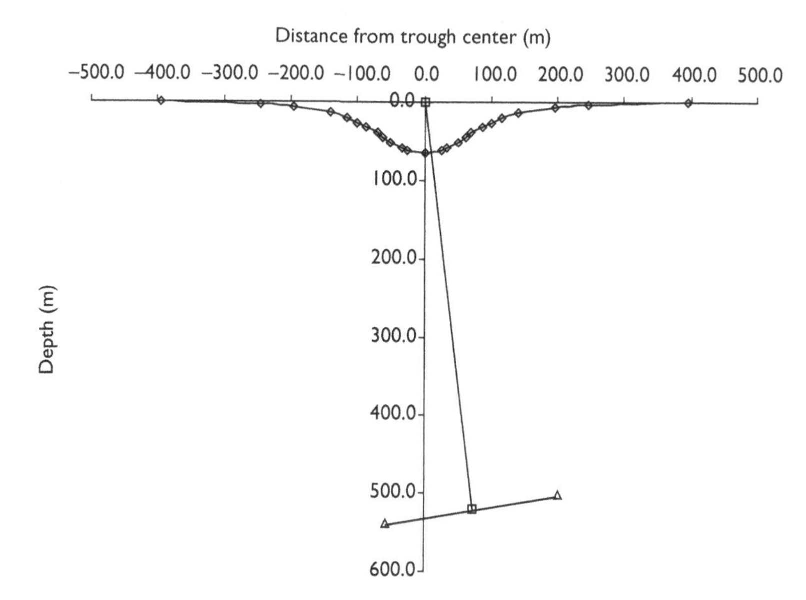

Sketch showing trough location relative to the mined panel. Subsidence scale is exaggerated by a factor of 50 relative to the distance scales.Vertical and horizontal distance scales are equal. Seam dip is 8?

Example 8.26 Consider the data in Example 8.24 and the flat seam case. Sketch the subsidence profile afer three panels are excavated side by side. Is excavation width now greater or less than critical width?

***Solution*:** The width to depth ratio of the three-panel excavation is (3)(260)/520 = 1.5. From the equation associated with Figure 8.15, S/h = 0.897, so the indicated maximum subsidence is (4.2)(0.897) = 3.77 m. However, the given subsidence factor is 0.67, so the actual maximum is (0.67/0.90)(3.77) = 2.80 m.

From Table 8.1, the relative trough data are

$s/S =$	0	0.05	0.1	0.2	0.3	0.4	0.5	0.6	0.7	0.8	0.9	0.95	1
$x/H = 1.4$	1.4	0.91	0.8	0.7	0.64	0.6	0.56	0.52	0.48	0.43	0.35	0.27	0
$x/H = 1.6$	1.5	1.01	0.9	0.8	0.74	0.7	0.66	0.62	0.58	0.53	0.45	0.37	0.05
$x/H = 1.5$	1.45	0.96	0.85	0.75	0.69	0.65	0.61	0.57	0.53	0.48	0.4	0.32	0.025

where a linear interpolation is used to obtain the third row.

The actual data after adjustment for angle of draw and subsidence factor, Equations (8.37) and (8.38) are

$s =$	0	0.14	0.28	0.56	0.84	1.12	1.4	1.68	1.96	2.24	2.52	2.66	2.8	2.8
$x =$	654.9	433.6	383.9	338.7	311.6	293.6	275.5	257.4	239.4	216.8	180.7	144.5	11.3	0.0

where a point at the trough center is added.

Because maximum subsidence is reached away from the trough center (at $x = +11.3$ m), the trough develops a flat area near the center of the three-panel excavation. This observation indicates that the excavation is now supercritical. A computation of the critical width is $W_c = 2H\tan(\delta) = (2)(520)\tan(27) = 530$ m also indicates the excavation is supercritical.

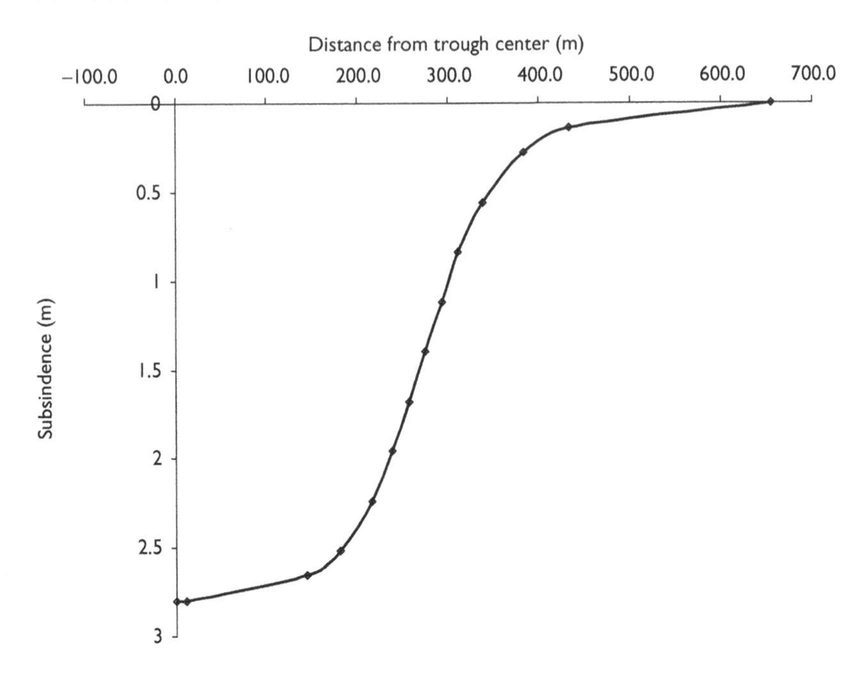

Sketch of one-half of the three-panel subsidence profile

NCB strain profile

Figure 8.19 illustrates a horizontal surface strain profile over an excavated panel. Ground outside the surface projection of a panel is stretched in tension that increases from zero at the lateral limits to subsidence to a tensile strain peak $+E$ as points toward the center of the subsidence profile are considered. Tensile strain decreases past the peak tension and becomes compressive with increase in distance from the lateral subsidence limit. A compressive strain peak $-E$ is eventually reached. This peak may occur at the center of the subsidence profile.

However, if the panel width is large relative to seam depth, then the compressive strain decreases past the peak and may reach a secondary peak at the trough center or, in case of wide troughs, decrease to zero. In case of very wide troughs, a segment of the subsidence profile near the trough center will be "flat." The corresponding rib-side strain profile segment will have a zero strain interval, as shown in Figure 8.19. A face-side strain profile may also be developed that is identical to the rib-side strain profile.

Table 8.2, adopted from the SHB, gives normalized values of strain at discrete distances from a panel center for various width to depth ratios. The normalized strain is e/E and is divided into tensile strain that is normalized by $+E$ and compressive strain that is normalized by $-E$. Distance from the panel center to a point of the table is normalized by panel depth.

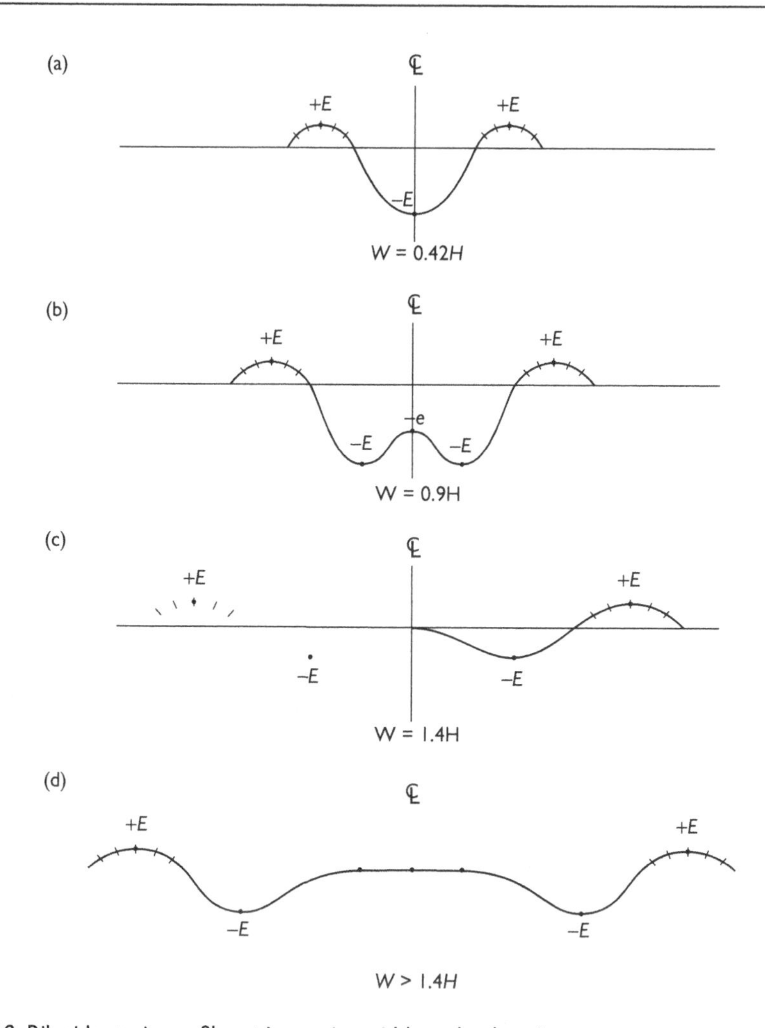

Figure 8.19 Rib-side strain profiles at increasing width to depth ratios.

Table construction is thus similar to that for the subsidence profile, although two normalizing factors ($+E$ and $-E$) are used instead of one (S).

The peak tensile and compressive strains are obtained from Table 8.3 that is also adopted from the SHB. Table 8.3 shows that the surface strains are proportional to the ratio of maximum subsidence to depth, that is, to S/H. The values in Table 8.3 are selected values from plots reproduced in Figure 8.20. An interesting feature of Figure 8.20 is the increase of compressive strain with depth at fixed panel width.

If the width to depth ratio is 0.5, then the peak tensile and compressive strains are $0.8(S/H)$ and $1.35(S/H)$, respectively. If further, $S = 12$ ft and $H = 1,200$ ft, then $E_t = (E+)(S/H) = (0.80)(0.010) = 0.008(0.8\%)$ and $E_c = (E-)(S/H) = (1.35)(0.010) = 0.0135(1.35\%)$. Strains of these magnitudes are significant in consideration of uniaxial strain to failure, say, of 0.1 to 1% for many geologic and manufactured materials.

Table 8.2 Strain profile points at various width to depth ratios (from SHB, 1975)

e/E	*0.00*	0.20	*0.40*	*0.60*	*0.80*	*1.00*	*0.80*	*0.00*	*0.20*	*0.40*	*0.60*	*0.80*	*1.00*	*0.80*	*0.60*	*0.40*	0.20	0.00
W/H									*X/H*									
0.20	0.80	0.74	0.69	0.66	0.61	0.49	0.42	0.32	0.27	0.23	0.18	0.12	0.00	0.00	0	0	0	0
0.24	0.82	0.70	0.63	0.60	0.54	0.44	0.37	0.28	0.23	0.20	0.15	0.10	0.00	0.00	0	0	0	0
0.28	0.84	0.66	0.58	0.54	0.49	0.39	0.33	0.24	0.21	0.17	0.13	0.08	0.00	0.00	0	0	0	0
0.32	0.86	0.63	0.55	0.49	0.45	0.35	0.30	0.22	0.19	0.16	0.12	0.07	0.00	0.00	0	0	0	0
0.36	0.88	0.62	0.53	0.46	0.42	0.33	0.29	0.21	0.18	0.15	0.11	0.07	0.00	0.00	0	0	0	0
0.40	0.90	0.61	0.52	0.45	0.40	0.32	0.28	0.21	0.18	0.15	0.11	0.07	0.00	0.00	0	0	0	0
0.44	0.92	0.60	0.51	0.43	0.39	0.31	0.27	0.20	0.18	0.15	0.11	0.07	0.01	0.00	0	0	0	0
0.48	0.94	0.60	0.51	0.43	0.38	0.31	0.27	0.20	0.18	0.15	0.12	0.08	0.01	0.00	0	0	0	0
0.52	0.96	0.60	0.51	0.43	0.38	0.32	0.27	0.21	0.18	0.16	0.12	0.09	0.02	0.00	0	0	0	0
0.56	0.98	0.61	0.51	0.44	0.39	0.33	0.28	0.22	0.19	0.17	0.13	0.10	0.03	0.00	0	0	0	0
0.60	1.00	0.62	0.52	0.45	0.40	0.34	0.29	0.22	0.20	0.18	0.14	0.11	0.05	0.00	0	0	0	0
0.64	1.02	0.63	0.53	0.46	0.41	0.35	0.31	0.23	0.21	0.19	0.15	0.12	0.06	0.00	0	0	0	0
0.68	1.04	0.64	0.54	0.47	0.43	0.37	0.32	0.24	0.22	0.20	0.16	0.13	0.07	0.00	0	0	0	0
0.72	1.06	0.66	0.55	0.49	0.44	0.38	0.34	0.26	0.24	0.21	0.17	0.15	0.09	0.00	0	0	0	0
0.76	1.08	0.67	0.57	0.51	0.46	0.40	0.36	0.27	0.25	0.22	0.19	0.16	0.10	0.01	0	0	0	0
0.80	1.10	0.69	0.58	0.53	0.48	0.42	0.37	0.29	0.26	0.24	0.20	0.17	0.11	0.02	0	0	0	0
0.84	1.12	0.71	0.60	0.54	0.49	0.44	0.39	0.30	0.28	0.25	0.22	0.19	0.13	0.04	0	0	0	0
0.88	1.14	0.73	0.62	0.56	0.51	0.46	0.41	0.32	0.29	0.27	0.24	0.21	0.15	0.05	0.01	0	0	0
0.92	1.16	0.75	0.64	0.58	0.53	0.47	0.43	0.34	0.31	0.29	0.25	0.22	0.16	0.07	0.02	0	0	0
0.96	1.18	0.77	0.66	0.60	0.55	0.49	0.45	0.35	0.33	0.30	0.27	0.24	0.18	0.09	0.04	0	0	0
1.00	1.20	0.79	0.68	0.62	0.57	0.51	0.47	0.37	0.35	0.32	0.29	0.26	0.20	0.10	0.05	0	0	0
1.20	1.30	0.88	0.77	0.71	0.66	0.61	0.56	0.46	0.44	0.41	0.38	0.35	0.29	0.20	0.13	0.07	0.02	0
1.40	1.40	0.98	0.87	0.81	0.75	0.70	0.66	0.56	0.53	0.51	0.48	0.45	0.39	0.30	0.23	0.17	0.1	0
1.60	1.50	1.08	0.97	0.91	0.85	0.80	0.76	0.66	0.63	0.61	0.58	0.55	0.49	0.40	0.33	0.27	0.2	0.03
1.80	1.60	1.17	1.07	1.01	0.95	0.90	0.86	0.76	0.73	0.71	0.68	0.65	0.59	0.50	0.43	0.37	0.3	0.1
2.00	1.70	1.28	1.17	1.11	1.05	1.00	0.96	0.86	0.84	0.81	0.78	0.75	0.69	0.60	0.53	0.47	0.4	0.2
2.20	1.80	1.38	1.27	1.21	1.15	1.10	1.06	0.96	0.94	0.91	0.88	0.85	0.79	0.70	0.63	0.57	0.5	0.3
2.40	1.90	1.48	1.37	1.31	1.25	1.20	1.16	1.06	1.04	1.01	0.98	0.95	0.89	0.80	0.73	0.67	0.6	0.4
2.60	2.00	1.58	1.47	1.41	1.36	1.30	1.26	1.16	1.14	LI I	1.08	1.05	0.99	0.90	0.83	0.77	0.7	0.5

Table 8.3 Peak strain multipliers[a]

W/H	*E+*	*E-*
0.10	0.25	2.40
0.20	0.52	2.25
0.30	0.70	2.00
0.40	0.77	1.73
0.50	0.80	1.35
0.60	0.75	1.06
0.70	0.68	0.86
0.80	0.65	0.70
0.90	0.65	0.60
1.00	0.65	0.55
1.10	0.65	0.54
1.20	0.65	0.53
1.30	0.65	0.52
1.40	0.65	0.51

Note
a Tension (+),compression (–).

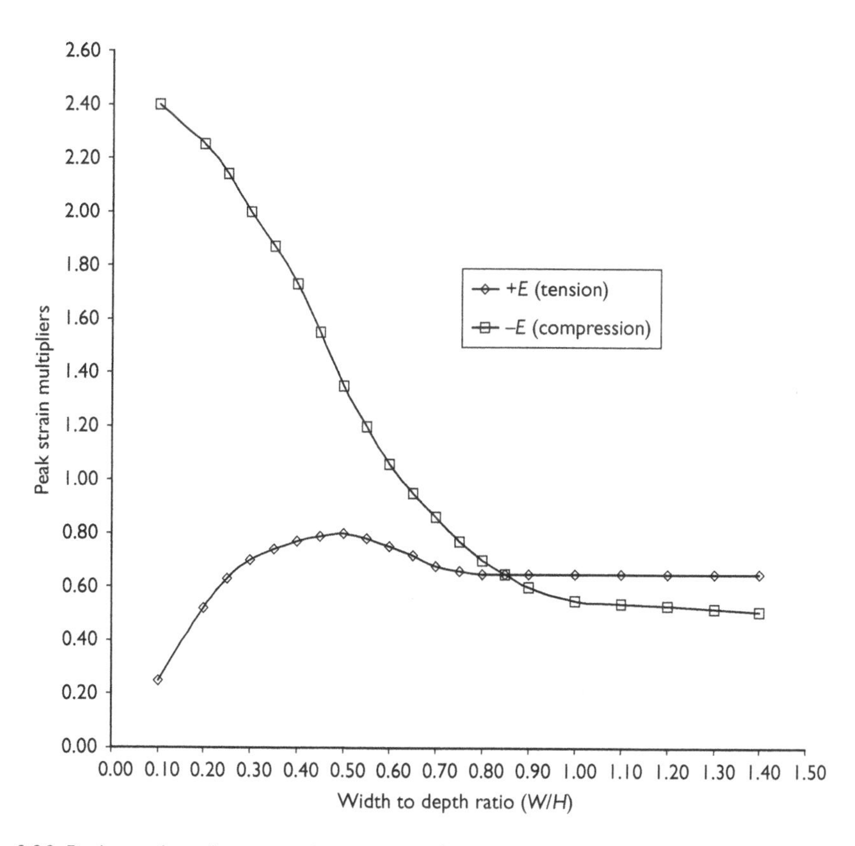

Figure 8.20 Peak tensile and compressive strains as functions of width to depth ratio.

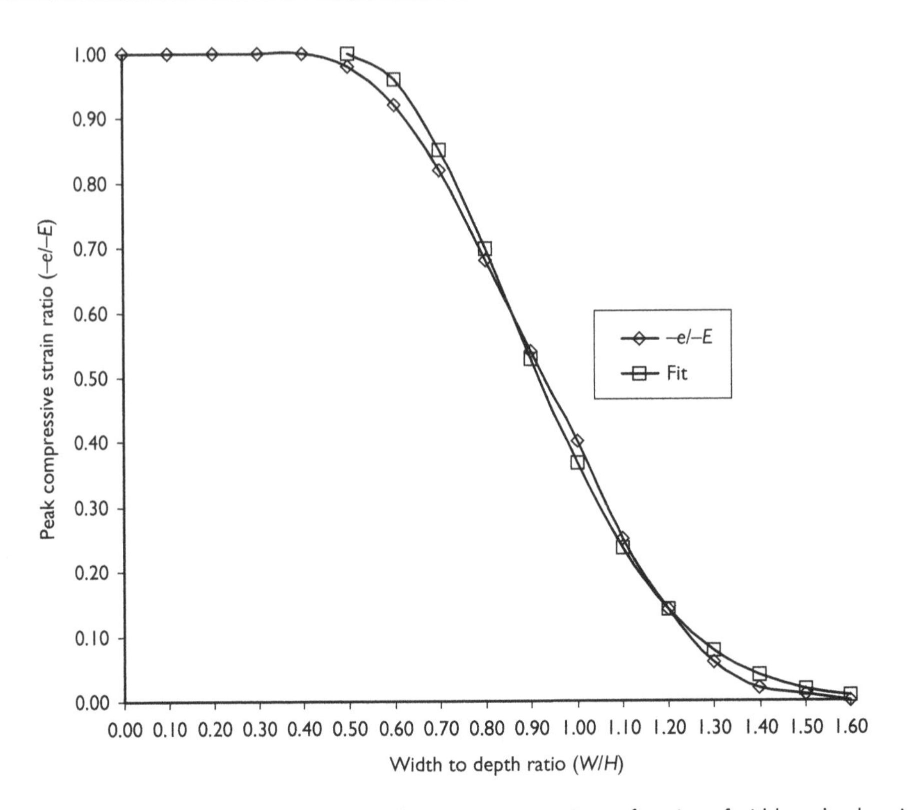

Figure 8.21 Ratio of secondary to primary peak compressive strain as a function of width to depth ratio.

The secondary compressive strain maximum that may occur at the center of the strain profile is related to the peak compressive strain. This relationship is given in Figure 8.21. There is fitted curve also shown in Figure 8.21 that has the form

$$\frac{-e}{-E} = \exp\left[-4\left(\frac{W}{H} - 0.5\right)^2\right]$$

where the minus sign indicates compression. The secondary compression peak $-e$ occurs at the trough center.

Steps to constructing a horizontal surface strain profile are:

1. Determine e/E along the strain profile from Table 8.2;
2. Determine $+E$ and $-E$ from Table 8.3 after finding S/H;
3. Find the secondary compressive strain maximum from Figure 8.21;
4. Calculate actual values of e and x and plot;
5. Draw a smooth curve through the plotted points.

An exaggerated vertical scale for strain e is needed relative to distance scale for x to present the data in useful form.

Example 8.27 A mine in a flat seam at a depth of 520 m excavates panels with a face width of 260 m and a length of 2,500 m. Mining is full seam height at 4.2 m. The angle of draw and subsidence factor are estimated to be 37° and 0.9, respectively. Determine peak tensile and compressive strains after completion of a single panel. Also determine the secondary compressive strain maximum that may occur at the trough center.

***Solution*:** The given angle of draw and subsidence factor correspond to UK conditions, so data from Tables 8.1, 8.2, and 8.3 can be used directly without adjustment. The width to depth ratio is 0.5, so from Example 8.24, maximum subsidence is 1.74 m; the subsidence depth ratio (S/H) is then $1.74/520 = 3.346(10^{-3})$. According to Table 8.3, $E+ = 0.80$ and $E- = 1.35$. Hence the peak tensile and compressive strains are

$$E_t = (E+)(S/H) = (0.80)(3.346)(10^{-3}) = 2.677(10^{-3})$$

$$E_c = (E-)(S/H) = (1.35)(3.346)(10^{-3}) = 4.517(10^{-3})$$

From Figure 8.21, there is indeed a secondary compressive strain peak e_c at the trough center. The magnitude of this peak may be obtained from

$$\frac{-e}{-E} = \exp\left[-4\left(\frac{W}{H} - 0.5\right)^2\right] = \exp[-4(0.5 - 0.5)^2] = 1.0$$

$$\therefore$$

$$e_c = 4.517(10^{-3})$$

which is compressive.

Example 8.28 Given the data in Example 8.27, determine the peak tensile and compressive strains after mining three panels side-by-side. Also determine the magnitude of a possible secondary compressive peak strain at the trough center.

***Solution*:** The given conditions correspond to UK conditions and allow for direct use of Table 8.1, 8.2, and 8.3 data. In this configuration, the total panel width to depth ratio is (3)(260)/520 = 1.5 and maximum subsidence, with aid of Figure 8.15, is (0.9)(4.2) = 3.78 m. The subsidence to depth ratio is $3.78/520 = 7.269(10^{-3})$ From Table 8.3, $E+$ and $E-$ are 0.65 and 0.51, respectively. Hence the peak tensile and compressive strains are

$$E_t = (E+)(S/H) = (0.65)(7.269)(10^{-3}) = 4.725(10^{-3})$$

$$E_c = (E-)(S/H) = (0.51)(7.269)(10^{-3}) = 3.707(10^{-3})$$

The panel width is supercritical so no secondary peak of compression is expected at the trough center, although Figure 8.21 suggests a small peak, about 0.02 times E_c.

Example 8.29 A mine in a flat seam at a depth of 520 m excavates panels with a face width of 260 m and a length of 2,500 m. Mining is full seam height at 4.2 m. The angle of draw and subsidence factor are estimated to be 27° and 0.67, respectively. Determine peak tensile and compressive strains after completion of a single panel. Also determine the secondary compressive strain maximum that may occur at the trough center.

***Solution*:** These are the conditions of Example 8.27, although with a different angle of draw and subsidence factor. The width to depth ratio is again 0.5 for a single panel and from Example 8.24, the maximum subsidence is 1.30 m. From Table 8.3, $E+ = 0.80$ and $E- = 1.35$. Peak tensile and compressive strains are then

$$E_t = (E+)(S/H) = (0.80)(1.30/520) = 2.000(10^{-3})$$

$$E_c = (E-)(S/H) = (1.35)(1.30/520) = 3.375(10^{-3})$$

These strains are somewhat less than those in Example 8.27 for UK conditions. The main reason is the lesser subsidence. Indeed, if no subsidence occurred, then no strains would be induced.

From Figure 8.21, there is indeed a secondary compressive strain peak e_c at the trough center. The magnitude of this peak may be obtained from the ratio

$$\frac{-e}{-E} = \exp\left[-4\left(\frac{W}{H} - 0.5\right)^2\right] = \exp[-4(0.5 - 0.5)^2] = 1.0$$

$$\therefore$$

$$e_c = 3.375(10^{-3})$$

which is compressive.

Example 8.30 A mine in a flat seam at a depth of 520 m excavates panels with a face width of 260 m and a length of 2,500 m. Mining is full seam height at 4.2 m. Compare strain profiles when the angle of draw and subsidence factor are 37° and 0.9, respectively, with the case when they are 27° and 0.67. In particular, locate the peak tensile and compressive strains in relation to the trough center in both cases.

***Solution*:** These are the conditions in Examples 8.27 and 8.29 (UK and US conditions, respectively). In the first case (UK case), the relative strains from Table 8.2 are

						E_t						E_c						
e/E	0.00	0.20	0.40	0.60	0.80	1.00	0.80	0.00	0.20	0.40	0.60	0.80	1.00	0.80	0.60	0.40	0.20	0.00
$W/H = 0.48$	0.94	0.60	0.51	0.43	0.38	0.31	0.27	0.20	0.18	0.15	0.12	0.08	0.01	0.00	0.00	0.00	0.00	0.00
$W/H = 0.52$	0.96	0.60	0.51	0.43	0.38	0.32	0.27	0.21	0.18	0.16	0.12	0.09	0.02	0.00	0.00	0.00	0.00	0.00
$W/H = 0.50$	0.95	0.60	0.51	0.43	0.38	0.32	0.27	0.21	0.18	0.16	0.12	0.09	0.02	0.00	0.00	0.00	0.00	0.00

where the last row is obtained by linear interpolation. Inspection of the compressive strain data shows the peak compression is reached very close to the trough center ($x/H = 0.02$). Distance entries after that point are zero indicating a sudden decrease of strain from near peak to zero at the trough center. However, this interpretation of the table data is incorrect. In fact, the strain at the trough center is only slightly less than the peak compression. Indeed, from Figure 8.21, one sees that a secondary peak of compression occurs at the trough center that is only slightly less than the peak strain.

The actual strain and distance data are

$e =$	0.000	0.533	1.067	1.600	2.134	2.667	2.134	0.000	−0.903	−1.807	−2.710	−3.614	−4.517	−4.000
$X =$	494	312	265.2	223.6	197.6	163.8	140.4	106.6	93.6	80.6	62.4	44.2	7.8	0

where the strains should be divided by 1,000 and the distance is in meters. Again, tension is positive and compression is negative.

In the US case, the data are

$e =$	0	0.4	0.8	1.2	1.6	2	1.6	0	−0.675	−1.35	−2.025	−2.7	−3.375	−3
$x =$	397.8	251.3	213.6	180.1	159.1	131.9	113.1	85.8	75.4	64.9	50.3	35.6	6.3	0.0

where the strains should be divided by 1,000 and the distance is in meters.

The peak strain data and locations are summarized in the table.

Table for Example 8.30.

Case	Data			
	E_t	E_c	X_t (m)	X_c(m)
UK	2.667 (10^{-3})	4.517(10^{-3})	164	7.8
US	2.000 (10^{-3})	3.375 (10^{-3})	132	6.3

Plots of data from both cases are shown in the sketch.

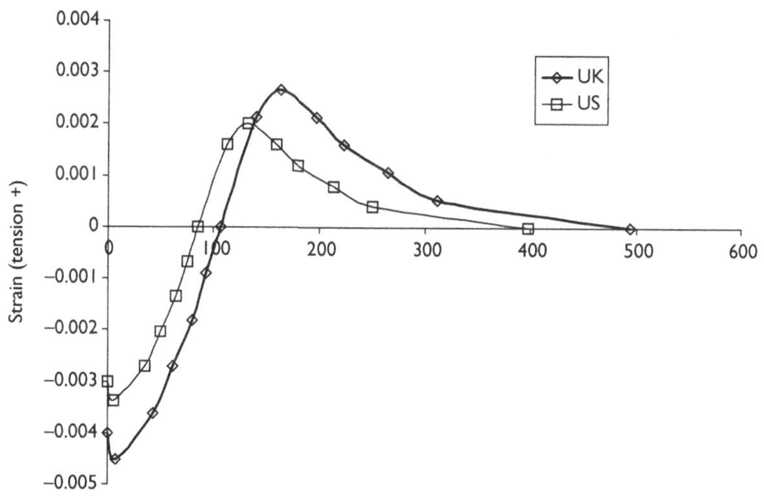

Plot of strain data for Example 8.30 (tension is positive). Panel half width is 130 m.

Surface damage

Damage to surface structures from mine subsidence depends on the structure type and the motion transmitted to the structure. Structures include residential buildings, multi-story office buildings, warehouses, factories, roads and bridges, electrical transmission towers, smoke stacks, water towers, water and gas pipelines, storm and sanitary sewers, water retention dams, canals, head frames, mine tailings impoundments, and so forth. A related damage question concerns aquifers in the overburden and whether subsidence may cause fracturing allowing water into the mine and lowering the water table that may support water wells and springs. No one criterion suffices for the great variety of structures that may be at risk.

Residences constructed with brick work may accommodate elongation and compression more readily than continuous walls. Relative displacement between walls is the product of average horizontal strain and distance between walls. As a reminder, the primitive definition of normal strain is $\varepsilon = \Delta L/L_o$. A plot of horizontal surface strain ε (y-axis) as a function of structure length L_o (x-axis, distance between walls) therefore has the shape of a hyperbola for a fixed tolerance or damage criterion, ΔL. The hyperbola moves away from the origin with increasing severity of damage (ΔL). Figure 8.22(a) shows these relationships, while Figure 8.22(b) shows the same strain as a function of the reciprocal of structure length with damage again as a parameter. The degrees of damage severity are described in SHB. For example, "slight" includes change in structure length between 0.03 and 0.06 m and implies: "Several slight fractures inside the building. Doors and windows may stick slightly. Repairs to decoration probably necessary." A change between 0.06 and 0.12 m indicates slight to appreciable damage, while a change between 0.12 and 0.18 is indicative of appreciable to severe damage.

An important question related to surface damage is when to begin repairs, if needed. If mining reaches the reserve limit and comes to a permanent halt, then repairs could begin almost immediately. Otherwise, some assurance that further subsidence will not occur must be obtained before repairs should begin. In this regard, the limit to subsidence ahead of a face or beyond a panel edge is defined by the angle of draw. This same limit extends behind the face, as shown in Figure 8.23. Thus, when the face has traveled a horizontal surface distance of (H) tan(δ) past a structure of interest, subsidence is essentially complete and repairs may start.

Example 8.31 Structures that are shown in the sketch have rectangular foundations 50 ft × 125 ft. Panel width is 260 m; panel length is 2,500 m. The angle of draw and subsidence factor are 27° and 0.67, respectively. Estimate the damage potential for each site after completion of a single panel.

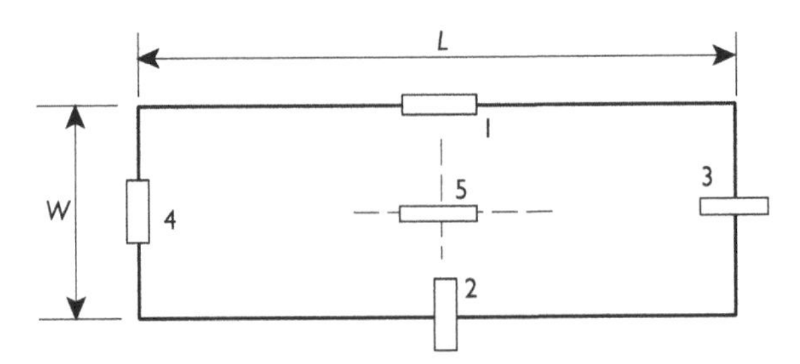

Solution: Sites 1, 2, 3, and 4 experience strain directly above the panel boundaries. Sites 1 and 2 experience strain from rib-side profiles, while sites 3 and 4 experience face-side profile strains. Site 5 over the panel center experiences the full history of the face-side profile as excavation proceeds, say, from left to right across the full width of the panel. If one assumes

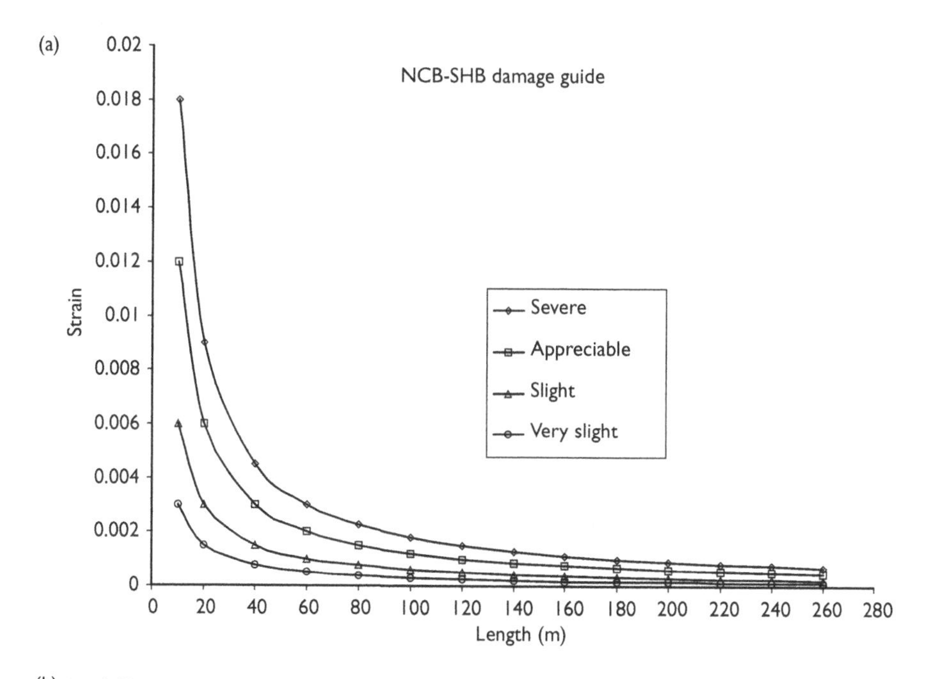

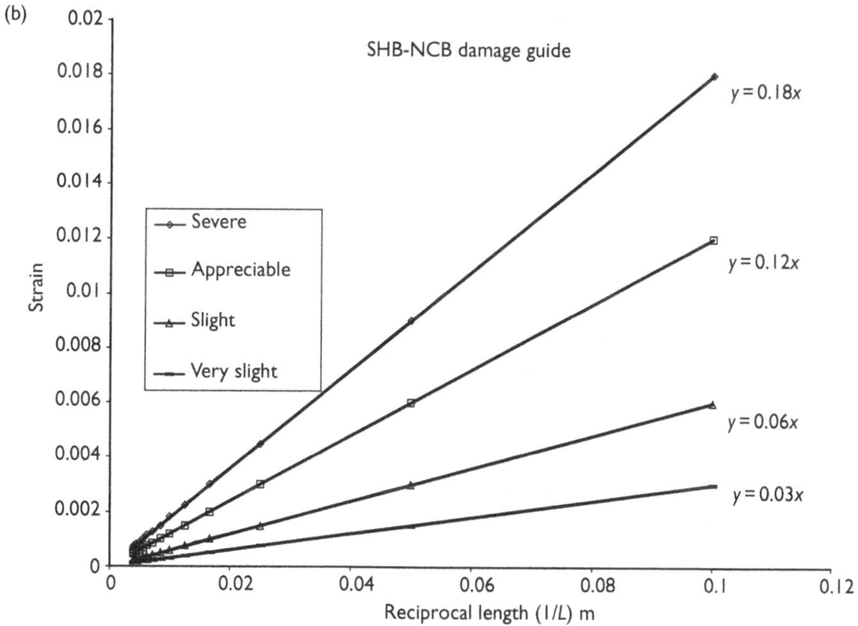

Figure 8.22 Damage severity as a function of surface strain and structure length (NCB): (a) strain versus length and (b) strain versus reciprocal length.

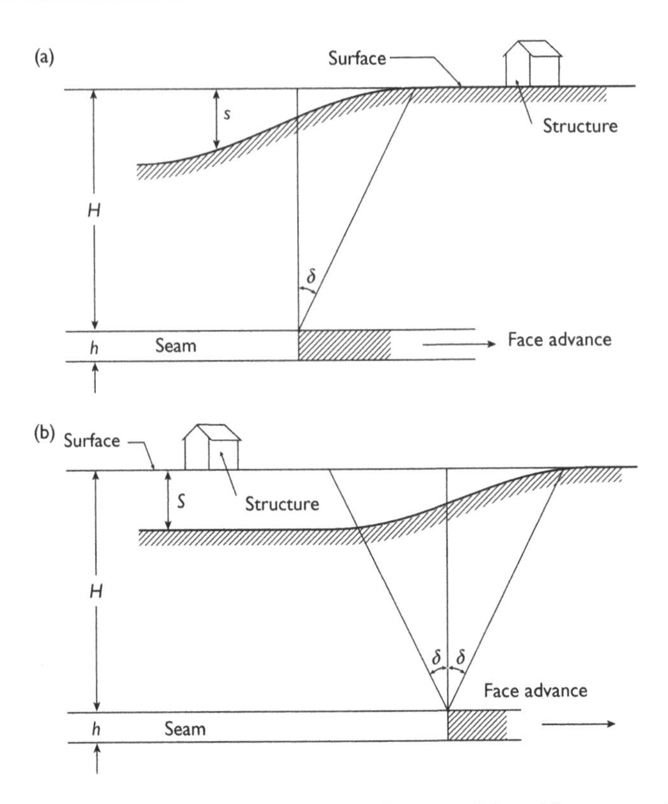

Figure 8.23 Zone of subsidence influence ahead and behind panel face: (a) structure beyond limit of subsidence from advancing face and (b) structure beyond influence of advancing face, repairs may begin.

the horizontal surface strains are transmitted with 100% efficiency to the structures, then changes in structure lengths are at Sites 1, 4: $\Delta L(\text{ft}) = \varepsilon(50)$, and at Sites 2, 3: $\Delta L(\text{ft}) = \varepsilon(125)$.

Strains directly over the panel boundaries occur at 130 m ($W/2$) from the center of the trough profile. The conditions are those of Example 8.30. Strains of 0.002 and 0.0016 occur at 110.8 m and 133.6 m from the trough center. Interpolation to 130 m gives a tensile strain of 0.0017. Changes in structure lengths are at Sites 1, 4: $\Delta L(\text{ft}) = (0.0017)(50) = 0.085$ ft (0.026 m), and at Sites 2, 3: $\Delta L(\text{ft}) = (0.0017)(125) = 0.2125$ ft (0.065 m). At site 5, the structure will experience peak compressive and tensile strain. Under peak tensile strain: $\Delta L(\text{ft}) = (0.002)(125) = 0.250$ft (0.076 m). These length changes are in the range of very slight to slight damage as seen in Figure 8.22. The worst threat is to the structure in the panel center. Length is 38 m (reciprocal = 0.026 1/m) and strain is 0.002. Potential damage to this structure is "slight."

In consideration of peak compressive strain, $\Delta L(\text{ft}) = (0.003375)(125) = 0.422$ft (0.129 m) which is in the range of appreciable to severe damage. If the foundation of a structure were a concrete slab, then the compressive strength of the foundation would not be adequate to prevent damage. The peak compressive strain is 0.003375 that when linked to an elastic modulus, say, of 4 million psi, generates a compressive stress of 13,500 psi (93 MPa), well above allowable stress for concrete. If the floor of a structure were a concrete slab, it would certainly crack in tension, too. Obviously, type of structure needs to be considered when details are needed.

Multipanel, multiseam subsidence

Panels are generally mined one after the other with only a row or so of small pillars left in between. These pillars generally are not mined, but are crushed and lost in the course of mining. Indeed, the high loads on tail gate entries in longwall panels often require additional support for successful ground control. The subsidence trough formed by adjacent panels may be estimated as a single panel of great width or from superposition of troughs from the several panels. With neglect of small pillars between panels, adjacent panels create an opening that is much wider than a single panel. Use of the combined panel dimensions allows for estimation of subsidence and strain profiles as for a single panel. An alternative *superposition* is shown in Figure 8.24(a).

In case of multi-seam mining, as shown in Figure 8.24(b), subsidence and strain profiles must be superposed. Although subsidence is certainly not an elastic phenomenon and therefore, strictly speaking, superposition does not apply, as a practical matter, combining effects of separate panels by adding the individual subsidence contributions may be satisfactory for estimation. The same practical considerations hold for strain profiles. Both are based on the empirical SHB approach. Of course, locally determined subsidence factors and angles of draw should be used, in any case.

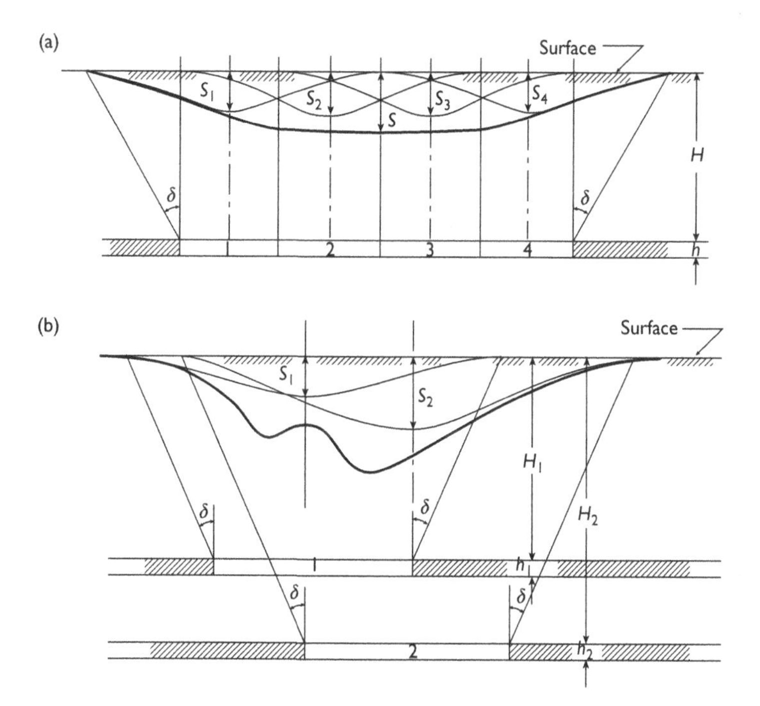

Figure 8.24 Superposition of subsidence profiles: (a) panels side by side with superposed subsidence profiles and (b) multi-seam superposition of subsidence profiles.

Example 8.32 Two seam longwall mining occurs in panels 280 m wide by 2,600 m long. The top seam is at a depth of 300 m and is 3.8 m thick; the bottom seam is at a depth of 320 m and is 4.5 m thick. The angle of draw and subsidence factors are 28° and 0.68. Estimate maximum subsidence after mining a top seam panel and then a bottom seam panel directly beneath the top seam panel. Determine width of the subsidence trough and sketch the subsidence profile.

Solution: From Figure 8.15 and the associated equation estimates of maximum subsidence under UK conditions are

$$\left(\frac{S}{h}\right) = 0.9\left\{1 - \exp\left[\frac{1}{2}\left(\frac{280/300}{0.45}\right)^2\right]\right\} = 0.795$$

$\therefore$

$$S = (3.8)(0.795) = 3.02\text{m (top)}$$

$$\left(\frac{S}{h}\right) = 0.9\left\{1 - \exp\left[\frac{1}{2}\left(\frac{280/320}{0.45}\right)^2\right]\right\} = 0.764$$

$\therefore$

$$S = (4.5)(0.764) = 3.44\text{m (bottom)}$$

Superposing the maximums is permissible because the panels are aligned horizontally. Thus, $S(\max) = 3.02 + 3.44 = 6.46$ m under UK conditions. Under the given conditions

$$S(\text{top}) = (3.02)\left(\frac{0.68}{0.9}\right) = 2.28 \text{ m}$$

$$S(\text{bot}) = (3.44)\left(\frac{0.68}{0.9}\right) = 2.60 \text{ m}$$

$$S(\max) = (6.46)\left(\frac{0.68}{0.9}\right) = 4.48 \text{ m}$$

The relative subsidence data for each panel is obtained from Table 8.1 and interpolation. Thus,

Table of relative subsidence and subsidence points for the top and bottom seam profiles

$s/S=$	0	0.05	0.1	0.2	0.3	0.4	0.5	0.6	0.7	0.8	0.9	0.95	1
$W/H=$													
0.84	1.12	0.65	0.54	0.44	0.38	0.34	0.3	0.26	0.22	0.18	0.12	0.09	0
0.88	1.14	0.67	0.56	0.45	0.4	0.36	0.32	0.28	0.24	0.2	0.13	0.1	0
0.875	1.14	0.67	0.56	0.45	0.40	0.36	0.32	0.28	0.24	0.20	0.13	0.10	0.00
0.9	1.15	0.68	0.57	0.46	0.4	0.36	0.32	0.29	0.25	0.2	0.14	0.1	0
0.94	1.17	0.69	0.58	0.48	0.42	0.38	0.34	0.31	0.26	0.22	0.16	0.11	0
0.933	1.17	0.69	0.58	0.48	0.42	0.38	0.34	0.31	0.26	0.22	0.16	0.11	0.00
s(top)	0.0	0.1	0.2	0.5	0.7	0.9	1.1	1.4	1.6	1.8	2.1	2.2	2.3
x(top)	286	169	142	117	102	92	83	75	63	53	38	27	0
s(bot)	0.0	0.1	0.3	0.5	0.8	1.0	1.3	1.6	1.8	2.1	2.3	2.5	2.6
x(bot)	297	174	145	117	104	93	83	73	62	52	34	26	0

To obtain the total subsidence trough interpolation and superposition are required. The data are

s' (top)	0.0	0.1	0.2	0.5	0.7	0.9	1.1	1.4	1.6	1.9	2.1	2.2	2.3
s(tot) =	0.0	0.2	0.5	1.0	1.4	1.9	2.4	3.0	3.4	3.9	4.5	4.6	4.9
x(tot) =	297	174	145	117	104	93	83	73	62	52	34	26	0

where the top row is the top seam subsidence at the x(tot) point. This amount is added to the bottom seam subsidence to obtain the total subsidence. Interpolation of top seam subsidence to the bottom seam points (x) is through use of the linear interpolation formula

$$s\prime(\text{top}) = s_1 + (s_2 - s_1)\left[\frac{x_1(\text{bot}) - x_1}{x_2 - x_1}\right]$$

where the subscripts refer to the top seam values except the point labeled bot. Point 1 is toward the trough center, while point 2 is farther away. A rough check on the calculation is seen at the trough center and end where no interpolation is needed for superposition. The plot shows the results.

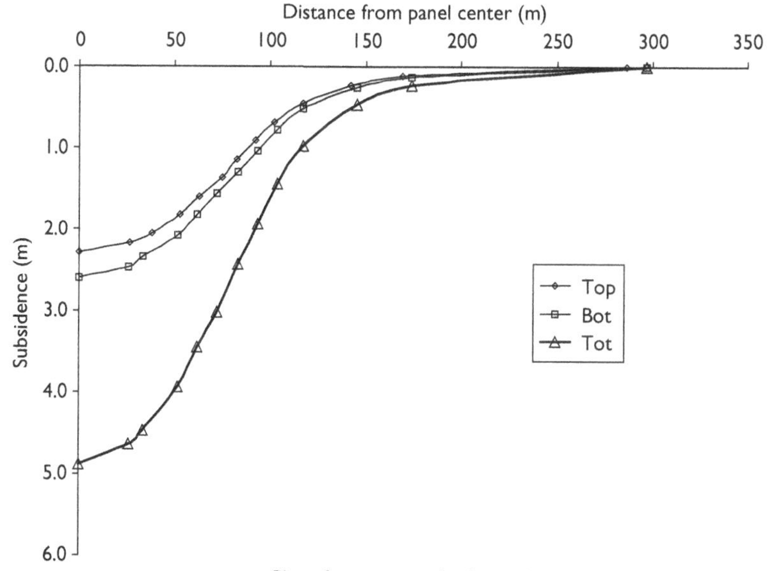

Plot of two seam subsidence data

From the tabulated data, width of the subsidence trough is (2) (297) = 594 m. From the formula $W' = W + 2H\tan(\delta) = 280 + (2)(320)\tan(28) = 620$. The difference of about 4% between the two estimates lies in the empirical nature of the SHB data, but is not a serious matter for subsidence estimation.

Example 8.33 Consider the data given in Example 8.32 and further suppose two top seam panels are mined side by side but only one bottom seam panel is mined directly below the first top seam panel. Estimate maximum subsidence after mining the three panels, then determine width of the subsidence trough and sketch the subsidence profile.

***Solution*:** From Figure 8.15 and the associated equation estimates of maximum subsidence under UK conditions are

$$\left(\frac{S}{h}\right) = 0.9\left\{1 - \exp\left[\frac{1}{2}\left(\frac{560/300}{0.45}\right)^2\right]\right\} = 0.900$$

$\therefore$

$$S = (3.8)(0.900) = 3.42\text{m (top)}$$

$$\left(\frac{S}{h}\right) = 0.9\left\{1 - \exp\left[\frac{1}{2}\left(\frac{280/320}{0.45}\right)^2\right]\right\} = 0.764$$

$\therefore$

$$S = (4.5)(0.764) = 3.44\text{m (bottom)}$$

Superposing the maximums is *not* permissible in this case because the excavation centers are not aligned horizontally. However, under the given conditions

$$S(\text{top}) = (3.42)\left(\frac{0.68}{0.9}\right) = 2.58 \text{ m}$$

$$S(\text{bot}) = (3.44)\left(\frac{0.68}{0.9}\right) = 2.60 \text{ m}$$

The relative subsidence data for each panel is obtained from Table 8.1 and interpolation. Thus, for the two-panel top seam excavation

$s/S=$	0	0.05	0.1	0.2	0.3	0.4	0.5	0.6	0.7	0.8	0.9	0.95	1
$W/H = 1.8$	1.6	1.11	1	0.9	0.84	0.8	0.76	0.72	0.68	0.63	0.55	0.47	0.1
$W/H = 2$	1.7	1.21	1.09	0.99	0.94	0.9	0.86	0.82	0.78	0.73	0.65	0.57	0.16
$W/H = 1.87$	1.63	1.14	1.03	0.93	0.87	0.83	0.79	0.75	0.71	0.66	0.58	0.50	0.12

so that

s(top)	0.00	0.13	0.26	0.52	0.77	1.03	1.29	1.55	1.81	2.06	2.32	2.45	2.58	2.58
x(top)	426	298	268	242	228	217	207	196	186	173	152	131	31	0

and for the bottom seam

s (bot) =	0.0	0.1	0.3	0.5	0.8	1.0	1.3	1.6	1.8	2.1	2.3	2.5	2.6
x(bot) =	297	174	145	117	104	93	83	73	62	52	34	26	0
x′(bot) =	157	34	5	−23	−36	−47	−57	−67	−78	−88	−106	−114	−140

where the distance measure is adjusted to give the bottom trough the same origin of coordinates as the top trough. For the composite subsidence trough interpolation is necessary when a bottom point falls between coordinates of a top point. For instance, the bottom point with $x' = 157$ falls between top points with $x = 173$ and $x = 152$. A plot of the troughs and the composite subsidence profile is shown in the sketch.

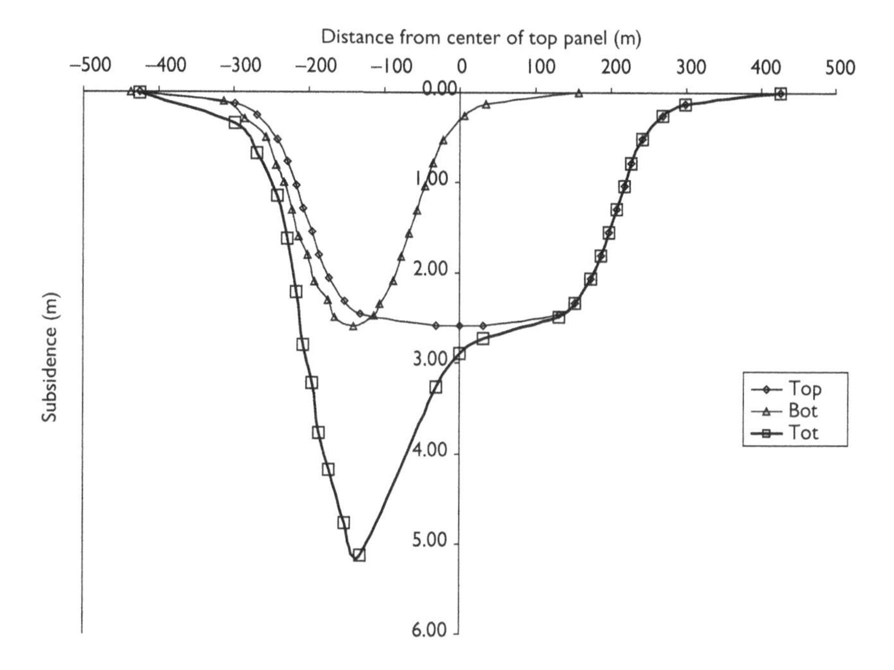

Plot of top and bottom troughs and composite

Example 8.34 Consider the data given in Examples 8.32 and 8.33 where two top seam panels are mined side by side but only one bottom seam panel is mined directly below the first top seam panel and individual panels are 280 m wide by 2,600 m long. The top seam is at a depth of 300 m and is 3.8 m thick; the bottom seam is at a depth of 320 m and is 4.5 m thick, and the angle of draw and subsidence factors are 28° and 0.68. Plot the strain profile and then determine the magnitude and location of peak tensile and compressive strains.

***Solution*:** The composite strain profile may be obtained by superposition of strain profiles induced by top and bottom seam mining. Thus, strain profiles for the top and bottom seam need to be obtained before superposing them to obtain the composite profile. From Examples 8.32 and 8.33 and Table 8.3 data, top seam subsidence is 2.58 m and bottom seam subsidence is 2.60. Hence,

Data	W/H	S/H	$E+$	$E-$	E_t	E_c	$-e/-E$
Top	1.876	8.600 (10^{-3})	0.65	0.51	5.59 (10^{-3})	4.386 (10^{-3})	0.0
Bottom	0.875	8.125 (10^{-3})	0.65	0.625	5.281 (10^{-3})	5.079 (10^{-3})	0.57

From Table 8.2 for the top seam

e(top)	0.00	1.12	2.24	3.35	4.47	5.59	4.47	0.00
x(top)	443	334	305	289	274	261	250	224

–0.88	–1.75	–2.63	–3.51	–4.39	–3.51	–2.63	–1.75	–0.88	0.00	0.00
219	211	203	195	180	156	138	123	104	52	0

where the strains are multiplied by 1,000; tension is in the upper part and compression is in the lower part of the tabulation.

For the bottom seam

e(bot)	0.00	1.06	2.11	3.17	4.22	5.28	4.22	0.00
x(bot)	289	185	157	142	129	116	104	81
–1.16	–2.32	–3.47	–4.63	–5.79	–4.63	–3.47	–3.30	
73	68	60	53	38	12	2	0	

that is also tabulated into tension and compression segments. In both tabulations, distance is from the center of the considered panel. Superposition requires a common coordinate system, of course.

For the composite

e(tot)	0.00	1.12	2.24	3.35	4.47	5.59	4.47	0.00
x(tot)	443	334	305	289	274	261	250	224
–0.88	–1.75	–2.63	–3.51	–4.39	–3.51	–2.52	–1.48	–0.42
219	211	203	195	180	156	138	123	104

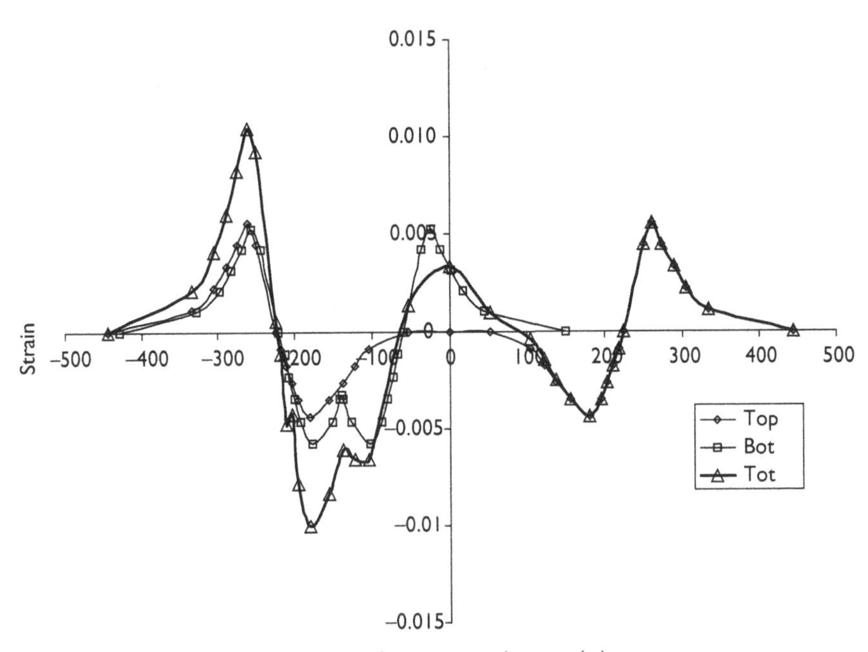

Distance from top panel center (m)
Plot of three strain profiles

0.98	3.32	1.33	–6.60	–6.59	–6.08	–8.32	–10.03	–7.81	–4.25	–4.79	–1.17
52	0	–52	–104	–123	–138	–156	–180	–195	–203	–211	–219
0.55	9.23	10.46	8.29	6.03	4.08	2.11	0.00				
–224	–250	–261	–274	–289	–305	–334	–443				

The three strain profiles are shown in the plot.

Superposition near the center ($x = 0$) is faulty because the sharp peak in tensile stress is obscured by the widely spaced interpolation points. This defect could be corrected by more closely spaced points, but in this example, the peak tension is elsewhere so the consequences are not serious.

Alternative approaches to subsidence

More accurate subsidence trough calculations may be obtained with the aid of computer programs designed for a particular coal field and calibrated against surface surveys over mines in the field. These programs make use of simple mathematical formulas that contain parameters determined from subsidence measurements. The formulas fall into two categories, profile functions and influence functions. They are many such formulas available and because they are essentially fitted to subsidence data, the choice between them is largely a matter of personal taste.

A profile function, as the name suggests, describes a surface subsidence trough in vertical section and has the simple form y = f(x) where y is surface subsidence at the point x. Alternatively, a "profile" is a curve formed by intersection of a vertical section with a subsidence trough and is clearly a two-dimensional object. Azimuth of the profile is the azimuth of the section. The sketch in Example 8.22 is a plot of a subsidence profile done from tables but could have easily be done using a profile function, if one were available. Subsidence profiles generally have an "S" shape when subsidence is continuous and not stepped and faulted. Accordingly, profile functions are constrained to produce an S-shape. There are several simple functions that do so: (1) $y = \tan^{-1}(x)$, (2) $y = \exp(-x^2)$, (3) $y = \tanh(x)$, (4)$y = (2/\sqrt{\pi})\int_0^x \phi(x)dx$ where $\phi(\mathrm{x}) = \exp(-\mathrm{x}^2)$. The last is the error function (erf) and is closely related to standard normal cumulative distribution function. The interest here is only in the geometric properties of the function, that is, the S-shaped curve that appears when plotted. The error function has the property: erf (−x) = −erf(x). The inverse tangent function also has this property. A complimentary error function is defined as: erfc = 1−erf and $(2/\sqrt{\pi})\int_x^\infty \phi(x)dx$, also useful for developing profile functions. Constants are introduced into profile functions to give a reasonable fit to subsidence measurements. Maximum subsidence, S_{max}, is one such constant. Another constant is subsidence observed at the inflection point of these S-curves. The inflection point is often placed over the edge (rib) of an excavated panel and assigned $S_{max}/2$. Subsidence trough width is yet another parameter that is often used in calibrating profile functions. Profile functions used in many European countries and examples of profile function use are reviewed in some detail by Brauner (1973) who also discusses influence functions and their relationship to profile functions.

Numerical models such as finite element or boundary element models that are based on first principles including physical laws, kinematics, and material laws offer an attractive alternative to empirical methods. The reason is they allow for inclusion of mine geology, mine geometry, topography, water effects, and other physical complications. Rock properties of the strata in the region of interest may be included, major faults can be taken into account and a more accurate description of changing mine geometry can be done. Aquifers and effects of water pressure and flow can also be computed. These features allow for more realistic simulation of the mining sequence and consequence for subsidence. Complex material behavior that includes inelasticity and time-dependency is possible. However, special expertise is required and the computational effort is relatively large and therefore more costly than simple SHB subsidence estimation.

Example 8.35 Show that the four example profile functions mentioned in the text can be used to represent subsidence trough profiles. The functions are:

$$y = \tan^{-1}(x), \tag{8.39}$$

$$y = \exp\left(-x^2\right), \tag{8.40}$$

$$y = \tanh(x), \tag{8.41}$$

$$y = \left(2/\sqrt{\pi}\right)\int_0^x \phi(x)dx \tag{8.42}$$

***Solution*:** First plot the functions as is and then adjust to subsidence trough conditions with appropriate constants. Primitive plots from spreadsheet computations are:

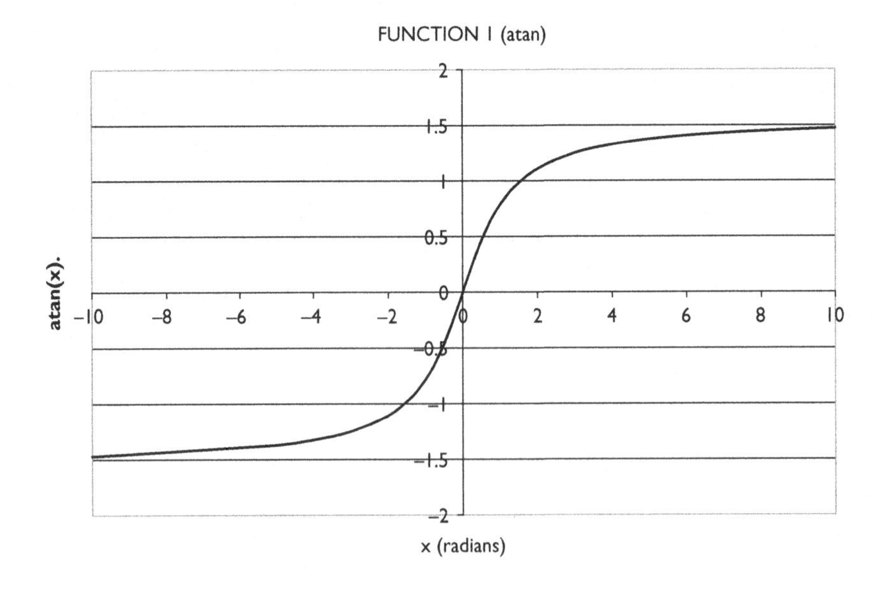

The inverse tangent function has potential as a profile function but needs to be extended to account for an entire trough.

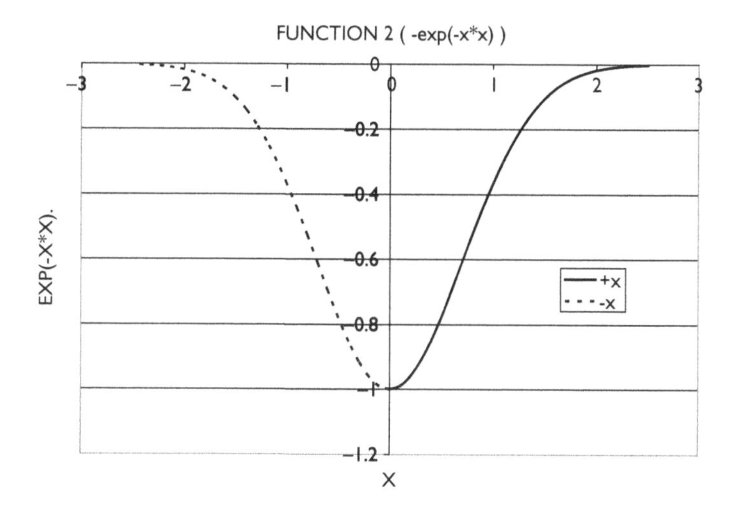

The exponential function plot is inverted by −1 multiplication and has two branches. An advantage of this function is the symmetry and the fact that the slope is zero at the origin.

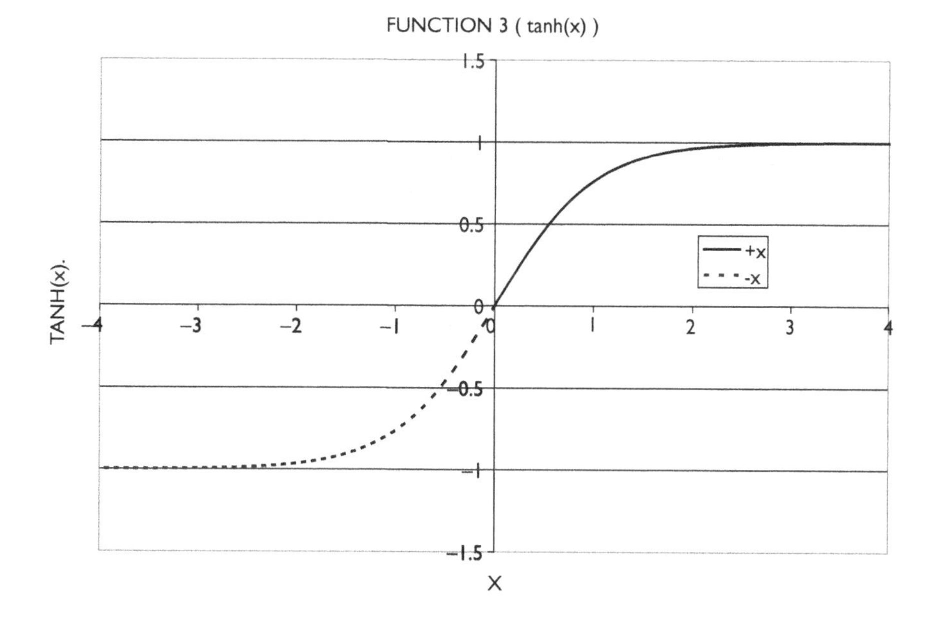

The hyperbolic tangent function also needs adjustment before use a profile function.

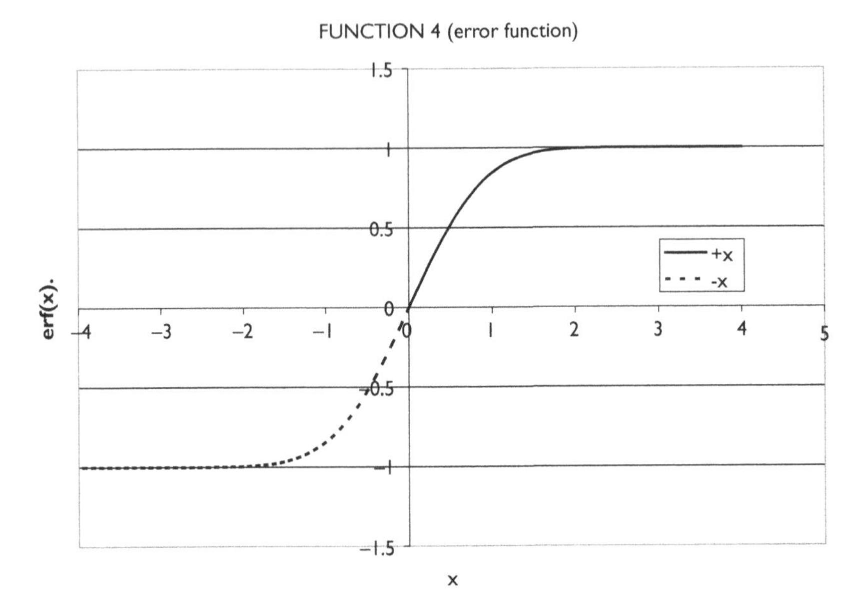

Again, adjustment is needed before using as a profile function. Functions 2, 3, and 4 have positive and negative branches as seen in the plots. Except for the exponential function, the origin is at an inflection point. Origin of the exponential function is above the peak value that occurs in the center of the plot.

Example 8.36 Consider the functions and plots in Example 8.35, then adjust Function (2), the exponential function, so the origin is above the center of a panel in vertical cross-section (and at ground surface before any subsidence has occurred). This procedure places the profile in relation to the excavation below. Also adjust the profile so the subsidence at the trough center is S, a maximum for a given width-depth ratio of a sub-critical width panel. Ensure subsidence is near zero at the edge of a trough estimated with a 35° angel of draw. Then evaluate and plot for a mining height of 10 ft, depth of 900 ft, panel width of 900 ft, and a subsidence factor of 0.9.

***Solution*:** From the text, the maximum subsidence is given by

$$(S/h) = (S_f)\left\{1 - \exp[-(1/2)\left(\frac{W/H}{S_f/2}\right)^2]\right\}$$

where h, S_f, W, and H are mining height, subsidence factor, panel width and panel depth, respectively. Hence, $s = (S)[\exp(-x^2)]$ where s is the surface subsidence at x units from the panel center measured along the original (and flat) ground surface. This equation is not dimensionally correct; the argument of the exponential function should be dimensionless and thus a "scaled" distance. Some additional adjustment is needed. Examination of the plot data shows that 2% of the maximum subsidence occurs at a dimensionless distance of 2. At the trough limits the distance from the trough center is $x' = (W/2) + H\tan(\delta) = W_s/2$ where the prime is added to indicate actual ft and not the dimensionless x in the exponential function argument. The notation W_s indicates trough width. The profile function now has the form

$$s = (S)\left\{\exp[-\left(\frac{x}{W_s/4}\right)^2]\right\}$$

that gives maximum subsidence (8.24 ft) at the trough center and near zero (2%) subsidence at the trough half width (1,080 ft) as seen in the plot. The panel is represented by the black rectangle at the bottom of the plot. The vertical scale is greatly exaggerated relative to the horizontal scale. Note that panel width is subcritical.

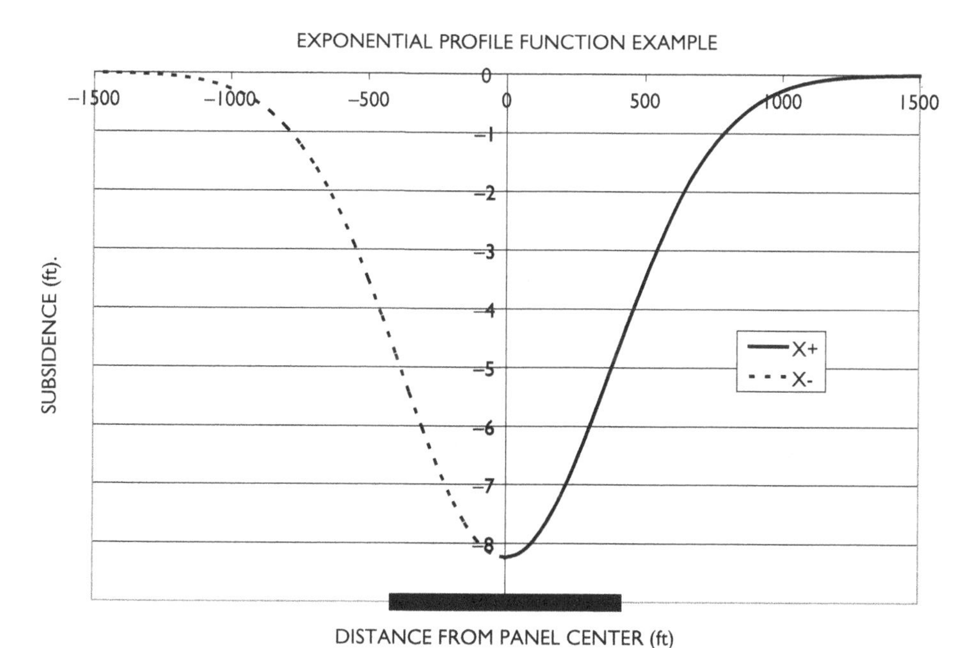

Plot for Example 8.36 using a calibrated exponential function.

Example 8.37 Given the data and results of Example 8.36, compare a subsidence profile using the function $(4) y = (2/\sqrt{\pi})\int_0^x \phi(x)dx$ with the exponential function.

***Solution*:** The process proceeds in much the same way as in Example 3.36. In this case, the scaled distance $x' = W_s/8$ and the arguments (limits) of the error function are (0, x"/x') where x" is the distance from the panel edge. The plotting distance x is the physical distance from the panel *center*. The panel edge is 450 ft from the panel center and the limit of subsidence is half trough width from the panel center (1,080 ft). The maximum subsidence occurs at the plot center and is 8.24 ft as in example 8.36. The left side of the plot is obtained by reflecting the right side about the y-axis. The operation produces a symmetric trough, but also produces a vertex at the trough bottom. Dotted lines extending horizontally across the bottom are plot portions that are cut-off during reflection to obtain the full subsidence trough. The vertex also occurs when using tanh(x) as a profile function. Thus, while the limit to subsidence is within 2% at the trough limit, the slope of the trough bottom is not zero *when the panel width is sub-critical*. For this reason, one may prefer the exponential function for use as a profile function to describe sub-critical trough subsidence.

When the panel is super-critical the flat across the bottom that is cut-off in this example may be retained and the error function and hyperbolic tangent function may be used as profile functions. In any case, some numerical experiments are necessary to obtain a reasonable profile that gives maximum subsidence and very small subsidence outside of the subsidence trough limits computed through application of the angle of draw concept.

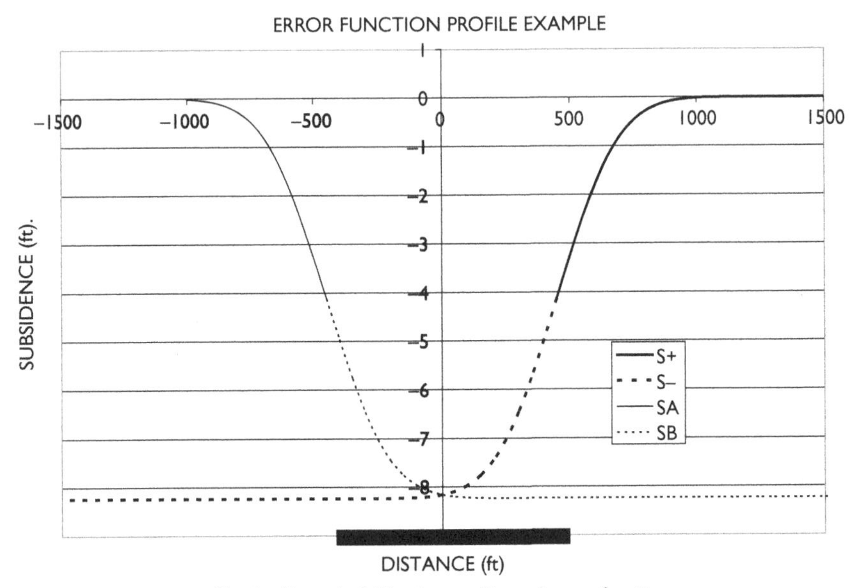

Plot for Example 8.37 using a calibrated error function.

Example 8.38 Inflection points play an important role in profile functions. Determine the inflection points for the functions: (1) $y = \tan^{-1}(x)$, (2) $y = \exp(-x^2)$, (3) $y = \tanh(x)$, (4)$y = (2/\sqrt{\pi})\int_0^x \phi(x)dx$

Solution: Inflection points occur where the slope changes sign with increasing or decreasing x. These points are evident in the plots, but need verification. The first derivative should not be zero at an inflection point. The functions, first derivatives, and second derivatives are given in the tabulation that shows where the inflection points are and verifies the constraint on the first derivatives. For convenience, the origin is often placed at the inflection point in those cases where the inflection point occurs at $x = 0$. This point is often considered a half-subsidence point, where the subsidence is one half the maximum subsidence.

$$
\begin{array}{llll}
y = \tan^{-1}(x), & dy/dx = 1/(1+x^2), & d^2y/dx^2 = -(1+x^2)^{-3}(2x), & x = 0 \\
y = -\exp(-x^2), & dy/dx = (2x)\exp(-x^2), & d^2y/dx^2 = (-4x^2+2)\exp(-x^2), & x = \pm 1\sqrt{2} \\
y = \tanh(x), & dy/dx = 1/\cosh^2(x), & d^2y/dx^2 = (-2)(\cosh^{-3})\sinh(x), & x = 0 \\
y = erf(x) & dy/dx = (2/\sqrt{\pi})\exp(-x^2), & d^2y/dx^2 = (2/\sqrt{\pi})(-2x)\exp(-x^2), & x = 0
\end{array}
$$

Example 8.39 Influence functions are used to compute subsidence troughs over tabular excavations by an integration process. A small element mined at seam level makes a small contribution to surface subsidence. This contribution diminishes with distance away from the mined element. If a point on the surface directly over the mined element is an origin of coordinates, then distance from the mined element can be measured by a radius from the origin. Excavation of an adjacent element creates a similar contribution to subsidence at the surface. The net subsidence is the sum of the two contributions to a given surface point. Excavation over an area A involves summation over all small elements. Given the exponential function as an influence function, derive the related profile function.

Solution: The exponential function in primitive form is $\exp(-x^2)$ and in consideration of the definition of an influence function, x is replaced by r, the radius at the surface from the center of the small excavation element at seam level. In fact, a scaled distance should be used, so one has $\exp[-(\frac{r}{W_s/2})^2]$ where the scale is the trough half-width. Again, by definition, the influence is nil beyond a limiting radius. Integration over a specified excavated area A at seam level gives the subsidence. Symbolically then, $s(x,y) = \int_A \exp[-(\frac{r}{W_s/2})^2]dA$. In more detail $s(x,y) = \int_x \int_y \exp[-(\frac{\sqrt{a^2+b^2}}{W_s/2})]dbda$ where a and b are dummy variables for integration. In case of a profile function coincident with the x-axis, $s(x) = \int_x \exp[-(\frac{a}{W_s/2})^2]da \int_\infty \exp[-(\frac{b}{W_s/2})^2]db$. The last integral on the right has $-/+$ infinity limits and a value of $(W_s/2\sqrt{\pi})$. Hence, $s(x) = \sqrt{\pi}(W_s/2)\int_x \exp[-(\frac{a}{W_s/2})^2]da$ which is now a line integral along the x-axis. This integral can also be written using the error function. Thus, $s(x) = (\text{constant})[1 - erf(\frac{x}{W_s/2})]$ where the constant needs adjustment for physical constraints. This result is the companion *profile* function to the exponential *influence* function. The constant can be evaluated using subsidence S at the trough center where $x = 0$. Finally,

$$s(x) = (S)[1 - erf(\frac{x}{W_s/2})]$$

Note: An influence function is in a sense a derivative of a profile function.

Example 8.40 Area integration of an influence function to obtain a surface subsidence trough is symbolically $s(x,y) = \int_A f(a,b;x,y)dA$ where a and b are variables of integration. This concept has interesting mathematics that become evident when explored in depth. However, implementation of the concept almost certainly requires recasting into a form suitable for numerical evaluation. Recast the symbolic concept of influence functions into a form that is amenable to numerical evaluation.

Solution: The main effort is to simply replace integration by summation. Consider a single seam element such that $s(a_m, b_n) = f(a_m, b_n)\Delta A$ where $s(a_m, b_m)$ are influence coefficients. These coefficients can be compactly represented in matrix form. Thus, $[s_{mn}]_k = f(a_m, b_n)\Delta A_k$ where the range of rows and columns (m, n) is (M, N) and the subscript k indicates the k-th seam element. In view of the radial form of influence functions, one would expect a square matrix ($M = N$). The size of the matrix is determined by the number of subdivisions one expects for accuracy and efficiency of computation, say, 10. The physical range of influence $R = H\tan(\delta)$, a distance that also gives the extent of subsidence beyond the edge of a panel. The magnitude of the influence coefficients is simply the product of the influence function by the area element. If a surface point comes under the influence of all the influence points associated with a given area element, then the resulting total for the point is the sum of all the elements of the matrix$[s_{mn}]_k$. Dividing the individual matrix elements by this total "normalizes" the matrix members, so the total becomes unity. A constant multiplier may then be used to scale the computed subsidence to what is expected, say, expected maximum subsidence.

Individual contributions from each seam area element to the points within the range of influence must be accumulated to obtain the total subsidence at each and every surface point. A summation of contributions is symbolically$[S_{ij}] = \Sigma_k [s_{mn}]_k$. This summation process is one of assembly where the smaller influence matrices $[s]$ are assembled into a global subsidence matrix $[S]$ (and is similar to assembly in finite element analysis where many thousands of points can be accommodated in seconds). Assuming points are spaced on a regular grid, location of a surface point (i, j) is $(x = i\Delta x, y = j\Delta y)$ where the subsidence is S_{ij}. This simple approach does not appear to have been published in the technical literature.

8.3 Problems

Chimney caving

8.1 Consider a vertical shrinkage stope W (ft, m) wide, H (ft, m) high filled with muck having a specific weight γ along a strike length of L (ft, m). Further suppose wall friction angle ϕ' and angle of internal friction ϕ apply. Cohesion of the muck and adhesion at the stope walls is negligible. (a) Derive a Janssen-type formula for this specific situation per unit of strike length. (b) If the stope is 35 ft (10.7 m) width, 120 ft (36.6 m) high, and 165 ft (56.3 m) along the strike, $\phi' = 37°$, $\phi = 30°$, $\gamma = 105$ pcf (16.6 kN/m^3), estimate the vertical stress, horizontal stress, and shear stress at the bottom of the muck. (c) If water infiltrates the stope, modify the "dry" formula from (a) and recalculate the bottom stresses in (b).

8.2 Derive a formula for the chimney cave height H shown in the sketch. Note that the initial void is roughly circular in plan and has a radius of a (ft, m). The circular region is h (ft, m) high. An elliptical vault forms in the back and extends b ft into the back at its highest point. Chimney caving begins from this configuration.

8.3 If the excavation width a in Problem 8.2 is 105 ft (32 m), the initial height of the circular region is 14 ft (4.3 m); the arch extends 35 ft (10.7 m) above the floor. What bulking porosity is indicated if the cave just reaches the surface from a seam depth (surface to seam top) of 1,150 ft (350 m)?

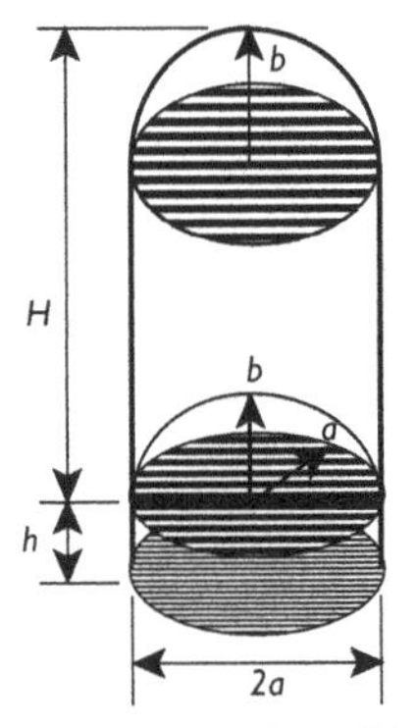

Sketch for problems 8.2 and 8.3

8.4 Consider the stope in the inclined vein shown in the sketch. If the stope height measured along the dip is 115 ft (35.1 m), the true width of the vein is 35 ft (10.7 m), and the bulking porosity, as determined from past observations at the mine, is 0.072, what cave height is expected if the caving is in the form of a vertical chimney over the stope and the stope strike length is 105 ft (32.0 m)? If the stope strike length is 205 ft (61.5 m), what cave height is expected?

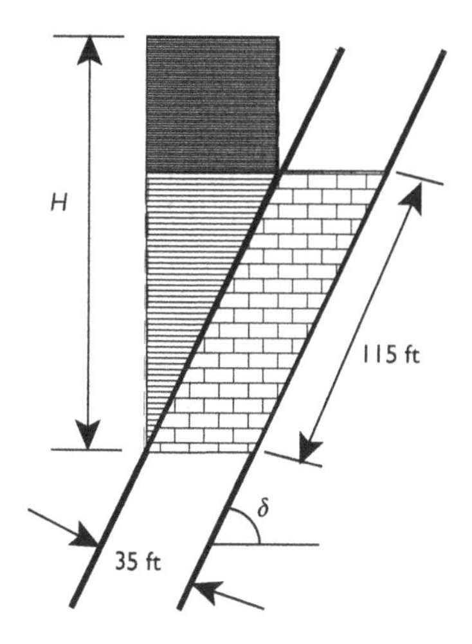

Note: Stope depth H is 4,650 ft (1,417 m) and the angle of repose of the caved ground is 38°.

8.5 A panel block caving system operates at a depth of 4,250 ft (1,417 m) as shown in the sketch. Dimensions of the panels are 150 ft by 300 ft (45.7 × 91.4 m) in plan. The initial undercut is 21 ft (6.4 m) high. Initial bulking porosity at the start of the cave, which forms rubble, is 0.37. Estimate the cave height possible before the swell must be drawn in order to maintain caving.

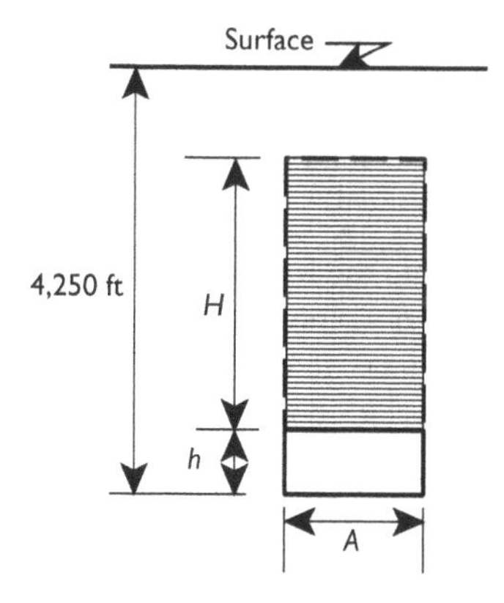

8.6 With reference to Problem 8.5, the caved ground has a specific weight of 105 pcf (16.6 kN/m^3); the angle of friction between muck and rock is 23°, while the angle of internal friction of the muck is 33°. The ratio of horizontal to vertical stress in the caved ground is estimated to be 0.32.

(a) Find the average vertical stress σ_v at the bottom of the cave as a function of the cave height z measured from the bottom of the undercut.
(b) With the origin in the upper left hand corner, plot vertical stress horizontally as a function of cave depth (height) positive down.
(c) What values of vertical, horizontal, and shear stresses are realized at the undercut level when the cave height reaches 300 ft (90 m)? 600 ft (180 m)?

8.7 A panel block caving system operates at a depth of 1,295 m. Dimensions of the panels are 46 m by 92 m ft in plan. The undercut is 6.4 m high. Initial bulking porosity at the start of the cave, which forms rubble, is 0.25. Solid ground *specific gravity* is 2.60; caved ground has a *specific gravity* of 1.73; the angle of friction between muck and rock is 23°, and the angle of internal friction of the muck is 33°. The ratio of horizontal to vertical stress in the caved ground is estimated to be 0.32.

(a) Estimate the cave height possible before the swell must be drawn in order to maintain caving.
(b) What values of vertical, horizontal, and shear stresses are realized at the undercut level when the cave height reaches 183 m?
(c) What bulking porosity is indicated when the cave just reaches the surface after drawing the necessary swell?

8.8 A panel block caving system operates at a depth of 4,250 ft. Dimensions of the panels are 150 ft by 300 ft in plan. The undercut is 21 ft high. Initial bulking porosity at the start of the cave, which forms rubble, is 0.25. The caved ground has a specific weight of 108 pcf; the angle of friction between muck and rock is 23°, and the angle of internal friction of the muck is 33°. The ratio of horizontal to vertical stress in the caved ground is estimated to be 0.32.

(a) Estimate the cave height possible before the swell must be drawn in order to maintain caving.
(b) What values of vertical, horizontal, and shear stresses are realized at the undercut level when the cave height reaches 600 ft?
(c) What bulking porosity is indicated when the cave just reaches the surface after drawing the necessary swell?

8.9 Block caving studies indicate the rock mass has an associated bulking porosity of 0.015. A caving panel is developed 1,570 ft (479 m) below ground surface. Suppose the panel is A square feet in plan view, say, A = 20, 000 square feet (1,858 m^2). How much ore ('solid rock') may be removed before subsidence just reaches the surface?

8.10 A circular ore pass 13 ft (4 m) in diameter has a muck cushion that extends 9 ft (9 m) above the bottom of the ore pass. The ore pass is vertical and extends 150 ft from level to level. The muck has a specific weight of 105 pcf (16.6 kN/m^3), is considered cohesionless with an angle of internal friction of 38°, and has a coefficient of friction against rock of 0.57. The ratio of horizontal to vertical normal stress in the ore pass is estimated at 0.33. Three tons of ore are dumped into the ore pass. Analysis of the ore fall indicates that on impact, the muck decelerates at 4g. Thus, the impact force on the muck cushion top is four times the muck weight. Consider this force to be a static surcharge, then determine the associated increase in vertical stress (psf, kPa) at the bottom of the muck cushion.

8.11 A proposal for panel caving a massive ore zone using undercuts 250 ft by 600 ft (76.2 × 183 m) in plan and 14 ft (5.4 m) high is made. If the bulking porosity is 0.075 and caving occurs directly over the undercut, how high above the undercut will the cave zone extend? Draw of ore removes solid material and creates additional void space. If caving extends to the surface, estimate the tons of ore drawn (12 ft^3/ton, 0.34 m^3/ton). If the draw rate is 50,000 tpd, what is the elapsed time from undercut completion until the draw appears at the surface?

Combination support

8.12 With reference to the paper by F. Kendorski (chapter 4, 1977) and the example calculations for Mine C (which relates to Mine B), a wall shear check is made using the equation $F_D = F_f + F_r + F_{ri}$. The numbers are given as

$$(48,400)(0.707) = (10)(144)(5)(1.414) + (48,000)(0.707)(0.21) \\ +(0.5)(0.60)(60,000) + (126)(144)A_{ri}$$

Explain the meaning of each number, where it is obtained and its physical meaning. Are the numbers correct?

8.13 With reference to Problems 8.5 and 8.6, 9 ft wide by 12 ft high grizzly drifts below the undercut level are first supported with 3/4 in diameter, grade 55 resin bolts 6 ft long (grouted full length) and wire mesh. A continuous concrete liner may be added later. Compressive strength of concrete is 5,500 psi. Unit weight of adjacent rock is 180 pcf, angle of internal friction is 42° and cohesion is 3,450 psi. Three sets of nearly orthogonal joints transect the rock mass. Joint friction angles vary between 27° and 38°, while joint cohesion varies from 7 to 70 psi. Fracture persistence determined from trace length scans is 87%, so only 13% is in intact rock. What thickness of liner should be used if a rock mass safety factor of 1.5 and a concrete liner safety factor of 1.4 are specified when the abutment load factor is 185%?

8.14 With reference to Problems 8.5 and 8.6, 2.7 m wide by 3.7 m high grizzly drifts below the undercut level are first supported with 3/4 in diameter, grade 55 (379 MPa yield strength) resin bolts 1.8 m long (grouted full length), and wire mesh. A continuous concrete liner may be added later. Compressive strength of concrete is 37.9 MPa. Unit weight of adjacent rock is 28.5 kN/m^3, angle of internal friction is 42° and cohesion is 23.8 MPa. Three sets of nearly orthogonal joints transect the rock mass. Joint friction angles vary between 27° and 38°, while joint cohesion varies from 48 to 480 kPa. Fracture persistence determined from trace length scans is 87%, so only 13% is in intact rock. What thickness of liner should be used if a rock mass safety factor of 1.5 and a concrete liner safety factor of 1.4 are specified when the abutment load factor is 185%?

8.15 A semicircular arched back heading 14 ft (4.3 m) wide and 17 ft (5.2 m) high is driven due north. Two joints are present; both strike due north. Set 1 dips 60° east; Set 2 dips 60° west. No other data are available. Yet, some support planning is required. Would conventional rock bolting be adequate in your estimation or would cable bolts,

Dywidag, or Swellex bolts need to be considered? Explain your choice with the aid of sketches.

8.16 A caving operation is conducted in a relatively weak rock mass where the joint persistence is 88%. Laboratory testing of intact core shows that $C_o = 12{,}750$ psi, and $\phi = 29°$. Specific weight of rock is 157 pcf. Joint cohesion is 75 psi and joint friction angle is 20°. Depth of the undercut is 3,750 ft. Development drifts are 13 ft wide by 18 ft high. Loading is expected to increase 100% after caving has reached the surface. If the rock mass is not self-supporting with a safety factor of 1.5, then steel sets will be used, possibly in conjunction with rock bolts and shotcrete. In this regard, spot bolting with 6 ft, 3/4 in. diameter, high strength bolts is standard procedure in any case. Determine a combination support system that will provide a safety factor of 1.5 in any steel support and other reinforcement.

8.17 A caving operation is conducted in a relatively weak rock mass where the joint persistence is 88%. Laboratory testing of intact core shows that $C_o = 88$ MPa, and $\phi = 29°$. Specific weight of rock is 24.8 kN/m^3. Joint cohesion is 517 kPa and joint friction angle is 20°. Depth of the undercut is 1,143 m. Development drifts are 4 m wide by 5.5 m high. Loading is expected to increase 100% after caving has reached the surface. If the rock mass is not self-supporting with a safety factor of 1.5, then steel sets will be used, possibly in conjunction with rock bolts and shotcrete. In this regard, spot bolting with 1.8 m, 1.9 cm diameter, high strength bolts is standard procedure in any case. Determine a combination support system that will provide a safety factor of 1.5 in any steel support and other reinforcement.

8.18 With reference to the data below, an open pit mine has reached an economic limit (3,000 ft, 914 m depth), so underground mining is being considered using a block caving method. A mainline ramp 16 ft (4.9 m) wide with semicircular backs 16 ft (4.9 m) high is planned from the pit bottom to spiral down to 5,000 ft (1,524 m) below the original ground surface. The naturally supported ramp is examined as if stress concentrations were those of a rectangular opening. No stress measurements have been made at this early stage of planning. (a) Estimate the minimum local safety factors (FS_c, FS_t) for the naturally supported ramp, then (b) consider a combination support system that may be required after caving is fully developed. In this latter circumstance, use bolts no more than one-half opening width and require a rock mass safety factor of 1.8. Specify details of the considered support system.

Open pit mine data: A Mohr–Coulomb failure criteria apply, the clay-filled joints constitute 83% of the potential shear failure surface, no tension cracks have yet appeared, current slide block weight is 9.891(10^7) lbf per foot (1.45 GN/m) of thickness and:

1	slope height	$H = 1{,}540$ ft(469m)
2	failure surface angle	$\alpha = 39°$
3	slope angle	$\beta = 55°$
4	friction angle (rock)	$\phi_r = 36°$
5	cohesion (rock)	$c_r = 1{,}870$ psi(12.9MPa)
6	friction angle (joint)	$\phi_j = 27°$
7	cohesion (joint)	$c_j = 17.0$ psi (117 kPa)
8	specific weight	$\gamma = 156$ pcf (24.7 kN/m^3)

9	tension crack depth	$h_c = 0.0$ ft (0.0 m)
10	water table depth	$h_w = 103$ft(31.4m)
11	seismic coefficient	$a_o = 0.10$
12	surcharge	$\sigma = 0.0$ psf (0.0 kPa)

Ultimately, the pit is planned to a depth of 3,000 ft (914 m).

8.19 A caving operation is conducted in a relatively weak rock mass where the joint persistence is 88%. Laboratory testing of intact core shows that $C_o = 12{,}750$ psi, and $\phi = 29°$. Specific weight of rock is 157 pcf. Joint cohesion is 75 psi and joint friction angle is 20°. Depth of the undercut is 3,750 ft. Development drifts are 13 ft wide by 18 ft high. Loading is expected to increase 100% after caving has reached the surface. If the rock mass is not self-supporting with a safety factor of 1.5, then steel sets will be used, possibly in conjunction with rock bolts and shotcrete. Determine a combination support system that will provide a safety factor of 1.5 in any steel support and other reinforcement.

8.20 A caving operation is conducted in a relatively weak rock mass where the joint persistence is 88%. Laboratory testing of intact core shows that $C_o = 87.9$ MPa, and $\phi = 29°$. Specific weight of rock is 24.8 kN/m^3. Joint cohesion is 517 kPa and joint friction angle is 20°. Depth of the undercut is 1,143 m. Development drifts are 4 m wide by 5.5 m high. Loading is expected to increase 100% after caving has reached the surface. If the rock mass is not self-supporting with a safety factor of 1.5, then steel sets will be used, possibly in conjunction with rock bolts and shotcrete. Determine a combination support system that will provide a safety factor of 1.5 in any steel support and other reinforcement.

Subsidence troughs

8.21 A longwall panel 6,200 ft long with a face length of 820 ft is planned for a seam 1,300 ft deep and 16.5 ft thick. Mining is full seam height. Assume UK conditions apply. Find:

1. The maximum subsidence that will develop during the life of the panel.
2. The maximum subsidence after the face has progressed 520 ft.
3. The surface subsidence profile when fully developed.
4. The surface strain profile; include specific values for E_t, E_c, e_c.
5. The width of a barrier pillar needed to protect a ventilation shaft that serves the mine.
6. The time required to mine the panel at 7,600 tons per shift, 2 shifts per day, 250 days per year.
7. The average rate of advance per day.
8. The maximum subsidence after mining four identical panels adjacent to the first panel and trough width.
9. The maximum subsidence and trough width after mining an identical panel 80 ft directly below the first panel.
10. Assuming US conditions with a subsidence factor of 0.65 and an angle of draw of 28°, find the maximum subsidence and trough width for a single panel and for four adjacent panels.

11 Estimate the maximum tensile and compressive strains for a single panel assuming US conditions (part 10).

8.22 Alongwall panel 1,900 m long with a face length of 250 m is planned for a seam 396 m deep and 5.0 m thick. Mining is full seam height. Assume UK conditions apply. Find (in m when):

1 The maximum subsidence that will develop during the life of the panel.
2 The maximum subsidence after the face has progressed 158 m.
3 Peak strains: $E+$, $E-$, e.
4 The width of a barrier pillar needed to protect a ventilation shaft that serves the mine.
5 The maximum subsidence after mining four identical panels adjacent to the first panel and trough width.
6 The maximum subsidence and trough width after mining an identical panel 25 m directly below the first panel.
7 Assuming US conditions with a subsidence factor of 0.65 and an angle of draw of 28°, find the maximum subsidence and trough width for a single panel.

8.23 A longwall panel 6,200 ft long with a face length of 820 ft is planned for a seam 1,300 ft deep and 16.5 ft thick. Mining is full seam height. Assume UK conditions apply. Find:

1 The maximum subsidence that will develop during the life of the panel.
2 The maximum subsidence after the face has progressed 520 ft.
3 Peak strains: E_t, E_c, e_c.
4 The width of a barrier pillar needed to protect a ventilation shaft that serves the mine.
5 The maximum subsidence after mining four identical panels adjacent to the first panel and trough width.
6 The maximum subsidence and trough width after mining an identical panel 80 ft directly below the first panel.
7 Assuming Utah conditions with a subsidence factor of 0.65 and an angle of draw of 28°, find the maximum subsidence and trough width for a single panel.

8.24 Consider a single longwall panel in a seam mined full height of 5 m at a depth of 400 m. The panel has a face width of 300 m; panel length will eventually reach 3,000 m. The seam is flat. Mining occurs at a rate of 8,000 tons per shift (1 ton = 2,000 lbs), two production and one maintenance shift per day, 5 days per week. Assume UK conditions, then determine for a single panel:

1 maximum subsidence expected
2 critical area
3 subsidence trough width
4 maximum tensile strain
5 maximum compressive strain
6 severity of damage (25 m structure width)
7 compressive strain at the center of the rib-side subsidence profile
8 maximum subsidence when the panel has advanced only 200 m
9 time required to mine the panel

10 maximum subsidence after mining four panels side by side and
11 maximum subsidence after mining a single panel at 400 m and an identical panel 420 m depth directly below.

8.25 Coal is mined 15 ft high underground in flat strata by the longwall method at a depth of 1,500 ft. Panels are 750 ft wide and 7,500 ft long. Assume US conditions, then considering trough subsidence, estimate (in ft):

1 maximum subsidence over a single panel (ft),
2 critical width (ft),
3 maximum subsidence over an area undermined by six adjacent panels (ft),
4 maximum tensile strain over a single panel,
5 maximum compressive strain over an area undermined by six adjacent panels.

8.26 A solution mine for nacholite that is processed into soda ash is developed by wells 1,800 ft below the surface in the Piceance basin of Colorado. The mine will produce about 1 million short tons per year initially. Assume the "pay zone" is 90 ft thick, dips 5° and that dissolvable nacholite is 20% of the zone material, then (1) specify an appropriate subsidence factor and angle of draw; (2) estimate the maximum subsidence; (3) critical width, and (4) trough width at the time mining extends to the critical area limit.

8.27 A solution mine for nacholite that is processed into soda ash is developed by wells 549 m below the surface in the Piceance basin of Colorado. The mine will produce about 1 million short tons per year initially. Assume the "pay zone" is 27 m thick, dips 5° and that dissolvable nacholite is 20% of the zone material, then (1) specify an appropriate subsidence factor and angle of draw; (2) estimate the maximum subsidence; (3) critical width; and (4) trough width at the time mining extends to the critical area limit.

Profile and influence functions

8.28 Show by plots that the profile functions:

$$(1)y = \tan^{-1}(x),\ (2)y = \exp(-x^2),\ (3)y = \tanh(x),\ (4)y = (2/\sqrt{\pi})\int_0^x \phi(x)dx$$

can be extended to represent a full, symmetric subsidence profiles. Center plots at panel centers.

8.29 Consider the functions and plots in Problem 8.28, then adjust Function (2), the exponential function, so the origin is above the center of a panel in vertical cross-section (and at ground surface before any subsidence has occurred). This procedure places the profile in relation to the excavation below. Also adjust the profile so the subsidence at the trough center is S, a maximum for a given width-depth ratio of a sub-critical width panel. Ensure subsidence is near zero at the edge of a trough estimated with a 35° angel of draw. Evaluate and plot for a mining height of 3 m, depth of 274 m, panel width of 274, and a subsidence factor of 0.9.

8.30 Given the data and results of Problem 8.29, compare subsidence profile using the function $(4)y = (2/\sqrt{\pi})\int_0^x \phi(x)dx$ with the exponential function.

8.31 Inflection points play an important role in profile functions. Determine the inflection points for the functions: (1) $y = \tan^{-1}(x)$, (2) $y = \exp(-x^2)$, (3) $y = \tanh(x)$, (4)$y = (2/\sqrt{\pi}) \int_0^x \phi(x)dx$

8.32 Influence functions are used to compute subsidence troughs over tabular excavations by an integration process. A small element mined at seam level makes a small contribution to surface subsidence. This contribution diminishes with distance away from the mined element. If a point on the surface directly over the mined element is an origin of coordinates, then distance from the mined element can be measured by a radius from the origin. Excavation of an adjacent element creates a similar contribution to subsidence at the surface. The net subsidence is the sum of the two contributions to a given surface point. Excavation over an area A involves summation over all small elements. Given the exponential function as an influence function, derive the related profile function.

8.33 Area integration of an influence function to obtain a surface subsidence trough is symbolically $s(x,y) = \int_A f(a,b;x,y)dA$ where a and b are variables of integration. This concept has interesting mathematics that become evident when explored in depth. However, implementation of the concept almost certainly requires recasting into a form suitable for numerical evaluation. Recast the symbolic concept of influence functions into a form that is amenable to numerical evaluation.

8.34 Consider a tabular excavation at a depth of 330 m in a seam 4 m high with a panel width of 280 m. Seam and surface are flat. Estimate maximum subsidence and trough width, then plot the subsidence profile using as profile functions the exponential and error function. Use a subsidence factor of 0.65 and an angle of draw of 28°.

8.35 Consider a tabular excavation at a depth of 330 m in a seam 4 m high with a panel width of 280 m and a second excavation at a depth of 360 m with the same width and excavation height directly below the top excavation. Seams and surface are flat. Compute maximum subsidence and trough width for each excavation. Estimate maximum total subsidence and composite trough width, then plot the subsidence profile using the exponential function as a profile function. Use a subsidence factor of 0.65 and an angle of draw of 28°.

8.36 Consider a tabular excavation at a depth of 330 m in a seam 4 m high with a panel width of 280 m and a second excavation at a depth of 360 m with the same width and excavation height. The bottom panel center is 140 m to the right of the top panel center as seen in vertical section. Seams and surface are flat. Estimate maximum subsidence and trough width of the composite subsidence profile. Use the exponential function as a profile function. Also use a subsidence factor of 0.65 and an angle of draw of 28°. Plot the composite subsidence profile with the top panel center as the origin.

8.37 Consider a tabular excavation at a depth of 2,100 ft in a seam 15 ft high with a panel width of 900 ft. Seam and surface are flat. Estimate maximum subsidence and trough width. Use the error function and the exponential function as a profile functions. Also use a subsidence factor of 0.7 and an angle of draw of 26°. Compare the subsidence profiles by plotting the troughs with panel center as origin of coordinates.

8.38 Consider two tabular excavations at a depth of 2,100 ft in a seam 15 ft high with a panel widths of 900 ft each excavated side by side. Seam and surface are flat. Estimate maximum subsidence and trough width for each panel and for the composite subsidence profile. Use the exponential function as a profile function. Also use a subsidence factor of 0.7 and an angle of draw of 26°. Plot the composite trough using the center of the first panel on the left as the origin.

8.39 Consider a tabular excavation at a depth of 640 m in a seam 4.57 m high with a panel width of 274 m. Seam and surface are flat. Estimate maximum subsidence and trough width. Use the error function and the exponential function as a profile functions. Also use a subsidence factor of 0.7 and an angle of draw of 26°. Compare the subsidence profiles by plotting the troughs with panel center as origin of coordinates.

8.40 Consider two tabular excavations at a depth of 640 m in a seam 4.57 m with panel widths of 274 m each excavated side by side. Seam and surface are flat. Estimate maximum subsidence and trough width for each panel and for the composite subsidence profile. Use the exponential function as a profile function. Also use a subsidence factor of 0.7 and an angle of draw of 26°. Plot the composite trough using the center of the first panel on the left as the origin.

Chapter 9

Dynamic Phenomena

Some physical phenomena cannot be understood on the basis of statics alone; consideration of dynamics is essential in many instances. An example is the magician's feat of pulling a table cloth from underneath a table set with fragile dinnerware and heavy silverware. If the table cloth is pulled slowly in quasi-static fashion, the table setting follows the slow motion of the table cloth to the floor. However, if the table cloth is snapped dynamically, the table setting remains in place even as the cloth is removed. The snap generates a pulse in the cloth that momentarily casts plates, glasses, knives, and forks, and so forth, upwards even as the cloth moves from underneath. With a proper snap, the cloth can be removed cleanly leaving the setting seemingly undisturbed.

Dynamic phenomena bring inertial forces into design analyses. Instead of equations of equilibrium, one must consider equations of motion. In the simplest and quite fundamental form, Newton's second law, which expresses the balance of linear momentum, is an equation of motion (for a particle). For bodies of finite size, the requisite equations of motion express balances of linear and angular momentum separately, and the conservation of mass. Thus, $F = \dot{P}$ and $L = \dot{H}$ with $\dot{M} = 0$ where F, P, L, H, and M are resultant of the external forces, linear momentum, resultant of external moments, angular momentum, and mass, respectively, and the dot denotes time rate of change. The first equation is Newton's second law. In the static case, the time rates of change of momenta are nil. Detonation of explosives in a blast hole is an example of a process that is certainly dynamic. The extremely rapid increase in pressure against the wall of a blast hole and the subsequent propagation of a stress pulse into the adjacent rock mass are certainly dynamic phenomena. Consideration of inertial forces is then necessary in the description of the associated waves that are propagated through the rock mass. Earthquakes that occur when stress exceeds strength on a fault also propagate stress or strain pulses or waves into the adjacent rock. The associated dynamic phenomena should be taken into account in design of surface structures such as buildings and rock excavations as well.

Stress or strain waves that are generated by blasting and earthquakes and subsequently propagated through the adjacent rock masses are generally within the elastic range of deformation, so the relationship between stress and strain is given by Hooke's law as in the static case. These waves are known by different names: sound waves, seismic waves, and "shock" waves. The latter is a misnomer in the context of mechanics where "shock" implies possible discontinuities in stress, strain, temperature, and density. Explosive detonation is accompanied by a true shock wave, but dynamic disturbances, pulses or waves in the rock mass beyond a small damage zone about a blast hole are in the elastic range of deformation. Shock discontinuities do not occur in the elastic range of stress and strain. While large earthquakes attract much public attention, they occur relatively infrequently. Small earthquakes that generally go unnoticed occur with much greater frequency. These *micro-seismic* events are often recorded in the vicinity of rock excavations using a network of sensitive *geophones*. Very small seismic events may be monitored during laboratory testing for rock properties on centimeter-scale test cylinders. Monitoring of blast "vibrations" in the field is also commonly done. In this regard, vibrations of structural elements such as beams and columns represent a special branch of dynamics of much interest to structural design engineers. Impact loading of rock by a drill bit is yet another example of an action associated with dynamic phenomena. In fact, impact of a drill hammer on a drill rod shank is dynamic. Energy of the impact is transferred along the drill rod to the drill bit by waves. As in blasting, elastic waves are subsequently propagated beyond a localized zone of damage near the bit into the surrounding rock mass.

Wave propagation constitutes a fascinating subject worthy of comprehensive study. Reference books on rock mechanics fundamentals contain much additional discussion of rock dynamics as do books about the mathematical theory of elasticity. A number of such references may be found in Appendix A. Only a brief outline of wave propagation in solids is given here with the limited objective of understanding several important dynamic phenomena, including *rock bursts* and *bumps*, which are related to design analyses in rock mechanics.

9.1 Fundamentals of wave propagation

Fundamentals of wave propagation include derivation of simple wave propagation equations, reflection and transmission of plane waves at normal incidence to a free or fixed face, reflection and transmission at normal incidence to an interface between two materials and oblique incidence to interfaces between materials.

Simple wave propagation models

Two simple models involving normal and shear stress wave propagation in bars illustrates several important features of wave or pulse propagation. Consider a simple, one-dimensional situation in a long bar of length L and cross-sectional area A that is being traversed by a disturbance or pulse as shown in Figure 9.1. According to Newton's second law

$$\begin{aligned} F &= \dot{P} \\ (\sigma + \Delta\sigma)A - \sigma A &= \frac{\partial(mv)}{\partial t} \\ \left(\frac{\partial \sigma}{\partial x} dx\right) A &= m\frac{\partial(v)}{\partial t} \\ \left(\frac{\partial \sigma}{\partial x}\right) dxA &= \rho A dx \frac{\partial(v)}{\partial t} \\ \left(\frac{\partial \sigma}{\partial x}\right) &= \rho \frac{\partial(v)}{\partial t} \end{aligned} \tag{9.1}$$

where m is mass, v is velocity, and ρ is mass density. Because mass of the small element considered is constant, it does not vary with time and $\dot{m} = 0$. The increment or change in stress from one side of the element to the other in the x-direction is simply the rate of change

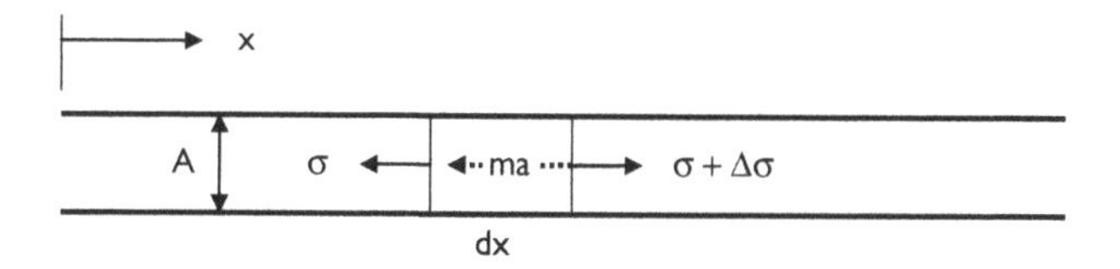

Figure 9.1 One-dimensional dynamic normal stress pulse propagation in a bar.

times the distance over which the rate acts. The inertia force is represented by the dotted arrow and opposes the motion. With this force, included in the sum of forces, one may obtain an expression describing dynamic equilibrium (9.1).

Because the deformation is elastic, Hooke's law applies. In one-dimensional form: $\sigma = E\varepsilon$ where E is Young's modulus and ε is strain. Thus,

$$\left(\frac{\partial E\varepsilon}{\partial x}\right) = \rho\frac{\partial(v)}{\partial t}$$
$$E\left(\frac{\partial \varepsilon}{\partial x}\right) = \rho\frac{\partial(v)}{\partial t} \tag{9.2}$$

According to the definition of strain and the strain-displacement relations $\varepsilon = \partial u/\partial x$ where u is the displacement in the x-direction. Another familiar kinematic relationship is between displacement and velocity such that $v = \partial u/\partial t$. After substitution in (9.2), one obtains

$$E\left(\frac{\partial^2 u}{\partial x^2}\right) = \rho\frac{\partial^2 u}{\partial t^2}$$
$$\left(\frac{\partial^2 u}{\partial t^2}\right) = (E/\rho)\frac{\partial^2 u}{\partial x^2} \tag{9.3}$$

The last equation is recognized as a *wave* equation with solutions in the form

$$u = f(x \mp ct) \tag{9.4}$$

where $c = \sqrt{E/\rho}$ which has the interpretation of *bar wave speed.* The minus sign relates to a wave moving in the positive x-direction; the plus sign relates to a wave moving in the negative x-direction. To see that (9.4) is a solution to (9.3), one simply differentiates (9.4) twice with respect to x and t and then substitutes into (9.3). Thus, $\partial^2 u/\partial x^2 = f''$ and $\partial^2 u/\partial t^2 = (f'')(c^2)$ where a prime indicates differentiation with respect to the function argument, for example, $f' = df/d(x - ct)$. The double prime is a second derivative. Hence,

$$\left(\frac{\partial^2 u}{\partial t^2}\right) = (E/\rho)\frac{\partial^2 u}{\partial x^2} \text{ and } c^2 f'' = (E/\rho)f'' \text{ as claimed.}$$

Strain and stress also follow an equation like (9.3) with solutions of the form (9.4). After differentiating (9.3) with respect to x and again noting that $\varepsilon = \partial u/\partial x$, one obtains

$$\frac{\partial^2 \varepsilon}{\partial t^2} = (E/\rho)\frac{\partial^2 \varepsilon}{\partial x^2} \text{ and since } \varepsilon = \frac{\sigma}{E} \text{ one also has } \frac{\partial^2 \sigma}{\partial t^2} = (E\rho)\frac{\partial^2 \sigma}{\partial x^2}.$$

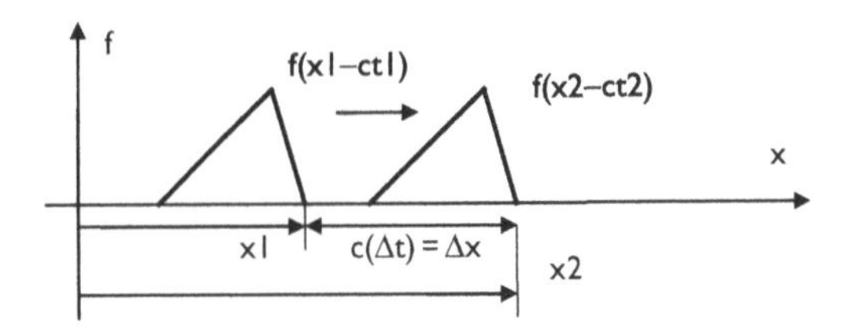

Figure 9.2 A wave traveling along the x-axis in the positive x-direction.

Thus, stress and strain are propagated dynamically as waves or pulses similar to displacement.

While the disturbance or wave moves along the bar with speed c, particles influenced by the pulse acquire a velocity v that differs from the wave velocity. Particle velocity $v = \partial u/\partial t = (\mp c)(f')$. But also $f' = \partial u/\partial x = \varepsilon = \sigma/E$. Thus, $v = (\mp c)(\sigma/E)$. Hence,

$$\sigma = \rho c v \tag{9.5}$$

where for simplicity the −/+ sign has been dropped. Equation (9.5) applies to tensile and compressive stresses and is a fundamental wave relationship that relates stress, density, wave speed, and particle velocity in one dimension, that is, for bar waves. The product ρc is often referred to as *acoustic impedance* in allusion to Ohm's law (voltage) = (impedance)(flow).

To see that (9.4) describes a wave, one may consider a plot at different times, say at time t1 when the wave front is at x1 and a later time t2 when the front is at x2 as shown in Figure 9.2. The wave front moves a distance x2 − x1=Δx in time t2−t1=Δt. At time $t1$, u = f (x1 − ct1) and at time t2 u = f (x2−ct2) = f(x1+Δ x−c(t1+Δ t)) = f(x1−ct1). Thus, the shape of the pulse at time t2 is the same as at time t1 while the wave front has moved from x1 to x2. A similar analyses holds for u = f (x + ct) and shows a wave moving in the negative x-direction. The considered wave may have a more complicated shape and have tensile and compressive parts.

Particle motion under a tensile pulse is opposite (anti-parallel) to the wave motion. An analogy is the start up of a train. As the engine moves ahead to the left; the first car behind the engine begins to move and then the next car moves in turn on so on. The cars are being pulled in tension and begin to slowly move to the left in the direction of the engine even as the tensile pulse travels rapidly to the right towards the rear of the train. If the engine is at the rear of the train and begins to push the cars to the left in compression, the cars again begin to move to the left slowly while the compressive pulse travels rapidly to the left towards to the front of the train. In the compressive case, particle velocity and wave travel are in the same (parallel) direction. The pulse in Figure 9.2 is entirely tensile. Pulses that have tensile and compressive parts move particles forward and back as the wave speeds on its way.

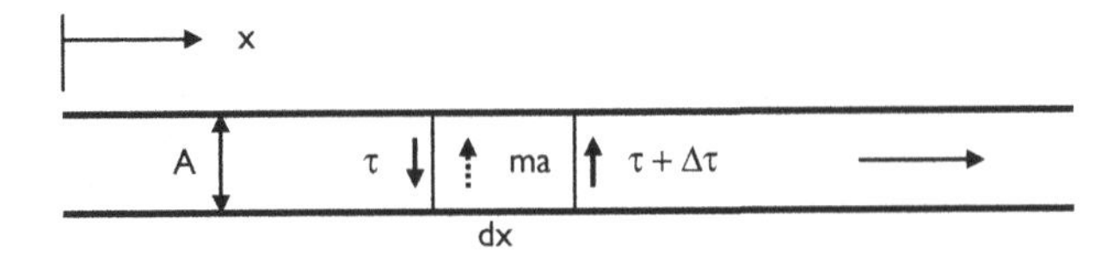

Figure 9.3 One-dimensional dynamic shear pulse propagation in a bar.

In case of shear stress propagation along a bar, a wave equation results from much the same analysis as in the case of normal stress propagation. However, in this case the particle motion is transverse to the direction of wave travel as shown in Figure 9.3.

The equation of motion is the same, Newton's second law. Thus,

$$F = \dot{P}$$

$$(\tau + \Delta\tau)A - \tau A = \frac{\partial(mv_y)}{\partial t}$$

$$\left(\frac{\partial\tau}{\partial x}dx\right)A = m\frac{\partial(v_y)}{\partial t}$$

$$\left(\frac{\partial\tau}{\partial x}\right)dxA = \rho A dx\frac{\partial(v_y)}{\partial t}$$

$$\left(\frac{\partial\tau}{\partial x}\right) = \rho\frac{\partial(v_y)}{\partial t} \tag{9.6}$$

$$\left(G\frac{\partial\gamma}{\partial x}\right) = \rho\frac{\partial^2(v)}{\partial t^2}$$

$$\left(G\frac{\partial^2 v}{\partial x^2}\right) = \rho\frac{\partial^2(v)}{\partial t^2}$$

$$\left(\frac{\partial^2 v}{\partial t^2}\right) = (G/\rho)\frac{\partial^2(v)}{\partial x^2}$$

where from Hooke's law $\tau = G_\gamma$, G=shear modulus, γ =shear strain given by the definition of engineering shear strain $\gamma = (\partial u/\partial y + \partial v/\partial x) = \partial v/\partial x$; no variation occurs in the y- direction. In this case, the displacement in the y-direction is v and should not be confused with velocity. Again, solution of the wave equation takes the form

$$v = g(x \mp ct) \tag{9.7}$$

where the wave is traveling in the x-direction and has a speed given by $c = \sqrt{G/\rho}$. To distinguish between the two wave speeds, a new notation is needed. In case of a normal stress wave in a bar, let the wave speed be denoted as a bar wave speed $c_B = \sqrt{E/\rho}$ and in case of shear wave propagation let the wave speed be denoted by $c_S = \sqrt{G/\rho}$. In consideration of the fact that $G < E$, then $c_S < c_B$. Shear strain and stress are also propagated as waves or pulses with a description like (9.7).

Example 9.1 Show that $\tau = \rho c_S V$ in case of a shear wave where ρ is mass density, V is particle velocity (perpendicular to the direction of wave front propagation), and c_S is shear wave speed ($c_S = \sqrt{G/\rho}$).

***Solution*:** According to (9.7), $v = g(x \mp c_S t)$ where v is shear displacement. Particle velocity by differentiation with respect to time is $\partial v/\partial t = \mp c_S g'$ and shear strain is $\partial v/\partial x = g'$ because there is no variation of the x-direction displacement in the y-direction. Here, the x-direction is the direction of wave travel. Solving for g' shows that $\mp V/c_S = \gamma = \tau/G$ where V, c_S, γ, τ, and G are particle velocity, shear wave speed, shear strain, shear stress, and shear modulus, respectively. Note: $\partial v/\partial x = g' = \gamma$ in consideration of bar shear strain. Thus, $\tau = \rho c_S V$ where $c_S = \sqrt{G/\rho}$.

Example 9.2 Show that shear strain and stress are propagated as waves similar to shear displacement.

***Solution*:** According to (9.6), $\frac{\partial^2 v}{\partial t^2} = (G/\rho)\frac{\partial^2 v}{\partial x^2}$ and the expression for bar wave shear strain $\partial v/\partial x = g' = \gamma$, one obtains. $\frac{\partial^2 v}{\partial t^2} = (G/\rho)\frac{\partial^2 \gamma}{\partial x^2}$. Differentiation of this last expression with respect to x and use of the fact that partial differentiation is commutative shows that $\frac{\partial^2 \gamma}{\partial t^2} = (G/\rho)\frac{\partial^2 \gamma}{\partial x^2}$. Hence, shear strain follows a wave equation. According to Hooke's law $\gamma = \tau/G$, so $\frac{\partial^2 \tau}{\partial t^2} = (G/\rho)\frac{\partial^2 \tau}{\partial x^2}$ that shows shear stress also follows a wave equation.

When waves travel in an extended, three-dimensional solid, the situation is not so simple. A detailed theoretical analysis in the context of dynamic elasticity shows that there are still two types of waves that travel in linear, homogeneous, isotropic elastic solids where Hooke's law applies. The two types of waves have various names. A shear wave is also known as an equi-voluminal wave because the wave propagates without volume change. The speed of this wave is the same as before: $c_S = \sqrt{G/\rho}$. The second type of wave is known as a dilatational or irrotational wave and, as the name suggests, is associated with volume change that, in turn, is associated with normal stresses and strains. This type of wave is also known as a longitudinal wave. The speed of a longitudinal wave is $c_L = \sqrt{(E/\rho)(1-v)/(1-v-2v^2)}$ where v is Poisson's ratio. This wave speed formula shows that longitudinal waves travel faster than bar waves in the same material with the exception of the case where Poisson's ratio is nil. Longitudinal and shear waves are also known as P- and S-waves (P for "primary", S for "secondary"). Also, $c_S < c_L$.

Some ranges of wave speeds for several rock types, ice and water are given in Table 9.1. The ranges in Table 9.1 are approximate and intended as guides only. A particular granite, for example, may have wave speeds outside the given range. In any case, the ratio of longitudinal to shear wave speed is constrained by Poisson's ratio in the range (0.0, 0.5) according to

Table 9.1 Wave speeds and density of several rock types (after Jaeger et al., 2007)

Rock Type	c_L *(m/s)*	c_S *(m/s)*	ρ *(kg/m³)*
Shales and clays*	1100–2500	200–800	2000–2400
Marls	2000–3000	750–1500	2100–2400
Sandstones*	2000–3500	800–1800	2100–2400
Limestones	3500–6000	2000–3300	2400–2700
Chalk	2300–2600	1100–1300	1800–2300
Salt	4500–5500	2500–3100	2100–2300
Anhydrite	4000–5500	2200–3100	2900–300
Dolomite	3500–6500	1900–3600	2500–2900
Granite	4500–6000	2500–3000	2500–2700
Basalt	5000–6000	2800–3400	2700–3100
Gneiss	4400–5200	2700–3200	2500–2700
Coal	2200–2700	1000–1400	1300–1800
Ice	3400–3800	1700–1900	900
Water	1450–1500	–	1000

* saturated

$$\nu = \frac{(1/2)(c_L/c_S)^2 - 1}{(c_L/c_S)^2 - 1}.$$

This constraint is also helpful in laboratory measurements of wave speeds. Wave speeds in porous media are influenced by the pore fluid and the amount of fluid present, that is, the degree of saturation. Moreover, detailed analysis of poroelastic dynamic phenomena shows that two longitudinal waves may occur; one is "fast" and the other "slow".

In case of gneiss and coal and other rock types that certainly have directional characteristics, the ranges in Table 9.1 should not obscure the fact that anisotropic rock has wave speeds that are different in different directions. In fact, there may be three longitudinal and three shear wave speeds that are noticeably different in anisotropic rock. As a rough guide, wave speeds parallel and perpendicular to stratification in laminated or foliated rock or rock with flow structure, are often in a ratio less than two.

Example 9.3 Determine whether the range of wave speeds for granite in Table 9.1 meet the constraint imposed by Poisson's ratio.

***Solution*:** The ranges of c_L and c_S for granite in Table 9.1 are (4500, 6000) m/s and (2500, 3000) m/s, respectively. The ratio of c_L/c_S ranges from 4500/2500 to 6000/2500 to 4500/3000 to 6000/3000, that is, over values (1.80, 2.40, 1.50, 2.00). The ratio range is therefore (1.5, 2.4). The corresponding range of Poisson's ratio is determined by $\nu = \frac{(1/2)(c_L/c_S)^2 - 1}{(c_L/c_S)^2 - 1}$. Hence, $0.10 \leq \nu \leq 0.39$, so the range of wave speeds for granite in Table 9.1 meet the constraint imposed by Poisson's ratio.

Bar waves suggest a simple method for determining elastic moduli in a laboratory test by measuring wave speeds in a long rock "bar" or diamond drill core. The drill core length should be greater than five times core diameter. A common test setup consists of two piezoelectric crystals attached to the ends of a drill core. One crystal is a transmitter; the other is a receiver. Upon energizing, fast (normal stress) and slow (shear stress) pulses are propagated down the drill core. Travel times are recorded with the aid of an oscilloscope and velocities computed simply as core length divided by travel times. Young's modulus and the shear modulus are then calculated after measuring drill core mass density. The slow or second arrival of the shear wave is often difficult to pick accurately, so a check is made to insure an acceptable Poisson's ratio by using the well-known formula $G = E/[2(1+\nu)]$, that is, $\nu = (E/2G) - 1$. Poisson's ratio should be within the range (0, 0.5). If not, then a second look at the data and pick of arrival times from the oscilloscope traces is needed. A common result of measuring elastic moduli dynamically is that they differ noticeably from the same moduli determined on the same core in static fashion.

Example 9.4 Show that the ratio of bar wave speed to shear wave speed lies in the interval $(\sqrt{2}, \sqrt{3})$.

***Solution*:** The ratio of bar wave to shear wave speed is $c_B/c_S = \sqrt{E/\rho}/\sqrt{G/\rho}$. Using the formula $G = \frac{E}{2(1+\nu)}$ one has $\frac{E}{G} = 2(1+\nu)$. As Poisson's ratio varies over the interval (0.0,0.5), the ratio $\frac{E}{G}$ varies over the interval (2, 3). Hence, $\sqrt{2} \leq c_B/c_S \leq \sqrt{3}$.

Drill core "bars" can also be made to "vibrate" in laboratory testing. Vibrations may be induced by cyclic loading of one end of the drill core while recording the induced periodic motion at the other end of the core. When the frequency of the periodic motion results in a peak in amplitude of the induced vibration, a "resonance" condition is achieved. The resonance frequency is related to the bar wave velocity. Thus, $c_B = 2f_BL = 2L/T$ where f_B, L, and T are resonant frequency, drill core length, and vibration period, respectively. The drill core is just one-half the length of the driving wave at resonance. Because the resonance method requires relatively long drill core for test measurements, the method is not used to any great extent. The pulse method is by far the more popular method of determining dynamic elastic moduli.

Kinetic energy and strain energy are attributes of a wave or pulse. The kinetic energy ke is readily calculated from the usual formula $ke = mv^2/2V = \rho v^2/2$ per unit volume. The strain energy or potential energy of deformation is $pe = \sigma\varepsilon/2 = E\varepsilon^2/2 = \sigma^2/2E$ per unit volume. The total energy per unit volume is the sum $U = ke + pe = (\rho v^2/2) + (\sigma\varepsilon/2)$. In the elastic case, this energy is conserved. As velocity varies so do stress and strain. A change in velocity must be accompanied by a compensating change in stress and strain. Interestingly, the partition of energy between kinetic and potential energies is an equal partition. Using (9.4), one has $ke = \rho v^2/2 = \rho(-cf')^2 = E(f')^2/2$ and $pe = \sigma\varepsilon/2 = E\varepsilon^2/2 = E(f')^2/2$.

Example 9.5 Derive expressions for shear wave total, kinetic, and potential energies.

Solution: Total energy U per unit volume of material is the sum of kinetic energy per unit volume ke and potential energy per unit volume pe. The latter is the elastic strain energy density given by the usual formulas: $pe = \tau\gamma/2 = G\gamma^2/2 = \tau^2/2G$. Kinetic energy is by definition $ke = (1/2)mV^2/volume = \rho V^2/2$ where ρ and V are mass density and particle shear velocity, respectively. Therefore, $U = ke + pe = (\rho V^2/2) + (\tau\gamma/2)$.

The energy of a wave at a spherical wave front spreading from a point source diminishes with radial distance r from the source. If U is the total energy and u is the energy density per unit area of the wave front, then $U = uA$ where A is the wave front area. In case of a spherical front,

$$U = u(r)A(r) = u(r)(4\pi r^2) \text{ and}$$

$$U = u(r)A(r) = u(r)(2\pi r)(L) \text{ in case of a cylindrical source.}$$

Thus, the energy density at a spherical wave front diminishes as $1/r^2$. In case of a cylindrical wave front, the energy density diminishes as $1/r$, while plane wave energy propagates undiminished according to the elastic model.

Wave energy density is proportional to stress squared, so in case of a spherical front $u(r) = \frac{U}{4\pi r^2} = \frac{\sigma^2}{\text{constant}}$. Hence, $\sigma^2 r^2 = \text{constant}$. Thus, $\sigma(r') = \sigma(r)(\frac{r}{r'})$. This results shows that stress decreases as $1/r$ as a spherical front propagates outward from a point source. In case of a cylindrical front, a similar analysis shows that $\sigma(r') = \sigma(r)\sqrt{(\frac{r}{r'})}$. Thus, diminution of stress from a cylindrical front is more gradual than from a spherical front.

Example 9.6 Show that a compressional, spherical wave converging towards a point generates extremely high stress and is thus may be enormously compactive.

Solution: From the text, the stress at a spherical wave front varies according to $\sigma(r') = \sigma(r)(\frac{r}{r'})$. Clearly, as the radius from the wave front to the considered point decreases, the stress increases, and as r' becomes very small, $\sigma(r')$ becomes very large. In this regard, fast compaction of porous materials may be done explosively with a suitable arrangement of explosives and material.

Example 9.7 A cylindrical compressive pulse is emitted from a vertical blast hole that has a diameter of 0.3 m. The stress at the hole wall at the time of the blast is a normal stress of magnitude σ_o. Show in a plot of stress versus radial distance from the hole center how the stress diminishes with distance from the hole. At what distance from the hole center is the magnitude of the outgoing stress pulse equal to dynamic tensile strength σ_t that is 1/20th of the dynamic compressive strength σ_c?

Solution: From the text $\sigma(r') = \sigma(r)\sqrt{\left(\frac{r}{r'}\right)}$, so $\sigma(r) = \sigma_0\sqrt{\left(\frac{0.3}{r}\right)}$ where r is distance in meters from the hole center. A plot of the ratio $\left(\frac{\sigma}{\sigma_0}\right) = \sqrt{\left(\frac{0.3}{r}\right)}$ shows the stress magnitude σ as a fraction of the blast hole wall stress versus distance from hole center.

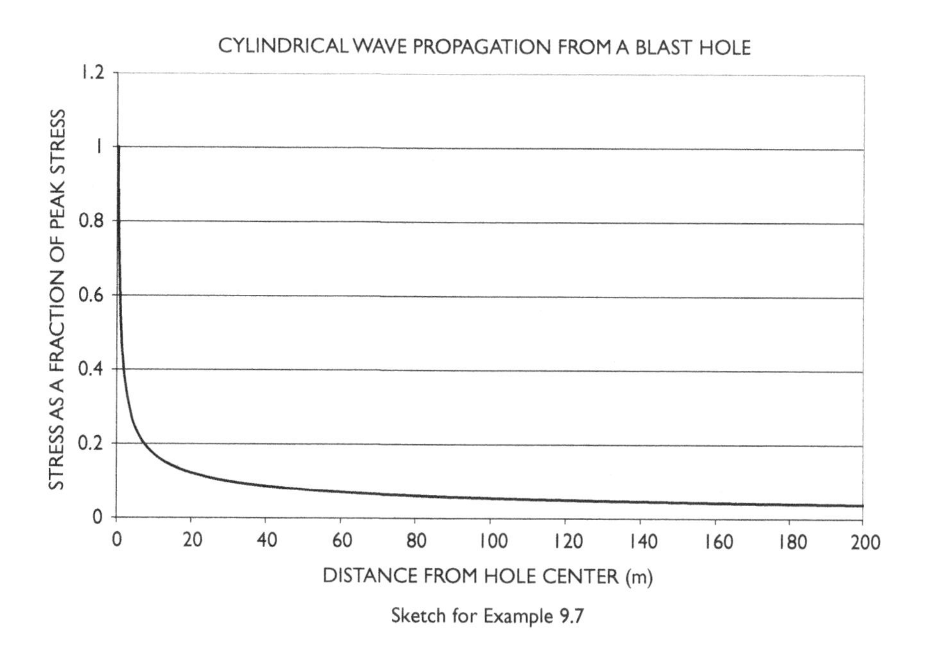

Sketch for Example 9.7

The stress at the blast hole wall is just at the dynamic compressive strength of the material as the wave is propagated *elastically* away from the hole wall. Because the dynamic tensile strength is just 1/20th of the compressive strength, the magnitude of the pulse equals the magnitude of tensile strength at a ratio of 0.05. A calculation using the plotted function shows the corresponding distance to be 120 m.

Example 9.8 Derive expressions for shear stress at the fronts of spreading spherical and cylindrical waves.

Solution: Energy of the wave may be computed as the energy per unit area of wave front multiplied by energy per unit area. Thus, $u(r) = \frac{U}{4\pi r^2} = \frac{\tau^2}{\text{constant}}$ where the proportionality of energy density to stress squared is used. Hence, $\tau^2 r^2 = \text{constant}$. Thus, $\tau(r') = \tau(r)\left(\frac{r}{r'}\right)$. This results shows that shear stress decreases as $1/r$ as a spherical front propagates outward from a point source. In case of a cylindrical front, a similar analysis shows that $\tau(r') = \tau(r)\sqrt{\left(\frac{r}{r'}\right)}$.

Thus, diminution of shear stress from a cylindrical front is more gradual than from a spherical front. These results are similar to those obtained in the text for normal stress propagation.

Reflection at free and fixed faces under normal incidence

A wave pulse traveling down a bar that arrives at a stress-free end must change to satisfy the given end condition. Suppose the wave is a compressive stress pulse. A free-end condition requires an equal but opposite stress to appear as time passes. This stress can be represented by an associated tensile pulse traveling in the opposite direction into the bar from the free end. Superposition of the two pulses is allowed by linear elasticity and creates a stress free end condition. Figure 9.4 illustrates the situation where a plot above the line is compression and below the line is tension. At the early time t1 the compressive pulse is traveling towards the free face along the solid line perpendicular to the face. A *virtual* tensile pulse is shown traveling towards the free face along the dotted line. The two pulses add at the interface with the resultant pulse shown at a time t2 in bold line. Amplitudes of the compressive and tensile pulses at the free face are of the same magnitude but opposite in sense. The result at the free face is zero stress as required. At time t2, part of the compressive pulse is still in the solid while part of the tensile pulse is still in air. The process amounts to a reflection of the compressive pulse at the free face back into the solid as a tensile pulse. The reflected pulse is shown as a tensile pulse at some time t3 after the reflection is complete. A similar diagram would depict reflection of a tensile pulse at a free face back into the solid as a compressive pulse. Mathematically, the two pulses have stresses that may be obtained from by (9.4). Thus, with subscripts c and t for compression and tension, $\sigma_c = E\varepsilon_c = E(-\partial u/\partial x)$ and $\sigma_t = E\varepsilon_t = E(\partial u/\partial x)$. Addition gives a net stress of zero at the free face for all time.

The free end moves during the reflection process with a velocity that is also given by superposition of the effects of the *two* pulses. Particle velocity associated with the compressive wave is in the same direction as the pulse moves while particle velocity associated with the tensile wave is opposite the direction of propagation. Thus, peak particle velocity doubles at the free face while the potential energy (strain energy) goes to zero, as it should with the reduction of stress to zero.

Example 9.9 A shear wave is propagated vertically from below upwards towards the ground surface. Show that the horizontal velocity of the surface upon wave impingement doubles while the shear stress falls to zero. Note: This wave is known as an SH stress wave.

***Solution*:** The situation is entirely analogous to free face reflection of a normal stress wave. However, in case of shear stress, particle velocity is perpendicular to the propagation

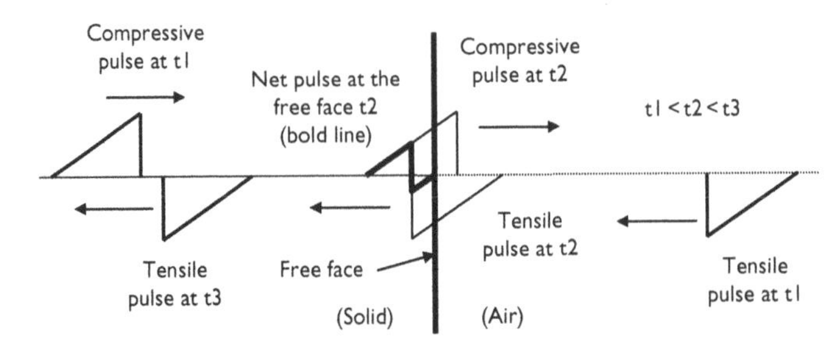

Figure 9.4 Reflection of a plane triangular pulse at a free face under normal incidence.

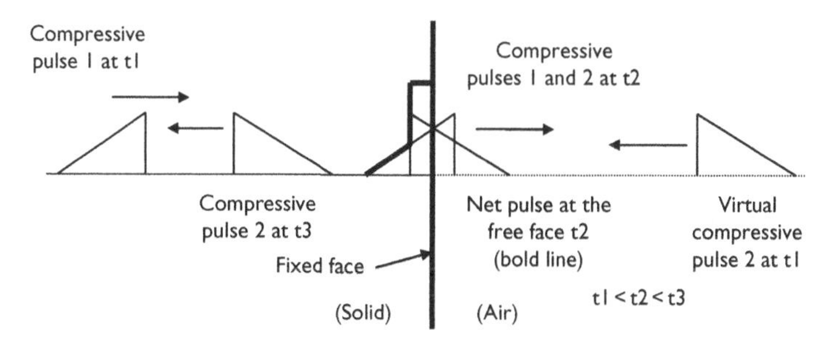

Figure 9.5 Reflection of a plane triangular pulse at a fixed face under normal incidence.

direction and is therefore parallel to the ground surface. Maintenance of a stress-free face is accomplished by introduction of a wave of opposite sense traveling into the ground. The energy of each wave is the sum of equal kinetic and potential energy parts. Potential energy vanishes at the stress-free face, while kinetic energy doubles in each wave and quadruples in total. Hence, the free-face velocity, which is parallel to the ground surface, doubles in magnitude.

If the plane of reflection is fixed rather than free, then the velocity of a particle must go to zero. The incoming virtual pulse must then be a compressive pulse as shown in Figure 9.5. Kinetic energy at the fixed face is zero, while peak stress during reflection doubles. The triangular pulses in Figure 9.4 and 9.5 are sharp-fronted and have very fast (infinite) rise times, but have the advantage of simplicity for illustrative purposes. Stresses associated with these triangular pulses have the simple expression $\sigma = \sigma_0 (1 - x/L)$ where σ_o is the peak stress, L is pulse width (base length of triangle), and x is measured along a triangle base. Strain is simply $\varepsilon = \sigma/E$. A similar equation can be written for a shear wave. Thus, $\tau = \tau_0 (1 - x/L)$ where τ_o is peak shear stress. Shear strain is $\gamma = \tau/G$.

Reflection and transmission at an interface under normal incidence

Reflection and transmission of planes waves at normal incidence to an interface between two different materials must be conserve mass, momenta, and energy. These requirements translate to a requirement for dynamic equilibrium that is consistent with elastic behavior. Equality of particle velocities (and displacements) at the interface is also required to avoid separation of the materials and physically impossible overlap. Thus,

$$\begin{aligned} \sigma_I + \sigma_R &= \sigma_T \\ V_I - V_R &= V_T \end{aligned} \tag{9.8}$$

where the subscripts I, R, and T denote incident, reflected, and transmitted pulses. Use of (9.5) in (9.8) leads to

$$\begin{aligned} \sigma_R &= \left(\frac{\rho_2 c_2 - \rho_1 c_1}{\rho_2 c_2 + \rho_1 c_1}\right)\sigma_I \\ \sigma_T &= \left(\frac{2\rho_2 c_2}{\rho_2 c_2 + \rho_1 c_1}\right)\sigma_I \\ \frac{\sigma_R}{\sigma_T} &= \left(\frac{\rho_2 c_2 - \rho_1 c_1}{\rho_2 c_2 + \rho_1 c_1}\right) \end{aligned} \tag{9.9}$$

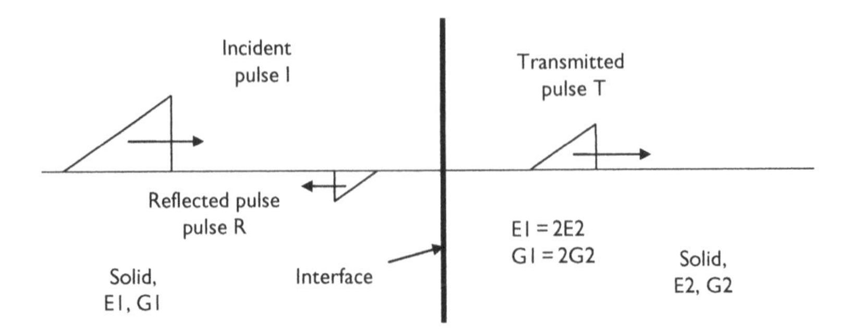

Figure 9.6 Normal incidence of a plane wave at an interface between two different materials.

Figure 9.6 illustrates the situation. The coefficients in the first two of (9.9) are reflection and transmission coefficients that give the magnitude and sign of reflected and transmitted pulses in terms of the incident pulse. The last of (9.9) shows that if the properties of the two materials are nearly the same, then almost all of the incident pulse is transmitted; very little is reflected. These relations also show that the reflected pulse will be in the opposite sense of the incident pulse when the second material has lower acoustic impedance (ρc product). Indeed in case of a free face, the second material is "air" with a negligible modulus relative to rock and reflection is complete while transmission is nil. When the acoustic impedances of the two materials are matched, no reflected wave is generated and maximum transmission of energy occurs from the first to the second material. Results similar to (9.9) hold in case of shear stress.

Example 9.10 A shear wave is propagated vertically from below upwards towards the ground surface. Note: This wave is known as an SH stress wave. A large structure is founded on the ground surface where the wave will arrive. A one-dimensional model of the situation may be used to make a rough estimate of the effect of the wave on the structure in terms of stress and velocity transferred to the structure. Derive expressions for the transmitted and reflected shear stresses and associated velocities using such a model.

***Solution*:** Equilibrium and compatibility at interface between ground surface and structure requires equality of stress and displacement or, equivalently, velocity. Thus, $\tau_I + \tau_R = \tau_T$ and $V_I - V_R = V_T$ where subscripts *I*, *R*, and *T* denote incident, reflected, transmitted quantities, respectively. Using the relationship $\tau = \rho c_S V$ in these two requirements, one obtains

$$\tau_I + \tau_R = \tau_T$$

$$\frac{\tau_I}{\rho c_1} - \frac{\tau_R}{\rho c_1} = \frac{\tau_T}{\rho c_2}$$

that lead to

$$\tau_R = \left(\frac{\rho_2 c_2 - \rho_1 c_1}{\rho_2 c_2 + \rho_1 c_1}\right)\tau_I$$

$$\tau_T = \left(\frac{2\rho_2 c_2}{\rho_2 c_2 + \rho_1 c_1}\right)\tau_I$$

$$\frac{\tau_R}{\tau_T} = \left(\frac{\rho_2 c_2 - \rho_1 c_1}{\rho_2 c_2 + \rho_1 c_1}\right)$$

where the wave speeds are shear wave speeds. The incident pulse, wave speeds, and densities are considered known quantities. Thus, one is able to estimate the shear stress magnitudes of the reflection and transmitted waves. In turn, one may estimate particle velocities using $\tau = \rho c_S V$.

If the acoustic impedance of the structure matches the impedance of the ground, then no refection occurs and all the wave energy is transmitted to the structure and the wave continues to propagate upwards to the top of the building where essentially a free-face reflection occurs with a doubling of horizontal particle velocity. This phenomenon is observed in the "whiplash" of tall buildings during earthquakes.

Reflection and transmission at an interface under oblique incidence

When an incident plane wave strikes an interface between two materials at an oblique angle, the situation is much more complicated than at normal incidence. Details may be found in the references. However, the requirements for dynamic equilibrium and maintenance of the interface without separation or overlap are similar to the normal incidence case, although shear stresses and velocities or displacements must now be considered as well as normal stresses and velocities or displacements. The main result is that an incident wave generally produces a transmitted normal and shear stress wave and a reflected normal and shear stress wave, that is, the reflected and transmitted waves are split into two waves each. Thus, five waves are involved at oblique incidence. Figure 9.7 illustrates this situation where displacement magnitudes are A's and B's with associated angles α's and β's measured from the perpendicular to the interface. A's correspond to normal stress waves and B's correspond

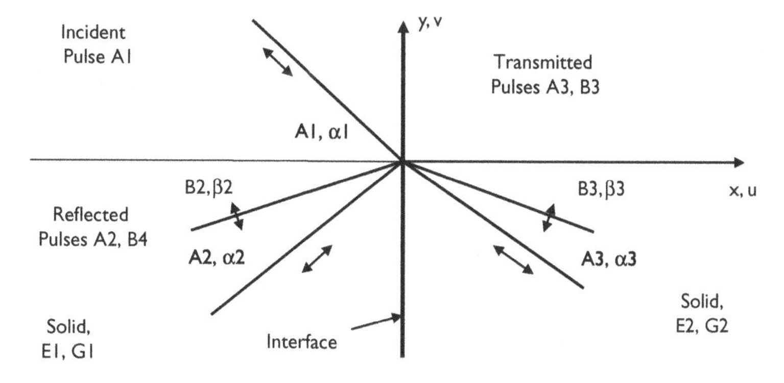

Figure 9.7 Oblique incidence of a plane wave at an interface between two different materials.

to shear stress waves as indicated by the arrows showing particle displacement. Propagation paths are the rays drawn from the origin. Displacements in the x- and y-directions at the interface must be equal, while the same is true of normal and shear stress. Thus,

$$\begin{gathered}\sum u(1) = \sum u(2) \\ \sum \sigma_{xx}(1) = \sum \sigma_{xx}(2) \\ \sum v(1) = \sum v(2) \\ \sum \tau_{xy}(1) = \sum \tau_{xy}(2)\end{gathered} \tag{9.10}$$

where (1) and (2) refer to the materials on the left and right sides of the interface, respectively. The oblique incident wave is considered known as in normal incidence. Equations (9.10) then provide four equations for the four unknown wave amplitudes. At normal incident, the last two of (9.10) are absent.

With reference to Figure 9.7, a detailed analysis shows that

$$\frac{\sin(\alpha 1)}{c_{L1}} = \frac{\sin(\alpha 2)}{c_{L1}} = \frac{\sin(\alpha 3)}{c_{L2}} = \frac{\sin(\beta 2)}{c_{S1}} = \frac{\sin(\beta 3)}{c_{S2}} \tag{9.11}$$

where subscripts L and S refer to longitudinal and shear wave speeds and subscripts *1* and *2* refer to the first and second materials in Figure 9.7. The first of (9.11) states that *the angle of reflection is equal to the angle of incidence* of the incident wave. The second of (9.11) is *Snell's law* that may be put in the form

$$\frac{\sin(\alpha 1)}{\sin(\alpha 3)} = \frac{c_{L1}}{c_{L2}} \tag{9.12}$$

and solved for the transmitted wave angle. Thus,

$$\sin(\alpha 3) = \sin(\alpha 1)\left(\frac{c_{L2}}{c_{L1}}\right) \tag{9.13}$$

Example 9.11 Consider a waves traveling towards a free surface from a source at a depth h below the surface as shown in the sketch. If longitudinal wave speed $c_L = 4,500$ m/s and shear wave speed c_S = 2,250 m/s, determine the angles of reflection of the longitudinal and shear waves at the surface when the angle of incidence of the longitudinal wave at the surface is 23°.

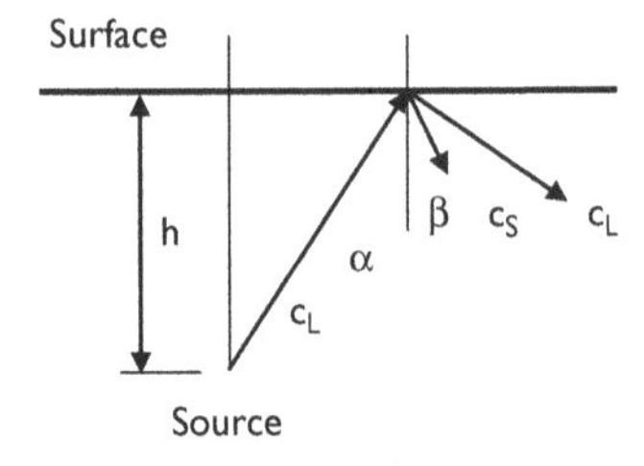

Sketch for Example 9.11

Solution: According to (9.11) $\sin(\alpha 2) = \sin(\alpha 1)[c_L(1)/c_L(2)]$, that is, the angle of reflection is equal to the angle of incidence of the source wave. Thus, α=23°. The angle the shear wave makes with the normal to the surface at the considered point is also given by (9.11).

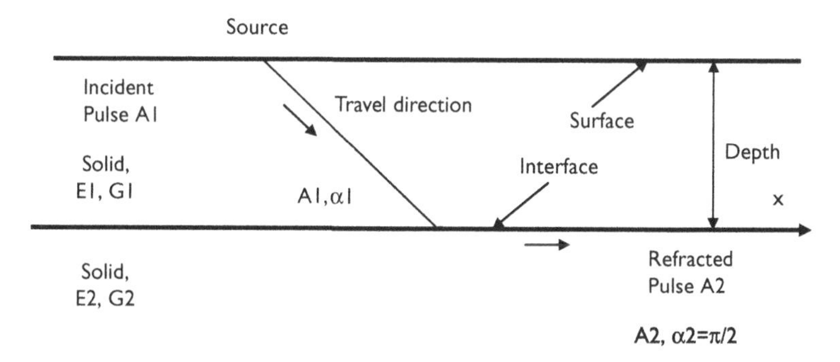

Figure 9.8 Grazing incidence of a plane wave at a horizontal interface.

Hence, $\sin(\beta 2) = \sin(\alpha 1)[c_S(1)/c_L(2)]$. Therefore $\sin(\beta 2) = \sin(23)[2, 2500/4500]$ and $\beta = 11.3°$.

If the wave speed in the second material is much greater than in the first, then the right side of (9.13) may be greater than one. In such a case there is no solution, that is, there is no transmitted wave. This possibility presents an interesting phenomenon known as "grazing" incidence. Reflection is total when the incident angle is at the critical value for grazing incidence. The transmitted wave of the same type as the incident wave is also known as a *refracted* wave. When refraction (transmission) occurs, the refracted wave may be at a lower or higher angle than the incident way depending on whether the wave speed is lower or higher in the second material than the first. Figure 9.8 depicts grazing incidence and refraction at a horizontal interface.

A diagram similar to Figure 9.7 applies in case of an obliquely incident shear stress. Requirements (9.10) still apply in case of shear stress, although at normal incidence the first two of (9.10) are absent. However, there are additional possibilities in case of shear waves that depend on Poisson's ratio, the angle of incidence and the particle motion relative to the interface. One possibility is complete reflection of the incident shear wave without generation of a longitudinal wave.

Example 9.12 Consider a longitudinal wave propagating from a surface source towards bedrock below a soil layer as shown in the sketch. Assume the longitudinal wave speed in bedrock is four times that in soil. At what angle of incidence will a longitudinal wave develop in the bedrock that propagates parallel to the horizontal contact between soil and bedrock?

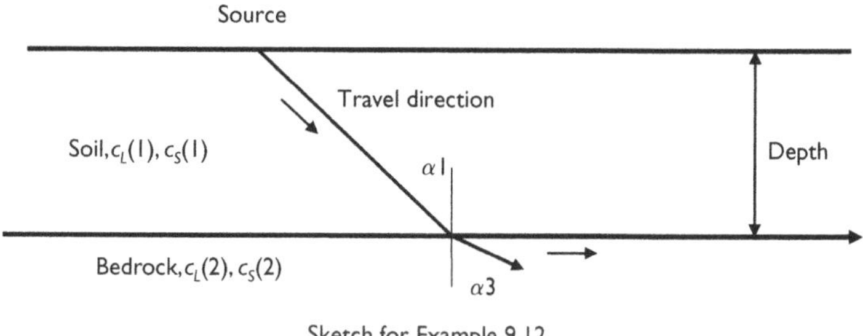

Sketch for Example 9.12

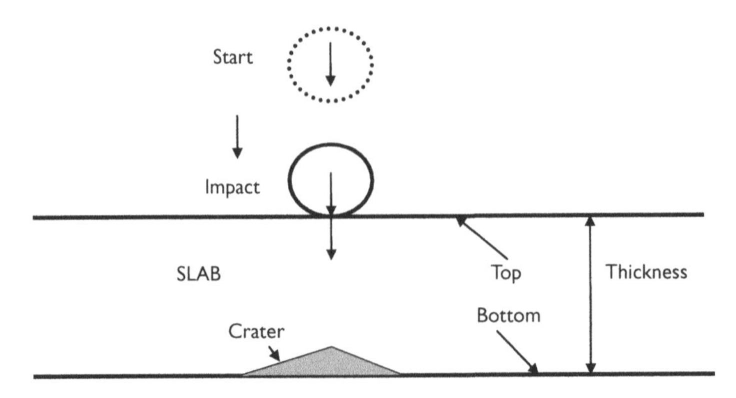

Figure 9.9 Crater formation at slab bottom as a consequence of top impact by a heavy object.

Solution: According to (9.11) $\sin(\alpha 3) = \sin(\alpha 1)[c_L(2)/c_L(1)]$. When $\alpha 3 = \pi/2$, the transmitted longitudinal wave runs horizontally. The incident angle is then $\sin(\alpha) = c_L(1)/c_L(2)$. Thus, $\sin(\alpha 1) = c_L(1)/c_L(2) = 1/4$ and $\alpha 1 = 14.5°$.

There is much more to dynamic phenomena associated with waves traveling in elastic solids. For example, waves traveling in solids may show *dispersion* or wave speed dependent on wave frequency. Several waves traveling out of phase may interfere and produce an envelope that leads to a distinction of *phase* velocity and *group* velocity. Moreover, there are other solutions to the wave equation besides the forms (9.4) and (9.7) that describe waves traveling in the interior of homogeneous solids. Another type of wave that travels along the surface of a solid with amplitude that decreases with distance into the solid is a *Rayleigh* wave. Rayleigh wave speed is slightly less than shear wave speed. Propagation of fractures that involves surface generation is considered limited by the Rayleigh wave speed. Wave phenomena are not limited to the elastic range of deformation. Indeed, *plastic* waves may also be generated in solids. Still, the fundamentals presented here should allow one to develop a basic understanding of observations of rock response to load that are not readily explained statically. As an example, consider a slab of rock in the form of a thick plate as shown in Figure 9.9 and suppose a heavy, round ball (perhaps a bowling ball) is dropped on the slab with the result that a crater forms on the bottom side of the slab opposite the point of impact on top. Surely a slow placement of the ball on the slab would not lead to formation of the crater. A crack might form if the slab deformed in bending and the bending stress exceeded tensile strength, but suppose this was not the case. How does one explain the crater? The explanation is straightforward in that the compressive wave formed at the impact point was reflected in tension at the slab bottom. Although the rock slab is strong in compression and no damage occurred at the impact point, the rock is weak in tension. When the reflected tensile wave reached tensile strength, failure occurred. The lateral extent and depth of the crater are consequences of reflection dynamics.

9.2 Rock bursts and bumps

Rock bursts and bumps are sudden, violent failures of rock that range in size from less than a cubic meter to thousands of cubic meters. They occur suddenly, often with little or no warning, and may cause serious injury to personnel in the vicinity and result in considerable damage to an excavation. Rock bursts and bumps occur at all depths in surface and underground excavations in hardrock and in softrock formations. However, bursts in hardrock excavations tend to occur more often in massive, brittle rock at great depth. A fascinating account of hardrock mine rock burst case histories in North America may be found in the references (Blake and Hedley 2003). Bumps are bursts in softrock formations, notably coal, and also tend to occur more often in deeper excavations. Periodic reviews may be found on the website of the Pittsburgh Research Laboratory of the National Institute of Occupational Health and Safety (NIOSH), a former US Bureau of Mines research center. The Spokane Research Laboratory (NIOSH) has done extensive research studies on bursts in hardrock mines, especially in the Coeur d'Alene mining district of northern Idaho where some of the most burst-prone hardrock mines in the world are located. More recently focus has been on wide area mine collapses and related dynamic phenomena. These studies and many related publications may be found on the web site www.cdc.gov/NIOSH/mining/topics ... where searches under "bumps" or "bursts" are helpful. Publications by *Whyatt* and colleagues are especially pertinent. International conferences on mine seismicity are also sources of case histories, data, and research efforts on bursts and bumps. In this regard, early optimism for burst and bump prediction has proven unfounded as was the case for earthquake prediction. In both areas of study, emphasis is now on understanding as research continues to reveal much more complexity than originally anticipated.

Bursts and bumps are defined in several ways depending on the organization concerned. For example, the U.S. Mine and Health Administration gives the definition: **"Burst** – An explosive breaking of coal or rock in a mine due to pressure; the sudden and violent failure of overstressed rock resulting in the instantaneous release of large amounts of accumulated energy where coal or rock is suddenly expelled from failed pillars. In coal mines they may or may not be accompanied by a copious discharge of methane, carbon dioxide, or coal dust; also called outburst; bounce; bump; rock burst." The presence of methane and similar strata gases complicates the mechanics of bursts and bumps through the concept of effective stress that governs strength of permeable media. An "outburst" is a violent failure associated with a sudden expulsion of coal from an excavation wall that is accompanied by release of large amounts of methane. Similar events occur in some salt mines with the production of "pop corn" salt, a highly granulated form of failed material (salt).

Some classification schemes distinguish bursts on the basis of damage; others classify bursts according to cause: strain bursts, pillar bursts, and fault-slip bursts. A strain burst is a consequence of high stress concentration at the periphery of an excavation. A strain burst could also be referred to as a stress burst. A pillar burst is also a result of high stress in a large portion of a pillar. Fault-slip bursts, as the label suggests, results from shear stress exceeding strength on a structural discontinuity. Strain and pillar bursts are associated with fast, violent release of strain (stress) energy stored in the rockmass. Stored strain energy is elastic strain energy that would be recoverable upon controlled release of load. Bursts caused by fault-slip and associated damage to excavations are a consequence of wave propagation and stress redistribution in the vicinity of slip. These types of bursts are illustrated in Figure 9.10

Excavation geometry and amount of extraction, preexcavation stress, and rock mass quality all influence the mode of failure and whether bursting failure may be expected. Thin, steeply dipping vein deposits under the influence of high horizontal stress are likely to be burstprone. Flat, tabular deposits in hard, brittle rock at depth are also likely to be burst-prone. Failure is always a question of strength versus stress, and even weak rock at shallow depth

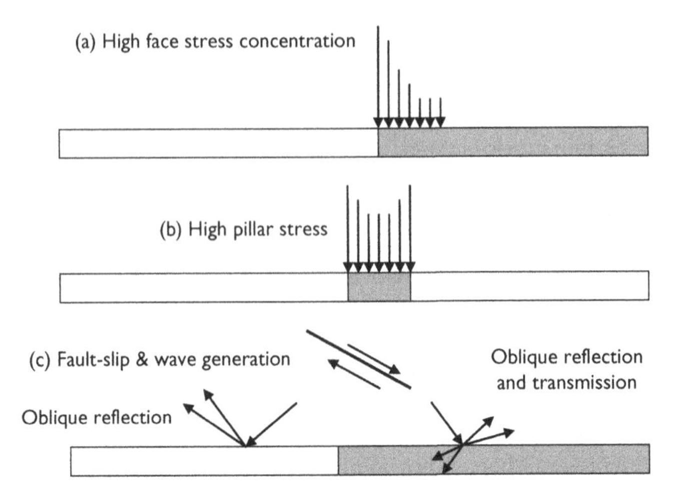

Figure 9.10 Schematic illustration of types of bursts and bumps.

may be burst-prone. Quantitative details are not easy to assess. Pillars in tabular excavations where extraction is high may be especially vulnerable. Perhaps the best indicator of burst or bump potential is the history of a mining or tunneling site. New excavations in districts with histories of bursts or bumps can be expected to pose burst and bump hazards.

The energy released in a burst may be seismically comparable to an earthquake and recorded by seismographs many miles away from the source. While large rock bursts are unusual, small seismic events abound about mining operations. Most small events go unnoticed, although such events may be recorded by an array of geophones at the surface or underground in the form of a micro-seismic network. Examination of recordings shows that the rock mass "talks" continuously in response to excavation. However, periods of relative inactivity occur when slippage on numerous discontinuities in the rock mass ceases and new fractures and cracks are no longer being generated. There are two situations that lead to quiescent periods. One is simply cessation of operations. The other occurs when the rock mass "locks" prior to failure. Acoustic emission during laboratory testing of rock cylinders in compression often follows a pattern of low emissions during initial application of load, increased emissions as load is increased, then decreased emissions, and ultimately failure. Such a pattern is explained by seating of the test cylinder, increased micro-cracking under increased load, clamping of micro-cracks ("locking") and further strain energy accumulation, and finally failure. In an active underground or surface operation, quiescence may be a precursor to a burst or simply a response to an ever changing stress distribution about an excavation. For this reason, microseismic monitoring is only a guide to rock mass stability and is not generally used as an early warning device to signal time to evacuate personnel.

Face bursts and bumps

Stress concentration at the operating face of a longwall panel grows as the face is advanced and the width to height ratio increases. The increase is linear with respect to the width to height ratio that easily obtains values well in excess of 100. Panel widths and lengths are measured in thousand of feet or hundreds of meters, while mining heights are of the order of two to six meters as shown in Figure 9.11. One reason mining is even possible using

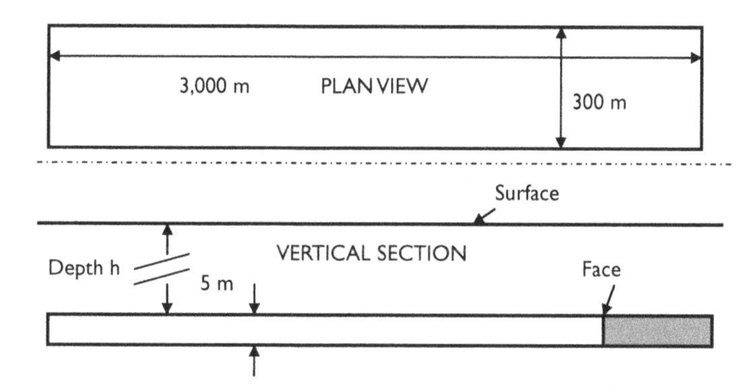

Figure 9.11 Schematic of a longwall panel layout showing the high width to height ratio that evolves as the face is advanced.

the longwall method is because the immediate roof stratum over the panel fails as the span becomes great enough to induce caving. Caving is desirable in longwall mining and produces many seismic events as roof and floor strata fracture in the normal course of mining. There may be a periodicity in the seismic event record as roof strata periodical reach a critical span only to fail by fast fracture and settle onto caved ground below.

In this regard, bursts and bumps are by definition dynamic and therefore associated with dynamic phenomena and seismic events. However, not all seismic events are bursts or bumps nor are all seismic events necessarily harmful. Indeed, seismic activity is beneficial to the extent that associated releases of strain energy occur without damage to excavation walls or harm to personnel. Dissipation of energy in numerous small events is helpful in reducing the number of large events.

Example 9.13 Suppose a longwall panel is developed at a depth of 460 m (1,510 ft) in a seam that is 3.1 m thick (10 ft). Assume elastic conditions, so no failure at the face occurs. Estimate the stress at the face when the panel has advanced 31 m (100 ft), 310 m (1,000 ft)?

***Solution*:** According to the regression data in Figure 3.9, the stress concentration is given by $K\,(\max) = a_1 + a_2 k + a_3 /k + a_4 M$ where k and M are aspect ratio and principal stress ratio, respectively. In the reasonable case of $M = 1/3$, $K\,(\max) = 0.69 + 1.27\,k + 0.68/k + 3.83\,M$. Hence, for $k = 10$, $K\,(\max) = 0.69 + 1.27(10) + 0.68/10 + 3.83(1/3) = 14.7$ and for $k = 100$, $K\,(\max) = 129$. An estimate of the vertical stress at 23 kPa/m of depth leads to a face stress of (14.7)(23)(460) = 156 MPa (22,551 psi) when the face has advanced 31 m, and 1,365 MPa (198,000 psi) when the face has advanced 310 m.
Comment: A strong coal may have an unconfined compressive strength of 28 MPa (4,100 psi); a strong quartzite may have an unconfined compressive strength of 200 MPa (29,000 psi). Thus, a face in a well-advanced longwall panel can be expected to reach the elastic limit and fail whether in hardrock of softrock.

If the immediate roof stratum is thick, massive sandstone, as is the case in some softrock mining districts, then caving may not occur in the desired manner. Rather, a long cantilever roof beam may form and exert unusually high force on face supports and the face proper. In combination with strong coal, a tendency towards bursting, bumping, and bouncing develops. A remedial measure is to reduce the stress concentration at the face by bringing down

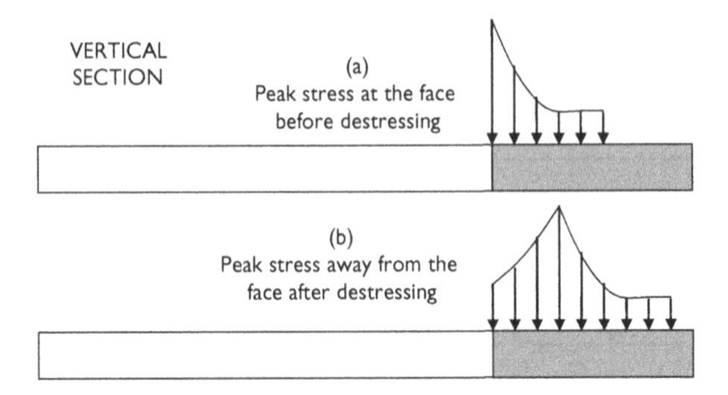

Figure 9.12 Effect of face destressing on peak stress position near a longwall face.

(shortening) the cantilever roof beam. Blast holes are drilled upwards into the roof between face supports (chocks, frames, shields) for this purpose.

Even when caving does occur, stress concentration at a longwall face can be extreme that, again, in combination with strong coal, leads to bursting. In this regard, anticipation of face bump potential in softrock mines is sometimes done by examining the volume of cuttings produce by a short drill hole into the face. Relief of stress concentration at the face is still required. In this case, stress relief may be done by "volley firing" that involves drilling blast holes into the face and shooting. The aim is to damage the coal at the face, reduce strength and stiffness, and thus allow the face to yield rather than to fail by fast, violent fracture.

A consequence of destressing the face is to move the peak stress away from the face into the solid as shown in Figure 9.12. The more distant the peak stress is away from the face the greater is the benefit because of the larger cushion of yielding ground between the face and potential burst location. The same technique is used in longwall methods for mining tabular ore bodies in hardrock. Notable are the much discussed deep gold mines in South Africa. Of importance is dissipation of strain energy in creation of additional surface during the fracturing process about the face and during shear slip on existing fractures and faults in the vicinity of the face. Reduction of the stored strain energy density then reduces burst potential.

Yet another method of destressing the face by reducing strength and stiffness is water infusion. In this method bore holes are drilled from panel sides into the solid ahead of the face and then pressurized with water. As the water pressure increases, the effective stress in the pores and micro-cracks of the coal is decreased, and consequently strength is decreased. Recall that shear strength according to the well-known Mohr-Coulomb criterion is given by $\tau = \sigma' \tan(\phi) + c = (\sigma - p)\tan(\phi) + c$ where σ', σ, p, ϕ, c are effective stress, total stress, pore pressure, angle of internal friction, and cohesion, respectively. Infusion not only reduces the frictional component of strength but also cohesion as the microstructure of the coal is progressively damaged.

Pillar bursts and bumps

Strain and pillar bursts release strain energy accumulated in the loaded volume of the burst. The total strain energy $U = uV$ where u and V are strain energy density (per unit volume) and volume, respectively. In the elastic range of deformation before failure, the strain energy density is $u = \sigma\varepsilon/2 = \sigma^2/2E = E\varepsilon^2/2$ under uniaxial stress or strain conditions. In three-dimensions, all stresses and strains need to be taken into account. Thus,

$$u = (\sigma_{xx}\varepsilon_{xx} + \sigma_{yy}\varepsilon_{yy} + \sigma_{zz}\varepsilon_{zz} + \tau_{yz}\gamma_{yz} + \tau_{zx}\gamma_{zx} + \tau_{xy}\gamma_{xy})/2$$

$$= \left(\frac{1}{2E}\right)\left(\sigma_{xx}^2 + \sigma_{yy}^2 + \sigma_{zz}^2\right) - \left(\frac{\nu}{E}\right)\left(\sigma_{yy}\sigma_{zz} + \sigma_{zz}\sigma_{xx} + \sigma_{xx}\sigma_{yy}\right)$$

$$+\left(\frac{1}{2G}\right)\left(\tau_{yz}^2 + \tau_{zx}^2 + \tau_{xy}^2\right)$$

$$u = \left[\frac{(1-\nu)G}{(1-2\nu)}\right]\left(\varepsilon_{xx}^2 + \varepsilon_{yy}^2 + \varepsilon_{zz}^2\right) + (\lambda)\left(\varepsilon_{yy}\varepsilon_{zz} + \varepsilon_{zz}\varepsilon_{yy} + \varepsilon_{xx}\varepsilon_{yy}\right)$$

$$+(G/2)\left(\gamma_{yz}^2 + \gamma_{zx}^2 + \gamma_{xy}^2\right) \tag{9.14}$$

where $\lambda = \frac{\nu E}{(1+\nu)(1-2\nu)}$, one of the Lame constants (G is the other Lame constant).

In the case that the stress is hydrostatic, so the three normal stresses are equal and the three shear stresses vanish, then

$$u = \left(\frac{3}{2E}\right)(\sigma^2) - \left(\frac{3\nu}{E}\right)(\sigma^2) = \left[\frac{3(1-2\nu)}{2E}\right]\sigma^2 = \left(\frac{1}{2K}\right)\sigma^2$$

where σ is the hydrostatic stress assumed and K is the bulk modulus of the rockmass. When Poisson's ratio is 1/4, $K = (2/3)E$. If $E = 1.5(10^6)$ psi, then $K = 1.0(10^6)$ psi which may be used to make a rough estimate of strain energy density as a function of depth. Assuming 1 psi/foot of depth, $u = (h)^2 (10^{-6})$ in − lbf/in^3 where h is depth in *feet*. At a depth of 1,000 ft, $u = 1$ in − lbf /in^3 or 144 ft-lbf/ft^3 (6.89 kN-m/m^3). Figure 9.13 shows strain energy as a function of rock mass volume with depth as a parameter. The scales are logarithmic in the figure. Also shown in the figure are earthquake magnitudes M from 2 to 5 and associated with energies obtained from the relationship $\log(U) = 9.4 + 2.14M$ where energy U is in ergs. Note: U (ft-lbf)=7.374(10^{-8})U (ergs). Very large rock bursts have magnitudes near 4 (strong). The largest burst in North America recorded to date had magnitude 5.2. This burst occurred in 1995 in a trona mine. Trona is mined in the U.S. in the state of Wyoming in stratified ground at depths of 1,500 ft more or less. Collapse extended over a 1000x2000 m area that was being mined by the room and pillar method.

Example 9.14 Seismic energy associated with a wide area collapse may be related to the potential energy given up by subsidence of the collapse cylinder above the mining horizon. Suppose an area 300 ft by 300 ft collapses suddenly at a depth of 1800 ft where the mining is full seam height of 18 ft. Estimate the potential energy of the collapse if the surface subsidence is 50% of seam thickness. Also estimate the magnitude of the associated seismic event.

***Solution*:** Potential energy in the gravity field is the product of weight by height above the chosen datum. Thus, $PE = Wz$ and the change because of collapse is $\Delta PE = W\,\Delta h$ or $\Delta PE = \gamma$ (300)(300)(1800)(0.5)(18) where γ is specific weight of the overburden. A reasonable assumption is $\gamma = 144$ pcf, so $\Delta PE = 2.1(10^{11})$ ft − lbf. According to the data in Figure 9.13, the associated seismic event would have a magnitude between 3 and 4. The event would be "strong".

Comment: Not all the potential energy would be converted to kinetic energy and transported from the collapse region by elastic waves. Much would be dissipated in frictional slip, fracture

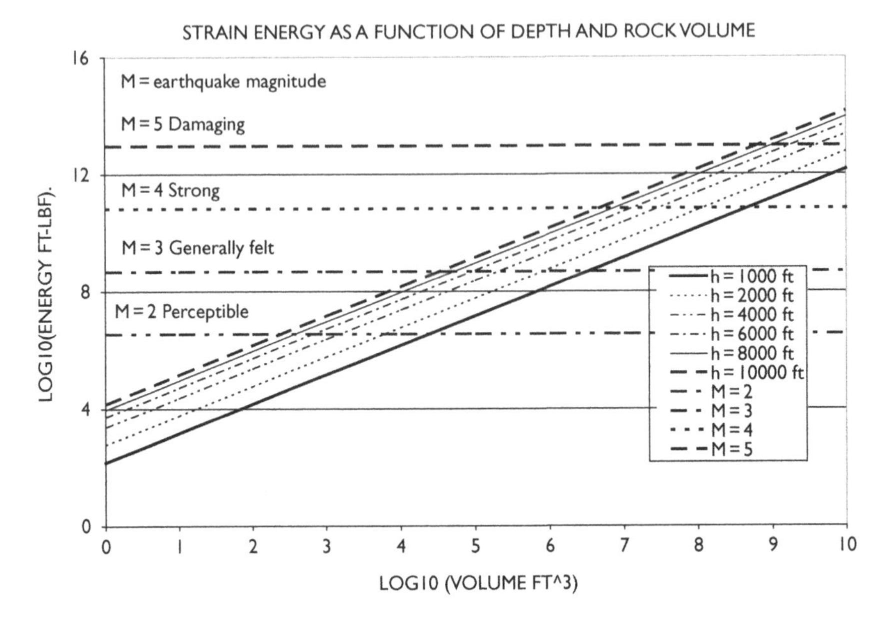

Figure 9.13 Log energy versus log volume with depth as a parameter. Note: m^3 = 0.0283 ft^3, N - m = 1.356 ft-lbf. (after Obert and Duvall).

generation, and other inelastic processes. Because of dissipation, the event magnitude would be smaller.

The data in Figure 9.13 are based on preexcavation stress caused by gravity loading alone. Pillars created in the course of excavation are stressed much higher. On average according to the tributary area model of a pillar in a large array of identical pillars, the average vertical stress in a pillar in flat strata is given by $S_p = S_v/(1 - R)$ where S_v is the overburden stress (unit weight times depth) and R is the area extraction ratio. At high extraction ratios, the pillar stress increases rapidly as shown in Figure 9.14. At 80% extraction, the pillar stress is five time the overburden stress. An increase of extraction by 10% to 90% doubles the pillar stress to 10 times overburden stress. If pillar stress is thought of in terms of stress concentration relative to overburden stress, then $S_p = KS_v$ then $K = 1 / (1 - R)$ where K is the stress concentration factor.

The situation is actually worse than the average vertical pillar stress indicates for two reasons. The first is the higher stress concentration that occurs at the pillars walls. The second reason is the width to height ratio of the pillar is likely to be relatively small when the extraction ratio is high. A consequence of small width is less horizontal stress build-up towards the center of the pillar and therefore lower confining pressure and less pillar core strength. Both features increase the risk of fast pillar failure. However, even relatively wide pillars may burst because of high stress concentration near pillar walls. Small pillar widths are generally avoided during first mining in room and pillar mining where development of panels results in extraction of about 35%. However, during retreat mining with pillar extraction aiming towards 100% recovery of the resource, small pillar remnants may be created along the pillar retreat line. These remnants are prone to violent failure and pose a substantial risk of "bumping".

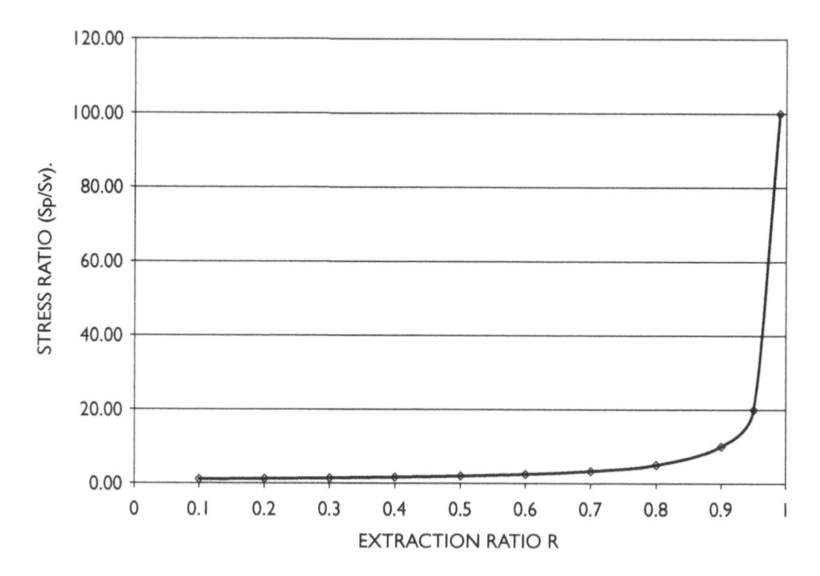

Figure 9.14 Ratio of average vertical pillar stress to overburden stress versus extraction ratio.

Fast failure of a pillar may be associated with *wide area* failures where an extensively mined room and pillar region exists only to collapse in a spectacular, damaging and often fatal manner. One hypothesis is that failure of a single pillar sheds load to adjacent pillars that then fail under the additional loading. As the process continues very wide areas of a mine may experience rapid failure with associated dynamic phenomena. Seismic signatures from fast, wide area failures are often large events with magnitudes of four or so. These events are distinguished at recording stations from fault-slip earthquake events by the seismic signatures observed. A mine collapse is an "implosion" associated with an inward motion of the adjacent rock mass. An explosion is the opposite and is associated with an outward motion. Signatures of fault slip depend on whether the faulting is normal, reversed or thrusting. Fast, sequential pillar failure is also known as a "cascade" failure. Although plausible, simple quantitative analysis casts doubt on the hypothesis. For example, if a pillar with a relatively low safety factor of 1.5 collapses and sheds load to the nearest four adjacent pillars, then these pillars see a stress increase of $S_p/4$ and safety factor reduced to 1.2 from 1.5. Cascade failure under static conditions thus requires very low pillar safety factors over a wide area prior to catastrophic failure.

Example 9.15 A large array of identical pillars are formed in a room and pillar mine. The pillars have an average safety factor based on the tributary area formula of 1.6. Two pillars side by side fail and the load is transferred according to a nearest neighbor load sharing rule as shown in the sketch. Determine the factor of safety of the nearest neighbor pillars after the two-pillar failure.

***Solution*:** The load transferred from the two failed pillars is $2Sp$. This load is distributed amongst the six nearest neighbor pillars, so each of these pillars now must support and additional $2Sp/6$ load. The resulting safety factor is $FS = \frac{Cp}{Sp+(1/3)Sp} = \frac{3Cp}{4Sp}$, that is, 75% of the original safety factor or 1.2. Cp is pillar compressive strength.

Comment: As the number of failed pillars increases, so does the load transferred to the nearest neighbor pillars. If the number of initial failures in a cluster is large enough, a cascade failure becomes possible.

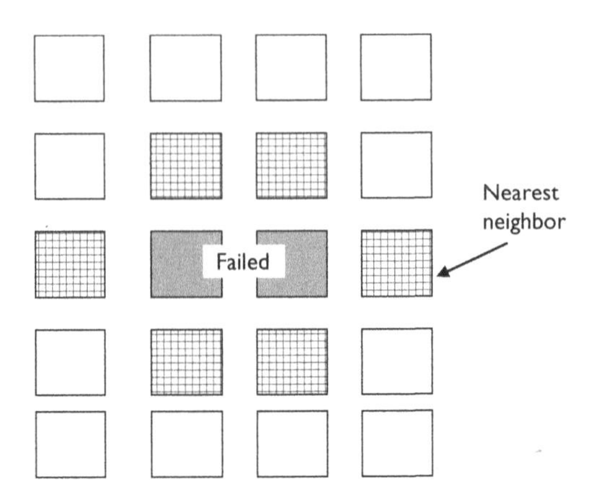

Sketch for Example 9.15

Because of the questionable static mechanism of wide area pillar collapse, a dynamic mechanism for increased loading of pillars may be involved. One simple mechanism is *sudden* application of overburden weight to pillars. Sudden application doubles the static, slow application of pillar load. Consider a dead weight block just in contact with a pillar block below. If the weight is allowed to settle on the pillar gradually, then stress and strain increase gradually until the full block weight is supported by the pillar. Static equilibrium is simply $F = W$ where F, and W are pillar force and block weight, respectively.

However, if the block is released suddenly or dropped, then an inertial force arises with deceleration of the block in accordance with the resistance of the pillar much like dynamic compression of an elastic spring. A model of the system is a block falling from a specified height onto an elastic spring as shown in Figure 9.15. The potential energy change in the fall of the block of weight W through height h is simply Wh. This energy is transformed to an equal amount of kinetic energy that, in turn, is transformed into potential energy of spring deformation $kx^2/2$. At the time of maximum spring displacement $W(h+\delta) = (1/2)(k)(\delta)^2$ that may be solved for the maximum displacement δ. Thus,

$$\delta = \left(\frac{W}{k}\right) + \sqrt{\left(\frac{W}{k}\right)^2 + 2\left(\frac{W}{k}\right)h} \tag{9.15}$$

Under slow application of load, the static displacement is W/k. From (9.15) one sees that when the drop height h is zero, the sudden application of load results in a momentary displacement twice the static displacement. Consequently, the associated force is twice the static load. Rebound and reloading occur in an oscillatory fashion that would continue indefinitely if the system were truly elastic. However, some dissipation of energy is expected and eventually the system will reach a static equilibrium condition.

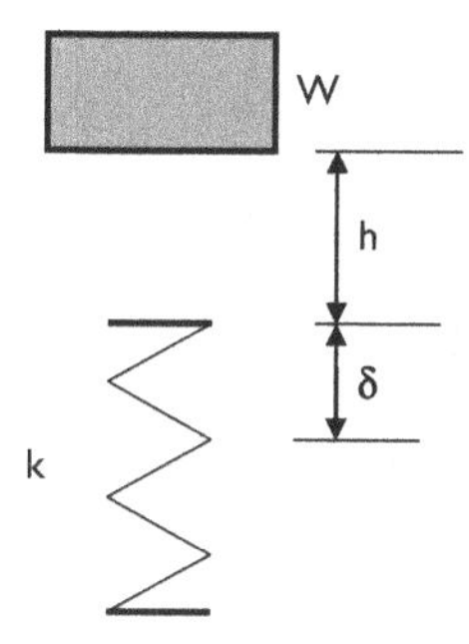

Figure 9.15 Dynamic pillar (spring) loading model.

Example 9.16 Consider a wooden support post 8 inches (20 cm) in diameter and 7 ft (2.13 m) high that has an elastic modulus of 1.6 (10^6) psi (11.0 GPa) and compressive strength of 4,000 psi (27.6 MPa) parallel to the grain. Estimate the shortening the post experiences at failure under static and dynamic loads.

***Solution*:** Under axial load, the relative displacement between top and bottom of the post is given by the formula $\delta(\text{static}) = PL/AE$ and according to (9.15), the dynamic deflection is twice the static deflection under the same load. The difference is in the load application, of course. In the static case, $\delta(\text{static}) = (P/A)(L/E) = (4000)(7)(12)/(1.6)(10^6) = 0.21$ inches, and in the dynamic case the $\delta(\text{dynamic}) = 0.421$ inches (0.533 and 1.06 cm, respectively).

Studies of wide area collapses often indicate the presence of a strong, stiff stratum in the overburden that may fail suddenly to form a large subsidence cylinder over the collapsed portion of a mine. Such an event may be a mechanism for dynamic pillar loading and cascade failure that otherwise seems precluded by static equilibrium requirements on which the tributary area method of pillar stress calculation is based. As the region below the flexed stratum is mined to a high extraction ratio over a large area, induced tensile bending stress at the abutments of the mined region may exceed tensile strength producing fractures that disrupt continuity of the stratum and allow it to move downwards should the support action of pillars below be insufficient. Alternatively, the pillars may be strong enough not to fail under the increased loading, while a soft floor under the pillars may yield or a soft roof may be penetrated. Both situations would allow collapse.

A situation occurs in steep vein mining that is similar in some respects to room and pillar mining in softrock. Figure 9.16 illustrates the situation in a simplified diagram of a vertical vein. As mining advances upwards, the remaining rock becomes progressively smaller and under greater stress. If one neglects the support action of the fill, then the stress in the pillar follows an extraction ratio formula: $S_p = S_h/(1 - R)$ where S_h is the horizontal preexcavation stress. In many mining districts, the horizontal stress is greater than the vertical preexcavation stress as is the case in the Coeur d'Alene mining district of northern Idaho, USA. As the pillar size diminishes, a condition is reached where bursting is highly probable. As mining progresses past this condition, the pillar will begin to crush and become less likely to burst. The obvious solution to pillar bursting is to eliminate the pillar. Indeed, this was done at one of the district mines where the direction of mining was changed from up to down, although the use of fill was retained (underhand cut and fill method). The number of large, damaging

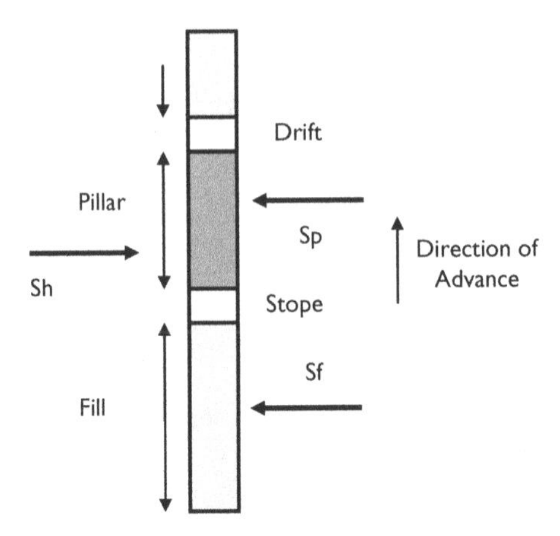

Figure 9.16 Mining upwards (overhand stoping) in a vertical vein of ore using sand as fill.

bursts was reduced, while the number of less consequential small bursts increased as a consequence of the change.

The same lesson of reducing bursts or bumps by eliminating pillars was demonstrated clearly in an underground coal mine in central Utah, USA, in a region known as the Book Cliffs. Mines in this region are developed from outcrops and experience increasing bumps as the depth of cover increases. A traditional method of developing longwall panels in the district was to use multiple entries (four or more) that also formed multiple pillars. Reduction of the number of entries to just two reduced the number of pillars and the number of damaging pillar bursts. Two entries require a regulatory exception in the US. Most longwall mines in the US use three entries, although some longwall mines elsewhere in the world use a single entry. In fact a study of advantages and disadvantages of a single entry system versus a double entry system was done at the mine in question. The main advantage from the rock mechanics view was elimination of the pillar in the two entry system. A potential economic advantage was in the elimination of an additional entry and the crosscuts between. However, in the overall scheme of operations, the single entry system was not sufficiently advantageous to justify a change.

Fault slip

Slip on faults remote from excavation generates waves that may be damaging and produce bursts in hardrock environments and bumps in softrock environments. Normal faulting, reverse faulting and thrust faulting generate longitudinal and shear waves in different proportions that are reflected and refracted at contacts between different formations. The most important reflections occur at the excavation boundaries where fast spalling, slip on nearby discontinuities, and rock-support interactions occur. Wave interactions with underground excavations have produced some surprising phenomena, some akin to the magician's trick of snapping a table cloth from underneath a dinnerware setting. Great lateral movement of the table cloth occurs while the dinnerware appears to remain in place. If one were moving with the table cloth, the dinnerware would appear to have moved horizontally. Figure 9.17 is an example of such phenomena that occur in practice. The first motion of an excavation may be either dilatational in the sense that the relative distance between roof and floor increases

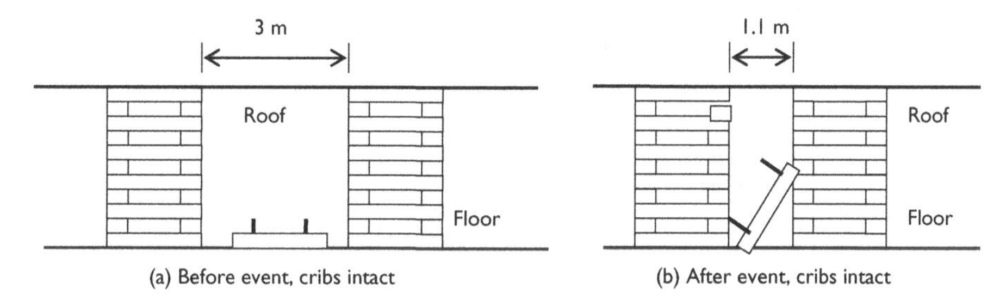

Figure 9.17 Displacement of timber cribs and rail track during a bump (after Whyatt, 2009).

or compressional with loading of pillars and standing support. Both are usually accompanied by lateral shearing action and sometimes large horizontal displacement as well. Motion of equipment, standing support such as timber cribs, and adjacent pillars will be affected in different ways depending on the first motion. In any case, violent ejection of rock into an excavation is almost certain to occur without adequate protection against such events. Although details of bursts and bump phenomena are not easy to quantify, there are design guides for mitigation of burst potential and control of damage.

Efforts to mitigate damage from fault-slip bursts and bumps include operational practices, mining methods, and dynamic support design. The symposium proceedings edited by Kaiser & McCreath (1992) contains much relevant information on the subject. Each mine is different so only the most general guidelines can be given. An example of an operational practice is to blast going off-shift to relieve strain energy accumulation in a timely manner when few personnel are underground. Mining methods include those that minimize pillar formation and avoid mining towards highly stressed regions. Mining away from regions of high stress moves operations in a favorable direction and is clearly a desirable practice. Mine fill also assists in ground control, especially when fill is placed alongside pillars that must be left in place. The confinement provided by fill augments pillar strength to some degree and, importantly, prevents large displacements of loose rock that may develop at pillar, hanging, and footwalls of mined areas. Indeed, mine fill is essential for overall mine stability, unless the mining method involves deliberate caving of the rock mass.

Guides to hazard mitigation and damage control include reduction of stress and use of support and reinforcement specifically intended for dynamic loading. The main support measure consists of rock bolts and screen that contains loose rock that may form in time or suddenly in the event of a burst or bump. Bolts and screen such as wire mesh may be used as a matter of course for reinforcement of the opening walls and to prevent small rock fragments from spilling into the opening. When considered for dynamic effects, additional anchorage of the screen may be necessary, say, in the form of cables laced in a criss-cross pattern along the screened walls and anchored by additional bolts.

An important feature of support designed to resist dynamic loading is the capacity to yield or telescope extremely fast thus avoiding damage that would otherwise occur. Bolts that are grouted in the bolt hole the full length of the bolt are prone to breakage under dynamic loading as are friction bolts that make contact along the full length of the bolt hole. Point anchored mechanical rock and roof bolts respond to dynamic loading in much better fashion. The reason is in the length of the "unbonded" portion of a bolt. In essence, long bolts are better able to absorb dynamic loading than short bolts. Consider a bolt under axial load P with Young's modulus E and cross-sectional area A. According to Hooke's law, the axial displacement is $u = PL/AE$ where L is bolt length. Work done on the bolt by the axial load is stored as strain energy. Thus, $U = \int_L Pdu = \int_L (AE/L)udu = (\sigma^2/2E)(AL)$. The last term is the product of strain energy per unit volume by volume of the bolt. Thus, a long bolt is able to absorb more strain energy than a short bolt other factors being equal. A simpler way of comparing long and short bolts if to compute the stretch to failure δ, that is, $\delta = T_o L/E$ where T_o is tensile strength. A long bolt can therefore accommodate greater inward movement of excavation walls than a short bolt before failing. A compromise that is sometimes used is to cement a bolt only over a portion of the bolt length while leaving a suitable open length between the bolt collar and the cemented segmented that runs to the hole end. The cemented segmented provides superior anchorage to point anchorage, while the cement-free portion allows for superior energy absorption to resist dynamic loading. Tailoring of the cemented portion to allow yielding also improves bolt resistance to failure under dynamic loading. In burst-prone mines, the additional expense of special bolts may be justified by reduction of burst damage and savings in repairs.

Example 9.17 Consider a point anchored, 3/4-inch diameter, grade 75, rock bolt installed in the wall of a tunnel. If Young's modulus of the bolt is $30(10^6)$ psi, determine the bolt length necessary to withstand a dynamic inward wall movement of 0.20 inches.

Solution: The axial stretch of the bolt is given $\delta = PL/AE$ and the stretch per unit length of bolt is $\delta/L = (25,\ 100)/(0.334)(30.0)(10^6) = 0.0025$ in./in where the bolt specifications are from Table 3.4. Thus, the bolt length should be 0.20/0.025 or 80 inches, that is, 6.7 ft (2.0 m).

9.3 Event location

Location of bumps, bursts, and related events is important to operations and understanding of the excavation environment. Spatial arrays of geophones serve the purpose. The basic principle is simply (distance)=(velocity)(time). Suppose an event occurs at (x_s, y_s, z_s) at time (t_s) in a rock mass where wave velocity is v. A geophone or receiver "i" senses the wave from the source "s" at time t_i. The distance from source to receiver is traveled by the wave according to the basic principle during the intervening time. Thus,

$$[(x_i - x_s)^2 + (y_i - y_s)^2 + (z_i - z_s)^2]^{1/2} = v(t_i - t_s) \tag{9.16}$$

relative to a rectangular coordinate system and master clock that keeps the same time at all receivers. There are four unknowns in (9.16), i.e., (x_s, y_s, z_s, t_s). However, (9.16) is not a linear expression, so some thought is required to compute event locations, especially when there are many geophones in an array. In this regard, event location in one, two, and three spatial dimensions is instructive. For details one may consult Hardy (2003) where the focus is on laboratory measurements. Applications of microseismic monitoring in underground mining are described by Blake, Leighton, and Duvall (1974).

Source and receivers on a line are shown in Figure 9.17.

Equation (9.16) has a simple, linear expression in this case. Thus,

$$(x_s - x_1) = v(t_1 - t_s)$$
$$(x_2 - x_s) = v(t_2 - t_s)$$

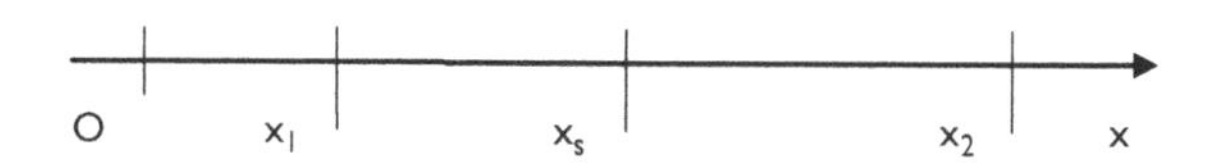

Figure 9.17 Source and receivers on a line with source between receivers.

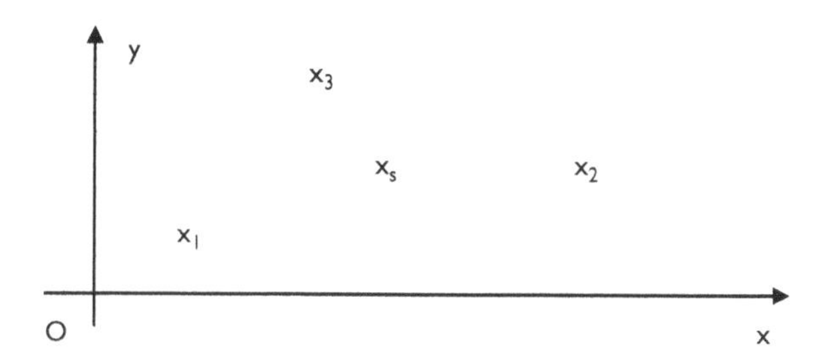

Figure 9.18 Event and receivers in a plane with event source inside the receiver triangle.

that can be brought into the form

$$x_s + vt_s = vt_1 + x_1$$
$$x_s + vt_s = vt_2 + x_2.$$

Alternatively in matrix form,

$$\begin{bmatrix} 1, & +v \\ 1, & -v \end{bmatrix} \begin{Bmatrix} x_s \\ t_s \end{Bmatrix} = \begin{Bmatrix} vt_1 + x_1 \\ -vt_2 + x_2 \end{Bmatrix}$$

that is readily inverted to

$$\begin{Bmatrix} x_s \\ t_s \end{Bmatrix} = \begin{bmatrix} -v, & -v \\ -1, & +1 \end{bmatrix} \begin{Bmatrix} vt_1 + x_1 \\ -vt_2 + x_2 \end{Bmatrix} /(-2v)$$

Thus,

$$\begin{aligned} x_s &= (x_1 + x_2)/2 - v(t_2 - t_1)/2 \\ t_s &= (t_1 + t_2)/2 - (x_2 - x_1)/2v \end{aligned} \tag{9.17}$$

This solution could be obtained directly at the start, but the reformulation into matrix form is helpful in locating an event on a plane.

A feature that is important to note is that the event is assumed to occur between receivers. When the event occurs outside the receivers, the event cannot be located. However, wave velocity can be determined from the observations.

An event on a plane is shown in relationship to three receivers in Figure 9.18.

Application of (9.16) gives three equations in the three unknowns (x_s, y_s, t_s). Thus,

$$\begin{aligned} [(x_1 - x_s)^2 + (y_1 - y_s)^2]^{1/2} &= v(t_1 - t_s) \\ [(x_2 - x_s)^2 + (y_2 - y_s)^2]^{1/2} &= v(t_2 - t_s) \\ [(x_3 - x_s)^2 + (y_3 - y_s)^2]^{1/2} &= v(t_3 - t_s) \end{aligned} \qquad (a, b, c)$$

This system can be reduced to a linear system by squaring each equation in turn and then subtracting a from b, b from c, and c from a. This procedure eliminates the squared terms (x_s^2, y_s^2, t_s^2) with the result

$$\begin{bmatrix} 2x_{12}, & 2y_{12}, & -2v^2t_{12} \\ 2x_{23}, & 2y_{23}, & -2v^2t_{23} \\ 2x_{31}, & 2y_{31}, & -2v^2t_{31} \end{bmatrix} \begin{Bmatrix} x_s \\ y_s \\ t_s \end{Bmatrix} = \begin{Bmatrix} v^2(t_2^2 - t_1^2) - (x_2^2 - x_1^2) - (y_2^2 - y_1^2) \\ v^2(t_3^2 - t_2^2) - (x_3^2 - x_2^2) - (y_3^2 - y_2^2) \\ v^2(t_1^2 - t_3^2) - (x_1^2 - x_3^2) - (y_1^2 - y_3^2) \end{Bmatrix} \quad (9.18)$$

The notation in the left side coefficient matrix indicates, e.g., $x_{23} = x_2 - x_3$, and $t_{31} = t_3 - t_1$. The order of the subscripts in this notation is important. Examination of (9.18) shows that the equation is dimensionally consistent in units of distance-squared. Inversion or solution of (9.18) may be done in several ways including the method of determinants as used in the two by two system for locating events on a line.

Extension of the plane procedure to three-space is straightforward with necessary additions of z-terms to columns and rows in (9.18). A minimum of four well-positioned receivers that box in the event is now required.

As with any system of equations, ill-conditioning is always a possibility. Another consideration in practice is the presence of many geophones. The system is then over determined by more observations (equations) than unknowns. These complications increase the mathematical complexity of source location technology; although the basic concept is simple and leads to a system that in matrix form is similar to (9.18). In case of 12 receivers, one has $[A]_{12x4}\{x\}_{4x1} = \{b\}_{12x1}$. Because $[A]$ is not square, inversion is problematic. However, the transformation obtained by multiplication using the transpose of $[A]$ offers some hope of solution. Thus, $[A]^T_{4x12}[A]_{12x4}\{x\}_{4x1} = [A]^T_{4x12}\{b\}_{12x1}$ may be invertible, perhaps in a least-squares sense or by some other mathematical technique.

Example 9.18 Consider an event source on a line passing through receivers and source. Determine from the basic principle of (distance)=(velocity)(time), the source location and time of event when locations are ordered such that $x_1 < x_2 < x_s$ and when $x_s < x_1 < x_2$, that is, when the source is on the same side of the receivers rather than in between.

***Solution*:** The basic principle states that

$$x_1 - x_s = v(t_1 - t_s)$$
$$x_2 - x_s = v(t_2 - t_s)$$

Solving these equations for the source location shows that

$$x_s = (x_1 - vt_1) + vt_s$$
$$x_s = (x_2 - vt_2) + vt_s$$

These last two equations are straight lines when plotted in an $t_s\ x_s$ plane, as shown in the sketch.

In fact the two lines are the same as can be shown by subtracting the last two equations to obtain

$$(x_2 - x_1) = v(t_2 - t_1)$$

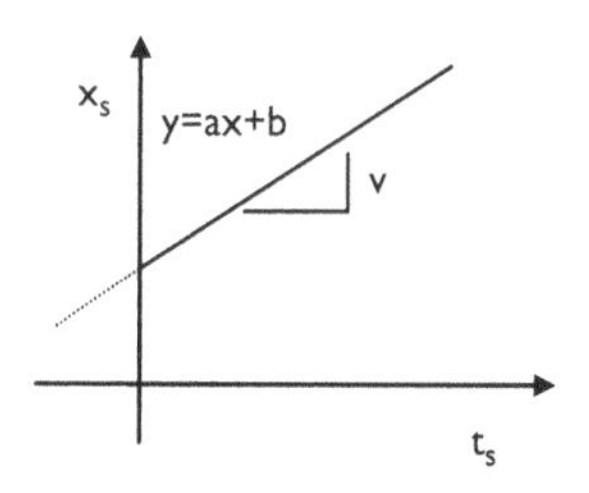

Sketch for Example 9.18

After substitution of this result into the second equation for source location, one has

$$\begin{aligned} x_s &= (x_2 - vt_2) + vt_s \\ &= [x_1 + v(t_2 - t_1) - vt_2] + vt_s \\ x_s &= (x_1 - vt_1) + vt_s \end{aligned}$$

Thus, one can determine wave velocity from the observations in this case, but the source location cannot be determined because there is only one independent equation but two unknowns.

Example 9.19 Consider an event on a line such that the event occurs at the location of one of the receivers. Determine whether this event can be located from receiver recordings.

***Solution*:** The given data are $x_s = x_1 < x_2$, $t_s = t_1$ or $x_1 < x_2 = x_s$, $t_s = t_2$. From the basic principle (9.16) $(x_1 - x_s) = v(t_1 - t_s) = 0$ or $(x_2 - x_s) = v(t_2 - t_s) = 0$. In the first case $x_1 = x_s$ and in the second case $x_2 = x_s$. Thus, in both cases the event location is determined.

Example 9.20 Consider a receiver array on a plane as shown in Figure 9.18 and suppose an event occurs at one of the receivers, say, x_2. Show that the event can be located by receiver data.

***Solution*:** In this case, the basic principle (9.16) shows that

$$(x_2 - x_s)^2 + (y_2 - y_s)^2 = v^2(t_2 - t_s)^2$$

The arrival time at the receiver $t_2 = t_s$ according to the given conditions. Therefore

$$(x_2 - x_s)^2 = -(y_2 - y_s)^2$$

This result requires both sides of the equation to equal zero (all numbers are real). Thus, $x_2 = x_s$ and $y_2 = y_s$ and the event can be located by the receiver data.

Example 9.21 Suppose planar array of receivers detects an event that occurs on a line between two receivers, say, between receivers 1 and 2. Determine whether the event can be located from the receiver data.

***Solution*:** The sketch shows the situation.

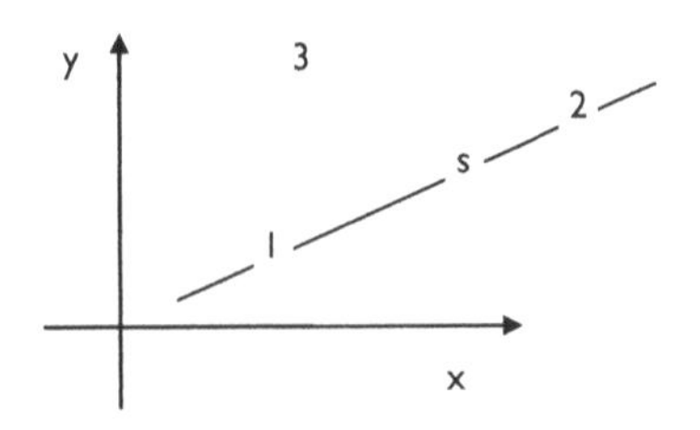

Sketch for Example 9.21

Application of (9.16) to receivers 1 and 2 gives

$$(x_1 - x_s)^2 + (y_1 - y_s)^2 = v^2(t_1 - t_s)^2$$
$$(x_2 - x_s)^2 + (y_2 - y_s)^2 = v^2(t_2 - t_s)^2$$

Rotation of the axes to bring the new x-axis on line with receivers 1 and 2 transforms these two equations into

$$(x_s - x'_1) = v(t_1 - t_s)$$
$$(x'_2 - x_s) = v(t_2 - t_s)$$

with proper choice of the square root operation. This system is just the system for locating an event between two receivers on a line described in the text. Therefore the event can be located, provided it occurs between receivers.

Example 9.22 Consider a planar array of three receivers located at the vertices of a right triangle and suppose arrival times are the same at each receiver. Determine the event location and time.

***Solution*:** The sketch shows the receiver array in a plane with origin fixed at the 1-receiver. From (9.16)

$$(x_1 - x_s)^2 + (y_1 - y_s)^2 = v^2(t_1 - t_s)^2$$
$$(x_2 - x_s)^2 + (y_2 - y_s)^2 = v^2(t_2 - t_s)^2$$
$$(x_3 - x_s)^2 + (y_3 - y_s)^2 = v^2(t_3 - t_s)^2$$

Because the arrival times are the same, the distance between receivers and source must also be the same. In consideration of the receiver locations then,

$$(-x_s)^2 + (-y_s)^2 = (x_2 - x_s)^2 + (-y_s)^2 = (-x_s)^2 + (y_3 - y_s)^2$$

Solution of the first two equations shows that $x_s = x_2/2$. Similarly, solution of the first and third equations shows that $y_s = y_2/2$. The event time may be obtained by back substitution. Thus, $t_s = t_1 + [(x_2/2)^2 + (y_2/2)^2]^{1/2}/v$.

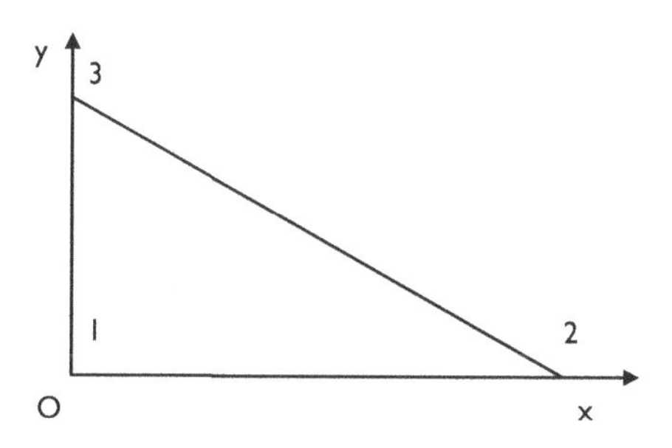

Sketch for Example 9.22

Example 9.23 Show that when two receivers detect an event at the same time, the source must lie on a perpendicular bisector of the line passing through the two receivers.

***Solution*:** From 9.16, distance travel - time relationships for the two receivers are

$$(x_1 - x_s)^2 + (y_1 - y_s)^2 + (z_1 + z_s)^2 = v^2(t_1 - t_s)^2$$
$$(x_2 - x_s)^2 + (y_2 - y_s)^2 + (z_2 + z_s)^2 = v^2(t_2 - t_s)^2$$

Rotation of axes to bring the new x-axis on line through the two receivers shows that

$$(-x_s)^2 + (-y_s)^2 + (-z_s)^2 = (x_2 - x_s)^2 + (-y_s)^2 + (-z_s)^2$$

in consideration of equal travel times where the new origin is positioned at receiver 1. The solution to this equation for x_s is $x_s = x_2/2$. This bisects the line passing through the two receivers. The source must lie in a plane perpendicular to this line and must contain this point. If the source were out of the bisecting plane, the travel times would differ contrary to the given conditions of the problem. Alternatively, $(-x_2/2)^2 + d_1^2 = (x_2 - x_2/2)^2 + d_2^2$ where the d 's are the yz distances between source and receivers. These distances must be equal to satisfy the basic, physical distance - time relationship. In essence, the event source must lie on the intersection of two spheres of equal radii emanating from the considered receivers. The intersection is a circle perpendicular to the line of receivers and the circle radius is a perpendicular bisector of the line.

Example 9.24 An event occurs in 3-space and is simultaneously recorded by four receivers positioned at the vertices of a tetrahedron as shown in the sketch. Determine the event location and time

***Solution*:** By the guiding principle (9.16)

$$(0 - x_s)^2 + (0 - y_s)^2 + (0 + z_s)^2 = v^2(t_o - t_s)^2$$
$$(x_2 - x_s)^2 + (0 - y_s)^2 + (0 + z_s)^2 = v^2(t_o - t_s)^2$$
$$(0 - x_s)^2 + (y_3 - y_s)^2 + (0 + z_s)^2 = v^2(t_o - t_s)^2$$
$$(0 - x_s)^2 + (0 - y_s)^2 + (z_4 + z_s)^2 = v^2(t_o - t_s)^2$$

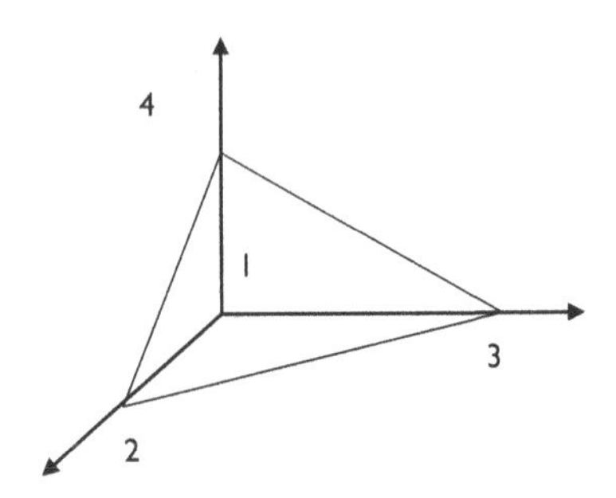

Sketch for Example 9.24

where the receivers sense the event at the same time t_o. The solution to this system for the source location using the first and second, then the first and third, and finally the first and fourth equations is $x_s = x_2/2, y_s = y_3/2, z_s = z_4/2$. Substitution of this result back into the first equation gives $t_s = t_0-(1/4\ v)[(x_2)^2 + (y_3)^2 + (z_4)^2]^{1/2}$.

Example 9.25 Three receivers on a line detect an event that occurs between receivers 2 and 3. Determine the event location and time.

***Solution*:** The receivers are located on a line as shown in the sketch.

—O——— 1 ——— 2 —— s ——— 3 ——→ x

Sketch for Example 9.25

From the basic distance - time principle

$$\begin{aligned} x_s - x_1 &= v(t_1 - t_s) & \qquad x_s + vt_s &= x_1 + vt_1 \\ x_s - x_2 &= v(t_2 - t_s) \quad \text{or} & \qquad x_s + vt_s &= x_2 + vt_2 \\ x_3 - x_s &= v(t_3 - t_s) & \qquad x_s - vt_s &= x_3 - vt_3 \end{aligned}$$

This system is over determined because there are three equations in only two unknowns. In matrix form the system is

$$\begin{bmatrix} 1 & v \\ 1 & v \\ 1 & -v \end{bmatrix} \begin{Bmatrix} x_s \\ t_s \end{Bmatrix} = \begin{Bmatrix} x_1 + vt_1 \\ x_2 + vt_2 \\ x_3 + vt_3 \end{Bmatrix}$$

Premultiplying by the transpose of the rectangular matrix on the left side gives

$$\begin{bmatrix} 1 & 1 & 1 \\ v & v & -v \end{bmatrix} \begin{bmatrix} 1 & v \\ 1 & v \\ 1 & -v \end{bmatrix} \begin{Bmatrix} x_s \\ t_s \end{Bmatrix} = \begin{bmatrix} 1 & 1 & 1 \\ v & v & -v \end{bmatrix} \begin{Bmatrix} x_1 + vt_1 \\ x_2 + vt_2 \\ x_3 - vt_3 \end{Bmatrix}, \text{so}$$

$$\begin{bmatrix} 3 & v \\ v & 3v^2 \end{bmatrix} \begin{Bmatrix} x_s \\ t_s \end{Bmatrix} = \begin{Bmatrix} (x_1 + vt_1) + (x_2 + vt_2) + (x_3 - vt_3) \\ v(x_1 + vt_1) + v(x_2 + vt_2) - v(x_3 - vt_3) \end{Bmatrix}$$

The inverse of the 2x2 coefficient matrix on the left is

$$\begin{bmatrix} 3v^2 & -v \\ -v & 3 \end{bmatrix} / (8v^2).$$

Hence

$$\begin{Bmatrix} x_s \\ t_s \end{Bmatrix} = (1/8v^2) \begin{bmatrix} 3v^2 & -v \\ -v & 3 \end{bmatrix} \begin{Bmatrix} (x_1 + vt_1) + (x_2 + vt_2) + (x_3 - vt_3) \\ v(x_1 + vt_1) + v(x_2 + vt_2) - v(x_3 - vt_3) \end{Bmatrix}$$

After multiplying the right side, the result is

$$\begin{Bmatrix} x_s \\ t_s \end{Bmatrix} = \begin{Bmatrix} (1/4)(x_1 + x_2) + (1/2)x_3 - (v/4)(t_1 + t_2) - (v/2)t_3 \\ (1/4v)(x_1 + x_2) - (1/2v)x_3 + (1/4)(t_1 + t_2) + (1/2)t_3 \end{Bmatrix}$$

Comment: In this simple case, the over determined system is readily solved by the procedure of "squaring the system". When many receivers are considered and the event may not be boxed in by the array, the situation is not so easily resolved, although techniques are available for event location in such circumstances.

9.4 Problems

9.1 Consider a long steel bar of diameter D and length L (10 ft) with bar wave speed $c_B = 20,000$ ft/s. Modulus of the steel is $30(10^6)$ psi (E) and Poisson's ratio is 0.25 (v). Impact at one end of the bar generates a square compressive stress pulse of amplitude σ and length l. Derive a formula for the time of travel to the far end of the bar and back to the struck end.

9.2 If the far end of the bar in Problem 9.1 is free, determine (a) the particle velocity as the pulse is traveling along the bar, and (b) the peak particle velocity during reflection from this end, when the pulse amplitude is 90,000 psi. Assume a steel specific weight of 0.284 lbf/in^3.

9.3 Consider a long steel bar of diameter D and length L (3.1 m) with wave speed $c_L = 6,100$ m/s. Modulus of the steel is 207 GPa (E) and Poisson's ratio is 0.25 (v). Impact at one end of the bar generates a square compressive stress pulse of amplitude σ and length l. Derive a formula for the time of travel to the far end of the bar and back to the struck end.

9.4 If the far end of the bar in Problem 9.3 is free, determine (a) the particle velocity as the pulse is traveling along the bar and (b) the peak particle velocity during reflection from this end, when the pulse amplitude is 621 MPa. Assume a steel specific weight of 77.5 kN/m^3.

9.5 A sharp-fronted triangular wave ("saw tooth") is reflected upon normal incidence at a free face. Peak stress is σ, longitudinal wave speed is c_L ft/s, pulse width is L ft. Dynamic tensile strength of rock is T_o that is 1/20th of dynamic compressive strength. Consequently, a free face reflection will cause tensile failure and form a spall. Derive an expression for spall thickness.

9.6 A sharp-fronted triangular wave ("saw tooth") is reflected upon normal incidence at a free face. Peak stress is σ (138 MPa), longitudinal wave speed is c_L (6,100) m/s, pulse

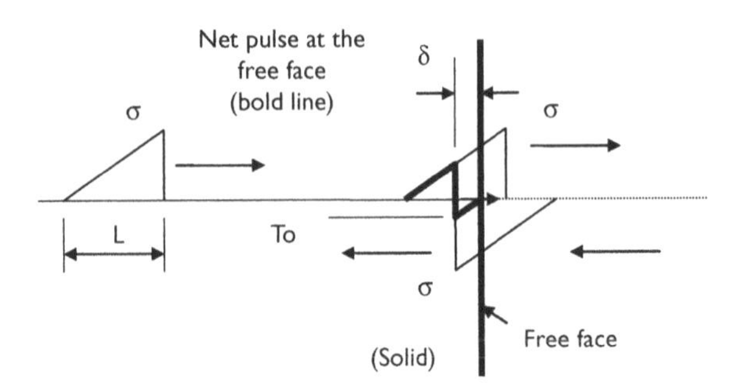

Sketch for Problems 9.5-9.10.

width is L (12 m). Dynamic tensile strength of rock is T_o that is 1/20th of dynamic compressive strength. Consequently, a free face reflection will cause tensile failure and form a spall. Determine the spall thickness.

9.7 A sharp-fronted triangular wave ("saw tooth") is reflected upon normal incidence at a free face. Peak stress is σ (138 MPa), longitudinal wave speed is c_L (6,100) m/s, pulse width is L (12 m) and specific weight γ =25 kN/m^3. Dynamic tensile strength of rock is T_o that is 1/20th of dynamic compressive strength. Consequently, a free face reflection will cause tensile failure and form a spall. Spall thickness is given by $\delta = T_o\, L/2\sigma$. Derive a formula for spall velocity and then evaluate the formula with the given data. Hint: consider the momentum trapped in the spall per square meter of face.

9.8 A sharp-fronted triangular wave ("saw tooth") is reflected upon normal incidence at a free face. Peak stress is σ (20,000 psi), longitudinal wave speed is c_L (20,000) ft/s, pulse width is L (40 ft) and specific weight γ =158 lbf/ft^3. Dynamic tensile strength of rock is T_o that is 1/20th of dynamic compressive strength. Consequently, a free face reflection will cause tensile failure and form a spall. Spall thickness is given by $\delta = T_o\, L/2\sigma$. Derive a formula for spall velocity and then evaluate the formula with the given data. Hint: consider the momentum trapped in the spall per square foot of face.

9.9 A sharp-fronted triangular wave ("saw tooth") is reflected upon normal incidence at a free face after traveling along a cylindrical wave front from a blast hole 5 m from the face. Peak stress is σ (138 MPa) at the blast hole that has a diameter of 0.3 m, longitudinal wave speed is c_L (6,100) m/s, pulse width is L (12 m) and specific weight γ =25 kN/m^3. Dynamic tensile strength of rock is T_o that is 1/20th of dynamic compressive strength. Consequently, a free face reflection will cause tensile failure and form a spall. Spall thickness is given by $\delta = T_o\, L/2\sigma$. Derive a formula for spall velocity and then evaluate the formula with the given data. Hint: consider the momentum trapped in the spall per square meter of face.

9.10 A sharp-fronted triangular wave ("saw tooth") is reflected upon normal incidence at a free face after traveling along a cylindrical wave front from a blast hole 15 ft from the face. Peak stress is σ (20,000 psi) at the blast hole that has a diameter of 1.0 ft, longitudinal wave speed is c_L (20,000) ft/s, pulse width is L (40 ft) and specific weight γ = 158 lbf/ft^3. Dynamic tensile strength of rock is T_o that is 1/20th of dynamic compressive strength. Consequently, a free face reflection will cause tensile failure and form a spall. Spall thickness is given by $\delta = T_o\ L/2\sigma$. Derive a formula for spall velocity and then evaluate the formula with the given data. Hint: consider the momentum trapped in the spall per square foot of face.

9.11 Angle hole drilling is used in bench blasting in quarry as shown in the sketch. A blast in a cylindrical hole that is parallel to the free face produces a triangular stress pulse with a magnitude just at the compressive strength of the material (σ = 30,000 psi, 206.9 MPa). A wave front travels from the blast hole wall to the free face. Dynamic tensile strength (T_o) is just 1/20th of the compressive strength. The blast hole is 1 ft (0.3048 m) in diameter. The distances h1, h2, and h3, are 7 ft, 5 ft, and 3 ft (2.13, 1.52, and 0.91 m), respectively. Specific weights are: $\gamma 1$ = 158 lbf/ft^3, $\gamma 2$ = 168 lbf/ft^3, γ 3 = 148 lbf/ft^3. (25, 26.6, 23.4 kN/m^3) Longitudinal wave speeds in materials 1,2 and 3 are 15,000 fps, 18,000 fps, and 12,000 fps (4,572, 5,486, 3,658 m/s), respectively. Determine the sign and amplitude of the reflected wave at the face that follows a ray path perpendicular to the face as shown in the sketch.

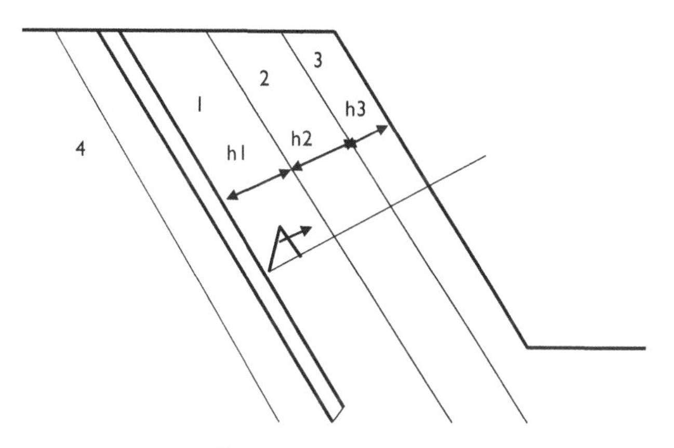

Sketch for Problem 9.11

9.12 Consider the data and sketch given in Problem 9.11 where the transmission coefficients for longitudinal waves are R12=1.121 and R23=0.74 as obtained from (9.9) in the text. Suppose the wave front is cylindrical. Blast hole diameter is 1 ft (0.3 m). Estimate the amplitude of the reflected wave at the free face of the bench.

9.13 A spherical source with a diameter of 1 ft (0.3 m) at ground surface emits longitudinal and shear waves of amplitudes 40,000 and 20,000 psi (276 and 138 MPa), respectively. An underground opening exists below the source in stratified ground as shown in the sketch. Properties of the five strata in the sketch are given in the table. Calculate the wave speeds c_L, c_S (m/s) and then determine the magnitude of the reflected longitudinal and shear waves upon normal incidence at the excavation roof.

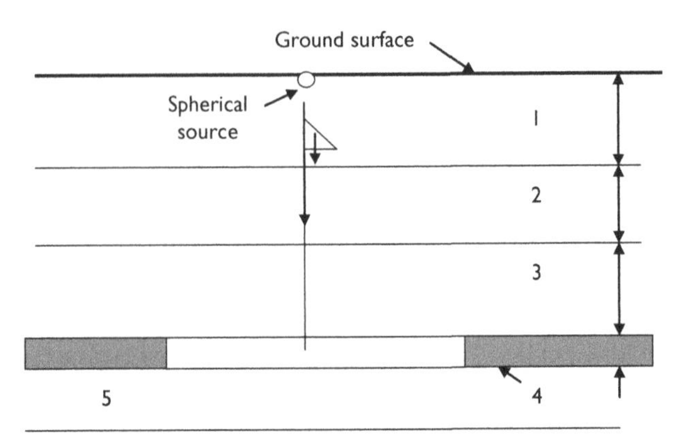

Sketch for Problem 9.13

Table of properties for Problem 9.13.

Property Material	E (*GPa*)	ν	Co (*MPa*)	To (*MPa*)	(*kN/m*^{3})	h(*m*)
1	10.0	0.20	50.0	5.0	18.9	230
2	30.0	0.30	100.0	6.7	22.1	176
3	20.0	0.25	150.0	7.5	25.3	84
4	2.0	0.15	20.0	1.3	14.2	4.0
5	25.0	0.35	175.0	8.7	26.6	12.3

9.14 Given the data in Problem 9.13, determine the travel times of the longitudinal and shear waves along a vertical ray path to the top of the excavation from the source point.

9.15 Consider the situation shown in the sketch where a source and receiver are positioned a variable distance $d1$ apart on the surface. Depth to bedrock is h.

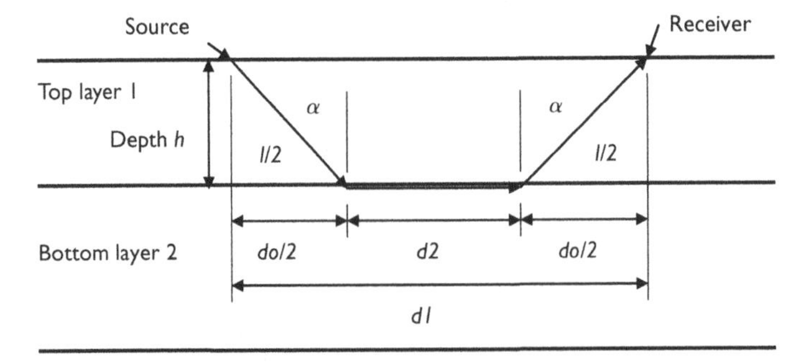

Sketch for Problem 9.15 illustrating the refraction method for determining depth to bedrock

As the receiver is positioned farther from the source, the arrival time $t1$ of the first wave increases. The received wave travels through the top layer of material at a speed

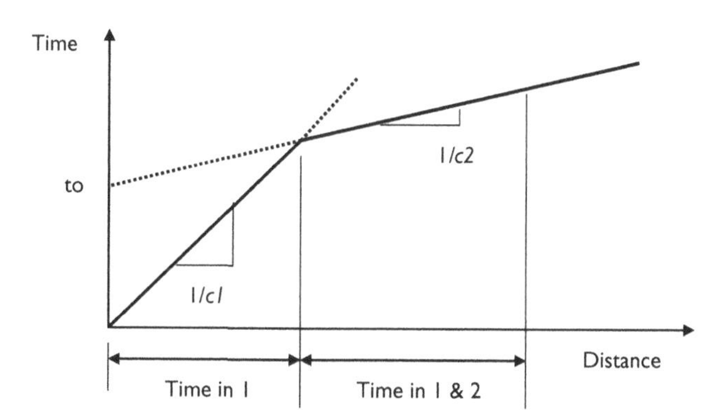

Sketch showing a plot of time versus distance between source and receiver for Problem 9.15.

of $c1$. As the receiver is positioned even farther from the source, the first arrival will be a wave that traveled downwards through the top layer and then sped along the bottom layer all the while emitting waves upwards towards the surface. Wave speed in the bottom layer is $c2$ and is greater than $c1$ for this event to occur. Show that data from a plot of arrival time t versus distance d allows one to determine the depth to bedrock. Such a time-distance plot is shown in the second sketch.

9.16 A longwall panel is excavated full height in a softrock seam 4 m thick at a depth of 500 m. The panel is 300 m wide and will eventually be 3,000 m long. Preexcavation stress is caused by gravity along. If the seam material has an unconfined compressive strength of 50 MPa, determine how far the face can be advanced before stress concentration at the face raises the stress to failure.

9.17 A longwall panel is excavated full height in a softrock seam 12 ft thick at a depth of 1,500 ft. The panel is 900 ft wide and will eventually be 9,000 ft long. Preexcavation stress is caused by gravity along. The seam material has an unconfined compressive strength of 7,250 psi. The face is advanced 360 ft giving the panel a width to height ratio of 30. Suppose the stress concentration at the face is inversely proportional to the square of distance from the face, that is, $S' = So/(1+x)^2 + Sv$ where So, Sv, and S' are starting stress, vertical premining stress, and stress at x ft from the face along the seam. Stresses are in psi. Determine the distance along the seam that the vertical stress exceeds unconfined compressive strength.

9.18 A longwall panel is excavated full height in a softrock seam 12 ft (4 m) thick at a depth of 1,500 ft (500 m). The panel is 900 ft (300 m) wide and will eventually be 9,000 ft (3000 m) long. Preexcavation stress is caused by gravity along. The seam material has an unconfined compressive strength of 7,250 psi (50 MPa), Young's modulus is $7.25(10^6)$ psi (50 GPa) and Poisson's ratio ν=0.2. The face is advanced 360 ft (120 m) giving the panel a width to height ratio of 30. Suppose the stress concentration at the face is inversely proportional to the square of distance from the face, that is, $S' = So/(1+x)^2 + Sv$ where So, Sv, and S' are starting stress, vertical premining stress, and stress at x ft from the face along the seam. Stresses are in psi (MPa). Determine the distance along the seam that the vertical stress exceeds unconfined compressive strength and then compute the energy in excess of the energy associated with stress at the unconfined compressive strength value.

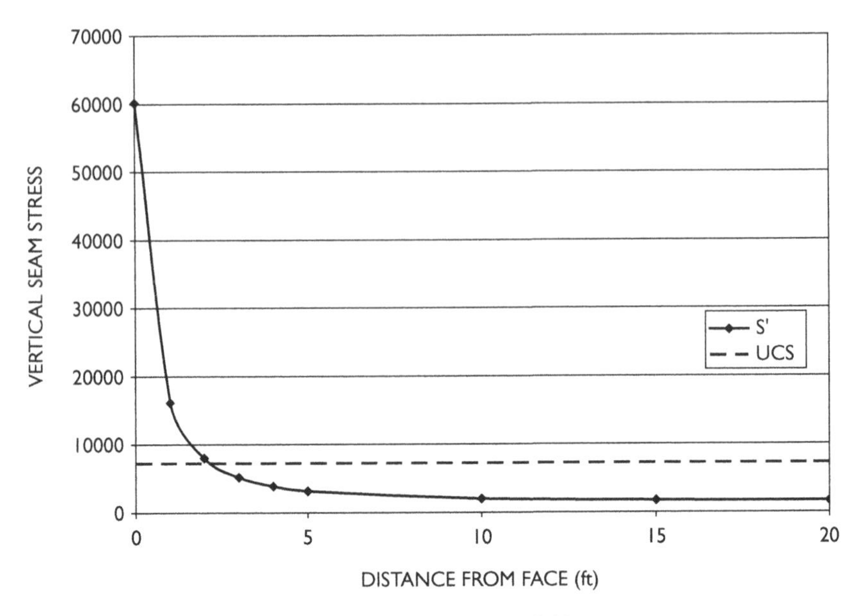

Sketch for Problem 9.17 - 9.19.

Note: This energy is excess, so to speak, and would preferably be released by yielding or possibly in a violent manner by a bump. The estimate may be based on a one-dimensional calculation.

9.19 Suppose the stress concentration at the a pillar wall is inversely proportional to the square of distance from the face, that is, $S' = So / (1 + x)^2 + Sv$ where *So, Sv*, and S' are starting stress, vertical premining stress, and stress at x ft from the wall into the pillar. Stresses are in psi. Further suppose that pillar strength unconfined compressive strength is 2,500 psi, seam depth is 1,500 ft and the area extraction ratio is 75% and the peak stress at the pillar wall is 2.5 times the average vertical stress in the pillar. Determine the distance along the seam that the vertical stress exceeds unconfined compressive strength and plot.

9.20 Confinement increases rock strength and suggests a way to mitigate hazards associated with violent pillar spalls is by bolting with screen between bolts. Derive a formula

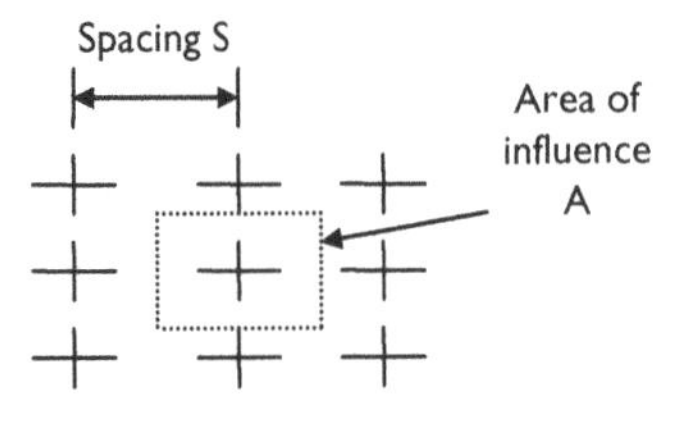

Sketch for Problem 9.20

that relates confinement to bolt force assuming a regular square bolting pattern, then evaluate for bolting pressure and compressive strength increase at a pillar wall.

Assume spacing is 2 m (6 ft), bolt force is 89.6 kN (20,000 lbf), and the rock follows a Mohr-Coulomb failure criterion and has a ratio of compressive to tensile strength of 15.

9.21 Mining proceeds overhand in a narrow vertical vein as shown in the sketch. As pillar dimensions are diminished, the peak stress in the pillar approaches the average stress and the danger of sudden failure of the pillar in burst fashion increases. Determine the pillar height at the point where average pillar stress approaches peak stress at the pillar walls. Neglect the effect of fill. Vein width is 5 m and level interval is 50 m. Depth is 1500 m where the horizontal stress is twice the vertical stress caused by gravity alone. Pillar unconfined compressive strength is 175 MPa. Hint: consider the excavation a rectangular section in a two-dimensional view and the pillar long into the page.

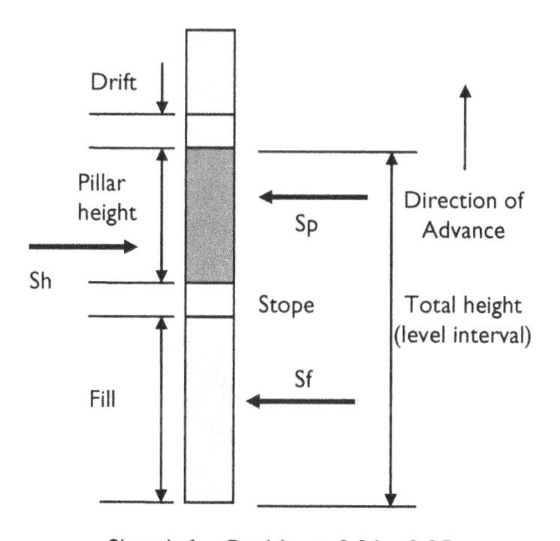

Sketch for Problems 9.21 - 9.25

9.22 Consider the pillar in that meets the conditions of Problem 9.21 when the pillar height (measured along the vein) is 3.45 m (11.3 ft). Determine the pillar stress and safety factor.

9.23 Mining proceeds overhand in a narrow vertical vein as shown in the sketch. As pillar dimensions are diminished, the average pillar approaches pillar strength and the danger of sudden failure of the pillar in burst fashion increases. Neglect any size effect on strength, then determine the pillar height at the point where the pillar safety factor is reduced to one. Neglect the effect of fill. Vein width is 15 ft and level interval is 150 ft. Depth is 5,000 ft where the horizontal stress is twice the vertical stress caused by gravity alone. Pillar unconfined compressive strength is 25,000 psi.

9.24 Consider the data given for Problem 9.23 and the result that at a pillar height measured along the vein, the pillar safety factor is one. Determine the stress concentrated at the pillar walls at this point.

9.25 Suppose a pillar is formed according to the conditions of Problem 9.23 where the pillar height measured along the vein is 66 ft and the pillar safety factor with respect to average pillar stress is one. Compute the strain energy contained in the pillar per foot into the page assuming Young's modulus is $7.5(10^6)$ psi and Poisson's ratio is

0.25. Use a one-dimensional pillar stress model. Estimate the magnitude of seismic event that would be generated if the contained energy were released as a burst.

9.26 A pillar has Mohr-Coulomb strength properties, c =2,500 psi, $\phi = 52^o$, and elastic properties, $E = 5(10^6)$ psi, $v = 0.25$. Specific weight $\gamma = 110$ pcf. The pillar is 12 ft high, 60 ft wide, and 90 ft long in a flat stratum. A spall extends 4 ft into a rib. Compute the strain energy in the spall per ft along the pillar length. If this energy is converted into kinetic energy, what is the velocity of the spall assuming a horizontal direction?

9.27 A pillar has Mohr-Coulomb strength properties, c =17.2 MPa, $\phi = 52^o$, and elastic properties, $E = 34.5$ GPa, $v = 0.25$. Specific weight $\gamma = 17.4$ kN/m^3. The pillar is 3.7 m high, 18.3 m wide, and 27.4 m long in a flat stratum. A spall extends 1.2 m into a rib. Compute the strain energy in the spall per m along the pillar length. If this energy is converted into kinetic energy, what is the velocity of the spall assuming a horizontal direction?

9.28 A pillar is formed in hard rock where the horizontal stress is 1.5 times the vertical preexcavation stress at a depth of 5,500 ft, as shown in the sketch. Unconfined compressive and tensile strengths are 32,400 psi and 1,850 psi, respectively. Young's modulus and shear modulus are $12(10^6)$ psi and $5(10^6)$ psi, respectively. Height of the excavation L is 150 ft and width Hp is 20 ft. Specific weight of rock is 162 pcf. Determine pillar width Wp when the pillar is on the verge of failure, then estimate the stored strain energy per ft of length along the pillar.

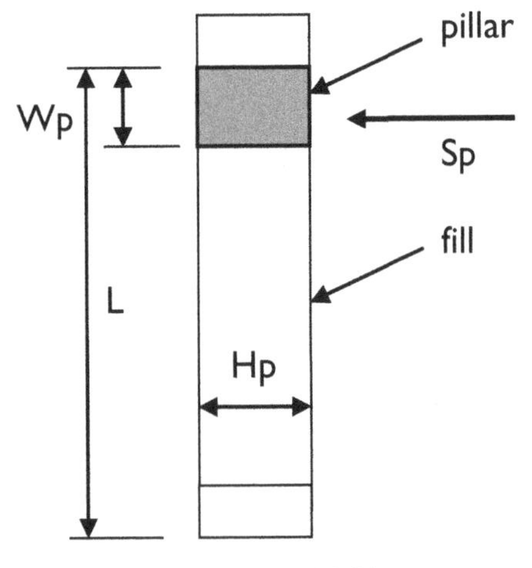

Sketch for 9.28-9.30.

9.29 A pillar is formed in hard rock where the horizontal stress is 1.5 times the vertical preexcavation stress at a depth of 1,676 m, as shown in the sketch. Unconfined compressive and tensile strengths are 223 MPa and 12.8 MPa, respectively. Young's modulus and shear modulus are 82.8 GPa and 34.5 GPa, respectively. Height of the excavation L is 45.7 m and width Hp is 6.1 m. Specific weight of rock is 26 kN/m^3. Determine pillar width Wp when the pillar is on the verge of failure, then estimate the stored strain energy per ft of length along the pillar.

9.30 Consider the pillar in Problem 9.28 or 9.29 and suppose a spall h ft or m thick forms at the pillar bottom, which is H high, is ejected downwards against the fill. The spall energy is "absorbed" by the fill as work is done at the spall bottom that is in contact with the fill top. Derive algebraically the displacement of the fill top relative to the fill bottom a distance L units below the top.

9.31 Two geophones are located along a tunnel at 100 ft (G1)and 1100 ft (G2) from the tunnel entrance. Longitudinal wave velocity in the adjacent rock mass is 10,000 ft/s. An event arrival is detected at G1 at 10:31.20.06 (hr, min, sec). An event arrival is also detected at G2 10:31:20.04. Determine if the event is locatable and if so, the location and time of the event.

9.32 Two geophones are located along a tunnel at 100 ft (G1)and 1100 ft (G2) from the tunnel entrance. Longitudinal wave velocity in the adjacent rock mass is 10,000 ft/s. An event arrival is detected at G1 at 10:31.20.06 (hr, min, sec). An event arrival is also detected at G2 10:31:20.16. Determine if the event is locatable and if so, the location and time of the event.

9.33 Two geophones are located along a tunnel at 10 m (G1)and 310 m (G2) from the tunnel entrance. Longitudinal wave velocity in the adjacent rock mass is 3,000 m/s. An event arrival is detected at G1 at 10:31.20.06 (hr, min, sec). An event arrival is also detected at G2 10:31:20.04. Determine if the event is locatable and if so, the location and time of the event.

9.34 Three geophones are located along a tunnel at 10 m (G1), 310 m (G2), and 640 m (G3) from the tunnel entrance. Longitudinal wave velocity in the adjacent rock mass is 3,000 m/s. An event arrival is detected at G1 at 10:31.20.06 (hr, min, sec). Event arrivals are also detected at G2 at 10:31:20.04 and at G3 at 10:31:15. Determine if the event is locatable and if so, best estimates of location and time of the event.

9.35 Three geophones are located along a tunnel at 10 m (G1), 310 m (G2), and 640 m (G3) from the tunnel entrance. Longitudinal wave velocity in the adjacent rock mass is 3,000 m/s. An event arrival is detected at G1 at 10:31.20.06 (hr, min, sec). Event arrivals are also detected at G2 at 10:31:20.16 and at G3 at 10:31:27. Determine if the event is locatable and if so, best estimates of location and time of the event.

9.36 Geophone pairs are installed in two drifts along a narrow, steeply dipping vein as shown in the sketch. An event occurs that may be a burst in the hanging wall. If this event is locatable, determine the source location (x_s, y_s) and time of the event(t_s). Assume the origin of coordinates is at G1 (0,0) and the coordinates of the other geophones are: G2(0,200), G3(200, 150), G4(0,150), so the geophones are at the corners of a plane rectangle. Units are ft. Relative to an arbitrary zero of time, arrival times in milliseconds are: G1(20.0), G2(32.36), G3(35.50) and G4(25.81). Wave speed for first arrivals (P-waves) is 7,071 ft/s.

Comment: The location requires extreme accuracy in event times. The problem was setup with the source at (x=50.0,y= 50.0,t= 0.010).

9.37 Geophone pairs are installed in two drifts along a narrow, steeply dipping vein as shown in the sketch. An event occurs that may be a burst in the hanging wall. If this event is locatable, determine the source location (x_s, y_s) and time of the event (t_s). Assume the origin of coordinates is at G1 (0,0) and the coordinates of the other geophones are: G2 (0,200), G3(200, 150), G4(0,150), so the geophones are at the corners of a plane rectangle. Units are meters. Relative to an arbitrary zero of time, arrival times in

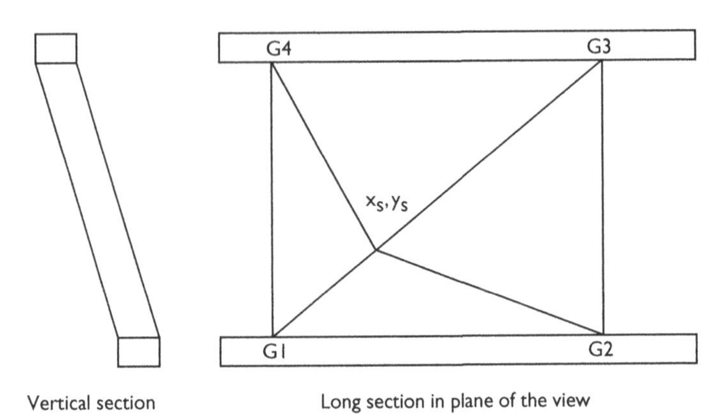

Sketch for Problem 9.36-9.37

seconds are: G1(0.05281), G2(0.09336), G3(0.10366) and G4(0.07187). Wave speed for first arrivals (P-waves) is 2,155 m/s.

9.38 Geophones are installed in four drifts along a wide but steeply dipping vein as shown in the sketch. An event is recorded at eight geophones. If this event is locatable, determine the source location (x_S, y_S, z_S) and time of the event (t_S). Assume the origin of coordinates is at G1 (0,0,0) and the coordinates of geophones: G2(0,200,0), G3(0,200,150), G4(0,0,150),G5(300,0,0) G6(300,200,0), G7(300,200,150), G8(300,0,150),so the geophones are at the corners of a box. Units are ft. Relative to an arbitrary zero of time, arrival times in milliseconds are: G1(42.25), G2(53.45), G3(56.46), G4(47.32), G5(66.74), G6(71.83), G7(73.59),and G8(68.73). Wave speed for first arrivals (P-waves) is 7,071 ft/s.

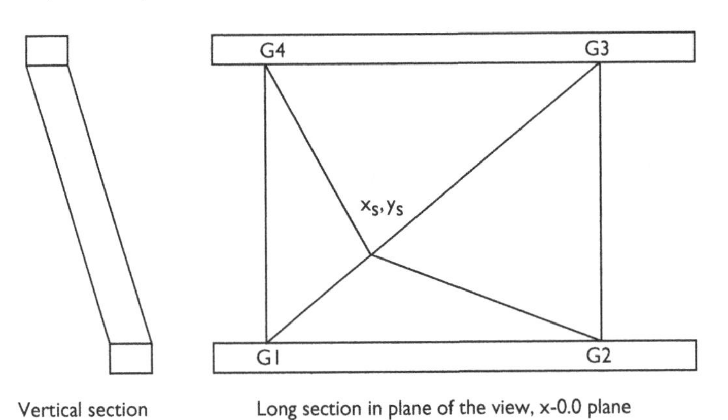

Sketch for Problems 9.38-9.40

9.39 Geophones are installed in four drifts along a wide but steeply dipping vein as shown in the sketch. An event is recorded at eight geophones. If this event is locatable, determine the source location (x_S, y_S, z_S) and time of the event (t_S). Assume the origin of coordinates is at G1 (0,0,0) and the coordinates of geophones: G2(0,200,0), G3(0,200,150),

G4(0,0,150),G5(300,0,0) G6(300,200,0), G7(300,200,150), G8(300,0,150),so the geophones are at the corners of a box. Units are m. Relative to an arbitrary zero of time, arrival times in seconds are: G1(0.0702), G2(0.1070), G3(0.1168), G4(0.0868), G5 (0.1506), G6(0.1673), G7(0.1730),and G8(0.1571). Wave speed for first arrivals (P-waves) is 2,155 m/s.

9.40 Consider the geophone layout and data in Problem 9.39 where they are installed in four drifts along a wide but steeply dipping vein as shown in the sketch. An event is recorded at eight geophones. If this event is locatable, determine the source location (x_s, y_s, z_s) and time of the event (t_s). Although the geophone layout is the same as in 9.39, the arrival times are different. Relative to an arbitrary zero of time, arrival times in seconds are: G1(0.0702), G2(0.1070), G3(0.1168), G4(0.0868), G5(0.1957), G6(0.2082), G7(0.2127),and G8(0.2005). Wave speed for first arrivals (P-waves) is 2,155 m/s.

Chapter 10

Foundations on jointed rock

This chapter briefly compares and contrasts foundation design on rock and soil. A traditional soil foundation approach based on plane plastic strain and limit load theorems is outlined. Demonstration of the validity and accuracy of numerical modeling of foundations using the popular finite element method is next. Guidelines for foundation design on jointed rock in case of square, rectangular and strip footings follow.

Foundation designs on rock and soil have some similarities but also important differences. Both share elasticity theory as a starting point for rational design and both often show nonlinearities under load. Numerical computations are therefore necessary for modern design analyses. In this regard, a foundation load – settlement plot captures the essential information needed for design. To be sure, much of traditional foundation design is based on limit loads determined from plane plastic strain analyses of "sliplines". These patterns are the basis for upper bound solutions to foundation load. An upper bound implies a collapse load is not greater than the bound. A lower bound would be more useful in that the foundation collapse load would not be less than the bound.

A most important feature of a load – settlement ($P - \delta$) plot is the plot slope K that characterizes the foundation stiffness. Here, the load P is the force per unit area of the foundation and δ is the average vertical displacement of the foundation surface. In a linear range, $P = K\delta$ where K is foundation rock stiffness, a constant with units of kpsi/inch (MPa/mm). Alternatively, $\delta = CP$ where C is foundation rock compliance, the inverse of K. Departure from linearity indicates yielding of the foundation rock. While site-specific design analysis is desirable, general load – settlement guidelines based on parametric studies may be useful in preliminary stages of design analyses. Such guidelines are presented in the sequel.

Although designs of foundations on rock and soil have similarities, there are also significant differences to consider, mainly in the differences between the engineering properties of rock and soil *in situ*. In both cases, there are two central questions: one of bearing capacity and one of settlement. *Bearing capacity* is somewhat ambiguous because it may refer to a limiting strength or a limiting displacement depending on the structure to be supported.

There may be an intermediate structure for transmitting load from above to support below. An example is a concrete footing. Piers and end-bearing piles are other examples. Of course, such structures must be designed to transmit the anticipated load or range of loads with an adequate margin of safety. Friction piles and tendons could also be included in an extensive treatment of foundation design. Although design of structural elements of foundations is an important consideration, the concern here is not with design of structural elements, but rather with ways that rock supporting a foundation may fail and therefore with the bearing capacity and settlement of *jointed rock*.

Soil is a particulate material with engineering properties largely determined by grain size that ranges from gravel to sand, silt and clay. Because soil particle size is orders of magnitude less than foundation size, a continuum mechanics view is applicable for engineering analyses. Indeed, soil and rock are often treated as elastic materials. In fact, elastic moduli and soil strengths can be determined by laboratory testing of samples acquired at a site of interest and by a variety of field tests. Soils generally have much lower strengths and elastic moduli than rock. Interestingly, strengths of both are often described by the famous Mohr-Coulomb failure criterion. However, rock under low confinement often fails by fracture, while soils often fail by flow, which may be localized in a zone of intense shearing. In any case, lateral dimensions of foundations are generally large compared with soil grain dimensions and elasticity theory provides a quantitative approach to design, that is, to the estimate of bearing capacity and settlement.

Design of foundations on soils is highly developed relative to rock foundation design. Foundations on soils are described in handbooks and design is often embedded in building codes. Some codes provide guidance for rock foundation design, but generally much is left to judgment and experience of the engineer. Almost any textbook on soil mechanics has a foundations section and is a logical starting place for the subject.

Rock is a quite variable material that ranges from soft, highly deformable evaporates such as salt, gypsum and potash strata, to porous, fractured sandstones and limestones, to strong, stiff granitic rocks and to highly directional schists and gneisses. Near surface rock masses tend to have weathered joints and to form blocky rock masses. While, intact rock between joints is usually of high strength, often exceeding that of concrete, joint strength is usually much lower. Slip and separation of joints thus become primary concerns in estimating bearing capacity of rock masses at the surface and the design of foundations on rock. Despite the presence of joints, a continuum view of intact rock *between* joints is still applicable in consideration of grain sizes of intact rock. A continuum approach is also still possible when joints are present by using an *equivalent* properties approach. Equivalent properties take into account intact rock and joint properties and geometry in a systematic and technically sound way that is consistent with physical laws, material laws, and kinematics.

Again, in case of rock *in situ*, the presence of joints forms a discontinuous material, often with joint spacing more or less the same as foundation dimensions. Intact rock between joints can be sampled, usually by core drilling, and tested in the laboratory for elastic and strength properties. Important properties are elastic moduli and strengths. Unconfined compressive and tensile strengths are especially important as is the effect of confining pressure on compressive strength. Joints pose a more difficult sampling procedure but can also be tested in the laboratory for stiffness and strength, usually in direct shear, once the necessary samples are acquired. Three especially useful references for rock foundations may be found in the reference list for this chapter (Goodman, Wyllie, and Kulhawy).

In both cases, a useful field test for directly determining bearing capacity is a plate bearing test. In this test, a stiff plate is pressed against the surface of interest while recording the applied load and measuring the associated displacement. This test has the advantage of being done on a sizeable, undisturbed sample of foundation material. This test is also used in the design of highways and runways. Another very useful test for determining foundation material properties is a pressure meter test in case of soils and a Goodman jack test in case of rock. Both give estimates of a "deformation" modulus that is related to Young's modulus of the material. However, the scale of these field tests while large compared with laboratory size test specimens, may be too small to adequately sample a jointed rock mass. The reason is joint spacing that may be much larger than the bearing plate or jack length, both a fraction of a meter.

A thorough geological site investigation is essential to successful design of foundations on rock (and soil). Rock contacts, strata dip and dip direction, strata thicknesses and properties (especially elastic moduli and strengths) should be determined. Joint geometry, continuity and properties (joint stiffness and strength) should also be determined. If water is present, then permeability or hydraulic conductivity and pressure are needed for rock and joints. Presence of faults and caverns (natural and man-made) should be known in advance of design analysis, of course.

10.1 Plane plastic strain

In two dimensional analyses (plane strain), the stress equations of equilibrium and a failure criterion for isotropic media form a system of three equations in three unknown stresses (Sxx, Syy, Txy) and thus allow for solution to problems of "limiting" equilibrium. Transformation of this system of partial differential equations to a system of ordinary equations allows for solution in special cases that are associated with patterns of "sliplines" or characteristic curves. Two patterns of special interest define regions of constant state and radial shear where yielding or failure is occurring. These patterns are illustrated in Figure 2.32 in the discussion of base failures of slopes. The discussion also applies to foundations. In Regions I and III the characteristic lines are two families of straight lines. In Region II one family of characteristic lines are straight lines or rays; the second family of lines are exponential (logarithmic) spirals. In both regions the characteristic lines form an acute angle $(\pi/2 - \phi)$ with the major principal compression (S1) where ϕ is an angle of internal friction.

Integration of the normal stress over the width of the foundation determines the vertical foundation force (per unit of distance in the z-direction (normal to the page) associated with the given pattern of sliplines. In case of oblique loading, integration of the shear stress over the foundation width gives the horizontal load (per unit of distance in the z-direction. In any case, slipline solutions are upper bound solutions and thus indicate the maximum load possible at collapse. Such solutions need to be used with a judicious factor of safety with respect to collapse (*FS*) because the collapse load may be less. In this regard, collapse occurs when the plastic or yielding region is no longer constrained by adjacent elastic ground (as in Figure 2.32). More useful solutions are lower bound solutions that indicate the collapse load is no less than a given number. For example, in case of a Mohr-Coulomb material with an angle of internal friction $\phi = 30^\circ$ an upper bound to a uniformly applied stress is $30.1k$ where k is cohesion. A lower bound is $13.9k$. Thus, the applied stress at collapse is greater than $13.9k$ but no greater than $30.1k$. A smaller spread between bounds is desirable, of course. If the angle of internal friction is zero, then upper and lower bounds are $5.14k$ and $4.0k$, respectively. Much ingenuity has been devoted to the finding upper and lower bounds to collapse loads in three dimensions as well as in two dimensions prior to the advent of the digital computer. Bounds now serve to provide checks on computer solutions and thus on programs for doing plasticity analyses. One should also note the cohesion k is several orders of magnitude greater in case of intact rock when compared with even a highly cohesive soil. Rock cohesion if often measured in psi (MPa) while soil cohesion is often measured in psf (kPa). Accordingly, allowable foundation loads on rock are usually much greater than on soil.

10.2 Uniformly loaded strip

A uniformly loaded strip is often the beginning of foundation calculations for bearing capacity and illustrates the development of numerical guidelines in the form of load – settlement plots. A strip foundation is likely to be used to support a wall. Rectangular, square and circular slabs or footings are shapes likely to support columns.

The strip problem has an analytical (closed form) solution that allows for comparison with numerical calculations which may then be used in cases where analytical solutions are not available. The problem is illustrated in Figure 10.1 that locates the origin at the center (bottom) of a slab of breadth B under a load per unit area P that is uniformly distributed. The slab may be very *flexible* or very *stiff* compared with the rock below. The problem is two-dimensional (plane strain) in the *yz* plane where the *x*-axis is out of the page, so the strip is long in the x-direction compared with width B. Material behavior is assumed to be purely elastic and isotropic. Numerical results relating to *stress* and *displacement* were obtained using the popular finite element method. Related results include a *pressure bulb* concept and a local *factor of safety* concept. Dimensions of the finite element model are given as multiples of the loaded strip width B and are more than sufficient to nullify any boundary effects on stresses and displacements in the vicinity of the loaded strip.

Surface displacements from two finite element models are illustrated in Figure 10.2. The slab is 55 units wide (B=55). Each square in the figure is 5×5 units. Maximum vertical displacement in the figure centers is roughly 1/6 units in both cases. While the stiff slab is 100 times stiffer than the flexible slab, deformation of the slab still occurs. In the flexible slab case, slab deformation conforms to the surface displacement.

Stress

Comparison of analytical with finite element results in the case of a very *flexible* slab is shown in Figure 10.3 with plots of vertical stress versus depth below the center of the loaded strip (y=0). The vertical stress from the analytical solution plots almost directly over the finite

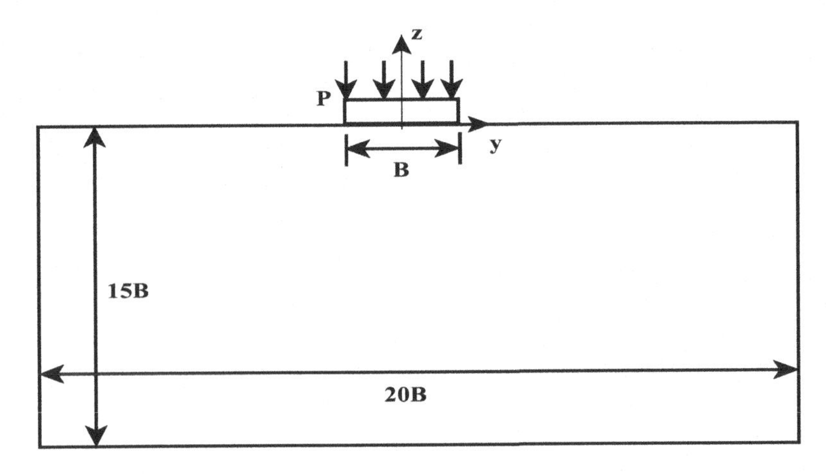

Figure 10.1 Uniformly distributed strip load and region of interest.

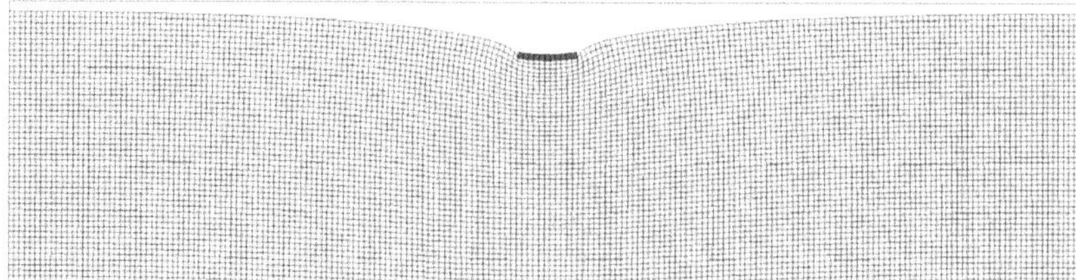

(a) Surface displacement under a uniformly loaded, but stiff slab (dark elements at the top center).

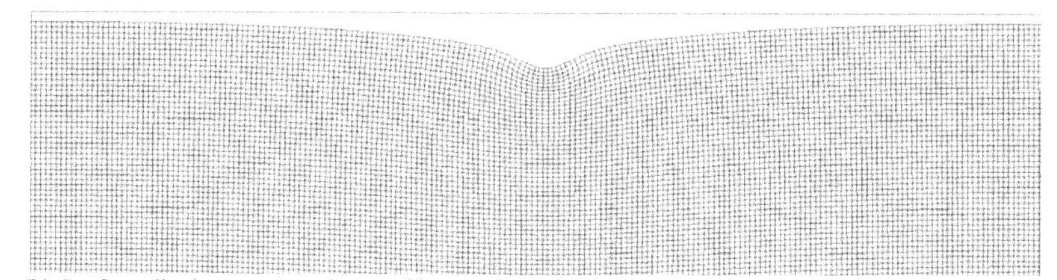

(b) Surface displacement under a uniformly loaded but flexible slab. Load width is the same as in the stiff slab case.

Figure 10.2 Surface displacements in two load cases. Width of the figures is 1005 units. Vertical displacements are exaggerated. Depth of the figures is 355 units.

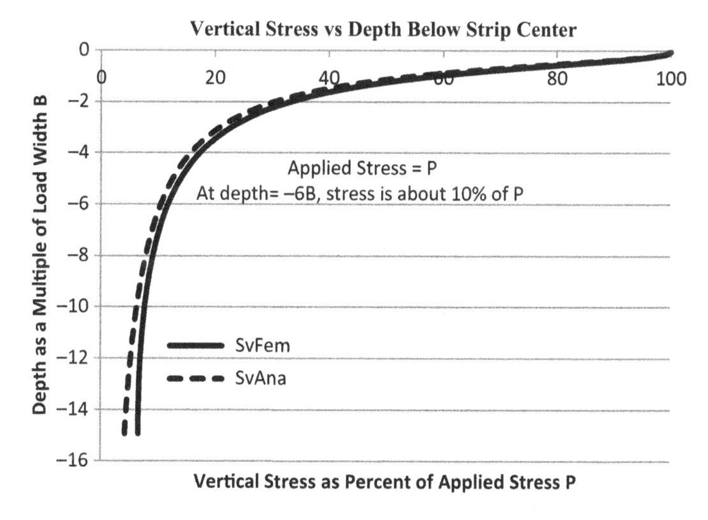

Figure 10.3 Vertical stress Sv versus depth from finite element (Fem) and analytical (Ana) solutions. Compression is positive. Flexible slab case.

element results. Importantly, high stresses near the loaded surface are in very close agreement. The analytical, theoretical solution is for the lower half-plane that has infinite extent. The finite element solution is necessarily for a finite region (the region in Figure 10.1) and is a numerical approximation, so some differences are expected. In this case, the differences are small and of no importance to the response of the foundation rock.

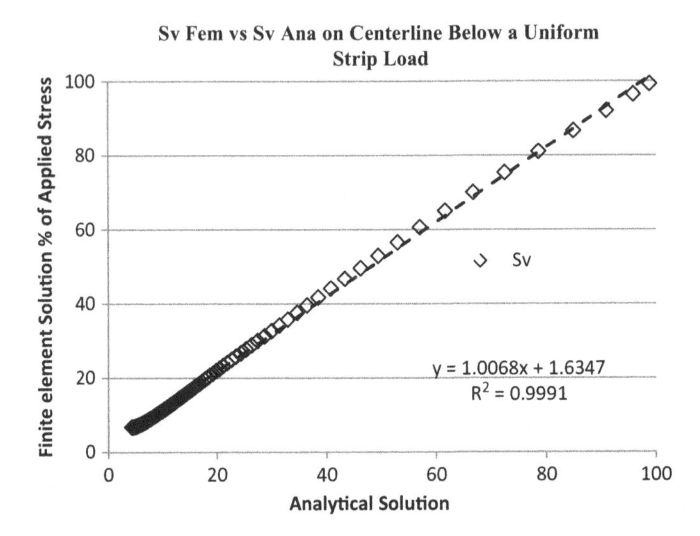

Figure 10.4 Regression analysis of finite element on analytical vertical stresses from Figure 10.2 in the flexible slab case.

In Figure 10.4, the analytical solution for vertical stress along the centerline is plotted along the x-axis as a percentage of the applied stress; the corresponding finite element stress is plotted on the y-axis. The correlation between analytical and numerical results is excellent in consideration of the high correlation coefficient and regression line slope, both near one.

Displacement

Because the slab is flexible, the surface displacements of the rock form a curved trough as seen in Figure 10.5. Two sets of data are plotted in the figure. One set shows the vertical surface displacements computed from the finite element solution (Uver). The second set shows the vertical surface displacements from the analytical solution (UverRel). Finite element displacements are always relative to some fixed point in the domain of analysis. In this case, the domain of analysis is shown in Figure 10.1 where vertical displacements along the bottom of the domain are fixed at zero. Surface displacements from the analytical solution are those induced in the lower half-plane that extends to an indefinitely large depth. To compare with finite element displacements, displacements from the analytical solution need to be made relative to the depth of the finite element domain. These displacements are plotted as UverRel in the figure. There is very little difference in the displacements near the loaded strip where the largest displacements occur.

Surface displacements in the stiff slab case are also shown in Figure 10.5 (UverStiff). The displacement trough is somewhat wider than in the flexible slab case, while the maximum displacement is somewhat less, about ten percent less than in the case of a flexible slab. Both maximums occur directly below the slab center. Figure 10.5 shows that surface displacements are nearly less than 15 percent of maximum displacement beyond 6B away from the slab center.

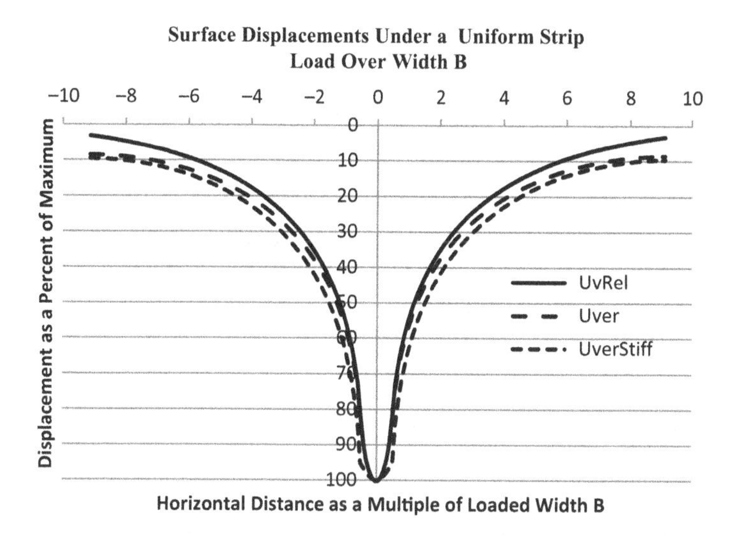

Figure 10.5 Vertical surface displacements in case of flexible and stiff slabs.

A quantitative comparison of displacements in the flexible case is shown in Figure 10.6 where a high correlation between finite element and analytical results is evident.

The comparisons of stress and displacement results clearly and quantitatively demonstrate the applicability of the finite element method to calculations of stress and displacement in the

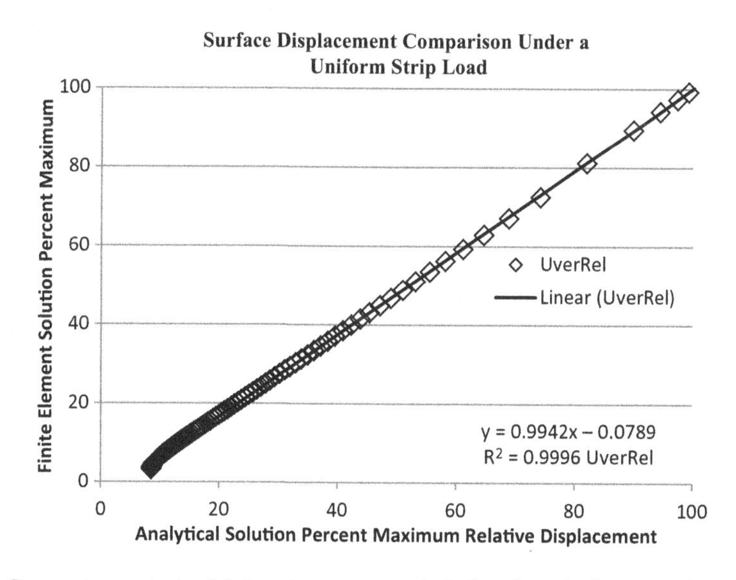

Figure 10.6 Regression analysis of finite element on analytical surface displacement from Figure 10.4 (flexible slab case).

case of a uniformly distributed strip load applied to a rock foundation through very flexible and very stiff slabs.

Distributions of displacements are of critical importance, of course, when design criteria are given in terms of limiting displacements. However, when strength criteria are given, then stresses are central to design. Some inelasticity is often tolerable under strength design criteria. The reason is that zones of inelasticity, yielding or plastic zones, when small are restrained by adjacent material that is within the elastic limit. Consequently, displacements in yielding zones are of the same magnitude as those in the adjacent elastic regions. If yielding zones grow to undermine a slab or loaded area, then displacements would no longer be restrained and collapse would be imminent.

Pressure bulb and factor of safety

The distribution of stress in the vicinity of a slab foundation under a uniform strip load is often described as a *pressure bulb*. Distribution of local safety factor *fs* is also bulb-like. A local safety factor is a ratio of strength to stress at a point: *fs* = "*strength*"/"*stress*" where appropriate measures of *strength* and *stress* are used. For example strength may be computed using the well-known Mohr-Coulomb or Drucker-Prager yield condition and a corresponding stress from stress analysis is implied. If the state of stress at a point is in the elastic domain, then *fs*>1. When *fs*=1 yielding is imminent. Figure 10.7 is a contour plot of a local safety factor distribution in case of a uniformly loaded strip or slab, the same two cases as before. The plots show similar distributions away from the slab, but noticeable differences immediate below the slab. The lowest safety factor (2.0) occurs at the slab edges in the stiff slab case where stress concentration is high. The lowest safety factor in the flexible slab case is 2.5 below the slab center.

Load – settlement

Examples of load – settlement plots associated with a uniformly loaded strip illustrate the design guidance that is inherent in such plots. Two cases of joint free, isotropic rock are shown in Figure 10.8. In the first case (cf=1.0), full rock strength is used; in the second case (cf=0.025), rock strength is only 0.025 times full strength. The plots are made dimensionless by dividing the applied force (stress by area) by a Mohr-Coulomb cohesion (*k* by area) and by dividing the vertical displacement of the footing centerline by the footing width (B). The cohesion is given as $k = \sqrt{C_o T_o / 4}$ where C_o and T_o are unconfined compressive and tensile strengths, respectively. The load is applied in 10 increments in both cases. No failures occur in the full strength case and the plot is a simple straight line passing through the origin. In the low strength case, noticeable departure from linearity is evident at the end of the fourth load increment that continues through application of full load. The displacement is almost four times the maximum displacement in the full strength case as seen in the figure. Indeed, as loading continues in the low strength case, each increment of load induces an even larger displacement increment suggesting collapse is imminent.

Figure 10.9 shows the element safety factor distributions for three load – settlement examples obtained at full strength and reduced strengths. No element failures occur at full strength and only minor failures occur near the slab edges at 0.2x strength. However, collapse is indicated at 0.1x strength as evident in the large "bubble" within the local factor of safety (*fs*) contour of one.

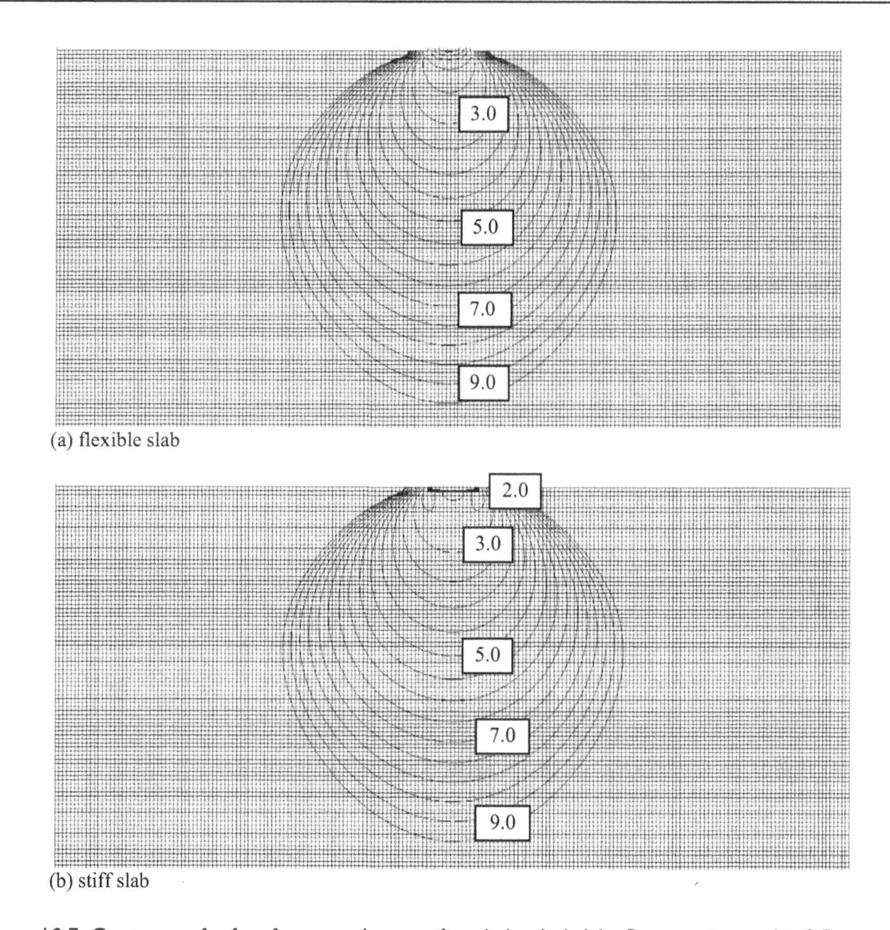

Figure 10.7 Contours of safety factor under a uniformly loaded slab. Contour interval is 0.5

Joints make rock *in situ* more compliant and weaker at the scale of most engineering works including foundations. Figure 10.10 shows the effect of a joint set on otherwise intact rock. The joints in each set dip 45 degrees and strike parallel to the long axis of the strip (into the page). Joint spacing in the moderately jointed case (S/B=0.36) allow for no more than two joints to intersect the surface immediately below the wide strip. In the case of closely spaced joints (S/B=0.09) seven joints could intersect the surface below the strip. The joints have moduli a mere $1/100^{th}$ of intact rock between joints. Because of the high intact rock strength, very few element and joint failures occur and the response in both cases is essentially elastic as seen in the linearity of the plots. However, in case of closely spaced joints, the response is quasi-linear and displacement is much larger than either of the other two cases. The effect of the joints is to induce a pronounced *anisotropy* with a noticeably reduced set of elastic moduli. The displacement at full load in case of moderately jointed rock is 34 percent greater than in the no-joints or isotropic case as seen in the plots. In case of closely jointed rock, the displacement at full load is several hundred percent times the no joint displacement. As a rough guide to a limiting settlement, a 1-inch (25 mm) line is also plotted in the figure. Displacement at maximum load exceeds this limit in the three example cases.

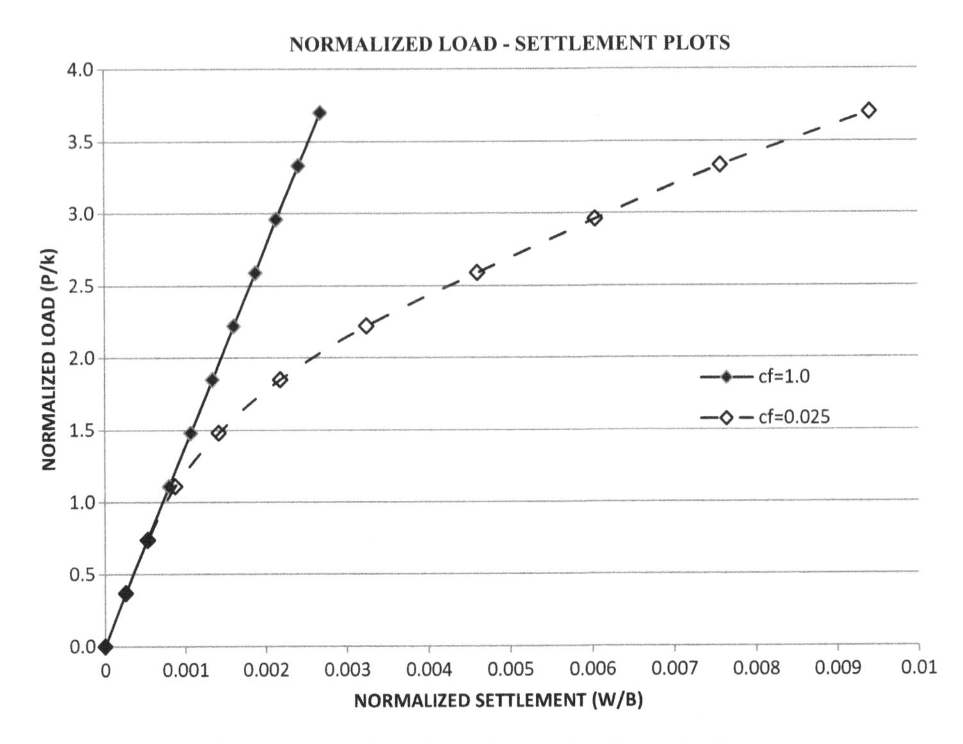

Figure 10.8 High and low rock strength load – settlement plots (normalized).

The surface profile or settlement curve in the vicinity of the strip is shown in the three cases: (1) without joints, (2) moderately jointed, and (3) closely jointed in Figure 10.11. Without joints, the rock is isotropic; with joints the rock is anisotropic. Without joints, settlement is uniform below the strip, but when joints are present the settlement becomes irregular as seen in the figure. Also evident in the figure is the greater settlement along the entire profile in the jointed rock case. In the case without joints, the settlement decreases to a small value, less than the 1- inch (25 mm) guide line at a distance of 5B from the strip center. The same occurs in case of moderately jointed rock as seen in the figure. In case of closely jointed rock the settlement at 5B from the strip center is only slightly less than the 1-inch (25 mm) limit. The irregularity of settlement under the strip in the jointed rock case occurs because the joints transect the slab in this example. A properly reinforced slab would address any structural design challenge for the strip caused by jointing and the profile would be smooth. But in each of the three example cases, the strip settlement is much greater than the 1-inch (25 mm) guide line.

Figure 10.12 shows safety factor contours that define a "pressure bulb" in the two cases: no joints and moderately jointed. The effect of joints on the distribution of the local factor of safety (*fs*) is clearly evident. As a reminder, the joints dip 45 degrees to the right and strike parallel to the long axis of the strip (into the page). Joint spacing is relatively wide at almost 0.4B; no more than two joints intersect the surface below the strip. Close inspection shows a short, narrow band of failures under the strip and also local failures at the outer edges of the strip where *fs*=1. Contours on the right hand side of the figure showing joint effects weave up and down the joints. A surface effect is also present where the joints intersect the ground

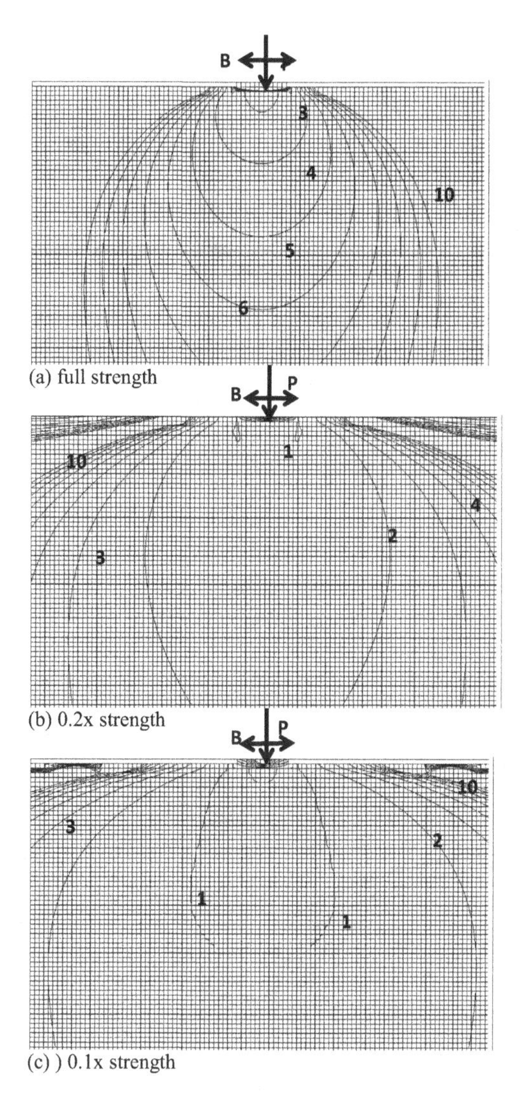

Figure 10.9 Safety factor contours near a strip footing at three strengths. B=slab breadth. P=uniformly distributed load. Numbers are safety factors for the nearby contour.

surface; the effect is asymmetric as seen in the figure where joint contours are visible on the right hand side but not the left hand side. Elastic ground constrains failure and thus prevents large, inelastic strains. However, the large zone directly below the strip defined by the 1.5 safety factor contour indicates caution and perhaps a smaller maximum load in both cases.

Figure 10.13 shows safety factor contours in case of closely spaced joints. Contours are again asymmetric. In fact, there is a significant surface effect adjacent to the right hand side of the strip where a sizeable zone of failing elements is present within the $fs=1$ contour. There are also zones of failure at the edges of the strip. However, confining pressure below the strip center strengthens the rock and joints in the vicinity and the safety factor increases to 1.5. The

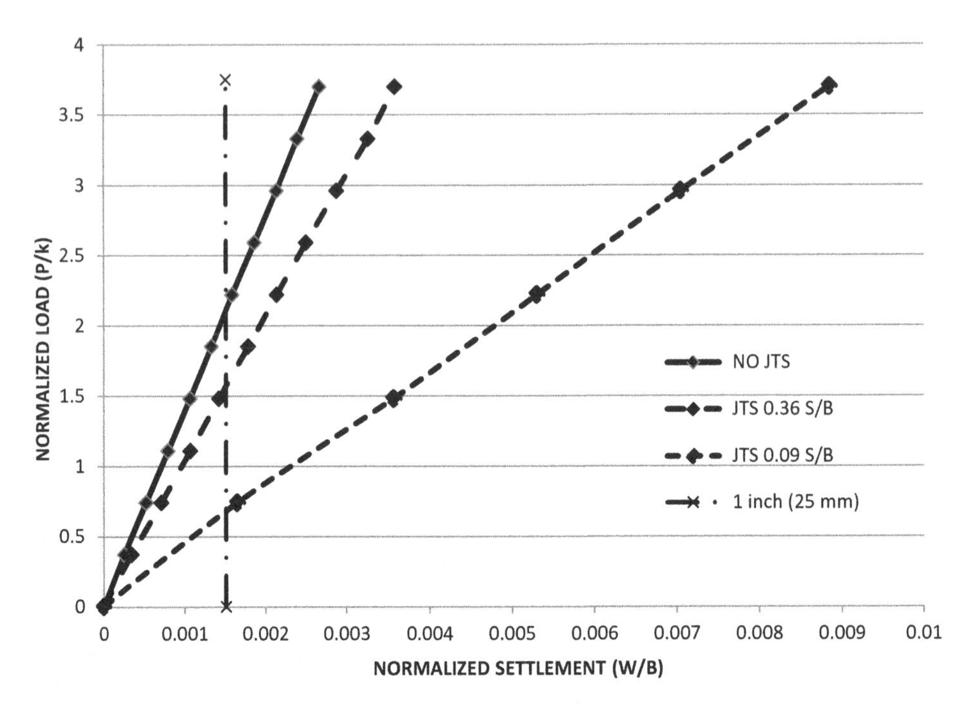

Figure 10.10 Example of effect of joints on strip load – settlement response.

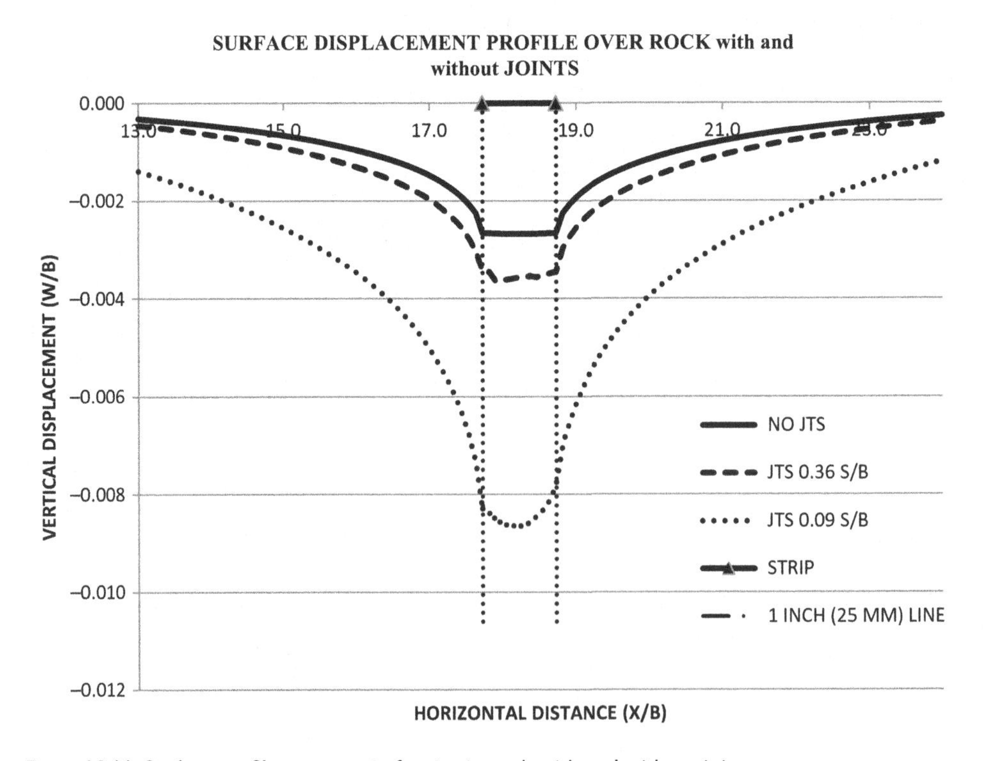

Figure 10.11 Surface profile near a strip footing in rock with and without joints.

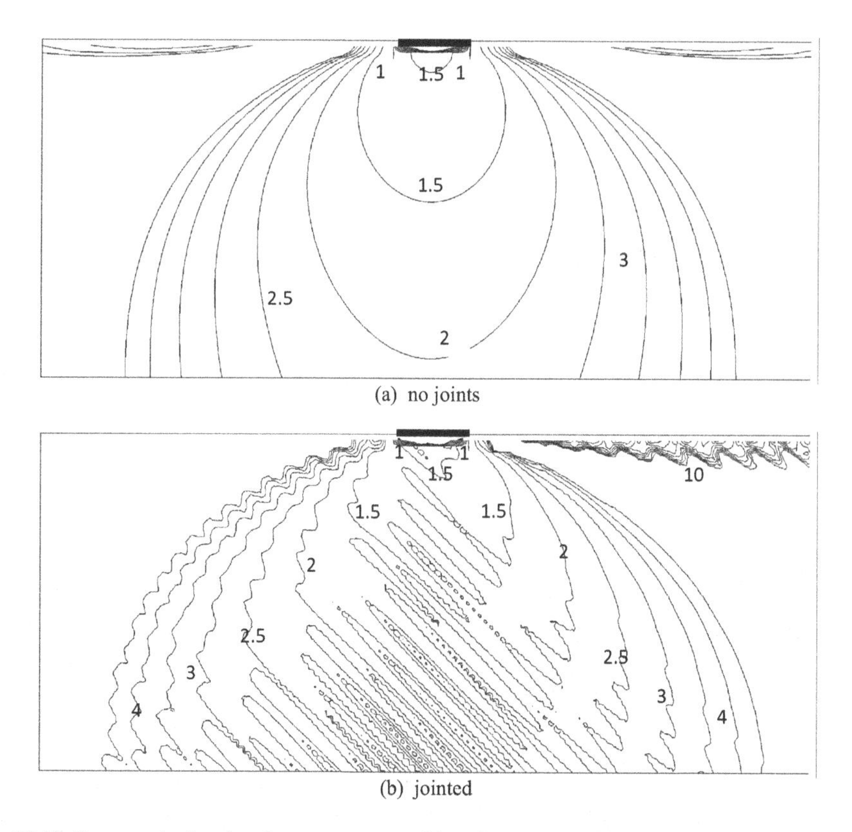

Figure 10.12 Pressure bulb safety factor contours: (a) without joints, (b) with moderately spaced joints at S/B=0.36. Dark line at the top center is the strip B units wide. View is 5B each side of the strip center. Depth is approximately 5B.

large region beyond that also shows a safety factor of 1.5 and indicates caution from the view of high stress as a limiting factor. In fact, settlement or displacement is likely to be the limiting design factor in this example. The ripples in the contours are caused by joints that dip 45 deg to the right and strike parallel to the long axis of the strip (normal to the page).

Summary

Design analysis for foundations on jointed rock requires a numerical approach to the solution of the governing system of equations generated from physical laws, kinematics and material behavior. The popular and well known finite element method is well suited to the task. Comparisons of stress and displacement results from numerical analyses with results from closed form solutions in case of a uniformly loaded strip founded on joint free, isotropic rock show close agreement and lend confidence to the use of the method.

While much traditional foundation analysis in soil mechanics is based on plane strain analysis of limiting equilibrium, the results are mainly upper bound estimates of collapse loads. Lower bounds to collapse loads may also be obtained within the theoretical framework of ideal plasticity where neither strain hardening or softening follow deformation beyond the

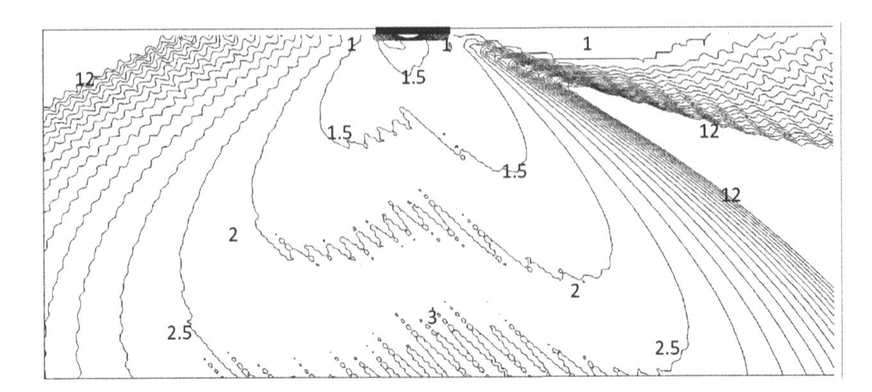

Figure 10.13 Pressure bulb safety factor contours with closely spaced joints (S/B=0.09). A large failure zone exists to the right of the strip where the safety factor contour value is 1.

elastic limit. Displacements usually are not considered in this approach and a safety factor against collapse must be used.

Design of foundations on jointed rock requires a different approach. An approach based on elasticity theory with the elastic limit constrained by strength of intact rock and joints is needed. Elastic-plastic analysis of a strip footing under a uniformly distributed load leads to a distribution of stress below the strip in the shape of a "pressure bulb". The same shape describes the distribution of a local safety factor (*fs*), defined as the ratio of strength to stress at a point, as contours of *fs* show in example problems of strip footings. Yielding at the elastic limit as loading continues is certainly possible and is indicated by *fs* contours of one.

Plots of applied load as a function of vertical displacement of a strip center contain much of the essential information needed for design. In this regard, design based on bearing capacity is ambiguous because reference may be to limiting strength or limiting displacement. A load – settlement plot shows both and also indicates onset of extensive yielding with the appearance of nonlinearity. Examples illustrate this fundamental concept in case of isotropic rock without joints and jointed rock. The presence of joints is shown to increase settlement and induce anisotropy as seen in asymmetric safety factor distributions. Examples also show that decreasing joint spacing increases settlement and decreases safety, as one might expect.

What is needed in lieu of specialized finite element software and expertise for doing site-specific analysis are broad guidelines to foundation design on jointed rock in the form of load – settlement plots that span a range of joint geometries. Strip footings are of interest, but rectangular footings are perhaps even more important, especially square footings. Circular footings are not considered but may not be so different from square footings on jointed rock. The number of combinations of joint sets, rock properties and slab size and shape is endless, so some selection of conditions for practical guideline development is necessary however limiting such may be.

10.3 Bearing capacity near the surface

A critical feature of rock *in situ* is the presence of joints. No analytical solution exists for estimating the bearing capacity of jointed rock masses, so numerical solution is required. Taking joints into account is possible by computing *equivalent* rock properties, elastic moduli certainly and possibly strengths. However, if joints are explicitly represented, then equivalent properties are not needed. A numerical solution then allows deformation and possible yielding of joints and intact rock between joints as individual materials. In practice, yielding should be quite limited and the load – settlement relationship should be well within the range of linearity.

Sedimentary rock often has three joint sets present, two vertical joint sets at right angles and a third set of bedding plane joints at a right angle to the vertical joint sets. These three mutually orthogonal joint sets influence the response of the composite material to foundation load, mainly through induced anisotropy and reduction in elastic moduli, but also through possible yielding of joints even when the intact rock between joints remains elastic. Figure 10.14a illustrates three joint sets within a sample cube where the bedding is flat. Figure 10.14b illustrates the case of three joint sets where the bedding dips so the flat joints are inclined to the horizontal at the dip angle and one set of vertical joints is rotated to the complimentary angle. The second set of vertical joins is not rotated in this example illustration.

Much depends on joint spacing relative to footing dimensions. If the footing is square with edge length B and joint spacing is S, then the number of vertical joints along a footing edge is approximately B/S when the strata are flat. The exact number depends on footing placement. In any case, quantitative analysis via the finite element method allows for the establishment of guidelines over a range of joint spacing and dip. A *wide* spacing would be the case if no more than, say, two joints occurred beneath a square footing. Thus, B/S=2 in case of joints relatively widely spaced. *Medium* spacing would perhaps be five joints more or less, i.e., B/S=5. *Closely* spaced joints would be B/S=10 and *very closely* spaced joints would be B/S=25. This classification is subjective but helpful in defining terms and developing guidelines that span a range of joint spacing.

Footing shapes may be characterized by width B and length L. Square footings are such that $B=L$; rectangular footings are characterized by $L>B$ and strip footings are such that L becomes indefinitely large. Circular footings could be considered with diameter D. However, attention here is on square footings, rectangular footings and strip footings.

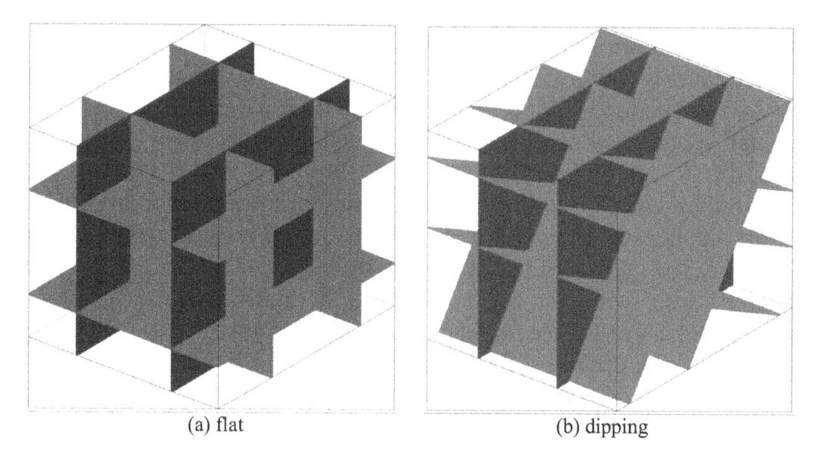

Figure 10.14 Three mutually orthogonal joint sets in a sample cube.

Table 10.1 Properties for guideline computations.

PROPERTY MATERIAL	E (10^6psi/GPa)	ν (–)	C_o (psi/MPa)	T_o (psi/MPa)
ROCK	4.8/33.1	0.20	9,600/66.2	800/5.52
JOINTS	0.048/0.33	0.20	480/3.3	4.8/0.033
FOOTING	4.8/33.1	0.20	4,800/33.1	400/2.76

E = Young's modulus, v = Poisson's ratio, C_o = unconfined compressive strength, T_o = tensile strength.

For all shapes, a load – settlement plot is simply $P = K\delta$ where P is load in units of force per unit area (stress), K is stiffness of jointed rock supporting a foundation footing (stress/length), and δ is settlement, downward vertical displacement in units of length. Alternatively, $K^{-1}P = \delta$ or $\delta{=}CP$ where C is foundation rock compliance or flexibility with units of length/stress, the inverse of foundation rock stiffness.

Properties for all guideline computations for footings on jointed, sedimentary rock are given in Table 10.1. The joints induce anisotropy and make the composite rock mass more compliant and weaker. The finite element method used for computations automatically takes the joints into account with the computation of equivalent properties and by tracking stress states in joints and intact rock separately. Joint modulus is a mere $1/100^{\text{th}}$ of the rock modulus as seen in the table. Rock strain to failure in uniaxial compression is 0.2 percent; joint strain to failure is 0.1 percent. These are reasonable values in consideration of rock and joint properties and reinforced concrete properties from handbooks where a considerable range of values are found, for example, in Appendix B.

Square footings

Figure 10.15 shows four *dimensionless* load – settlement plots in the case of a *square* footing of edge length B founded on a flat sedimentary rock containing three orthogonal joint sets. The ratio $S_v/C_o\%$ is a percentage ratio of applied vertical stress to unconfined compressive strength of intact rock. The ratio δ/B is a percentage ratio of vertical displacement to footing edge length. Joint spacing varies from *wide* to *medium* to *close* to *very close* in the plots. Also shown in the figure is a vertical line denoting 1 inch (25 mm) of displacement. In this regard, stress and displacement are averages over the footing area. The footing deflects only slightly and the applied stress is uniform, so the averages are representative of the footing response. The plots show some nonlinearity that increases with a decrease in joint spacing. At close and very close joint spacing and maximum load, foundation displacement exceeds the 1 inch (25 mm) line.

Figure 10.16 shows much the same information but in dimensional form with regression lines that pass through the origin. The shapes are similar because rescaling induces no important change. The regression line slopes are stiffnesses of sorts; they tend to be larger than tangent or secant stiffnesses but do not differ by much from the table values. The regression lines also show that the plots are only weakly nonlinear. The data show that the closely spaced joint case exhibits stiffness less than one-half the widely spaced joint case in this example of flat strata. The very closely spaced joint case shows an even greater reduction in stiffness relative to the widely spaced joint case.

Figure 10.17 shows load – settlement plots when two sets of joints dip at 45 degrees in opposite directions and a third set of joints is vertical. Slopes of the regression lines in case of

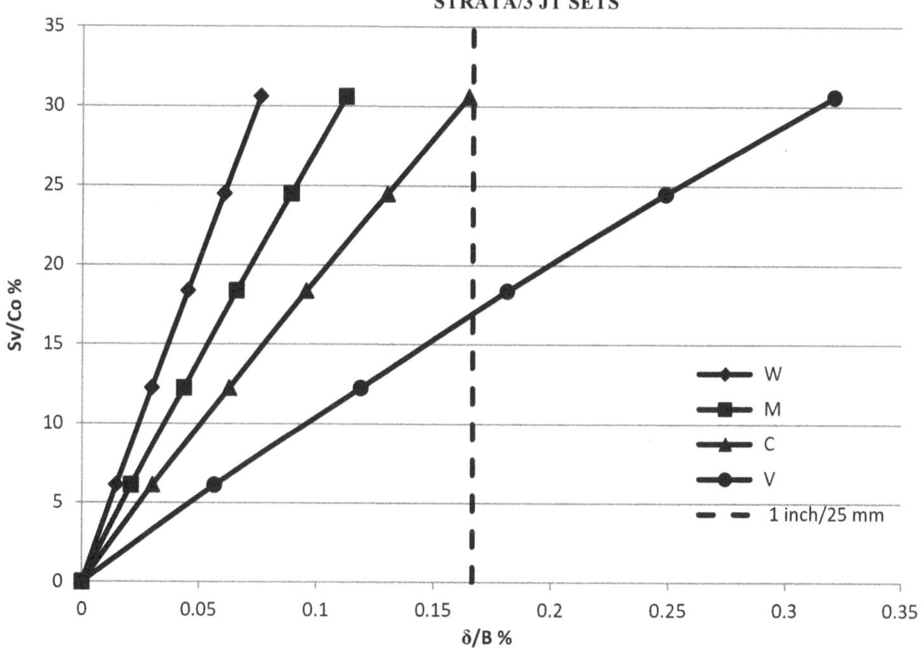

Figure 10.15 Dimensionless load – settlement plots in case of a square footing on flat strata with three joint sets. W = wide spacing, M = medium spacing, C = close spacing, V = very close spacing.

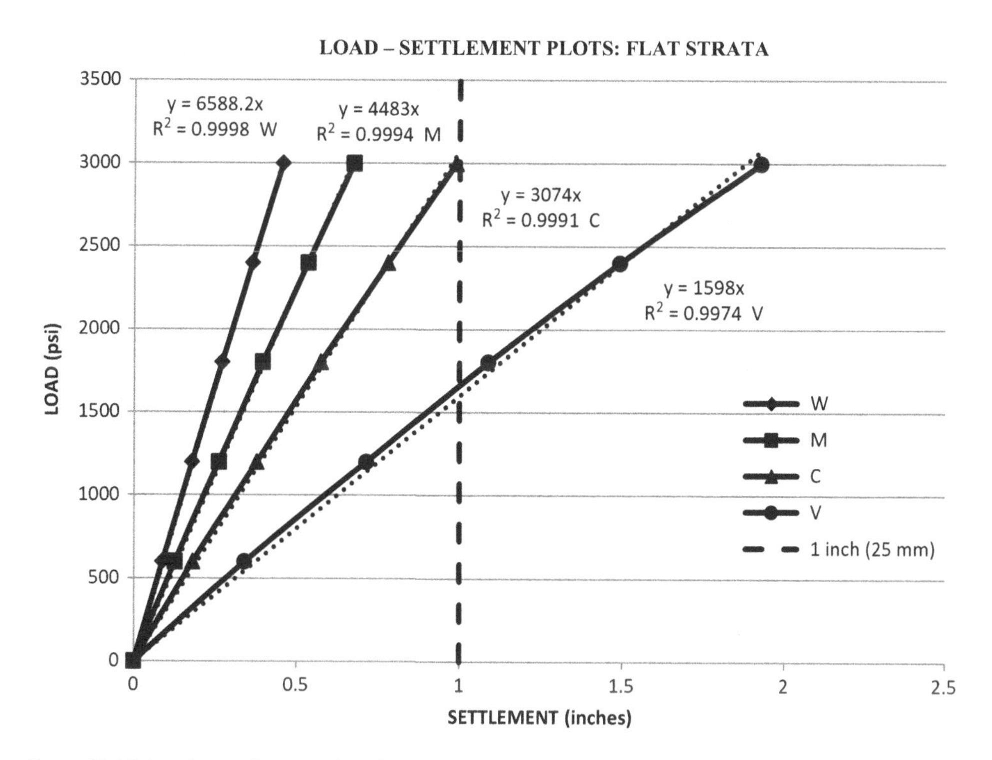

Figure 10.16 Load – settlement plots for a square footing on flat strata with three joint sets.

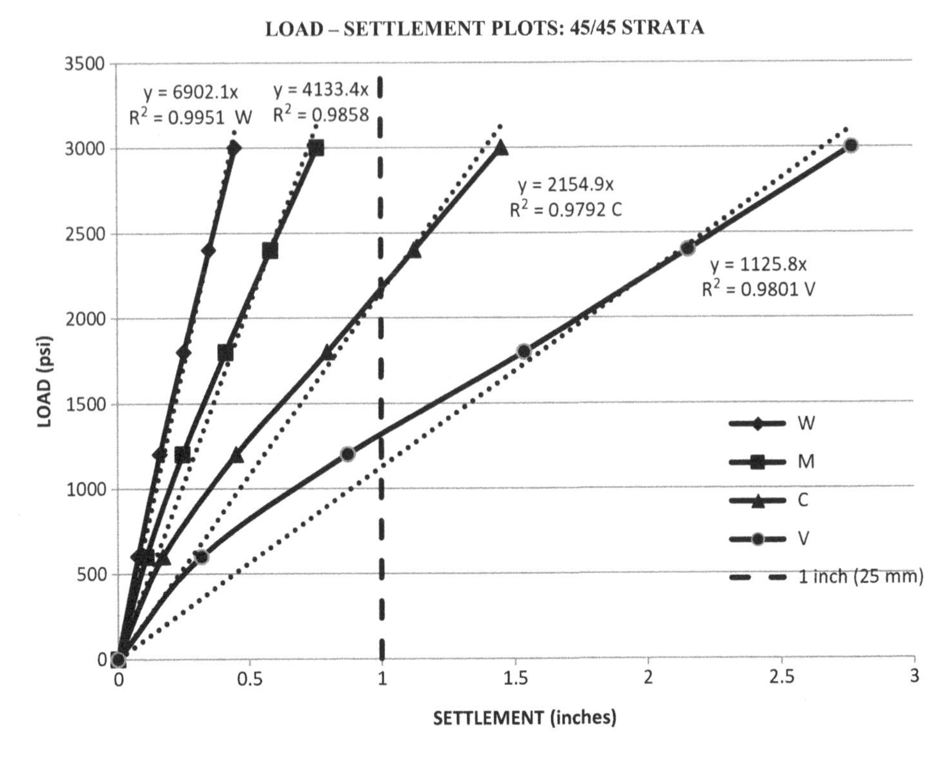

Figure 10.17 Load – settlement plots for a square footing on dipping strata with three joint sets.

wide and medium joint spacing are greater than the case of flat strata (Figure 10.16), but when joints are closely and very closely spaced, the dipping strata slopes are less than in the flat strata case. In all cases, the plot slopes decrease with decreasing joint spacing. All plots are only weakly nonlinear as the relatively high correlation coefficients indicate.

A range of dips as well as a range of joint spacing is needed for guidelines in sedimentary rock. Table 10.2 shows the combination of circumstances for which load – settlement plots are generated. Table entries are foundation rock stiffnesses, that is, the slopes K of load – settlement $P-\delta$ plots for the various combinations of spacing and dip indicated in the row and column headings. Because of nonlinearity, a choice of a *tangent* modulus at 60 percent of full load is used to compute the table values. These tangent stiffness values are within a few

Table 10.2 Square foundation stiffness for various combinations of joint spacing and strata dip*.

SPACING *DIP (deg)*	*WIDE* *B/S=2*	*MEDIUM* *B/S=5*	*CLOSE* *B/S=10*	*VERY CLOSE* *B/S=25*
0/90	6.48/1.76	4.37/1.19	2.98/0.81	1.54/0.42
15/75	5.68/1.54	3.50/0.95	2.05/0.56	1.06/0.29
30/60	6.04/1.64	3.76/1.02	2.00/0.54	1.06/0.29
45/45	6.37/1.73	3.56/0.97	1.78/0.48	0.94/0.25

Notes
* Units of stiffness are kpsi/inch / MPa/mm. Conversion: (MPa/mm) = 0.272(kpsi/inch).

percent of *secant* moduli. A secant modulus is the slope of a straight line to full load from the origin. Units are kpsi/inch and MPa/mm. Thus, the numerical value of stiffness K is just load P that results in one inch (25 mm) of settlement.

In all cases, flat and dipping strata, the more closely joints are spaced the lower the stiffness is of the composite foundation rock. The trend is evident in Table 10.2 data. In this regard, there are over one million joint segments in the 45 degree dipping strata case with closely spaced joints and over three million in the very closely spaced case. More than one half of these joint segments fail at full load.

The data in Table 10.2 are presented in graphical form in Figures 10.18a that show foundation rock *compliance* as a function of relative joint spacing. Recall compliance relates settlement to load by $\delta = CP$. The regression lines and equations indicate compliances are quasi-linear functions of relative joint spacing at all dip angles. Compliance increases with decreasing joint spacing at all dips.

Example 10.1 Consider a square footing on a jointed rock mass that contains three orthogonal joint sets. The strata dip 30 degrees, although the foundation surface is flat. A footing 3 × 3 ft square is subject to a load of 1,000 psi. Estimate the settlement expected: (a) when joint spacing is 6 ft, and (b) when joint spacing is 0.6 ft. Young's modulus is $4.8(10^6)$ psi. Poisson's ratio is 0.2.

Solution: Settlement follows a load – settlement relationship $P = K\delta$, so

$$\delta = P/K$$

Properties are Table 10.1 values. Hence, from Table 10.2

$$\delta = 1,000/6.04(10^3) = 0.17 \text{ inches (a) widely spaced joints (B/S} < 2)$$

$$\delta = 1,000/3.76(10^3) = 0.27 \text{ inches (b) medium spaced joints (B/S} = 5)$$

Example 10.2 Consider a square footing on a jointed rock mass that contains three orthogonal joint sets. The strata dip 15 degrees, although the foundation surface is flat. A footing 1.5 × 1.5 m square is subject to a load of 10 MPa. Estimate the settlement expected: (a) when joint spacing is 2 m, and (b) when joint spacing is 0.3 m. Young's modulus is 33.1 GPa. Poisson's ratio is 0.2.

Solution: Settlement follows a load – settlement relationship $P = K\delta$, so

$$\delta = P/K$$

Properties are Table 10.1 values. Hence, from Table 10.2

$$\delta = 10,000/1.54(10^3) = 6.5 \text{ mm (a) widely spaced joints (B/S} < 2)$$

$$\delta = 10,000/0.95(10^3) = 10.5 \text{ mm (b) medium spaced joints (B/S} = 5)$$

Rock properties may differ from those in Table 10.1 and thus the stiffnesses in Table 10.2 may also differ. A simple adjustment is possible in recognition of the way that Young's modulus enters calculations. In a one-dimensional case of a column under load P, Hooke's law states $P = E\varepsilon = E\Delta L/L = (E/L)\delta$ where P=load, E=Young's modulus, δ=settlement and

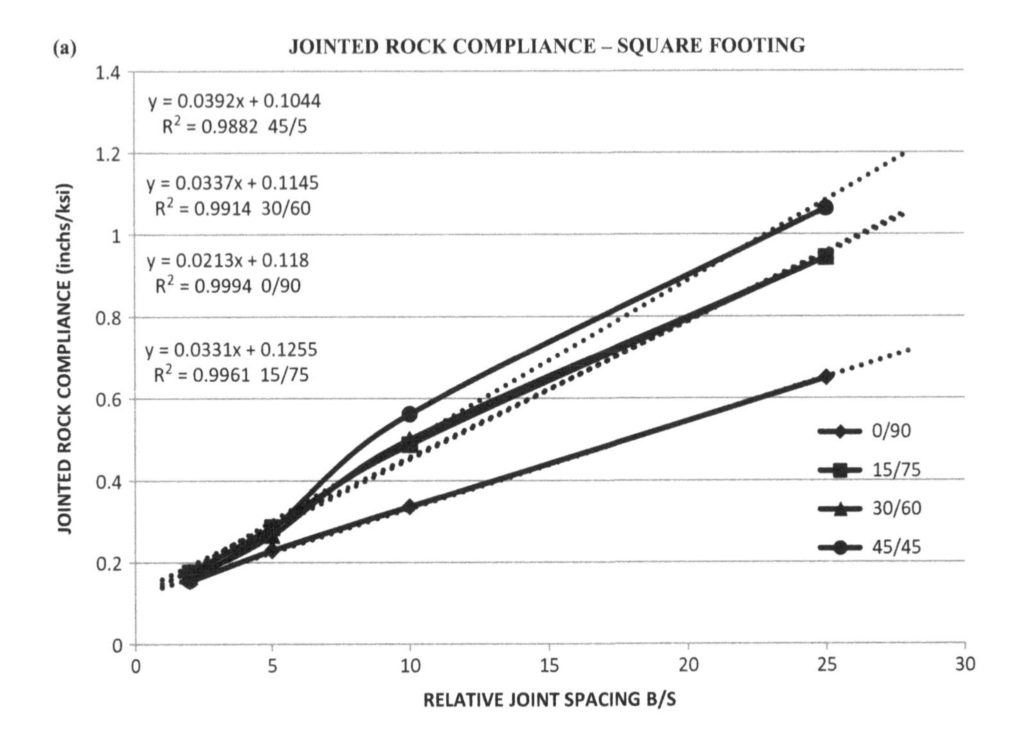

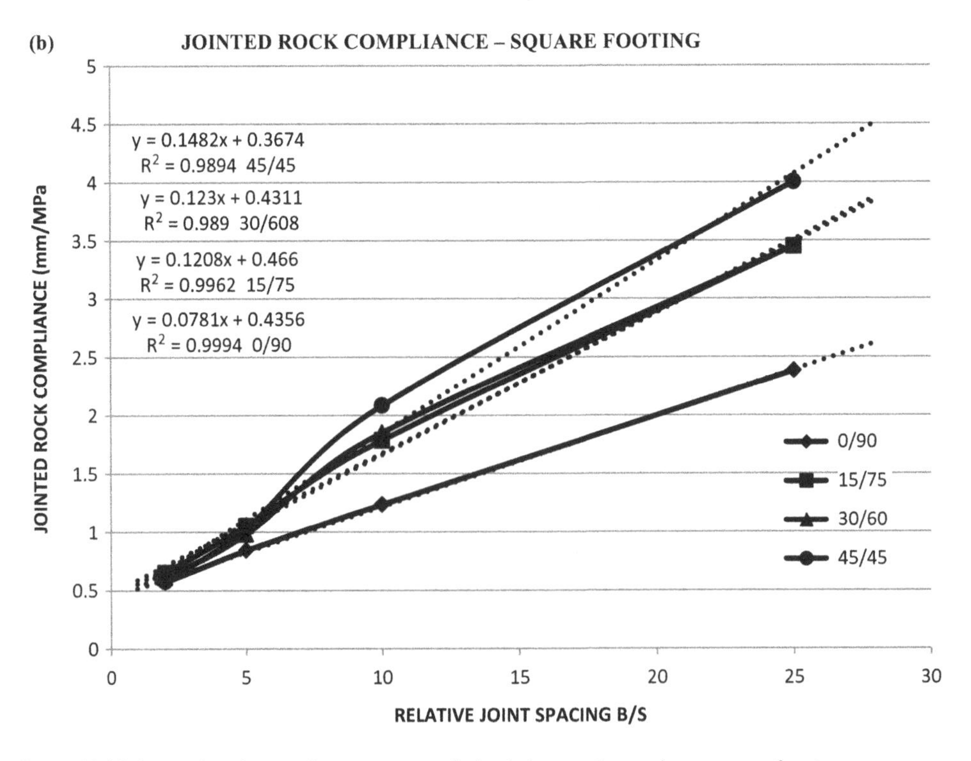

Figure 10.18 Jointed rock compliance versus relative joint spacing under a square footing.

L=column height. This observation suggests that rescaling the stiffnesses in Table 10.2 may provide a reasonable estimate of foundation stiffness in case of a very different Young's modulus of intact rock. Thus, $K' = K(E'/E)$ is an estimate of stiffness for rock with Young's modulus E' from Table 10.1 and Table 10.2 values of E and K.

Example 10.3 Consider a square footing on a jointed rock mass that contains three orthogonal joint sets. The strata dip 15 degrees, although the foundation surface is flat. A footing 1.5 × 1.5 m square is subject to a load of 10 MPa. Estimate the settlement expected: (a) when joint spacing is 2 m, and (b) when joint spacing is 0.3 m. Young's modulus is 43.1 GPa. Poisson's ratio is 0.2.

Solution: Settlement follows a load – settlement relationship $P = K\delta$, so

$$\delta = P/K$$

Properties are *not* Table 10.1 values. Hence, $K' = K(E'/E)$

$$\delta = 10,000/[1.54(10^3)(43.1/33.1)] = 5.0 \text{ mm (a) widely spaced joints (B/S} < 2)$$

$$\delta = 10,000/[0.95(10^3)(43.1/33.1)] = 8.1 \text{ mm (b) medium spaced joints (B/S} = 5)$$

Comment: The smaller settlement values are consistent with a greater Young's modulus and stiffer rock below the footing.

An alternative adjustment to foundation rock stiffness is one based on uniaxial strain rather than uniaxial stress. According to Hooke's law $P = E(\frac{1-v}{(1-2v)(1+v)})\varepsilon = E(1.11)\delta$ for Poisson's ratio equal to 0.2. An increase of Young's modulus of 11 percent is indicated. A reduction in settlement of approximately 11 percent would be expected. Ignoring a Poisson's ratio effect on foundation rock stiffness is therefore conservative in the sense that settlement would be somewhat less if a Poisson's ratio effect were taken into account. If Poisson's ratio were 0.25 rather than the Table 10.1 value, then $P = E(1.25)\delta$ and the adjustment to E is greater; settlement is less. At the extreme of an incompressible material, Young's modulus becomes indeterminate and indeed the guidance of uniaxial stress and strain becomes untenable. The reason is that foundation settlement is simply not that of a column of rock as recall of the pressure bulb concept shows. As a practical matter, foundation rock stiffness adjustment is still warranted based on Young's modulus, but Poisson's ratio effect may be ignored in this context.

Example 10.4 Consider a square footing on a jointed rock mass that contains three orthogonal joint sets. The strata dip 15 degrees, although the foundation surface is flat. A footing 1.5 × 1.5 m square is subject to a load of 10 MPa. Estimate the settlement expected: (a) when joint spacing is 2 m, and (b) when joint spacing is 0.3 m. Young's modulus is 43.1 GPa. Poisson's ratio is 0.25.

Solution: Settlement follows a load – settlement relationship $P = K\delta$, so

$$\delta = P/K$$

Properties are *not* Table 10.1 values. Hence, $K' = K(E'/E)$ and in consideration of Poisson's ratio $E' = E(\frac{1-v}{(1-2v)(1+v)}) = (43.1)(1.11) = 47.8$

$$\delta = 10,000/[1.54(10^3)(47.8/33.1)] = 4.5 \text{ mm (a) widely spaced joints } (B/S < 2)$$

$$\delta = 10,000/[0.95(10^3)(47.8/33.1)] = 7.3 \text{ mm (b) medium spaced joints } (B/S = 5)$$

Comment: The smaller settlement values are consistent with a greater Young's modulus and a Poisson's ratio effect that induce stiffer rock below the footing. The settlement values in this example are slightly less than in Example. 4.3 as they should be.

Example 10.5 Consider a square footing on a jointed rock mass that contains three orthogonal joint sets. The strata dip 45 degrees, although the foundation surface is flat. A footing 5 × 5 ft square is subject to a load of 1,500 psi. Estimate the settlement expected: (a) when joint spacing is 5 ft, and (b) when joint spacing is 0.5 ft. Young's modulus is $4.8(10^6)$ psi. Poisson's ratio is 0.2.

Solution: Settlement follows a load – settlement relationship $P = K\delta$, so

$$\delta = P/K$$

Properties are Table 10.1 values. Hence,

$$\delta = 1,500/6.37(10^3) = 0.24 \text{ inches (a) widely spaced joints } (B/S = 1)$$

$$\delta = 1,500/1.78(10^3) = 0.84 \text{ inches (b) closely spaced joints } (B/S = 10)$$

Comment: In the case of closely spaced joints, settlement is near the 1-inch guideline for acceptance leaving little room for error in properties that are likely to vary *in situ*.

Example 10.6 Consider a square footing on a jointed rock mass that contains three orthogonal joint sets. The strata dip 20 degrees, although the foundation surface is flat. A footing 5 × 5 ft square is subject to a load of 1,500 psi. Estimate the settlement expected: (a) when joint spacing is 5 ft, and (b) when joint spacing is 0.5 ft. Young's modulus is $4.8(10^6)$ psi. Poisson's ratio is 0.2.

Solution: Settlement follows a load – settlement relationship $P = K\delta$, so

$$\delta = P/K$$

Properties are Table 10.1 values, but the dip is in between Table 10.2 values of 15/75 and 30/60. Interpolation is indicated. Assuming linearity, at a wide spacing (B/S=1), K is between 5.62 and 5.92 psi/inch. Hence, $K = (\frac{6.04-5.68}{30-15})(25-15) + 5.68 = 5.92$.

$$\delta = 1,500/5.92(10^3) = 0.25 \text{ inches (a) widely spaced joints } (B/S = 1)$$

For closely spaced joints (B/s=10), $K = (\frac{2.00-2.05}{30-15})(25-15) + 2.00 = 1.97$

$$\delta = 1,500/1.74(10^3) = 0.86 \text{ inches (b) closely spaced joints } (B/S = 10)$$

Example 10.7 Consider a square footing on flat sedimentary rock containing three orthogonal joint sets that are widely spaced. If the footing is 2 m on edge, is a load equal to one half the unconfined compressive strength of intact rock below the footing acceptable, that is, within a 25 mm settlement limit? Properties are given in Table 10.1.

Solution:

$$\delta = P/K = (0.5 * 66.2\text{MPa})/1.76(\text{kPa}) = 18.8\text{mm}$$

which is slightly less than the 25 mm limit.

Example 10.8 Consider the footing in Example 10.7 with properties given in Table 10.1 but with joints closely spaced. Determine the load that causes settlement to reach the 25 mm limit.

Solution:

$$\begin{aligned} P &= K\delta \\ &= (0.81 \text{ MPa/mm})(25 \text{ mm}) \\ P &= 20.25 \text{ MPa} \end{aligned}$$

which is well below the unconfined compressive strength of the rock.

Example 10.9 A square footing 7 by 7 ft is founded on jointed rock dipping 30 degrees. Three sets of orthogonal joints are present. Material properties are given in Table 10.1. Joint spacing leads to value of B/S=7. Estimate the settlement under an applied load of 2,300 psi.

Solution. Applied load and settlement are related by the stiffness equation $P = K\delta$, so

$$\delta = P/K$$

Properties are consistent with Table 10.2 stiffnesses, but the given joint spacing is between table values, so interpolation is required. Thus,

$$K = (\frac{2.00 - 3.76}{10 - 5})(7 - 5) + 3.76 = 3.06$$

$$\delta = (2,300 \text{ psi})/[2.8(10^3) \text{ psi/inch}] = 0.82 \text{ inches}$$

Rectangular footings

Rock, joint and footing properties for rectangular footing computations are the same as for square footings and are given in Table 10.1. Tables 10.3 and 10.4 give stiffness of jointed rock in two cases of rectangular footings: (1) footings with L/B=2 and (2) footings with L/B=4, respectively. Again, three sets of mutually orthogonal joints are embedded in the finite

Table 10.3 Rectangular L/B=2 foundation rock stiffness*.

SPACING DIP (deg)	*WIDE B/S=2*	*MEDIUM B/S=5*	*CLOSE B/S=10*	*VERY CLOSE B/S=25*
0/90	5.14/1.40	3.37/0.91	2.14/0.58	1.24/0.34
15/75	4.60/1.25	2.93/0.80	1.70/0.46	0.88/0.24
30/60	4.70/1.28	3.08/0.84	1.62/0.44	0.84/0.23
45/45	4.67/1.27	2.94/0.80	1.43/0.34	0.72/0.20

Notes
* Units of stiffness are kpsi/inch / MPa/mm. Conversion: (MPa/mm)=0.272(kpsi/inch).

element meshes used to compute table entries. Joint spacing ranges from wide to very closely spaced as before. Also, the range of dips is the same: 0/90, 15/75, 30/60, 45/45 degrees. Table 10.3 units are (psi/inch)(10^3) = kpsi/inch and MPa/mm.

Comparison of Table 10.3 entries for a rectangular footing with Table 10.2 entries for a square footing shows that jointed rock stiffness in case of the rectangular footing is less than the square footing case for all combinations of joint spacing and strata dip. Somewhat similar trends in jointed rock stiffness with respect to joint spacing and strata dip are present: decreasing stiffness with decreasing joint spacing and increasing stiffness with strata dip except for the closely and very closely spaced joint cases.

The data in Table 10.3 are presented in graphical form in Figure 10.19 that shows foundation rock *compliance* as a function of relative joint spacing. Recall compliance relates settlement to load by $\delta = CP$. The regression lines and equations indicate compliances are quasi-linear functions of relative joint spacing at all dip angles.

A decrease in foundation rock stiffness implies an increase in settlement under the same load. If this trend continues, then foundation rock stiffness in case of a longer rectangular footing should decrease and similarly for a strip footing. Inspection of entries in Table 10.4 in case of a rectangular footing with L/B of 4 shows this trend to be the case. An explanation for this trend from square to strip footings is in the trend from three-dimensional to two-dimensional load distribution. Square footings distribute load to four sides while strip footings distribute load to just two sides. For the same applied stress, the rock load is greater

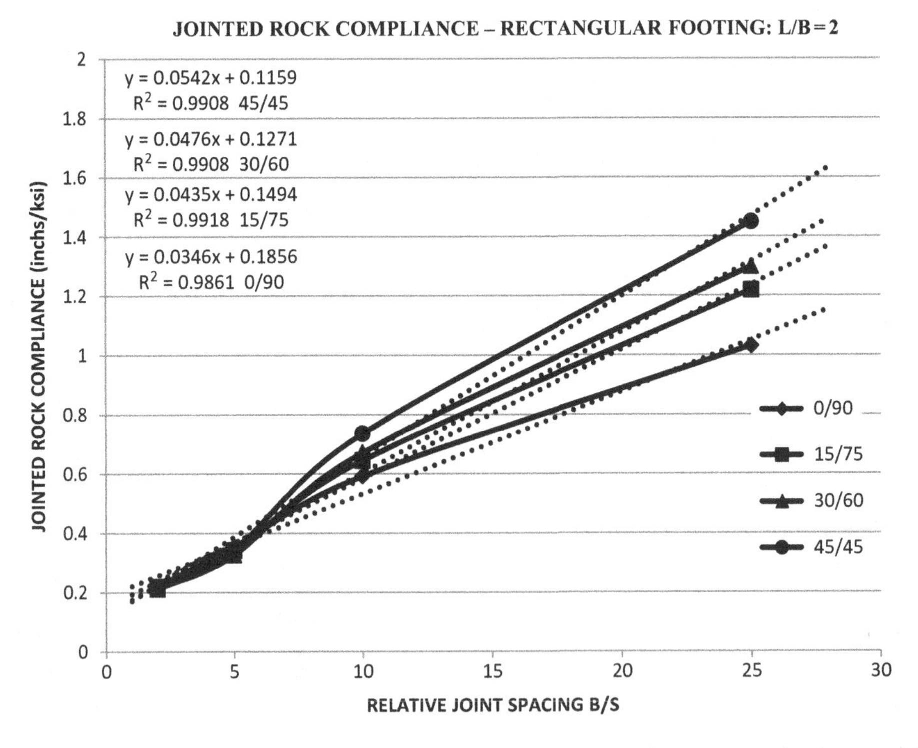

Figure 10.19 Jointed rock compliance versus relative joint spacing under a rectangular footing with L/B=2.

Table 10.4 Rectangular L/B=4 foundation rock stiffness.

SPACING *DIP (deg)*	*WIDE* *B/S=2*	*MEDIUM* *B/S=5*	*CLOSE* *B/S=10*	*VERY CLOSE* *B/S=25*
0/90	4.51/1.23	2.99/0.81	1.89/0.51	0.99/0.27
15/75	4.24/1.15	2.74/0.74	1.56/0.43	0.83/0.23
30/60	4.55/1.24	2.88/0.78	1.56/0.42	0.81/0.22
45/45	4.36/1.18	2.73/0.74	1.34/0.36	0.67/0.18

Notes

* Units of stiffness are kpsi/inch / MPa/mm. Conversion: (MPa/mm)=0.272(kpsi/inch).

below strip footings and so is the settlement. Consequently, foundation rock stiffness is less for strip footings. A similar phenomenon occurs in comparison with stress concentration about spherical excavations with long, cylindrical (circular) excavations. Stress concentration about a spherical excavation is much less than about a circular excavation of the same diameter. Table 10.4 units are (psi/inch)(10^3) = kpsi/inch and MPa/mm.

The data in Table 10.4 are presented in graphical form in Figure 10.20 that shows foundation rock *compliance* as a function of relative joint spacing. Recall compliance relates settlement to load by $\delta = CP$. The regression lines and equations indicate compliances are quasi-linear functions of relative joint spacing at all dip angles. The scales in Figure 10.20 are

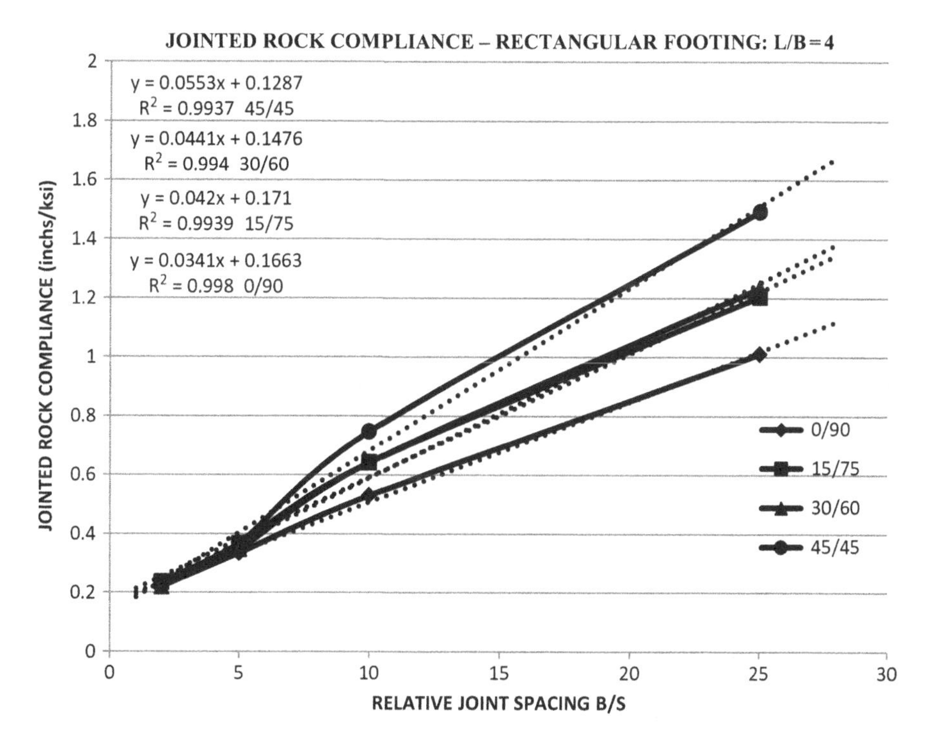

Figure 10.20 Jointed rock compliance versus relative joint spacing under a rectangular footing with L/B=4.

the same as in Figures 10.18 and 10.19 for ease of comparison. In this regard, as compliances increase, stiffnesses decrease and therefore settlement increases for a given load.

Example 10.10 Estimate the change in load at 25.4 mm of settlement when a 1×1 m footing is changed to a 1 × 2 m footing on rock containing three orthogonal joints at a medium spacing with flat bedding plane joints.

Solution: The load – settlement equations from Table 10.3 and Table 10.4 for square and rectangle footings (L/B=2) for medium joints spacing are:

$$P = K\delta,$$
$$P(square) = 1.19(MPa/mm)(25.4\ mm) = 30.2\quad MPa$$
$$P(rec2) = 0.91(MPa/mm)(25.4\ mm) = 23.1\quad MPa$$

Thus, the change in load is a decrease of 7.10 MPa (1,030 psi).

Example 10.11 Estimate the change in load at 1 inch of settlement when a 0.3 × 0.3 ft footing is changed to an 0.3 × 0.6 footing on rock containing three orthogonal joints at a medium spacing with flat bedding plane joints.

Solution: The load – settlement equations from Table 10.3 and Table 10.4 for square and rectangle footings (L/B=2) for medium joints spacing are:

$$P = K\delta,$$
$$P(square) = 4.37(kpsi/inch)(1.0\ \ inches) = 4,370\ \ psi$$
$$P(rec2) == 3.37(ksi/inch)(1.0\ inch) = 3,370\ \ psi$$

Thus, the change in load to reach a one inch settlement is a decrease of 1,000 psi (6.90 MPa)

Note: This is the same problem as Example 10.10. The numerical difference is in round-off of units, somewhat less than three percent.

Strip footings

Rock, joint and footing properties for strip footing computations are the same as for square and rectangular footings and are given in Table 10.1. Table 10.5 gives stiffness of jointed rock under strip loading where strip length L becomes indeterminately large. Again, three sets of mutually orthogonal joints are embedded in the finite element meshes used to compute table

Table 10.5 Strip footing foundation rock stiffness*.

SPACING DIP (deg)	*WIDE B/S=2*	*MEDIUM B/S=5*	*CLOSE B/S=10*	*VERY CLOSE B/S=25*
0/90	3.30/0.90	2.09/057	1.31/0.35	0.65/0.18
15/75	3.60/0.98	2.39/0.65	1.42/0.39	0.72/0.20
30/60	3.69/1.00	2.48/0.67	1.33/0.36	0.66/0.18
45/45	3.73/1.01	2.26/0.61	1.08/0.29	0.52/0.14

Notes
* Units of stiffness are kpsi/inch / MPa/mm. Conversion: (MPa/mm)=0.272(kpsi/inch).

entries. Joint spacing ranges from wide to medium to very closely spaced as before. The range of dips is also the same: 0/90, 15/75, 30/60, 45/45 degrees. Table 10.5 units are (psi/inch)(10^3) = kpsi/inch and MPa/mm. The overall trend of decreasing jointed rock stiffness with increasing length of footing is again evident in the table values. A decrease in stiffness implies an increase in settlement at a given load.

The data in Table 10.5 are presented in graphical form in Figures 10.21a.b that show foundation rock *compliance* as a function of relative joint spacing. Recall compliance relates settlement to load by $\delta = CP$. The regression lines and equations indicate compliances are quasi-linear functions of relative joint spacing at all dip angles. The scales in Figure 10.21 are the same as in Figures 10.19 and 10.20.

Example 10.12 A column exerts a load of 1,000 psi on a square footing 4×4 ft. If the same load is exerted by a wall on a strip footing, what is the increase in displacement when the foundation rock is dipping 30 deg and joints of a three joint set are closely spaced?

Solution: The load – settlement curve is the usual $P = K\delta$ so $P/K = \delta$. Hence, from Tables 10.2 and 10.5, respectively,

$$(1,000)/(2.00\ kpsi/inch) = 0.50\ inches\ (square)$$
$$(1,000)/(1.33\ kpsi/inch) = 0.75\ inches\ (strip)$$

Thus, the difference is 0.25 inches.

Note: The increase from square to strip footings is a substantial 50 percent.

Example 10.13 A column exerts a load of 6.9 MPa on a square footing 1.22×1.22 m. If the same load is exerted by a wall on a strip footing, what is the increase in displacement when the foundation rock is dipping 30 deg and joints of a three joint set are closely spaced?

Solution: The load – settlement curve is the usual $P = K\delta$ so $P/K = \delta$. Hence, from Tables 10.2 and 10.5, respectively,

$$(6.9MPa)/(0.54MPa/mm) = 12.8\ mm\ (square)$$
$$(6.9MPa)/(0.36\ MPa/mm) = 19.2\ mm\ (strip)$$

Thus, the difference is 6.4mm.

Note: The increase from square to strip footings is a substantial 50 percent.

Example 10.14 Explain the increase in settlement in Example 10.13 where the same load is applied to a square and strip footing.

Solution: The explanation is in the dimensionality of the footings. In the square footing case, the applied load is distributed to all four sides. In the strip footing case, essentially two-dimensional, load is distributed to two sides only and therefore increases the stress on the supporting rock mass that in turn increases settlement.

Example 10.15 Estimate the settlement that occurs when the load reaches one-third the unconfined compressive strength of a jointed rock mass containing three sets of orthogonal joints with bedding plane joints that are flat and closely spaced below a strip footing.

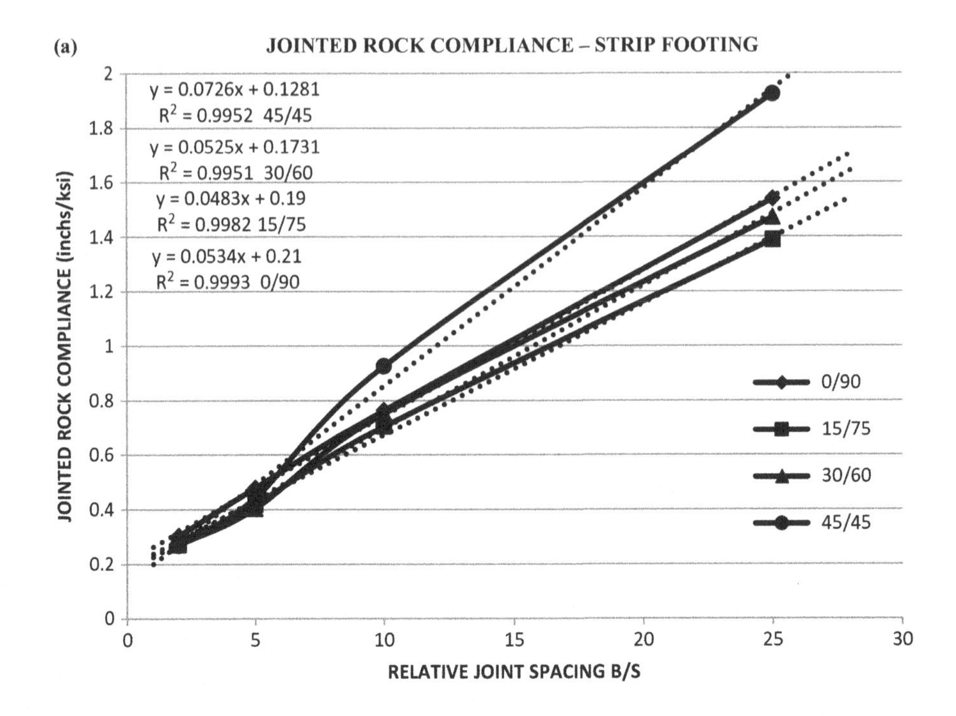

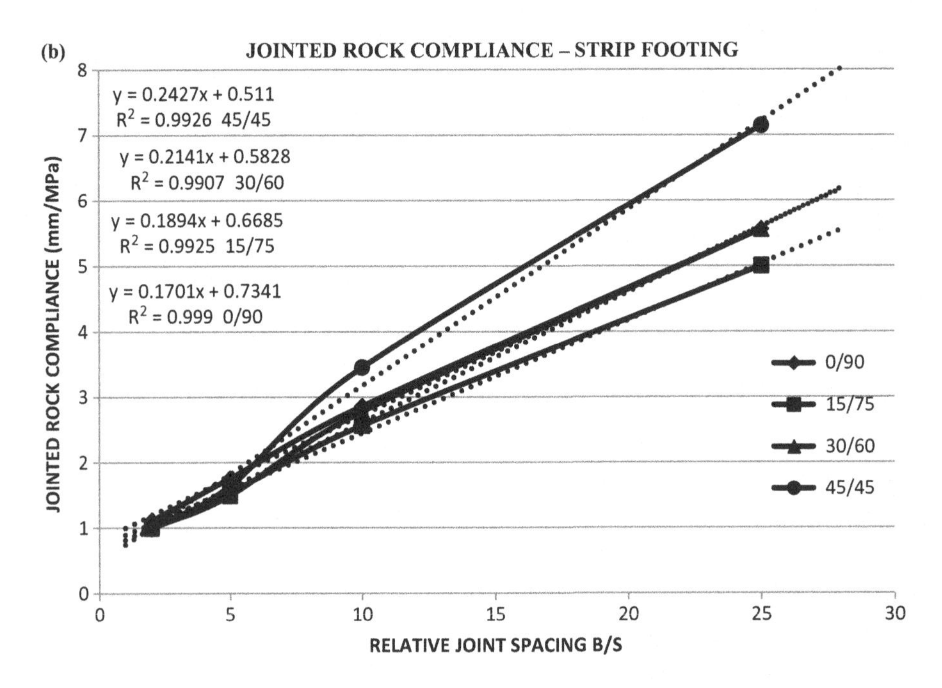

Figure 10.21 Jointed rock compliance versus relative joint spacing under a strip footing.

Solution: Stiffness tables are based on properties in Table 10.1 that shows rock unconfined compressive strength is 9,600 psi (66.2 MPa). Hence, from Table 10.5

$$P = K\delta \text{ so } P/K = \delta = (9,600/3)/1,310 = 2.44 \ inches \ (62 \ mm)$$

that is well beyond the 1 inch (25 mm) guideline for limiting settlement even though the load is only one third of the rock unconfined compressive strength while the foundation rock is confined and even stronger. The large settlement is a consequence of jointing, of course, and indeed joints are apt to fail in large numbers under this relatively large load.

Comment

Nothing has been said about wet ground and the role of water in the vicinity of a footing. In this regard, water pressure in jointed rock near the surface is quite low relative to bearing stress exerted by a footing on the rock below. The reason is the presence of joints tends to make the foundation rock free draining thereby eliminating excess pore fluid pressure. In effect, the total stresses and effective stresses are equal. This situation would not be the case if the rock foundation were under deep water. In case of a foundation on soil, the consolidating effect of foundation load would be of concern, especially on soils of low permeability.

Another possible concern is the slope of the foundation rock. If the foundation rock is inclined and a notch cut into the slope to secure a flat footing, there is a potential for slope instability. However, such a notch is likely to be small, of the order of a few feet or meters and thus not an issue provided any potential blocks formed by jointing are secured. In case of a foundation on sloping soil, adjustments to bearing capacity in terms of load would be necessary.

10.4 Problems

Note: All problems are based on Table 10.1 rock properties data and three sets of mutually orthogonal joints with equal spacing unless otherwise stated.

10.1 Estimate the load required to cause a settlement of 25 mm on a square footing on rock containing three sets of mutually orthogonal joints including bedding plane joints that are flat and at a medium spacing. Intact rock modulus is 33.1 GPa; unconfined compressive strength is 66.2 MPa.

10.2 Estimate the load required to cause a settlement of 1 inch on a square footing on rock containing three sets of mutually orthogonal joints including bedding plane joints that are flat and at a medium spacing. Intact rock modulus is 4.8 (10^6) psi; unconfined compressive strength is 9,600 psi.

10.3 Settlement is restricted to 1 inch in rock with three orthogonal joints that are spaced 2 ft apart. The bedding plane joints dip 15 deg and the companion joints dip in the opposite direction at 75 deg. The third joint set is vertical. A square footing 4×4 ft is being considered to support a load of 1250 psi. Determine whether the settlement expected is within the 1 inch limit.

10.4 Settlement is restricted to 25 mm in rock with three orthogonal joints that are spaced 0. 6 m apart. The bedding plane joints dip 15 deg and the companion joints dip in the opposite direction at 75 deg. The third joint set is vertical. A square footing

1.2×1.2 m is being considered to support a load of 8.62 MPa. Determine whether the settlement expected is within the 25 mm limit.

10.5 Bedding plane joints have a dip direction of 90 deg (East) and dip 30 deg and a companion set of joints has a dip direction of 180 deg (West) and dip 30 deg. Joints in a third set dip 90 deg and bear due north. Joints in each set are closely spaced. Laboratory testing indicates Young's modulus of intact rock is 3.2 million psi. Estimate the foundation rock stiffness where square footings are anticipated.

10.6 Bedding plane joints have a dip direction of 90 deg (East) and dip 30 deg and a companion set of joints has a dip direction of 180 deg (West) and dip 30 deg. Joints in a third set dip 90 deg and bear due north. Joints in each set are closely spaced. Laboratory testing indicates Young's modulus of intact rock is 33.1 GPa. Estimate the foundation rock stiffness where square footings are anticipated.

10.7 A rectangular footing with L/B=2 is proposed on rock with flat bedding plane joints that are closely spaced. Rock and joint properties are given in Table 10.1. Determine the load that will just cause settlement of 25 mm. Suppose the footing has L/B=4, and the same load is applied, what is the settlement? Briefly explain the difference.

10.8 A rectangular footing with L/B=2 is proposed on rock with flat bedding plane joints that are closely spaced. Rock and joint properties are given in Table 10.1. Determine the load that will just cause settlement of 1 inch. Suppose the footing has L/B=4, and the same load is applied, what is the settlement? Briefly explain the difference.

10.9 Consider a rectangular footing 6×12 ft on a laminated rock with laminations on 4 inch intervals. The laminations are in effect bedding plane joints. The bedding dips 15 degrees. Estimate the bearing capacity under a 1 inch settlement limit.

10.10 Consider a rectangular footing 2×4 m on a laminated rock with laminations on 100 mm intervals. The laminations are in effect bedding plane joints. The bedding dips 15 degrees. Estimate the bearing capacity under a 25 mm settlement limit.

10.11 Consider a strip footing on jointed rock where bedding plane joints that are closely spaced dip 30 degrees. Young's modulus of intact rock between joints is $1.2(10^6)$ psi. A load of 5 tons per square foot is anticipated. Estimate the expected settlement in inches.

10.12 Consider a square footing on jointed rock where bedding plane joints that are closely spaced dip 30 degrees. Young's modulus of intact rock between joints is 8.281 GPa. A load of 0.5 MPa is anticipated. Estimate the expected settlement in inches.

10.13 An alternative to a series of square footings for support of columns in a line is a strip footing. The square footings would be 5×5 ft on rock where joints are spaced 1.0 ft apart and dip 45 deg. The square footings would be spaced on 15 ft centers. There are nine square footings in the series. A strip footing would extend to the end of the square footing series. Intact rock Young's modulus is $2.4(10^6)$ psi. Settlement is limited to 0.5 inches in any case. Estimate the load per square footing allowed and then estimate the width of a strip footing to support the same total load under the same settlement limit.

10.14 An alternative to a series of square footings for support of columns in a line is a strip footing. The square footings would be 1.5×1.5 m on rock where joints are spaced 0.3 m apart and dip 45 deg. The square footings would be spaced on 4.5 m centers. There are nine square footings in the series. A strip footing would extend to the end of the square footing series. Intact rock Young's modulus is 16.6 GPa. Settlement is limited to 12 mm in any case. Estimate the load per square footing allowed and then estimate the width of a strip footing to support the same total load under the same settlement limit.

10.15 Consider a factor of safety with respect to unconfined compressive strength of intact rock between joints associated with a strip footing 3.5 ft wide on flat and very closely jointed rock. If a safety factor of three is required, what settlement is indicated? If settlement is greater than a one inch limit, what limit load is indicated?

10.16 Consider a factor of safety with respect to unconfined compressive strength of intact rock between joints associated with a strip footing 1.1 m wide on flat and very closely jointed rock. If a safety factor of three is required, what settlement is indicated? If settlement is greater than a 25 mm limit, what limit load is indicated?

10.17 Consider a rectangular footing with L/B=4. Table 10.4 containing foundation rock stiffnesses is based on the assumption that the long dimension of the footing is parallel to the strike of dipping joints. How would the Table 10.4 values change if the footing was oriented perpendicular to the strike of dipping joints? Would stiffness increase, decrease or remain the same? Explain your response.

10.18 Suppose a strip footing on a flat but jointed rock mass where the intact rock is permeable as are the joints is saturated. If a the footing is 4 ft wide and subject to a 1,230 psi load, estimate water pressure, vertical total stress and vertical effective stress at a depth of 24 ft.

10.19 Suppose a strip footing on a flat but jointed rock mass where the intact rock is permeable as are the joints is saturated. If a the footing is 1.2 m wide and subject to a 8.48 MPa load, estimate water pressure, vertical total stress and vertical effective stress at a depth of 7.2 m.

10.20 Consider as square footing on jointed rock dipping 45 degrees where joint spacing is medium. Estimate the maximum error in specifying foundation rock compliance using linear regression equations.

Appendix A

Background literature

Background literature includes texts that develop fundamentals of strength of materials, mechanics of materials, elasticity theory, and continuum mechanics. Also included are reference works in rock mechanics. Rock properties data necessary for engineering design analysis may be found in numerous technical articles and handbooks.

A.1 Books about fundamentals of mechanics

1. Popov, E. P. (1957) *Mechanics of Materials*. Prentice-Hall, 441 pp.
   Comment: An early undergraduate text in mechanics of materials. One that first introduced the author to the subject. Contains illustrative material and examples.
2. Popov, E. P. and T. A. Balan (1991) *Engineering Mechanics of Solids* (2nd ed). Prentice-Hall, 864 pp.
   Comment: An expanded edition with updated example problems and discussion.
3. Hibbeler, R. C. (2003) *Mechanics of Materials* (5th ed). Prentice-Hall, 843 pp.
   Comment: An undergraduate text with numerous examples and problems for homework assignments. Includes chapters on two- and three-dimensional stress and strain, as do most recent mechanics of materials text books.

   More advanced treatments of mechanics including elasticity, plasticity, fracture, and general continuum theory that the author has personally consulted on a number of occasions include:
4. Fung, Y. C. (1965) *Foundations of Solid Mechanics*. Prentice-Hall, 525 pp.
5. Hill, R. (1950) *The Mathematical Theory of Plasticity*. Oxford, 355 pp.
6. Jaunzemis, W. (1967) *Continuum Mechanics*. The Macmillan Co., 604 pp.
7. Knott, J. F. (1973) *Fundamentals of Fracture Mechanics*. Butterworths, 273 pp.
8. Lekhnitskii, S. G. (1963) *Theory of Elasticity of an Anisotropic Elastic Body*. Translation by P. Fern. Holden-Day, 404 pp.
9. Lawn, B. R. and T. R. Wilshaw (1975) *Fracture of Brittle Solids*. Cambridge University Press, 204 pp.
10. Love, A. E. H. (1944) *A Treatise on the Mathematical Theory of Elasticity* (4th ed). Dover, 643 pp.
11. Malvern, L. E. (1969) *Introduction to the Mechanics of a Continuous Medium*. Prentice-Hall, 713 pp.
12. Nadai, A. (1963) *Theory of Flow and Fracture of Solids*. Vol. II, McGraw-Hill, 705 pp.
13. Prager, W. (1961) *Introduction to Continuum Mechanics*. Dover, 230 pp.

14. Savin, G. N. (1961) *Stress Concentration around Holes*. Translation Ed. – W. Johnson. Pergamon, 430 pp.
15. Sokolnikoff, I. S. (1956) *Mathematical Theory of Elasticity*. McGraw-Hill, 476 pp.
16. Sokolovski, V. V. (1960) *Statics of Soil Media*. Translation by D. H. Jones and A. N. Schofield. Butterworths, 237 pp.
17. Thomas, T. Y. (1961) *Plastic Flow and Fracture in Solids*. Academic Press, 267 pp.
18. Timoshenko, S. and J. N. Goodier (1951) *Theory of Elasticity*. McGraw-Hill, 506 pp.

A.2 Books about rock mechanics

1. Obert, L. and W. I. Duvall (1967) *Rock Mechanics and the Design of Structures in Rock*. John Wiley & Sons, 650 pp.
   Comment: One of the first references about rock mechanics from stress analysis to laboratory testing to applications to excavations in massive rock and in stratified ground. With the benefit of hindsight, treatment of joints is conspicuous by its absence. Now out of print, but perhaps available in libraries.
2. Coates, D. F. (1970) *Rock Mechanics Principles*. Canadian Mines Branch Monograph 874 (Revised 1970, first edition 1967).
   Comment: Another early work in rock mechanics that includes chapters on foundations and rock dynamics. Contains useful appendices including the subjects of stress concentration factors, beam bending, and stereonets. Also out of print.
3. Jaeger, J. C. and N. G. W. Cook (1969) *Fundamentals of Rock Mechanics*. Methuen, 513 pp.
   Comment: A reference work by two pioneering authorities in rock mechanics. Contains important chapters on rock friction and the effect of fractures on rock mass elastic moduli. Discusses excavation of tabular caverns and the role of strain energy release rate. Contains an extensive list of references, not undergraduate fare. An expanded 4th edition was extensively updated and augmented by R.W. Zimmermann in 2009.
4. Hoek, E. and J. Bray (1977) *Rock Slope Engineering* (2nd ed). Institution of Mining and Metallurgy, London, 402 pp.
   Comment: A practical guide to rock slope stability analysis by two pioneers in rock engineering applications at Imperial College. Computer programs are now available for planar block slides and wedge failures.
5. Barry Voight (ed.) (1978) *Rockslides and Avalanches – Natural Phenomena* Vol. 1. Elsevier, 833 pp.
6. Barry Voight (ed.) (1978) *Rockslides and Avalanches – Engineering Sites* Vol. 2. 850 pp.
   Comment: This two volume set is a remarkable collection of historic developments, observations of natural phenomena and engineering case histories that is an outstanding interdisciplinary merger of geology and engineering.
7. Peng, S. S. (1978) *Coal Mine Ground Control*. John Wiley & Sons, 450 pp.
   Comment: An excellent reference for practical approaches to coal mine ground control by an outstanding educator and contributor in mining engineering. Contains details about longwall supports such as shields not found elsewhere. An expanded 3rd edition came out in 2008.

8. Hoek, E. and E. T. Brown (1980) *Underground Excavations in Rock*. Institution of Mining and Metallurgy, London, 527 pp.
Comment: A guide to underground opening analysis by two internationally recognized authorities in rock mechanics applications to mine engineering and civil works. A successful sequel to the surface mining effort by the same rock mechanics group at Imperial College. Contains important chapters on data collection preceding design analysis and includes discussion of the role of rock mass classification systems for engineering purposes. References abound following each chapter including those on blasting and instrumentation.
9. Goodman, R. E. (1980) *Introduction to Rock Mechanics*. John Wiley & Sons, 478 pp.
Comment: An introduction to rock mechanics from the geological – civil engineering perspective by an outstanding educator and practioner in rock mechanics. Contains an excellent chapter on foundations on rock. Second edition published in 1989 (562 pp).
10. Brady, B. H. G. and E. T. Brown (1980) *Rock Mechanics for Underground Mining* (2nd ed). Chapman-Hall, 571 pp.
Comment: A well-illustrated tour de force that ranges from the theoretical to the empirical, from numerical analysis to support specification from rock mass classification. Some problems and examples are included, but still not an undergraduate text. A recent revision toward computer modeling is available.
11. Bieniawski, Z. T. (1989) *Engineering Rock Mass Classifications*. John Wiley & Sons, 251 pp.
Comment: "A complete manual for engineers and geologists in mining, civil and petroleum engineering" to quote the description following the title. Indeed, this is a handy reference to the subject of rock mass classification schemes for engineering purposes that includes the most important schemes to date beginning with the Terzaghi scheme for estimating tunnel supports developed in the late 1930's, early 1940's.
12. Hudson, J. A. (ed.) (1993) *Comprehensive Rock Engineering*. Pergamon Press, (in 5 volumes).
Comment: No list of rock mechanics references would be complete without mention of this huge work in five volumes by an international group of eminent rock mechanics specialists. Vol. 1 = Fundamentals, Vol. 2 = Analysis and Design Methods, Vol. 3 = Rock Testing and Site Characterization, Vol. 4 = Excavation, Support and Monitoring, Vol. 5 = Surface and Underground Project Case Histories.
13. Hoek, E., P. K. Kaiser, and W. F. Bawden (1995) *Support of Underground Excavations in Hard Rock*. Balkema, 215 pp.
Comment: An authoritative treatment of support for underground excavations in rock for civil and mining purposes. Presented with a preference for rock mass classification schemes, the Hoek–Brown failure criterion, and points toward computer programs for detailed stress analysis. Well illustrated with figures and photos.
14. Harrison, J.P. and J.A. Hudson 2000. *Engineering Rock Mechanics: Part 2 Illustrative Worked Examples*. Elsevier, 506 pp.
Comment: A work that only recently came to my attention, but one that would be especially useful to beginners and practicing professionals alike in the broad area of geotechnical engineering. The unusual question and answer format seems especially suited for self-study. Part 1 (1997) of this two-part series deals with principles of rock mechanics and has the authorship reversed.

A.3 Books containing rock properties

The most important rock properties for engineering design are elastic moduli and strengths of intact rock and joint stiffnesses and strengths. Other properties such as thermal conductivity may be important in special cases, of course. All the books about rock mechanics mentioned previously have some tables of rock properties; some also have joint properties.

1. Vutukuri, V. S., R. D. Lama, and S. S. Saluja (1974) *Handbook on Mechanical Properties of Rock* (Testing Techniques and Results) Vol. I. Trans Tech Publications, 280 pp.
2. Lama, R. D. and V. S. Vutukuri (1978) *Handbook on Mechanical Properties of Rock* (Testing Techniques and Results) Vol. II. Trans Tech Publications, 481 pp.
3. Lama, R. D. and V. S. Vutukuri (1978) *Handbook on Mechanical Properties of Rock* (Testing Techniques and Results) Vol. III. Trans Tech Publications, 406 pp.
4. Lama, R. D. and V. S. Vutukuri (1978) *Handbook on Mechanical Properties of Rock* (Testing Techniques and Results) Vol. IV. Trans Tech Publications, 515 pp.
   Comment: This enormous work consists of four volumes that contain an extensive tabulation of mechanical properties test data and measurements from the international literature in rock mechanics. Descriptions of test and measurement procedures complement the data.
5. Hartman, H. L. (ed.) (1992) *SME Mining Engineering Handbook*, Vol. 2 (Appendix), Society for Mining, Metallurgy and Exploration.
   Comment: The second volume of this large, two-volume set contains extensive tables of data in the appendix to Vol. 2 where references to the data sources are also cited. The data relate mainly to tests on intact laboratory test pieces. The references cited also point to sources of other data relevant to geophysics, for example. Unfortunately, one will need to look hard for joint data; "joint" is not in the index.
6. Kulhawy, F. H. (1975) Stress deformation properties of rock and rock discontinuities. *Engineering Geology*, Vol. 9, pp. 327–350.
7. Kulhawy, F. H. (1978) Geomechanical model for rock foundation settlement. *Journal of the Geotechnical Engineering Division*, ASCE, Vol. 104 (GT2), pp. 211–225.
   Comment: The two articles by Professor Kulhawy are cited repeatedly in the literature because of the large amount of data on rock and joints that are tabulated in these very helpful contributions to rock mechanics.

A.4 General sources of rock mechanics information

There is an enormous accumulation of technical literature concerning rock mechanics. Growth has been especially rapid since the late 1960s with the proliferation of journals and symposia. Two journals that have stood the test of time are:

1. *International Journal of Rock Mechanics and Mining Sciences* (now published by Elsevier). First issue published January, 1964, and has a web site.
2. *Rock Mechanics and Rock Engineering* (now published by Springer Wien). Available on line. Antecedent issue published as *Rock Mechanics*, Vol. 1, 1969, "continuing the tradition of Felsmechanik und Ingenierugeologie."

There are also two series of symposia of interest that are rich sources of rock mechanics theories and practices from around the world:

1. The U.S. Rock Mechanics Symposium that began in 1956 at the Colorado School of Mines and continued almost annually to the 34th Symposium in 1993 when it began alternating with a North American Rock Mechanics Symposium.
2. The International Congress of the International Society for Mechanics series that began in Lisbon, 1967, and has continued meeting at four year intervals.

Appendix B

Mechanical properties of intact rock and joints

Mechanical properties of rock that are of most importance to engineering design are elastic properties and strengths. In this regard, field-scale rock masses are composite materials that are composed of numerous joints and blocks of intact rock between joints. Deformation of this composite under changing loads is largely determined by: (1) intact rock elastic moduli and strengths and (2) stiffnesses and strengths of joints. These properties may be determined by laboratory tests on relatively small samples that have dimensions of the order of inches or centimeters. Confining pressure, rate of loading, temperature, time, and peculiarities of testing apparatus are often observed to affect test results on strength but have only a minor influence on elastic moduli. Within the engineering design domain and range of environmental variables, the effect of confining pressure on strength is most significant. Rock includes an enormous variety of materials from volcanic glass to reef coral, from fresh granite to welded tuff, from pegmatite to porphyry, limestone to marble, shale to slate, sandstone to quartzite, peat to coal, and so on. Not too surprisingly, there are exceptions and departures from the norms.

Rock types based on genetic or geological classification systems seldom relate to engineering properties. Although one may intuitively suppose that there is an association of properties values with rock types, there is often little correlation. Some sedimentary rocks have high strength, while some igneous rocks have low strength, and so on. However, there are several rock classification systems available for engineering purposes that allow for preliminary estimates of excavation stability or safety. The most popular engineering classification systems are the rock mass rating (RMR) and quality (Q) systems. These classification systems characterize rock masses by a single number based on a combination of laboratory measurements and field observations, that is, on intact rock properties, joint properties, joint spacing, and whether water is present. The purpose of classification schemes is to assist in determining tunnel support requirements in advance of excavation; several are discussed in Appendix C. While rock and joint properties determined by laboratory testing do not automatically lead to elastic moduli and strengths of field-scale rock masses, they form an essential part of any site-specific database needed for design, even at the earliest stages of site evaluation.

The American Society for Testing and Materials (ASTM) and the International Society for Rock Mechanics (ISRM) have published standards for conducting a variety of laboratory tests. ISRM also has published suggested standards for field tests and measurements *in situ.* These standards give meaning to rock properties test results and allow for comparisons amongst rock types and testing laboratories. Various tabulations of rock properties data may be found in the references.

Table B.1 Specific gravity and porosity of rock (after Kulhawy, 1975)

Property	*Rock type*					
	Igneous		*Metamorphic*		*Sedimentary*	
	Plutonic	*Volcanic*	*Non-foliated*	*Foliated*	*Clastic*	*Chemical*
(SG)						
Number	22	10	4	21	17	20
Maximum	3.04	3.00	2.82	2.86	2.72	2.90
Minimum	2.50	1.45	2.63	2.18	2.32	1.79
Average	2.68	2.61	2.72	2.66	2.53	2.56
(n%)						
Number	22	10	4	21	20	17
Maximum	9.6	42.5	1.9	22.4	21.4	36.0
Minimum	0.3	2.7	0.9	0.4	1.8	0.3
Average	2.3	10.6	1.3	4.5	11.3	8.1

Rock specific gravity and porosity are also of interest. Table B.1 shows ranges and averages of specific gravity (SG) and porosity (n) according to the three major categories of rock, igneous, metamorphic, and sedimentary. This table was adopted from a survey of laboratory test data by Kulhawy (1975). Data from Table B.1 show a range of specific weights (densities) from 91 lbf/ft^3 (1.45 g/cm^3) to 190 lbf/ft^3 (3.04 g/cm^3). Some rock may be much lighter, for example, pumice stone that floats on water, or heavier, for example, galena-rich lead ore. Porosity may be high, especially in some volcanic rock.

B.1 Elastic moduli of intact rock

Elastic moduli of isotropic media commonly determined in laboratory testing are Young's modulus E and Poisson's ratio v. Shear modulus G is usually computed from E and v. These moduli may be determined statically and dynamically, although static elastic moduli are determined most often. Measurements are usually done on cylindrical test specimens prepared from diamond drill core obtained during site exploration. An additional measurement of lateral displacement allows for the determination of Poisson's ratio. Figure B.1 is a schematic of a laboratory test for Young's modulus.

Young's modulus

Cores prepared with smooth ends and with a length to diameter ratio (L/D) of two are loaded axially in compression. A cycle of loading and unloading is done first to ensure proper functioning of the system and to seat the apparatus. Subsequent cycles of loading and unloading show greater reproducibility, narrower loops, and steeper plots of axial force versus displacement. Data are obtained after seating the apparatus and test cylinder. If strain gages are attached to the test specimen, then a plot of axial stress as a function of axial strain allows for the determination of Young's modulus. Strain gages have the advantage of reacting directly to the rock but are more expensive than is the use of displacement transducers. Figure B.2 presents results of laboratory test for Young's modulus for a variety of rock types. The range of test

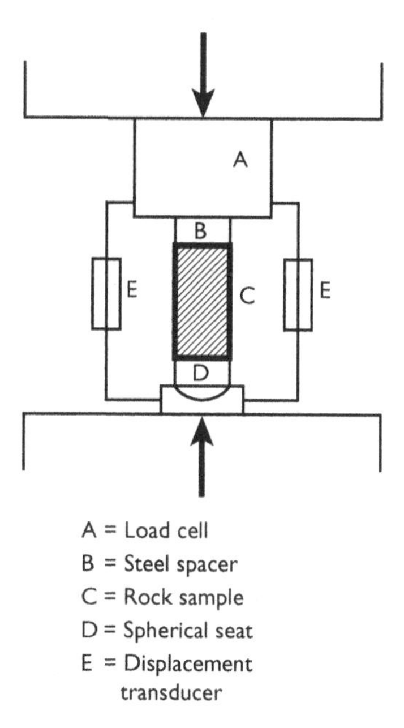

Figure B.1 Schematic of apparatus for measuring Young's modulus and Poisson's ratio on an intact rock test cylinder.

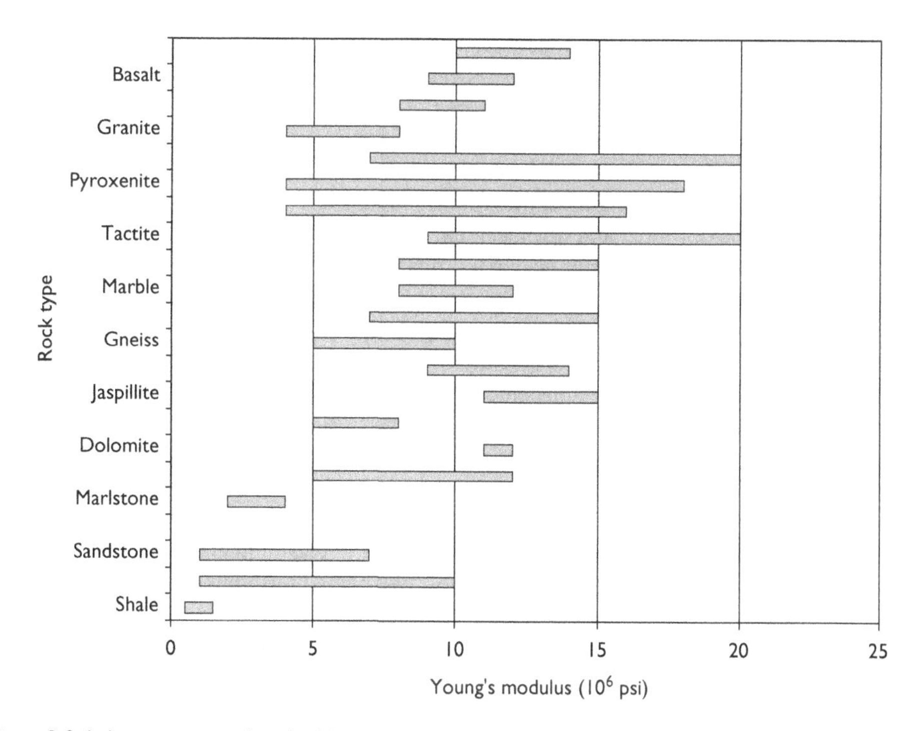

Figure B.2 Laboratory test data for Young's modulus (after Windes and Blair, 1949).

Table B.2 Young's modulus by rock type (after Kulhawy, 1975)

Young's modulus E 10^6 psi (GN/m²)	*Rock type*					
	Igneous		*Metamorphic*		*Sedimentary*	
	Plutonic	*Volcanic*	*Non-foliated*	*Foliated*	*Clastic*	*Chemical*
Number	40	17	12	29	30	163
Maximum	14.4 (99.4)	12.2 (83.8)	12.8 (88.4)	11.9 (81.7)	13.5 (90.0)	14.4 (99.4)
Minimum	11.3 (7.8)	0.17 (1.2)	5.21 (35.9)	8.6 (5.9)	0.67 (4.6)	0.17 (1.2)
Average	8.21 (56.6)	5.53 (38.1)	8.6 (59.6)	6.8 (47.0)	6.8 (47.0)	6.3 (43.4)

results indicates that variability is to be expected. Table B.2 shows ranges of Young's modulus according to rock type and indicates a wide range for a given rock type and no correlation of modulus with rock type.

Coal is a rock type that is in a class by itself because of its organic constitution and commercial importance. Vutukuri and Lama (Vol. II, 1978a) report a wide range for Young's modulus from 1 to over 51 GN/m^2 (0.15 to 7.4 × 10^6 psi). This large range is caused, in part, by directional dependencies (anisotropy), but mainly by coal rank that ranges from sub-bituminous to anthracite. A likely range of Young's modulus for most bituminous coal is perhaps 2 to 5 GN/m^2 (0.29 to 0.73 × 10^6 psi). However, recognition of anisotropy may be important at any particular site.

The theory behind the test for Young's modulus is Hooke's law that in consideration of the uniaxial loading reduces to $\sigma_a = E\varepsilon_a$ for the axial direction. Thus, a plot of axial stress versus axial strain does indeed allow Young's modulus to be determined by simply measuring the slope of the plot. The plot is usually curved, as shown in Figure B.3 where compression is positive. Slope of a tangent to the stress–strain curve defines a *tangent modulus* E_t. A line drawn from the origin to a point on the curve defines a *secant modulus* E_s. Measurement of the tangent modulus at 50% of the unconfined compressive strength defines Young's modulus for engineering design.

Measurements of axial force and displacement using displacement transducers instead of strain gages allows for a more economical determination of Young's modulus. However, care must be taken to account for displacement of steel spacers, spherical seat, and other material in series with the test specimen. This accounting is easily done graphically, as shown in Figure B.4, where a total force–displacement plot and a force–displacement plot without the test specimen are presented. The displacement of the test specimen is the difference between the two plots at the considered level of force. According to Hooke's law $\sigma_a = F/A = E\varepsilon_a = EU/L$ where F, U, A, and L are force, displacement, cross-sectional area of the test specimen, and test specimen length, respectively. Again, data reduction is done at about 50% of the unconfined compressive strength of the material.

The equation of a *force–displacement* plot for a linearly elastic material is $F = KU$; K is obviously the slope of the plot. The core, steel spacers, and so forth may be analyzed as two springs in series, as shown in Figure B.5. Each linear, elastic spring experiences the same force, while the total displacement is the sum of the displacements of spring and steel. Thus,

$$F = K_s U_s, \quad F = K_r U_r, \quad F = KU, \quad U = U_s + U_r$$

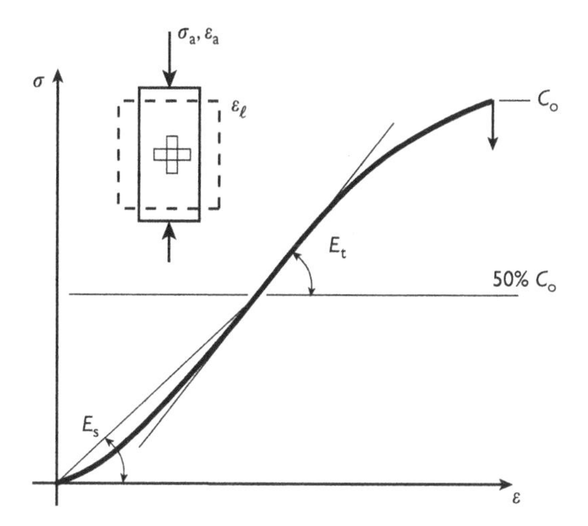

Figure B.3 An axial stress–strain plot for Young's modulus.

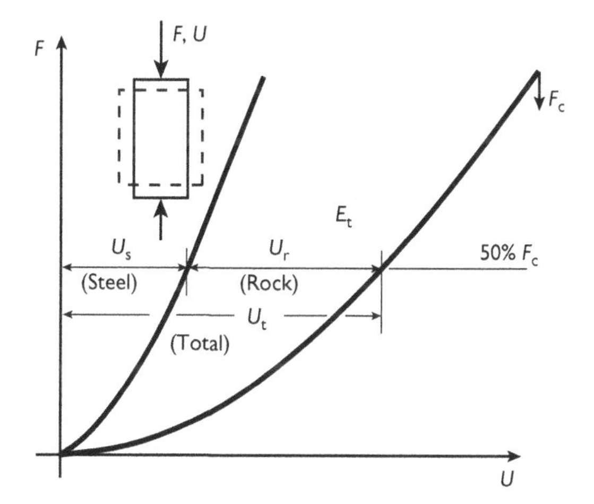

Figure B.4 Force–displacements plots for steel and steel plus rock total.

where subscripts s and r stand for steel and rock. The combined system stiffness (no subscript) is

$$\frac{1}{K} = \frac{1}{K_s} + \frac{1}{K_r}$$

and $Kr = E/AL$ for the rock core, so one may also determine the rock modulus E from the slopes of the experimental "total" and "steel" force–displacement curves.

Poisson's ratio

Hooke's law also indicates the measurements needed to determine Poisson's ratio v. Under axial loading, the lateral strain is $\varepsilon_l = -v\sigma_a/E$. Thus, $-E/v$ is the slope of a plot of axial stress

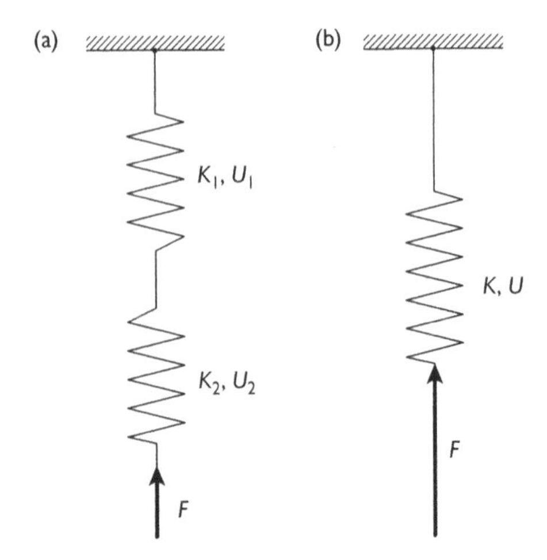

Figure B.5 Springs in series representing steel and rock in a laboratory test.

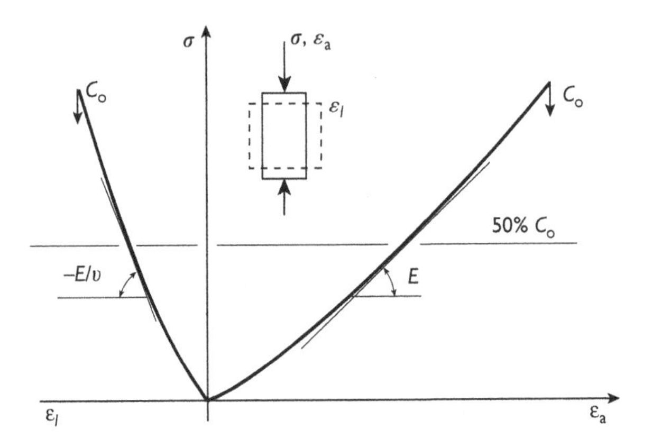

Figure B.6 Axial stress versus axial and lateral strain for Poisson's ratio.

versus lateral strain obtained during a uniaxial compression test, as shown in Figure B.6. Determination of Young's modulus from an axial stress–strain plot and measurement of lateral strain then allows for determination of Poisson's ratio. Alternatively, under uniaxial stress, one may compute Poisson's ratio from $v = |\varepsilon_1/\varepsilon_a|$. The algebraic signs of axial and lateral strains are always opposite whether compression or tension is considered positive; the use of the absolute value signs prevents error. Negative values of Poisson's ratio and values greater than 0.5 are suspect and likely indicate measurement error or loading beyond the elastic limit. A range of 0.02 to 0.46 for Poisson's ratio is reported by Kulhawy (1975); the average of the data reviewed is a reasonable 0.20. Table B.3 shows Poisson's ratio for the main rock types and the range observed.

Table B.3 Poisson's ratio by rock type (after Kulhawy, 1975)

Poisson's ratio ν	*Rock type*					
	Igneous		*Metamorphic*		*Sedimentary*	
	Plutonic	*Volcanic*	*Non-foliated*	*Foliated*	*Clastic*	*Chemical*
Number	36	17	12	25	26	141
Maximum	0.39	0.32	0.40	0.46	0.73[a]	0.73[a]
Minimum	0.05	0.09	0.02	0.03	0.04	0.02
Average	0.20	0.20	0.21	0.17	0.26	0.20

Note

a Dilatant values, not elastic.

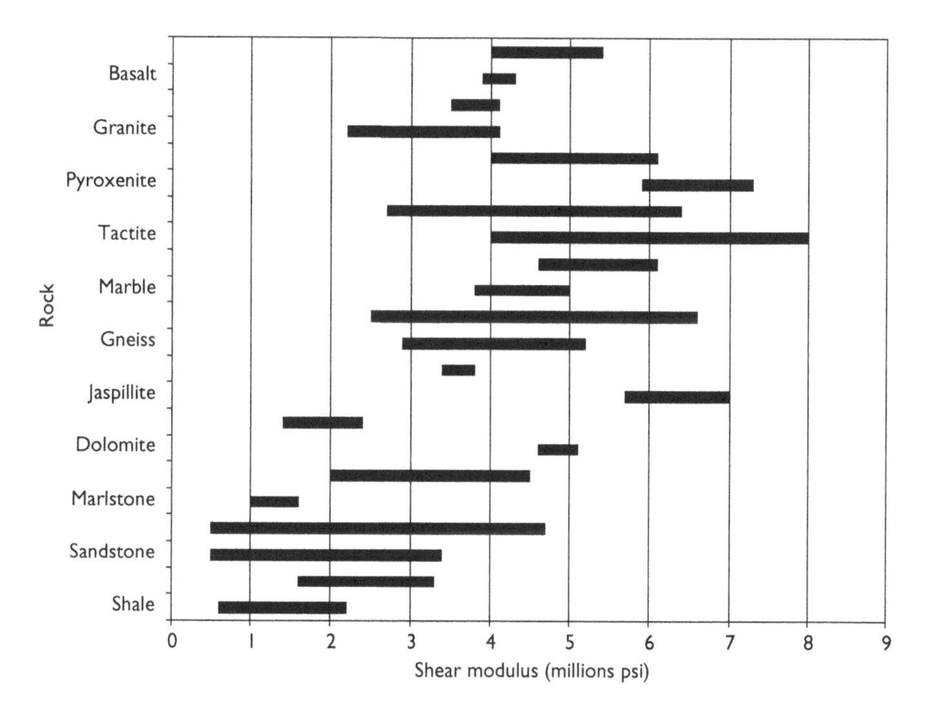

Figure B.7 Shear modulus for a variety of rock types (after Windes and Blair, 1949).

Shear modulus

Direct experimental measurement for shear modulus G would require torsion and is rarely done. Usually, G is computed from Young's modulus and Poisson's ratio by the formula $G = E/2(1 + \nu)$under the assumption of isotropy. A maximum G corresponds to a minimum ν for a given E, so that G ranges from $0.50E$ down to $0.33E$, corresponding to ν ranging from 0.0 up to 0.5. A reasonable assumption of $\nu = 0.20$ gives a shear modulus $G = E/2.4 = 0.417E$. Figure B.7 presents values of shear modulus for a variety of rock types. Again, variability is the rule rather than the exception.

Anisotropy

Many rock types have directional features such as bedding, foliation, and flow structures that are reflected in anisotropic elastic properties. For example, Young's modulus parallel to bedding of a laminated sandstone is often greater than perpendicular to the bedding. Sedimentary rocks may therefore exhibit anisotropy in the form of transversely isotropic elastic properties. Other rock types may also be transversely isotropic. Transversely isotropic rock requires five elastic moduli for characterization. If v is vertical and perpendicular to the bedding and h is horizontal, then core from vertical and horizontal holes allow for the determination of E_v, E_h, G_v, G_h, ν_v, ν_h. The horizontal direction is in a plane of isotropy, so $G_h = E_h/2(1 + \nu_h)$, but a similar relationship does not apply to the vertical direction. The five independent elastic constants in this case are: E_v, E_h, G_v, G_h, and ν_v.

Metamorphic rocks, such as gneisses, and other rock types, such as rhyolite with flow structure, may have three distinct directions of anisotropy (orthotropic) and thus require nine elastic moduli for characterization. The principal axes of anisotropy are preferred material directions, say, *abc* where a is down dip, b is parallel to strike, and c is normal to the plane of *ab* (foliation) of folds that are not plunging. Oriented core from holes drilled parallel to these directions when tested under uniaxial stress and torsion with suitably oriented strain gages allow for determining Young's moduli, Poisson's ratio, and shear moduli: E_a, E_b, E_c, ν_{ab}, ν_{bc}, ν_{ca}, G_a, G_b, G_c. Symmetry requirements for a generalized (anisotropic) Hooke's law restrict ν_{ba}, ν_{cb}, ν_{ac}, for example $\nu_{ac}E_c = \nu_{ca}E_a$.

Laboratory tests aimed at determining anisotropy require a much greater effort than testing under the assumption of isotropy. Relatively few results reported in the technical literature indicate that Young's modulus parallel to bedding is often no more than about twice Young's modulus perpendicular to the bedding. Table B.4 presents a sample of laboratory test data for Young's modulus perpendicular to the bedding and in two orthogonal directions parallel to the bedding (orthotropy).

In a detailed study of coal from the Pittsburgh seam, Ko and Gerstle (1976) showed the coal was orthotropic with preferred directions normal to bedding (c), parallel to a set of main

Table B.4 Anisotropic Young's moduli (after Obert and Duvall, 1967)

Rock type	*Young's modulus 10^6 psi (GN/m^2)*						
	Perpendicular		Parallel-1		Parallel-2		(Emax/Emin)
Marble (MD)	7.15	(49.3)	9.15	(63.1)	10.4	(71.7)	1.45
Limestone (IN)	4.84	(33.4)	5.94	(41.0)	5.39	(37.2)	1.23
Limestone (OH)	9.93	(68.5)	8.19	(56.5)	8.96	(61.8)	1.21
Granite (VM)	4.41	(30.4)	3.97	(27.4)	6.41	(44.2)	1.61
Granite (NC)	—	—	3.28	(22.6)	4.39	(30.3)	—
Slate (PA)	—	—	13.6	(93.8)	12.1	(83.4)	—
Granite (CO)	5.41	(37.3)	6.13	(42.3)	—	—	1.13
Sandstone (OH)	0.87	(6.0)	0.97	(6.7)	1.28	(8.8)	1.47
Sandstone (OH)	1.03	(7.1)	1.54	(1.1)	1.63	(11.2)	1.58
Sandstone (UT)	1.39	(9.6)	1.53	(1.1)	—	—	1.10
Gneiss (GA)	2.70	(18.6)	3.35	(23.1)	1.80	(12.4)	1.86
Oil Shale (CO)	1.80	(12.4)	3.10	(21.4)	—	—	1.72
Oil Shale (CO)	3.06	(21.1)	4.82	(33.2)	—	—	1.58

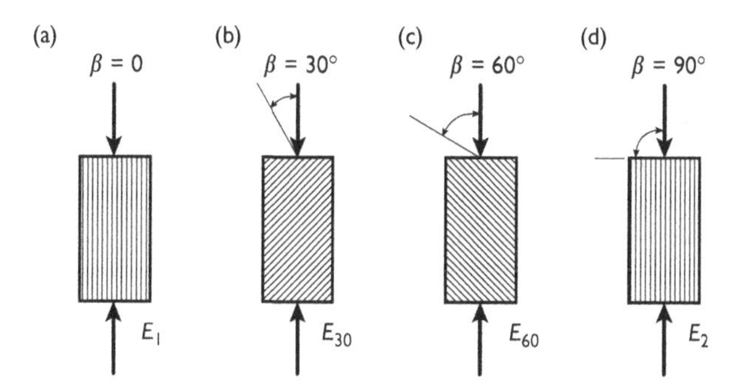

Figure B.8 Testing anisotropic rock at an angle to bedding or foliation.

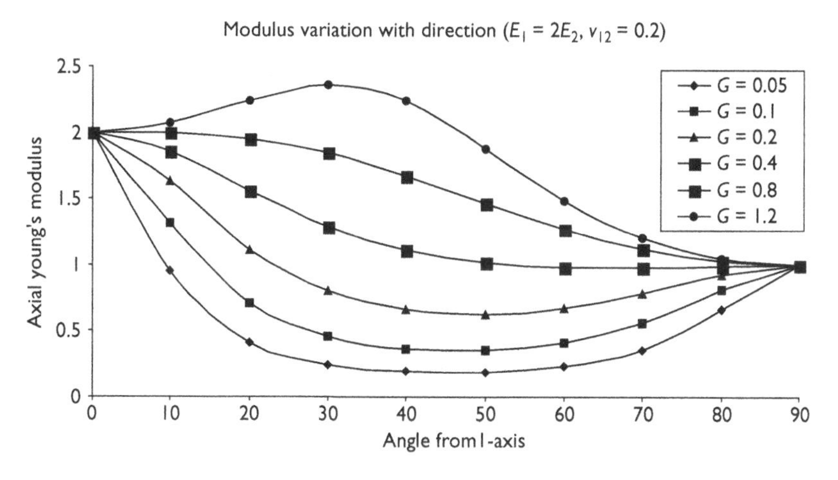

Figure B.9 Young's modulus at an angle to bedding with shear modulus as a parameter.

cleats (b), and to a set of minor cleats (a), all mutually orthogonal. In this case: $E_a = 0.48$, $E_b = 0.47$, $E_c = 0.38$, $G_a = 0.10$, $G_b = 0.05$, $G_c = 0.11 (\times 10^6$ psi). Shear moduli were much less than an isotropic estimate. Poisson's ratios ranged from 0.33 to 0.53.

Testing of core at an angle to the bedding is an alternative to torsion for the determination of shear moduli. With Young's moduli determined parallel and perpendicular to the bedding and also the relevant Poisson's ratio (E_1, E_2, and ν_{12}), additional experimental measurements of Young's modulus E at an angle to the bedding allows for the determination of the relevant shear modulus by

$$\frac{1}{E} = \frac{\cos^2(\beta)}{E_1} + \left(\frac{1}{G_{12}} - \frac{2\nu_{12}}{E_1}\right)\sin^2(\beta)\cos^2(\beta) + \frac{\sin^4(\beta)}{E_2}$$

where β is the angle between bedding and axial load, as shown in Figure B.8. In this regard, one may suppose that Young's modulus at an angle to the bedding varies between the moduli values parallel and perpendicular to the bedding. However, this supposition is not at all the case. A plot of Young's modulus at an angle to the bedding assuming Young's modulus parallel to the bedding is twice the value of Young's modulus perpendicular to the bedding and a Poisson's ratio of 0.20 is shown in Figure B.9 for different values of the shear modulus.

Clearly, Young's modulus at angle to the bedding may be much less or even greater than the moduli parallel and perpendicular to the bedding. In this regard, shear moduli of anisotropic rock have great influence on design analysis, for example, in estimation of surface subsidence of stratified ground. For this reason, shear moduli of anisotropic rock should be measured rather than estimated, as is too often the case.

B.2 Strength of intact rock

Strength properties are constants that appear in failure criteria. Isotropic media require only two independent strength constants such as tensile strength (T_o) and unconfined or uniaxial compressive strength (C_o). Tensile strength is usually assumed to be independent of confining pressure, while compressive strength under confining pressure (C_p) varies but may be expressed in terms of C_o and T_o. Alternatively, cohesion c and angle of internal friction φ may be used. Any one strength constant can be expressed in terms of two others. If test data indicate nonlinearity of a failure criterion, then the degree of nonlinearity must also be specified. For example, if the failure envelope $\tau = f(\sigma)$ has the form $\tau^n = \sigma \tan(\phi) + c$, then the value of the exponent n must be specified even though the material is isotropic. Strength tests according to standards, are carried out in a few minutes at loading rates of 10 to 1,000 or so psi/s (0.07 MPa/s to 7.0 MPa/s); results are considered to be independent of time or loading rate. Dynamic or impact loading giving rise to wave propagation and inertia effects, say, of the order of 10,000 psi/s (70 MPa/s), is impossible with conventional testing machines.

Tensile strength

Tensile strength may be determined by direct pull testing, bending, and indirectly by the popular "Brazil" test, also known as a "splitting" test. Point load tests also indicate tensile strength. Figure B.10 shows various laboratory testing configurations for tensile strength. Direct pull testing, while the most natural test configuration, is usually avoided because of difficulties associated with end attachments, the brittle nature of many rocks in tension and low tensile strength that makes the overall procedure relatively costly, delicate, and uncertain. Bending test specimens need to be long relative to thickness and are thus difficult to prepare and rather costly, especially when samples are taken from densely jointed rock sites. Tensile

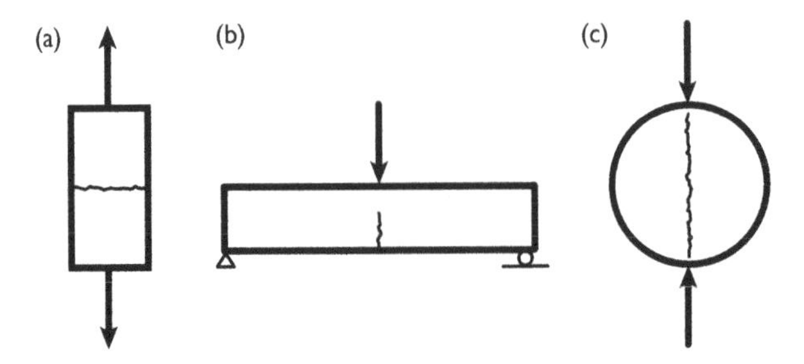

Figure B.10 Tensile test configurations: (a) direct pull, (b) bending, and (c) indirect.

strength in bending is also known as the modulus of rupture. Point load testing in various ways requires knowledge of empirical constants that relate the breaking load to tensile strength. By far the most common tensile strength test is the Brazil test.

The Brazil test involves compression of an intact rock disk between opposing forces across a disk diameter. Test specimen preparation is simple and inexpensive. The ends of the disk do not require a smooth finish, only careful saw cuts from a cylinder, usually a diamond drill core. Typically, an NX-core (2-1/8 in. diameter, about 50 mm) is sawed into a disk about 3/4 in. (20 mm) thick. Thickness should be large relative to grain size, but certainly no greater than disk radius, so the disk is "thin". Rock grain size is of the order of a millimeter in coarse grained rocks where grains are easily distinguished by the eye. Grain size in many rocks is much less. Disk thickness of 3/4 in. (2 cm) is usually satisfactory. Interpretation of test results comes from a detailed analysis of stress in a disk that is assumed to be linearly elastic, homogeneous, and isotropic. The main result is that an almost uniform horizontal tension is induced in a disk loaded vertically, although tension at the disk center is slightly elevated. In the immediate vicinity of the load points, the stress state is highly compressive. Fracture is expected to propagate from the disk center towards the load points at the disk perimeter. The test is carried out in a few minutes or less and the results are considered time-independent.

Tensile strength from the Brazil test is given by the formula

$$T_o = \frac{2F_c}{\pi Dh} \tag{B.1}$$

where F_c, D, and h are load at fracture, diameter, and thickness, respectively. Table B.5 shows a sampling of tensile strength from Brazil test results. The values in psi are rounded off to the nearest 10 psi; the derived statistics are based on MPa values. Table B.5 suggests that tensile strength of rock is no more than a few thousand psi or tens of MPa and often less. Of interest is the large variation in results that is characterized by the coefficient of variation (standard deviation as a percentage of the mean). A coefficient of variation of 40% is not unusual and seems to be an intrinsic characteristic of rock. The standard deviation of the mean is obtained by division of the data standard deviation by the square root of the number of tests. To reduce the standard deviation associated with tensile strength, say, by 50%, a quadrupling of test numbers would be required, other factors remaining the same. Because of the relatively large

Table B.5 Brazil test tensile strengths (after Singh, 1989)

Rock type	*Mean tensile strength To psi (MPa)*	*Standard deviation psi (MPa)*	*Number of tests (n)*	*Std. dev. of the mean psi (MPa)*	*Coeff. of variation (%)*
Dolomite	1,260 (8.7)	480 (3.3)	13	130 (0.92)	37.9
Granite (Barre)	2,000 (13.8)	300 (2.1)	150	25 (0.17)	13.3
Limestone (Bedford)	1,080 (7.5)	520 (3.6)	34	90 (0.62)	48.0
Limestone (Indiana)	1,320 (9.1)	550 (3.8)	17	130 (0.92)	41.8
Magnetite silica	1,810 (12.5)	250 (1.7)	12	70 (0.49)	13.6
Rhyolite porphyry	2,090 (14.4)	260 (1.8)	9	90 (0.60)	12.5
Sandstone	1,120 (7.7)	260 (1.8)	21	60 (0.39)	23.4
Sandstone (Berea)	1,030 (7.1)	750 (5.2)	33	130 (0.91)	67.5
Sandstone (Berea)	1,480 (10.2)	830 (5.7)	17	200 (1.38)	55.9
Shale	1,460 (10.1)	280 (1.9)	24	60 (0.39)	18.8

Table B.6 Anisotropic tensile strength data (after Chenevert and Gatlin, 1965)

Strength (psi)	*Rock type*								
	Arkansas sandstone			*Green River shale*			*Permian shale*		
	Mean	*Std. dev.*	*No.*	*Mean*	*Std. dev.*	*No.*	*Mean*	*Std. dev.*	*No.*
Perpendicular to bedding	1,700	250	10	3,140	400	10	2,500	35	3
Parallel to bedding	1,390	180	4	1.970	450	6	1,660	380	3

Notes
Mean = average, Std. dev. = standard deviation, No. = number tested. Rounded to 10 psi.

increase in sample tests and costs and in view of the intrinsic variability of test results, some caution should be exercised before recommending additional testing for the purpose of increasing confidence in test results (by reducing the standard deviation of the mean).

Rocks with pronounced directional features such as lamination, foliation, schistosity, and flow structure have anisotropic strength characteristics. Usually tensile strength perpendicular to bedding is less than tensile strength parallel to the bedding. When the bedding is inclined to the load axis (and the disk axis is a principal material direction), fracture between load points is uncertain. However, when fracture occurs between load points in the usual manner, then a measure of tensile strength at an angle to the bedding is obtained. There is no guarantee that tensile strength at an angle to the bedding falls between strengths parallel and perpendicular to the bedding, as in the case of Young's modulus of anisotropic rock. Table B.6 presents anisotropic strength data for three rock types. These are Brazil test data and show tensile strength ratios parallel and perpendicular to bedding of less than two and coefficients of variation less than 24%.

Figure B.11 shows a hypothetical variation of tensile strength with bedding plane angle in anisotropic rock. The angle is from the bedding (parallel to the bedding) to the load axis which would be parallel to a drill core tested in direct pull. With the 1-direction parallel to the bedding and the 2-direction perpendicular to the bedding, the strength parameters are $T_1 = (1/9)C_1 = 1$, $T_2 = (1/19)C_2$, $T_3 = (1/9)C_3$, $C_2 = 2C_1$, $C_3 = C_1$. Thus, T_1 is the arbitrary unit of the plot. Strength in pure shear about the 3-axis, R_{12}, is a parameter (R) in the plot. Because $T_2 = (2/19)C_1 = (18/19)T_1 = 0.95T_1$, relative minima and maxima tend to be near 45°, although the critical point tends to migrate from slightly greater than 45° to slightly less than 45° as the shear strength is increased. If the material were isotropic, the shear strength would correspond to $R = 0.115$, and the plot would simply be a horizontal line at one.

Figure B.11 reveals the importance of shear strength along the bedding in anisotropic rock and shows why simply averaging strengths parallel and perpendicular to the bedding may be a serious mistake. In this example, at low shear strength relative to an isotropic value, tensile strength at even a small angle to the bedding is only about 20% of tensile strength parallel to the bedding. At a modestly elevated bedding plane shear strength, there is little difference in tensile strength regardless of bedding plane angle. Although not shown in the plot, at relatively high bedding plane shear strengths, tensile strength at a bedding plane angle near 45° increases greatly. These data were constructed according to theory (Pariseau, 1972).

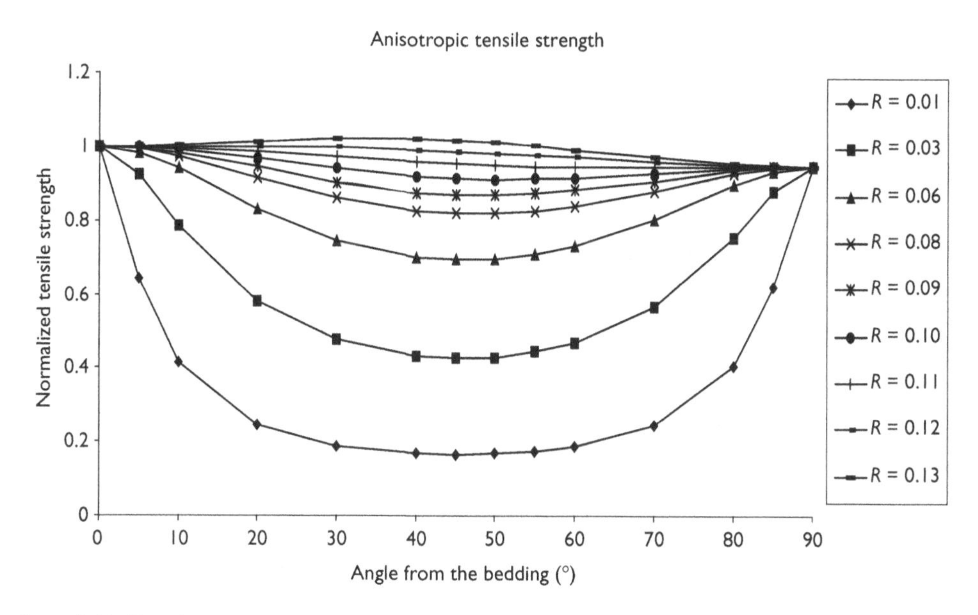

Figure B.11 Tensile strength as a function of bedding plane angle in anisotropic rock.

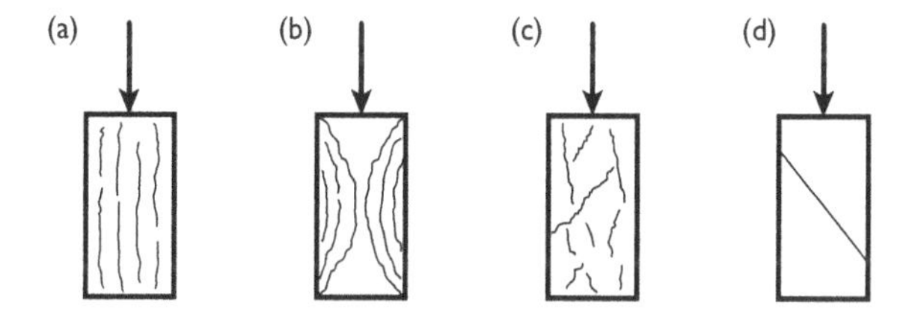

Figure B.12 Several modes of compressive test failure. (a) Axial splitting, (b) spalling, (c) multiple fractures, and (d) single shear fracture.

Unconfined compressive strength

Unconfined compressive strength (C_o or UCS) is perhaps the most widely used rock property for design. The reason is that compressive stresses are the rule. An important exception is beam action of roof rock in stratified ground. Naturally supported excavation surfaces are free of normal and shear forces. Consequently, one of the principal stresses at a point on such a surface is zero. The state of stress at the considered point is therefore characterized by a lack of confining pressure; the corresponding Mohr's circle passes through the origin in a normal stress–shear stress plot. Thus, regardless of depth of excavation, minimum stress at a naturally supported excavation wall is often zero, while the maximum compression is limited by the unconfined compressive strength.

Unconfined compression test cylinders usually fail by fracture in the form of: (1) axial splitting, (2) spalling or hour-glassing, (3) single shear fracturing, and (4) multiple fractures that defy easy description. Figure B.12 illustrates these modes. Smooth, well-lubricated test cylinder ends favor axial splitting because of the elimination of lateral restraint. High end

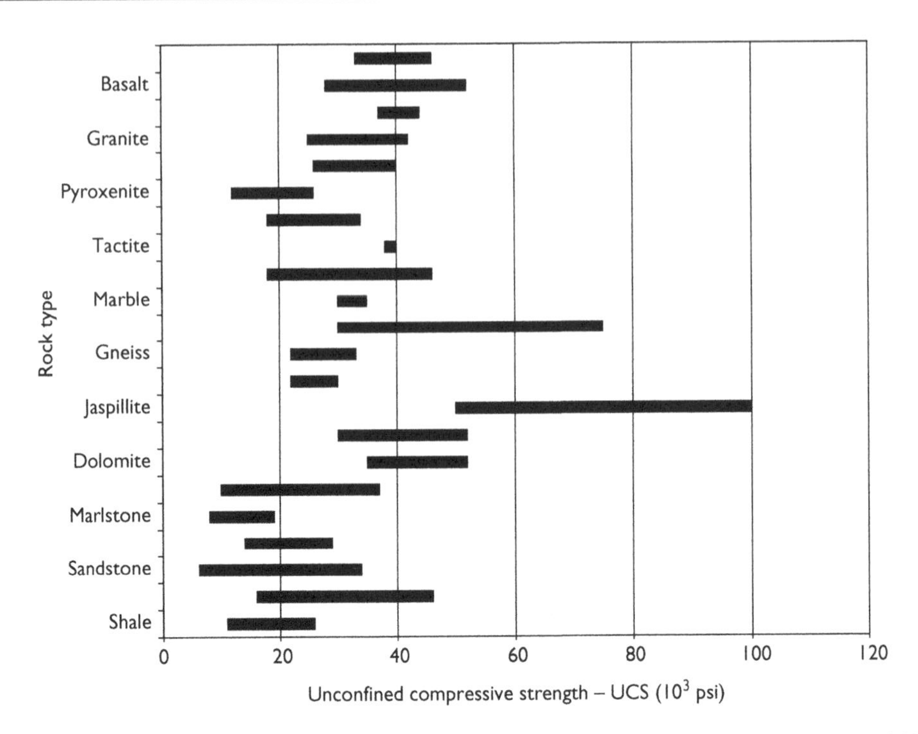

Figure B.13 Unconfined compressive strength of a variety of rock types (after Windes and Blair, 1949).

friction favors spalling and the formation of an hourglass shape. Single shear fracturing is unusual in unconfined test cylinders; multiple fractures are more likely in routine unconfined compressive strength tests.

Figure B.13 lists a range of unconfined compressive strength test results for a variety of rock types. Again, there is little association between strength and rock type. Additional statistical data for unconfined compressive strength are presented in Table B.7. There is considerable scatter measured by the coefficient of variation, even when relatively large numbers are tested. The reason is the intrinsic microstructural variability of test specimens. Testing ever greater numbers does reduce the standard deviation of the mean value distribution and thus may improve confidence in using mean values for design. The intrinsic variability of even intact rock indicates a need for safety factors higher than those used in the design of conventional structures.

Coal strength varies considerably with coal rank just as the elastic moduli do. Vutukuri and Lama (Vol. II, 1978a) report a wide range of unconfined compressive strength. A range from 200 to 20,000 psi (1.4 to 140 MPa) is possible; however, a practical range excluding extremes would be 2,000 to 6,000 psi (14 to 41 MPa). Strengths parallel and perpendicular to bedding may differ by a factor of two. Tensile strengths may be one-tenth to one-twentieth of unconfined compressive strength as is often the case for other rock types. Test on very large samples of coal in-seam, some up to nine feet in height, show considerably lower strengths than conventional, laboratory size test cylinders that lack defects in the field-scale test volumes.

Table B.7 Unconfined compressive strength statistics (after Singh, 1989)

Rock type	*Unconfined compressive strength C_o (MPa)*	*Number tested (n)*	*Standard deviation (MPa)*	*Coeff. of variation (%)*	*Standard deviation of the mean (MPa)*
Limestone	75.3	17	36.0	48	8.7
Limestone	37.7	11	8.7	23	2.6
Limestone	103.8	15	26.2	25	6.8
Magnetite	136.6	15	23.7	18	6.1
Marble	98.6	14	22.6	23	6.0
Marble	136.8	20	75.6	55	6.9
Marble	123.3	35	58.1	47	9.8
Qtz. monzonite	177.1	29	41.0	23	7.6
Dacite porphyry	101.4	55	31.8	31	4.3
Rhyolite porphyry	226.0	14	62.3	28	4.5
Potash	32.8	56	12.9	39	1.7
Quartzite	143.7	20	50.4	35	4.4
Quartzite	112.4	55	37.3	33	5.0
Rhyodacite	200.5	14	45.4	32	12.1
Rock Salt	23.7	36	2.8	12	0.5
Sandstone	24.2	104	11.0	45	1.1
Sandstone	59.8	69	32.0	54	3.9
Shale	22.8	14	15.5	68	4.1
Shale	116.4	59	29.8	25	3.9
Tonalite	85.1	33	36.7	43	6.4
Tuff	16.6	24	6.9	42	1.4
Tuff	65.3	19	50.1	77	11.5
Tuff	52.4	44	19.8	38	3.0

A "size" effect is often associated with compressive strength. This effect is discerned from laboratory test data where cylinders of varying L/D ratio are tested for compressive strength. The results generally show a trend of decreasing strength with increasing L/D ratio. The label "size" effects is thus a misnomer because of the changing shape. At a constant L/D ratio, a true size effects might be observed, or when testing cubes. When the L/D ratio of a cylinder is varied, the decrease of strength is more aptly called a "shape" effect. There are two difficulties with this label: (1) the data are sparse and are mostly in the range of 0.5 to 2.0, so the specimens tend to become stubby and (2) the effect of end friction becomes more important for stubby specimens. The best description of the apparent decrease of strength with increasing L/D ratio in uniaxial compression testing is therefore "end" effects. A formula of longstanding that describes "size," "shape," or "end" effects is

$$C = C_1 \left[0.78 + 0.22 \frac{1}{L/D} \right] \tag{B.2}$$

where C_1 is the unconfined compressive strength of a cylinder with an $L/D = 1$ and C is unconfined compressive strength at some other ratio. A plot of this hyperbolic equation is shown in Figure B.14. A minimum strength occurs with the horizontal asymptote or of $0.78C_1$. Strength increases indefinitely as L/D as the test specimen becomes very stubby.

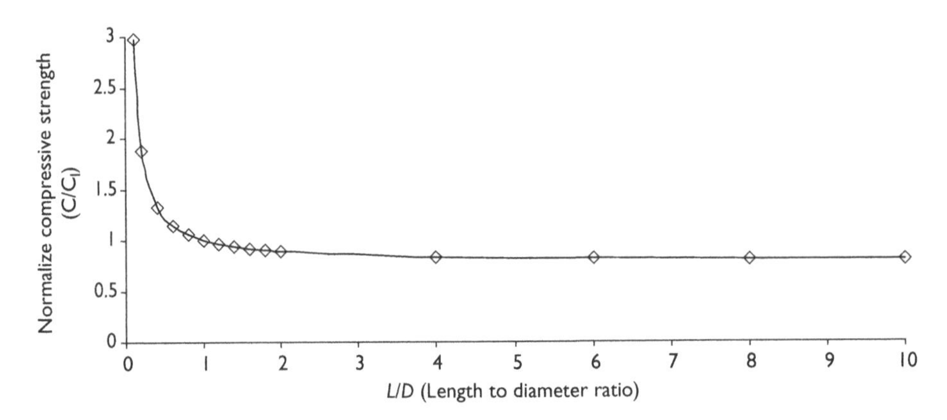

Figure B.14 Size, shape, or end effect on uniaxial compressive strength of rock cylinders.

Alternatively, the classic size effects formula may be written in straight-line form:

$$C = C_1 \left[0.78 + 0.22 \left(\frac{D}{L} \right) \right] \tag{B.3}$$

which has strength axis intercept $0.78C_1$ where the ratio D/L tends to zero. More generally, one may simply suppose that

$$C = B + A \left(\frac{D}{L} \right) \tag{B.4}$$

where A and B are slope and intercept, respectively. Figure B.15 from a report by Pariseau and others (1977) shows the ultimate, unconfined compressive strength of a Utah *coal* (Beehive Mine) as a function of D/L for tests of cylinders having diameters of 1, 2, 4, 6, 8, and 12 inches and length to diameter ratios of 1/2, 1, 3/2, and 2. The range of strength at each L/D ratio is considerable and is associated with test cylinder diameter; the larger diameter cylinders tend to be weaker. Also shown in Figure B.15 is a linear regression line of C on D/L for all 371 tests that shows the mean strength steadily increases with increasing D/L ratio, as expected. A much closer fit may be obtained by using averages for each D/L ratio, of course. This procedure masks the scatter, however.

Table B.8 shows linear regression statistics for each diameter tested and for the total data set. The last column in Table B.8 shows that the correlation coefficient is relatively high for each diameter, but drops noticeably when the data are pooled. Mean strength increases with increasing D/L ratio at all diameters. The coefficient of variation (standard deviation as a percentage of the mean strength) is roughly 33% for diameters up to eight inches, but drops in the case of the 12 inch diameter cylinders, although at 25%, is high by standards for manufactured materials. These values are within the range of coefficients of variation associated with other rock types such as those given in Table B.7.

A natural expectation for a given suite of rock test specimens is that higher Young's modulus is associated with higher strength. Thus, measurements on the same test cylinder are expected to be correlated. Figure B.16 shows a plot of Young's modulus as a function of

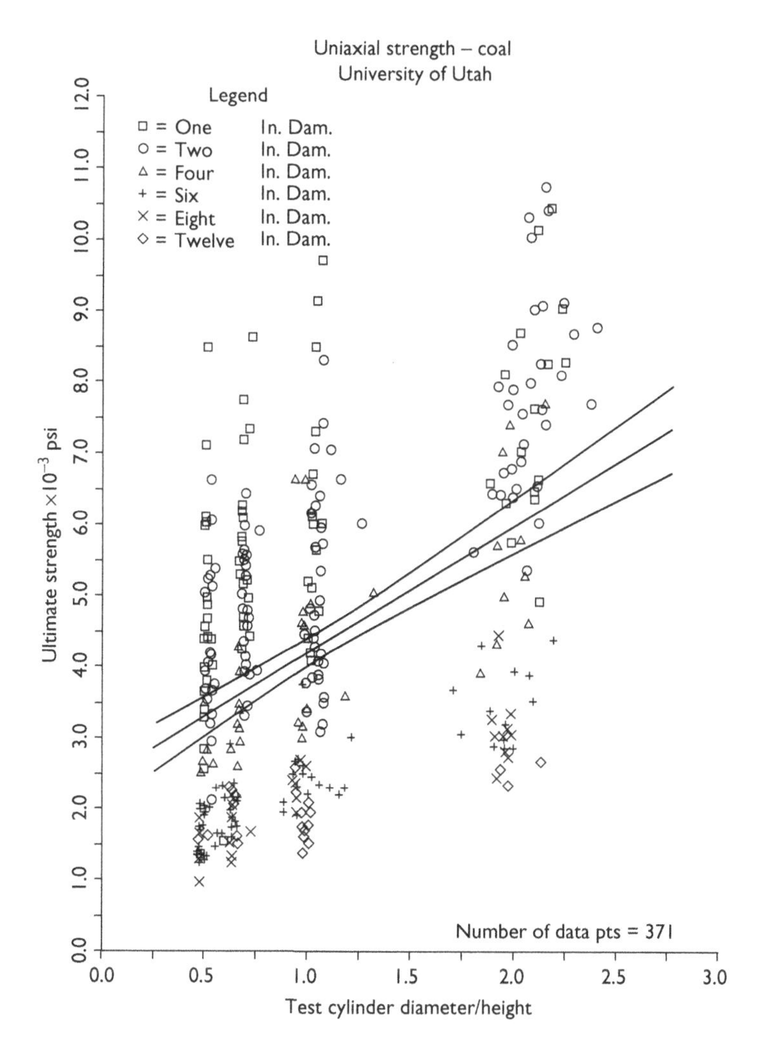

Figure B.15 Linear regression and 95% confidence limits of ultimate strength on diameter to height ratio for all.

Table B.8 Linear regression statistics for unconfined compressive strength as a function diameter to length ratio *D/L* (after Pariseau and others, 1977)

D	*Ave D/L*	*SD(D/L)*	*Ave C*	*SD(C)*	*CV%*	*n*	*A*	*B*	*r*
1.00	0.933	0.599	5,625	1793	35.1	91	1,965	3,792	0.61
2.125	1.180	0.609	5,623	1871	33.3	112	2,335	2,867	0.76
4.00	1.051	0.540	4,122	1427	34.6	45	1,903	2,121	0.72
5.75	1.068	0.579	2,413	780	32.3	59	1,163	1,171	0.86
8.00	1.022	0.577	2,252	752	33.4	39	1,071	1,158	0.82
12.00	1.062	0.500	1,959	425	21.7	25	589	1,327	0.69
All	1.062	0.577	4,330	2141	49.4	371	1,780	2,439	0.48

Notes

D = test cylinder (inches), Ave *D/L* = average ratio, SD(*D/L*) = standard deviation of *D/L*, Ave *C* = average unconfined compressive strength (psi), SD(C) = standard deviation of *C*, CV% = coefficient of variation as a percentage, *n* = number tested, *A* = regression line slope, *B* = regression line intercept, *r* = correlation coefficient.

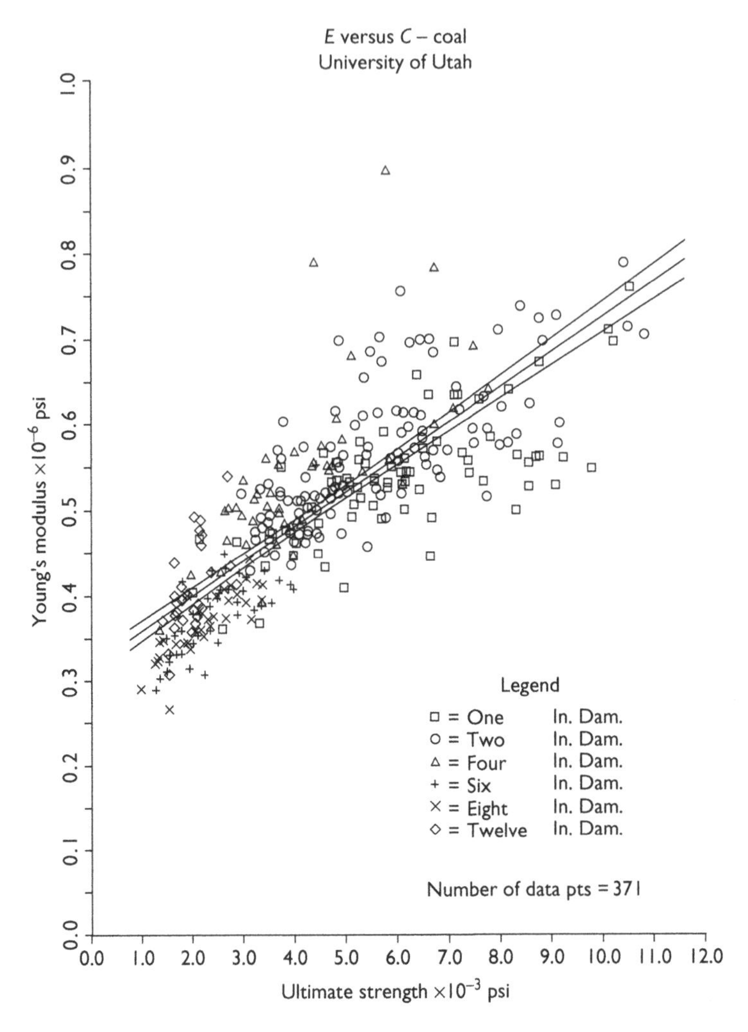

Figure B.16 Linear regression and 95% confidence limits of Young's modulus on strength for all diameters.

unconfined compressive strength for the same Utah coal data presented in Figure B.15 and Table B.8. The expectation of higher Young's modulus with higher strength is realized in Figure B.16. Test statistics are presented in Table B.9. These data show that scatter in Young's modulus tends to be much less than scatter in strength. This phenomenon is often observed in other rock types and testing programs.

The scatter in modulus and strength test data poses important design questions. One question is whether to adopt a probabilistic design approach or to use a more conventional safety factor criterion for design. The first creates considerable mathematical complexities while the latter calls for specification of appropriate safety factors that lead to safe, stable excavations. There is no consensus about safety factors, although Obert and Duvall (1967)

Table B.9 Linear regression statistics for Young's modulus on unconfined compressive strength (after Pariseau and others, 1977)

D	*Ave E*	*SD(E)*	*CV(E)%*	*Ave C*	*n*	*A*	*B*	*r*
1.00	0.5249	0.0663	12.6	5.625	91	0.0262	0.3776	0.71
2.125	0.5698	0.0814	14.3	5.623	112	0.0312	0.3946	0.72
4.00	0.5493	0.0982	17.9	4.122	45	0.0489	0.3479	0.71
5.75	0.3824	0.0491	12.8	2.413	59	0.0501	0.2615	0.80
8.00	0.3723	0.0438	11.8	2.252	39	0.0493	0.2612	0.85
12.00	0.4080	0.0566	13.9	1.959	25	0.0803	0.2508	0.60
All	0.4984	0.1067	21.6	4.330	371	0.0463	0.3189	0.82

Notes

D = test cylinder (inches), Ave *E* = average Young's modulus in millions of psi SD(*E*) = standard deviation of *E* in millions of psi, CV(*E*)% = coefficient of variation %, Ave *C* = average unconfined compressive strength in thousands of psi, *n* = number tested, *A* = regression line slope, *B* = regression line intercept, *r* = correlation coefficient.

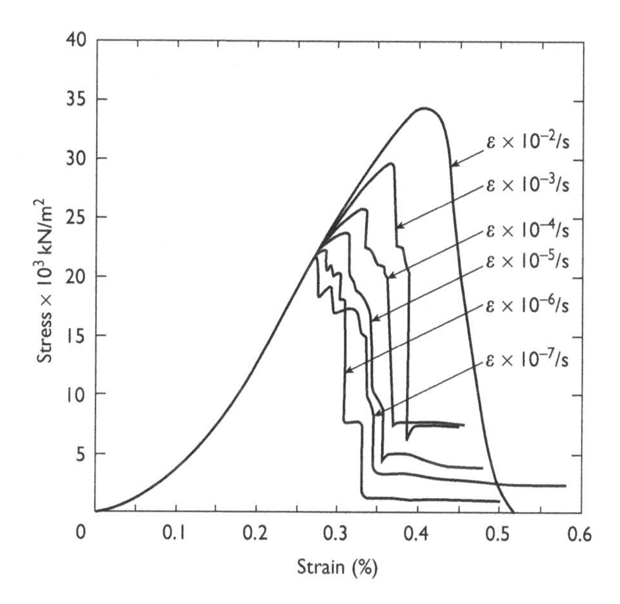

Figure B.17 Complete stress–strain curves for tuff at various strain rates (after Peng and Podnieks, 1973).

suggest the use of safety factors of two to four in compression and four to eight in tension. Lower safety factors may be acceptable when the design life is short.

The effect of strain rate in the quasi-static range, say, from, 10^{-7} to 10^{-2}/s is to raise the elastic limit, that is, to increase the unconfined compressive strength. At these rates, 10^5 s (28 h) to 1 s are required to reach a strain of 1% which is a rough strain to failure in uniaxial compression. The increase depends on the rock and other variables such as temperature. For example, a test of a tuff (Peng and Podnieks, 1972) showed an increase over 50% in compressive strength as the strain rate increased from 10^{-7} to 10^{-2}/s, as shown in Figure B.17. Peng (1973) further found that the unconfined compressive strength of Charcoal granite, Tennessee marble, and Berea sandstone increased about 25%, 21%, and 33% with an increase in the strain rate from 10^{-8} to 10^{-4}/s. Greater increases would be expected at greater strain rates. However, decreasing strain rates would not seem to decrease strength indefinitely to a limit such that at zero strain rate strength is zero. Thus, if strength were a linear function of strain rate, then

$$C_o(\dot{\varepsilon}) = a\dot{\varepsilon} + b \tag{B.5}$$

where a is a constant and b is a lower limit or long-term strength C_o in case of vanishing strain rate. Different functions could be fitted to data with greater accuracy, of course. This form simply indicates an observed trend of increasing strength with strain rate. Routine tests for C_o are conducted at a strain rate of about 10^{-5}/s and require just a few minutes to complete.

Whether the effect of strain rate on strength is important may be measured against the usual scatter in test data where the standard deviation may be more than 35% of the mean. The effect of strain rate on strength over several orders of magnitude may thus be masked in ordinary test data variability.

The usual range of temperature in engineering design is unlikely to have a noticeable effect on strength. Generally, temperature rises less than about one-third the melting point are not important; higher temperatures will decrease strength, while very low temperatures may increase strength. However, the presence of water at below freezing temperatures may decrease strength. Figure B.18 shows a stress–strain curve and strength change over a large range of elevated temperatures. Although the effect of temperature on the initial yield point or unconfined compressive strength is small in Figure B.18, the effect on ductility, indicated by strain after reaching the elastic limit, is substantial. Figure B.19 shows a slowly increasing strength with temperature decrease below room temperature. For this particular test on sandstone, the change in strength from 40 to −80°C even in the water saturated case is less than 10%. Also shown in Figure B.19 is the effect of cold temperatures on Young's modulus

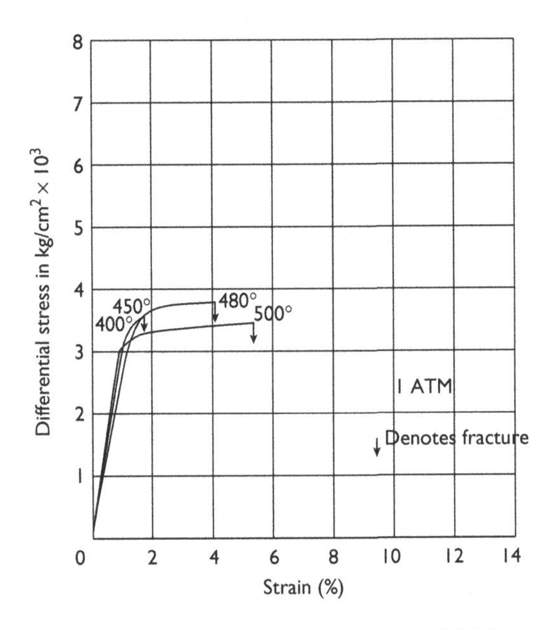

Figure B.18 Stress–strain for Solenhofen limestone. Temperature in °C (after Heard, 1960).

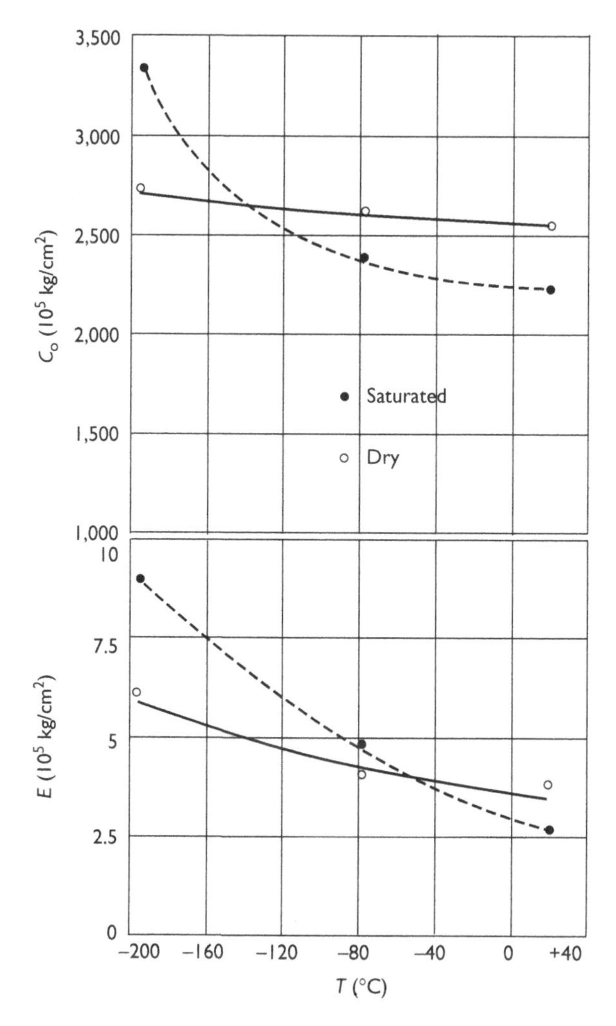

Figure B.19 Effect of temperature on strength and Young's modulus of a sandstone (after Brighenti, 1970).

which also increases somewhat with decreasing temperature. These data suggest a simple strength formula for temperature effect:

$$C_o(T) = a - bT \tag{B.6}$$

where a is room temperature C_o, b is a small empirical constant, and T is temperature. More complicated formulas could be postulated, of course.

Directional dependency of unconfined compressive strength is expected for laminated and foliated rocks in much the same manner as anisotropic elastic moduli and rock that has directional dependent tensile strength. Again, there is no guarantee that compressive strength at an angle to the bedding falls between values measured parallel and perpendicular to the

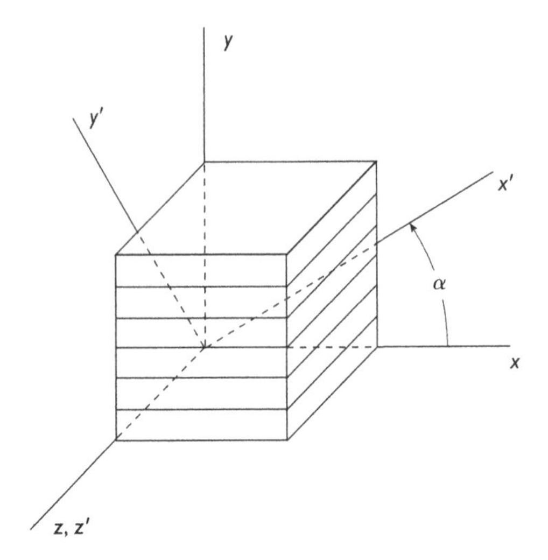

Figure B.20 Bedding plane angle definition.

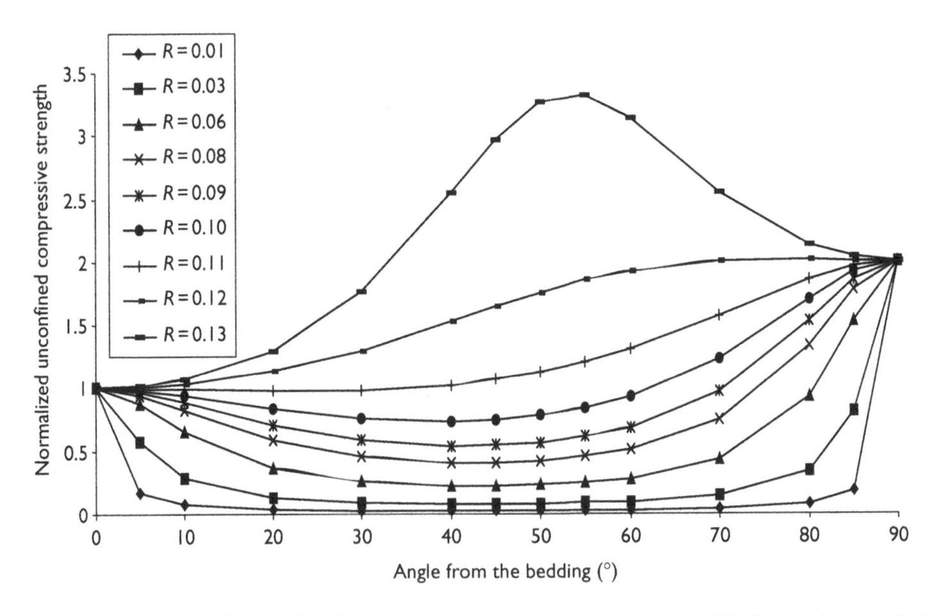

Figure B.21 Hypothetical unconfined compressive strength as a function of bedding plane angle in anisotropic rock.

bedding. Indeed, much data suggest that a minimum strength is often observed when the bedding forms a small angle with the applied load. A bedding plane angle is shown in Figure B.20.

Figure B.21 shows a hypothetical variation of unconfined compressive strength with bedding plane angle (Pariseau, 1972). Strength parameters of this plot are the same as those used to demonstrate effect of anisotropy on tensile strength ($C_2 = 2C_1$, $C_3 = C_1$, etc.). This plot is in units of C_1 (unconfined compressive strength parallel to bedding). Unconfined compressive strength perpendicular to bedding is C_2. The importance of bedding plane shear strength (R) in this example is even greater than for tensile strength. At bedding plane angles from 5° to 95°, compressive strength is only a small fraction of strengths parallel or perpendicular to bedding and suggests that failure would be characterized by shear along bedding planes, when shear strength is low relative to an isotropic value of $R = 0.115$. This phenomenon reveals the danger in assuming that the unconfined compressive strength at an angle to the bedding lies between strengths measured parallel and perpendicular to bedding. At 0° or 90°, compressive strength failure is associated with formation of fault planes that necessarily results in shear across bedding. As shear strength increases, the lowest strength minimum, initially at an angle less than 45°, diminishes and eventually vanishes. Relative maxima form at relatively higher shear strengths at angles greater than 45°.

Compressive strength under confining pressure

Excavation support and reinforcement provides confinement to excavation walls, so stress at supported or reinforced walls is limited by compressive strength under confining pressure. Confining pressure is also provided for rock away from excavation walls by adjacent rock. There are exceptions, for example, walls of water tunnels that are under fluid pressure which may induce tensile stress. The deepest of excavations are about 10,000 ft or 3,300 m below ground surface. At approximately 1.1 psi per foot of depth (2.3 kPa/m), the vertical stress is about 11,000 psi (75.9 MPa). Horizontal stresses may be more or less. Thus, a practical engineering limit to consideration of confining pressure effect on strength is roughly 11,000 psi (75.9 MPa), less than a kilobar. Some laboratory investigations have examined confining pressure effects on strength into the kilobar range and temperature effects to hundreds of degrees Celsius because of interest in the behavior of deeply buried crustal rock. However, such extremes would be exceptional in engineering design. Unconfined compressive strength may be more or less than 11,000 psi (75.9 MPa); a practical range of confining pressure would then be no more than C_o .

Strength determined by uniaxial tension, compression, or possibly torsion is simply a number, T_o , C_o , or R_o , as the case may be. These numbers mark the limit to a purely elastic response to load. Strength under biaxial and triaxial stress states may also be defined as the limit to a purely elastic response. However, a single number is then no longer adequate to describe strength. In three dimensions, strength is a function, not a single number. Instead of $\sigma - T_o = 0$, $\sigma - C_o = 0$, or $\tau - R_o = 0$, one has $F(\sigma) = 0$ where σ stands for components of stress at the elastic limit.

Brittle materials, by definition, fail by fracture at the elastic limit with negligible inelastic deformation. Generally, tensile fracture, fracture under compressional load, that is, shear fracture and fracture under torsional stress (sometimes referred to as diagonal tension) lead to different strengths. Thus, even under uniaxial stress, brittle materials require a functional description. Ductile materials fail by yielding and experience plastic deformation as loading is continued beyond the elastic limit. This deformation is seen as a permanent "set" upon release of load.

Ductile materials generally have equal tensile and compressive strengths, so strength in pure shear is just one half strength under normal stress failure. In three dimensions, the mode of failure may change. Materials that are brittle under uniaxial load may become ductile under triaxial loading and thus experience a brittle–ductile transition. Materials that are ductile at room temperature may become brittle under extreme cold. One concludes that a functional description of strength is needed in any case.

General three-dimensional stress states may be visualized in terms of the principal stresses acting (σ_1 , σ_2 , σ_3) on the faces of a small cube (principal planes). The principal directions define a mutually orthogonal triple of axes along which the principal stresses may be plotted. In such a plot, σ_1 (principal stress acting parallel to the 1-direction) is measured in the positive or negative 1-direction depending on algebraic sign. The same is true of σ_2 and σ_3 . As the stresses are varied, a set of points that define the limit to elasticity is generated. These points define a surface $F(\sigma_1, \sigma_2, \sigma_3) = 0$ that represents a relationship amongst the stresses at failure. The function F is thus a failure criterion.

In consideration of Hooke's law and linearly elastic behavior, the uniaxial stress failure criteria, $\sigma - T_o = 0$, $\sigma - C_o = 0$, or $\tau - R_o = 0$, are also uniaxial strain failure criteria. More generally, any state of stress at the limit to elasticity is associated with a companion state of strain through Hooke's law. For this reason, there is no substantial difference between stress and strain failure criterion. One implies the other when strength is defined as stress at the elastic limit.

Energy may also be considered as a criterion for failure. For example, strain energy density U at the elastic limit may be postulated as a failure criterion. Under uniaxial stress, strain energy per unit volume of material $U = \sigma\varepsilon/2$ in case of normal stress; $U = \tau\gamma/2$ in case of shear stress. Alternatively, $U = \sigma^2/2E = E\varepsilon^2/2$ where E is Young's modulus and σ is To or Co (ε is T_o/E or C_o/E), or $U = \tau^2/2G = U = G\gamma^2/2$ where G is a shear modulus and $\tau = Ro$ (γ is τ/G). Because strains and strain energies can be calculated from the stresses at failure through Hooke's law, no significant differences exist between stress, strain, or strain energy criteria for failure when strength is defined as stress at the elastic limit. The same is true in three dimensions.

The most popular functions that describe strength in rock mechanics have names:

1 Mohr–Coulomb (MC)
2 Hoek–Brown (HB)
3 Drucker–Prager (DP).

The functions these names imply are failure criteria, also referred to as yield functions. They apply to isotropic materials that lack directional features. Materials such as laminated sandstones that have different strengths parallel and perpendicular to the laminations are anisotropic and require more complex failure criteria than MC, HB, or DP. Because strength of isotropic materials is independent of direction, the orientation of the principal stresses is not important. Hence, instead of (σ_{xx} , σ_{yy} , σ_{zz} , τ_{xy} , τ_{yz} , τ_{zx}) in F or (σ_1 , σ_2 , σ_3 , θ_x , θ_y , θ_z) where σ_1 , σ_2 , σ_3 are now the major, intermediate, and minor principal stresses and the angles specify the direction of the 1-axis relative to xyz, one only needs σ_1 , σ_2 , σ_3 to describe the strength of isotropic materials. Figure B.22 shows MC and DP in principal stress space.

Most strength functions are simple polynomials in stress, even in case of anisotropy. However, not all functions are candidates for the description of material strength; there are restrictions. One obvious restriction is that any such function gives the appropriate strength under uniaxial tensile, compressive, or shear loading. Another requirement is symmetry with respect to the normal stress axis when plotted in a normal stress, shear stress plane.

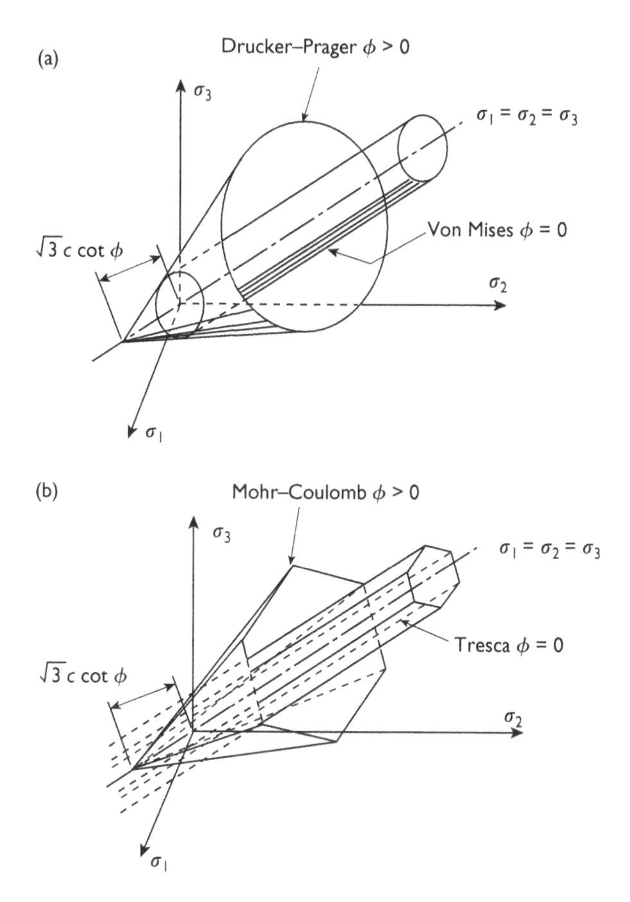

Figure B.22 (a) Drucker–Prager and Von Mises failure criteria in principal stress space. (b) Mohr–Coulomb and Tresca failure criteria in principal stress space. Compression is positive in both (a) and (b) (after Zienkiewicz, 1977).

Mohr–Coulomb strength

The Mohr–Coulomb failure criterion is

$$|\tau| = \sigma \tan(\phi) + c \tag{B.7}$$

where τ and σ are shear and normal stresses at a point on a potential failure surface and compression is positive. The angle φ is the *angle of internal friction*; c is the *cohesion*. These properties characterize the strength of the material. In this regard, the algebraic sign of the shear stress is not physically significant, so the absolute value sign is used in (B.7). Mohr strength theory postulates a functional relationship between τ and σ , that is, $\tau = f(\sigma)$. The linear form (B.7) is associated with Coulomb. This criterion forms a pair of straight lines when plotted in a normal stress (x-axis)–shear stress (y-axis) plane. The slopes of these lines are $\pm \tan(\varphi)$; they have τ-axis intercepts $\pm c$. The function (B.7) is thus symmetric with respect to the σ-axis and is tangent to Mohr circles representing stress states at failure. Function (B.7) is sometimes referred to as the Mohr envelope.

The MC criterion may also be expressed in terms of the major and minor principal stresses. Thus,

$$\left|\left(\frac{\sigma_1 - \sigma_3}{2}\right)\right| = \left(\frac{\sigma_1 + \sigma_3}{2}\right)\sin(\phi) + (c)cos(\phi) \tag{B.8}$$

Alternatively,

$$|\tau_{\mathrm{m}}| = \sigma_{\mathrm{m}}\sin(\phi) + (c)cos(\phi) \tag{B.9}$$

where τ_{m} is the maximum shear stress (also the numerical value of the minimum shear stress), and σ_{m} is the mean normal stress in the plane of $\sigma_1 - \sigma_3$. The function (B.8) is also a pair of straight lines in the normal stress–shear stress plane. These lines pass through the tops and bottoms of the Mohr circles representing stress states at failure. Function (B.9) may be unlinked from the failure envelope (B.7) by the simple generalization

$$|\tau_{\mathrm{m}}| = \sigma_{\mathrm{m}}\tan(\psi) + k \tag{B.10}$$

where the slope tan(ψ) may now exceed the limits of ± 1·set by sin(ϕ) in (B.9). Thus, in the unusual case that a failure envelope does not exist, (B.10) may be used. An envelope exists whenever ψ is less than $\pi/2$.

When the angle of internal friction is zero, MC reduces to a criterion associated with the name Tresca. The Tresca or maximum shear stress criterion is often used for ductile metals. This criterion may also be considered a maximum shear strain criterion in consideration of Hooke's law and linearly elastic behavior up to the elastic limit.

Rearrangement of MC shows that

$$\sigma_1 = \sigma_3\left(\frac{1 + \sin(\phi)}{1 - \sin(\phi)}\right) + \frac{2c\cos(\phi)}{1 - \sin(\phi)} \tag{B.11}$$

Alternatively,

$$\sigma_1 = C_o + \frac{C_o}{T_o}\sigma_3 \tag{B.12}$$

Hence, compressive strength C_{p} under confining pressure p according to MC is

$$C_{\mathrm{p}} = C_o + \frac{C_o}{T_o}p \tag{B.13}$$

with rate of increase of strength given by

$$\frac{\mathrm{d}C_p}{\mathrm{d}p} = \frac{C_o}{T_o} \tag{B.14}$$

Hoek–Brown strength

The HB failure criterion in original form may be stated as

$$\sigma_1 = \sigma_3 + \sqrt{a\sigma_3 + b^2} \tag{B.15}$$

where a and b are strength properties of the material. This criterion is obviously nonlinear. In terms of τ_m and σ_m , (B.15) is

$$\tau_m = \left(\frac{1}{2}\right)[a(\sigma_m - \tau_m) + b^2]^{(1/2)} \tag{B.16}$$

or

$$\tau_m^2 = (a/4)(\sigma_m - \tau_m) + b^2/4 \tag{B.17}$$

that after solving for τ_m gives

$$\tau_m = -\frac{a}{8} \pm \frac{1}{8}\sqrt{a^2 + 16(a\sigma_m + b^2)} \tag{B.18}$$

which is strange because two substantially different values of strength in pure shear are implied, that is,

$$\tau_m = -\frac{a}{8} \pm \left(\frac{1}{8}\right)\sqrt{a^2 + 16b^2} \tag{B.19}$$

Neither (B.16), (B.17) nor (B.18) is symmetric about the normal stress axis in the normal stress–shear stress plane and thus violate this symmetry requirement.

Compressive strength under confining according to HB is

$$C_p = p + \sqrt{\left(\frac{C_o^2 - T_o^2}{T_o}\right)p + C_o^2} \tag{B.20}$$

with rate of increase approximately

$$\frac{dC_p}{dp} \approx \frac{1}{2}\frac{C_o}{T_o} + 1 \tag{B.21}$$

which is less than the rate associated with Mohr–Coulomb strength, a consequence of the nonlinearity of HB.

Drucker–Prager strength

The DP criterion expressed in terms of the principal stresses is

$$\left(\sqrt{\frac{2}{3}}\right)\left[\left(\frac{\sigma_1 - \sigma_2}{2}\right)^2 + \left(\frac{\sigma_2 - \sigma_3}{2}\right)^2 + \left(\frac{\sigma_3 - \sigma_1}{2}\right)^2\right]^{(1/2)} = A(\sigma_1 + \sigma_2 + \sigma_3) + B \tag{B.22}$$

where A and B are strength properties of the material. The term on the left of (B.22) is $\sqrt{2}$ times the root-mean-square value of the principal shears because the terms under the square root sign are the squares of the principal shear stresses (relative maxima and minima). The first term in parentheses on the right is just three times the three-dimensional mean normal stress. This criterion is often written in abbreviated form:

$$J_2^{1/2} = AI_1 + B \tag{B.23}$$

where J_2 is the second principal invariant of deviatoric stress and I_1 is the first principal invariant of total stress. The DP criterion is a cone centered on the space diagonal when plotted with the principal stresses $(\sigma_1, \sigma_2, \sigma_3)$ as coordinates. When the strength parameter $A =$ 0, DP reduces to a criterion associated with Von Mises that is widely used for ductile metals. This criterion is also closely associated with an "octahedral" shear stress criterion (τ_{oct}) because $\sqrt{J_2}$ is a constant times the octahedral shear stress $(\tau_{oct} = \sqrt{2/3J_2})$. The Von Mises criterion, $\sqrt{J_2} = B$, is a cylinder centered on the space diagonal in principal stress space; the octahedral plane is normal to the space diagonal.

The DP strength criterion differs from MC and HB by inclusion of the *intermediate* principal stress, although it is linear in stress. The MC and HB criteria imply that the intermediate principal stress has only a negligible influence on strength. The MC criterion forms a pyramidal surface in principal stress space with flat sides.

Compressive strength under confining pressure according to DP is

$$C_p = C_o + \left(\frac{3}{2}\frac{C_o}{T_o} - \frac{1}{2}\right)p \tag{B.24}$$

with rate of increase

$$\frac{dC_p}{dp} = \frac{3}{2}\frac{C_o}{T_o} - \frac{1}{2} \tag{B.25}$$

The ratio of unconfined compressive to tensile strength for rock is typically in the range of 10 to 20. In this range, HB shows an increase of compressive strength with confining pressure about 50% less than MC, while DP shows an increase of about 50% more. These forecasts seem to have never been tested experimentally.

All of the strength parameters introduced with MC, HB, and DP can be expressed in terms of the unconfined or uniaxial tensile and compressive strengths, T_o and C_o, respectively. Because the considered materials are isotropic, shear strength R_o may also be expressed in terms of T_o and C_o. Thus,

$$\begin{aligned} \sin(\phi) &= \frac{C_o - T_o}{C_o + T_o} & C_o &= \frac{2c\cos(\phi)}{1 - \sin(\phi)} \\ c &= \sqrt{C_o T_o / 4} & T_o &= \frac{2c\cos(\phi)}{1 + \sin(\phi)} \\ a &= \frac{C_o^2 - T_o^2}{T_o} & C_o &= b \\ b &= C_o & T_o &= -\frac{a}{2} + \frac{1}{2}\sqrt{a^2 + 4b^2} \end{aligned} \tag{B.26}$$

$$A = \frac{1}{\sqrt{3}}\left(\frac{C_o - T_o}{C_o + T_o}\right) \quad C_o = \frac{\sqrt{3}B}{1 - \sqrt{3}A}$$
$$B = \frac{2}{\sqrt{3}}\left(\frac{C_o T_o}{C_o + T_o}\right) \quad T_o = \frac{\sqrt{3}B}{1 + \sqrt{3}A}$$

Nonlinear n-type strength

Nonlinear forms of MC, HB, and DP are possible. Thus,

$$\begin{aligned} \tau^n &= \sigma \tan(\phi) + c \\ \tau_m^n &= \sigma_m \tan(\psi) + k \\ \sigma_1^n &= \sigma_3 + \sqrt{a\sigma_3 + b^2} \\ (J_2^{1/2})^n &= AI_1 + B \end{aligned} \qquad \text{(B.27a – d)}$$

where n is some positive number, not necessarily an integer. Equations (B.27a,b) are nonlinear forms that reduce to MC and extended MC, respectively; (B.27c) reduces to HB, while the last reduces to DP when the exponent $n = 1$. In principle, all the constants in (B.27) can also be expressed in terms of T_o and C_o. Figure B.23 shows fits to sandstone test data including three values of n in (B.27b); $n = 1.34$ provides the best fit of all (B.27).

Compressive strength test data

Compressive strength under confining pressure is determined in the laboratory by enclosing the test cylinder in an impermeable jacket that allows transmission of an external fluid

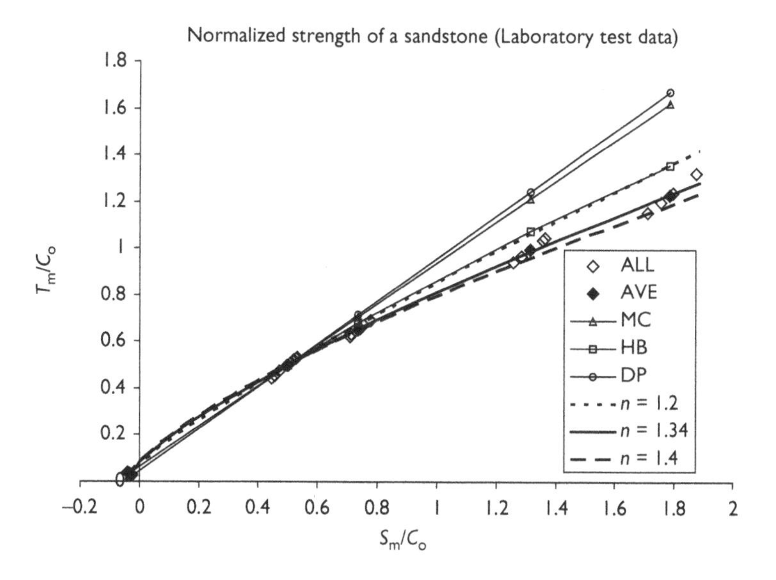

Figure B.23 Normalized sandstone compressive strength data and various fits of n-type criteria. Four tests were done at each confining pressure (ALL). Averaged data = (AVE).

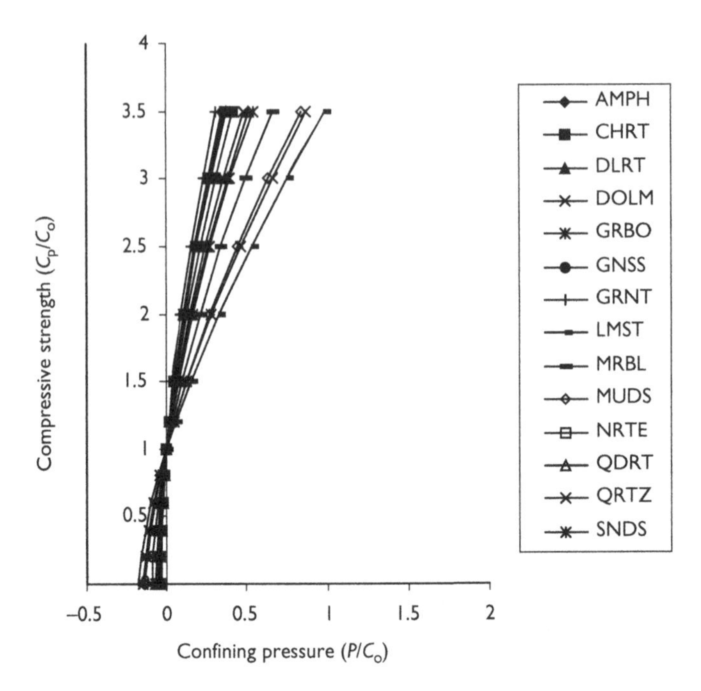

Figure B.24 Compressive strength versus confining pressure (after Hoek and Brown, 1980).

pressure p to the lateral surface of the test cylinder. When compression is positive, the axial load is associated with the major principal stress; the minor principal stress is provided by confining pressure in the radial direction. With the assumption of Hooke's law, the circumferential stress is equal to the radial stress. An impermeable membrane prevents confining pressure from forcing fluid into a test cylinder. Testing of saturated porous, fractured rock may involve separate control of pore fluid pressure. In this case, effective stress, total stress minus pore fluid pressure, determines rock strength. Of course, in a dry test, the effective and total stresses are the same.

Figure B.24 shows a plot of compressive strength as a function of confining pressure for a variety of rock types. The data plotted in Figure B.24 were generated from average m values reported by Hoek and Brown (1980) who did an extensive literature survey and fitted the data to the HB criterion in dimensionless form:

$$\frac{C_p}{C_o} = \frac{p}{C_o} + \sqrt{m\frac{p}{C_o} + 1} \tag{B.28}$$

where C_p, C_o, and p are compressive strength under confining pressure, unconfined compressive strength, and confining pressure, respectively. Because the data are normalized by C_o, all plots necessarily pass through the point (0,1). The range of confining pressure is the maximum C_o of the data, about 50,000 psi (345 MPa).

Inspection of the plot shows a rapid increase in compressive strength with confining pressure for all rock types, more than threefold at confining pressures less than the unconfined compressive strength. The rate of increase is approximately equal to $m/2$ near the zero of

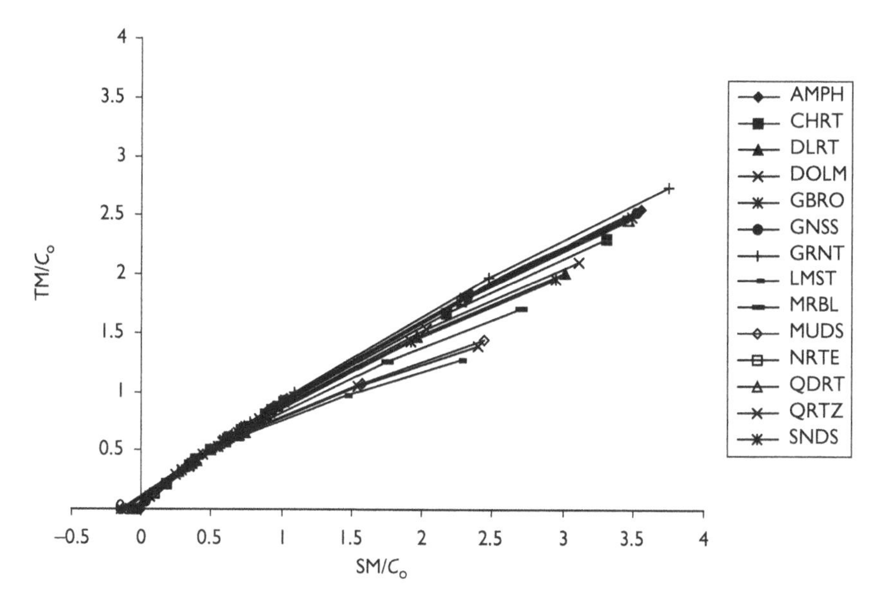

Figure B.25 Shear strength as a function of mean normal stress in units of unconfined compressive strength for a variety of rock types (data from Hoek and Brown, 1980).

confining pressure and gradually lessens with increasing confining pressure. The constant m is given by

$$m = \frac{C_o}{T_o} - \frac{T_o}{C_o} \approx \frac{C_o}{T_o}$$

where the approximation holds for many suites of intact rock tests because the ratio C_o/T_o is often greater than 10 and sometimes greater than 20; the slope near the strength axis intercept thus ranges from 5 to 10 or so. The HB criterion is actually a parabola that is symmetric with respect to a 45° line, so a restriction is necessary – normalized confining pressure must be no less than $-1/m$. Negative confining pressure allows for tensile stresses.

One may also consider the effect of confining pressure on compressive strength as the effect of the mean normal stress on shear strength. Figure B.25 shows data plotted in terms of of maximum shear stress, $\mathrm{TM} = \tau_\mathrm{m} = (\sigma_1 - \sigma_3)/2$, and mean normal stress, $\mathrm{SM} = \sigma_\mathrm{m} = (\sigma_1 + \sigma_3)/2$, after a change of variables in the HB criterion. Because the plots are in units of unconfined compressive strength, all plots pass through the point (0.5, 0.5). A complete plot of HB after transformation to normal stress–shear stress variables would be a parabola described by

$$\left(\tau_\mathrm{m} + \frac{m}{8}\right)^2 = \left(\frac{m}{4}\right)\sigma_\mathrm{m} + \left(\frac{m}{8}\right)^2 + \left(\frac{1}{4}\right)$$

where the stress variables are normalized by division by C_o as before. This transformed HB parabola is offset to the negative side of the shear stress axis and is therefore not symmetrical with respect to the normal stress axis. Thus, the full transformed HB parabola is physically unacceptable. However, by rewriting this parabola in the form

$$\left|\tau_{\rm m}\right| = \left(\frac{-m}{8}\right) + \left[\frac{m^2}{8} + \left(\frac{1}{4}\right) + \left(\frac{m}{4}\right)\sigma_{\rm m}\right]^{1/2} \tag{B.29}$$

and then applying the restriction: $\sigma_{\rm m} \geq -1/m$ symmetry is obtained. The graphical result is that the plot in Figure B.24 is reflected across the normal stress axis. The nose of the plot is not rounded, but then neither is the nose of a Mohr–Coulomb plot.

If one starts with a parabola in normal stress shear stress variables such that

$$\tau_{\rm m}^2 = a\sigma_{\rm m} + b,$$

where a and b are strength constants that can be determined in terms of tensile and unconfined compressive strength, and transforms the criterion to a plot of major principal stress versus minor principal stress, the result is unexpected. Thus,

$$\sigma_1 - a = \sigma_3 \pm \left[4a\sigma_3 + (1-a)^2\right]^{1/2}$$

which is a parabola that is offset from both axes and symmetric with respect to the line $\sigma_1 = \sigma_3 + a$. Again some restriction is necessary for physical meaning: $\sigma_3 \geq -b/a$, which is also necessary for $\sigma_{\rm m}$ in the original parabola in normal stress–shear stress form.

A parabola may show that the decreasing rate of strength increase with normal stress is too rapid, so some adjustment is needed. This adjustment is to simply replace the exponent with a more suitable number. Thus,

$$\tau_{\rm m}^n = a\sigma_{\rm m} + b$$

where a and b are strength constants and n is some number that results in an improved fit to the data, for example, $n = 1.8$. This form is a category of n-type criteria (Pariseau, 1967, 1972). A similar approach may be used in plots of $\sigma_1 = f(\sigma_3)$, for example,

$$\sigma_1^n = a\sigma_3 + b$$

where a and b are different strength constants. This equation also provides good fits to the data in Figure B.24 for the same m values.

Data in Figure B.25 are plotted to much higher confining pressures than in Figure B.24 and show near linearity of strength as a function of confining pressure over a limited range. This observation suggests a Mohr–Coulomb (MC) criterion would fit the data well at relatively high or low confining pressures, but with different slopes and intercepts. The high confining pressures in Figure B.24 are beyond the range of engineering interest for the most part. Recall, slope, tan ψ, and strength axis intercept, k, are related to cohesion, c, and angle of internal friction, ϕ: $k = c\cos(\phi)$ and $\tan\psi = \sin(\phi)$. Of interest are data near the origin of strength plots where confining pressure is low and the mean normal stress may even be tensile. In the tensile region, strength plots steepen rapidly. Consequently, backward extrapolation of high confining pressure strength test data to the tensile stress region tends to overestimate tensile strengths, as shown in Figure B.26, which also shows that backward extrapolation tends to overestimate cohesion. These observations suggest that, when only compressive stress states are anticipated in a design analysis, cohesion c, and angle of internal

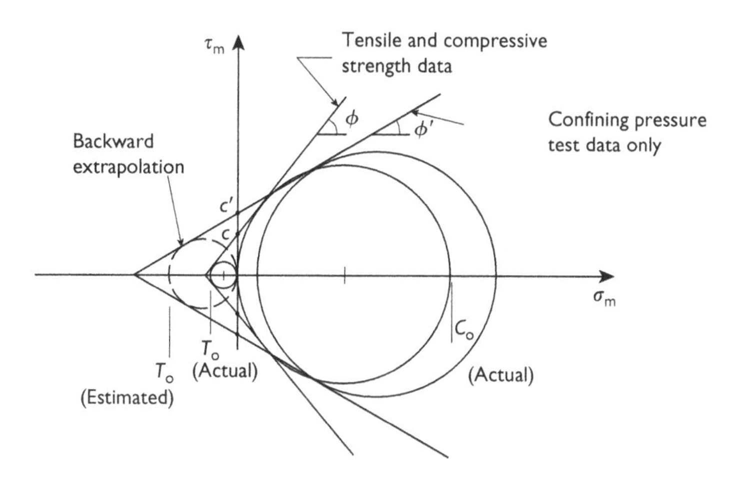

Figure B.26 Mohr circles and linear envelope from uniaxial and confining pressure test data showing difficulties with extrapolation.

Table B.10 Cohesion and angle of internal friction (after Kulhawy, 1975)

Property		*Rock type*					
		Igneous		*Metamorphic*		*Sedimentary*	
		Plutonic	*Volcanic*	*Non-foliated*	*Foliated*	*Clastic*	*Chemical*
Cohesion c (MN/m^2)	No.	12	17	8	14	35	22
	Max.	176.0	77.4	70.6	70.3	73.1	96.0
	Min.	16.5	0.0	0.0	14.8	0.0	0.0
	Ave.	56.1	32.2	22.9	45.7	31.7	26.3
Angle ϕ of internal friction	No.	12	17	8	14	35	22
	Max.	56.0	64.0	60.0	47.6	55.5	61.0
	Min.	23.8	0.0	25.3	15.0	7.5	7.0
	Ave.	45.6	24.7	36.6	27.3	29.2	35.9

friction ϕ are the most useful strength parameters to account for the effect of confining pressure rather than T_o and C_o.

Data for the three basic rock types from the survey by Kulhawy (1975) are given in Table B.10 in the form of cohesion and angle of internal friction in the range of confining pressures of engineering interest. These show that in some instances, rock is purely cohesive with zero angle of internal friction, and occasionally, purely frictional with zero cohesion. Such test results are rare, however.

Anisotropy may be important just as it was in consideration of directional features of unconfined compressive strength test cylinders. Figure B.27 shows how compressive strength of a slate is influenced by confining pressure and bedding plane angle. The confining pressure p is given in ksi (1,000s psi). These data clearly show that compressive strength does not simply vary monotonically between strength parallel to bedding and perpendicular to bedding.

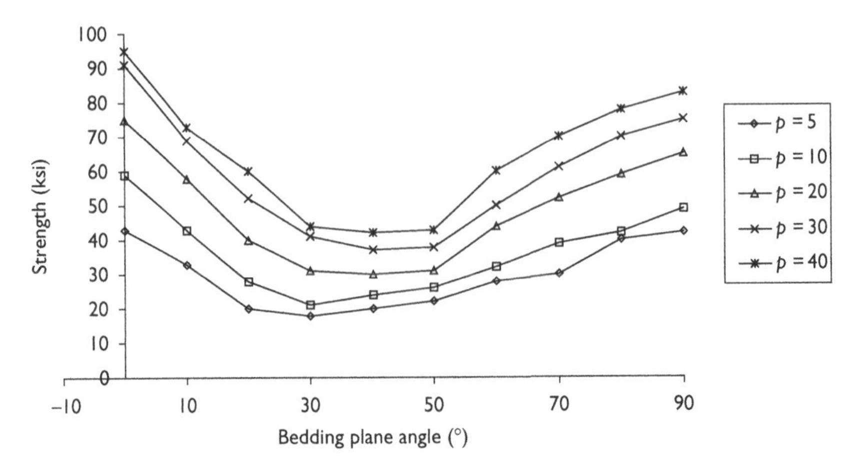

Figure B.27 Variation of compressive strength with confining pressure and bedding plane angle for a slate (after McLamore and Gray, 1967).

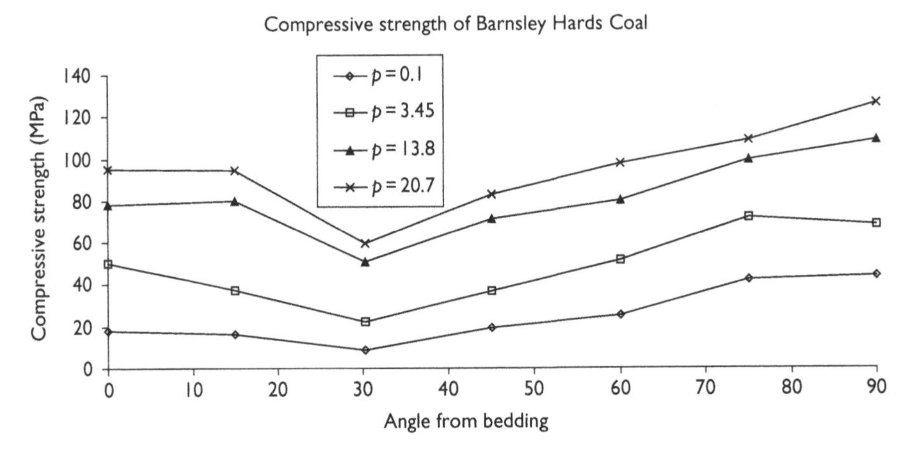

Figure B.28 Variation of compressive strength of Barnsley Hards Coal with confining pressure and bedding plane angle (after Pomeroy *et al.*, 1971).

They show a minimum strength at an angle to bedding and a migration of the minimum toward a larger angle with increasing confining pressure, as forecast by theory (Pariseau, 1972). These results also indicate that for this slate, compressive strength parallel to the bedding is greater than perpendicular to the bedding regardless of confining pressure. Maximum compressive strength (parallel to bedding) for this rock is about twice the minimum strength (at an angle to bedding) at all confining pressures. Similar results were obtained for two types of an oil shale (Green River Shale).

Coal shows a similar effect of confining pressure on strength and strength variation with direction. Figure B.28 shows how strength of a bituminous coal (Barnsley Hards) varies with bedding plane angle and confining pressure. This coal has a main set of cleats and a minor seat of cleats that are normal to the bedding and to each other, so an orthotropic form of anisotropy is expected. The data in Figure B.28 were from tests on specimens in the plane of the main cleats but at an angle to the bedding.

B.3 Joint stiffness

Joint stiffnesses, k_n and k_s, relate joint stress to relative displacement between opposing points on the two surfaces that define a joint. Thus,

$$\sigma = k_n u \quad \text{and} \quad \tau = k_s v$$

where σ, τ, u, and v are joint normal stress, joint shear stress, relative displacement across a joint, and relative displacement along a joint, respectively. The units of these joint stiffnesses are stress over displacement units, for example, psi/in. or GPa/m. In analogy to Hooke's law, one may compute a joint Young's modulus E_j, and shear modulus G_j as $E_j = k_n h$ and $G_j = k_s h$ where h is joint "thickness." Poisson's ratio for a joint is then defined by the usual isotropic relationship $v_j = (E/2G) - 1$, which is dimensionless.

Closure and shear along a joint tend to be highly nonlinear, so any joint stress displacement relationship should be considered incrementally, that is, in differential form. Thus,

$$d\sigma = k_n du, \quad \text{and} \quad d\tau = k_s dv$$

defines joints stiffnesses more realistically as local slopes of stress displacement plots.

If a thickness h is assigned to a joint, then one may define joint normal and shear strains as

$$\varepsilon = \frac{u}{h} \quad and \quad \gamma = \frac{v}{h}$$

Joint thickness may be related to asperity height or a filling thickness, if present. In this way, joint stress–strain relations become

$$\sigma = k_{nn}\varepsilon + k_{ns}\gamma$$
$$\tau = k_{sn}\varepsilon + k_{ss}\gamma$$

where coupling between normal and shear effects is introduced by new, additional joint stiffnesses and $k_{ns} = k_{sn}$, that is, symmetry is assumed. The units of these joint stiffness are the units of stress. Again, a differential formulation would be a more physically accurate description of joint behavior.

One may also consider a joint as a thin layer of material that differs from intact material away from the joint. Thickness of such a layer may be related to asperity height, joint roughness, filling thickness, or similar physical feature associated with a joint. Once a joint or interface between blocks of intact rock is considered to be a thin layer of different material than the intact rock, the full form of Hooke's law becomes available for the description of joint stress–strain relations. In this regard, assignment of a thickness to a joint is problematic, but can be avoided by direct experimental determination of joint stiffnesses or moduli. Triaxial tests on cylinders with joints and direct shear tests on joint samples allow for the determination of joint stiffnesses and strengths.

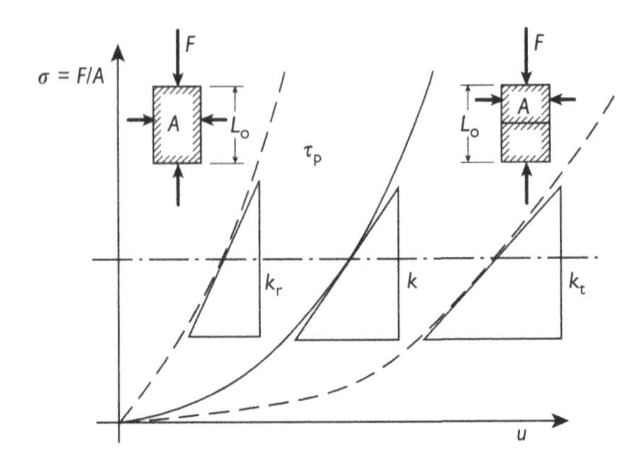

Figure B.29 Testing for joint normal stiffness using intact and jointed test pieces.

Normal stiffness

Consider a test cylinder loaded axially in compression that contains a joint perpendicular to the load axis and the corresponding force displacement curve, as shown in Figure B.29. Normal stress σ acting across the joint is simply axial load F divided by joint area A. Relative displacement between sample ends is a total u_t that is the sum of displacement of intact rock u_r and joint closure u. Thus,

$$u_t = u + u_r = \frac{F}{Ak_t} = \frac{F}{Ak} + \frac{F}{Ak_r}$$

Hence, joint stiffness may be obtained from slopes of force displacement curves of intact rock only and intact rock with a joint. Thus,

$$\frac{1}{k_n} = \frac{1}{k_t} - \frac{1}{k_r}$$

Alternatively, a jointed sample normal strain ε_t may be obtained by dividing the relative displacement between sample ends, ΔL or u_t, by original sample length L_o . This is a total strain that includes deformation of the intact rock above and below the joint as well as closure of the joint. A test of the same sample before introduction of the joint under the same load allows for estimating strain of intact rock ε_r in the jointed test cylinder. The joint normal strain is $\varepsilon = \varepsilon_t - \varepsilon_r$ and the joint Young's modulus is $E = \sigma/\varepsilon$. A joint thickness may be computed: $h = k_n/E$, if desired, although there is no need after having determined E directly. If the experimental measurements correspond to a particular point on a joint force displacement curve, then joint normal stiffness and Young's modulus are tangential values suitable for use in the differential form of the joint normal stress displacement or stress–strain relations.

Force displacement plots from joint testing are usually far from linear. A *normal* force displacement plot generally shows an increasingly steep slope, but eventually reaches a constant value that may well reflect the stiffness of intact rock with effectively full closure of the joint. Such behavior is sometimes referred to as "locking."

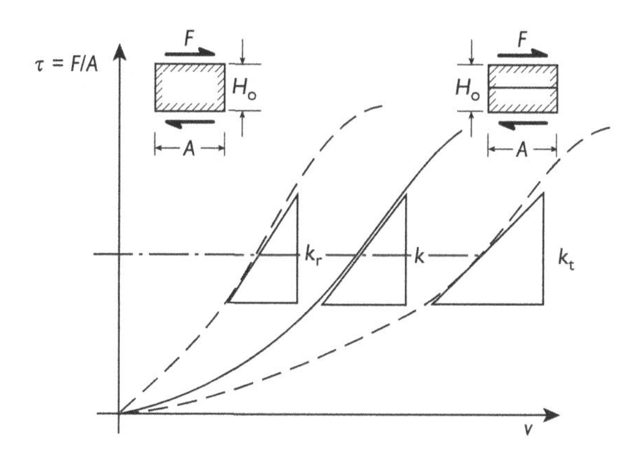

Figure B.30 Testing for joint shear stiffness using intact and jointed test pieces.

Shear stiffness

Shear stiffness of a joint is obtained much the same way as normal stiffness is determined. However, testing is often done in direct shear as illustrated in Figure B.30. Again, displacement of the intact rock is determined from a test on a sample without a joint and subtracted from the total displacement of the joint sample to obtain joint displacement v. Shear force divided by area is shear stress τ. Joint shear stiffness is then

$$\frac{1}{k_s} = \frac{1}{k'_t} - \frac{1}{k'_r}$$

where the primes indicate a shear test.

Alternatively, a total shear strain γ_t may be computed as a total horizontal displacement, ΔL or u, divided by original sample *height* Ho and similarly for an intact sample γ_r. The joint strain is then the difference, that is, $\gamma = \gamma_t - \gamma_r$, which is related to joint shear stress τ by Hooke's law. Thus, the joint shear modulus is $G = \tau/\gamma$. Again, a thickness may be computed $h = k_s/G$ which may be different than a thickness computed from normal stiffness and Young's modulus.

Table B.11 shows a range of single joint stiffness reported in the technical literature from several sources. In this regard, measurements on test cylinders containing inclined joints allow for simultaneous determination of normal and shear stiffness, although such tests are more difficult to do. The results in Table B.11 range over an order of magnitude for individual rock types and give the impression that normal stiffness is usually, but not always, greater than shear stiffness. When all rock types are considered, the range in joint stiffnesses is perhaps four orders of magnitude or more.

B.4 Joint strength

Shear force displacement plots are more likely to steepen and then flatten as slip begins. Indeed, the slope of a shear force–displacement curve may turn negative after reaching an initial peak and then flatten. Such behavior is often referred to as a peak residual response and

Table B.11 Joint normal and shear stiffness ranges (after Moon, 1987)[a]

Rock type	*Normal stiffness* k_n *(10⁵ psi/in.)*	*Shear stiffness* k_s *(10⁵ psi/in.)*	*Source*
Sandstone	0.58–24.8	0.10–11.7	Brechtel, 1978
Shale	0.31–3.72	0.44–3.50	Rosso, 1976
Basalt	10.7–19.6	8.79–16.2	Hart *et al.*, 1985
Granite,limestone schist, gneiss, slate, shale, and sandstone	0.01–2.49	0.0004–1.16	Goodman, 1968; Kulhawy, 1975

Note
a 1 GPa/m = 3,687 psi/in.

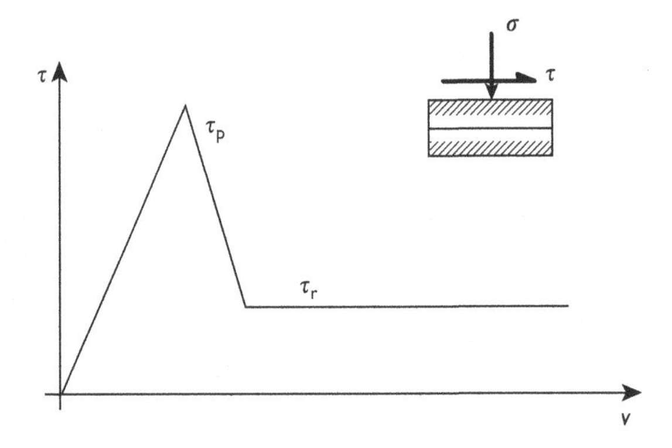

Figure B.31 Peak-residual shear stress–displacement behavior.

is characteristic of rough joints sheared under relatively low normal stress. Smooth joints usually do not exhibit peak residual behavior. Figure B.31 is an idealized plot of peak residual behavior of a shear joint that is initially rough. Figure B.32 shows data obtained from some experimental measurements on smooth, saw cut joints, and rough joints that were grooved. The rough joints show peak residual behavior while the smooth joints do not; both data sets depend on the applied normal stress. The peak of a joint shear stress displacement curve at the end of a nominally elastic response is peak shear strength. Shear stress on the flat portion of the curve following a peak is residual strength. Peak residual behavior is often observed when a fresh fracture is sheared for the first time. The stress displacement plot rises linearly and steeply to a peak when asperities on the joint begin to fail.

A volume expansion or *dilatation* occurs as the top set of asperities ride up and over the lower opposing set. Tips of the highest asperities begin to fail. Continued shearing causes additional asperity failure and decreasing resistance to shear. At the same time, failed material (gouge) begins to fill valleys between asperities. Eventually, the entire sample shear plane is smoothed by asperity shear and fill; residual strength is reached. Continued shearing then proceeds with no additional drop in resistance. An increased normal stress

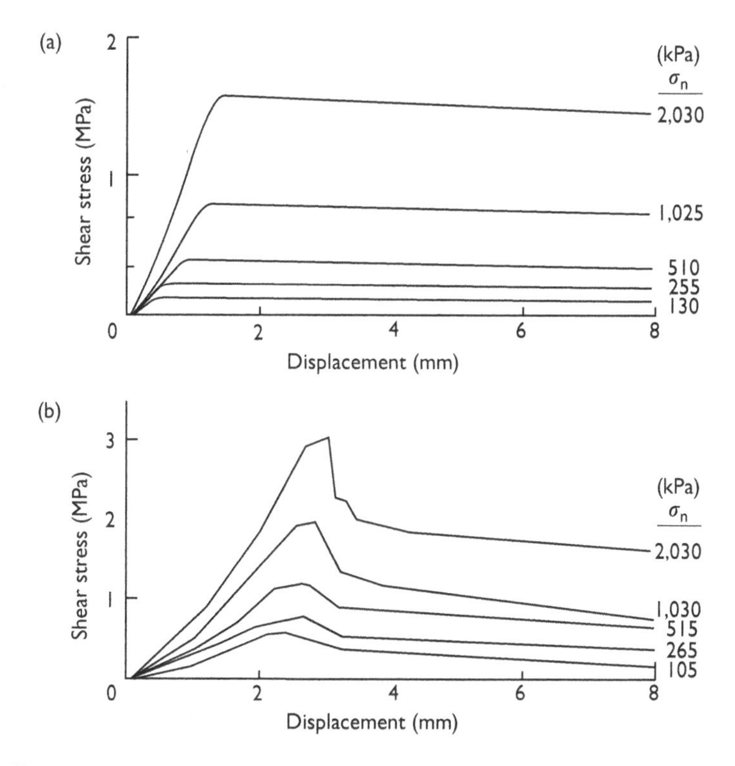

Figure B.32 Shear stress–displacement data for (a) smooth and (b) rough 4-groove surfaces (after Hassani and Scoble, 1985).

across a joint tends to decrease dilatation which may be suppressed entirely and very high normal stresses. At the same time, shear strength is increased with an increased normal stress.

A natural joint that has a shear history may not exhibit peak residual behavior because of smoothing action associated with prior shearing. However, joint cementation in time may erase past effects and thus appear to be fresh and exhibit peak residual behavior when shearing is reinitiated. A filled joint is also likely to have cohesive strength and present some resistance to separation as well to shear. This observation suggests that a cohesive frictional description of joint shear strength may be appropriate. Indeed, the most common model for joint shear strength is a Coulomb friction with cohesion description:

$$\tau = \sigma \tan(\phi_j) + c_j$$

where σ, τ, ϕ_j, and c_j are joint normal stress, shear stress, friction angle, and cohesion, respectively.

A smooth joint lacking filling would certainly be cohesionless; the same is true of a rough joint. However, rough joint data often fit the linear model very well, but such fits often produce a shear axis intercept indicating presence of cohesion. This pseudo-cohesion is an artifact of the data fit. An alternative, is a power law fit that uses an n-type failure criterion for cohesionless material. The n-type model is a special case because of the lack of cohesion and has the form

Table B.12 Ranges of cohesion c and friction angle ϕ for several types of rock joints

Joint rock type	*Joint property*	
	Joint cohesion psi (MPa)	*Joint friction angle (°)*
Sandstone, limestone granite, slate, schist, gneiss[a]	0–377 (0–2.69)	19–56
Granite, gabbro, trachyte, sandstone, marble[b]	40–160 (0.28–1.10)	27–37

Notes
a after Kulhawy, 1975.
b after Jaeger and Cook, 1969.

$$\tau^n = a\sigma \quad \text{or} \quad \tau = a'\sigma^m$$

where $a' = a^{1/n}$ and $m = 1/n$. The value of m would be less than 1 because $n \geq 1$. Of course, if a joint has cohesion, then an appropriate constant must be added to the above criteria.

Figure B.33 shows actual experimental data for sandstone and mudstone and shows that the linear Coulomb friction model fits the data quite well, but implies a fictitious cohesion. The nonlinear n-type fits are superior for these cohesionless joints and have higher correlation coefficients. A range of cohesion and friction angle values for a number of different jointed rock types is presented in Table B.12. Not too surprisingly, considerable variation of joint properties within a given rock type occurs. Reasons include joint genesis, history, and test environments. Although every joint is unique, experimental observations indicate that joint friction angle seldom exceeds 45° and is often less.

If a joint is infiltrated with fluid (water), then the effect of fluid pressure on strength should be taken into account. There may also be important chemical effects on strength in the presence of joint fluids. Such effects may be temperature-dependent. Solution mining of fracture hosted ore bodies pose complex questions of joint behavior. Extraction of petroleum from fracture hosted reservoirs is another source of questions about complex joint behavior. Fluid pressure decreases the normal stress transmitted through solid–solid contacts between opposing joint surfaces and thus decreases the normal stress that is associated with shear strength. Hence, shear strength of joints is determined by effective stress, total normal stress less the pore fluid pressure. The situation is entirely analogous to the role of effective stress in intact porous rock. Effective stress determines strength, while total stress determines equilibrium. Thus, normal joint stress is always understood to be effective stress when computing joint shear strength.

B.5 Simple combinations of intact rock and joints

Field-scale rock masses are composed of rock joints and intact rock between. The composite rock mass properties therefore differ from those of rock and joints. Generally, joints reduce

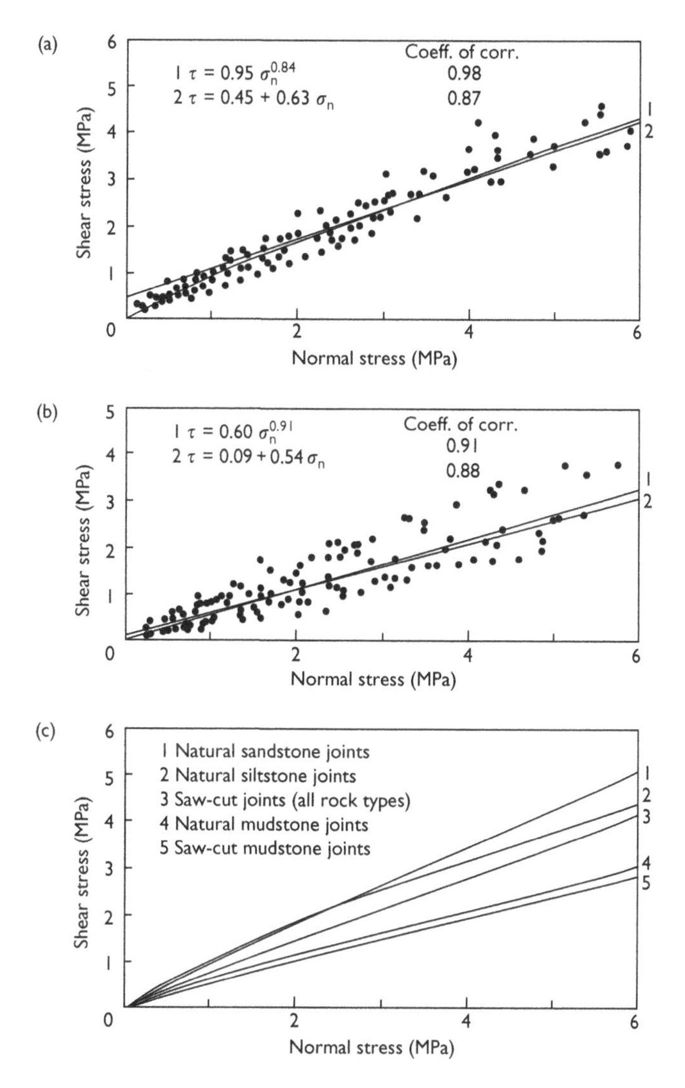

Figure B.33 Non-linear, cohesionless joint and linear, fictitious joint failure criteria (after Hassani and Scoble, 1985). (a) Shear failure envelopes for all natural sandstone joint data; (b) shear failure envelopes for all natural mudstone joints; and (c) comparison of general failure envelopes for different rock discontinuities.

rock mass elastic moduli and strengths from those of intact rock. Despite a number of equivalent properties of models of rock masses of varying degrees of complexity, there is no general consensus on how to compute rock mass properties based on laboratory test data from intact rock and joints. Certainly, the geometry of joint sets in the rock mass of interest are important. Joint direction, dip, and spacing must influence rock mass response to excavation.

Another joint set parameter, *persistence*, is also important. Persistence is a measure of joint continuity and may be defined in several ways. A useful definition is a linear one that defines persistence in a plane containing joints as the ratio of distance between joint areas to the total

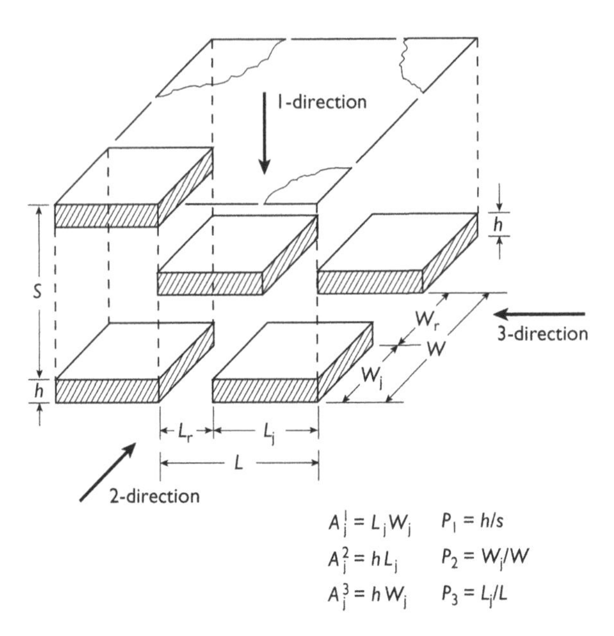

Figure B.34 Persistences of discontinuous joints of thickness *h*.

distance. With reference to Figure B.34 that shows a number of identical rectangular joint areas in a joint plane, persistence measured in a joint plane parallel to the longest joint dimension is $P_3 = L_j/L$, while persistence measured parallel to the shortest joint dimension is $P_2 = W_j/W$ where L_j and W_j are joint length and width, respectively. When joints are continuous in a given direction, the corresponding persistence (in the 2- or 3-direction) is one. Spacing between joint planes is S; by another definition, $P_1 = h/S$.

Over a planar area A that is composed of joint area, $A_j = L_j W_j$. Area of intact rock is A_r, so $A = A_j + A_r$. *Joint area ratio* R is then defined simply as the ratio of intact rock bridge area to total area, that is, $R = A_r/A$, which is analogous to area extraction ratio used in the tributary area approach to pillar design. This area ratio is associated with a normal direction, the 1-direction in Figure B.34, and is therefore $R_1 = L_j W_j/LW = P_2 P_3$ and is perhaps the most important geometric feature of a joint set. Other joint area ratios associated with the 2- and 3-directions are $R_2 = hL_j/LS = P_3 P_1$ and $R_3 = hW_j/WS = P_1 P_2$, respectively. These joint persistences and area parameters are useful in consideration of simple models of discontinuously jointed rock masses.

Continuously jointed rock mass moduli

Again consider a joint under normal load as in Figure B.35a and suppose joint spacing is S and joint thickness is h, as before, and the joint is *continuous* across the test piece (persistences P_1 and P_2 are one). Average overall normal strain, rock strain, and joint strain, are

$$\varepsilon = \frac{u}{S}, \quad \varepsilon_r = \frac{u_r}{S-h}, \quad \varepsilon_j = \frac{u_j}{h}$$

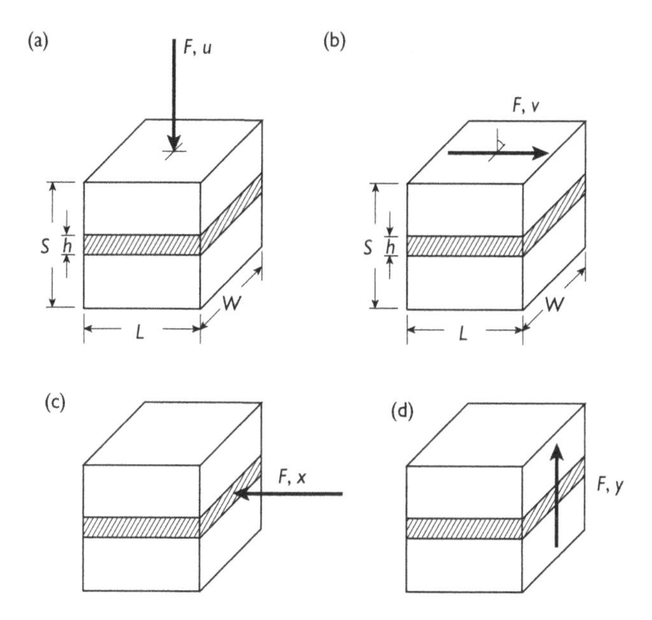

Figure B.35 Normal and shear loads applied to a continuous joint sample. (a) Load normal to joint; (b) shear load parallel to joint; (c) normal load parallel to joint; and (d) shear load across joint.

where u, u_r, and u_j are total, rock, and joint normal displacements, respectively, and $u = u_r + u_j$. Corresponding normal stresses are equal, that is, $\sigma = \sigma_r = \sigma_j$. In consideration of Hooke's law for this one-dimensional load case, $\sigma = E\varepsilon$ overall and also for rock and joint.

Thus,

$$S\varepsilon = (S-h)\varepsilon_r + h\varepsilon_j = S\frac{\sigma}{E} = (S-h)\frac{\sigma_r}{E_r} + h\frac{\sigma_j}{E_j}$$

Hence

$$\frac{1}{E_n} = \frac{1-(h/S)}{E_r} + \frac{(h/S)}{E_j} = \frac{1-P_1}{E_r} + \frac{P_1}{E_j} \approx \frac{1}{E_r} + \frac{1}{Sk_n}$$

When h/S is small relative to one and $k_n = hE_j$ is joint normal stiffness, as before, the approximation on the right-hand side is reasonable. Indeed, this formulation is frequently mentioned in the technical literature and is one of the earliest models of joint effects on rock mass moduli. Young's modulus E_n is an equivalent modulus for the combination of joint and intact rock and applies to loading perpendicular to a continuous joint.

A similar analysis applies in shear. With reference to Figure B.35b, the shear strains are assumed small, so

$$\gamma = \frac{v}{S}, \quad \gamma_r = \frac{v_r}{S-h}, \quad \gamma_j = \frac{v_j}{h}$$

where v, v_r, and v_j are total, rock, and joint shear displacement, respectively. Again, equilibrium requires equality of shear stresses, so $\tau = \tau_r = \tau_j$. In consideration of Hooke's law, $\tau = G\gamma$ for each component and the composite. Thus,

$$\frac{1}{G_p} = \frac{1 - h/S}{G_r} + \frac{h/S}{G_j} = \frac{1 - P_1}{G_r} + \frac{P_1}{G_j} \approx \frac{1}{G_r} + \frac{1}{Sk_s}$$

where G_r and G_j are shear moduli of composite, rock and joint, respectively, and $k_s = hG_j$ is joint shear stiffness. The shear modulus G_p applies to a continuous joint loaded by a shear stress acting parallel to the joint. Again, the approximation applies when h/S is small relative to one.

Young's modulus and shear modulus could be used to compute a Poisson's ratio assuming isotropy. However, the assumption of isotropy is not justified because of the directional character joints impart to a rock mass. Indeed, loading parallel to the joint leads to a Young's modulus that differs from the one obtained under a normal load analysis. In fact, normal loading parallel to a joint and shear loading normal to a joint allow for estimating two moduli in each case.

In case of normal loading parallel to a joint that completely transects the test piece, as shown in Figure B.35c, a "compatible" assumption may be made that imposes equal strains (or displacements) in the load direction, that is, $\varepsilon = \varepsilon_r = \varepsilon_j$. A volume weighted average stress is then

$$\sigma = (1 - h/S)\sigma_r + (h/S)\sigma_j$$

According to Hooke's law in this one-dimensional loading case, $\sigma = E\varepsilon$ for composite and components. Hence,

$$E_p = (1 - h/S)E_r + (h/S)E_j = (1 - P_1)E_r + P_1 E_j$$

where E_p applies to a joint rock combination that has a normal stress applied parallel to the joint.

Alternatively, if an "equilibrium" or uniform stress assumption may be made, so $\sigma = \sigma_r = \sigma_j$. A volume weighted average strain is then

$$\varepsilon = (1 - h/S)\varepsilon_r + (h/S)\varepsilon_j$$

After introduction of Hooke's law, the result is

$$\frac{1}{E'_p} = (1 - h/S)\frac{1}{E_r} + (h/S)\frac{1}{E_j}$$

which is the same expression derived for Young's modulus E_n under equilibrium normal loading perpendicular to the joint. In fact, a second "compatible"estimate for E_n could be obtained that would be equal to E_p (no prime). These moduli, E_n and E_p, actually provide lower and upper bounds to Young's modulus ($E_n < E_p$) when the rock mass is considered isotropic. Obviously a jointed rock mass has directional features and would therefore be anisotropic. Hence, a reasonable approach would be to use both, applying E_n for normal loading perpendicular to the joint and E_p for normal loading parallel to the joint.

Similar analyses using compatible and equilibrium assumptions for shear moduli lead to similar results with the subscripts interchanged. Thus,

$$G'_n = (1 - P_1)G_r + P_1 G_j$$

The suggestion is then that G_n and G_p apply for shearing normal or across the joint and parallel to the joint.

To summarize the continuously jointed rock case, suggested moduli are those given by

$$\frac{1}{E_n} = (1 - P_1)\frac{1}{E_r} + (P_1)\frac{1}{E_j}$$

$$E_p = (1 - P_1)E_r + (P_1)E_j$$

$$G_p = (1 - P_1)G_r + (P_1)G_j$$

$$\frac{1}{G_n} = (1 - P_1)\frac{1}{G_r} + (P_1)\frac{1}{G_j}$$

Unfortunately, questions concerning Poisson's ratio(s) are not easily answered even with these relatively simple thought experiments. The reason relates to fundamental requirements that appear in consideration of an inclined joint. When the joint is flat, an equilibrium requirement is obvious and dictates equality of stresses; when the joint is vertical, a reasonable assumption is equality of strains. Consideration of normal and shear loading a test volume containing an inclined joint is more difficult because of simultaneous requirements of equilibrium and continuity across the joint. These difficulties have been addressed in detail. Direct physical verification of theory is not practical because of the nature of the problem and, in particular, the enormous sample size that would be required.

However, numerical experiments that simulate laboratory-like tests on samples of arbitrary size allow for highly controlled testing of theory. As an example, consider the cubical test volume shown in Figure B.36a that contains a single joint at various angles of inclination. The joint thickness to spacing ratio is about 0.014. If the joint were 1.4 in. thick, joint spacing would be 100 in., which is a relatively wide spacing in some respects. Rock masses could be much more densely jointed. The ratio of intact rock modulus to joint modulus is 100. If Young's modulus of intact is 2.4×10^6 psi, the joint Young's modulus is 0.024×10^6 psi and the joint normal stiffness is 0.17×10^5 psi/in.

Figure B.36(b) and (c) shows comparisons between theory and (numerical) experiment for Young's modulus and shear modulus; the agreement is excellent at all angles of inclination. These results are normalized in that the jointed sample modulus is divided by the intact rock modulus at all angles ranging from parallel loading (zero degrees) to loading perpendicular to the joint. Although the joint is thin relative to spacing, the effect on the moduli is substantial and results in a reduction of $E = E_n$ to 42% of the intact rock modulus. The modulus for normal loading parallel to the joint $E = E_p$ is reduced slightly, to about 99% of the intact rock modulus, as might be expected for this thin joint example. When the joint is flat and loaded in shear parallel to the joint, the shear modulus $G = G_p$ is reduced to about 42% of the intact shear modulus. When the joint is vertical, the reduction is about 58% and is influenced by the diagonal direction of the joint relative to the applied shear stresses. Shearing with the joint so that $G = G_n$ results in only a small decrease of shear modulus, about 99%. These results verify the elementary formulas derived for Young's moduli and shear moduli under loads parallel

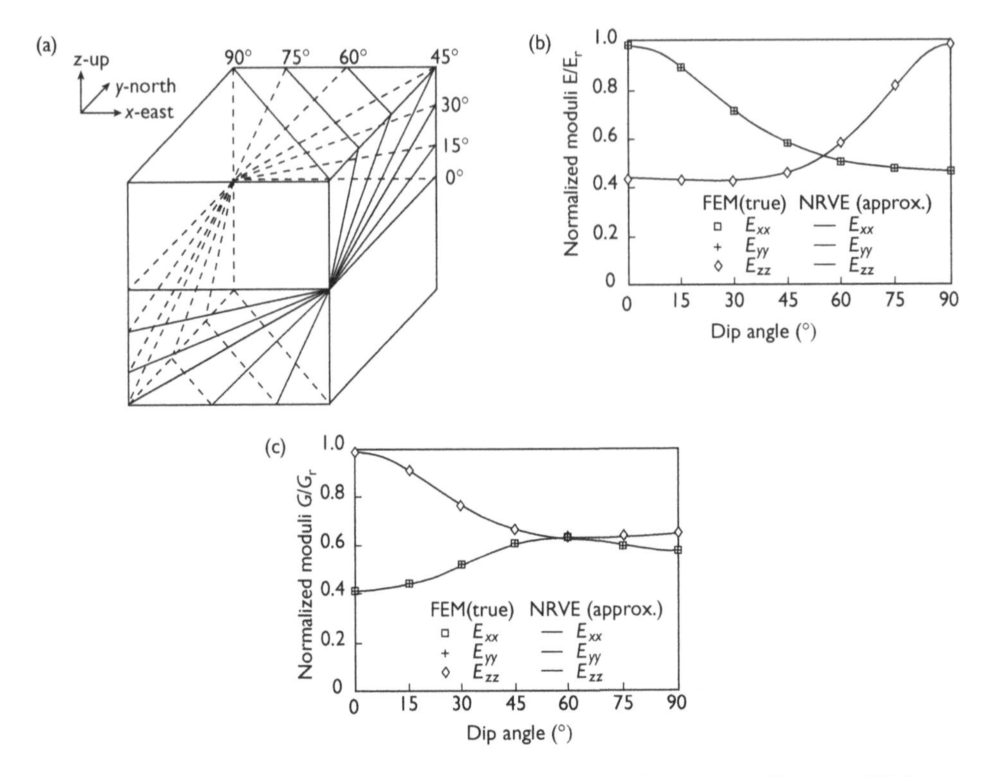

Figure B.36 Jointed test cube geometry and results from numerical experiment (Pariseau, 1994).

and perpendicular to a joint. Addition of similar joints to the test volume has the same effect as decreasing joint spacing S and does not change the nature of the results. Although, these results are limited to linear elasticity, they further underscore the importance of joints to rock mass deformation and indicate restrictions to the elementary formula for design applications where joints and joint sets may be numerous and oriented at arbitrary angles to stress directions that vary with excavation.

Discontinuously jointed rock mass moduli

When joints are discontinuous, only part of the plane containing joints is occupied by joints; the equivalent rock mass moduli formulas then change in a systematic way. Under normal loading perpendicular to the plane containing joints and the equilibrium assumption of uniform stress and Hooke's law, one has using the notation in Figure B.34,

$$\frac{1}{E_1} = (1 - R_1 P_1)\frac{1}{E_{\mathrm{r}}^1} + (R_1 P_1)\frac{1}{E_{\mathrm{j}}^1}$$

where E_1 is an estimate of the jointed rock mass modulus in the 1-direction; rock and joint moduli are also in the 1-direction, indicated by a superscript, which may be different in the 2- and 3-directions. When the jointing is continuous, $R_1 = 1$ and the estimate reduces to the previous formula derived for a continuous joint.

The compatible assumption leads to more useful estimates for normal loadings parallel to the joint:

$$E_2 = (1 - R_2P_2)E_r^2 + (R_2P_2)E_j^2$$
$$E_3 = (1 - R_3P_3)E_r^3 + (R_3P_3)E_j^3$$

where the superscripts indicate direction. These formulas also reduce to the previous formulas for normal loading parallel to a joint when the joint is continuous $(R_2 = R_3 = R_1 = 1)$ and also allowance is made for rock and joint anisotropy.

The alternative assumption of compatible strains leads to the estimate which is not particularly useful.

$$E'_1 = (1 - R_1P_1)E_r^1 + (R_1P_1)E_j^1$$

When normal loading is parallel to the joint in the 2-direction, the equilibrium assumption of uniform stress leads to

$$\frac{1}{E'_2} = (1 - R_2P_2)\frac{1}{E_r^2} + (R_2P_2)\frac{1}{E_j^2}$$

and similarly for normal loading parallel to the 3-direction,

$$\frac{1}{E'_3} = (1 - R_3P_3)\frac{1}{E_r^3} + (R_3P_3)\frac{1}{E_j^3}$$

which are also not particularly useful.

Analysis of shear moduli for discontinuously jointed rock masses proceeds in a manner similar to that for Young's modulus. Two estimates are available from the equilibrium and compatible assumptions. Figure B.37 defines the subscript notation used to identify the shear moduli. For example, the 1-subscript indicates shear seen looking down the 1-axis at the 2-3 plane. Shear in this case amounts to shearing two adjacent intact rock slabs separated by a thin

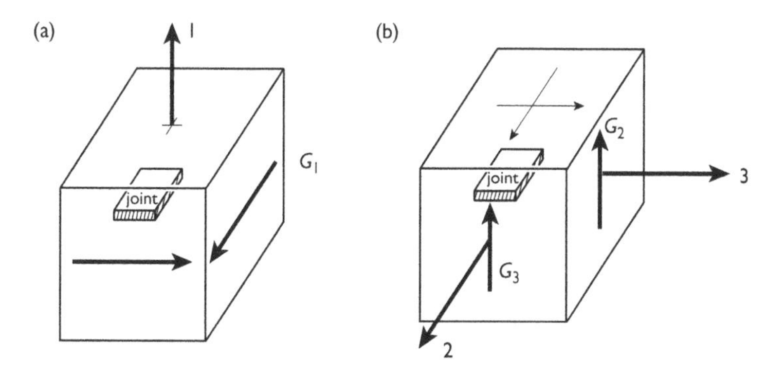

Figure B.37 Notation for shear analysis of a discontinuously jointed test cube.

joint layer. Resistance to shear is almost entirely intact rock. Shear about the 2- and 3-axes is akin to shearing a layered structure that has much less shear resistance because of the joint, despite the thinness of the joint.

To summarize results for the discontinuously jointed rock mass case, the recommended expressions for Young's modulus and shear modulus in the principal 1-, 2-, and 3-directions are

$$\frac{1}{E_1} = (1 - R_1 P_1)\frac{1}{E_r^1} + (R_1 P_1)\frac{1}{E_j^1}$$

$$E_2 = (1 - R_2 P_2)E_r^2 + (R_2 P_2)E_j^2$$

$$E_3 = (1 - R_3 P_3)E_r^3 + (R_3 P_3)E_j^3$$

$$G_1 = (1 - R_1 P_1)G_r^1 + (R_1 P_1)G_j^1$$

$$\frac{1}{G_1} = (1 - R_1 P_1)\frac{1}{G_r^1} + (R_1 P_1)\frac{1}{G_j^1}$$

$$G_2 = (1 - R_2 P_2)G_r^2 + (R_2 P_2)G_j^2$$

$$G_3 = (1 - R_3 P_3)E_r^3 + (R_3 P_3)E_j^3$$

where the superscripts indicate the mutually orthogonal directions relative to the joint plane. For example, the 1-direction is normal to the joint plane.

Continuously jointed rock mass strengths

Strength of intact rock and joints are often described by Mohr–Coulomb criteria. The form is the same, $\tau = \sigma \tan(\phi) + c$, but the cohesions are much different and so are the friction angles. Usually the joint strength properties are less than those for intact rock. When a joint transects a test volume, as shown in Figure B.38, three failure modes are possible: (1) joint failure; (2) intact rock failure, and (3) failure of both. Tensile, compressive, and shear stresses may be present at failure. Whether the joint fails depends on orientation measured by joint dip δ.

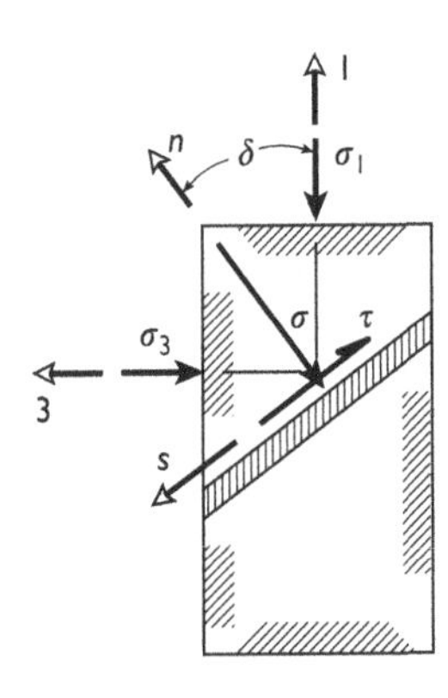

Figure B.38 Notation and sign convention for joint analysis.

Under the simplifying assumption of uniform stresses in the test volume, joint failure occurs provided the maximum shear stress exceeds the value given by

$$\tau_m = \frac{\sigma_m \sin(\phi_j) + c_j \cos(\phi_j)}{\sin(2\delta - \phi_j)} = \sigma_m a + b$$

where the subscript j indicates a joint value and σ_m is the mean normal stress in the view of interest. However, the maximum shear stress cannot exceed intact rock strength which is given by

$$\tau_m = \sigma_m \sin(\phi_r) + c_r \cos(\phi_r)$$

where the subscript r indicates intact rock.

Figure B.39 shows Mohr circles representing the extremes of first joint failure and intact rock failure when the mean normal stress is held constant while the maximum shear stress (deviatoric stress) is increased. Joint failure first becomes possible provided the joint dip $\delta = \pi/4 + \phi_j/2$. If the joint dips at some other angle, joint failure does not occur. However, by increasing shear stress at constant mean normal stress, joints at other dips may fail according to the maximum shear stress criterion. Ultimately joint dips that fall within the arc *AB* in Figure B.39 ($2\delta_A$, $2\delta_B$) may fail.

Alternatively, one may recast the thought experiment in terms of compressive strength under confining pressure. Thus, failure is possible provided the major principal compression exceeds the value

$$\sigma_1 = \left(\frac{1+a}{1-a}\right)\sigma_3 + b = C_p = \left(\frac{1+a}{1-a}\right)p + b$$

where *a* and *b* are the parameters in the maximum shear stress formula for joint failure. Figure B.40 shows the compressive strength of a continuously jointed test volume. The curves in Figure B.40 are in ascending order of confining pressure. The lowermost curve relates to an unconfined compressive strength test. The concept leading to Figure B.40 is also known as a single plane of weakness model.

Laboratory test data are generally not represented very well by this model, but the considerable weakening effect joints may have on rock masses is clearly revealed. The

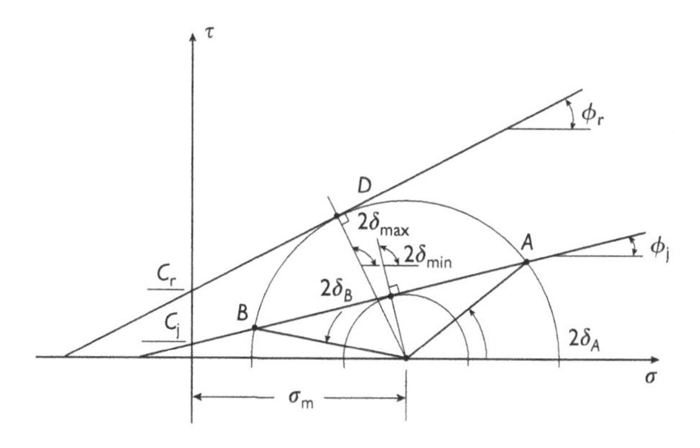

Figure B.39 Intact rock and joint failure criteria.

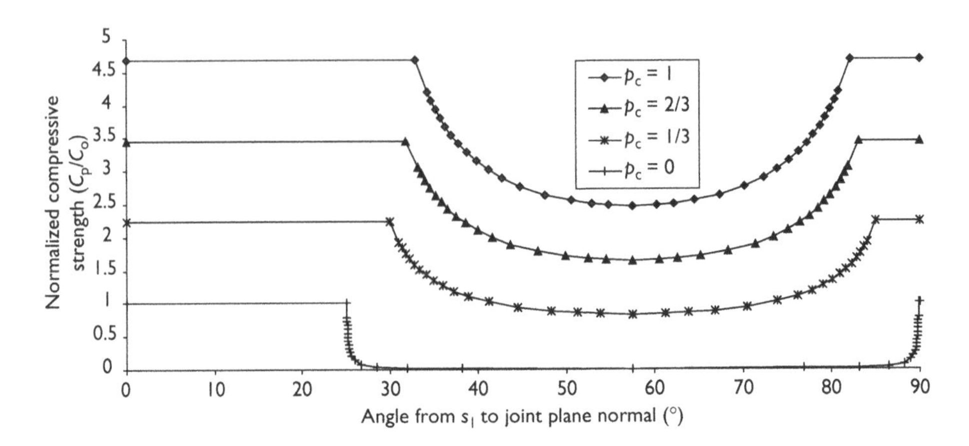

Figure B.40 Compressive strength under confining pressure of a test cylinder containing a single joint plane. Strength and confining pressure are in units of unconfined compressive rock strength. Joint friction angle is 25°; rock angle of internal friction is 35°. Rock cohesion is about 0.26 unconfined compressive strength of rock and joint cohesion is 0.01 of intact rock.

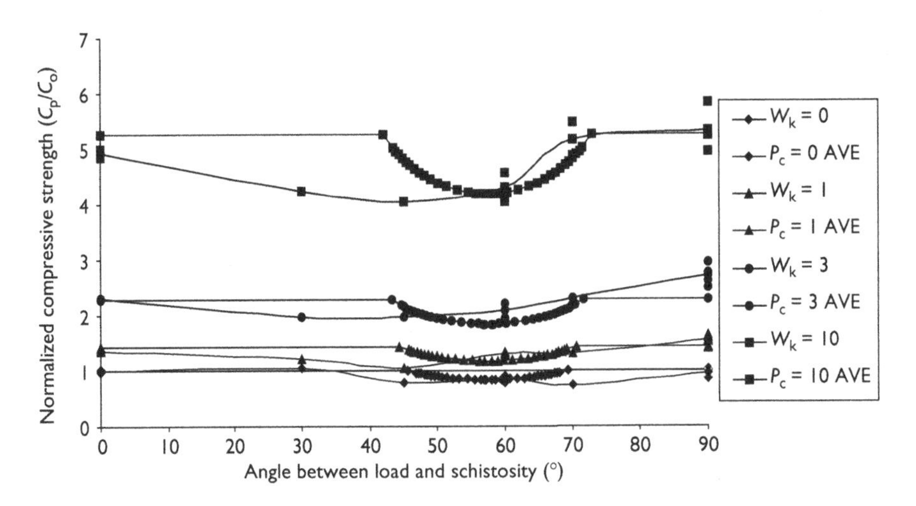

Figure B.41 Plane of weakness model test against laboratory test data on gneiss (after Deklotz and Brown, 1967). P_c = confining pressure (ksi) and experimental data with a smoothed line drawn through average values at each inclination, W_k = single plane of weakness model results at the same confining pressure using cohesion and friction angles from experimental data for initialization.

model has also been used to describe anisotropic rock, but without much success as the data in Figure B.41 illustrate. In fact, there is no reason to expect any agreement because of the physical differences between anisotropic rock and intact (anisotropic) rock containing a plane of weakness (a joint) which has entirely different properties.

Discontinuously jointed rock mass strengths

When joints are discontinuous, then shear failure requires shearing along joints and through intact rock between joints. If the total shear resistance over the joint plane is T, then $T = T_r + T_j$ where the subscripts r and j signify rock and joint, respectively. The resisting shear forces are products of average shear stresses and areas. Hence,

$$\tau = \tau_r\left(\frac{A_r}{A}\right) + \tau_j\left(\frac{A_j}{A}\right)$$

where $A = LW$, $A_j = L_jW_j$, and $A_r = 1 - A$ as in Figure B.34. Similar analysis for shear across the joint plane leads to expressions for shear strength on planes normal to the 1-, 2-, and 3-directions. Thus, the principal shear strengths of the discontinuously jointed shear plane are

$$\tau_1 = (1 - R_1)\tau_r^1 + (R_1)\tau_j^1$$
$$\tau_2 = (1 - R_2)\tau_r^2 + (R_2)\tau_j^2$$
$$\tau_3 = (1 - R_3)\tau_r^3 + (R_3)\tau_j^3$$

where the superscripts indicate direction normal to the considered shear plane. For example, the 1-direction is normal to the 2-3 plane containing the joint areas. Also, $R_1 = L_jW_j$, $R_2 = hL_j$, $R_3 = hW_j$, h is joint thickness and S is spacing between joint planes, as in the discussion of jointed rock mass moduli.

If one invokes Mohr–Coulomb criterion for intact rock and joints, then

$$\tau = \sigma[(1 - R)\tan(\phi_r) + (R)\tan(\phi_j)] + [(1 - R)c_r(R)c_j]$$

which may be written in more compact form

$$\tau = \sigma\tan(\phi') + c'$$

where the primes indicate *effective* quantities, that is,

$$\tan(\phi') = (1 - R)\tan(\phi_r) + (R)\tan(\phi_j)$$
$$c' = (1 - R)c_r + (R)c_j$$

Thus, under the assumption of Mohr–Coulomb criteria, possibly different in different directions, shear strengths for a discontinuously jointed test volume are given by:

$$\tau_1 = \sigma\ \tan({\phi'}_1) + {c'}_1$$
$$\tau_2 = \sigma\ \tan({\phi'}_2) + {c'}_2$$
$$\tau_3 = \sigma\ \tan({\phi'}_3) + {c'}_3$$

where the subscripts indicate direction normal to the considered plane of shear and the effective properties are

$$\tan({\phi'}_1) = (1 - R_1)\tan(\phi_r^1) + (R_1)\tan(\phi_j^1)$$
$$\tan({\phi'}_2) = (1 - R_2)\tan(\phi_r^2) + (R_2)\tan(\phi_j^2)$$

$$\tan(\phi'_3) = (1 - R_3)\tan(\phi^3_\mathrm{r}) + (R_3)\tan(\phi^3_\mathrm{J})$$
$$c'_1 = (1 - R_1)c^1_\mathrm{r} + (R_1)c^1_\mathrm{j}$$
$$c'_2 = (1 - R_2)c^2_\mathrm{r} + (R_2)c^2_\mathrm{j}$$
$$c'_3 = (1 - R_3)c^3_\mathrm{r} + (R_3)c^3_\mathrm{j}$$

and, for example, c^1_r is intact rock cohesion in a shear plane with a 1-direction normal.

According to a model proposed by Terzaghi (1962), shear strength of a (jointed) rock mass is

$$\tau = \sigma\tan(\phi) + c_\mathrm{r}\left(\frac{A_\mathrm{r}}{A}\right)$$

where τ , σ , ϕ, and c_r are rock mass shear strength, normal stress acting on a potential shear failure surface, friction angle, and intact rock cohesion. This model is actually a special case of a more general model. The Terzaghi model follows from assuming equal friction angles ($\phi_\mathrm{r} = \phi_\mathrm{j} = \phi'$) and negligible joint cohesion ($c_\mathrm{j} = 0$). The last term in the Terzaghi model is $c'_1 = (1 - R_1 c^1_\mathrm{r})$

The Mohr–Coulomb form allows for calculation of unconfined compressive and tensile strengths in terms of effective cohesion and angle of internal friction. Thus,

$$\left.\begin{matrix} C'_\mathrm{o} \\ T'_\mathrm{o} \end{matrix}\right\} = \frac{2c'\cos(\phi')}{1 \mp \sin(\phi')}$$

where the primes indicate composite rock mass properties. Directional features, if present in joints and intact rock could be incorporated into these formulas which apply strictly to isotropic materials.

The general model cannot be expected to hold for relatively large displacements on a composite rock mass shear failure plane because of progressive diminution of intact rock bridges between adjacent joint areas. Progressive shearing of rock bridges would also be accompanied by destruction of joint cohesion. A peak resistance would be rapidly exceeded and followed by a sharp reduction to residual strength which would be essentially frictional. In essence, the rock mass strain softens. The residual friction angle could be different from either of the initial friction angles because of asperity shear and gouge formation. However, design on the basis of peak strength is reasonable in view of the fact that residual strength is only approached after peak strength is exceeded.

There is no direct experimental confirmation of either jointed rock mass shear strength model because of the impracticality of testing the very large samples that would be required, samples that could be many feet or meters in linear dimensions. Indeed, this situation is precisely why a model or theory is needed. However, there are some laboratory test data on low strength, rock-like mixtures of sand and cement that offer some guidance toward the potential utility of such models. Figure B.42 shows data from Lajtai (1969) who tested a number of jointed plaster blocks in direct shear. Joints in all tests were cohesionless and the ratio of intact rock to total area (persistence) was 0.50. Figure B.42(a) shows direct shear results on a single joint and indicates clearly the lack of cohesion and a friction angle of

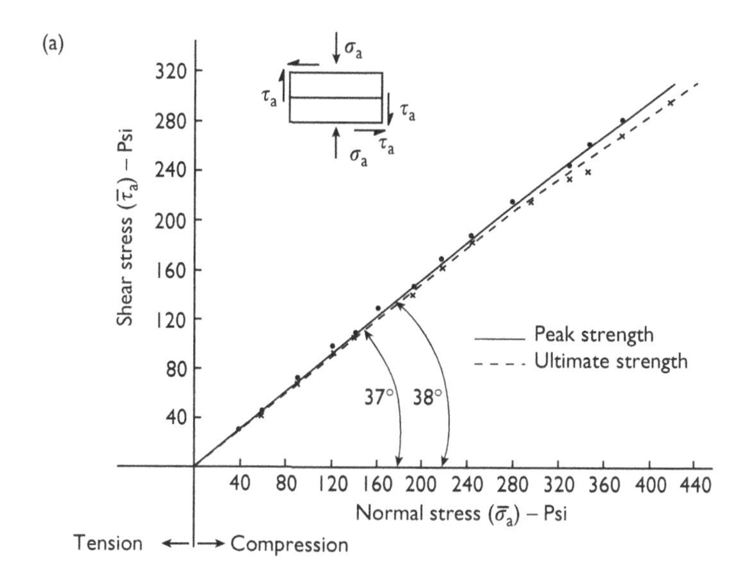

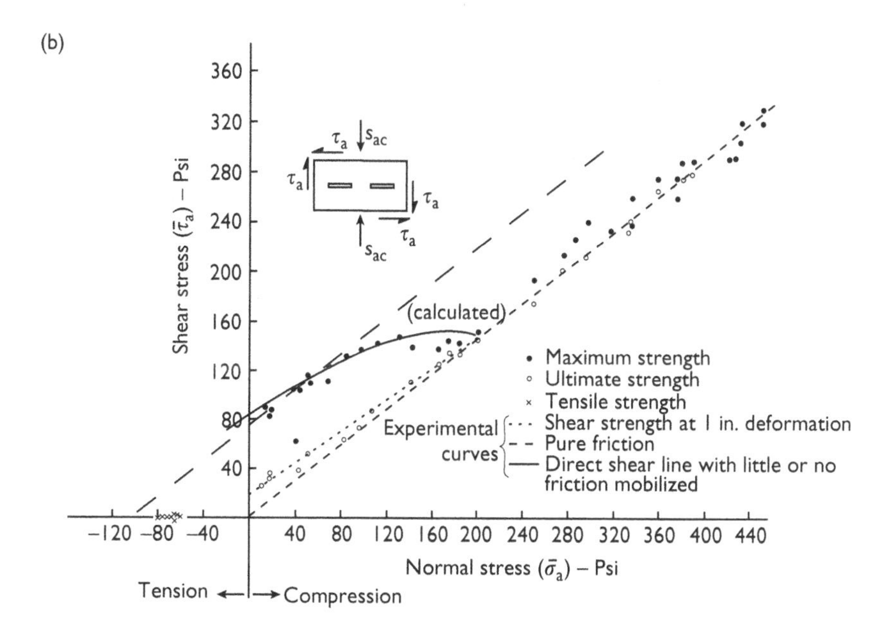

Figure B.42 Direct shear test results on continuous and discontinuous cast joints (after Lajtai, 1969). (a) Continuous joint, (b) open, discontinuous joints, (c) interlocked joints, and (d) shear stress–displacement curves.

(c)

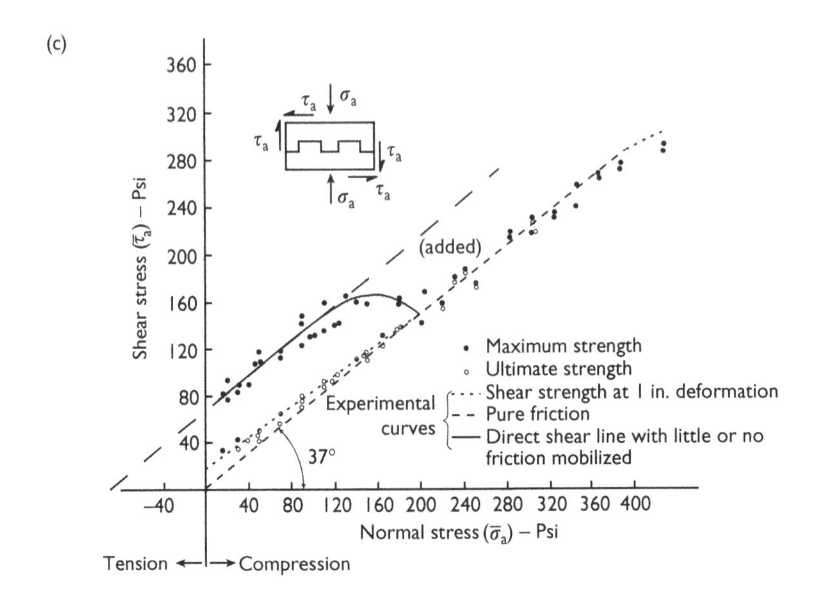

(d)

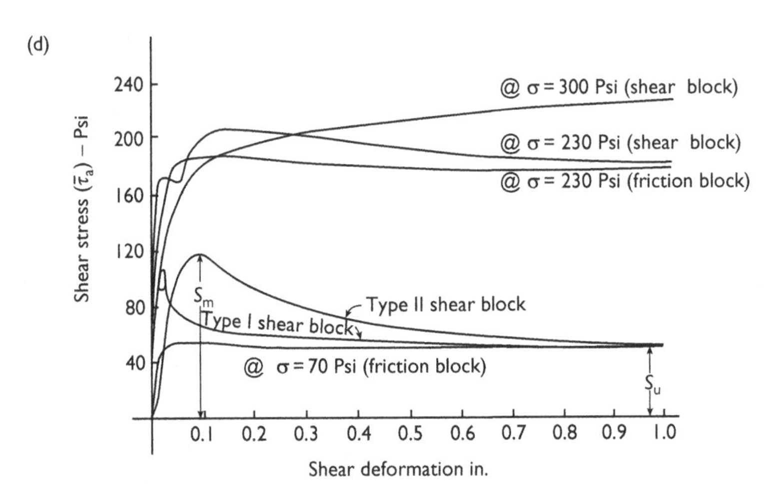

Figure B.42 Continued.

37–38° with little difference between peak and residual angles, as one might expect. Additional tests show the plaster blocks have an unconfined compressive strength of about 600 psi and tensile strength between 120 and 160 psi. These data correspond to intact block cohesion of about 145 psi and an angle of internal friction of 38°, according to the Mohr–Coulomb strength criterion.

Figure B.42(b) shows results of a block with open joints. Shear strength is initially high and falls on the added peak strength line but undergoes a transition to the residual strength line that passes through the origin. The results in Figure B.42(c) on interlocked blocks shows similar behavior. Shear stress as a function of shear displacement is shown in Figure B.42(d) and reveals peak residual shear strength behavior at relatively low normal stress in both types of jointed test blocks. This behavior is suppressed at higher normal stress. Because the joints are cohesionless and persistence is 0.5, the effective cohesion is simply one-half the intact rock cohesion, about 78 psi. Inspection of Figures B.42(b) and (c) shows shear axes intercepts of about the same amount. The friction angles of joint and intact "rock" are nearly equal. These data are consistent with the general model of jointed rock mass shear strength and therefore also with the Terzaghi model *at normal (compressive) stress less than tensile strength of the intact "rock"* (about 145 psi).

At normal stresses higher than tensile strength, shear strength of the jointed blocks is entirely frictional which suggests that intact rock cohesion is destroyed prior to reaching peak strength; peak strength is diminished to residual strength. In effect, persistence is reduced to zero and the shear plane becomes entirely jointed and, in this case, entirely frictional. This result is not anticipated in supposing that intact rock bridges between joints fail in shear.

Appendix C

Rock mass classification schemes for engineering

Rock mass classification schemes are used mainly to characterize a rock mass according to qualities that influence support requirements in tunnels. The Terzaghi scheme in Table 4.2 was an early effort to describe a rock mass with the objective of determining support requirements. In this scheme, a rock arch of height H_p above the tunnel top comes to bear on fixed steel sets inside the tunnel. A subsequent effort at improvement by Deere (1963) eventually led to the introduction of Rock Quality Designation, RQD (Deere, *et al.*, 1967), a modified diamond drill core recovery measurement. A more definitive rock mass classification scheme was introduced as a Rock Structure Rating (RSR) which was followed by Rock Mass Rating (RMR) and Quality (Q) classification schemes. RMR and Q are statistically correlated as one might expect from physical considerations. The original Terzaghi scheme was also modified during the development of other classification schemes. All these schemes provide estimates of rock arch height H_p and may be compared. A different approach to rock mass classification is found in a Geological Strength Index (GSI) that is used to estimate strength and deformation modulus of excavation scale rock masses. These estimates may then be used as input for numerical analyses of excavation safety and possible support requirements.

C.1 Rock quality designation

Rock quality designation is determined by only counting core pieces longer than 4 in. (100 mm) in a run, say, of 5 ft or 2 m of core (Deere, *et al.*, 1967). By comparison, ordinary drill core recovery is the percentage of material brought to the surface in a core barrel and placed in a core box. A 100% core recovery implies the core volume is equal to the volume of the core barrel. In RQD, the core is assumed to be NX-size, that is, 2.12 in. in diameter. The hole is 3.0 in. in diameter. Diamond drill core today would almost certainly be NQ-size (wireline series) rather than NX. Wireline apparatus allows retrieval of the core barrel without the need to break down the drill pipe; the core is pulled to the surface inside the drill pipe. NQ core diameter is about 1.87 in. The same length to diameter ratio used for RQD based on NX core should be preserved when other core sizes are used, whether smaller, when deep exploration holes are needed, or larger, when soft rock such as coal is cored. RQD is a number that is qualitatively associated with five divisions of rock mass quality as indicated in Table C.1.

C.2 Terzaghi modified scheme

The Terzaghi scheme was modified by Rose (1982) using RQD. In the modified scheme, a reduction was made mainly because the effect of water was much less than supposed in

Table C.1 Rock mass quality and RQD (after Bieniawski, 1989).

Rock mass quality	*RQD %*
I. Excellent	91–100
II. Good	76–90
III. Fair	51–75
IV. Poor	26–50
V. Very Poor	<25

Table C.2 Revision of the Terzaghi Classification Scheme for H_p (after Rose, 1982)

Rock condition	*RQD (%)*	*Rock load H_P (ft)*	*Original[a] H_P (ft)*
1 Hard and intact	95–100	Nil	No change
2 Hard stratified, schistose	90–99	0.0–0.5B	No change
3 Massive, some jointing	85–95	0.0–0.25B	No change
4 Moderately blocky, seamy	75–85	0.25B–0.20(B + H_t)	0.25B to 0.35(B + H_t)
5 Very blocky and seamy	30–75	(0.20–0.60)(B + H_t)	(0.35 to 1.10)(B + H_t)
6 Completely crushed	3–30	(0.60–1.10)(B + H_t)	(1.10)(B + H_t)
6a Sand and gravel	0–3	(1.10–1.40)(B + H_t)	—
7 Squeezing rock, moderate depth	NA	(1.10–2.10)(B + H_t)	No change
8 Squeezing rock, great depth	NA	(2.10–4.50)(B + H_t)	No change
9 Swelling rock	NA	Up to 250 ft independent of B, H_t	No change

Note
a Original is the same value as in Table 4.2 (after Terzaghi).

the Terzaghi scheme which suggested, in effect, doubling the dry rock load in wet ground. Observations indicated that water had little effect on rock load H_p. The modification of the Terzaghi scheme by Rose is shown in Table C.2. Rock conditions 7, 8, and 9 in Table C.2 (and in Table 4.2) are contrary to the concept of a rock arch at the top of a tunnel the weight of which comes to bear on steel sets. Instead, stress drives squeezing ground all around the tunnel and encloses the tunnel in a yielding zone that surrounds the excavation. Invert braces or full-circle ring sets are likely necessary. Squeezing and swelling rock in these categories are also not amenable to diamond drill coring, so RQD is not applicable either. As in the original scheme, in dry ground only one-half the values of H_p should be used.

C.3 RSR, RMR, and Q

Two popular rock mass classification schemes that are in current use are the RMR scheme and the Quality (Q) scheme. These schemes and several modifications for purposes other than tunnel support load estimation are described in detail by Bieniawaski (1989). The ABC's of the RMR scheme quite literally began with a RSR scheme that was developed through a US Bureau of Mines contract (Wickham and Tidemann, 1974; Wickham et al., 1974). In fact, RSR = A + B + C where A ranges from 6 to 30% and characterizes intact rock type and hardness and geologic structure; B ranges from 7 to 45% and depends on joint attitude and spacing, and C depends on water flow and ranges to 25%. Thus, RSR ranges up to 100%. The higher is the rating the better is the rock mass condition and the less support needed. The most important factor in RSR according to the allocated percentage is jointing. These same factors, intact rock, joints, and water, are also the main factors of RMR and Q as indeed they are in the original Terzaghi classification scheme. RSR is an improvement over the Terzaghi scheme because of a more focused identification of rock mass features that affect support load requirements.

RMR also ranges to 100%. In RMR, A is essentially uniaxial compressive strength with 0–15% allocation; B is allocated up to 70% and is a union of joint spacing, orientation, condition, and RQD, while C is related to groundwater and is rated to 15%. Joints dominate RMR in view of the up to 70% influence that features associated with joints exert on the rock mass rating. Although there are a number of adjustments in the scheme that take details of the rock mass into account, in the end RMR = A + B + C.

In Q, A is a composite of rock strength and stress; B is composed of RQD, number of joint sets, joint roughness, and degree of alteration, while C is related to groundwater. The ABC's of this system are multiplied to obtain Q which ranges over many orders of magnitude, for example, from 10^{-3} to 10^{+3} .Specifically,

$$Q = \left(\frac{1}{\mathrm{SRF}}\right)\left(\frac{\mathrm{RQD}}{J_n}\frac{J_r}{J_a}\right)(J_w) = (A)(B)(C) \tag{C.1}$$

where RQD = rock quality designation, J_n = joint set number, J_r = joint roughness number, J_a = joint alteration number, SRF = stress reduction factor.

C.4 Comparisons of H_p estimates

Not too surprisingly, RMR and Q values are statistically correlated. When applied to the same tunnel site, the values obtained should lead to similar tunnel support recommendations. One correlation (Bieniawski 1989) is

$$\mathrm{RMR} = 9\ln Q + 454, \quad Q = \exp[(RMR - 44)/9] \tag{C.2}$$

For example, when RMR = 98%, $Q = \exp[6] = 403$, "very good" rock. When RMR = 44%, Q = 1, only "fair" rock.

Another important relationship associated with RMR is (Bieniawski 1989)

$$H_p = (1 - RMR)B \tag{C.3}$$

where RMR is expressed in decimal form rather than as a percentage. This relationship shows a linear dependency of H_P on RMR at fixed tunnel width B and is plotted in Figure C.1 assuming that $H_t = B$, a reasonable assumption. The association of RQD with H_P in Table C.2 also allows plotting H_P as a function of RQD. There are two such plots possible, the original Terzaghi H_P and the modified H_P by Rose. Both are shown in Table C.2 and plotted in Figure C.1. Partitioning of Figure C.1 is done according to rock quality for RMR and RQD labels. The Rose reduction of H_P from the Terzaghi values is clear for moderate RQD values. Also evident is the independence of H_P on tunnel dimensions for "very poor" rock in the Terzaghi scheme. Values of H_P from RMR are the lowest of the three schemes.

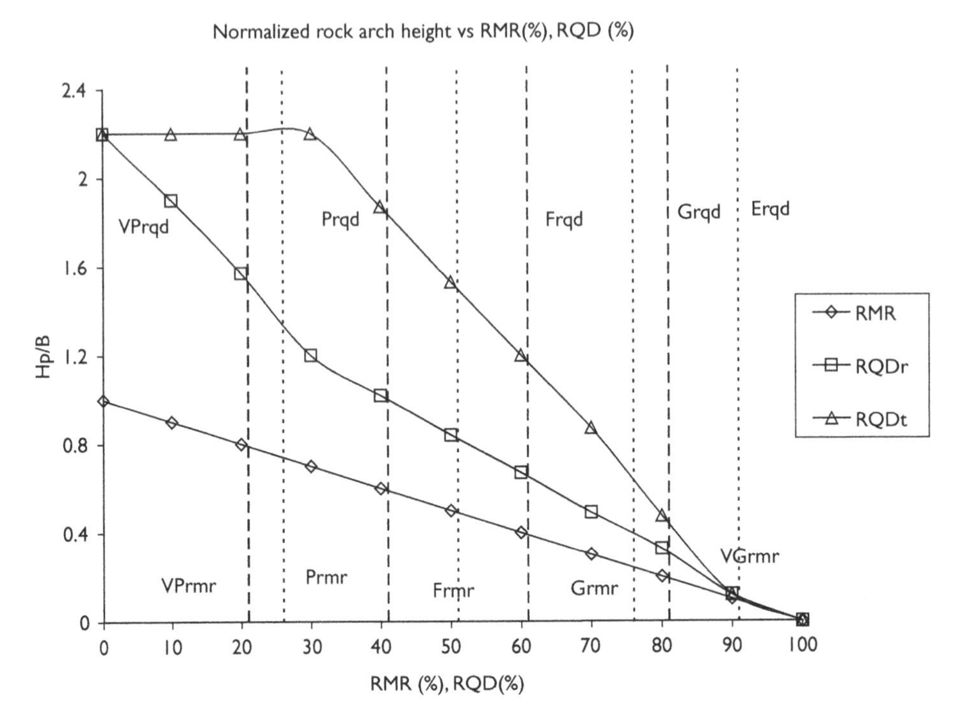

Figure C.1 Normalized rock arch height as a function of RMR and RQD when $B = H_t$. RQDr = Rose values, RQDt =Terzaghi values. E = excellent, VG = very good, G = good, F = fair, P = poor, VP = very poor. - - - - = partition for RMR rock quality, ... = partition for RQD rock quality.

The disparate values of H_P are considerable and raise a question concerning reliability of any of the three estimates. Detailed support estimation using fixed steel sets requires H_P, so a choice must be made. However, in the general case where alternative support systems are considered, for example, rock bolts with shotcrete, then H_P is not important. Indeed, the main use of rock mass classification schemes in tunneling is to decide what type of support may be needed. Thus, the utility of classification schemes is in classifying the rock mass according to rock quality (..."good", " fair"...) as determined from the ABC's (intact rock, joints, water) and with quality thus specified, the anticipated support system. Once support systems are standardized for a particular tunnel site or mine, then identification of rock type (e.g., I, II, etc., or Excellent, Very Good, etc.) determines the support plan as one from the list of standard support designs. In this regard, mapping of rock mass quality at a mine can be a valuable aid to safety and stability.

C.5 GSI

GSI stands for Geological Strength Index, an empirical measurement of rock quality that was promulgated for use in estimating strength and deformation modulus of very low quality rock (Hoek and Brown, 1997). GSI is intimately linked to the popular Hoek-Brown (HB) failure criterion. In this context, "deformation modulus" is synonymous with Young's modulus. GSI is particularly favored in Canada and is growing in popularity elsewhere. GSI differs from RSR, RMR and Q which were derived from tunneling observations and were intended to estimate tunnel support requirements. However, the correlation between RMR and Q and a formula for estimating deformation modulus from RMR links the three indices. Indeed, the original GSI formulation related GSI to RMR by the simple equality GSI=RMR_{76} where the subscript stands for 1976 version of RMR.

All rock mass classification schemes tend to evolve over time. GSI and companion HB are no exceptions (Hoek and others, 2002). A generalized HB has the form

$$\sigma_1 = \sigma_3 + (C_0)[m\sigma_3/C_o + s]^a \tag{C.4}$$

where σ_1, σ_3, m, C_0, s, and a are major and minor principal (effective) stresses at failure, a material constant, unconfined compressive strength of intact rock, a scale factor, and an exponent, respectively. Here, $m = C_o/T_0 - T_o/C_o \cong C_o/T_0$ for intact rock; T_o is tensile strength. Generalized HB is independent of the intermediate principal stress and does not lead to a symmetric function in the normal stress shear stress plane and therefore requires qualification or restriction, as is the case for the original form. (Unfortunately, such restriction is seemingly ignored in almost all usage.) In case of intact rock tested at the laboratory scale, $s = 1$, and in the original formulation, $a = 0.5$. In case of a rock mass, unconfined compressive strength $C_{om} = C_0\ [s]^a$. The scale factor s ranges on the approximate interval (0,1] and a ranges between about 1/3 and 1/2 , so compressive strength is "scaled" by a factor also on the interval (0,1]. However, a and s are not independent because both are functions of GSI. If they were independent, different combinations could lead to the same rock mass compressive strength. For example, the combination $a = 1/2$ and $s = 1/4$ and the combination $a = 1/3$ and $s = 1/8$ give the same rock unconfined compressive strength as just 1/2 of the intact unconfined compressive strength.

GSI is used to compute the scale factor s, the exponent a, the constant m, and a deformation modulus E according to empirical formulas based on physical reasoning and many excavation scale observations. A relatively recent complication enters these formulas in the form of a damage factor D that is intended to account for blast damage at excavation walls and ranges on the interval [0,1]. Values of GSI and D are essentially table look-ups based of qualitative descriptions of rock mass geology and structure and are intended for application to isotropic rock only. Joint sets and other features such as schistosity that impart directional character to rock masses rule out application of GSI as remarked in a cautionary note (V. Marinos, P. Marinos, and E. Hoek, 2005). Once values of GSI and D are decided upon, corresponding values of the various parameters in HB may be evaluated and thus rock mass strength. Another formula is used for deformation modulus E, that is,

$$E(GPa) = \left(1 - \frac{D}{2}\right)\left(\sqrt{C_0(MPa)/100}\right)\left\{10^{[(GSI-10)/40]}\right\} \tag{C.5}$$

which is not based on mechanics but rather is derived from empirical curve fitting efforts.

C.6 Comment

An often used alternative approach to empirical estimation of rock mass properties is simple scaling, that is, multiplying intact rock strengths and moduli by "scale factors" on the interval (0,1]. Scale factors are generally attributed to "size effects" in passing from the laboratory scale of testing to the scale of excavation (Heuze, 1980). The transition between scales is logically hazardous in consideration of the scale of testing, joint spacing and the scale of excavation. In any case, the challenge is to select applicable scale factors for the problem at hand. In many instances, back-analysis that compares computed with measured displacements induced by excavation leads to appropriate scale factors. Back-analysis is essentially an elaborate curve fitting scheme but one based on mechanics and one that falls within the mathematical discipline of the *inverse problem*. Back-analysis for scale factors may be considered the state of engineering practice (2011).

Another alternative to empirical rock mass classification and to back analysis for estimating rock mass properties are *equivalent properties models* of jointed rock masses. Equivalent properties models based on fundamental principles represent a return to basics from empirical schemes. To be sure, there are empirical schemes that are also labeled as equivalent properties models. The basic approach enables computation of composite jointed rock mass properties in advance of excavation from properties of joint sets and intact rock between joints using fundamental principles (e.g., Pariseau, 1999, Mars Ivars *et al*, 2011).

Appendix D

Some useful formulas

A brief outline of several formulas that are helpful in analysis of stress, strain, and stress–strain relationships (Hooke's law) in elasticity is given in this appendix. Derivation details can be found in mechanics of materials texts and books about elasticity and continuum mechanics. A few are cited in Appendix A.

D.1 Stress

Rotation of axes from (x, y) to (a, b) about z, c:

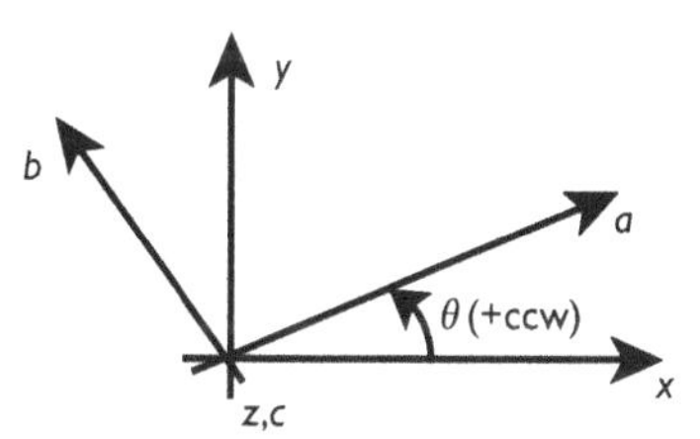

Sketch for rotation about the z-axis

$$
\begin{aligned}
\sigma_{aa} &= (1/2)(\sigma_{xx} + \sigma_{yy}) + (1/2)(\sigma_{xx} - \sigma_{yy})\cos(2\theta) + \tau_{xy}\sin(2\theta) \\
\sigma_{bb} &= (1/2)(\sigma_{xx} + \sigma_{yy}) - (1/2)(\sigma_{xx} - \sigma_{yy})\cos(2\theta) - \tau_{xy}\sin(2\theta) \\
\tau_{ab} &= \tau_{ba} = (1/2)(\sigma_{xx} - \sigma_{yy})\sin(2\theta) + \tau_{xy}\cos(2\theta) \\
\sigma_{cc} &= \sigma_{zz} \\
\tau_{ca} &= \tau_{xz}\cos(\theta) + \tau_{yz}\sin(\theta) \\
\tau_{cb} &= -\tau_{xz}\sin(\theta) + \tau_{yz}\cos(\theta)
\end{aligned}
\tag{D.1}
$$

The first 3 of (D.1) is the two-dimensional (xy, ab) or plane part of the rotation; the second 3 of (D.1) is the three-dimensional part (z,c).

General rotation in matrix, that is, table form:

Table of direction angles.

New axes	Old axes		
	x	y	z
a	ax	ay	az
b	bx	by	bz
c	cx	cy	cz

The normal direction ($c_x\, c_y\, c_z$) is sometimes written as ($n_x\, n_y\, n_z$), "n" for normal to a plane of interest.

Rotation matrix R is given by

$$[R] = \begin{bmatrix} \cos(ax) & \cos(ay) & \cos(az) \\ \cos(bx) & \cos(by) & \cos(bz) \\ \cos(cx) & \cos(cy) & \cos(cz) \end{bmatrix} = \begin{bmatrix} l_1 & m_1 & n_1 \\ l_2 & m_2 & n_2 \\ l_3 & m_3 & n_3 \end{bmatrix} \tag{D.2}$$

where ($lm\, n$) are direction cosines of the *abc* (or 123) directions.

The known stress state is given in the array

$$[\sigma(xyz)] = \begin{bmatrix} \sigma_{xx} & \tau_{xy} & \tau_{xz} \\ \tau_{yx} & \sigma_{yy} & \tau_{yz} \\ \tau_{zx} & \tau_{zy} & \sigma_{zz} \end{bmatrix} \tag{D.3}$$

The same stress state referred to the rotated system is

$$[\sigma(abc)] = [R][\sigma(xyz)][R]^t \tag{D.4}$$

where the superscript t means transpose.

Rotation about the z-axis counterclockwise through an angle θ in matrix form is done using

$$[R] = \begin{bmatrix} \cos(\theta) & \sin(\theta) & 0 \\ -\sin(\theta) & \cos(\theta) & 0 \\ 0 & 0 & 1 \end{bmatrix} \tag{D.5}$$

Rotation of axes from compass coordinates (xyz, x = east, y = north, z = up) to joint plane coordinates (abc, a = down dip, b = strike = dip direction less 90°, c = joint plane normal) is

$$[R] = \begin{bmatrix} \cos(\delta)\sin(\alpha) & \cos(\delta)\cos(\alpha) & -\sin(\delta) \\ -\cos(\alpha) & \sin(\alpha) & 0 \\ \sin(\delta)\sin(\alpha) & \sin(\delta)\cos(\alpha) & \cos(\delta) \end{bmatrix} \tag{D.6}$$

where α is the azimuth of the dip direction measured positive clockwise from north (0–360°), as is customary, and δ is the dip, always positive and down (0–90°). If the azimuth is measured counterclockwise from north, then α is negative. This result may be obtained by two successive rotations, first counterclockwise about the z-axis by the magnitude α' (90° less the dip direction α) to the dip direction, say, from (xyz) to ($x'y'z'$), then counterclockwise through the angle about the y' axis to the dip or a-axis, from ($x'y'z'$) to (abc). Thus,

$$[R] = [Ry][Rz] = \begin{bmatrix} \cos(\delta) & 0 & -\sin(\delta) \\ 0 & 1 & 0 \\ \sin(\delta) & 0 & \cos(\delta) \end{bmatrix} \begin{bmatrix} \cos(\alpha') & \sin(\alpha') & 0 \\ -\sin(\alpha') & \cos(\alpha') & 0 \\ 0 & 0 & 1 \end{bmatrix} = [R]$$

where the last is obtained after the substitutions $\cos(\alpha') = \sin(\alpha)$, $\sin(\alpha') = \cos(\alpha)$.

Normal and shear stress on a plane

Calculation of normal and shear stress on a plane, perhaps a fault or joint plane, is a special case of axes rotation because only part of the full three-dimensional rotation, say, from compass coordinates (xyz) to joint plane coordinates (abc) is needed. With c the direction normal to the joint plane, the tractions acting on the joint plane referred to xyz are

$$\begin{aligned} T_x &= \sigma_{xx}\cos(cx) + \tau_{xy}\cos(cy) + \tau_{xz}\cos(cz) \\ T_y &= \tau_{yx}\cos(cx) + \sigma_{yy}\cos(cy) + \tau_{yz}\cos(cz) \\ T_z &= \tau_{zx}\cos(cx) + \tau_{zy}\cos(cy) + \sigma_{zz}\cos(cz) \end{aligned} \tag{D.7}$$

where the direction cosines may be obtained from the last row of the rotation (D.6). The magnitude of the traction acting on the considered plane is simply

$$T = \sqrt{T_x^2 + T_y^2 + T_z^2}$$

while the normal component which is the normal stress acting on the considered plane is

$$N = T_x\cos(cx) + T_y\cos(cy) + T_z\cos(cz) = \sigma_{cc} = \sigma_p \tag{D.8}$$

The magnitude of the shear component of traction, the shear stress, on the considered plane is

$$S = \sqrt{T^2 - N^2} = \tau_p \tag{D.9}$$

The direction of S is shown in the sketch where

$$\tan(\theta) = \frac{\tau_{ca}}{\tau_{cb}} \tag{D.10}$$

that requires the calculation of the c-direction shear stresses. Because S can also be calculated by

$$S = \tau_p = \sqrt{\tau_{ca}^2 + \tau_{cb}^2}$$

One needs only three stresses or the third row of the final stress rotation array (D.4) in the abc system.

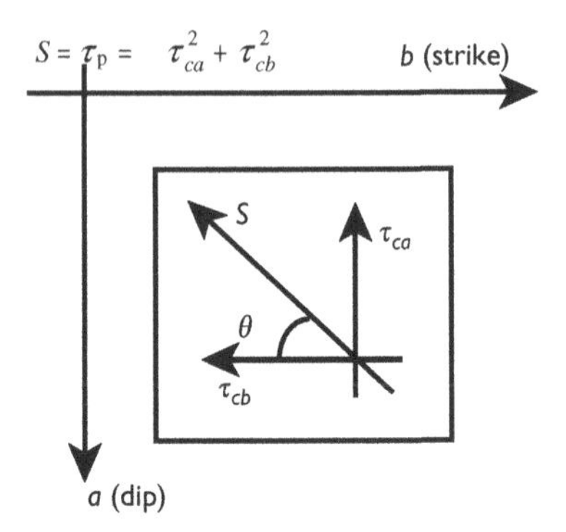

Sketch showing resultant shear stress on a plane

Principal (normal) stresses

Generally, there are three mutually orthogonal directions 123 that define directions of *principal stress*. These directions are normal to planes that are entirely free of shear stress. Associated normal stresses are the major, intermediate, and minor principal stresses, $(\sigma_1, \sigma_2, \sigma_3)$ where $\sigma_1 > \sigma_2 > \sigma_3$. The ordering is algebraic, so when compression is positive σ_1 is the greatest compression at the considered point (or the least tension when all principal stresses are tensile).

The principal stresses and directions are not always unique. For example, in case of hydrostatic stress, the principal stresses are equal and every direction is a principal direction.

Any plane free of shear stress is a principal plane with a normal that defines a principal direction. The other two principal directions must lie in the shear free plane. The stress acting normal to the considered plane is one of the three principal stresses.

When the z-axis is parallel to the intermediate principal stress direction, as shown in the sketch,

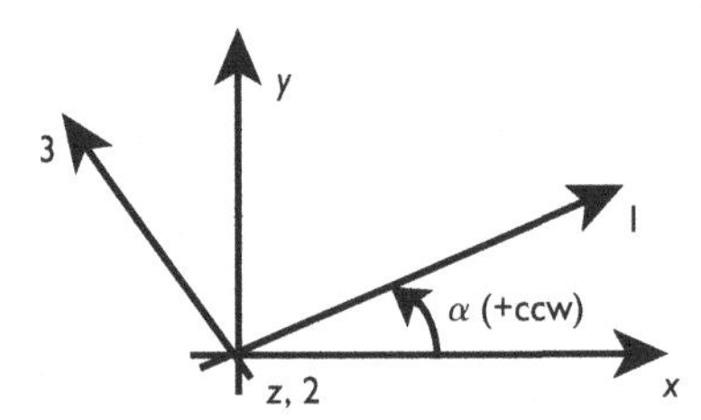

Sketch for principal axes orientation

then the principal directions seen in the xy-plane may be computed from the formulas

$$
\begin{aligned}
\tan(2\alpha) &= \frac{\tau_{xy}}{(1/2)(\sigma_{xx} - \sigma_{yy})} \\
R &= \sqrt{[(\sigma_{xx} - \sigma_{yy})/2]^2 + (\tau_{xy})^2} \\
\sin(2\alpha) &= \frac{\tau_{xy}}{R} \\
\cos(2\alpha) &= \frac{(\sigma_{xx} - \sigma_{yy})/2}{R}
\end{aligned}
\tag{D.11}
$$

There are two solutions to the first of (D.11) that are 90° apart. The angle $\alpha = \alpha_1$ is measured positive counterclockwise from either the x- or y-axis (to the direction of σ_1), whichever is associated with the largest normal stress (σ_{xx} or σ_{yy}). This rule may be used in conjunction with a hand calculator that gives positive or negative solutions to the *atan* function between −90° and +90°.

Magnitudes of the principal stresses seen in the xy-plane are

$$\sigma_1 = \frac{\sigma_{xx} + \sigma_{yy}}{2} + R$$
$$\sigma_3 = \frac{\sigma_{xx} + \sigma_{yy}}{2} - R \tag{D.12}$$

and by assumption, $\sigma_2 = \sigma z$. An interesting observation of (D.12) is that the sum of the normal stresses in the considered plane is constant, that is, invariant with respect to the orientation of the reference axes (xy, ab, 13). Thus by addition of (D.12),

$$\sigma_{xx} + \sigma_{yy} = \sigma_1 + \sigma_3 = \sigma_{aa} + \sigma_{bb} \tag{D.13}$$

There is always a rotation about an axis that causes the shear stress in the plane of interest to vanish. However, shear stresses normal to the considered plane may still be present and the z-direction may not coincide with the direction of the intermediate principal stress. In this case, the principal stresses given by (D.12) with directions from (D.11) are *secondary principal stresses*.

Determination of true principal stresses in the general three-dimensional case requires solution of a cubic equation and consideration of several special cases where directions may not be unique. The task is algebraically straightforward but lengthy and is thus better left to a computer program. In essence, bringing the real, symmetric 3 × 3 array (D.3) of stress to diagonal form solves the problem of determining principal stress magnitudes (eigenvalues) and directions (eigenvectors).

Principal shear stresses

An inquiry concerning whether shear stress at a point reaches a maximum as the reference axes are rotated leads to the six principal shear stresses:

$$\pm\frac{\sigma_1 - \sigma_3}{2}, \quad \pm\frac{\sigma_1 - \sigma_2}{2}, \quad \pm\frac{\sigma_2 - \sigma_3}{2} \tag{D.14}$$

which are invariant with respect to orientation of the reference axes. The most interesting principal shear stresses are the maximum and minimum shear stresses, the first of (D.14), which are equal in magnitude but opposite in sign. Thus,

$$\tau(\max) = \frac{\sigma_1 - \sigma_3}{2}, \quad \tau(\min) = -\frac{\sigma_2 - \sigma_3}{2} \tag{D.15}$$

The directions of the principal shears bisect the directions of principal stress in the plane considered (13, 12, 23).

Mohr's circle

The formulas for referring the state of stress at a point to a set of axis rotated in some arbitrary fashion from the given axes have a graphical representation in the form of Mohr circles. Consider the two-dimensional part of (D.1) and the accompanying sketch.

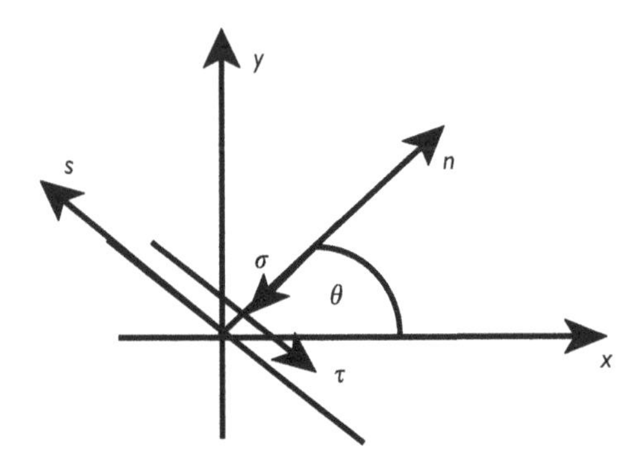

Sketch for Mohr's circle

The normal and shear stress associated with the rotated (*ns*) axes are

$$\sigma = \frac{\sigma_{xx} + \sigma_{yy}}{2} + \frac{\sigma_{xx} - \sigma_{yy}}{2}\cos(2\theta) + \tau_{xy}\sin(2\theta)$$

$$\tau = -\frac{\sigma_{xx} - \sigma_{yy}}{2}\sin(2\theta) + \tau_{xy}\cos(2\theta)$$

where compression is considered positive. After rearrangement

$$\sigma - \frac{\sigma_{xx} + \sigma_{yy}}{2} = +\frac{\sigma_{xx} - \sigma_{yy}}{2}\cos(2\theta) + \tau_{xy}\sin(2\theta)$$

$$\tau = \frac{\sigma_{xx} - \sigma_{yy}}{2}\sin(2\theta) + \tau_{xy}\cos(2\theta)$$

that after squaring and adding gives

$$\left(\sigma - \frac{\sigma_{xx} + \sigma_{yy}}{2}\right)^2 + \tau^2 = \left(\frac{\sigma_{xx} - \sigma_{yy}}{2}\right)^2 + (\tau_{xy})^2$$

$$\therefore$$

$$(\sigma - \sigma_{\mathrm{m}})^2 + \tau^2 = \tau_{\mathrm{m}}^2 \tag{D.16abc}$$

$$\sigma_{\mathrm{m}} = \frac{\sigma_{xx} + \sigma_{yy}}{2} = \frac{\sigma_1 + \sigma_3}{2}, \quad \tau_{\mathrm{m}} = \sqrt{\left(\frac{\sigma_{xx} - \sigma_{yy}}{2}\right)^2 + (\tau_{xy})^2} = \left(\frac{\sigma_1 - \sigma_3}{2}\right)$$

where the invariance of the mean normal stress (σ_{m})and maximum shear (τ_{m}) stress in the considered plane is indicated.

Equations (D.16ab) define a circle in a normal stress (*x*-axis)- shear stress (*y*-axis) plane with a center along the normal stress axis at the mean normal stress. The circle radius is just the maximum shear stress as shown in the sketch.

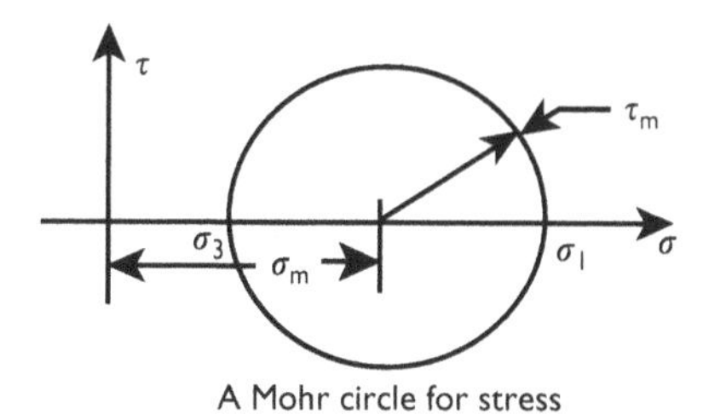

A Mohr circle for stress

D.2 Strain

The entire discussion of principal values and so on for stress also applies to strain. The mathematics is the same. Of course, the physical meaning is quite different. For example, the maximum shear strain is $(\varepsilon_1 - \varepsilon_3)/2$. There is a potential for ambiguity in discussion of strain because of the presence of two shear strains, a "mathematical" or tensorial shear strain ε_{xy} and an "engineering" shear strain γ_{xy}. The two are related by $\gamma_{xy} = 2\varepsilon_{xy}$, and similarly for the other two shear strains. Thus, $\gamma_{max} = 2(\varepsilon_1 - \varepsilon_3)/2 = (\varepsilon_1 - \varepsilon_3)$. In Hooke's law, for example, $\tau xy = G\gamma xy = 2G\varepsilon xy$.

Strain rosettes

Strain rosettes allow for the determination of the state of strain at a measurement site and with the aid of Hooke's law the state of stress. Three strain gauges are usually required in a rosette. Each gauge responds to normal strain as shown in the sketch.

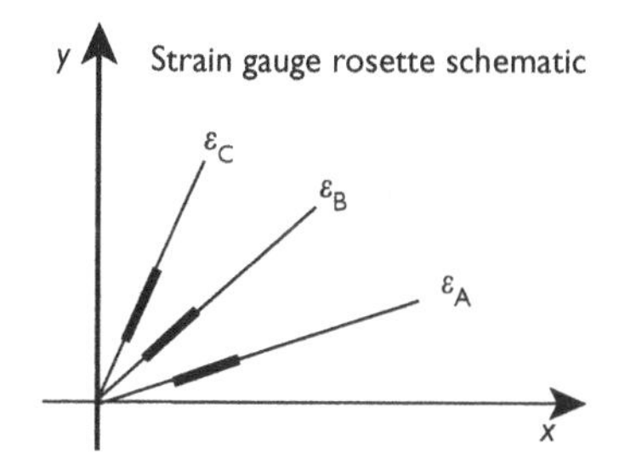

According to the equations of transformation of reference axes

$$\begin{aligned}
\varepsilon_A &= \frac{\varepsilon_{xx} + \varepsilon_{yy}}{2} + \frac{\varepsilon_{xx} - \varepsilon_{yy}}{2}\cos(2\theta_A) + \varepsilon_{xy}\sin(2\theta_A) \\
\varepsilon_B &= \frac{\varepsilon_{xx} + \varepsilon_{yy}}{2} + \frac{\varepsilon_{xx} - \varepsilon_{yy}}{2}\cos(2\theta_B) + \varepsilon_{xy}\sin(2\theta_B) \\
\varepsilon_C &= \frac{\varepsilon_{xx} + \varepsilon_{yy}}{2} + \frac{\varepsilon_{xx} - \varepsilon_{yy}}{2}\cos(2\theta_C) + \varepsilon_{xy}\sin(2\theta_C)
\end{aligned} \tag{D.17}$$

where the angles are measured from the x-axis to the gauge axes. The left side of (D.17) and the angles are known from measurement. Thus, equations (D.17) constitute a system of three equations in the three unknown Cartesian strains. With a change in notation,

$$\varepsilon_{\mathrm{A}} = S + D\cos(2\theta_{\mathrm{A}}) + T\sin(2\theta_{\mathrm{A}})$$
$$\varepsilon_{\mathrm{B}} = S + D\cos(2\theta_{\mathrm{B}}) + T\sin(2\theta_{\mathrm{B}})$$
$$\varepsilon_{\mathrm{C}} = S + D\cos(2\theta_{\mathrm{C}}) + T\sin(2\theta_{\mathrm{C}})$$

that treats the unknowns as a sum term S, a difference term D and a shear term T. In matrix notation

$$\begin{Bmatrix} \varepsilon_{\mathrm{A}} \\ \varepsilon_{\mathrm{B}} \\ \varepsilon_{\mathrm{C}} \end{Bmatrix} = \begin{bmatrix} 1 & CA & SA \\ 1 & CB & SB \\ 1 & CC & SC \end{bmatrix} \begin{Bmatrix} S \\ D \\ T \end{Bmatrix}$$

where, for example, $CA = \cos(2\theta_{\mathrm{A}})$. The gauge angles are a matter of choice and are chosen to make the computation efficient. For example, a choice of 0°, 45°, and 90° gives

$$\varepsilon_{\mathrm{A}} = S + D = \varepsilon_{xx}$$
$$\varepsilon_{\mathrm{B}} = S + T = (1/2)(\varepsilon_{xx} + \varepsilon_{yy}) + \varepsilon_{xy}$$
$$\varepsilon_{\mathrm{C}} = S - D = \varepsilon_{yy}$$

Hence, for a 0–45–90 rosette,

$$\begin{aligned} \varepsilon_{xx} &= \varepsilon_{\mathrm{A}} \\ \varepsilon_{yy} &= \varepsilon_{\mathrm{C}} \\ \gamma_{xy} &= -(\varepsilon_{\mathrm{A}} + \varepsilon_{\mathrm{C}}) + 2\varepsilon_{\mathrm{B}} \end{aligned} \tag{D.18}$$

where engineering shear strain γ is used instead of tensorial shear strain.

Because the rosette is bonded to a traction-free surface (with normal direction z), one has $\sigma_{zz} = \tau_{zx} = \tau_{zy} = 0$. Hooke's law for isotropic material implies the z-direction shear strains are also zero, that is, $2\varepsilon_{zx} = 2\varepsilon_{zy} = \gamma_{zx} = \gamma_{zy} = 0$ and that $\varepsilon_{zz} = -\nu(\sigma_{xx} + \sigma_{yy})/E$ while $\tau_{xy} = G\gamma_{xy}$ where ν and G are Poisson's ratio and shear modulus, respectively. At this juncture, there remains the determination of two normal stresses and one normal strain, σ_{xx}, σ_{yy}, ε_{zz}. Clearly once the two normal stresses are known, one can then compute the z-direction normal strain. From Hooke's law

$$\sigma_{xx} = \left(\frac{E}{1-\nu^2}\right)(\varepsilon_{xx} + \nu\varepsilon_{yy})$$

$$\sigma_{yy} = \left(\frac{E}{1-\nu^2}\right)(\varepsilon_{yy} + \nu\varepsilon_{xx})$$

Small strain–displacement relations

When the strains are small relative to 1, so squares and products are negligible, for example, $\varepsilon = 0.001$ and $\varepsilon^2 = 0.000001$, then

$$\begin{aligned}
&\varepsilon_{xx} = \frac{\partial u}{\partial x}, \quad \varepsilon_{yy} = \frac{\partial v}{\partial y}, \quad \varepsilon_{zz} = \frac{\partial w}{\partial z} \\
&\gamma_{xy} = \frac{\partial u}{\partial y} + \frac{\partial v}{\partial x}, \quad \gamma_{yz} = \frac{\partial v}{\partial z} + \frac{\partial w}{\partial y}, \quad \gamma_{zx} = \frac{\partial w}{\partial x} + \frac{\partial u}{\partial z}
\end{aligned} \tag{D.19}$$

where u, v, and w are displacements components in the x, y, and z directions, respectively.

D.3 Stress–strain relationships, Hooke's law

Hooke's law for linearly elastic, *isotropic* materials is of great importance. However, many materials show directional features and are *anisotropic*. For example, some sedimentary formations show pronounced differences between elastic moduli measured parallel and perpendicular to bedding. These materials are often characterized as transversely isotropic. Other formations, for example, some gneisses, show three distinct material directions. These materials are orthotropic. As the degree of anisotropy increases, so do the number of elastic constants needed to characterize the material. The most general anisotropic material requires 21 elastic constants to define the stress–strain law. Anisotropic stress–strain relations are sometimes referred to as "generalized" Hooke's law.

Hooke's law in one dimension – Young's modulus and shear modulus

If the constant of proportionality between normal stress and normal strain is E, then Hooke's law in one dimension is simply

$$\sigma = E\varepsilon \tag{D.20}$$

A plot of (D.20) with ε as abscissa and σ ordinate would obviously be a straight line passing through the origin, as shown in Figure D.1. The slope of the plot E is *Young's modulus*.

An important feature of linearity is *superposition*. If a given state of stress is $\sigma(1) = E\varepsilon(1)$ and a second state is $\sigma(2) = E\varepsilon(2)$, addition of strains $\varepsilon(1) + \varepsilon(2) = \varepsilon(3)$ implies addition of stresses such that $\sigma(1) + \sigma(2) = \sigma(3)$ because $\sigma(3) = E\varepsilon(1) + E\varepsilon(2)$. Superposition implies that states of stress and corresponding strain can simply be added to obtain the combined effects. Solutions to complicated problems may then be reduced to solution to several simpler problems. Linearity implies superposition, while nonlinearity precludes applicability of superposition.

In one dimension, the area under a stress–strain curve (line) represents the work done per unit volume of material during loading to the present state of strain. This work is stored in the body as *strain energy*. If A is the cross-sectional area of a prismatic bar and L is the length, then the average normal stress is $\sigma = F/A$ and the average normal strain is $\varepsilon = (L - L_o)/L_o$ as shown in Figure D.1. An element of area under the stress–strain curve dw is $\sigma\, d\varepsilon$, so

$$dw = \sigma d\varepsilon = \frac{F dL}{A L_o}$$

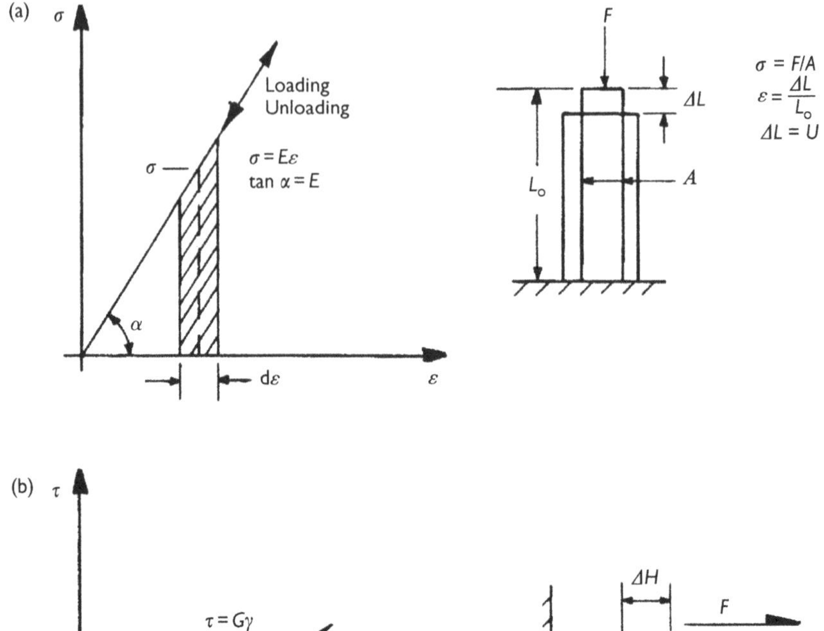

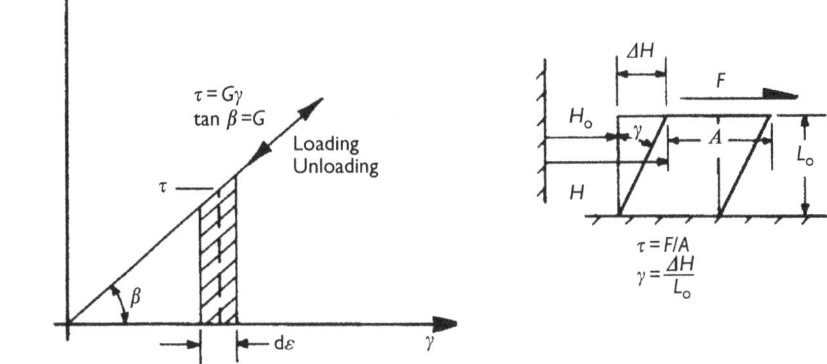

Figure D.1 One-dimensional linearly elastic normal stress–strain and shear stress–strain response: (a) one-dimensional compression and (b) one-dimensional shear.

Hence,

$$w = \int_L \sigma \, d\varepsilon = \frac{1}{AL_o} \int_L F dL$$

where ALo is the volume of the bar, the integral on the far right is the work done by the externally applied load and w is the area under the stress–strain curve. In fact w is the work per unit volume of the bar. Alternatively,

$$w = \int_L \sigma \, d\varepsilon = \int_L E\varepsilon \, d\varepsilon = \frac{E\varepsilon^2}{2} = \frac{\sigma\varepsilon}{2} = \frac{\sigma^2}{2E} \tag{D.21}$$

which expresses the external work in terms of the internal variables of stress and strain. When the work per unit volume done by the applied loads is expressed in terms of strain, the result is a *strain energy density.*

Young's modulus is closely associated with another elastic feature *normal stiffness*. However, stiffness is a structural feature and not an intrinsic material property. In an elastic structure composed of a number of connected elastic elements, an applied force is directly proportional to displacement; the constant of proportionality is a stiffness K_n . Thus, in one dimension, $F = K_n U$ where U is displacement in the direction of the applied force. If the structure is simply the prismatic bar considered previously, then the relevant displacement is the change in length of the sample, that is, $U = (L - L_o) = \varepsilon L_o$. Thus,

$$F = \sigma A = K_n U = K_n \varepsilon L_o = K_n \frac{\sigma}{E} L_o \tag{D.22}$$

which shows that

$$K_n = \frac{AE}{L_o} \tag{D.23}$$

Stiffness (normal stiffness) therefore depends not only on Young's modulus of the material but also on the geometry of the structure, which is quite simple in the case of a prismatic bar, but less so in the case of a wood crib, for example.

Application of load may be to a body that is already stressed or strained. Under such circumstances, change in stress and strain that follow Hooke's law are of interest. Final states of stress and strain are obtained by adding the stress and strain changes to the initial values. For example, if a strain εo is present in the unstressed state as shown in Figure D.2, then the stress in one dimension is given by

$$\sigma = E(\varepsilon - \varepsilon_o)$$

or if a stress σ_o is present in the unstrained state as shown in Figure D.2, then

$$(\sigma - \sigma_o) = E\varepsilon$$

Becau of linearity of the stress strain law, one can calculate a stress that corresponds to an initial strain and a strain that corresponds to an initial stress as shown in Figure D.2. Over a linear region shown in Figure D.2, one has

$$(\sigma - \sigma_o) = \Delta\sigma = E(\varepsilon - \varepsilon_o) = E\Delta\varepsilon$$

so the increments or changes of stress and strain are linearly related by Hooke's law in the elastic domain.

The incremental approach is often useful even in the nonlinear elastic case where Young's modulus depends on strain, that is, where $\sigma = E(\varepsilon)\varepsilon$. If at a given strain, the change in E is small, then the stress change for practical purposes is linearly related to the strain change at the given strain. A common occurrence of these conditions is in experimental determination of Young's modulus by wave propagation through a test specimen under static load.

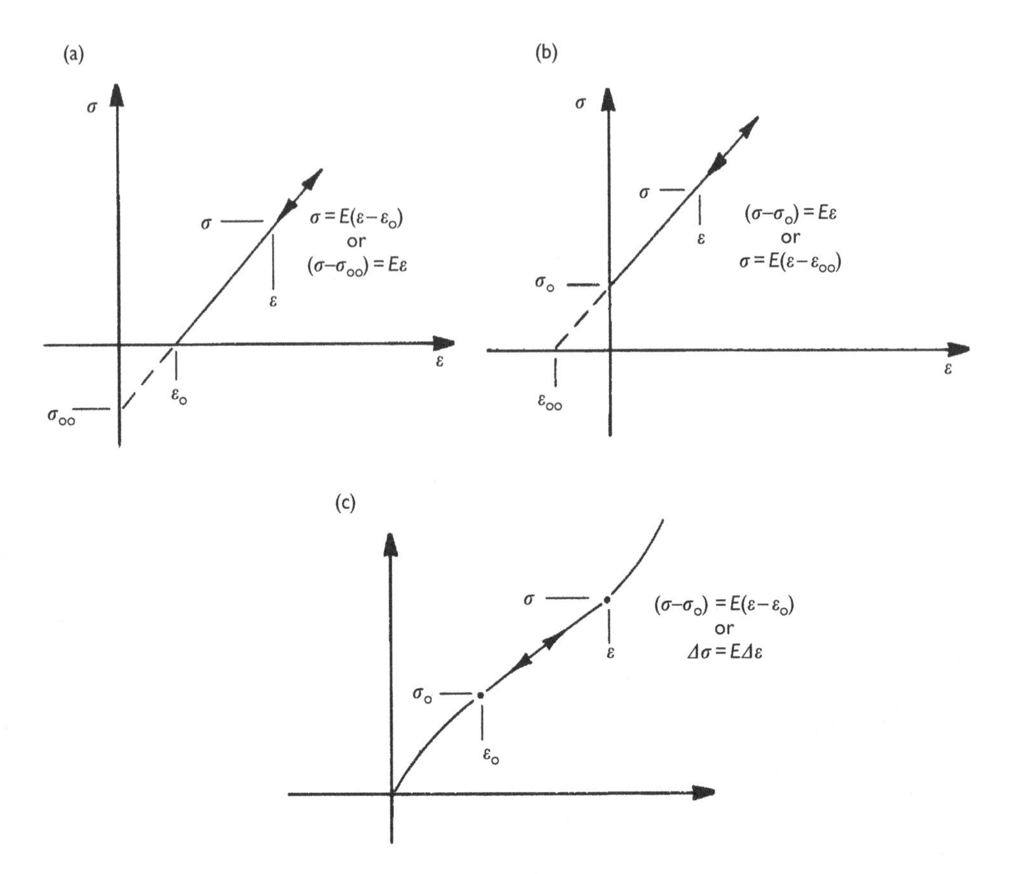

Figure D.2 (a) Initial strain, (b) initial stress, and (c) incremental elasticity concepts.

When the applied stress is a shear stress τ then the corresponding strain is a shear strain (engineering shear strain). The constant of proportionality between the two is a *shear modulus*. Hooke's law in the case of one-dimensional shear is simply

$$\tau = G\gamma \tag{D.24}$$

where G is the shear modulus of the considered material. The shear modulus is also known as the modulus of rigidity and is often denoted by the symbol μ. Numerical values of G generally range between one-third and one-half of E. A plot of as a function of is a straight line passing through the origin that has slope G as shown in Figure D.1.

Shear strain energy density w is the work per unit volume done by the applied shear force acting through the shear displacement and is graphically represented by the area under the shear stress shear strain curve. Thus,

$$w = \frac{G\gamma^2}{2} = \frac{\tau\gamma}{2} = \frac{\tau^2}{2G} \tag{D.25}$$

is the shear strain energy density. Shear stiffness K_s in contrast to shear modulus is

$$K_s = \frac{AG}{L_o} \tag{D.26}$$

as illustrated in Figure D.1. In structures such as wood cribs that have little resistance to side forces, shear stiffness may be only a small fraction of the of the wood shear modulus.

In the *isotropic* case, Hooke's law is

$$\begin{aligned}
\varepsilon_{xx} &= \frac{1}{E}\sigma_{xx} - \frac{v}{E}\sigma_{yy} - \frac{v}{E}\sigma_{zz}, \quad \gamma_{xy} = \frac{1}{G}\tau_{xy} \\
\varepsilon_{yy} &= -\frac{v}{E}\sigma_{xx} + \frac{1}{E}\sigma_{yy} - \frac{v}{E}\sigma_{zz}, \quad \gamma_{yz} = \frac{1}{G}\tau_{yz} \\
\varepsilon_{zz} &= -\frac{v}{E}\sigma_{xx} - \frac{v}{E}\sigma_{yy} + \frac{1}{E}\sigma_{zz}, \quad \gamma_{zx} = \frac{1}{G}\tau_{zx}
\end{aligned} \tag{D.27}$$

where v is *Poisson's ratio*, a new elastic constant. In cylindrical coordinates

$$\begin{aligned}
\varepsilon_{rr} &= \frac{1}{E}\sigma_{rr} - \frac{v}{E}\sigma_{\theta\theta} - \frac{v}{E}\sigma_{zz}, \quad \gamma_{r\theta} = \frac{1}{G}\tau_{r\theta} \\
\varepsilon_{\theta\theta} &= -\frac{v}{E}\sigma_{rr} + \frac{1}{E}\sigma_{\theta\theta} - \frac{v}{E}\sigma_{zz}, \quad \gamma_{\theta z} = \frac{1}{G}\tau_{\theta z} \\
\varepsilon_{zz} &= -\frac{v}{E}\sigma_{rr} - \frac{v}{E}\sigma_{\theta\theta} + \frac{1}{E}\sigma_{zz}, \quad \gamma_{zr} = \frac{1}{G}\tau_{zr}
\end{aligned}$$

which is obtained simply by replacing subscripts xyz by $r\theta z$. The same change is possible for any orthogonal coordinate system.

Because an isotropic material responds the same regardless of direction, rotation of the reference axis leaves (D.27) unchanged and leads to the relationships

$$v = \frac{E}{2G} - 1, \quad G = \frac{E}{2(1+v)}, \quad E = 2G(1+v) \tag{D.28}$$

so that any one of the three elastic constants can be obtained from the other two. Under uniaxial stress σ_a, as illustrated in Figure D.3, the corresponding strain is $\varepsilon_a = \sigma_a/E$ according to (D.28). The strain across the bar, the transverse strain, $\varepsilon_t = v\sigma_a/E$. Hence,

$$v = \left|\frac{\varepsilon_t}{\varepsilon_a}\right| \tag{D.29}$$

Transverse strain induced by load at right angles is often referred to as a Poisson's ratio effect. Under tension, the transverse strain is a contraction, but under compression a thickening occurs. Poisson's ratio practically varies between extremes of 0 and 0.5; values of about 0.15–0.35 are common.

Another useful elastic constant is defined in terms of a volumetric strain mean normal stress relation. Adding the first three of (D.27) to obtain the volumetric strain gives

$$(\varepsilon_{xx} + \varepsilon_{yy} + \varepsilon_{zz}) = \varepsilon_v = (\sigma_{xx} + \sigma_{yy} + \sigma_{zz})\frac{(1-2v)}{E} = 3\sigma_m\frac{(1-2v)}{E} = \frac{\sigma_m}{K} \tag{D.30}$$

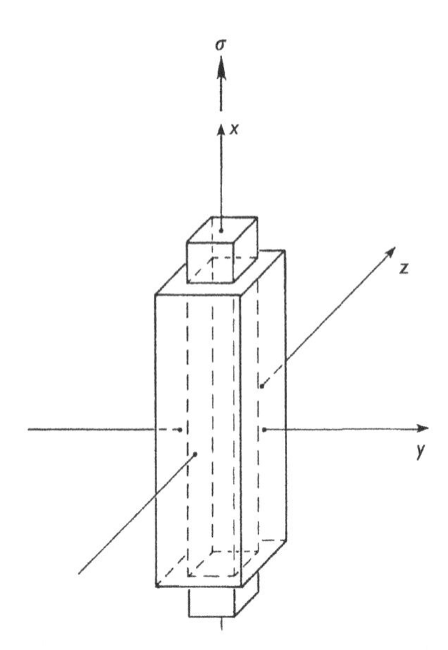

Figure D.3 Axial elongation and lateral contraction of a prism under axial load only.

where ε_v is volumetric strain, σ_m is the mean (average) normal stress and K is the *bulk modulus* of the material, $K = E/3(1-2\nu)$. A Poisson's ratio of 0.5 implies an incompressible material. Compressibility is the reciprocal of the bulk modulus and is zero when the material is incompressible.

Equations (D.27) in inverted form are

$$\begin{aligned}
\sigma_{xx} &= \left[\frac{E}{(1+\nu)(1-2\nu)}\right][(1-\nu)\varepsilon_{xx} + \nu\varepsilon_{yy} + \nu\varepsilon_{zz}], \quad \tau_{xy} = G\gamma_{xy} \\
\sigma_{yy} &= \left[\frac{E}{(1+\nu)(1-2\nu)}\right][\nu\varepsilon_{xx} + (1-\nu)\varepsilon_{yy} + \nu\varepsilon_{zz}], \quad \tau_{yz} = G\gamma_{yz} \\
\sigma_{zz} &= \left[\frac{E}{(1+\nu)(1-2\nu)}\right][\nu\varepsilon_{xx} + \nu\varepsilon_{yy} + (1-\nu)\varepsilon_{zz}], \quad \tau_{zx} = G\gamma_{zx}
\end{aligned} \tag{D.31}$$

An interesting feature of isotropic materials is the uncoupling of normal and shear components of stress and strain. Inspection of (D.27) and (D.31) shows that normal stresses are linked only to normal strains and that shear stresses are linked only to shear strain. This feature also holds for transversely isotropic and orthotropic materials when the reference axes coincide with the material axes.

Work done by externally applied forces during deformation that is stored in a body as strain energy is given by

$$w = \int \mathrm{d}w = \int_V (\sigma_{xx}\mathrm{d}\varepsilon_{xx} + \sigma_{yy}\mathrm{d}\varepsilon_{yy} + \sigma_{zz}\mathrm{d}\varepsilon_{zz} \\ \tau_{xy}\mathrm{d}\gamma_{xy} + \tau_{yz}\mathrm{d}\gamma_{yz} + \tau_{zx}\mathrm{d}\gamma_{zx})$$

Use of Hooke's law allows integration. Thus, the three-dimensional strain energy density as a quadratic form in strain is,

$$w = \frac{K(\varepsilon_v)^2}{2} + G\left[e_{xx}^2 + e_{yy}^2 + e_{zz}^2 + \left(\frac{1}{2}\right)(\gamma_{xy}^2 + \gamma_{yz}^2 + \gamma_{zx}^2)\right] \tag{D.32}$$

which shows volumetric and deviatoric strain contributions to the total strain energy density (strain energy per unit volume). Equation (D.32) can also be expressed in terms of mean normal stress and deviatoric stress, that is,

$$w = \frac{(\sigma_m)^2}{2} + \left(\frac{1}{2G}\right)\left[\tau_{xy}^2 + \tau_{yz}^2 + \tau_{zz}^2 + \left(\frac{1}{2}\right)(s_{xx}^2 + s_{yy}^2 + s_{zz}^2)\right] \tag{D.33}$$

where s is deviatoric stress, for example, $s_{xx} = \sigma_{xx} - \sigma_m$, and where σ_m is the mean normal stress. Equation (D.33) shows the contribution of the mean normal stress (hydrostatic part of stress) and the deviatoric stresses to the strain energy density.

In the *transversely isotropic* case one may consider bedding to be horizontal and the vertical axis to be the axis of rotational symmetry as shown in Figure D.4. Because transverse isotropy implies no distinction of direction in the horizontal plane, the material axes are simply vertical and horizontal, v and h. In this case, Hooke's law has the form

$$\begin{aligned}
\varepsilon_{xx} &= \frac{1}{E_h}\sigma_{xx} - \frac{\nu_h}{E_h}\sigma_{yy} - \frac{\nu_v}{E_h}\sigma_{zz}, \quad \gamma_{xy} = \frac{1}{G_h}\tau_{xy} \\
\varepsilon_{yy} &= -\frac{\nu_h}{E_h}\sigma_{xx} - \frac{1}{E_h}\sigma_{yy} - \frac{\nu_v}{E_h}\sigma_{zz}, \quad \gamma_{yz} = \frac{1}{G_h}\tau_{yz} \\
\varepsilon_{zz} &= -\frac{\nu_v}{E_h}\sigma_{xx} - \frac{\nu_h}{E_h}\sigma_{yy} - \frac{1}{E_h}\sigma_{zz}, \quad \gamma_{zx} = \frac{1}{G_h}\tau_{zx}
\end{aligned} \tag{D.34}$$

The physical meanings of the various elastic constants in (D.34) are illustrated in Figure D.4. Although there are 6 elastic constants in (D.34, 2 E's, 2 G's and 2 ν's, only 5 are independent. In fact, the in-plane properties with subscript h are related by the isotropic formula

$$G_h = \frac{E_h}{2(1 + \nu_h)} \tag{D.35}$$

The Poisson's ratio ν_h relates a horizontal strain at right angles to a horizontal stress; ν_v relates a vertical strain to a horizontal stress. However, ν_v is not the same as ν'_v which relates a horizontal stain to a vertical stress, although symmetry requires

$$\frac{\nu_v}{E_h} = \frac{\nu'_v}{E_v} \tag{D.36}$$

Moduli E_h and E_v relate horizontal and vertical stresses to corresponding strains; G_h and G_v, respectively, relate shear stress and strain parallel and perpendicular to the bedding, as illustrated in Figure D.4.

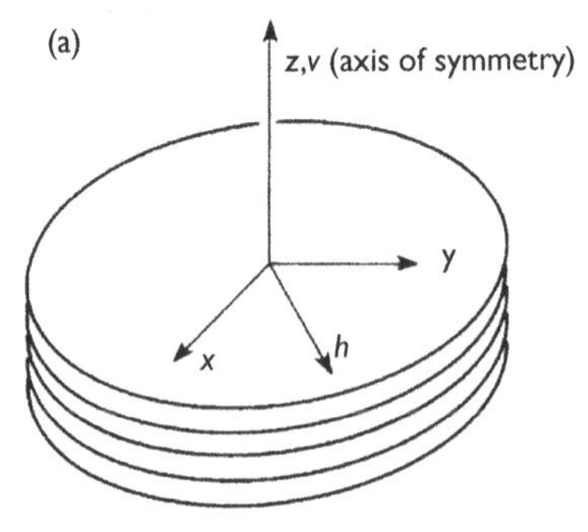

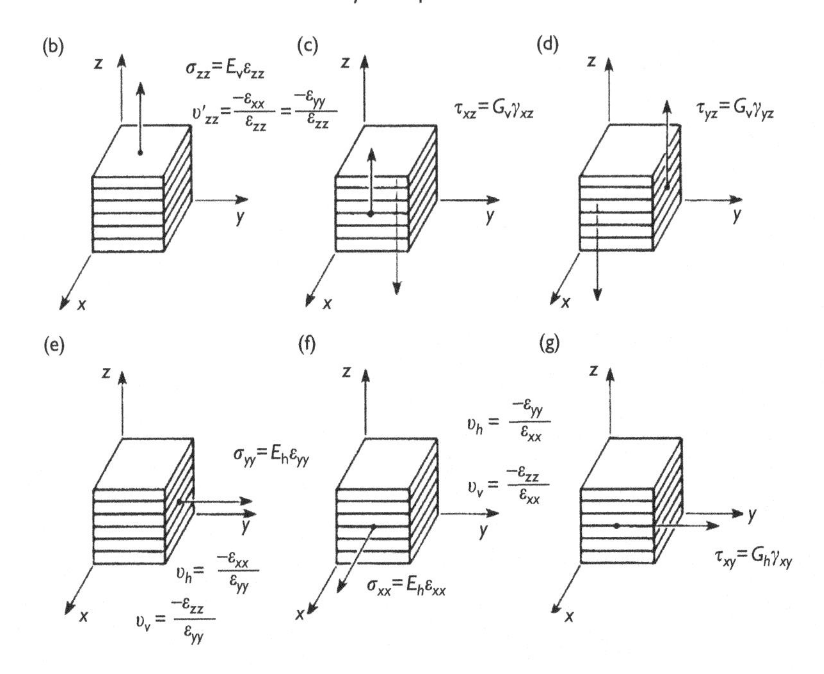

Figure D.4 Interpretation of transversely isotropic elastic constants.

In the *orthotropic* case when one uses the material axes a, b, c for reference, Hooke's law is

$$
\begin{aligned}
\varepsilon_{aa} &= \frac{1}{E_a}\sigma_{aa} - \frac{\nu_{ba}}{E_b}\sigma_{bb} - \frac{\nu_{ca}}{E_c}\sigma_{cc}, \quad \gamma_{ab} = \frac{1}{G_c}\tau_{ab} \\
\varepsilon_{bb} &= -\frac{\nu_{ab}}{E_a}\sigma_{aa} + \frac{1}{E_b}\sigma_{bb} - \frac{\nu_{cb}}{E_c}\sigma_{cc}, \quad \gamma_{bc} = \frac{1}{G_a}\tau_{bc} \\
\varepsilon_{cc} &= -\frac{\nu_{ac}}{E_a}\sigma_{aa} - \frac{\nu_{bc}}{E_b}\sigma_{bb} + \frac{1}{E_c}\sigma_{cc}, \quad \gamma_{ca} = \frac{1}{G_b}\tau_{ca}
\end{aligned}
\tag{D.37}
$$

Although there appear to be 12 elastic constants in (D.37), 3 E's, 3 G's and 6 v's, symmetry reduces the number to 9 (E_a, E_b, E_c, G_a, G_b, G_c, v_{ab}, v_{bc}, v_{ca}). The order of subscripts with the Poisson's ratios is important. For example, v_{ab} relates the effect of a stress in the a-direction to a strain in the b-direction, while v_{ba} relates a stress in the b-direction to a strain in the a-direction. The two are generally not equal. However, symmetry requires

$$\frac{v_{ba}}{E_b}=\frac{v_{ab}}{E_a},\quad \frac{v_{ca}}{E_c}=\frac{v_{ac}}{E_a},\quad \frac{v_{cb}}{E_c}=\frac{v_{bc}}{E_c} \tag{D.38}$$

which allows for computation of the other Poisson's ratios. Modulus E_a relates a stress applied in the a-direction to a strain in the same direction and similarly for E_b and E_c. A shear modulus G_c relates shear stress to shear strain in the ab-plane and similarly for G_a and G_b for shear in the bc- and ca-planes.

Hooke's law in two-dimensions – plane stress and plane strain

There are two important special cases of three-dimensional stress and strain that reduce to mainly two-dimensional considerations. The first is *plane stress*; the second is *plane strain*. Two-dimensional implies that variation in the third, say, z-direction are negligible. In one-dimensional analysis, five of the considered stresses were known to be zero. This fact then allowed for calculation of all the strains. Plane stress and plane strain analyses are more complex than one-dimensional analyses, but not as complex as the general three- dimensional case. Both are illustrated in Figure D.5. Many important practical problems are well approximated as plane stress or plane strain problems. Slabs and plates loaded on edge are examples of the former; tunnels and shafts are examples of the latter.

In *plane stress*, the three z-direction stresses are zero, while the three in-plane stresses remain to be determined. Thus, $\sigma_{zz}=\tau_{xz}=\tau_{yz}=0$ in plane stress, while σ_{xx}, σ_{yy}, and τ_{xy} are unknown. Vanishing of the z-direction shear stresses implies through Hooke's law vanishing of the z-direction shear strains, so $\gamma_{xz}=\gamma_{yz}=0$ and $E\varepsilon_{zz}=-v(\sigma_{xx}+\sigma_{yy})$. Solution of the two-dimensional problem for the stresses then allows for computing ε_{zz} and the in-plane strains ε_{xx}, ε_{yy}, and γ_{xy}. Hooke's law reduces to

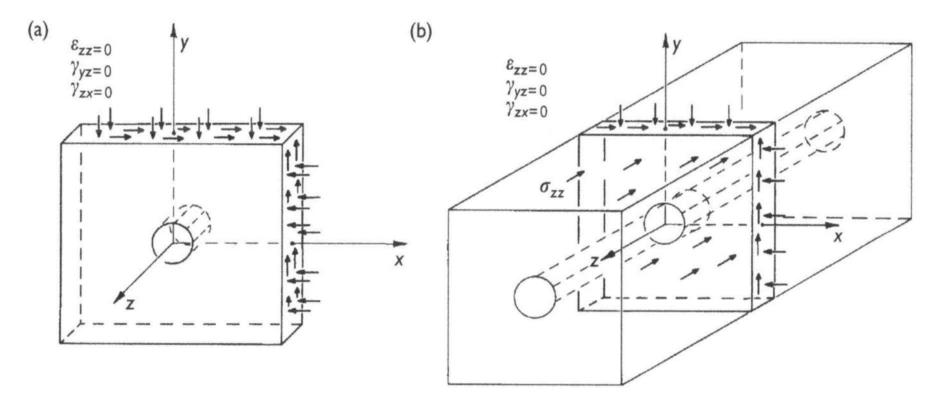

Figure D.5 Concepts of (a) edge-loaded slab: plane stress and (b) imaginary slab in a long block: plane strain.

$$\begin{aligned}
\varepsilon_{xx} &= \frac{1}{E}\sigma_{xx} - \frac{\nu}{E}\sigma_{yy} \\
\varepsilon_{yy} &= -\frac{\nu}{E}\sigma_{xx} + \frac{1}{E}\sigma_{yy} \\
\varepsilon_{xy} &= \frac{1}{G}\tau_{xy}
\end{aligned} \tag{D.39}$$

in plane stress. Excavation of a hole in an edge-loaded slab induces stress and strain changes that follow the plane stress idealization. Interestingly enough, the change in slab thickness varies from point to point and is largest at the hole rim where a "lip" forms.

Plane strain may be defined by setting the z-direction strains to zero, so $\varepsilon_{zz} = \gamma_{xz} = \gamma_{yz} = 0$. The z-direction shear strains imply through Hooke's law that the z-direction shear stresses are zero. Also, the z-direction normal strain implies that $\sigma_{zz} = \nu(\sigma_{xx} + \sigma_{yy})$. Again, solution for the in-plane stresses allows for determination of all stresses and strains. In the case of plane strain, Hooke's law has the form

$$\begin{aligned}
\varepsilon_{xx} &= \left(\frac{1-\nu^2}{E}\right)\sigma_{xx} - \left(\frac{\nu(1+\nu)}{E}\right)\sigma_{yy} \\
\varepsilon_{yy} &= -\left(\frac{\nu(1+\nu)}{E}\right)\sigma_{xx} + \left(\frac{1-\nu^2}{E}\right)\sigma_{yy} \\
\varepsilon_{xy} &= \frac{1}{G}\tau_{xy}
\end{aligned} \tag{D.40}$$

A thin imaginary slab perpendicular to the axis of a long tunnel located in a region where the pretunnel stress field shows no shear stress parallel to the tunnel axis is shown in Figure D.5. Excavation of the tunnel is reasonably idealized as a plane strain problem, that is, the changes in stress and strain induced by excavation satisfy plane strain conditions. Stresses and strains about the tunnel after excavation are obtained by addition of the changes to the pretunnel stresses and strains. Displacements induced by excavation are associated with the strain changes and are entirely in-plane. Thickness of the considered slab remains unchanged in a plane strain analysis.

In anisotropic rock where the tunnel axis is not parallel to a material axis, vanishing of shear strains (stresses) no longer implies vanishing of the corresponding shear stresses (strains). Additional considerations are then required for plane stress and plane strain analyses that involve subtle details of the problem.

References

2 Slope stability

Bishop, A. W. 1955. The use of the slip circle in the stability analysis of slopes. *Geotechnique*. Vol. 5, No. 1, pp. 7–17.

Chen, R. H. 1981. "Three-dimensional slope stability analysis," Joint Highway Research Project, Eng. Experiment station, Purdue University, Report JHRP–81–17.

Hamel, J. V. 1974. Rock strength from failure cases: left bank slope stability study, Libby Dam and Lake Koocanusa, Montana. *U.S. Army Corp of Engineers Technical Report MRD-1-74*. p. 239.

Janbu, N. 1957. Earth pressures and bearing capacity calculations by generalized procedure of slices. *Proc. 4th Intl Conf. on Soil Mechanics*. Vol. 1, pp. 207–221.

—— 1976. *The Pit Slope Manual* (R. Sage, ed.). Canada Centre for Mineral and Energy Technology (CANMET). In 10 separate chapters.

Pariseau, W. G. 1980. A simple mechanical model for rockslides and avalanches. *Engineering Geology*. Vol. 16, pp. 111–123.

—— Rocscience software. http://www.rocscience.com

—— Course downloads. http://www.utah.edu/mining/mgen5160.htm and mgen5150.htm

Seegmiller, B. L. 1974. How cable bolt stabilization may benefit open pit operations. *Mining Engineering*. Vol. 26, No. 12, p. 2934.

—— 1975. Cable bolts stabilize pit slopes, steepen walls to strip less waste. *World Mining*. July Issue.

3 Shafts

Cornish, E. 1967. Vertical shafts. *Mineral Industries Bulletin*. Vol. 10, No. 5.

Green, A. E. 1940. General biharmonic analysis for a plate containing circular holes. *Proc. Royal Soc. London*, Series A, Vol. 176, pp. 121–139.

Haddon, R. A. 1967. Stresses in an infinite plate with two unequal circular holes. *Q. J. Mech. 7 Appl. Math*., Vol. 20, pp. 277–291.

Heller, S. R., Jr, Brock, J. S., and R. Bart., 1958. The stresses around a rectangular opening with rounded corners in a uniformly loaded plate. *Proc. 3rd U.S. National Congress Applied Mechanics*. ASME, NY, pp. 357–368.

Howland, R. C. 1935. Stresses in a plate containing an infinite row of holes. *Proc. Royal Soc. London*, Series A, Vol. 148, pp. 471–491.

Howland, R. C. J. and R. C. Knight. 1939. Stress functions for a plate containing groups of circular holes. *Proc. Phil. Trans*. Vol. A238, pp. 357–392.

Jaeger, J. C. and N. G. W. Cook. 1969. *Fundamentals of Rock Mechanics*. Methuen & Co. Ltd. London, p. 513. Ellipse 253–260.

Ling, C.-H. 1948. On the stresses in a plate containing two circular holes. *J. of Applied Physics*. Vol. 19, pp. 77–82.

Love, A. E. H. 1944. *A Treatise on the Mathematical Theory of Elasticity*. Dover, NY, p. 643.
Manual of Steel Construction (7th ed.) AISC, NY, 1970.
National Design Specification for Wood Construction Supplement (1997 ed.) American Forest & Paper Association.
Obert, L. and W. I. Duvall. 1967. *Rock Mechanics and the Design of Structures in Rock*. John Wiley & Sons, Inc., NY, p. 650.
Obert, L., W. I. Duvall, and R. H. Merrill. 1960. Design of underground openings in competent rock. U.S. Bureau of Mines Bulletin 587. U.S. Government Printing Office, pp. 6–7.
Pariseau, W. G. 1977. Estimation of support Load Requirements for Underground Mine Openings by Computer Simulation of the Mining Sequence. *Transactions*. SME/AIME, Vol. 262, pp. 100–109.
Pariseau, W.G. 1987. An alternative solution for the in situ stress state inferred from borehole stress relief data. *Proc. 6th Intl. Congress on Rock Mechanics*. Vol. 2, Balkema, Rotterdam, pp 1201–1205.
Stillborg, B. 1986. *Professional Users Handbook for Rock Bolting*. Trans Tech Publications, Clausthal, p. 145.
—— ASTM Standard F 432-95 Standard Specification for Roof and Rock Bolts and Accessories. Copyright ASTM International, 100 Barr Harbor Drive, West Conshocken, PA 19428.
Terzaghi, K. 1962. Stability of steep slopes on hard unweathered rock. *Geotechnique*. Vol. 12, No. 4, pp. 251–270.
Wang, J., S. G. Mogilevskaya, and S. L. Crouch. 2001. A Galerkin Boundary Integral Method for Nonhomogeneous Materials with Cracks. *Proc. 38th U.S. Rock Mechanics Symp.* Vol. 2, Balkema, pp. 1453–1460.

4 Tunnels

Bischoff, J. A. and J. D. Smart. 1977. A method of computing a rock reinforcement system which is structurally equivalent to an internal support system. In: *Proc. 16th U.S. Symp. Rock Mechanics* (Fairhurst, C. and S. L. Crouch, eds). ASCE, NY, pp. 279–184.
Hoadley, A. 1964. *Essentials of Structural Design*. John Wiley & Sons, Inc. NY, p. 348.
Kendorski, F. S. 1977. Caving operations drift support design. In: *Proc. 16th U.S. Symp. Rock Mechanics* (Fairhurst, C. and S. L. Crouch, eds). ASCE, NY, pp. 277–286.
Proctor, R. V. and T. L. White. 1968. *Rock Tunneling with Steel Supports*. Commercial Shearing & Stamping Company. The Youngstown Printing Co, Youngstown, p. 291.
Timoshenko, S.P. and S. Woinoswaky-Krieger (1959) *Theory of Plates and Shells* (2nd ed.). McGraw-Hill, Tokyo, p17.

5 Entries in stratified ground

Gere, J. M. and S. P. Timoshenko. 1997. *Mechanics of Materials*. PWS Publishing Company, p. 912.
Hibbeler, R. C. 2003. *Mechanics of Materials*. Prentice-Hall, p. 848.
Popov, E. P. 1952. *Mechanics of Materials*. Prentice-Hall, p. 441.

6 Pillars in stratified ground

Pariseau, W.G. 1982. Shear stability of mine pillars in dipping seams. *Proc. 23rd U.S. Symp. Rock Mech.* SME/AIME, NY, pp. 1077–101090.
—— 1983. Ground Control in Multi-Level Room and Pillar Mines. U.S. Department of Interior, Bureau of Mines, Final Report. Grant No. G1115491, p. 111.

7 Three-dimensional excavations

ASTM. 1998. Standard specification for steel strand, uncoated seven-wire for prestressed concrete. American Society for Testing and Materials Standard A416/A 416M - 96. *Annual Book of Standards*. Vol. 01.04, pp. 213–215.

Coates, D. F. 1970. *Rock Mechanics Principles*. Canadian Mines Branch Monograph 874 (revised). Information Canada, Ottawa.

Landriault, D., D. Welch, and D. Morrison. 1996. *Mine tailings disposal as a paste backfill for underground mine backfill and surface waste deposition*. SME Short Course Notes, Little, Colorado.

Love, A. E. H. (1944) *A Treatise on the Mathematical Theory of Elasticity*. (4th ed.) Dover publication, NY

Hutchinson, D. J. and M.S. Diedrichs. 1996. *Cable Bolting in Underground Mines*. BiTech Publishers, Ltd., Richmond, B.C., Canada, p 407.

Pariseau, W. G. 2005. Preliminary design guidelines for large deep underground caverns. *Proc. 40th U.S. Symp. Rock Mech*. American Rock Mechanics Association. No. 853.

Pariseau, W. G., J. R. M. Hill, M. M. McDonald, and L. M. McNay. 1976 A support-performance prediction method for hydraulic backfill. U.S. Bureau of Mines R.I. 8161. p. 19.

Sadowsky, M. A. and E. Sternberg. 1949. Stress concentration around a triaxial ellipsoidal cavity. *J. Appl. Mech*. Trans. ASME, Series E, Vol. 71, pp. 141–157.

Schmuck, C. H. 1979. Cable bolting at the Homestake gold mine. *Mining Engineering*. Vol. 31, No. 12, pp. 1677–1681.

Terzaghi, K. and F. E. Richart, Jr. 1952. Stresses in rock about cavities. *Geotechnique*. Vol. 2, No. 3, pp. 57–90.

8 Subsidence

Brauner, G. 1973. Subsidence due to underground mining (In two parts) 1. Theory and practices in predicting surface deformation. U.S. Bureau of Mines I.C. 8571, p. 55. 2. Ground movements and mining damage. U.S. Bureau of Mines I.C. 8572, p. 53.

Janssen, H. A. 1896. On the pressure of grain in silos. *Minutes of Proc. of the Inst. Civ. Engr*. Vol. 124, pp. 553–555.

NCB. 1975. *Subsidence Engineer's Handbook*. National Coal Board, Mining Department, p. 111.

Peng, S. S. 1978. *Coal Mine Ground Control*. John Wiley & Sons, p. 450.

Voight, B. and W. G. Pariseau. 1970. State of predictive art in subsidence engineering. *J. Soil Mech. Fnd. Div*. ASCE, SM2, pp. 721–750.

9 Dynamic phenomena

Blake, W., F. Leighton and W. I. Duvall. 1974. Microseismic Techniques for Monitoring the Behavior of Rock Structures. *U.S. Bureau of Mines Bulletin 665*. pgs 65.

Blake, W. and D.G.F. Hedley (2003) *Rockbursts: Case studies from North American hard-rock mines*. Society for Mining, Metallurgy, and Exploration, Inc., Littleton, Colorado, USA. pgs 119.

Bollinger, G.A. (1971) *Blast vibration analysis*. Southern Illinois University Press, pgs 132.

Bourbie, T., O. Coussy and B. Zinszner (1987) *Acoustics of porous media*. Gulf Publishing Co., Houston. pgs 334.

Bullen, K.E. (1979) *An introduction to the theory of seismology* (3rd ed). Cambridge University Press, Cambridge, U.K. pgs 381.

Hardy, H. R., Jr. (2003). *Acoustic Emission/Microseismic Activity*. Vol. 1, A.A. Balkema, Lisse.

Jaeger, J.C., N.G.W. Cook, and R.W. Zimmerman (2007) *Fundamentals of Rock Mechanics* (4th edition), John Wiley & Sons, N.Y., pgs 500.

Kaiser, P. and D.R. McCreath (editiors). 1992. Rock Support. (*Proc. Intl. Symp. Rock Support.*) A.A. Balkema, Rotterdam, pp 593–691.
Kolsky, H. (1963) *Stress waves in solids*. Dover Publications, Inc. N.Y., pgs 213.
Obert, L. and W.I. Duvall (1967) *Rock mechanics and the design of structures in rock*. John Wiley and Sons, Inc., N.Y., pgs 650.
Rinehart, J. S. (1960) On fractures caused by explosions and impacts. *Quarterly of the Colorado School of Mines*, Vol. 55, No. 4. pgs 155.
Rinehart, J.S. (1975) *Stress transients in solids*. Hyper Dynamics, Santa Fe, New Mexico, pgs 230.
Whyatt, J.K. and M.C. Loken. (2009) Coal bumps and odd dynamic phenomena - a numerica investigation. *Proceedings 28th International Conference on Ground Control*. (Peng, S.S. et al: eds) Morgantown, West Virginia.

10 Foundations on jointed rock

Goodman, R. E. 1980. *Introduction to Rock Mechanics*. John Wiley & Sons, New York, pp 288–331.
Kulhawy, F. H. 2005. Analysis and Design of Foundations on and in Rock. Short Course Lecture Notes. 40th U.S. Symposium on Rock mechanics, Anchorage, Alaska.
Poulus, H. G. and E. H. Davis 1974. *Elastic Solutions for Soil and rock Mechanics*. John Wiley & Sons, Inc., New York, pgs 411.
Selvadurai, A,.P.S. and R. O. Davis. 1996. *Elasticity and Geomechanics*. Cambridge University Press, Cambridge, U.K., pgs 201.
Wyllie, D. C. 1999. *Foundations on Rock* (2nd ed). E & FN SPON, London, pgs 401.

Appendix B Mechanical properties of intact rock and joints

Balmer, G. G. 1953. Physical Properties of Some Typical Foundation Rocks. U.S. Department of Interior, Bureau of Reclamation, Concrete Laboratory Report No. SP–39.
Blair, B. E. 1955. Physical Properties of Mine Rock-III. U.S. Bureau of Mines Report of Investigations 5130.
—— 1956. Physical Properties of Mine Rock-IV. U.S. Bureau of Mines Report of Investigations 5244.
Brechtel, C. E. 1978. The Strength and Deformation of Singly and Multiply Jointed Sandstone Specimens. M.S. thesis, University of Utah, p. 112.
Brighenti, G. 1970. Influence of Cryogenic Temperatures on Mechanical Characteristics of Rocks. *Proc. 2nd Congress of the International Society for Rock Mechanics*. Belgrade, Vol. 1, Theme 2, Paper 2–27.
Chenevert, M. E. and C. Gatlin. 1965. Mechanical Anisotropies of Laminated Sedimentary Rocks. *Society of Petroleum Engineers Journal*. pp. 67–77.
Deere, D. U., A. J. Henderon, Jr, F. D. Patton, and E. J. Cording. (1967). Design of Surface and Near-Surface Construction in Rock. In: *Proc. 8th U.S. Symp. Mechanics*. AIME/SME, NY, 237–302.
Deklotz, E. J. and J. W. Brown. 1967. Tests for Strength Characteristics of a Schistose Gneiss. U.S. Army Corps of Engineers. Omaha, Technical Report No. 1–67.
Goodman, R. E. 1968. The Effects of Joints on the Strength of tunnels. U.S. Army Corps of Engineers. Omaha, Technical Report No. 5.
Hart, R. D., P. A. Cundall, and M. L. Cramer. 1985. Analysis of a Loading Test on a Large Basalt Block. *Proc. 26th U.S. Symposium on Rock Mechanics*. Vol. 2, Balkema, pp. 759–768.
Hassani, F. G. and J. J. Scoble. 1985. Frictional Mechanisms and Properties of rock Discontinuities. *Proc. Intl. Symp. on Fundamentals of Rock Joints*. Centek, Lulea, pp. 185–196.
Heard, H. C. 1960. Transition from Brittle Fracture to Ductile flow in Solenhofen Limestone as a Function of Temperature, Confining Pressure and Interstitial Fluid Pressure. In: *Rock Deformation* (eds: D. Griggs and J. Handin). The Geological Society of America Memoir 79, pp. 193–226.

Hoek, E. 1964. Fracture of Anisotropic Rock. *J. of the South African Institute of Mining and Metallurgy*. pp. 501–518.

Hoek, E. and E. T. Brown. 1980. *Underground Excavations in Rock*. The Institution of Mining and Metallurgy, p. 527.

Jaeger, J. C. and N. G. W. Cook. 1969. *Fundamentals of Rock Mechanics*. Methuen & Co., p. 513.

Ko, H.-Y. and K. H. Gerstle. 1976. Elastic Properties of Two Coals. *Int. J. Rock Mech. Min. Sci. & Geomech. Abstr.* Vol. 13, No. 3, pp. 81–90.

Kulhawy, F. L. 1975. Stress Deformation Properties of Rock and Rock Discontinuities. *Engineering Geology*. Vol. 9, pp. 327–350.

Lajtai, E. Z. 1969. Shear Strength of Weakness Planes in Rock. *Int. J. Rock Mech. Min. Sci. & Geomech. Abstr.* Vol. 6, No. 5, pp. 499–515.

Lama, R. D. and V. S. Vutukuri. 1978a. *Handbook on Mechanical Properties of Rocks*. Volume II – Testing Techniques and Results. Trans Tech Publications, pp. 481.

—— 1978a. *Handbook on Mechanical Properties of Rocks*. Volume III – Testing Techniques and Results. Trans Tech Publications, p. 406.

—— 1978b. *Handbook on Mechanical Properties of Rocks*. Volume IV – Testing Techniques and Results. Trans Tech Publications, p. 515.

McLamore, R. and K. E. Gray. 1967. The Mechanical Behavior of Anisotropic Sedimentary Rocks. *Journal of Engineering for Industry (ASME Transactions, Series B)*, Vol. 89, pp. 62–67.

Moon, H. 1987. Elastic Moduli of Well-Jointed Rock Masses. Ph.D. Dissertation, University of Utah, p. 284.

Obert, L. and W. I. Duvall. 1967. *Rock Mechanics and the Design of Structures in Rock*. John Wiley & Sons, p. 650.

Pariseau, W. G. 1967. *Post-yield Mechanics of Rock and Soil*. Mineral Industries. The Pennsylvania State University, College of Earth & Mineral Sciences, Vol. 36, No. 8, pp. 1–5.

—— 1972. Plasticity Theory for Anisotropic Rocks and Soils. *Proc. 10th Symposium on Rock Mechanics*. SME/AIME, pp. 267–295.

—— 1994. Design Cosiderations for Stopes in Wet Mines. In: *Proc. 12th Annual Workshop Generic Mineral Technology Center Mine System Design and Ground Control*, Department of Mining and Minerals Engineering, Virginia Polytechnic Institute and State University, Blacksburg, pp. 37–48.

Pariseau, W. G., W. A. Hustrulid, S. R. Swanson, and L. L. Van Sambeek. 1977. Coal Pillar Strength Study (the Design of Production Pillars in Coal Mines). U.S.B.M. Contract H0242059 Final Report. p. 417.

Peng, S. S. 1973. Time-dependent Aspects of Rock Behavior as Measured by a Servo-controlled Hydraulic Testing Machine. *Int. J. Rock Mech. Min. Sci. & Geomech. Abstr.* Vol. 10, No. 3, pp. 235–246.

Peng, S. S. and E. R. Podnieks. 1972. Relaxation and the Behavior of Failed Rock. *Int. J. Rock Mech. Min. Sci. & Geomech. Abstr.* Vol. 9, No. 6, pp. 669–712.

Pomeroy, C. D., D. W. Hobbs, and A. Mahmoud. 1971. The Effect of Weakness-plane Orientation on the Fracture of Barnsley Hards by Triaxial Compression. *Int. J. Rock Mech. Min. Sci. & Geomech. Abstr.* Vol. 8, No. 3, pp. 227–238.

Rosso, R. S. 1976. A Comparison of Joint Stiffness Measurements in Direct Shear, Triaxial Compression and In Situ. *Int. J. Rock Mech. Min. Sci. & Geomech. Abstr.* Vol. 13, No. 6, pp. 167–172.

Singh, M. M. 1989. Strength of Rock. In: *Physical Properties of Rocks and Minerals* (Y. S. Touloukian, W. R. Judd, and R. F. Roy, eds). Hemisphere Publishing Corp. pp. 83–121.

Terzaghi, K. 1962. Stability of Steep slopes on Hard Unweathered Rock. *Geotechnique*. Vol. 12, No. 4, pp. 251–270.

Vutukuri, V. S., R. D. Lama and S. S. Saluja. 1974. *Handbook on Mechanical Properties of Rocks*. Volume I – Testing Techniques and Results. Trans Tech Publications, p. 280.

Windes, S. L. 1949a. Physical Properties of Mine Rock-I. U.S. Bureau of Mines Report of Investigations 4459.

—— 1949b. Physical Properties of Mine Rock-II. U.S. Bureau of Mines Report of Investigations 4727.
Wuerker, R. G. 1956. Annotated Tables of Strength and Elastic Properties of Rocks. Petroleum Branch, AIME, Paper 663-G.
Zienkiewicz, O. C. (1977) *The Finite Element Method.* (3rd ed.). McGraw-Hill, London.

Appendix C Rock mass classification schemes for engineering

Bieniawaski, Z. T. 1989. *Engineering Rock Mass Classifications.* Wiley, NY, p. 251.
Deere, D. U. 1963. Technical description of rock core for engineering purposes. *Rock Mech. Eng. Geol.* Vol. 1, pp. 16–22.
Deere, D. U., A. J. Hendron, Jr, F. D. Patton, and E. J. Cording. 1967. Design of surface and near-surface construction in rock. *Proceedings of the 8th US Symposium on Rock Mechanics* SME/AIME, NY, pp. 237–302.
Heuze, F. 1980. Scale effects in the determination of rock mass strength and deformability. *Rock Mech. Rock Eng.* Vol. 12, pp 167–192.
Hoek, E., C. Carranza-Torres and B. Corkum. Hoek-Brown failure criterion – 2002 edition. 2002. *Proc. 5th North American Rock Mechanics Symposium.* U. of Toronto Press, Toronto, Vol. 1, pp 267–273.
Hoek, E. and M.S. Diederichs. 2006. Empirical estimation of rock mass modulus. *Intl. J. Rock Mech. Min. Sci.* Vol. 44, No. 2, pp 203–215.
Marinos, V., P. Marinos and E. Hoek. 2005. The geological strength index: applications and limitations. *Bull. Geol. Environ.* Vol. 64, pp 55–65.
Mars Ivars, D, M.E. Pierce, C. Darcel, J. Reyes-Montes, D. O. Potyondy, R. P. Young and P. A. Cundall. 2011. The synthetic rock mass approach for jointed rock mass modeling. *Intl. J. Rock Mech. Min. Sci.* Vol. 48, No. 2, pp 219–244.
Pariseau, W.G. 1999. Equivalent elastic plastic properties of rocks. *Intl. J. Rock Mech. Min. Sci.* Vol. 36, No. 7, pp 907–918.
Rose, D. 1982. Revising Terzaghi's Rock Load Coefficients. *Proceedings of the 23rd U.S. Symposium on Rock Mechanics* SME/AIME, NY, pp. 953–960.
Wickham, G. E. and H. R. Tidemann. 1974. Ground support prediction model (RSR Concept). U.S. Bureau of Mines Final Report. Contract No. Ho220075. p. 271.
Wickham, G. E., H. R. Tidemann, and E. H. Skinner. 1974. Ground support prediction model RSR concept. *Proceedings of the Rapid Tunneling and Excavation Conference.* SME/AIME, NY, Vol. 1, pp. 691–707.

Index